Coulson & Richards

CHEMICAL ENGINE

CU01082487

VOLUME 1

Fluid Flow, Heat Transfer and Mass Transfer

Coulson & Richardson's Chemical Engineering

Chemical Engineering, Volume 1, Fifth edition
Fluid Flow, Heat Transfer and Mass Transfer
J. M. Coulson and J. F. Richardson
with J. R. Backhurst and J. H. Harker

Chemical Engineering, Volume 2, Fourth edition
Particle Technology and Separation Processes
J. M. Coulson and J. F. Richardson
with J. R. Backhurst and J. H. Harker

Chemical Engineering, Volume 3, Third edition
Chemical & Biochemical Reactors & Process Control
Edited by J. F. Richardson and D. G. Peacock

Chemical Engineering, Volume 4, Second edition
Solutions to the Problems in Volume 1
J. R. Backhurst and J. H. Harker

Chemical Engineering, Volume 5, Second edition
Solutions to the Problems in Volumes 2 and 3
J. R. Backhurst and J. H. Harker

Chemical Engineering, Volume 6, Second edition
Chemical Engineering Design
R. K. Sinnott

Coulson & Richardson's

CHEMICAL ENGINEERING

VOLUME 1

FIFTH EDITION

Fluid Flow, Heat Transfer and Mass Transfer

J. M. COULSON

Late Emeritus Professor of Chemical Engineering
University of Newcastle-upon-Tyne

and

J. F. RICHARDSON

Department of Chemical Engineering
University of Wales, Swansea

WITH

J. R. BACKHURST and J. H. HARKER

Department of Chemical and Process Engineering
University of Newcastle-upon-Tyne

Butterworth-Heinemann
Linacre House, Jordan Hill, Oxford OX2 8DP
225 Wildwood Avenue, Woburn, MA 01801-2041
A division of Reed Educational and Professional Publishing Ltd

$\mathcal{R}$ A member of the Reed Elsevier plc group

OXFORD   BOSTON   JOHANNESBURG
MELBOURNE   NEW DELHI   SINGAPORE

First published by Pergamon Press 1954
Second edition 1964
Third edition 1977
Fourth edition 1990
Fifth edition 1996
Reprinted 1997, 1998

British Library Cataloguing in Publication Data
Coulson, J. M.
    Chemical Engineering.-Vol.1:Fluid Flow,
    Heat Transfer and Mass Transfer.-
    5Rev.edn
    I. Title II. Richardson, J. F.
    III. Backhurst, J. R. IV. Harker, J. H.
    660

ISBN 0 7506 2557 0

Library of Congress Cataloguing in Publication Data
Coulson, J. M. (John Metcalfe)
    Chemical engineering / J. M. Coulson and J. F. Richardson; with
    J. R. Backhurst and J. H. Harker.-5th edn
    p. cm.
    At head of title: Coulson and Richardson's.
    Includes bibliographical references and index.
    Contents: v. 1. Fluid flow, heat transfer and mass transfer
    ISBN 0 7506 2557 0 (v. 1.: pbk.)
    1. Chemical engineering. I. Richardson, J. F. (John Francis)
    II. Backhurst, J. R. III. Harker, J. H. (John Hadlett), 1937-
    IV. Title.
    TP155.C69 1995
    660-dc20
                                                95-23187
                                                CIP

Typeset by Laser Words, Madras, India
Printed in Great Britain by The Bath Press, Bath

Contents

13.   Humidification and Water Cooling   621

Appendix   673

Problems   711

Index   751

Professor J. M. Coulson

JOHN COULSON, who died on 6 January 1990 at the age of 79, came from a family with close involvement with education. Both he and his twin brother Charles (renowned physicist and mathematician) who died 20 years ago became professors. John did his undergraduate studies at Cambridge and then moved to Imperial College where he took the postgraduate course in chemical engineering — the normal way to qualify at that time — and then carried out research on the flow of fluids through packed beds. He then became an Assistant Lecturer at Imperial College and, after war-time service in the Royal Ordnance Factories, returned as Lecturer and was subsequently promoted to a Readership. At Imperial College he initially had to run the final year of the undergraduate course almost single-handed, a very demanding assignment. During this period he collaborated with Sir Frederick (Ned) Warner to write a model design exercise for the I. Chem. E. Home Paper on "The Manufacture of Nitrotoluene". He published research papers on heat transfer and evaporation, on distillation, and on liquid extraction, and co-authored his textbook of Chemical Engineering. He did valiant work for the Institution of Chemical Engineers which awarded him its Davis medal in 1973, and was also a member of the Advisory Board for what was then a new Pergamon journal, *Chemical Engineering Science.*

In 1954 he was appointed to the newly established Chair at Newcastle-upon-Tyne, where Chemical Engineering became a separate Department and independent of Mechanical Engineering of which it was formerly part, and remained there until his retirement in 1975. He took a period of secondment to Heriot Watt University where, following the splitting of the joint Department of Chemical Engineering with Edinburgh, he acted as adviser and *de facto* Head of Department. The Scottish university awarded him an Honorary D.Sc. in 1973.

John's first wife Dora sadly died in 1961; they had two sons, Anthony and Simon. He remarried in 1965 and is survived by Christine.

JFR

Preface to Fifth Edition

This textbook has been the subject of continual up-dating since it was first published in 1954. There have been numerous revised impressions and the opportunity has been taken on the occasion of each reprinting to make corrections and revisions, many of them in response to readers who have kindly pointed out errors or who have suggested modifications and additions. When the summation of the changes has reached a sufficiently high level, a new edition has been produced. We have now reached this point again and the fifth edition incorporates all the alterations in the 1993 revision of the fourth edition, together with new material, particularly on simultaneous mass transfer and chemical reaction for unsteady-state processes.

There have been changes in publisher too. Since the appearance of the fourth edition in 1990, Pergamon Press has become part of Elsevier Science and now, following a re-organisation in the Reed Elsevier group of companies, the responsibility for publishing the Chemical Engineering series has passed to Butterworth–Heinemann, another Reed Elsevier company.

We are grateful to our readers for their interest and very much hope they will continue to make suggestions for the improvement of the series.

JFR

Preface to Fourth Edition

THE First Edition of Volume 1 was published in 1954 and Volume 2 appeared a year later. In the intervening 35 years or so, there have been far-reaching developments in Chemical Engineering and the whole approach to the subject has undergone a number of fundamental changes. The question therefore arises as to whether it is feasible to update a textbook written to meet the needs of the final year students of an undergraduate course in the 1950's so that it can continue to fulfill a useful purpose in the last decade of the century. Perhaps it would have been better if a new textbook had been written by an entirely new set of authors. Although at one stage this had seemed likely through the sponsorship of the Institution of Chemical Engineers, there is now no sign of any such replacement book appearing in the United Kingdom.

In producing the Fourth Edition, it has been necessary to consider whether to start again with a clean sheet of paper — an impossibly daunting task — or whether to retain the original basic structure with relatively small modifications. In following the latter course, the authors were guided by the results of a questionnaire sent to a wide range of University (and Polytechnic) Departments throughout the English-speaking-world. The clear message which came back was "Do not tamper over-much with the devil we know, in spite of all his faults!"

It was in 1971 that Volume 3 was added to the series, essentially to make good some of the more glaring omissions in the earlier volumes. Volume 3 contains a series of seven specialist chapters written by members of the staff of the Chemical Engineering Department at the University College of Swansea, with Dr D. G. Peacock of the School of Pharmacy, London as a joint editor. In 1977/9, as well as contributing significantly to the new editions of Volumes 1 and 2, two colleagues at the University of Newcastle-upon-Tyne, Dr J. R. Backhurst and The Revd. Dr J. H. Harker, prepared Volumes 4 and 5, the solutions to the problems in Volumes 1 and 2, respectively. The final major development was the publication of Volume 6 on Chemical Engineering Design by Mr R. K. Sinnott in 1983. With the preparation of a Fourth Edition, the opportunity has presented itself for a degree of rationalisation, without introducing major changes to the structure. This has led to the following format:

Volume 1   Fluid Flow, Heat Transfer and Mass Transfer
Volume 2   Particle Technology and Separation Processes
Volume 3   Chemical and Biochemical Reactor Engineering and Control
Volume 4/5 Solutions to the Problems in Volumes 1, 2 and 3
Volume 6   Chemical Engineering Design

The details of this new arrangement are as follows:

1. Volume 1 has acquired an abbreviated treatment of non-Newtonian Flow, formerly in Volume 3.

2. Liquid Mixing appears as a new Chapter in Volume 1, which incorporates the relevant material formerly in Volumes 2 and 3.
3. Separate chapters now appear in Volume 1 on Compressible Flow and on Multiphase Flow, the latter absorbing material previously scattered between Volumes 1 and 2.
4. New chapters are added to Volume 2 to cover four separation processes of increasing importance — Adsorption (from Volume 3), Ion Exchange, Chromatographic Separations and Membrane Separations.
5. Volume 3 is now devoted to various aspects of Reaction Engineering and Control, material which is considerably expanded.
6. Some aspects of design, previously in the earlier volumes, are now transferred to a more appropriate home in Volume 6.

As far as Volume 1 is concerned, the opportunity has been taken to update existing material. The major changes in Fluid Flow include the incorporation of non-Newtonian Flow, an extensive revision of Compressible Flow and the new chapters on Multiphase Flow and Liquid Mixing. Material for this last chapter has been contributed by Dr R P Chhabra of the Indian Institute of Technology at Kanpur. There has also been a substantial revision of the presentation of material on Mass Transfer and Momentum, Heat and Mass Transfer. To the Appendix have been added the Tables of Laplace Transform and Error Functions which were formerly in Volume 3, and throughout this new edition, all the diagrams have been redrawn. Some further problems have been added at the end.

Sadly, John Coulson was not able to contribute as he had done previously and his death in January 1990 leaves us with a gap which is difficult to fill. John Backhurst and John Harker, who made a substantial contribution to the preparation of the Third Edition in 1977, have taken an increased share of the burden of revising the book and contributing new material, and have taken a special responsibility for those sections which originated from John Coulson, in addition to the special task of up-dating the illustrations. Without their continued support and willing co-operation there would have been no Fourth Edition.

Finally, we would all like to thank our many readers who have made such helpful suggestions in the past and have pointed out errors, many of which the authors would never have spotted. It is hoped that readers will continue to act in this way as unseen authors.

June 1990   JFR

NOTE TO REVISED REPRINT 1993

The reprint incorporates corrections and minor amendments, but the opportunity has been taken to effect some re-arrangements and additions, as follows:

Chapter 5 Multiphase Flow
   Revision of section on pneumatic conveying.

Chapter 9 Heat Transfer
   (i) Re-arrangement of material on plasmas and transfer to particles.
   (ii) Inclusion of an Example on lagging.

Chapter 10 Mass Transfer
   (i) Re-arrangement at the beginning to give a more logical sequence.
   (ii) Revision of final section on practical results.

Chapter 11 Boundary Layer
   Addition of section on flow with constant heat flux at surface.

Steam Tables
   These have been recast to facilitate interpolation.

Preface to Third Edition

THE introduction of the SI system of units by the United Kingdom and many other countries has itself necessitated the revision of this engineering text. This clear implementation of a single system of units will be welcomed not only by those already in the engineering profession, but even more so by those who are about to join. The system which is based on the c.g.s. and m.k.s. systems using length (L), mass (M), and time (T) as the three basic dimensions, as is the practice in the physical sciences, has the very great advantage that it removes any possible confusion between mass and force which arises in the engineering system from the common use of the term *pound* for both quantities. We have therefore presented the text, problems, and examples in the SI system, but have arranged the tables of physical data in the Appendix to include both SI and other systems wherever possible. This we regard as important because so many of the physical data have been published in c.g.s. units. For similar reasons, engineering units have been retained as an alternative where appropriate.

In addition to the change to the SI system of units, we have taken the opportunity to update and to clarify the text. A new section on the flow of two-phase gas–liquid mixtures has been added to reflect the increased interest in the gas and petroleum industries and in its application to the boiling of liquids in vertical tubes.

The chapter on Mass Transfer, the subject which is so central and specific to chemical engineering, has been considerably extended and modernised. Here we have thought it important in presenting some of the theoretical work to stress its tentative nature and to show that, although some of the theories may often lack a full scientific basis, they provide the basis of a workable technique for solving problems. In the discussion on Fluid Flow reference has been made to American methods, and the emphasis on Flow Measurement has been slanted more to the use of instruments as part of a control system. We have emphasised the importance of pipe-flow networks which represent a substantial cost item in modern large-scale enterprises.

This text covers the physical basis of the three major transfer operations of fluid flow, heat transfer, and mass transfer. We feel that it is necessary to provide a thorough grounding in these operations before introducing techniques which have been developed to give workable solutions in the most convenient manner for practical application. At the same time, we have directed the attention of the reader to such invaluable design codes as TEMA and the British Standards for heat exchanger design and to other manuals for pipe-flow systems.

It is important for designers always to have in their minds the need for reliability and safety: this is likely to follow from an understanding of the basic principles involved, many of which are brought out in the text.

We would like to thank our many friends from several countries who have written with suggestions, and it is our hope that this edition will help in furthering growth and interest in the profession. We should also like to thank a number of industrialists who have made available much useful information for incorporation in this edition; this help is acknowledged at the appropriate point. Our particular thanks are due to Dr B. Waldie for his contribution to the high temperature aspects of heat transfer and to the Kellogg International Corporation and Humphreys and Glasgow Limited for their help. In conclusion, we would like to thank Dr J. R. Backhurst and Dr J. H. Harker for their editorial work and for recalculating the problems in SI units and converting the charts and tables.

Since the publication of the Second Edition of this Volume, Volume 3 of *Chemical Engineering* has been published in order to give a more complete coverage of those areas of chemical engineering which are of importance in both universities and industry in the 1970's.

January 1976   JMC
                                                                                                 JFR

Preface to Second Edition

IN presenting this second edition, we should like to thank our many friends from various parts of the world who have so kindly made suggestions for clarifying parts of the text and for additions which they have felt to be important. During the last eight years there have been changes in the general approach to chemical engineering in the universities with a shift in emphasis towards the physical mechanisms of transport processes and with a greater interest in unsteady state conditions. We have taken this opportunity to strengthen those sections dealing with the mechanisms of processes, particularly in Chapter 7 on mass transfer and in the chapters on fluid mechanics where we have laid greater emphasis on the use of momentum exchange. Many chemical engineers are primarily concerned with the practical design of plant and we have tried to include a little more material of use in this field in Chapter 6 on heat transfer. An introductory section on dimensional analysis has been added but it has been possible to do no more than outline the possibilities opened up by the use of this technique. Small changes will be found throughout the text and we have tried to meet many readers' requests by adding some more worked examples and a further selection of problems for the student. The selection of material and its arrangement are becoming more difficult and must be to a great extent a matter of personal choice, but we hope that this new edition will provide a sound basis for the study of the fundamentals of the subject and will perhaps be of some value to practising engineers.

J. M. COULSON
J. F. RICHARDSON

Preface to First Edition

THE idea of treating the various processes of the chemical industry as a series of unit operations was first brought out as a basis for a new technology by Walker, Lewis and McAdams in their book in 1923. Before this, the engineering of chemical plants had been regarded as individual to an industry and there was little common ground between one industry and another. Since the early 1920's chemical engineering as a separate subject has been introduced into the universities of both America and England and has expanded considerably in recent years so that there are now a number of university courses in both countries. During the past twenty years the subject matter has been extensively increased by various researches described in a number of technical journals to which frequent reference is made in the present work.

Despite the increased attention given to the subject there are few general books, although there have been a number of specialised books on certain sections such as distillation, heat transfer, etc. It is the purpose of the present work to present to the student an account of the fundamentals of the subject. The physical basis of the mechanisms of many of the chemical engineering operations forms a major feature of chemical engineering technology. Before tackling the individual operations it is important to stress the general mechanisms which are found in so many of the operations. We have therefore divided the subject matter into two volumes, the first of which contains an account of these fundamentals — diffusion, fluid flow and heat transfer. In Volume 2 we shall show how these theoretical foundations are applied in the design of individual units such as distillation columns, filters, crystallisers, evaporators, etc.

Volume 1 is divided into four sections, fluid flow, heat transfer, mass transfer and humidification. Since the chemical engineer must handle fluids of all kinds, including compressible gases at high pressures, we believe that it is a good plan to consider the problem from a thermodynamic aspect and to derive general equations for flow which can be used in a wide range of circumstances. We have paid special attention to showing how the boundary layer is developed over plane surfaces and in pipes, since it is so important in controlling heat and mass transfer. At the same time we have included a chapter on pumping since chemical engineering is an essentially practical subject, and the normal engineering texts do not cover the problem as experienced in the chemical and petroleum industries.

The chapter on heat transfer contains an account of the generally accepted techniques for calculation of film transfer coefficients for a wide range of conditions, and includes a section on the general construction of tubular exchangers which form a major feature of many works. The possibilities of the newer plate type units are indicated.

In section three, the chapter on mass transfer introduces the mechanism of diffusion and this is followed by an account of the common relationships between heat, mass and momentum transfer and the elementary boundary layer theory. The final section includes the practical problem of humidification where both heat and mass transfer are taking place simultaneously.

It will be seen that in all chapters there are sections in small print. In a subject such as this, which ranges from very theoretical and idealised systems to the practical problems with empirical or experimentally determined relations, there is much to be said for omitting the more theoretical features in a first reading, and in fact this is frequently done in the more practical courses. For this reason the more difficult theoretical sections have been put in small print and the whole of Chapter 9 may be omitted by those who are more concerned with the practical utility of the subject.

In many of the derivations we have given the mathematical analysis in more detail than is customary. It is our experience that the mathematical treatment should be given in full and that the student should then apply similar analysis to a variety of problems.

We have introduced into each chapter a number of worked examples which we believe are essential to a proper understanding of the methods of treatment given in the text. It is very desirable for a student to understand a worked example before tackling fresh practical problems himself. Chemical engineering problems require a numerical answer and it is essential to become familiar with the different techniques so that the answer is obtained by systematic methods rather than by intuition.

In preparing this text we have been guided by courses of lectures which we have given over a period of years and have presented an account of the subject with the major emphasis on the theoretical side. With a subject that has grown so rapidly, and which extends from the physical sciences to practical techniques, the choice of material must be a matter of personal selection. It is, however, more important to give the principles than the practice, which is best acquired in the factory. We hope that the text may also prove useful to those in industry who, whilst perhaps successfully employing empirical relationships, feel that they would like to find the extent to which the fundamentals are of help.

We should like to take this opportunity of thanking a number of friends who have helped by their criticism and suggestions, amongst whom we are particularly indebted to Mr F. E. Warner, to Dr M. Guter, to Dr D. J. Rasbash and to Dr L. L. Katan. We are also indebted to a number of companies who have kindly permitted us to use illustrations of their equipment. We have given a number of references to technical journals and we are grateful to the publishers for permission to use illustrations from their works. In particular we would thank the Institution of Chemical Engineers, the American Institute of Chemical Engineers, the American Chemical Society, the Oxford University Press and the McGraw-Hill Book Company.

South Kensington,
London S.W.7
1953

Acknowledgements

THE authors and publishers acknowledge the kind assistance of the following organizations in providing illustrative material.

Figure 6.6*a* Gauges Bourdon (GB) Ltd, Guildford, Surrey.
Figures 6.7, 6.6*b* Foxboro Great Britain Ltd, Redhill, Surrey.
Figures 6.17, 6.26 Kent Meters Ltd.
Figure 6.18 KDG Mobrey Ltd, Burgess Hill, Sussex.
Figure 6.24 Baird and Tatlock Ltd.
Figures 7.26, 7.27, 7.28 Chemineer Ltd, Derby.
Figure 7.29 Sulzer Ltd, Switzerland.
Figure 8.1 Worthington-Simpson Ltd.
Figure 8.3 ECD Co Ltd.
Figure 8.5 Ibex Engineering Company Ltd, Hastings, Sussex.
Figure 8.6 Watson-Marlow Ltd, Falmouth, Cornwall.
Figures 8.7, 8.8 Mono Pumps Ltd, Manchester.
Figure 8.13 Weir Pumps Ltd, Glasgow.
Figure 8.22 Crane Packing Ltd, Wolverhampton, W. Midlands.
Figure 8.24 International Combustion Ltd, Derby.
Figures 8.27, 8.28, 8.30, 8.31 Reavell and Co Ltd.
Figure 8.29 Nash Engineering Co (GB) Ltd, Manchester.
Figures 8.36–8.40 Hick Hargreaves and Co Ltd.
Figure 9.62 Brown Fintube Co.
Figure 9.65 Alfa Laval Ltd. Brentford, Middlesex.
Figures 11.2, 11.3 Professor F. N. M. Brown, University of Notre Dame.
Figure 13.10 Casella London Ltd, Bedford.
Figures 13.12, 13.13 Visco Ltd, Croydon, Surrey.

CHAPTER 1

Units and Dimensions

1.1. INTRODUCTION

At an early stage students of chemical engineering will discover that the data which they use are expressed in a great variety of different units, so that they must convert quantities into a common system before proceeding with calculations. An attempt at standardisation has been made with the introduction of the *Système International d'Unités* (SI), discussed later, which is the main system used in this edition. This system is widely used on the continent of Europe and is now accepted in the United Kingdom and is likely to spread internationally. Most of the physical properties determined in the laboratory will have been expressed in the c.g.s. system, whereas the dimensions of the full-scale plant, its throughput, design, and operating characteristics will have appeared either in some form of general engineering units or in special units which have their origin in the history of the particular industry. This inconsistency is quite unavoidable and is a reflection of the fact that chemical engineering has in many cases developed as a synthesis of scientific knowledge and practical experience. Familiarity with the various systems of units and an ability to convert from one to another are therefore essential. In this chapter the main systems of units will be discussed and the importance of understanding dimensions emphasised. It will then be shown how dimensions can be used to help very considerably in the formulation of relationships between large numbers of parameters.

The magnitude of any physical quantity is expressed as the product of two quantities; one is the magnitude of the unit and the other is the number of those units. Thus the distance between two points may be expressed as 1 m or as 100 cm or as 3.28 ft. The metre, centimetre, and foot are respectively the size of the units, and 1, 100, and 3.28 are the corresponding numbers of units.

Since the physical properties of a system are interconnected by a series of mechanical and physical laws, it is possible to regard certain quantities as basic and other quantities as derived. The choice of basic dimensions varies from one system to another but it is usual to take length and time as fundamental. These quantities will be denoted by $\mathbf{L}$ and $\mathbf{T}$. The dimensions of velocity, which is a rate of increase of distance with time, can be written as $\mathbf{LT}^{-1}$, and those of acceleration, the rate of increase of velocity, are $\mathbf{LT}^{-2}$. An area has dimensions $\mathbf{L}^2$ and a volume has the dimensions $\mathbf{L}^3$.

The volume of a body does not completely define the amount of material which it contains, and therefore it is usual to define a third basic quantity, the amount of matter in the body, that is its mass $\mathbf{M}$. Thus the density of the material, its mass per unit volume, has the dimensions $\mathbf{ML}^{-3}$. Alternatively, the third basic quantity which is chosen may be force $\mathbf{F}$; in some cases, both force and mass can be used simultaneously as basic quantities as will be shown later.

Physical and mechanical laws provide a further set of relations between dimensions. The most important of these is that the force required to produce a given acceleration of a body is proportional to its mass and, similarly, the acceleration imparted to a body is proportional to the applied force.

Thus force is proportional to the product of mass and acceleration,

or:
$$\mathbf{F} = \text{const } \mathbf{M}(\mathbf{LT}^{-2})$$

The proportionality constant therefore has the dimensions:

$$\frac{\mathbf{F}}{\mathbf{M}(\mathbf{LT}^{-2})} = \mathbf{FM}^{-1}\mathbf{L}^{-1}\mathbf{T}^2$$

In any set of consistent units it is convenient to put the proportionality constant arbitrarily equal to unity and to define unit force as that which will impart unit acceleration to unit mass. Provided that no other relationship between force and mass is used, the constant may be arbitrarily regarded as dimensionless and the dimensional relationship:

$$\mathbf{F} = \mathbf{MLT}^{-2}$$

is obtained.

If, however, some other physical law were to be introduced so that, for instance, the attractive force between two bodies is proportional to the product of their masses, then this relation between **F** and **M** will no longer hold.

1.2. SYSTEMS OF UNITS

Although in scientific work mass is taken as the third fundamental quantity and in engineering force is sometimes used as mentioned above, four fundamental quantities **L**, **M**, **F**, **T** may be used.

The various systems of units and the basic quantities associated with them will now be considered. A summary is given in Table 1.1.

1.2.1. The centimetre–gram–second (c.g.s.) system

In this system the basic units are of length **L**, mass **M**, and time **T** with the nomenclature:

Length:	Dimension **L**:	Unit 1 centimetre	(1 cm)
Mass:	Dimension **M**:	Unit 1 gram	(1 g)
Time:	Dimension **T**:	Unit 1 second	(1 s)

The unit of force is that which will give a mass of 1 g an acceleration of 1 cm/s^2 and is known as the dyne:

Force: Dimension $\mathbf{F} = \mathbf{MLT}^{-2}$:   Unit 1 dyne (1 dyn)

Table 1.1.
Units

Quantity	c.g.s.	SI	f.p.s.	Engineering system	Dimensions in **M, L, T, θ**	Dimensions **F, L, T, θ**	Dimensions in **F, M, L, T, θ**
Mass	gram	kilogram	pound	slug	**M**	**FL⁻¹T²**	**M**
Length	centimetre	metre	foot	foot	**L**	**L**	**L**
Time	second	second	second	second	**T**	**T**	**T**
Force	dyne	newton	poundal	pound force	**MLT⁻²**	**F**	**F**
Energy	erg (= 10⁻⁷ joules)	joule	foot-poundal	foot-pound	**ML²T⁻²**	**FL**	**FL**
Pressure	dyne/square centimetre	newton/sq metre	poundal/square foot	pound force/square foot	**ML⁻¹T⁻²**	**FL⁻²**	**FL⁻²**
Power	erg/second	watt	foot-poundal/second	foot-pound/second	**ML²T⁻³**	**FLT⁻¹**	**FLT⁻¹**
Entropy per unit mass	erg/gram °C	joule/kilogram K	foot-poundal/pound °C	foot-pound/slug °F	**L²T⁻²θ⁻¹**	**L²T⁻²θ⁻¹**	**FM⁻¹Lθ⁻¹**
Universal gas constant	8.314 × 10⁷ erg/mole °C	8314 J/kmol K	8.94 ft-poundal/lb mol °C	4.96 × 10⁴ foot-pound/ slug mol °F	**MN⁻¹L²T⁻²θ⁻¹**	**MN⁻¹L²T⁻²θ⁻¹**	**FN⁻¹L θ⁻¹**

Heat units

Quantity	c.g.s.	SI	British/American engineering system	Dimensions in **M, L, T, θ**	Dimensions in **H, M, L, T, θ**
Temperature	degree centigrade	degree Kelvin	degree Fahrenheit	**θ**	**θ**
Thermal energy or heat	calorie	joule	British thermal unit (Btu)	**Mθ**	**H**
Entropy per unit mass, specific heat	calorie/gram-°C	joule/kilogram K	Btu/pound °F	—	**HM⁻¹θ⁻¹**
Mechanical equivalent of heat, *J*	4.18 × 10⁷ erg/gram-°C	1 J (heat energy) = 1 J (mechanical energy)	2.50 × 10⁴ foot-poundal/pound °F	**L²T⁻²θ⁻¹**	**H⁻¹ML²T⁻²**
Universal Gas Constant **R**	1.986 calorie/mole °C	8314 J/kmol K	1.986 Btu/lb-mol °F	**MN⁻¹L²T⁻²θ⁻¹**	**HN⁻¹θ⁻¹**

1.2.2. The système international d'unités (SI)

This system is a modification of the c.g.s. system but uses larger units. The basic dimensions are again of **L**, **M**, and **T**, but their values are different.

$$\begin{array}{llll}
\text{Length:} & \text{Dimension } \mathbf{L}: & \text{Unit 1 metre} & (1 \text{ m}) \\
\text{Mass:} & \text{Dimension } \mathbf{M}: & \text{Unit 1 kilogram} & (1 \text{ kg}) \\
\text{Time:} & \text{Dimension } \mathbf{T}: & \text{Unit 1 second} & (1 \text{ s})
\end{array}$$

The unit of force, known as the Newton, is that force which will give an acceleration of 1 m/s^2 to a mass of one kilogram. Thus $1 \text{ N} = 1 \text{ kg m/s}^2$ with dimensions $\mathbf{MLT}^{-2}$, and it will be seen that one Newton equals 10^5 dynes. The energy unit, the Newton metre, is 10^7 ergs and is called the Joule; and the power unit, equal to one Joule per second, is the Watt.

$$\begin{array}{llll}
\text{Thus:} & \text{Force:} & \text{Dimension } \mathbf{MLT}^{-2}: & \text{Unit 1 Newton (1 N)} & \text{or } 1 \text{ kg m/s}^2 \\
& \text{Energy:} & \text{Dimension } \mathbf{ML}^2\mathbf{T}^{-2}: & \text{Unit 1 Joule (1 J)} & \text{or } 1 \text{ kg m}^2/\text{s}^2 \\
& \text{Power:} & \text{Dimension } \mathbf{ML}^2\mathbf{T}^{-3}: & \text{Unit 1 Watt (1 W)} & \text{or } 1 \text{ kg m}^2/\text{s}^3
\end{array}$$

For many purposes, the chosen unit in the SI system will be either too large or too small for practical purposes. The following prefixes are adopted as standard. Multiples or sub-multiples in powers of 10^3 are preferred; thus, for instance, millimetre should always be used in preference to centimetre.

10^{18}	exa	(E)	10^{-1}	deci	(d)
10^{15}	peta	(P)	10^{-2}	centi	(c)
10^{12}	tera	(T)	10^{-3}	milli	(m)
10^{9}	giga	(G)	10^{-6}	micro	(μ)
10^{6}	mega	(M)	10^{-9}	nano	(n)
10^{3}	kilo	(k)	10^{-12}	pico	(p)
10^{2}	hecto	(h)	10^{-15}	femto	(f)
10^{1}	deca	(da)	10^{-18}	alto	(a)

The prefixes should be used with great care and be written immediately adjacent to the unit to be qualified; furthermore only one prefix should be used at a time to precede a given unit. Thus, for example, 10^{-3} metre, which is one millimetre, is written 1 mm. 10^3 kg is written as 1 Mg, not as 1 kkg — this shows immediately that the name *kilogram* is an unsuitable one for the basic unit of mass and a new name may well be given to it in the near future.

Some special terms are acceptable, and commonly used in the SI system. A mass of 10^3 kg (1 Mg) is called a *tonne* (t). A pressure of 100 kN/m^2 is called a *bar*.

1.2.3. The foot–pound–second (f.p.s.) system

The basic units in this system are as follows:

$$\begin{array}{llll}
\text{Length:} & \text{Dimension } \mathbf{L}: & \text{Unit 1 foot} & (1 \text{ ft}) \\
\text{Mass:} & \text{Dimension } \mathbf{M}: & \text{Unit 1 pound} & (1 \text{ lb}) \\
\text{Time:} & \text{Dimension } \mathbf{T}: & \text{Unit 1 second} & (1 \text{ s})
\end{array}$$

The unit of force is that which will give a mass of 1 lb an acceleration of 1 ft/s^2; it is known as the poundal (pdl).

In an alternative form of this system (*Engineering system*) the units of length (ft) and time (s) are unchanged, but the third fundamental is a unit of force ($\mathbf{F}$) instead of mass and is known as the pound force (lb$_f$). It is defined as the force which gives a mass of 1 lb an acceleration of 32.1740 ft/s^2 (the "standard" value of the acceleration due to gravity). It is therefore a fixed quantity and must not be confused with the pound weight which is the force exerted by the earth's gravitational field on a mass of one pound and which varies from place to place as g varies.

The unit of mass in this system is known as the slug, and is the mass which is given an acceleration of 1 ft/s^2 by a one pound force:

$$1 \text{ slug} = 1 \text{ lb}_f \text{ ft}^{-1}\text{s}^2$$

Misunderstanding often arises from the fact that the unit of mass in the one system has the same name as the unit of force in the other, that is the pound. To avoid confusion the pound mass should be written as lb or even lb$_m$ and the unit of force always as lb$_f$.

It will be noted that:

$$1 \text{ slug} = 32.1740 \text{ lb mass} \quad \text{and} \quad 1 \text{ lb}_f = 32.1740 \text{ pdl}$$

Many writers, particularly in America, use both the pound mass and pound force as basic units in the same equation because they are the units in common use. This is an essentially incoherent system and requires great care in use. In this system a proportionality factor between force and mass is defined as g_c given by:

$$\text{Force (in pounds force)} = \text{(mass in pounds)} \text{(acceleration in ft/s}^2)/g_c$$

Thus in terms of dimensions: $\quad \mathbf{F} = (\mathbf{M})(\mathbf{LT}^{-2})/g_c$

g_c has the dimensions $\mathbf{F}^{-1}\mathbf{MLT}^{-2}$ and therefore has a value of 32.1740 lb ft s^{-2}/lb$_f$. Similarly the units of gram force (g$_f$) and kilogram force (kg$_f$) are sometimes used. They are, respectively, the gravitational force acting on one gram mass or one kilogram mass at the "standard" value of the acceleration due to gravity (980.665 cm/s^2 or 9.80665 m/s^2). g_c then has values of 980.665 g cm s^{-2}/g$_f$ and 9.80665 kg m s^{-2}/kg$_f$. Using the relation $\mathbf{F} \equiv \mathbf{MLT}^{-2}$, g_c is dimensionless and numerically equal to the acceleration due to gravity expressed in the appropriate units.

1.2.4. Derived units

The three fundamental units of the SI and of the c.g.s. systems are length, mass, and time. It has been shown that force can be regarded as having the dimensions of $\mathbf{MLT}^{-2}$; the dimensions of many other parameters may be worked out in terms of the basic $\mathbf{MLT}$ system.

For example:

Energy is given by the product of force and distance with dimensions $\mathbf{ML}^2\mathbf{T}^{-2}$.
Pressure is the force per unit area with dimensions $\mathbf{ML}^{-1}\mathbf{T}^{-2}$.

Viscosity is defined as the shear stress per unit velocity gradient with dimensions $(\mathbf{MLT^{-2}/L^2})/(\mathbf{LT^{-1}/L}) = \mathbf{ML^{-1}T^{-1}}$.

Kinematic viscosity is the viscosity divided by the density with dimensions $\mathbf{ML^{-1}T^{-1}/ML^{-3}} = \mathbf{L^2T^{-1}}$.

The units, dimensions, and normal form of expression for these quantities in the SI system may now be stated.

Parameter	Unit	Dimensions	In terms of kg, m, s
Force	Newton	$\mathbf{MLT^{-2}}$	1 kg m/s^2
Energy or work	Joule	$\mathbf{ML^2T^{-2}}$	1 kg m^2/s^2 ($= 1$ N m)
Power	Watt	$\mathbf{ML^2T^{-3}}$	1 kg m^2/s^3 ($= 1$ J/s)
Pressure	Pascal	$\mathbf{ML^{-1}T^{-2}}$	1 kg/m s^2 ($= 1$ N/m^2)
Viscosity	Pascal-second	$\mathbf{ML^{-1}T^{-1}}$	1 kg/m s ($= 1$ N s/m^2)
Frequency	Hertz	$\mathbf{T^{-1}}$	1 s^{-1}

1.2.5. Thermal units

Since heat is a form of energy, it can be expressed in units with dimensions $\mathbf{ML^2T^{-2}}$, but it is not always convenient to express it in this way. If the concept of temperature is introduced with dimension θ then the thermal energy $\mathbf{H}$ of a body is conveniently expressed as $(\mathbf{H}) \propto$ mass $(\mathbf{M}) \times$ temperature (θ). The proportionality constant is the specific heat capacity which varies from material to material unlike the proportionality constant in Newton's law of motion. It is therefore necessary to define heat quantities in terms of a particular material, usually water at 15°C (298 K). If this is done using thermal units for heat energy the proportionality constant can be put equal to unity for water and, dimensionally:

$$H = M\theta$$

In the SI system heat is always expressed in Joules, in the same way as other forms of energy. In the c.g.s. system the unit of heat (the calorie) is defined as that quantity which will raise the temperature of one gram of water through 1°C at atmospheric pressure.

In the f.p.s. system the British thermal unit (Btu) is the quantity of heat required to raise one pound of water through 1°F (60–61°F) and the pound calorie (lb cal) or centigrade heat unit is that quantity required to raise it through 1°C. The Btu is too small a unit for practical purposes and the therm (10^5 Btu) is the usual practical unit, for example for the sale of gas for heating purposes.

The calorie and kilocalorie have now been redefined in terms of the Joule as exactly 4.1868 J and 4186.8 J respectively.

Expressing thermal energy in the same units as mechanical energy, specific heat has the dimensions $\mathbf{ML^2T^{-2}/M\theta} = \mathbf{L^2T^{-2}\theta^{-1}}$ and the specific heat of water becomes equal to the mechanical equivalent of heat (see below). Thus according to the system adopted, the specific heat of water will have the following values:

c.g.s. system	1 cal/g °C
Engineering units	1 Btu/lb °F
SI system	4186.8 J/kg K

It is necessary to be able to relate mechanical and thermal energy and this is conveniently done by introducing the term J as the mechanical equivalent of heat:

$$J = \frac{\text{Mechanical energy}}{\text{Thermal energy}} = \frac{\mathbf{LF}}{\mathbf{H}} = \frac{(\mathbf{L})(\mathbf{MLT}^{-2})}{\mathbf{M\theta}}$$

$$= \mathbf{L^2T^{-2}\theta^{-1}}$$

Thus where mechanical and thermal energies appear in the same equation, care must be taken to ensure that both are expressed in the same units. In the SI system, J is unity since all forms of energy are expressed in Joules.

1.2.6. Molar units

When working with ideal gases and systems in which a chemical reaction is taking place, it is usual to work in terms of *molar* units rather than mass. The *mole (mol)* is defined in the SI system as the quantity of material which contains as many entities (e.g. atoms, molecules or formula units) as there are in 12 g of carbon 12. It is more convenient, however, to work in terms of the *kilomole (kmol)* which relates to 12 kg of carbon 12, and the kilomole will be used exclusively in this book. The number of molar units will be denoted by dimensional symbol $\mathbf{N}$. The number of kilomoles of a substance $\mathbf{A}$ is obtained by dividing its mass in kilograms ($\mathbf{M}$) by its *molecular weight* M_A. M_A thus has the dimensions $\mathbf{MN}^{-1}$. The Royal Society recommends the use of the term *relative molecular mass* in place of *molecular weight*, but *molecular weight* will be used here because of its general adoption by chemical engineers.

1.2.7. Electrical units

Electrical current (I) has been chosen as the basic SI unit in terms of which all other electrical quantities are defined. Unit current, the *ampere* (A) (or amp), is defined in terms of the force exerted between two parallel conductors in which a current of 1 amp is flowing. Since the unit of power, the watt, is the product of current and potential difference, the *volt* (V) is defined as watts per amp and therefore has dimensions of $\mathbf{ML^2T^{-3}I^{-1}}$. From Ohm's law the unit of resistance, the *ohm*, is given by the ratio volts/amps and therefore has dimensions of $\mathbf{ML^2T^{-3}I^{-2}}$. A similar procedure can be followed for the evaluation of the dimensions of other electrical units.

1.3. CONVERSION OF UNITS

Conversion of units from one system to another is simply carried out if the quantities are expressed in terms of the fundamental units of mass, length, time, temperature. The conversion factors for the British and metric systems are, then:

$$\text{Mass} \quad 1 \text{ lb} = \frac{1}{32.2} \text{ slug} = 453.6 \text{ g} = 0.4536 \text{ kg}$$

$$\text{Length} \quad 1 \text{ ft} = 30.48 \text{ cm} = 0.3048 \text{ m}$$

$$\text{Time} \quad 1 \text{ s} = \frac{1}{3600} \text{ h}$$

Temperature $1°F = \left(\dfrac{1}{1.8}\right)°C = \left(\dfrac{1}{1.8}\right) K$
difference

Force   1 pound force = 32.2 poundal = 4.44×10^5 dyne = 4.44 N

If a viscosity is to be converted from poise (gram per centimetre-second) to a viscosity in British or in SI units, the procedure is as follows in Example 1.1.

Example 1.1

Convert 1 poise to British engineering units and SI units.

Solution

$$1 \text{ poise} = 1 \text{ g/cm s} = \frac{1 \text{ g}}{1 \text{ cm} \times 1 \text{ s}}$$

$$= \frac{(1/453.6) \text{ lb}}{(1/30.48) \text{ ft} \times 1 \text{ s}}$$

$$= 0.0672 \text{ lb/ft s}$$

$$= \underline{\underline{242 \text{ lb/ft h}}}$$

$$1 \text{ poise} \qquad = 1 \text{ g/cm s} = \frac{1 \text{ g}}{1 \text{ cm} \times 1 \text{ s}}$$

$$= \frac{(1/1000) \text{ kg}}{(1/100) \text{ m} \times 1 \text{ s}}$$

$$= 0.1 \text{ kg/m s}$$

$$= \underline{\underline{0.1 \text{ N s/m}^2}} \, [(\text{kg m/s}^2)\text{s/m}^2]$$

The conversion of power in SI units to horse power (h.p.) is illustrated in Example 1.2.

Example 1.2

Convert 1 kW to h.p.

Solution

$$1 \text{ kW} = 10^3 \text{ W} = 10^3 \text{ J/s}$$

$$= 10^3 \times \frac{1 \text{ kg} \times 1 \text{ m}^2}{1 \text{ s}^3}$$

$$= \frac{10^3 \times (1/0.4536) \text{ lb} \times (1/0.3048)^2 \text{ ft}^2}{1 \text{ s}^3}$$

$$= 23{,}730 \text{ lb ft}^2/\text{s}^3$$

$$= \frac{23{,}730}{32.2} = 737 \text{ slug ft}^2/\text{s}^3$$

$$= 737 \text{ lb}_f \text{ ft/s}$$

$$= \frac{737}{550} = \underline{\underline{1.34 \text{ h.p.}}}$$

or: $1 \text{ h.p.} = 0.746 \text{ kW.}$

Table 1.2 provides factors to enable conversions to be made to SI units.

Table 1.2.   Factors for conversion to SI units

mass		pressure	
1 lb	0.454 kg	1 lb_f/in^2	6.895 kN/m^2
1 ton	1016 kg	1 standard atm	101.3 kN/m^2
		1 bar	100 kN/m^2
length		1 ft water	2.99 kN/m^2
1 in	25.4 mm	1 in water	249 N/m^2
1 ft	0.305 m	1 in Hg	3.39 kN/m^2
1 mile	1.609 km	1 mm Hg	133 N/m^2
time		*viscosity*	
1 min	60 s	1 P	0.1 N s/m^2
1 h	3.6 ks	1 lb/ft h	0.414 mN s/m^2
1 day	86.4 ks	1 stoke	10^{-4} m^2/s
1 year	31.5 Ms	1 ft^2/h	0.258 cm^2/s
area		*mass flow*	
1 in^2	645.2 mm^2	1 lb/h	0.126 g/s
1 ft^2	0.093 m^2	1 ton/h	0.282 kg/s
		1 lb/h ft^2	1.356 g/s m^2
volume		*thermal*	
1 in^3	16,387.1 mm^3	1 Btu/h ft^2	3.155 W/m^2
1 ft^3	0.0283 m^3	1 Btu/h ft^2 °F	5.678 W/m^2 K
1 Imperial gal	4546 cm^3	1 Btu/lb	2.326 kJ/kg
1 US gal	3786 cm^3	1 Btu/lb °F	4.187 kJ/kg K
1 US gal	$\frac{5}{6}$ Imperial gal	1 Btu/h ft °F	1.731 W/m K
force		*energy*	
1 pdl	0.138 N	1 kWh	3.6 MJ
1 lb_f	4.45 N	1 therm	105.5 MJ
1 dyne	10^{-5} N		
energy			
1 ft lb_f	1.36 J	*calorific value*	
1 cal	4.187 J	1 Btu/ft^3	37.26 kJ/m^3
1 erg	10^{-7} J	1 Btu/lb	2.326 kJ/kg
1 Btu	1.055 kJ		
1 therm	1.055×10^5 kJ	*density*	
		1 lb/ft^3	16.02 kg/m^3
power			
1 h.p.	746 W		
1 Btu/h	0.293 W		

Conversion factors are also listed in Appendix A2.

1.4. DIMENSIONAL ANALYSIS

Dimensional analysis depends upon the fundamental principle that any equation or relation between variables must be *dimensionally consistent*; that is to say that each term in the relationship must have the same dimensions. Thus in the simple application of the principle, an equation may consist of a number of terms, each representing, and therefore having, the dimensions of length. It is not permissible to add, say, lengths and velocities in an algebraic equation because they are quantities of different characters. The corollary of this principle is that if the whole equation is divided through by any one of the terms, each remaining term in the equation must be dimensionless. The use of these *dimensionless groups*, or *dimensionless numbers* as they are called, is of considerable value in developing relationships in chemical engineering.

The requirement of dimensional consistency places a number of constraints on the form of the functional relation between variables in a problem and forms the basis of the technique of *dimensional analysis* which enables the variables in a problem to be grouped into the form of dimensionless groups. Since the dimensions of the physical quantities can be expressed in terms of a number of fundamentals, usually mass, length, and time, and sometimes temperature and thermal energy in addition, the requirement of dimensional consistency must be satisfied in respect of each of the fundamentals. Dimensional analysis will give no information about the form of the functions, nor will it provide any means of evaluating numerical proportionality constants.

The study of problems in fluid dynamics and in heat transfer is made difficult by the many parameters which appear to affect them. In most instances further study shows that the variables may be grouped together in dimensionless groups, thus reducing the effective number of variables. It is rarely possible, and certainly time consuming, to try to vary these many variables separately, and the method of dimensional analysis by providing a smaller number of independent groups is most helpful to the experimenter.

The application of the principles of dimensional analysis may best be understood by considering the following illustration.

It is found, as a result of experiment, that the pressure difference (ΔP) between two ends of a pipe in which a fluid is flowing is a function of the following variables: pipe diameter d, pipe length l, fluid velocity u, fluid density ρ, and fluid viscosity μ. It will now be shown by dimensional analysis how these variables are related.

The relationship may be written as:

$$\Delta P = f_1(d, l, u, \rho, \mu) \tag{1.1}$$

The form of the function is unknown, but since any function can be expanded as a power series, the function can be regarded as the sum of a number of terms each consisting of products of powers of the variables. The simplest form of relation will be where the function consists simply of a single term, when:

$$\Delta P = \text{const } d^{n_1} l^{n_2} u^{n_3} \rho^{n_4} \mu^{n_5} \tag{1.2}$$

The requirement of dimensional consistency is that the combined term on the right-hand side will have the same dimensions as that on the left; that is, it must have the dimensions of pressure.

Each of the variables in equation 1.2 can be expressed in terms of mass, length, and time. Thus, dimensionally:

$$\Delta P \equiv \mathbf{ML^{-1}T^{-2}} \qquad u \equiv \mathbf{LT^{-1}}$$

$$d \equiv \mathbf{L} \qquad \rho \equiv \mathbf{ML^{-3}}$$

$$l \equiv \mathbf{L} \qquad \mu \equiv \mathbf{ML^{-1}T^{-1}}$$

and: $$\mathbf{ML^{-1}T^{-2}} \equiv \mathbf{L}^{n_1}\mathbf{L}^{n_2}(\mathbf{LT^{-1}})^{n_3}(\mathbf{ML^{-3}})^{n_4}(\mathbf{ML^{-1}T^{-1}})^{n_5}$$

The conditions of dimensional consistency must be met for each of the fundamentals of $\mathbf{M}$, $\mathbf{L}$, and $\mathbf{T}$ and the indices of each of these variables can be equated. Thus:

$$\mathbf{M} \qquad 1 = n_4 + n_5$$

$$\mathbf{L} \qquad -1 = n_1 + n_2 + n_3 - 3n_4 - n_5$$

$$\mathbf{T} \qquad -2 = -n_3 - n_5$$

Thus three equations and five unknowns result and the equations may be solved in terms of any two unknowns. Solving in terms of n_2 and n_5:

$$n_4 = 1 - n_5 \text{ (from the equation in } \mathbf{M})$$

$$n_3 = 2 - n_5 \text{ (from the equation in } \mathbf{T})$$

Substituting in the $\mathbf{L}$ equation:

$$-1 = n_1 + n_2 + (2 - n_5) - 3(1 - n_5) - n_5$$

i.e.: $$0 = n_1 + n_2 + n_5$$

$\therefore$ $$n_1 = -n_2 - n_5$$

Thus, substituting into equation 1.2:

$$\Delta P = \text{const } d^{-n_2 - n_5} l^{n_2} u^{2 - n_5} \rho^{1 - n_5} \mu^{n_5}$$

i.e.: $$\frac{\Delta P}{\rho u^2} = \text{const } \left(\frac{l}{d}\right)^{n_2} \left(\frac{\mu}{d u \rho}\right)^{n_5} \tag{1.3}$$

Since n_2 and n_5 are arbitrary constants, the above equation can only be satisfied if each of the terms $\Delta P / \rho u^2$, l/d, and $\mu / d u \rho$ is dimensionless. As a check, if the dimension of each group is evaluated, it will be found to be dimensionless.

The group $u d \rho / \mu$, known as the *Reynolds number*, is one which will frequently arise in the study of fluid flow and affords a criterion by which the type of flow in a given geometry may be characterised. Equation 1.3 involves the reciprocal of the Reynolds number, but can be rewritten as:

$$\frac{\Delta P}{\rho u^2} = \text{const } \left(\frac{l}{d}\right)^{n_2} \left(\frac{u d \rho}{\mu}\right)^{-n_5} \tag{1.4}$$

The right-hand side of equation 1.4 is a typical term in the function for $\Delta P / \rho u^2$. More generally:

$$\frac{\Delta P}{\rho u^2} = f_2 \left(\frac{l}{d}, \frac{u d \rho}{\mu}\right) \tag{1.5}$$

Comparing equations 1.1 and 1.5, it is seen that a relationship between six variables has been reduced to a relationship between three dimensionless groups. Later, this statement will be generalised to show that the number of dimensionless groups is normally the number of variables less the number of fundamentals.

A number of important points emerge from a consideration of the preceding example:

(1) If the index of a particular variable is found to be zero, this indicates that this variable is not of significance in the problem.
(2) If two of the fundamentals always appear in the same combination, e.g. if $\mathbf{L}$ and $\mathbf{T}$ always occur as powers of $\mathbf{LT}^{-1}$, then the same equation for the indices will

be obtained for both $\mathbf{L}$ and $\mathbf{T}$ and the number of effective fundamentals is thus reduced by one.

(3) The form of the final solution will depend upon the method of solution of the simultaneous equations. If the equations had been solved, say, in terms of n_3 and n_4 instead of n_2 and n_5, the resulting dimensionless groups would have been different. However, these new groups would simply have been products of powers of the original groups. Any number of fresh groups can be formed in this way.

Clearly the maximum degree of simplification of the problem is achieved by using the greatest possible number of fundamentals since each yields a simultaneous equation of its own. In certain problems, force may be used as a fundamental in addition to mass, length, and time, provided that at no stage in the problem is force defined in terms of mass and acceleration. In heat transfer problems, temperature is usually an additional fundamental, and heat can also be used as a fundamental provided it is not defined in terms of mass and temperature and provided that the equivalence of mechanical and thermal energy is not utilised. Considerable experience is needed in the proper use of dimensional analysis, and its application in a number of areas of fluid flow and heat transfer will be seen in the relevant chapters of this book.

The choice of physical variables to be included in the dimensional analysis must be based on an understanding of the nature of the phenomenon being studied. Nevertheless, on occasions there may be some doubt as to whether a particular quantity is relevant or not.

If a variable is included which does not exert a significant influence on the problem, the value of the dimensionless group in which it appears will have little effect on the final numerical solution of the problem, and therefore the exponent of that group must approach zero. This presupposes that the dimensionless groups are so constituted that the variable in question appears in only one of them. If an important variable is omitted, on the other hand, it will be found that there is no unique relationship between the dimensionless groups.

Chemical engineering analysis requires the formulation of relationships which will apply over a wide range of size of the individual items of a plant. This problem of scale up is vital and can be much helped by dimensional analysis.

Since linear size is included among the variables, the influence of scale, which may be regarded as the influence of linear size without change of shape or other variables, has been introduced. Thus in the viscous flow past an object, a change in linear dimension $\mathbf{L}$ will alter the Reynolds number and therefore the flow pattern around the solid, though if the change in scale is accompanied by a change in any other variable in such a way that the Reynolds number remains unchanged, then the flow pattern around the solid will not be altered. This ability to change scale but still maintain a design relationship is one of the many attractions of dimensional analysis.

It should be noted that it is permissible to take a function, such as a logarithm, only of a dimensionless quantity. Where this principle appears to have been invalidated, it is generally because the equation includes a further term, such as an integration constant, which will restore the requirements of dimensional consistency when combined with the (logarithmic) term. It is easy to appreciate this argument when account is taken of the fact that any function (such as a logarithm) can be expanded as a power series, each term of which must have the same dimensions.

1.5. BUCKINGHAM'S Π THEOREM

The need for dimensional consistency imposes a restraint in respect of each of the fundamentals involved in the dimensions of the variables. This was apparent during the previous discussion in which a series of simultaneous equations was solved, one equation for each of the fundamentals. A generalisation of this statement is provided in Buckingham's Π theorem which states that the number of dimensionless groups is equal to the number of variables minus the number of fundamental dimensions (but see below). In mathematical terms, this can be expressed as follows:

If there are n variables, $Q_1, Q_2, \ldots, Q_n$, the functional relationship between them may be written:

$$f_3(Q_1, Q_2, \ldots, Q_n) = 0 \qquad (1.6)$$

If there are m fundamental dimensions, there will be $(n - m)$ dimensionless groups $(\Pi_1, \Pi_2, \ldots, \Pi_{n-m})$ and the functional relationship between them may be written:

$$f_4(\Pi_1, \Pi_2, \ldots, \Pi_{n-m}) = 0 \qquad (1.7)$$

The groups Π_1, Π_2, etc., must be independent of one another, and no one group should be capable of being formed by multiplying together powers of the other groups.

By making use of the theorem it is possible to obtain the dimensionless groups more simply than by solving the simultaneous equations for the indices. Furthermore, the functional relationship can often be obtained in a form which is of more immediate use.

The method involves choosing m of the original variables to form what is called a *recurring set*. Any set m of the variables may be chosen with the following two provisions:

(a) Each of the fundamentals must appear in at least one of the m variables.
(b) It must not be possible to form a dimensionless group from some or all of the variables within the recurring set. If it were so possible, this dimensionless group would, of course, be one of the Π terms.

The procedure is then to take each of the remaining $(n - m)$ variables on its own and to form it into a dimensionless group by combining it with one or more members of the recurring set. In this way the $(n - m)$ Π groups are formed, the only variables appearing in more than one group being those that constitute the recurring set. Thus, if it is desired to obtain an explicit functional relation for one particular variable, that variable should not be included in the recurring set.

In some cases, the number of dimensionless groups will be greater than predicted by the Π theorem. Thus, for instance, if two of the fundamentals always occur in the same combination; for example, length and time always as $\mathbf{LT}^{-1}$, they will constitute a single fundamental instead of two fundamentals. By referring back to the method of equating indices, it is seen that each of the two fundamentals gives the same equation, and therefore only a single constraint is placed on the relationship by considering the two variables. Thus, although m is normally the number of fundamentals, it is more strictly defined as the *maximum number of variables from which a dimensionless group cannot be formed*.

The procedure is more readily understood by consideration of the illustration given before. The relationship between the variables affecting the pressure drop for flow of fluid in a pipe can be written:

$$f_5(\Delta P, d, l, \rho, \mu, u) = 0 \tag{1.8}$$

Equation 1.8 includes six variables, and three fundamental quantities (mass, length, and time) are involved.

$$\text{Number of groups} = 6 - 3 = 3$$

The recurring set must contain three variables that cannot themselves be formed into a dimensionless group. This imposes the following two restrictions:

(a) Both l and d cannot be chosen as they can be formed into the dimensionless group l/d.

(b) ΔP, ρ and u cannot be used since $\Delta P/\rho u^2$ is dimensionless.

Outside these constraints, any three variables can be chosen. However, it should be remembered that the variables forming the recurring set are liable to appear in all the dimensionless groups. As one is concerned in this problem with the effect of conditions on the pressure difference ΔP, it is convenient if ΔP appears in only one group and therefore it is preferable not to include it in the recurring set.

If the variables d, u, ρ are chosen as the recurring set, it fulfils all the above conditions. Dimensionally:

$$d = \mathbf{L}$$

$$u = \mathbf{LT}^{-1}$$

$$\rho = \mathbf{ML}^{-3}$$

Each of the dimensions $\mathbf{M}$, $\mathbf{L}$, $\mathbf{T}$ can then be obtained explicitly in terms of the variables d, u, ρ. Thus:

$$\mathbf{L} = d$$

$$\mathbf{M} = \rho d^3$$

$$\mathbf{T} = du^{-1}$$

The three dimensionless groups are thus obtained by taking each of the remaining variables ΔP, l, and μ in turn.

ΔP has dimensions $\mathbf{ML}^{-1}\mathbf{T}^{-2}$.
$\Delta P\mathbf{M}^{-1}\mathbf{LT}^2$ is therefore dimensionless.

Group Π_1 is, therefore, $\Delta P(\rho d^3)^{-1}(d)(du^{-1})^2 = \dfrac{\Delta P}{\rho u^2}$

l has dimensions $\mathbf{L}$.
$l\mathbf{L}^{-1}$ is therefore dimensionless.

Group Π_2 is, therefore, $l(d^{-1}) = \dfrac{l}{d}$

μ has dimensions $\mathbf{ML}^{-1}\mathbf{T}^{-1}$.
$\mu\mathbf{M}^{-1}\mathbf{LT}$ is therefore dimensionless.

Table 1.3. Some important dimensionless groups

Symbol	Name of group	In terms of other groups	Definition	Application
Ar	Archimedes	Ga	$\dfrac{\rho(\rho_s - \rho)gd^3}{\mu^2}$	Gravitational settling of particle in fluid
Db	Deborah		$\dfrac{t_P}{t_F}$	Flow of viscoelastic fluid
Eu	Euler		$\dfrac{P}{\rho u^2}$	Pressure and momentum in fluid
Fo	Fourier		$\dfrac{D_H t}{l^2}, \dfrac{Dt}{l^2}$	Unsteady state heat transfer/mass transfer
Fr	Froude		$\dfrac{u^2}{gl}$	Fluid flow with free surface
Ga	Galileo	Ar	$\dfrac{\rho(\rho_s - \rho)gd^3}{\mu^2}$	Gravitational settling of particle in fluid
Gr	Grashof		$\dfrac{l^3 \rho^2 \beta g \Delta T}{\mu^2}$	Heat transfer by natural convection
Gz	Graetz		$\dfrac{G C_p}{kl}$	Heat transfer to fluid in tube
He	Hedström		$\dfrac{R_Y \rho d^2}{\mu_p^2}$	Flow of fluid exhibiting yield stress
Le	Lewis	$Sc \cdot Pr^{-1}$	$\dfrac{k}{C_p \rho D} = \dfrac{D_H}{D}$	Simultaneous heat and mass transfer
Ma	Mach		$\dfrac{u}{u_w}$	Gas flow at high velocity
Nu	Nusselt		$\dfrac{hl}{k}$	Heat transfer in fluid
Pe	Peclet	$Re \cdot Pr$	$\dfrac{ul}{D_H}$	Fluid flow and heat transfer
		$Re \cdot Sc$	$\dfrac{ul}{D}$	Fluid flow and mass transfer
Pr	Prandtl		$\dfrac{C_p \mu}{k}$	Heat transfer in flowing fluid
Re	Reynolds		$\dfrac{ul\rho}{\mu}$	Fluid flow involving viscous and inertial forces
Sc	Schmidt		$\dfrac{\mu}{\rho D}$	Mass transfer in flowing fluid
Sh	Sherwood		$\dfrac{h_D l}{D}$	Mass transfer in fluid
St	Stanton	$Nu \cdot Pr^{-1} \cdot Re^{-1}$	$\dfrac{h}{C_p \rho u}$	Heat transfer in flowing fluid
We	Weber		$\dfrac{\rho u^2 l}{\sigma}$	Fluid flow with interfacial forces
ϕ	Friction factor		$\dfrac{R}{\rho u^2}$	Fluid drag at surface
$\mathbf{N}_p$	Power number		$\dfrac{\mathbf{P}}{\rho N^3 d^5}$	Power consumption for mixers

Group Π_3 is, therefore, $\mu(\rho d^3)^{-1}(d)(du^{-1}) = \dfrac{\mu}{du\rho}$.

Thus: $f_6\left(\dfrac{\Delta P}{\rho u^2}, \dfrac{l}{d}, \dfrac{\mu}{ud\rho}\right) = 0$ or $\dfrac{\Delta P}{\rho u^2} = f_7\left(\dfrac{l}{d}, \dfrac{ud\rho}{\mu}\right)$

$\mu/ud\rho$ is arbitrarily inverted because the Reynolds number is usually expressed in the form $ud\rho/\mu$.

Dimensionless analysis is a useful tool in chemical engineering in that it shows the possible ways in which the variables involved can be grouped. It is a dangerous tool in that it can give incorrect results if the physical nature of the problem is not understood and can lead to false conclusions if a significant variable is omitted from the problem. It should therefore be used with caution.

1.6. FURTHER READING

BLACKMAN, D. R.: *SI Units in Engineering* (Macmillan, 1969).
BRIDGMAN, P. W.: *Dimensional Analysis* (Yale University Press, 1931).
FOCKEN, C. M.: *Dimensional Methods and their Applications* (Edward Arnold, London, 1953).
IPSEN, D. C.: *Units, Dimensions, and Dimensionless Numbers* (McGraw-Hill, 1960).
JOHNSTONE, R. E. and THRING, M. W.: *Pilot Plants, Models and Scale-up in Chemical Engineering* (McGraw-Hill, 1957).
KLINKENBERG, A. and MOOY, H. H.: *Chem. Eng. Prog.* **44** (1948) 17. Dimensionless groups in fluid friction, heat and material transfer.
MASSEY, B. S.: *Units, Dimensional Analysis and Physical Similarity* (Van Nostrand Reinhold, 1971).
MULLIN, J. W.: *The Chemical Engineer* (London) No. 211 (Sept. 1967) 176. SI units in chemical engineering.
MULLIN, J. W.: *The Chemical Engineer* (London) No. 254 (1971) 352. Recent developments in the change-over to the international system of units (SI).
BS 3763: 1964: British Standard (British Standards Institution, London) *The International System (SI) Units*.
British Standards Institution Publication PD 5686 (1967). *The Use of SI Units*.
Quantities, Units and Symbols. (The Symbols Committee of the Royal Society, 1971).
SI. The International System of Units (HMSO, 1970).

1.7. NOMENCLATURE

		Units in SI system	Dimensions in M, L, T, θ, I
C_p	Specific heat capacity at constant pressure	J/kg K	$L^2T^{-2}\theta^{-1}$
D	Diffusion coefficient, molecular diffusivity	m²/s	L^2T^{-1}
D_H	Thermal diffusivity $k/C_p\rho$	m²/s	L^2T^{-1}
d	Diameter	m	L
f	A function	—	—
G	Mass rate of flow	kg/s	MT^{-1}
g	Acceleration due to gravity	m/s²	LT^{-2}
g_c	Numerical constant equal to standard value of "g"	—	—
h	Heat transfer coefficient	W/m²K	$MT^{-3}\theta^{-1}$
h_D	Mass transfer coefficient	m/s	LT^{-1}
I	Electric current	A	I
J	Mechanical equivalent of heat	—	—
k	Thermal conductivity	W/m K	$MLT^{-3}\theta^{-1}$
l	Characteristic length	m	L
M_A	Molecular weight (relative molecular mass) of A	kg/kmol	MN^{-1}
m	Number of fundamental dimensions	—	—

		Units in SI system	Dimensions in $\mathbf{M, L, T, \theta, I}$
N	Rotational speed	s^{-1}	$\mathbf{T}^{-1}$
n	Number of variables	—	—
$\mathbf{P}$	Power	W	$\mathbf{ML^2T^{-3}}$
P	Pressure	N/m^2	$\mathbf{ML^{-1}T^{-2}}$
ΔP	Pressure difference	N/m^2	$\mathbf{ML^{-1}T^{-2}}$
Q	Physical quantity	—	
R	Shear stress	N/m^2	$\mathbf{ML^{-1}T^{-2}}$
R_y	Yield stress	N/m^2	$\mathbf{ML^{-1}T^{-2}}$
$\mathbf{R}$	Universal gas constant	(8314)J/kmol K	$\mathbf{MN^{-1}L^2T^{-2}\theta^{-1}}$
ΔT	Temperature difference	K	$\mathbf{\theta}$
t	Time	s	$\mathbf{T}$
t_F	Characteristic time for fluid	s	$\mathbf{T}$
t_P	Characteristic time for process	s	$\mathbf{T}$
u	Velocity	m/s	$\mathbf{LT^{-1}}$
u_w	Velocity of a pressure wave	m/s	$\mathbf{LT^{-1}}$
V	Potential difference	V	$\mathbf{ML^2T^{-3}I^{-1}}$
β	Coefficient of cubical expansion of fluid	K^{-1}	$\mathbf{\theta^{-1}}$
Π	A dimensionless group	—	—
μ	Viscosity	$N\ s/m^2$	$\mathbf{ML^{-1}T^{-1}}$
μ_p	Plastic viscosity	$N\ s/m^2$	$\mathbf{ML^{-1}T^{-1}}$
ρ	Density of fluid	kg/m^3	$\mathbf{ML^{-3}}$
ρ_s	Density of solid	kg/m^3	$\mathbf{ML^{-3}}$
σ	Surface or interfacial tension	N/m	$\mathbf{MT^{-2}}$
Ω	Electrical resistance	Ohm	$\mathbf{ML^2T^{-3}I^{-2}}$

Dimensions

$\mathbf{F}$	Force	$\mathbf{M}$	Mass
$\mathbf{H}$	Heat	$\mathbf{T}$	Time
$\mathbf{L}$	Length	$\mathbf{\theta}$	Temperature

CHAPTER 2

Flow of Fluids — Energy and Momentum Relationships

2.1. INTRODUCTION

Chemical engineers are interested in many aspects of the problems involved in the flow of fluids. In the first place, in common with many other engineers, they are concerned with the transport of fluids from one location to another through pipes or open ducts, which requires the determination of the pressure drops in the system, and hence of the power required for pumping, selection of the most suitable type of pump, and measurement of the flow rates. In many cases, the fluid contains solid particles in suspension and it is necessary to determine the effect of these particles on the flow characteristics of the fluid or, alternatively, the drag force exerted by the fluid on the particles. In some cases, such as filtration, the particles are in the form of a fairly stable bed and the fluid has to pass through the tortuous channels formed by the pore spaces. In other cases the shape of the boundary surfaces must be so arranged that a particular flow pattern is obtained: for example, when solids are maintained in suspension in a liquid by means of agitation, the desired effect can be obtained with the minimum expenditure of energy as the most suitable flow pattern is produced in the fluid. Further, in those processes where heat transfer or mass transfer to a flowing fluid occurs, the nature of the flow may have a profound effect on the transfer coefficient for the process.

It is necessary to be able to calculate the energy and momentum of a fluid at various positions in a flow system. It will be seen that energy occurs in a number of forms and that some of these are influenced by the motion of the fluid. In the first part of this chapter the thermodynamic properties of fluids will be discussed. It will then be seen how the thermodynamic relations are modified if the fluid is in motion. In later chapters, the effects of frictional forces will be considered, and the principal methods of measuring flow will be described.

2.2. INTERNAL ENERGY

When a fluid flows from one location to another, energy will, in general, be converted from one form to another. The energy which is attributable to the physical state of the fluid is known as internal energy; it is arbitrarily taken as zero at some reference state, such as the absolute zero of temperature or the melting point of ice at atmospheric pressure. A change in the physical state of a fluid will, in general, cause an alteration in the internal energy. An elementary reversible change results from an infinitesimal change in one of the intensive factors acting on the system; the change proceeds at an infinitesimal rate and a small change in the intensive factor in the opposite direction would have caused the process to take place in the reverse direction. Truly reversible changes never occur in practice but they provide a useful standard with which actual processes can be compared. In an irreversible process, changes are caused by a finite difference in the intensive factor and take place at a finite rate. In general the process will be accompanied by the conversion of electrical or mechanical energy into heat, or by the reduction of the temperature difference between different parts of the system.

For a stationary material the change in the internal energy is equal to the difference between the net amount of heat added to the system and the net amount of work done by the system on its surroundings. For an infinitesimal change:

$$dU = \delta q - \delta W \tag{2.1}$$

where dU is the small change in the internal energy, δq the small amount of heat added, and δW the net amount of work done on the surroundings.

In this expression consistent units must be used. In the SI system each of the terms in equation 2.1 is expressed in Joules per kilogram (J/kg). In other systems either heat units (e.g. cal/g) or mechanical energy units (e.g. erg/g) may be used. dU is a small change in the internal energy which is a property of the system; it is therefore a perfect differential. On the other hand, δq and δW are small quantities of heat and work; they are not properties of the system and their values depend on the manner in which the change is effected; they are, therefore, not perfect differentials. For a reversible process, however, both δq and δW can be expressed in terms of properties of the system. For convenience, reference will be made to systems of unit mass and the effects on the surroundings will be disregarded.

A property called entropy is defined by the relation:

$$dS = \frac{\delta q}{T} \tag{2.2}$$

where dS is the small change in entropy resulting from the addition of a small quantity of heat δq, at a temperature T, under reversible conditions. From the definition of the thermodynamic scale of temperature, $\oint \delta q/T = 0$ for a reversible cyclic process, and the net change in the entropy is also zero. Thus, for a particular condition of the system, the entropy has a definite value and must be a property of the system; dS is, therefore, a perfect differential.

For an irreversible process:

$$\frac{\delta q}{T} < dS = \frac{\delta q}{T} + \frac{\delta F}{T} \quad \text{(say)} \tag{2.3}$$

δF is then a measure of the degree of irreversibility of the process; it represents the amount of mechanical energy converted into heat or the conversion of heat energy at one temperature to heat energy at another temperature. For a finite process:

$$\int_{S_1}^{S_2} T \, dS = \Sigma \delta q + \Sigma \delta F = q + F \quad \text{(say)} \tag{2.4}$$

When a process is isentropic, $q = -F$; a reversible process is isentropic when $q = 0$, i.e. a reversible adiabatic process is isentropic.

The increase in the entropy of an irreversible process may be illustrated in the following manner. Consider the spontaneous transfer of a quantity of heat δq from one part of a system at a temperature T_1 to another part at a temperature T_2; the net change in the entropy of the system as a whole is then:

$$dS = \frac{\delta q}{T_2} - \frac{\delta q}{T_1}$$

T_1 must be greater than T_2 and dS is therefore positive. If the process had been carried out reversibly, there would have been an infinitesimal difference between T_1 and T_2 and the change in entropy would have been zero.

The change in the internal energy can be expressed in terms of properties of the system itself. For a reversible process:

$$\delta q = T\,dS \quad \text{(from equation 2.2)} \quad \text{and} \quad \delta W = P\,dv$$

if the only work done is that resulting from a change in volume, dv.

Thus, from equation 2.1:

$$dU = T\,dS - P\,dv \tag{2.5}$$

Since this relation is in terms of properties of the system, it must also apply to a system in motion and to irreversible changes where the only work done is the result of change of volume.

Thus, in an irreversible process, for a stationary system:

from equations 2.1 and 2.2: $dU = \delta q - \delta W = T\,dS - P\,dv$

and from equation 2.3: $\delta q + \delta F = T\,dS$

$\therefore$ $\delta W = P\,dv - \delta F \tag{2.6}$

that is, the useful work performed by the system is less than $P\,dv$ by an amount δF, which represents the amount of mechanical energy converted into heat energy.

The relation between the internal energy and the temperature of a fluid will now be considered. In a system consisting of unit mass of material and where the only work done is that resulting from volume change, the change in internal energy after a reversible change is given by:

$$dU = \delta q - P\,dv \quad \text{(from equation 2.1)}$$

If there is no volume change:

$$dU = \delta q = C_v\,dT \tag{2.7}$$

where C_v is the specific heat at constant volume.

As this relation is in terms of properties of the system, it must be applicable to all changes at constant volume.

In an irreversible process:

$$dU = \delta q - (P\,dv - \delta F) \quad \text{(from equations 2.1 and 2.6)} \tag{2.8}$$

$$= \delta q + \delta F \qquad \text{(under conditions of constant volume)}$$

This quantity δF thus represents the mechanical energy which has been converted into heat and which is therefore available for increasing the temperature.

Thus:
$$\delta q + \delta F = C_v \, dT = dU. \tag{2.9}$$

For changes that take place under conditions of constant pressure, it is more satisfactory to consider variations in the enthalpy H. The enthalpy is defined by the relation:

$$H = U + Pv. \tag{2.10}$$

Thus:
$$dH = dU + P \, dv + v \, dP$$

$$= \delta q - P \, dv + \delta F + P \, dv + v \, dP \quad \text{(from equation 2.8)}$$

for an irreversible process: (For a reversible process $\delta F = 0$)

$$\therefore \qquad dH = \delta q + \delta F + v \, dP \tag{2.11}$$

$$= \delta q + \delta F \quad \text{(at constant pressure)}$$

$$= C_p \, dT \tag{2.12}$$

where C_p is the specific heat at constant pressure.

No assumptions have been made concerning the properties of the system and, therefore, the following relations apply to all fluids.

From equation 2.7:
$$\left(\frac{\partial U}{\partial T} \right)_v = C_v \tag{2.13}$$

From equation 2.12:
$$\left(\frac{\partial H}{\partial T} \right)_P = C_p \tag{2.14}$$

2.3. TYPES OF FLUID

Fluids may be classified in two different ways; either according to their behaviour under the action of externally applied pressure, or according to the effects produced by the action of a shear stress.

If the volume of an element of fluid is independent of its pressure and temperature, the fluid is said to be incompressible; if its volume changes it is said to be compressible. No real fluid is completely incompressible but liquids may generally be regarded as such when their flow is considered. Gases have a very much higher compressibility than liquids, and appreciable changes in volume may occur if the pressure or temperature is altered. However, if the percentage change in the pressure or in the absolute temperature is small, for practical purposes a gas may also be regarded as incompressible. Thus, in practice, volume changes are likely to be important only when the pressure or temperature of a gas changes by a large proportion. The relation between pressure, temperature, and volume of a real gas is generally complex, but except at very high pressures the behaviour of gases approximates to that of the ideal gas for which the volume of a given mass is inversely proportional to the pressure and directly proportional to the absolute temperature. At high pressures and when pressure changes are large, however, there may be appreciable deviations from this law and an approximate equation of state must then be used.

The behaviour of a fluid under the action of a shear stress is important in that it determines the way in which it will flow. The most important physical property affecting the stress distribution within the fluid is its viscosity. For a gas, the viscosity is low and even at high rates of shear, the viscous stresses are small. Under such conditions the gas approximates in its behaviour to an inviscid fluid. In many problems involving the flow of a gas or a liquid, the viscous stresses are important and give rise to appreciable velocity gradients within the fluid, and dissipation of energy occurs as a result of the frictional forces set up. In gases and in most pure liquids the ratio of the shear stress to the rate of shear is constant and equal to the viscosity of the fluid. These fluids are said to be *Newtonian* in their behaviour. However, in some liquids, particularly those containing a second phase in suspension, the ratio is not constant and the apparent viscosity of the fluid is a function of the rate of shear. The fluid is then said to be *non-Newtonian* and to exhibit rheological properties. The importance of the viscosity of the fluid in determining velocity profiles and friction losses is discussed in Chapter 3.

The effect of pressure on the properties of an incompressible fluid, an ideal gas, and a non-ideal gas will now be considered.

2.3.1. The incompressible fluid (liquid)

By definition, v is independent of P, so that $(\partial v/\partial P)_T = 0$. The internal energy will be a function of temperature but not a function of pressure.

2.3.2. The ideal gas

An ideal gas is defined as a gas whose properties obey the law:

$$PV = n\mathbf{R}T \qquad (2.15)$$

where V is the volume occupied by n molar units of the gas, $\mathbf{R}$ the universal gas constant, and T the absolute temperature. Here n will be expressed in kmol when using the SI system.

This law is closely obeyed by real gases under conditions where the actual volume of the molecules is small compared with the total volume, and where the molecules exert only a very small attractive force on one another. These conditions are met at very low pressures when the distance apart of the individual molecules is large. The value of $\mathbf{R}$ is then the same for all gases and in SI units has the value of 8.314 kJ/kmol K.

When the only external force on a gas is the fluid pressure, the equation of state is given by:

$$f(P, V, T, n) = 0$$

Any property can be expressed in terms of any three other properties. Consider the dependence of the internal energy on temperature and volume:

$$U = f(T, V, n)$$

For unit mass of gas:

$$U = f(T, v)$$

and:

$$Pv = \frac{\mathbf{R}T}{M} \tag{2.16}$$

where M is the molecular weight of the gas.
and v is the volume per unit mass.

so that:

$$dU = \left(\frac{\partial U}{\partial T}\right)_v dT + \left(\frac{\partial U}{\partial v}\right)_T dv \tag{2.17}$$

Now:

$$T\,dS = dU + P\,dv \quad \text{(from equation 2.5)}$$

$$\therefore \quad T\,dS = \left(\frac{\partial U}{\partial T}\right)_v dT + \left[P + \left(\frac{\partial U}{\partial v}\right)_T\right] dv$$

and:

$$dS = \left(\frac{\partial U}{\partial T}\right)_v \frac{dT}{T} + \frac{1}{T}\left[P + \left(\frac{\partial U}{\partial v}\right)_T\right] dv \tag{2.18}$$

Thus:

$$\left(\frac{\partial S}{\partial T}\right)_v = \frac{1}{T}\left(\frac{\partial U}{\partial T}\right)_v \tag{2.19}$$

and:

$$\left(\frac{\partial S}{\partial v}\right)_T = \frac{1}{T}\left[P + \left(\frac{\partial U}{\partial v}\right)_T\right] \tag{2.20}$$

Then differentiating equation 2.19 by v and equation 2.20 by T and equating:

$$\frac{1}{T}\frac{\partial^2 U}{\partial T \partial v} = \frac{1}{T}\left[\left(\frac{\partial P}{\partial T}\right)_v + \frac{\partial^2 U}{\partial v \partial T}\right] - \frac{1}{T^2}\left[P + \left(\frac{\partial U}{\partial v}\right)_T\right]$$

i.e.:

$$\left(\frac{\partial U}{\partial v}\right)_T = T\left(\frac{\partial P}{\partial T}\right)_v - P \tag{2.21}$$

This relation applies to any fluid. For the particular case of an ideal gas, since $Pv = \mathbf{R}T/M$ (equation 2.16):

$$T\left(\frac{\partial P}{\partial T}\right)_v = T\frac{\mathbf{R}}{Mv} = P$$

so that:

$$\left(\frac{\partial U}{\partial v}\right)_T = 0 \tag{2.22}$$

and:

$$\left(\frac{\partial U}{\partial P}\right)_T = \left(\frac{\partial U}{\partial v}\right)_T\left(\frac{\partial v}{\partial P}\right)_T = 0 \tag{2.23}$$

Thus the internal energy of an ideal gas is a function of temperature only. The variation of internal energy and enthalpy with temperature will now be calculated.

$$dU = \left(\frac{\partial U}{\partial T}\right)_v dT + \left(\frac{\partial U}{\partial v}\right)_T dv \qquad \text{(equation 2.17)}$$

$$= C_v\,dT \quad \text{(from equations 2.13 and 2.22)} \tag{2.24}$$

Thus for an ideal gas under all conditions:

$$\frac{\mathrm{d}U}{\mathrm{d}T} = C_v \tag{2.25}$$

In general, this relation applies only to changes at constant volume. For the particular case of the ideal gas, however, it applies under all circumstances.

Again, since $H = \mathrm{f}(T, P)$:

$$\mathrm{d}H = \left(\frac{\partial H}{\partial T}\right)_P \mathrm{d}T + \left(\frac{\partial H}{\partial P}\right)_T \mathrm{d}P$$

$$= C_p \mathrm{d}T + \left(\frac{\partial U}{\partial P}\right)_T \mathrm{d}P + \left(\frac{\partial (Pv)}{\partial P}\right)_T \mathrm{d}P \quad \text{(from equations 2.12 and 2.10)}$$

$$= C_p \mathrm{d}T$$

since $(\partial U/\partial P)_T = 0$ and $[\partial(Pv)/\partial P]_T = 0$ for an ideal gas.

Thus under all conditions for an ideal gas:

$$\frac{\mathrm{d}H}{\mathrm{d}T} = C_p \tag{2.26}$$

$$\therefore \qquad C_p - C_v = \frac{\mathrm{d}H}{\mathrm{d}T} - \frac{\mathrm{d}U}{\mathrm{d}T} = \frac{\mathrm{d}(Pv)}{\mathrm{d}T} = \frac{\mathbf{R}}{M} \tag{2.27}$$

Isothermal processes

In fluid flow it is important to know how the volume of a gas will vary as the pressure changes. Two important idealised conditions which are rarely obtained in practice are changes at constant temperature and changes at constant entropy. Although not actually reached, these conditions are approached in many flow problems.

For an *isothermal* change in an ideal gas, the product of pressure and volume is a constant. For unit mass of gas:

$$Pv = \frac{\mathbf{R}T}{M} = \text{constant} \tag{equation 2.16}$$

Isentropic processes

For an *isentropic* process the enthalpy can be expressed as a function of the pressure and volume:

$$H = \mathrm{f}(P, v)$$

$$\mathrm{d}H = \left(\frac{\partial H}{\partial P}\right)_v \mathrm{d}P + \left(\frac{\partial H}{\partial v}\right)_P \mathrm{d}v$$

$$= \left(\frac{\partial H}{\partial T}\right)_v \left(\frac{\partial T}{\partial P}\right)_v \mathrm{d}P + \left(\frac{\partial H}{\partial T}\right)_P \left(\frac{\partial T}{\partial v}\right)_P \mathrm{d}v$$

Now: $\qquad \left(\frac{\partial H}{\partial T}\right)_P = C_p \tag{equation 2.14}$

and:
$$\left(\frac{\partial H}{\partial T}\right)_v = \left(\frac{\partial U}{\partial T}\right)_v + \left(\frac{\partial (Pv)}{\partial T}\right)_v$$

$$= C_v + v\left(\frac{\partial P}{\partial T}\right)_v \quad \text{(from equation 2.13)}$$

Further:
$$dH = dU + P\,dv + v\,dP \quad \text{(from equation 2.10)}$$

$$= T\,dS - P\,dv + P\,dv + v\,dP \quad \text{(from equation 2.5)}$$

$$= T\,dS + v\,dP \tag{2.28}$$

$$= v\,dP \quad \text{(for an isentropic process)} \tag{2.29}$$

Thus, for an isentropic process:

$$v\,dP = \left[C_v + v\left(\frac{\partial P}{\partial T}\right)_v\right]\left(\frac{\partial T}{\partial P}\right)_v dP + C_p\left(\frac{\partial T}{\partial v}\right)_P dv$$

i.e.:
$$\left(\frac{\partial T}{\partial P}\right)_v dP + \frac{C_p}{C_v}\left(\frac{\partial T}{\partial v}\right)_P dv = 0$$

From the equation of state for an ideal gas (equation 2.15):

$$\left(\frac{\partial T}{\partial P}\right)_v = \frac{T}{P} \quad \text{and} \quad \left(\frac{\partial T}{\partial v}\right)_P = \frac{T}{v}$$

$$\therefore \qquad \left(\frac{dP}{P}\right) + \gamma\left(\frac{dv}{v}\right) = 0$$

where $\gamma = C_p/C_v$.

Integration gives:

$$\ln P + \gamma \ln v = \text{constant}$$

or:
$$Pv^\gamma = \text{constant} \tag{2.30}$$

This relation holds only approximately, even for an ideal gas, since γ has been taken as a constant in the integration. It does, however, vary somewhat with pressure.

2.3.3. The non-Ideal gas

For a non-ideal gas, equation 2.15 is modified by including a compressibility factor Z which is a function of both temperature and pressure:

$$PV = Zn\mathbf{R}T \tag{2.31}$$

At very low pressures, deviations from the ideal gas law are caused mainly by the attractive forces between the molecules and the compressibility factor has a value less than unity. At higher pressures, deviations are caused mainly by the fact that the volume of the molecules themselves, which can be regarded as incompressible, becomes significant compared with the total volume of the gas.

Many equations have been given to denote the approximate relation between the properties of a non-ideal gas. Of these the simplest, and probably the most commonly used, is van der Waals' equation:

$$\left(P + a\frac{n^2}{V^2}\right)(V - nb) = n\mathbf{R}T \tag{2.32}$$

where b is a quantity which is a function of the incompressible volume of the molecules themselves, and a/V^2 is a function of the attractive forces between the molecules. It is seen that as P approaches zero and V approaches infinity, this equation reduces to the equation of state for the ideal gas.

A chart which correlates experimental $P - V - T$ data for all gases is included as Figure 2.1 and is known as the generalised compressibility-factor chart.[1] Use is made of *reduced* coordinates where the *reduced temperature* T_R, the *reduced pressure* P_R, and the *reduced volume* V_R are defined as the ratio of the actual temperature, pressure, and volume of the gas to the corresponding values of these properties at the critical state. It is found that, at a given value of T_R and P_R, nearly all gases have the same molar volume, compressibility factor, and other thermodynamic properties. This empirical relationship applies to within about 2 per cent for most gases, and the most important exception to the rule is ammonia.

Figure 2.1.   Compressibility factors of gases and vapours

The generalised compressibility-factor chart is not to be regarded as a substitute for experimental $P - V - T$ data. If accurate data are available, as they are for some of the more common gases,[2] they should be used.

It will be noted from Figure 2.1 that Z approaches unity for all temperatures as the pressure approaches zero. This serves to confirm the statement made previously that all gases approach ideality as the pressure is reduced to zero. For most gases the critical pressure is 3 MN/m² or above. Thus at atmospheric pressure (101.3 kN/m²), P_R is 0.033 or less. At this pressure, for any temperature above the critical temperature ($T_R = 1$), it will be seen that Z deviates from unity by no more than 1 per cent. Thus at atmospheric pressure for temperatures greater than the critical temperature, the assumption that the ideal gas law is valid usually leads to errors of less than 1 per cent. It should also be noted that for reduced temperatures between 3 and 10 the compressibility factor is nearly unity for reduced pressures up to a value of 6. For very high temperatures the isotherms approach a horizontal line at $Z = 1$ for all pressures. Thus all gases tend towards ideality as the temperature approaches infinity.

Example 2.1

It is required to store 1 kmol of methane at 320 K and 60 MN/m^2. Using the following methods, estimate the volume of the vessel which must be provided:

    (a) ideal gas law;
    (b) van der Waals' equation;
    (c) generalised compressibility-factor chart;
    (d) experimental data from PERRY.[2]

Solution

For 1 kmol of methane,
    (a) $PV = \mathbf{R}T$, where $\mathbf{R} = 8.314$ kJ/kmol K.

In this case:
$$P = 60 \times 10^3 \text{ kN/m}^2; T = 320 \text{ K}$$

∴
$$V = 8.314 \times \frac{320}{(60 \times 10^3)} = \underline{\underline{0.0443 \text{ m}^3}}$$

    (b) In van der Waals' equation (2.32), the constants may be taken as:

$$a = \frac{27\mathbf{R}^2 T_c^2}{64 P_c}; \quad b = \frac{\mathbf{R}T_c}{8 P_c}$$

where the critical temperature $T_c = 191$ K and the critical pressure $P_c = 4640$ kN/m^2 for methane as shown in the Appendix tables.

∴
$$a = \frac{27 \times 8.314^2 \times 191^2}{(64 \times 4640)} = 229.3 \text{ (kN/m}^2)(\text{m}^3)^2/(\text{kmol})^2$$

and:
$$b = 8.314 \times \frac{191}{(8 \times 4640)} = 0.0427 \text{ m}^3/\text{kmol}$$

Thus in equation 2.32:

$$\left(60 \times 10^3 + 229.3 \times \frac{1}{V^2}\right)(V - 1 \times 0.0427) = 1 \times 8.314 \times 320$$

or:
$$60{,}000V^3 - 5233V^2 + 229.3V = 9.79$$

Solving by trial and error:
$$V = \underline{\underline{0.066 \text{ m}^3}}$$

(c)
$$T_r = \frac{T}{T_c} = \frac{320}{191} = 1.68$$

$$P_r = \frac{P}{P_c} = \frac{60 \times 10^3}{4640} = 12.93$$

Thus from Figure 2.1, $Z = 1.33$

and:
$$V = \frac{Zn\mathbf{R}T}{P} \quad \text{(from equation 2.31)}$$

$$= \frac{1.33 \times 1.0 \times 8.314 \times 320}{(60 \times 10)} = \underline{\underline{0.0589 \text{ m}^3}}$$

(d) PERRY[2] lists experimental values of A' for various gases, where $A' = PV/(PV)_0$, where $(PV)_0$ is the PV product at 273 K and 101.3 kN/m^2.

For methane at 320 K and 60 MN/m^2: $A' = 1.565$

$\therefore$
$$(PV)_0 = 101.3 \times 10^3 \times 22.4 = 2.27 \times 10^6 \, (\text{N/m}^2)(\text{m}^3/\text{kmol})$$

$$V = \frac{A'(PV)_0}{P} = \frac{1.565 \times 2.27 \times 10^6}{(60 \times 10^6)}$$

$$= \underline{\underline{0.0592 \text{ m}^3}}$$

Compared with the experimental value, the percentage errors in the other values are:

(a) ideal gas law -25.2%
(b) van der Waals' equation $+11.5\%$
(c) compressibility factor -0.5%

Joule-Thomson effect

It has already been shown that the change of internal energy of unit mass of fluid with volume at constant temperature is given by the relation:

$$\left(\frac{\partial U}{\partial v}\right)_T = T\left(\frac{\partial P}{\partial T}\right)_v - P \qquad \text{(equation 2.21)}$$

For a non-ideal gas:
$$T\left(\frac{\partial P}{\partial T}\right)_v \neq P$$

and therefore $(\partial U/\partial v)_T$ and $(\partial U/\partial P)_T$ are not equal to zero.

Thus the internal energy of the non-ideal gas is a function of pressure as well as temperature. As the gas is expanded, the molecules are separated from each other against the action of the attractive forces between them. Energy is therefore stored in the gas; this is released when the gas is compressed and the molecules are allowed to approach one another again.

A characteristic of the non-ideal gas is that it has a finite *Joule-Thomson effect*. This relates to the amount of heat which must be added during an expansion of a gas from a pressure P_1 to a pressure P_2 in order to maintain isothermal conditions. Imagine a gas flowing from a cylinder, fitted with a piston at a pressure P_1 to a second cylinder at a pressure P_2 (Figure 2.2).

Figure 2.2.   Joule-Thomson effect

The net work done by unit mass of gas on the surroundings in expanding from P_1 to P_2 is given by:

$$W = P_2 v_2 - P_1 v_1 \qquad (2.33)$$

A quantity of heat (q, say) is added during the expansion so as to maintain isothermal conditions. The change in the internal energy is therefore given by:

$$\Delta U = q - W \quad \text{(from equation 2.1)}$$

$$\therefore \qquad q = \Delta U - P_1 v_1 + P_2 v_2 \qquad (2.34)$$

For an ideal gas, under isothermal conditions, $\Delta U = 0$ and $P_2 v_2 = P_1 v_1$. Thus $q = 0$ and the ideal gas is said to have a zero Joule-Thomson effect. A non-ideal gas has a Joule-Thomson effect which may be either positive or negative.

2.4. THE FLUID IN MOTION

When a fluid flows through a duct or over a surface, the velocity over a plane at right angles to the stream is not normally uniform. The variation of velocity can be shown by the use of streamlines which are lines so drawn that the velocity vector is always tangential to them. The flowrate between any two streamlines is always the same. Constant velocity over a cross-section is shown by equidistant streamlines and an increase in velocity by closer spacing of the streamlines. There are two principal types of flow which are discussed in detail later, namely streamline and turbulent flow. In streamline flow, movement across streamlines occurs solely as the result of diffusion on a molecular scale and the flowrate is steady. In turbulent flow the presence of circulating current results in transference of fluid on a larger scale, and cyclic fluctuations occur in the flowrate, though the time-average rate remains constant.

A group of streamlines can be taken together to form a streamtube, and thus the whole area for flow can be regarded as being composed of bundles of streamtubes.

Figure 2.3.   Streamlines in a straight tube

Figure 2.4.   Streamlines in a constriction

Figure 2.5.   Streamlines for flow past an immersed object

Figures 2.3, 2.4, and 2.5 show the flow patterns in a straight tube, through a constriction and past an immersed object. In the first case, the streamlines are all parallel to one another, whereas in the other two cases the streamlines approach one another as the passage becomes constricted, indicating that the velocity is increasing.

2.4.1. Continuity

Considering the flow of a fluid through a streamtube, as shown in Figure 2.6, then equating the mass rates of flow at sections 1 and 2:

$$\mathrm{d}G = \rho_1 \dot{u}_1 \, \mathrm{d}A_1 = \rho_2 \dot{u}_2 \, \mathrm{d}A_2 \tag{2.35}$$

where ρ_1, ρ_2 are the densities; $\dot{u}_1$, $\dot{u}_2$ the velocities in the streamtube; and $\mathrm{d}A_1$, $\mathrm{d}A_2$ the flow areas at sections 1 and 2 respectively.

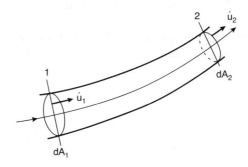

Figure 2.6.   Flow through a streamtube

On integration:

$$G = \int \rho_1 \dot{u}_1 \, \mathrm{d}A_1 = \int \rho_2 \dot{u}_2 \, \mathrm{d}A_2 = \rho_1 u_1 A_1 = \rho_2 u_2 A_2 \tag{2.36}$$

where u_1, u_2 are the average velocities (defined by the above equations) at the two sections. In many problems, the mass flowrate per unit area G' is the important quantity.

$$G' = \frac{G}{A} = \rho u \tag{2.37}$$

For an incompressible fluid (a liquid or a gas where the pressure changes are small):

$$u_1 A_1 = u_2 A_2 \qquad (2.38)$$

It will be seen that it is important to be able to determine the velocity profile so that the flowrate can be calculated. For streamline flow in a pipe the mean velocity is 0.5 times the maximum stream velocity which occurs at the axis. For turbulent flow, the profile is flatter and the ratio of the mean velocity to the maximum velocity is about 0.82.

2.4.2. Momentum changes in a fluid

As a fluid flows through a duct its momentum and pressure may change. The magnitude of the changes can be considered by applying the momentum equation (force equals rate of change of momentum) to the fluid in a streamtube and then integrating over the cross-section of the duct. The effect of frictional forces will be neglected at first and the relations thus obtained will strictly apply only to an inviscid (frictionless) fluid. Considering an element of length dl of a streamtube of cross-sectional area dA, increasing to $dA + (d(dA)/dl)dl$, as shown in Figure 2.7, then the upstream pressure $= P$ and force attributable to upstream pressure $= P\,dA$.

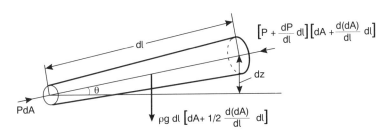

Figure 2.7.   Forces on fluid in a streamtube

The downstream pressure $= P + \dfrac{dP}{dl}dl$

This pressure acts over an area $dA + (d(dA)/dl)dl$ and gives rise to a total force of $-\{P + (dP/dl)dl\}\{dA + [d(dA)/dl]dl\}$.

In addition, the mean pressure of $P + \frac{1}{2}(dP/dl)dl$ acting on the sides of the streamtube will give rise to a force having a component $[P + \frac{1}{2}(dP/dl)dl][d(dA)/dl]dl$ along the streamtube. Thus, the net force along the streamtube due to the pressure gradient is:

$$P\,dA - \left(P + \frac{dP}{dl}dl\right)\left(dA + \frac{d(dA)}{dl}dl\right) + \left(P + \frac{1}{2}\frac{dP}{dl}dl\right)\frac{d(dA)}{dl}dl \approx -\frac{dP}{dl}dl\,dA$$

The other force acting is the weight of the fluid

$$= \rho g\,dl\left(dA + \frac{1}{2}\frac{d(dA)}{dl}dl\right)$$

The component of this force along the streamtube is

$$-\rho g\, dl \left(dA + \frac{1}{2}\frac{d(dA)}{dl}dl \right) \sin\theta$$

Neglecting second order terms and noting that $\sin\theta = dz/dl$:

$$\text{Total force on fluid} = -\frac{dP}{dl}dl\,dA - \frac{dz}{dl}\rho g\,dl\,dA \qquad (2.39)$$

The rate of change of momentum of the fluid along the streamtube

$$= (\rho \dot{u}\, dA)\left[\left(\dot{u} + \frac{d\dot{u}}{dl}dl\right) - \dot{u}\right]$$

$$= \rho \dot{u}\frac{d\dot{u}}{dl}dl\,dA \qquad (2.40)$$

Equating equations 2.39 and 2.40:

$$\rho \dot{u}\, dl\, dA \frac{d\dot{u}}{dl} = -dl\, dA \frac{dP}{dl} - \rho g\, dl\, dA \frac{dz}{dl}$$

$$\therefore \qquad \dot{u}\, d\dot{u} + \frac{dP}{\rho} + g\, dz = 0 \qquad (2.41)$$

On integration:

$$\frac{\dot{u}^2}{2} + \int \frac{dP}{\rho} + gz = \text{constant} \qquad (2.42)$$

For the simple case of the incompressible fluid, ρ is independent of pressure, and:

$$\frac{\dot{u}^2}{2} + \frac{P}{\rho} + gz = \text{constant} \qquad (2.43)$$

Equation 2.43 is known as Bernoulli's equation, which relates the pressure at a point in the fluid to its position and velocity. Each term in equation 2.43 represents energy per unit mass of fluid. Thus, if all the fluid is moving with a velocity u, the total energy per unit mass ψ is given by:

$$\psi = \frac{u^2}{2} + \frac{P}{\rho} + gz \qquad (2.44)$$

Dividing equation 2.44 by g:

$$\frac{u^2}{2g} + \frac{P}{\rho g} + z = \text{constant} \qquad (2.45)$$

In equation 2.45 each term represents energy per unit weight of fluid and has the dimensions of length and can be regarded as representing a contribution to the total fluid head.

Thus: $\dfrac{u^2}{2g}$ is the velocity head

$$\frac{P}{\rho g} \text{ is the pressure head}$$

and: z is the potential head

Equation 2.42 can also be obtained from consideration of the energy changes in the fluid.

Example 2.2

Water leaves the nozzle (25 mm diameter) of a fire hose at a velocity of 25 m/s. What will be the reaction force at the nozzle which the fireman will need to counterbalance?

Solution

$$
\begin{aligned}
\text{Mass rate of discharge of water, } G &= \rho u A \\
&= 1000 \times 25 \times \frac{\pi}{4}(0.025)^2 \\
&= 12.27 \text{ kg/s} \\
\text{Momentum of fluid per second} &= Gu \\
&= 12.27 \times 25 \\
&= 307 \text{ N}
\end{aligned}
$$

Reaction force = Rate of change of momentum = $\underline{\underline{307 \text{ N}}}$

Example 2.3

Water is flowing at 5 m/s in a 50 mm diameter pipe which incorporates a right-angle horizontal bend (Figure 2.8). What will be the force to which a retaining bracket will be subjected, as a result of the momentum changes in the liquid, if it is arranged symmetrically in the pipe bend?

Solution

Momentum per second of approaching liquid in Y-direction

$$
\begin{aligned}
&= \rho u^2 A \\
&= 1000 \times 25 \times \frac{\pi}{4}(0.050)^2 \\
&= 49.1 \text{ N}
\end{aligned}
$$

The pipe bracket must therefore exert a reaction force of -49.1 N in the Y-direction (in which the fluid is accelerating). Similarly, the force in the X-direction $= 49.1$ N

Resultant force in direction of arm of bracket $49.1 \cos 45° + 49.1 \sin 45°$

$$
= 49.1 \left(\frac{1}{\sqrt{2}} + \frac{1}{\sqrt{2}} \right)
$$

$$
= \underline{\underline{69.4 \text{ N}}}
$$

Note: This calculation takes no account of the reaction force due to the internal pressure in the pipe. If, for example, this were 5 bar (4 bar in excess of the external atmospheric pressure), the force in each of the principal (X, Y) directions would be $[4 \times 10^5] \times [\pi/4(0.050)^2] = 785$ N, considerably in excess of that due to the rate of change of momentum.

Figure 2.8.   Force on support for pipe bend

Water hammer

If the flowrate of a liquid in a pipeline is suddenly reduced, by rapid closure of a valve
for example, its rate of change of momentum can be sufficiently high for very large forces
to be set up which may cause damage to the installation. In a pipeline carrying water,
the resulting pressure wave may have a velocity as high as 1200 m/s. The behaviour of
the pipe network will be influenced by a large number of factors, including the density
and the bulk modulus of elasticity of the liquid, Young's modulus for the material of the
pipe, and the design and layout of the installation. The phenomenon is complex[3] and
reference should be made to one of the specialised texts, such as those by PARMAKIAN[4]
and SHARP[5] for a detailed analysis. The situation can arise with the flow of any liquid,
but it is usually referred to as *water hammer* on account of the characteristic sound arising
from water distribution systems.

2.4.3. Energy of a fluid in motion

The total energy of a fluid in motion is made up of a number of components. Reference
will be made to unit mass of fluid and changes in magnetic and electrical energy will be
neglected.

Internal energy U

This has already been discussed in Section 2.2.

Pressure energy

This represents the work which must be done in order to introduce the fluid, without
change in volume, into the system. It is therefore given by the product Pv, where P is
the pressure of the system and v is the volume of unit mass of fluid.

Potential energy

The potential energy of the fluid, due to its position in the earth's gravitational field, is
equal to the work which must be done on it in order to raise it to that position from some

arbitrarily chosen datum level at which the potential energy is taken as zero. Thus, if the fluid is situated at a height z above the datum level, the potential energy is zg, where g is the acceleration due to gravity which is taken as constant unless otherwise stated.

Kinetic energy

The fluid possesses kinetic energy by virtue of its motion with reference to some arbitrarily fixed body (normally taken as the earth). If the fluid is moving with a velocity u, the kinetic energy is $u^2/2$.

The total energy of unit mass of fluid is, therefore:

$$U + Pv + gz + \frac{u^2}{2} \tag{2.46}$$

If the fluid flows from section 1 to section 2 (where the values of the various quantities are denoted by suffixes 1 and 2 respectively) and q is the net heat absorbed from the surroundings and W_s is the net work done by the fluid on the surroundings, other than that done by the fluid in entering or leaving the section under consideration:

$$U_2 + P_2v_2 + gz_2 + \frac{u_2^2}{2} = U_1 + P_1v_1 + gz_1 + \frac{u_1^2}{2} + q - W_s \tag{2.47}$$

$$\Delta U + \Delta(Pv) + g\Delta z + \Delta \frac{u^2}{2} = q - W_s \tag{2.48}$$

where Δ denotes a finite change in the quantities.

Thus:
$$\Delta H + g\Delta z + \frac{\Delta u^2}{2} = q - W_s \tag{2.49}$$

It should be noted that the shaft work W_s is related to the total work W by the relation:

$$W = W_s + \Delta(Pv) \tag{2.50}$$

For a small change in the system:

$$dH + g\,dz + u\,du = \delta q - \delta W_s \tag{2.51}$$

For many purposes it is convenient to eliminate H by using equation 2.11:

$$dH = \delta q + \delta F + v\,dP \tag{equation 2.11}$$

Here δF represents the amount of mechanical energy irreversibly converted into heat.

Thus:
$$u\,du + g\,dz + v\,dP + \delta W_s + \delta F = 0 \tag{2.52}$$

When no work is done by the fluid on the surroundings and when friction can be neglected, it will be noted that equation 2.52 is identical to equation 2.41 derived from consideration of momentum, since:

$$v = \frac{1}{\rho}$$

Integrating the above equation for flow from section 1 to section 2 and summing the terms δW_s and δF:

$$\Delta \frac{\dot{u}^2}{2} + g\Delta z + \int_{P_1}^{P_2} v\,dP + W_s + F = 0 \tag{2.53}$$

Equations 2.41 to 2.53 are quite general and apply therefore to any type of fluid.

With incompressible fluids the energy F is either lost to the surroundings or causes a very small rise in temperature. If the fluid is compressible, however, the rise in temperature may result in an increase in the pressure energy and part of it may be available for doing useful work.

If the fluid is flowing through a channel or pipe a frictional drag arises in the region of the boundaries and gives rise to a velocity distribution across any section perpendicular to the direction of flow. For the unidirectional flow of fluid, the mean velocity of flow has been defined by equation 2.36 as the ratio of the volumetric flowrate to the cross-sectional area of the channel. When equation 2.52 is applied over the whole cross-section, therefore, allowance must be made for the fact that the mean square velocity is not equal to the square of the mean velocity, and a correction factor α must therefore be introduced into the kinetic energy term. Thus, considering the fluid over the whole cross-section, for small changes:

$$\frac{u\,du}{\alpha} + g\,dz + v\,dP + \delta W_s + \delta F = 0 \tag{2.54}$$

and for finite changes:

$$\Delta \left(\frac{u^2}{2\alpha}\right) + g\Delta z + \int_{P_1}^{P_2} v\,dP + W_s + F = 0 \tag{2.55}$$

Before equation 2.55 can be applied to any particular flow problem, the term $\int_{P_1}^{P_2} v\,dP$ must be evaluated.

Equation 2.50 becomes:

$$\Delta \left(\frac{u^2}{2\alpha}\right) + g\Delta z + \Delta H = q - W_s \tag{2.56}$$

For flow in a pipe of circular cross-section α will be shown to be exactly 0.5 for streamline flow and to approximate to unity for turbulent flow.

For turbulent flow, and where no external work is done, equation 2.54 becomes:

$$u\,du + g\,dz + v\,dP = 0 \tag{2.57}$$

if frictional effects can be neglected.

For horizontal flow, or where the effects of change of height can be neglected, as normally with gases, equation 2.57 simplifies to:

$$u\,du + v\,dP = 0 \tag{2.58}$$

2.4.4. Pressure and fluid head

In equation 2.54 each term represents energy per unit mass of fluid. If it is multiplied throughout by density ρ, each term has the dimensions of pressure and represents energy per unit volume of fluid:

$$\rho \frac{u\,du}{\alpha} + \rho g\,dz + dP + \rho \delta W_s + \rho \delta F = 0 \qquad (2.59)$$

If equation 2.54 is divided throughout by g, each term has the dimensions of length, and, as already mentioned, can be regarded as a component of the total head of the fluid and represents energy per unit weight:

$$\frac{1}{g} \frac{u\,du}{\alpha} + dz + v \frac{dP}{g} + \frac{\delta W_s}{g} + \frac{\delta F}{g} = 0 \qquad (2.60)$$

For an incompressible fluid flowing in a horizontal pipe of constant cross-section, in the absence of work being done by the fluid on the surroundings, the pressure change due to frictional effects is given by:

$$v \frac{dP_f}{g} + \frac{\delta F}{g} = 0$$

i.e.:

$$-dP_f = \frac{\delta F}{v} = \rho g\,dh_f \qquad (2.61)$$

where dh_f is the loss in head corresponding to a change in pressure due to friction of dP_f.

2.4.5. Constant flow per unit area

When the flow rate of the fluid per unit area G' is constant, equation 2.37 can be written:

$$\frac{G}{A} = G' = \frac{u_1}{v_1} = \frac{u_2}{v_2} = \frac{u}{v} \qquad (2.62)$$

or:

$$G' = u_1 \rho_1 = u_2 \rho_2 = u\rho \qquad (2.63)$$

Equation 2.58 is the momentum balance for horizontal turbulent flow:

$$u\,du + v\,dP = 0 \qquad (\text{equation } 2.58)$$

i.e.:

$$u \frac{du}{v} + dP = 0$$

Because u/v is constant, on integration this gives:

$$\frac{u_1(u_2 - u_1)}{v_1} + P_2 - P_1 = 0$$

i.e.:

$$\frac{u_1^2}{v_1} + P_1 = \frac{u_2^2}{v_2} + P_2 \qquad (2.64)$$

2.4.6. Separation

It should be noted that the energy and mass balance equations assume that the fluid is continuous. This is so in the case of a liquid, provided that the pressure does not fall to such a low value that boiling, or the evolution of dissolved gases, takes place. For water at normal temperatures the pressure should not be allowed to fall below the equivalent of a head of 1.2 m of liquid. With gases, there is no lower limit to the pressures at which the fluid remains continuous, but the various equations which are derived need modification if the pressures are so low that the linear dimensions of the channels become comparable with the mean free path of the molecules, that is when the so-called *molecular flow* sets in.

2.5. PRESSURE–VOLUME RELATIONSHIPS

2.5.1. Incompressible fluids

Here v is independent of pressure so that

$$\int_{P_1}^{P_2} v \, dP = (P_2 - P_1)v \tag{2.65}$$

Therefore equation 2.55 becomes:

$$\frac{u_1^2}{2\alpha_1} + gz_1 + P_1v = \frac{u_2^2}{2\alpha_2} + gz_2 + P_2v + W_s + F \cdot \tag{2.66}$$

or:

$$\Delta\frac{u^2}{2\alpha} + g\Delta z + v\Delta P + W_s + F = 0 \tag{2.67}$$

In a frictionless system in which the fluid does not work on the surroundings and α_1 and α_2 are taken as unity (turbulent flow)

$$\frac{u_1^2}{2} + gz_1 + P_1v = \frac{u_2^2}{2} + gz_2 + P_2v \tag{2.68}$$

Example 2.4

Water flows from a tap at a pressure of 250 kN/m^2 above atmospheric. What is the velocity of the jet if frictional effects are neglected?

Solution

From equation 2.68:

$$0.5(u_2^2 - u_1^2) = g(z_1 - z_2) + \frac{(P_1 - P_2)}{\rho}$$

Using suffix 1 to denote conditions in the pipe and suffix 2 to denote conditions in the jet and neglecting the velocity of approach in the pipe:

$$0.5(u_2^2 - 0) = 9.81 \times 0 + \frac{250 \times 10^3}{1000}$$

$$u_2 = 22.4 \text{ m/s}$$

2.5.2. Compressible fluids

For a gas, the mean value of the specific volume can be used unless the pressure or temperature change is very large, when the relation between pressure and volume must be taken into account.

The term $\int_{P_1}^{P_2} v\,dP$ will now be evaluated for the ideal gas under various conditions. In most cases the results so obtained can be applied to the non-ideal gas without introducing an error greater than is involved in estimating the other factors concerned in the process. The only common exception to this occurs in the flow of gases at very high pressures and for the flow of steam, when it is necessary to employ one of the approximate equations for the state of a non-ideal gas, in place of the equation for the ideal gas. Alternatively, one may use equation 2.56 and work in terms of changes in enthalpy. For a gas, the potential energy term is usually small compared with the other energy terms.

The relation between the pressure and the volume of an ideal gas depends on the rate of transfer of heat to the surroundings and the degree of irreversibility of the process. The following conditions will be considered.

(a) an isothermal process;
(b) an isentropic process;
(c) a reversible process which is neither isothermal nor adiabatic;
(d) an irreversible process which is not isothermal.

Isothermal process

For an isothermal process $Pv = \mathbf{R}T/M = P_1 v_1$, where the subscript 1 denotes the initial values and M is the molecular weight.

Thus
$$\int_{P_1}^{P_2} v\,dP = P_1 v_1 \int_{P_1}^{P_2} \frac{1}{P}\,dP = P_1 v_1 \ln \frac{P_2}{P_1} \tag{2.69}$$

Isentropic process

From equation 2.30, for an isentropic process:

$$Pv^{\gamma} = P_1 v_1^{\gamma} = \text{constant}$$

$$\int_{P_1}^{P_2} v\,dP = \int_{P_1}^{P_2} \left(\frac{P_1 v_1^{\gamma}}{P}\right)^{1/\gamma} dP$$

$$= P_1^{1/\gamma} v_1 \int_{P_1}^{P_2} P^{-1/\gamma}\,dP$$

$$= P_1^{1/\gamma} v_1 \frac{1}{1-(1/\gamma)} (P_2^{1-(1/\gamma)} - P_1^{1-(1/\gamma)})$$

$$= \frac{\gamma}{\gamma-1} P_1 v_1 \left[\left(\frac{P_2}{P_1}\right)^{(\gamma-1)/\gamma} - 1\right] \tag{2.70}$$

$$= \frac{\gamma}{\gamma-1} \left[P_1 \left(\frac{P_2}{P_1}\right)^{(\gamma-1)/\gamma} \left(\frac{P_2}{P_1}\right)^{1/\gamma} v_2 - P_1 v_1\right]$$

$$= \frac{\gamma}{\gamma-1} (P_2 v_2 - P_1 v_1) \tag{2.71}$$

Further, from equations 2.29 and 2.26, taking C_p as constant:

$$\int_{P_1}^{P_2} v \, dP = \int_{H_1}^{H_2} dH = C_p \Delta T \tag{2.72}$$

The above relations apply for an ideal gas to a reversible adiabatic process which, as already shown, is isentropic.

Reversible process — neither isothermal nor adiabatic

In general the conditions under which a change in state of a gas takes place are neither isothermal nor adiabatic and the relation between pressure and volume is approximately of the form $Pv^k = $ constant for a reversible process, where k is a numerical quantity whose value depends on the heat transfer between the gas and its surroundings. k usually lies between 1 and γ but may, under certain circumstances, lie outside these limits; it will have the same value for a reversible compression as for a reversible expansion under similar conditions. Under these conditions therefore, equation 2.70 becomes:

$$\int_{P_1}^{P_2} v \, dP = \frac{k}{k-1} P_1 v_1 \left[\left(\frac{P_2}{P_1} \right)^{(k-1)/k} - 1 \right] \tag{2.73}$$

Irreversible process

For an irreversible process it may not be possible to express the relation between pressure and volume as a continuous mathematical function but, by choosing a suitable value for the constant k, an equation of the form $Pv^k = $ constant may be used over a limited range of conditions. Equation 2.73 can then be used for the evaluation of $\int_{P_1}^{P_2} v \, dP$. It should be noted that, for an irreversible process, k will have different values for compression and expansion under otherwise similar conditions. Thus, for the irreversible adiabatic compression of a gas, k will be greater than γ: for the corresponding expansion k will be less than γ. This means that more energy has to be put into an irreversible compression than will be received back when the gas expands to its original condition.

2.6. ROTATIONAL OR VORTEX MOTION IN A FLUID

In many chemical engineering applications a liquid undergoes rotational motion, for example in a centrifugal pump, in a stirred vessel, in the basket of a centrifuge or in a cyclone-type separator. In the first instance, the effects of friction may be disregarded and consideration will be given to how the forces acting on the liquid determine the pressure distribution. Imagine the liquid to be rotating about a vertical axis. The fluid will then be subjected to vertical forces due to gravity and centrifugal forces in a horizontal plane. The total force on the liquid and the pressure distribution is then obtained by summing the two components. The vertical pressure gradient attributed to the force of gravity is given by:

$$\frac{\partial P}{\partial z} = -\rho g = -\frac{g}{v} \tag{2.74}$$

The centrifugal force acts in a horizontal plane and the resulting pressure gradient may be obtained by taking a force balance on a small element of liquid as shown in Figure 2.9.

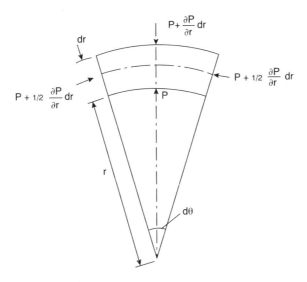

Figure 2.9.   Forces acting on element of fluid in a vortex

At radius r, the pressure is P.

At radius $r + dr$, the pressure is $P + (\partial P/\partial r)dr$.

For small values of dr, the pressure on the "cut" faces may be taken as $P + \frac{1}{2}(\partial P/\partial r)dr$.

Then, a force balance in the radial direction on an element of inner radius r, outer radius $r + dr$, depth dz and subtending a small angle $d\theta$ at the centre gives:

$$\left(P + \frac{\partial P}{\partial r}dr\right)(r + dr)d\theta\, dz - Pr\, d\theta\, dz - 2\left(P + \frac{1}{2}\frac{\partial P}{\partial r}dr\right)dr\, dz\, \sin\left(\frac{d\theta}{2}\right)$$

$$- r\, d\theta\, dr\, dz\, \rho r\omega^2 = 0$$

Simplifying and neglecting small quantities of second order and putting $\sin(d\theta/2)$ equal to $d\theta/2$ for a small angle:

$$\frac{\partial P}{\partial r} = \rho\omega^2 r = \frac{\rho u_t^2}{r} \tag{2.75}$$

where u_t is the tangential component of the liquid velocity at radius r.

Now:

$$dP = \frac{\partial P}{\partial z}dz + \frac{\partial P}{\partial r}dr \tag{2.76}$$

Substituting for $\partial P/\partial z$ and $\partial P/\partial r$ from equations 2.74 and 2.75:

$$dP = (-\rho g)dz + (r\rho\omega^2)\, dr \tag{2.77}$$

Equation 2.76 can be integrated provided that the relation between ω and r is specified. Two important cases will be considered:

(a) The *forced vortex* in which ω is constant and independent of r, and

(b) The *free vortex* in which the energy per unit mass of the liquid is constant.

2.6.1. The forced vortex

In a forced vortex the angular velocity of the liquid is maintained constant by mechanical means, such as by an agitator rotating in the liquid or by rotation in the basket of a centrifuge.

Thus, on integration of equation 2.76, for a constant value of ω:

$$P = -\rho gz + \frac{\rho \omega^2 r^2}{2} + \text{constant}$$

If the z-coordinate is z_a at the point on the axis of rotation which coincides with the free surface of the liquid (or the extension of the free surface), then the corresponding pressure P_0 must be that of the atmosphere in contact with the liquid.

i.e. when $r = 0$, $z = z_a$ and $P = P_0$. (See Figure 2.10.)

Then, on evaluation of the constant:

$$P - P_0 = \frac{\rho \omega^2 r^2}{2} - \rho g(z - z_a) \tag{2.78}$$

For any constant pressure P, equation 2.78 is the equation of a parabola, and therefore all surfaces of constant pressure are paraboloids of revolution. The free surface of the liquid is everywhere at the pressure P_0 of the surrounding atmosphere and therefore is itself a paraboloid of revolution. Putting $P = P_0$ in equation 2.77 for the free surface ($r = r_0, z = z_0$):

$$(z_0 - z_a) = \frac{\omega^2}{2g} r_0^2 \tag{2.79}$$

Differentiating equation 2.79:

$$\frac{dz_0}{dr_0} = \frac{r_0 \omega^2}{g} \tag{2.80}$$

Thus the greater the speed of rotation ω, the steeper is the slope. If $r_0\omega^2 \gg g$, $dz_0/dr_0 \to \infty$ and the surface is nearly vertical, and if $r_0\omega^2 \ll g$, $dz_0/dr_0 \to 0$ and the surface is almost horizontal.

The total energy ψ per unit mass of fluid is given by equation 2.44:

$$\psi = \frac{u_t^2}{2} + \frac{P}{\rho} + gz \tag{equation 2.44}$$

where u_t denotes the tangential velocity of the liquid.

Substituting $u_t = \omega r$ and for P/ρ from equation 2.78:

$$\psi = \frac{\omega^2 r^2}{2} + \left(\frac{P_0}{\rho} + \frac{\omega^2 r^2}{2} - g(z - z_a) \right) + gz$$

$$= \omega^2 r^2 + \frac{P_0}{\rho} + gz_a \tag{2.81}$$

Thus, the energy per unit mass increases with radius r and is independent of depth z. In the absence of an agitator or mechanical means of rotation energy transfer will take place to equalise ψ between all elements of fluid. Thus the *forced vortex* tends to decay into a *free vortex* (where energy per unit mass is independent of radius).

Figure 2.10.   Forced vortex

Application of the forced vortex—the centrifuge

Some of the important cases of forced vortexes are:

(a) The movement of liquid within the impeller of a centrifugal pump when there is no flow as, for example, when the outlet valve is closed.
(b) The rotation of liquid within the confines of a stirrer in an agitated tank.
(c) The rotation of liquid in the basket of a centrifuge. This application will now be considered. The operation of centrifuges is considered in detail in Volume 2, Chapter 9.

If liquid is contained in a cylindrical basket which is rotated about a vertical axis, the surfaces of constant pressure, including the free surface are paraboloids of revolution.

Thus, in general, the pressure at the walls of the basket is not constant, but varies with height. However, at normal operating speeds the centrifugal force will greatly exceed the gravitational force, and the inner surface of the liquid will be approximately vertical and the wall pressure will be nearly constant. At high operating speeds, where the gravitational force is relatively small, the functioning of the centrifuge is independent of the orientation of the axis of rotation. If mixtures of liquids or suspensions are to be separated in a centrifuge it is necessary to calculate the pressure at the walls arising from the rotation of the basket.

From equation 2.75:

$$\frac{\partial P}{\partial r} = \rho \omega^2 r \qquad \text{(equation 2.75)}$$

If it is assumed that there is no slip between the liquid and the basket, ω is constant and a forced vortex is created.

For a basket of radius R and with the radius of the inner surface of the liquid equal to r_0, the pressure P_R at the walls of the centrifuge is given by integration of equation 2.75

for a given value of z:

$$P_R - P_0 = \frac{\rho\omega^2}{2}(R^2 - r_0^2) \tag{2.82}$$

that is the pressure difference across the liquid at any horizontal level is

$$\frac{\rho\omega^2}{2}(R^2 - r_0^2). \tag{2.82a}$$

Example 2.5

Water is contained in the basket of a centrifuge of 0.5 m internal diameter, rotating at 50 revolutions per second. If the inner radius of the liquid is 0.15 m, what is the pressure at the walls of the basket?

Solution

Angular speed of rotation $= 2\pi \times 50 = 314$ rad/s
  The wall pressure is given by equation 2.82 as:

$$\frac{1000 \times 314^2}{2}(0.25^2 - 0.15^2)$$

$$= \underline{\underline{1.97 \times 10^6 \ \text{N/m}^2}}$$

2.6.2. The free vortex

In a free vortex the energy per unit mass of fluid is constant, and thus a free vortex is inherently stable. The variation of pressure with radius is obtained by differentiating equation 2.44 with respect to radius at constant depth z:

$$u_t \frac{\partial u_t}{\partial r} + \frac{1}{\rho}\frac{\partial P}{\partial r} = 0 \tag{2.83}$$

But:

$$\frac{\partial P}{\partial r} = \frac{\rho u_t^2}{r} \tag{equation 2.75}$$

$\therefore$

$$\frac{\partial u_t}{\partial r} + \frac{u_t}{r} = 0$$

and:

$$u_t r = \text{constant} = \kappa \tag{2.84}$$

Hence the angular momentum of the liquid is everywhere constant.

Thus:

$$\frac{\partial P}{\partial r} = \frac{\rho\kappa^2}{r^3} \tag{2.85}$$

Substituting from equations 2.74 and 2.85 into equation 2.76 and integrating:

$$P - P_\infty = (z_\infty - z)\rho g - \frac{\rho\kappa^2}{2r^2} \tag{2.86}$$

where P_∞ and z_∞ are the values of P and z at $r = \infty$.

Putting $P = P_\infty$ = atmospheric pressure:

$$z = z_\infty - \frac{\kappa^2}{2r^2 g} \tag{2.87}$$

Substituting into equation 2.44 gives:

$$\psi = \frac{P_\infty}{\rho} + g z_\infty \tag{2.88}$$

ψ is constant by definition and equal to the value at $r = \infty$ where $u = 0$.

A free vortex (Figure 2.11) exists:

(a) outside the impeller of a centrifugal pump;
(b) outside the region of the agitator in a stirred tank;
(c) in a cyclone separator or hydrocyclone;
(d) in the flow of liquid into a drain, as in a sink or bath;
(e) in liquid flowing round a bend in a pipe.

In all of these cases the free vortex may be modified by the frictional effect exerted by the external walls.

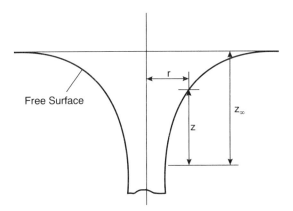

Figure 2.11.   Free vortex

2.7. FURTHER READING

DODGE, B. F.: *Chemical Engineering Thermodynamics* (McGraw-Hill, New York, 1944).
DE NEVERS, N.: *Fluid Mechanics for Chemical Engineers*, 2nd edn (McGraw-Hill, New York, 1970).
DOUGLAS, J. F.: *Solution of Problems in Fluid Mechanics* (Pitman, London, 1971).
MASSEY, B. S.: *Mechanics of Fluids*, 6th edn (Chapman and Hall, London, 1989).
MILNE-THOMSON, L. M.: *Theoretical Hydromechanics* (Macmillan, London, 1968).
SCHLICHTING, H.: *Boundary Layer Theory*, 5th edn (McGraw-Hill, New York, 1968).
SMITH, J. M. and VAN NESS, H. C.: *Introduction to Chemical Engineering Thermodynamics*, 2nd edn (McGraw-Hill, New York, 1959).

2.8. REFERENCES

1. HOUGEN, O. A. and WATSON, K. M.: *Chemical Process Principles* (Wiley, New York, 1964).
2. PERRY, R. H. and GREEN, D. W. (eds): *Perry's Chemical Engineers' Handbook* 6th edn (McGraw-Hill, New York, 1984).
3. CLARKE, D.: *The Chemical Engineer (London)* No.452 (1988) 34. Waterhammer: 1.
4. PARMAKIAN, J.: *Waterhammer Analysis* (Prentice-Hall, Englewood Cliffs, 1955).
5. SHARP, B. B.: *Water Hammer: Problems and Solutions* (Edward Arnold, London, 1981).

2.9. NOMENCLATURE

		Units in SI system	Dimensions in M, N, L, T, θ
A	Area perpendicular to direction of flow	m^2	$\mathbf{L}^2$
a	Coefficient in van der Waals' equation	(kN/m^2)(m^3)^2/(kmol)^2	$\mathbf{MN}^{-2}\mathbf{L}^5\mathbf{T}^{-2}$
b	Coefficient in van der Waals' equation	m^3/kmol	$\mathbf{N}^{-1}\mathbf{L}^3$
C_p	Specific heat at constant pressure per unit mass	J/kg K	$\mathbf{L}^2\mathbf{T}^{-2}\theta^{-1}$
C_v	Specific heat at constant volume per unit mass	J/kg K	$\mathbf{L}^2\mathbf{T}^{-2}\theta^{-1}$
F	Energy per unit mass degraded because of irreversibility of process	J/kg	$\mathbf{L}^2\mathbf{T}^{-2}$
G	Mass rate of flow	kg/s	$\mathbf{MT}^{-1}$
G'	Mass rate of flow per unit area	kg/m^{2}s	$\mathbf{ML}^{-2}\mathbf{T}^{-1}$
g	Acceleration due to gravity	9.81 m/s^2	$\mathbf{LT}^{-2}$
H	Enthalpy per unit mass	J/kg	$\mathbf{L}^2\mathbf{T}^{-2}$
h_f	Head lost due to friction	m	$\mathbf{L}$
k	Numerical constant used as index for compression	—	—
l	Length of streamtube	m	$\mathbf{L}$
M	Molecular weight	kg/kmol	$\mathbf{MN}^{-1}$
n	Number of molar units of fluid	kmol	$\mathbf{N}$
P	Pressure	N/m^2	$\mathbf{ML}^{-1}\mathbf{T}^{-2}$
P_B	Pressure at wall of centrifuge basket	N/m^2	$\mathbf{ML}^{-1}\mathbf{T}^{-2}$
P_R	Reduced pressure	—	—
q	Net heat flow into system	J/kg	$\mathbf{L}^2\mathbf{T}^{-2}$
R	Radius of centrifuge basket	m	$\mathbf{L}$
$\mathbf{R}$	Universal gas constant	(8314)J/kmol K	$\mathbf{MN}^{-1}\mathbf{L}^2\mathbf{T}^{-2}\theta^{-1}$
r	Radius	m	$\mathbf{L}$
S	Entropy per unit mass	J/kg K	$\mathbf{L}^2\mathbf{T}^{-2}\theta^{-1}$
T	Absolute temperature	K	θ
T_R	Reduced temperature	—	—
t	Time	s	$\mathbf{T}$
U	Internal energy per unit mass	J/kg	$\mathbf{L}^2\mathbf{T}^{-2}$
u	Mean velocity	m/s	$\mathbf{LT}^{-1}$
u_t	Tangential velocity	m/s	$\mathbf{LT}^{-1}$
$\dot{u}$	Velocity in streamtube	m/s	$\mathbf{LT}^{-1}$
V	Volume of fluid	m^3	$\mathbf{L}^3$
V_R	Reduced volume	—	—
v	Volume per unit mass of fluid	m^3/kg	$\mathbf{M}^{-1}\mathbf{L}^3$
W	Net work per unit mass done by system on surroundings	J/kg	$\mathbf{L}^2\mathbf{T}^{-2}$
W_s	Shaft work per unit mass	J/kg	$\mathbf{L}^2\mathbf{T}^{-2}$
Z	Compressibility factor for non-ideal gas	—	—
z	Distance in vertical direction	m	$\mathbf{L}$
z_a	Value of z_0 at $r_0 = 0$	m	$\mathbf{L}$
α	Constant in expression for kinetic energy of fluid	—	—
γ	Ratio of specific heats C_p/C_v	—	—
ψ	Total mechanical energy per unit mass of fluid	J/kg	$\mathbf{L}^2\mathbf{T}^{-2}$

		Units in SI system	Dimensions in **M, N, L, T, θ**
ρ	Density of fluid	kg/m^3	**ML^{-3}**
θ	Angle	—	—
ω	Angular velocity of rotation	rad/s	**T^{-1}**

Suffix

0	Value at free surface
∞	Value at $r = \infty$

CHAPTER 3

Flow in Pipes and Channels

3.1. INTRODUCTION

In the process industries it is often necessary to pump fluids over long distances from storage to processing units, and there may be a substantial drop in pressure in both the pipeline and in the individual units themselves. Many intermediate products are pumped from one factory site to another, and raw materials such as natural gas and petroleum products may be pumped very long distances to domestic or industrial consumers. It is necessary, therefore, to consider the problems concerned with calculating the power requirements for pumping, with designing the most suitable flow system, with estimating the most economical sizes of pipes, with measuring the rate of flow, and frequently with controlling this flow at a steady rate. The flow may take place at high pressures, for instance when nitrogen and hydrogen flow to an ammonia synthesis plant, or at low pressures when, for example, vapour leaves the top of a vacuum distillation plant. The fluid may consist of one or more phases, it may contain suspended solids, it may be near its boiling point, and it may have non-Newtonian properties, all of which complicate the analysis.

The design and layout of pipe systems is an important factor in the planning of modern plants and may represent a significant part of the total cost. In this chapter, the problem of determining the drop in pressure during flow will be discussed; in Chapter 6 the techniques of flow measurement, and in Chapter 8 the problem of selecting the most suitable pump for a given application will be considered.

Chemical engineering design frequently concerns equipment for the transfer of material or heat from one phase to another, and, in order to understand the mechanism of the transport process, the flow pattern of the fluid, and particularly the distribution of velocity near a surface, must be studied. It must be realised that when a fluid is flowing over a surface or through a pipe, the velocity at various points in a plane at right angles to the stream velocity is rarely uniform, and the rate of change of velocity with distance from the surface will exert a vital influence on the resistance to flow and on the rate of mass or heat transfer.

3.2. THE NATURE OF FLUID FLOW

When a fluid is flowing through a tube or over a surface, the pattern of the flow will vary with the velocity, the physical properties of the fluid, and the geometry of the surface. This problem was first examined by REYNOLDS[1] in 1883 using an apparatus shown in Figure 3.1. A glass tube with a flared entrance was immersed in a glass tank fed with water and, by means of the valve, the rate of flow from the tank through the glass

Figure 3.1.   Reynolds' apparatus for tracing flow patterns

tube could be controlled. By introducing a fine filament of coloured water from a small reservoir centrally into the flared entrance of the glass tube, the nature of the flow could be observed. At low rates of flow the coloured filament remained at the axis of the tube indicating that the flow was in the form of parallel streams which did not interfere with each other. Such flow is called *laminar* or *streamline* and is characterised by the absence of bulk movement at right angles to the main stream direction, though a small amount of radial dispersion will occur as a result of diffusion. As the flowrate was increased, a condition was reached in which oscillations appeared in the coloured filament which broke up into eddies causing dispersion across the tube section. This type of flow is known as *turbulent flow* and is characterised by the rapid movement of fluid as eddies in random directions across the tube. The general pattern is as shown in Figure 3.2. These experiments clearly showed the nature of the transition from streamline to turbulent flow. Below the critical velocity, oscillations in the flow were unstable and any disturbance would quickly disappear. At higher velocities, however, the oscillations were stable and increased in amplitude, causing a high degree of radial mixing. However, it was found that even when the main flow was turbulent there was a region near the wall (the *laminar sub-layer*) in which streamline flow persisted.

Figure 3.2.   Break-up of laminar thread in Reynolds' experiment

Only the problem of steady flow will be considered in which the time average velocity in the main stream direction X is constant and equal to u_x. In laminar flow, the instantaneous velocity at any point then has a steady value of u_x and does not fluctuate. In turbulent flow the instantaneous velocity at a point will vary about the mean value of u_x. It is convenient to consider the components of the eddy velocities in two directions — one along the main stream direction X and the other at right angles to the stream flow Y. Since the net flow in the X-direction is steady, the instantaneous velocity u_i can be imagined as being made up of a steady velocity u_x and a fluctuating velocity u_{Ex}, so that:

$$u_i = u_x + u_{Ex} \tag{3.1}$$

Since the average value of the main stream velocity is u_x, the average value of u_{Ex}, $\overline{u}_{Ex}$, will be zero, although the fluctuating component may at any instant amount to several per cent of the stream velocity. The fluctuating velocity in the Y-direction will also vary but, again, must have an average value of zero since there is no net flow at right angles to the stream flow. Turbulent flow is of great importance in chemical engineering because it causes rapid mixing of the fluid elements and is therefore responsible for promoting high rates of heat and mass transfer.

3.2.1. Flow over a surface

When a fluid flows over a surface the elements in contact with the surface will be brought to rest and the adjacent layers retarded by the viscous drag of the fluid. Thus the velocity in the neighbourhood of the surface will change with distance at right angles to the stream flow. It is important to realise that this change in velocity originates at the walls or surface. Suppose a fluid flowing with uniform velocity approaches a surface in the form of a plane as in Figure 3.3. When the fluid reaches the surface, a velocity gradient is set up at right angles to the surface because of the viscous forces acting within the fluid. The fluid in contact with the surface must be brought to rest as otherwise there would be an infinite velocity gradient at the wall, and an infinite stress to correspond. If u_x is the velocity in the X-direction at distance y from the surface, u_x will increase from zero at the surface $(y = 0)$ and will gradually approach the stream velocity u_s at some distance from the surface. Thus, if the values of u_x are measured, the velocity profile will be as shown in Figure 3.3. The velocity distributions are shown for three different distances downstream, and it is seen that in each case there is a rapid change in velocity near the wall and that the thickness of the layer in which the fluid is retarded becomes greater with distance in the direction of flow. The line 01 divides the stream into two sections; in the lower part the velocity is increasing away from the surface, whilst in the upper portion the velocity is approximately equal to u_s. This line indicates the thickness of the zone of retarded fluid which was termed the *boundary layer* by PRANDTL[2]. As will be shown in Chapter 11, the main stream velocity is approached asymptotically, and therefore the boundary layer strictly has no precise outer limit. However, it is convenient to define the boundary layer thickness such that the velocity at its outer edge equals 99 per cent of the stream velocity. Other definitions will be given later. Thus, by making certain assumptions covering the velocity profile, it will be shown in Chapter 11 that the boundary layer thickness δ at a distance x from the leading edge of a surface is dependent on the Reynolds number of flow.

Figure 3.3.   Development of boundary layer

Towards the leading edge of the surface, the flow in the boundary layer is laminar, but at a critical distance eddies start to form giving a turbulent boundary layer. In the turbulent layer there is a small region near the surface where the flow remains laminar; this is known as the *laminar sub-layer*. The change from laminar to turbulent flow in the boundary layer will occur at different distances downstream depending on the roughness of the surface and the physical properties of the fluid. This is discussed at length in Chapter 11.

3.2.2. Flow in a pipe

When a fluid with a uniform velocity enters a pipe, the layers of fluid adjacent to the walls are slowed down as on a plane surface and a boundary layer forms at the entrance. This builds up in thickness as the fluid passes into the pipe. At some distance downstream from the mouth, the boundary layers reach a thickness equal to the pipe radius and join at the axis, after which conditions remain constant and *fully developed flow* exists. If the flow in the boundary layers is streamline when they meet, laminar flow exists in the pipe. If the transition has already taken place before they meet, turbulent flow will persist in the region of fully developed flow. The region before the boundary layers join is known as the *entry length* and is discussed in greater detail in Chapter 11.

3.3. SHEARING CHARACTERISTICS OF A FLUID

As a fluid is deformed because of flow and applied external forces, frictional effects are exhibited by the motion of molecules relative to each other. The effects are encountered in all fluids and are due to their *viscosity*. Considering a thin layer of fluid between two parallel planes, distance dy apart as shown in Figure 3.4 with the lower plane fixed and a shearing force F applied to the other, since fluids deform continuously under shear, the upper plane moves at a steady velocity du_x relative to the fixed lower plane. When conditions are steady, the force F will be balanced by an internal force in the fluid due to its viscosity and the shear force per unit area is proportional to the velocity gradient in the fluid, i.e.:

$$\frac{F}{A} = R_y \propto \frac{du_x}{dy} \qquad (3.2)$$

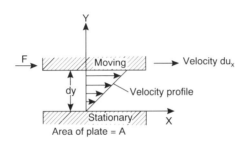

Figure 3.4.   Shear stress and velocity gradient in a fluid

where R is the shear stress in the fluid and du/dy is the velocity gradient or the rate of shear. The proportionality sign may be removed by the introduction of the proportionality factor μ, which is the coefficient of viscosity, to give:

$$R_y = \mu \frac{du_x}{dy} \tag{3.3}$$

A *Newtonian* fluid is one in which, provided that the temperature and pressure remain constant, the shear rate increases linearly with shear stress over a wide range of shear rate. As shear stresses within the fluid can act in both positive and negative directions, equation 3.3 should strictly be written with the incorporation of modulus signs to give:

$$\mu = \frac{|R_y|}{|du_x/dy|} \tag{3.4}$$

The viscosity strongly influences the shear stresses and hence the pressure drop for the flow. Viscosities for liquids are generally two orders of magnitude greater than gases at atmospheric pressure. For example, at 294 K, $\mu_{\text{water}} = 1.0 \times 10^{-3}$ N s/m^2 and $\mu_{\text{air}} = 1.8 \times 10^{-5}$ N s/m^2. Thus for a given shear rate, the shear stresses are considerably greater for liquids. It should be noted that with increase in temperature, the viscosity of a liquid decreases and that of a gas increases. At high pressures, especially near the critical point, the viscosity of a gas increases with increase in pressure. There are, however, a wide range of industrially important liquids such as high molecular weight polymers, colloids, suspensions and emulsions which do not exhibit a constant ratio of shear stress to shear rate. These are known as *non-Newtonian* fluids and are discussed further in Section 3.7 of this chapter.

3.4. THE DROP IN PRESSURE FOR FLOW THROUGH A TUBE

Experimental work by REYNOLDS[1], NIKURADSE[3], STANTON and PANNELL[4], MOODY[5], and others on the drop in pressure for flow through a pipe is most conveniently represented by plotting the head loss per unit length against the average velocity through the tube. The graph shown in Figure 3.5 is obtained in which $i = h_f/l$ is known as the hydraulic gradient. At low velocities the curve is linear showing that i is directly proportional to the velocity, but at higher velocities the pressure drop increases more rapidly. If a logarithmic plot is used, as in Figure 3.6, the results fall into three sections. Over the region of low velocity the line PB has a slope of unity but beyond this region over section BC there is instability with poorly defined points. At higher velocities, the line CQ has a slope of about

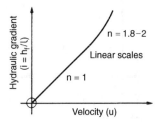

Figure 3.5.   Hydraulic gradient versus velocity

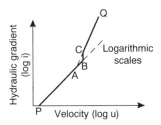

Figure 3.6. Critical velocities

1.8. If QC is produced backwards, it cuts PB at the point A, corresponding in Reynolds' earlier experiments to the change from laminar to turbulent flow and representing the critical velocity. Thus for streamline flow the pressure gradient is directly proportional to the velocity, and for turbulent flow it is proportional to the velocity raised to the power of 1.8.

The velocity corresponding to point A is taken as the lower critical velocity and to B as the higher critical velocity. Experiments with pipes of various sizes showed that the critical velocity was inversely proportional to the diameter, and that it was less at higher temperatures where the viscosity was lower. This led Reynolds to develop his own criterion based on the velocity of the fluid, the diameter of the tube, and the viscosity and density of the fluid. The dimensionless group $du\rho/\mu$ is termed the Reynolds number (Re) and is of vital importance in the study of fluid flow. It has been found that for values of Re less than about 2000 the flow is usually laminar and for values above 4000 the flow is usually turbulent. There is, however, no sharp change from laminar to turbulent flow but the transition occurs gradually as the velocity is increased in a given system. The precise velocity will depend on the geometry and on the pipe roughness. It is important to realise that there is no such thing as stable transitional flow. When the Reynolds number lies between 2000 and 4000 the flow is changing rapidly from laminar to turbulent and the portion of any cycle time for turbulent flow increases as Re approaches 4000.

If a turbulent fluid passes into a pipe so that the Reynolds number there is less than 2000, the flow pattern will change and the fluid will become streamline at some distance from the point of entry. On the other hand, if the fluid is initially streamline ($Re < 2000$), the diameter of the pipe can be gradually increased so that the Reynolds number exceeds 2000 and yet streamline flow will persist in the absence of any disturbance. Unstable streamline flow has been obtained in this manner at Reynolds numbers as high as 40,000. The setting up of turbulence requires a small force at right angles to the flow to initiate eddies.

The property of the fluid which appears in the Reynolds number is the kinematic viscosity μ/ρ. The kinematic viscosity of water at 294 K and atmospheric pressure is 10^{-6} m^2/s compared with 15.5×10^{-6} m^2/s for air. Thus, gases typically have higher kinematic viscosities than liquids at atmospheric pressure.

3.4.1. Shear stress in fluid

It is convenient to relate the pressure drop $-\Delta P_f$ to the shear stress R_0, at the walls of a pipe. If R_y is the shear stress at a distance y from the wall of the pipe, the corresponding value at the wall R_0 is given by:

$$R_0 = -\mu \left(\frac{du_x}{dy} \right)_{y=0} \qquad \text{(from equation 3.4)}$$

In this equation the negative sign is introduced in order to maintain a consistency of sign convention when shear stress is related to momentum transfer (see Chapter 11). Since $(du_x/dy)_{y=0}$ must be positive (velocity increases towards the pipe centre), R_0 is negative. It is therefore more convenient to work in terms of R, the shear stress exerted by the fluid on the surface ($= -R_0$) when calculating friction data.

Supposing a fluid is flowing through a length l of pipe of radius r (diameter d) over which the change in pressure due to friction is $-\Delta P_f$, then a force balance on the fluid in the pipe gives:

$$-\Delta P_f \pi r^2 = 2\pi r l (-R_0)$$

or:
$$R_0 = \Delta P_f \frac{r}{2l} \qquad (3.5)$$

or:
$$R = -\Delta P_f \frac{r}{2l} = -\Delta P_f \frac{d}{4l} \qquad (3.6)$$

If a force balance is taken over the central core of fluid of radius s:

$$-\Delta P_f \pi s^2 = 2\pi s l (-R_y)$$

i.e.:
$$-R_y = -\Delta P_f \frac{s}{2l} \qquad (3.7)$$

Thus from equations 3.5 and 3.7:

$$\frac{R_y}{R_0} = \frac{s}{r} = 1 - \frac{y}{r} \qquad (3.8)$$

Thus the shear stress increases linearly from zero at the centre of the pipe to a maximum at the walls:

$$\frac{|R_y|}{R} = 1 - \frac{y}{r} \qquad (3.9)$$

3.4.2. Resistance to flow in pipes

STANTON and PANNELL[4] measured the drop in pressure due to friction for a number of fluids flowing in pipes of various diameters and surface roughnesses. The results were expressed by using the concept of a friction factor, defined as the dimensionless group $R/\rho u^2$, which is plotted as a function of the Reynolds number; here $R(= -R_0)$ represents the resistance to flow per unit area of pipe surface. For a given pipe surface a single curve was found to express the results for all fluids, pipe diameters, and velocities. As with the results of Reynolds the curve was in three parts (Figure 3.7). At low values of Reynolds number ($Re < 2000$), $R/\rho u^2$ was independent of the surface roughness, but at high values ($Re > 2500$), $R/\rho u^2$ varied with the surface roughness. At very high Reynolds numbers the friction factor became independent of Re and a function of the surface roughness

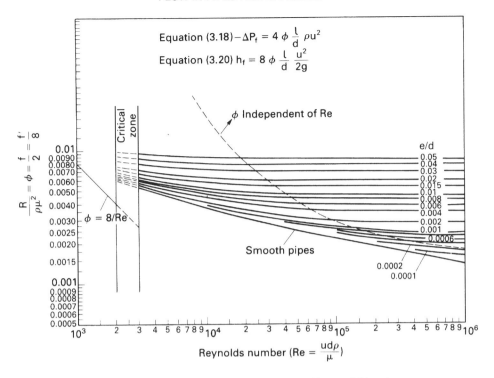

Figure 3.7. Pipe friction chart ϕ versus Re (also see fold-out)

only. Over the transition region of Re, from 2000 to 2500, $R/\rho u^2$ increased very rapidly, showing the great increase in friction as soon as turbulent motion commenced. This general relationship is one of the most widely used in all problems associated with fluid motion, heat transfer, and mass transfer. MOODY[5] worked in terms of a friction factor (here denoted by f') equal to $8R/\rho u^2$ and expressed this factor as a function of the two dimensionless terms Re and e/d where e is a length representing the magnitude of the surface roughness. This relationship can be obtained from dimensional analysis.

Thus if R is a function of $u, d, \rho, \mu, e,$
the analysis gives:

$$\frac{R}{\rho u^2} = \text{function of } \frac{ud\rho}{\mu} \text{ and } \frac{e}{d}$$

Thus a single curve will correlate the friction factor with the Reynolds group for all pipes with the same degree of roughness of e/d. This curve is of very great importance since it not only determines the pressure loss in the flow but can often be related to heat transfer or mass transfer (see Chapter 12). Such a series of curves for varying values of e/d is given in Figure 3.7 which shows the values of $R/\rho u^2$ and the values of the Moody factor f' related to the Reynolds group. Four separate regions may be distinguished:

Region 1 ($Re < 2000$) corresponds to streamline motion and a single curve represents all the data, irrespective of the roughness of the pipe surface. The equation of the curve is $R/\rho u^2 = 8/Re$.

Region 2 (2000 < *Re* < 3000) is a transition region between streamline and turbulent flow conditions. Reproducible values of pressure drop cannot be obtained in this region, but the value of $R/\rho u^2$ is considerably higher than that in the streamline region. If an unstable form of streamline flow does persist at *Re* greater than 2000, the frictional force will correspond to that given by the curve $R/\rho u^2 = 8/Re$, extrapolated to values of *Re* greater than 2000.

Region 3 (*Re* > 3000) corresponds to turbulent motion of the fluid and $R/\rho u^2$ is a function of both *Re* and e/d, with rough pipes giving high values of $R/\rho u^2$. For smooth pipes there is a lower limit below which $R/\rho u^2$ does not fall for any particular value of *Re*.

Region 4. Rough pipes at high *Re*. In this region the friction factor becomes independent of *Re* and depends only on (e/d) as shown below:

$$\frac{e}{d} = 0.05 \qquad Re > 1 \times 10^5 \qquad \frac{R}{\rho u^2} = 0.0087$$

$$\frac{e}{d} = 0.0075 \quad Re > 1 \times 10^5 \qquad \frac{R}{\rho u^2} = 0.0042$$

$$\frac{e}{d} = 0.001 \qquad Re > 1 \times 10^6 \qquad \frac{R}{\rho u^2} = 0.0024$$

A number of expressions have been proposed for calculating $R/\rho u^2 (= \phi)$ in terms of the Reynolds number and some of these are as follows:

Smooth pipes: $2.5 \times 10^3 < Re < 10^5$ $\qquad \phi = 0.0396 Re^{-0.25}$ $\qquad\qquad$ (3.10)

Smooth pipes: $2.5 \times 10^3 < Re < 10^7$ $\quad \phi^{-0.5} = 2.5 \ln(Re\phi^{0.5}) + 0.3$ (see equation 12.75)
$\qquad\qquad\qquad\qquad\qquad\qquad\qquad\qquad\qquad\qquad\qquad\qquad\qquad\qquad$ (3.11)

Rough pipes: $\qquad\qquad\qquad\qquad\qquad \phi^{-0.5} = -2.5 \ln \left(0.27 \, \frac{e}{d} + 0.885 Re^{-1} \phi^{-0.5} \right)$
$\qquad\qquad\qquad\qquad\qquad\qquad\qquad\qquad\qquad\qquad\qquad\qquad\qquad\qquad$ (3.12)

Rough pipes: $\dfrac{e}{d} Re\phi^{0.5} \gg 3.3$ $\qquad \phi^{-0.5} = 3.2 - 2.5 \ln \dfrac{e}{d}$ $\qquad\qquad$ (3.13)

Equation 3.10 is due to BLASIUS[6] and the others are derived from considerations of velocity profile. In addition to the Moody friction factor $f' = 8R/\rho u^2$, the Fanning or Darcy friction factor $f = 2R/\rho u^2$ is often used in other texts. It is extremely important therefore to be clear about the exact definition of the friction factor when using this term in calculating head losses due to friction.

3.4.3. Calculation of drop in pressure along a pipe

For the flow of a fluid in a short length of pipe dl of diameter d, the total frictional force at the walls is the product of the shear stress R and the surface area of the pipe $(R\pi d \, dl)$. This frictional force results in a change in pressure dP_f so that for a horizontal pipe:

$$R\pi d \, dl = -dP_f \pi \frac{d^2}{4}$$
$\qquad\qquad\qquad\qquad\qquad\qquad\qquad\qquad\qquad\qquad\qquad\qquad\qquad$ (3.14)

i.e.:
$$-\mathrm{d}P_f = 4R\frac{\mathrm{d}l}{d} = 4\frac{R}{\rho u^2}\frac{\mathrm{d}l}{d}\rho u^2 \tag{3.15}$$

and:
$$\frac{R}{\rho u^2} = \left(-\frac{\mathrm{d}P}{\mathrm{d}l}\right)\left(\frac{d}{4\rho u^2}\right) \tag{3.16}$$

The head lost due to friction:

$$\mathrm{d}h_f = -\frac{\mathrm{d}P_f}{\rho g} = 4\frac{R}{\rho u^2}\frac{\mathrm{d}l}{d}\frac{u^2}{g} \tag{3.17}$$

For an incompressible fluid flowing in a pipe of constant cross-section, u is not a function of pressure or length, and these equations can be integrated, over a length l to give:

$$-\Delta P_f = 4\frac{R}{\rho u^2}\frac{l}{d}\rho u^2 \tag{3.18}$$

The energy lost per unit mass F is then given by equation 3.19:

$$F = -\frac{\Delta P_f}{\rho} = 4\frac{R}{\rho u^2}\frac{l}{d}u^2 = 4\phi\frac{l}{d}u^2 \tag{3.19}$$

To obtain $-\Delta P_f$ it is therefore necessary to evaluate e/d and then from a knowledge of the value of Re the corresponding value of $R/\rho u^2$ is obtained. This is then used in equation 3.18 to give $-\Delta P_f$ or the head loss due to friction h_f as:

$$h_f = -\frac{\Delta P_f}{\rho g} = 8\phi\frac{l}{d}\frac{u^2}{2g} \tag{3.20}$$

Using the friction factors of Moody and Fanning, f' and f respectively, the head loss due to friction is obtained from the following equations:

Moody:
$$h_f = f'\frac{l}{d}\frac{u^2}{2g} \tag{3.21}$$

Fanning:
$$h_f = 4f\frac{l}{d}\frac{u^2}{2g} \tag{3.22}$$

The energy loss per unit mass because of the irreversibility of the process is given by $\delta F = -\mathrm{d}P_f/\rho$ or, over a length of pipe l, the energy loss is $F = 4\phi(l/d)u^2$, i.e. equation 3.19.

If it is required to calculate the flow in a pipe where the maximum pressure drop is specified, the velocity u is required but the Reynolds number is unknown, the above method cannot be used directly to give $R/\rho u^2$. One method of solution is to estimate the value of $R/\rho u^2$ and calculate the velocity and the value of Re. The value of $R/\rho u^2$ is then found and if different from the assumed value, a further trial becomes necessary.

An alternative solution to this problem is as follows:

$$\frac{R}{\rho u^2}\left(\frac{\rho u d}{\mu}\right)^2 = \phi Re^2 = \frac{Rd^2\rho}{\mu^2} = \frac{-\Delta P_f d^3\rho}{4l\mu^2} \tag{3.23}$$

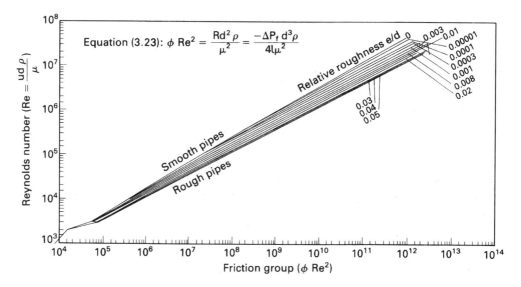

Figure 3.8.   Pipe friction chart ϕRe^2 versus Re for various values of e/d (also see fold-out)

If Re is plotted as a function of ϕRe^2 and e/d as shown in Figure 3.8, the group $(-\Delta P_f d^3 \rho/4l\mu^2) = \phi Re^2$ may be evaluated directly as it is independent of velocity. Hence Re may be found from the graph and the required velocity obtained. Similarly, if the diameter of pipe is required to transport fluid at a mass rate of flow G with a given fall in pressure, the following group, which is independent of d, is useful:

$$\left(\frac{R}{\rho u^2}\right)\left(\frac{ud\rho}{\mu}\right)^{-1} = \phi Re^{-1} = \frac{-\Delta P_f \mu}{4\rho^2 u^3 l} \tag{3.24}$$

3.4.4. Roughness of pipe surfaces

The estimation of the roughness of the surface of the pipe normally presents considerable difficulty. The use of an incorrect value, however, will not be serious even for turbulent flow at low Reynolds numbers because the pressure drop is not critically dependent on the roughness in this region. However, at high values of Reynolds number, the effect of pipe roughness is considerable, as can be seen from the graph of $R/\rho u^2$ against the Reynolds number shown in Figure 3.7. The values of the absolute roughness have been measured for a number of materials and examples are quoted in Table 3.1. Where the value for the pipe surface in question is not given, it is necessary to take an approximate value estimated from the known figures. Where pipes have become corroded, the value of the roughness will commonly be increased, up to tenfold.

Values of roughness applicable to materials used in the construction of open channels are also included in Table 3.1.

For practical installations it must be remembered that the frictional losses cannot be estimated with very great accuracy both because the roughness will change with use and the pumping unit must therefore always have ample excess capacity.

Table 3.1.   Values of absolute roughness e

	(ft)	(mm)
Drawn tubing	0.000005	0.0015
Commercial steel and wrought-iron	0.00015	0.046
Asphalted cast-iron	0.0004	0.12
Galvanised iron	0.0005	0.15
Cast-iron	0.00085	0.26
Wood stave	0.0006–0.003	0.18–0.9
Concrete	0.001–0.01	0.3–3.0
Riveted steel	0.003–0.03	0.9–9.0

Example 3.1

Ninety-eight per cent sulphuric acid is pumped at 4.5 t/h (1.25 kg/s) through a 25 mm diameter pipe, 30 m long, to a reservoir 12 m higher than the feed point. Calculate the pressure drop in the pipeline.

$$\text{Viscosity of acid } = 25 \text{ mN s/m}^2 \text{ or } 25 \times 10^{-3} \text{ N s/m}^2$$

$$\text{Density of acid } = 1840 \text{ kg/m}^3$$

Solution

Reynolds number:

$$Re = \frac{ud\rho}{\mu} = \frac{4G}{\pi\mu d}$$

$$= \frac{4 \times 1.25}{\pi \times 25 \times 10^{-3} \times 25 \times 10^{-3}}$$

$$= 2545$$

For a mild steel pipe, suitable for conveying the acid, the roughness e will be between 0.05 and 0.5 mm (0.00005 and 0.0005 m).

Relative roughness: $\dfrac{e}{d} = 0.002$ to 0.02

From Figure 3.7: $\dfrac{R}{\rho u^2} = 0.006$ over this range of $\dfrac{e}{d}$

Velocity: $u = \dfrac{G}{\rho A} = \dfrac{1.25}{1840 \times (\pi/4)(0.025)^2}$

$$= 1.38 \text{ m/s}$$

The kinetic energy attributable to this velocity will be dissipated when the liquid enters the reservoir. The pressure drop can now be calculated from the energy balance equation and equation 3.19. For turbulent flow of an incompressible fluid:

$$\Delta \frac{u^2}{2} + g\Delta z + v(P_2 - P_1) + 4\frac{R}{\rho u^2}\frac{l}{d}u^2 = 0 \quad \text{(from equation 2.67)}$$

$\therefore$
$$(P_1 - P_2) = \rho \left[0.5 + 4\frac{R}{\rho u^2}\frac{l}{d} \right] u^2 + g\Delta z$$

$$= 1840 \left\{ \left[0.5 + 4 \times 0.006\frac{30}{0.025} \right] 1.38^2 + 9.81 \times 12 \right\}$$

$$= 3.19 \times 10^5 \text{ N/m}^2$$

or:
$$\underline{\underline{320 \text{ kN/m}^2}}$$

Example 3.2

Water flows in a 50 mm pipe, 100 m long, whose roughness e is equal to 0.013 mm. If the pressure drop across this length of pipe is not to exceed 50 kN/m^2, what is the maximum allowable water velocity? The density and viscosity of water may be taken as 1000 kg/m^3 and 1.0 mN s/m^2 respectively.

Solution

From equation 3.23:

$$\frac{R}{\rho u^2} Re^2 = -\frac{\Delta P_f d^3 \rho}{4l\mu^2} = \phi Re^2$$

$$-\frac{\Delta P_f d^3 \rho}{4l\mu^2} = \frac{50{,}000(0.05)^3 1000}{4 \times 100(1 \times 10^{-3})^2}$$

$$= 1.56 \times 10^7$$

$$\frac{e}{d} = \frac{0.013}{50} = 0.00026$$

From Figure 3.8, when $\phi Re^2 = 1.56 \times 10^7$ and $(e/d) = 0.00026$:

$$Re = \frac{\rho u d}{\mu} = 7.9 \times 10^4$$

Hence:
$$u = \frac{7.9 \times 10^4 (1 \times 10^{-3})}{1000 \times 0.05}$$

$$= \underline{\underline{1.6 \text{ m/s}}}$$

Example 3.3

A cylindrical tank, 5 m in diameter, discharges through a horizontal mild steel pipe 100 m long and 225 mm in diameter connected to the base. What is the time taken for the water level in the tank to drop from 3 m to 0.3 m above the bottom? The viscosity may be taken as 1 mN s/m^2.

Solution

At time t let the liquid level be D m above the bottom of the tank. Designating point 1 as the liquid level and point 2 as the pipe outlet, then applying the energy balance equation for turbulent flow:

$$\Delta \frac{u^2}{2} + g\Delta z + v(P_2 - P_1) + F = 0$$

Now:
$$P_2 = P_1 = 101.3 \text{ kN/m}^2$$

$$\frac{u_1}{u_2} = \left(\frac{0.225}{5}\right)^2 = 0.0020, \quad \text{i.e. } u_1 \text{ can be neglected}$$

and:
$$\Delta z = -D$$

Thus:
$$\frac{u_2^2}{2} - Dg + 4\frac{R}{\rho u^2}\frac{l}{d}u_2^2 = 0$$

or:
$$u_2^2 - 19.62D + 8\frac{R}{\rho u^2}(444)u_2^2 = 0$$

$$\therefore \quad u_2 = \frac{4.43\sqrt{D}}{\sqrt{1 + 3552(R/\rho u^2)}}$$

As the level of liquid in the tank changes from D to $(D + dD)$, the quantity of water discharged $= (\pi/4)5^2(-dD) = -19.63\,dD$ m^3.

The time taken for the level to change by an amount dD is given by:

$$dt = \frac{-19.63\,dD}{(\pi/4)0.225^2 \times 4.43\sqrt{D}/\sqrt{[1 + 3552(R/\rho u^2)]}}$$

$$= -111.5\sqrt{\left[1 + 3552\frac{R}{\rho u^2}\right]}D^{-0.5}dD$$

Total time:

$$t = -\int_3^{0.3} 111.5\sqrt{\left[1 + 3552\frac{R}{\rho u^2}\right]}D^{-0.5}dD$$

Assuming that $R/\rho u^2$ is constant over the range of flow rates considered:

$$t = 264\sqrt{\left[1 + 3552\frac{R}{\rho u^2}\right]}\text{ s}$$

Assuming that the kinetic energy of the liquid is small compared with the frictional losses, an approximate value of $R/\rho u^2$ may be calculated.

Pressure drop along the pipe $= D\rho g = \dfrac{4Rl}{d}$:

$$\frac{R}{\rho u^2}Re^2 = \frac{Rd^2\rho}{\mu^2} = \frac{Dg\rho^2 d^3}{4l\mu^2}$$

$$= \frac{D \times 9.81 \times 1000^2 \times 0.225^3}{4 \times 100 \times 0.001^2}$$

$$= 2.79 \times 10^8 D$$

As D varies from 3 m to 0.3 m, $(R/\rho u^2)Re^2$ varies from 8.38×10^8 to 0.838×10^8. It is of interest to consider whether the difference in roughness of a new or old pipe has a significant effect at this stage.

For a mild steel pipe:

New: $e = 0.00005$ m, $\dfrac{e}{d} = 0.00022$, $Re = 7.0$ to 2.2×10^5 (from Figure 3.7)

Old: $e = 0.0005$ m, $\dfrac{e}{d} = 0.0022$, $Re = 6.0$ to 2.2×10^5 (from Figure 3.7)

For a new pipe $R/\rho u^2$ therefore varies from 0.0019 to 0.0020 and for an old pipe $R/\rho u^2 = 0.0029$ (from Figure 3.7).

Taking a value of 0.002 for a new pipe, assumed constant:

$$t = 264\sqrt{(1 + 3552 \times 0.002)} = 264\sqrt{8.1} = 750\text{ s}$$

Pressure drop due to friction is approximately $(7.1/8.1) = 0.88$ of the total pressure drop, that is the drop due to friction plus the change in kinetic energy.

Thus:

$$\frac{R}{\rho u^2}Re^2\text{ varies from about } 7.4 \times 10^8 \text{ to } 0.74 \times 10^8$$

$$Re\text{ varies from about } 6.2 \times 10^5 \text{ to } 1.9 \times 10^5$$

and:

$$\frac{R}{\rho u^2}\text{ varies from about } 0.0019 \text{ to } 0.0020$$

which is sufficiently close to the assumed value of $R/\rho u^2 = 0.002$.

The time taken for the level to fall is therefore about $\underline{\underline{750\text{ s}}} = \underline{\underline{12.5\text{ min}}}$.

Example 3.4

Two storage tanks, A and B, containing a petroleum product, discharge through pipes each 0.3 m in diameter and 1.5 km long to a junction at D. From D the liquid is passed through a 0.5 m diameter pipe to a third storage tank C, 0.75 km away as shown in Figure 3.9. The surface of the liquid in A is initially 10 m above that in C and the liquid level in B is 6 m higher than that in A. Calculate the initial rate of discharge of liquid into tank C assuming the pipes are mild steel. The density and viscosity of the liquid are 870 kg/m^3 and 0.7 mN s/m^2 respectively.

Figure 3.9.   Tank layout for Example 3.4

Solution

Because the pipes are long, the kinetic energy of the fluid and minor losses at the entry to the pipes may be neglected.

Assume, as a first approximation, that $R/\rho u^2$ is the same in each pipe and that the velocities in pipes AD, BD, and DC are u_1, u_2, and u_3 respectively. Let the pressure at D be P_D and let point D be z_d m above the datum for the calculation of potential energy, the liquid level in C.

Applying the energy balance equation between D and the liquid level in each of the tanks:

A-D
$$(z_d - 10)g + vP_D + 4\frac{R}{\rho u^2}\frac{1500}{0.3}u_1^2 = 0 \tag{1}$$

B-D
$$(z_d - 16)g + vP_D + 4\frac{R}{\rho u^2}\frac{1500}{0.3}u_2^2 = 0 \tag{2}$$

D-C
$$-z_d g - vP_D + 4\frac{R}{\rho u^2}\frac{750}{0.5}u_3^2 = 0 \tag{3}$$

From equations 1 and 2:
$$6g + 20{,}000\frac{R}{\rho u^2}(u_1^2 - u_2^2) = 0 \tag{4}$$

From equations 2 and 3:
$$-16g + 20{,}000\frac{R}{\rho u^2}(u_2^2 + 0.30u_3^2) = 0 \tag{5}$$

Taking the roughness of mild steel pipe e as 0.00005 m, therefore e/d varies from 0.0001 to 0.00017. As a first approximation, take $R/\rho u^2 = 0.002$ in each pipe. Substituting this value in equations 4 and 5:

$$58.9 + 40(u_1^2 - u_2^2) = 0 \tag{6}$$

$$-156.0 + 40(u_2^2 + 0.30u_3^2) = 0 \qquad (7)$$

The flowrate in DC is equal to the sum of the flowrates in AD and BD.

$$\therefore \qquad \frac{\pi}{4}0.3^2u_1 + \frac{\pi}{4}0.3^2u_2 = \frac{\pi}{4}0.5^2u_3$$

or: $$u_1 + u_2 = 2.78u_3 \qquad (8)$$

From equation 6: $$u_1^2 = u_2^2 - 1.47 \qquad (9)$$

From equations 7, 8, and 9:

$$-156.0 + 40\left\{ u_2^2 + 0.3 \times \left(\frac{1}{2.78}\right)^2 [u_2^2 + u_3^2 - 1.47 + 2u_2\sqrt{(u_2^2 - 1.47)}] \right\} = 0$$

$$\therefore \qquad u_2\sqrt{(u_2^2 - 1.47)} = 50.7 - 13.8u_2^2$$

$$\therefore \qquad u_2^4 - 7.38u_2^2 + 13.57 = 0$$

$$\therefore \qquad u_2^2 = 0.5[7.38 \pm \sqrt{(54.46 - 54.28)}] = 3.90 \text{ or } 3.48$$

$$\therefore \qquad u_2 = 1.975 \text{ or } 1.87 \text{ m/s}$$

Substituting in equation 9: $$u_1 = 1.56 \text{ or } 1.42 \text{ m/s}$$

Substituting in equation 8: $$u_3 = 1.30 \text{ or } 1.18 \text{ m/s}$$

When these values of u_1, u_2, and u_3 are substituted back in equation 7, the lower set of values satisfies the equation and the higher set, introduced as false roots during squaring do not.

Thus: $$u_1 = 1.42 \text{ m/s}$$
$$u_2 = 1.87 \text{ m/s}$$
$$u_3 = 1.18 \text{ m/s}$$

The assumed value of 0.002 for $R/\rho u^2$ can now be checked:

For pipe AD: $$Re = \frac{0.3 \times 1.42 \times 870}{0.7 \times 10^{-3}} = 5.3 \times 10^5$$

For pipe BD: $$Re = \frac{0.3 \times 1.87 \times 870}{0.7 \times 10^{-3}} = 6.9 \times 10^5$$

For pipe DC: $$Re = \frac{0.5 \times 1.18 \times 870}{0.7 \times 10^{-3}} = 7.3 \times 10^5$$

For $e/d = 0.0001$ to 0.00017 and this range of Re, $R/\rho u^2$ varies from 0.0019 to 0.0017. The assumed value of 0.002 is therefore sufficiently close.

The volumetric flowrate is, therefore:

$$\frac{\pi}{4} \times 0.5^2 \times 1.18 = \underline{\underline{0.23 \text{ m}^3/\text{s}}}$$

3.5. TYPES OF FLOW

3.5.1. Reynolds number, shear stress, and momentum transfer

For a fluid flowing through a pipe the momentum per unit cross-sectional area is given by ρu^2. This quantity, which is proportional to the inertia force per unit area, is the force required to counterbalance the momentum flux.

The ratio u/d represents the velocity gradient in the fluid, and thus the group $(\mu u/d)$ is proportional to the shear stress in the fluid, so that $(\rho u^2)/(\mu u/d) = (du\rho/\mu) = Re$ is proportional to the ratio of the inertia forces to the viscous forces. This is an important physical interpretation of the Reynolds number.

In turbulent flow with high values of Re, the inertia forces become predominant and the viscous shear stress becomes correspondingly less important.

Shear stress and viscosity

In steady streamline flow the direction and velocity of flow at any point remain constant and the shear stress R_y at a point where the velocity gradient at right angles to the direction of flow is $\partial u_x/\partial y$ and is given for a Newtonian fluid by the relation:

$$R_y = -\mu \frac{\partial u_x}{\partial y} = -\frac{\mu}{\rho} \frac{\partial(\rho u_x)}{\partial y} \tag{3.25}$$

showing the relation between shear stress and momentum per unit volume (ρu_x). (See equation 3.2.)

In turbulent motion, the presence of circulating or eddy currents brings about a much-increased exchange of momentum in all three directions of the stream flow, and these eddies are responsible for the random fluctuations in velocity u_E. The high rate of transfer in turbulent flow is accompanied by a much higher shear stress for a given velocity gradient.

Thus:
$$R_y = -\left(\frac{\mu}{\rho} + E\right) \frac{\partial(\rho u_x)}{\partial y} \tag{3.26}$$

where E is known as the *eddy kinematic viscosity* of the fluid, which will depend upon the degree of turbulence in the fluid, is not a physical property of the fluid and varies with position.

In streamline flow, E is very small and approaches zero, so that μ/ρ determines the shear stress. In turbulent flow, E is negligible at the wall and increases very rapidly with distance from the wall. LAUFER[7], using very small hot-wire anemometers, measured the velocity fluctuations and gave a valuable account of the structure of turbulent flow. In typical chemical engineering transport operations of mass, heat, and momentum transfer, the transfer has to be effected through the laminar layer near the wall, and it is here that there is the greatest resistance to transfer.

The Reynolds group will often be used where a moving fluid is concerned. Thus the drag produced as a fluid flows past a particle is related to the Reynolds number in which the diameter of the particle is used in place of the diameter of the pipe. Under these conditions the transition from streamline to turbulent flow occurs at a very much lower value. Again for the flow of fluid through a bed composed of granular particles, a mean dimension of the particles is used, and the velocity is usually calculated by dividing the flowrate by the total area of the bed. In this case there is no sharp transition from streamline to turbulent flow because the sizes of the individual flow passages vary.

If the surface over which the fluid is flowing contains a series of relatively large projections, turbulence may arise at a very low Reynolds number. Under these conditions, the frictional force will be increased but so will the coefficients for heat transfer and mass transfer, and therefore turbulence is often purposely induced by this method.

3.5.2. Velocity distributions, streamline flow

The distribution of velocity (in the stream direction) over the cross-section of a fluid flowing in a pipe is not uniform. Whilst this distribution in velocity over a diameter can be calculated for streamline flow this is not possible in the same basic manner for turbulent flow.

The pressure drop due to friction and the velocity distribution resulting from the shear stresses within a fluid in streamline Newtonian flow will be considered for three cases: (a) the flow through a pipe of circular cross-section, (b) the flow between two parallel plates, and (c) the flow through an annulus. The velocity at any distance from the boundary surfaces will be calculated and the mean velocity of the fluid will be related to the pressure gradient in the system. For flow through a circular pipe, the kinetic energy of the fluid will be calculated in terms of the mean velocity of flow.

Pipe of circular cross-section

A horizontal pipe with a concentric element marked *ABCD* is shown in Figure 3.10. Since the flow is steady, the net force on this element must be zero. The forces acting are the normal pressures over the ends and shear forces over the curved sides.

$$\text{Force over } AB = P\pi s^2$$

$$\text{Force over } CD = -\left(P + \frac{dP}{dl}\delta l\right)\pi s^2$$

$$\text{Force over curved surface } = 2\pi s\delta l R_y$$

$$\text{where the shear stress } R_y = \mu\frac{du_x}{ds}\left(= -\mu\frac{du_x}{dy}\right) \tag{3.27}$$

Figure 3.10. Flow through pipe

Taking a force balance:

$$P\pi s^2 - \left(P + \frac{dP}{dl}\delta l\right)\pi s^2 + 2\pi s\,\delta l\,\mu\frac{du_x}{ds} = 0$$

i.e.:

$$\left(-\frac{dP}{dl}\right)s + 2\mu\frac{du_x}{ds} = 0 \tag{3.28}$$

From equation 3.27:

$$\frac{du_x}{dy} = -\frac{du_x}{ds}$$

and hence in equation 3.28:

$$\frac{\mathrm{d}u_x}{\mathrm{d}y} = \left(-\frac{\mathrm{d}P}{\mathrm{d}l}\right)\frac{s}{2\mu}$$

and the shear stress at the wall is given by:

$$\left(\frac{\mathrm{d}u_x}{\mathrm{d}y}\right)_{y=0} = \left(-\frac{\mathrm{d}P}{\mathrm{d}l}\right)\frac{r}{2\mu} = \left(-\frac{\mathrm{d}P}{\mathrm{d}l}\right)\frac{d}{4\mu} \qquad (3.29)$$

The velocity at any distance s from the axis of the pipe can now be found by integrating equation 3.28.

Thus:

$$u_x = \frac{1}{2\mu}\frac{\mathrm{d}P}{\mathrm{d}l}\frac{s^2}{2} + \text{constant}$$

At the walls of the pipe (i.e. where $s = r$) the velocity u_x must be zero to satisfy the condition of zero wall slip. Substituting the value $u_x = 0$, when $s = r$:

$$\text{constant} = \frac{1}{2\mu}\left(-\frac{\mathrm{d}P}{\mathrm{d}l}\right)\frac{r^2}{2}$$

and, therefore:

$$u_x = \frac{1}{4\mu}\left(-\frac{\mathrm{d}P}{\mathrm{d}l}\right)(r^2 - s^2) \qquad (3.30)$$

Thus the velocity over the cross-section varies in a parabolic manner with the distance from the axis of the pipe. The velocity of flow is seen to be a maximum when $s = 0$, i.e. at the pipe axis.

Thus the maximum velocity, at the pipe axis, is given by u_{CL} where:

$$u_{max} = u_{CL} = \frac{1}{4\mu}\left(-\frac{\mathrm{d}P}{\mathrm{d}l}\right)r^2 = \left(-\frac{\mathrm{d}P}{\mathrm{d}l}\right)\frac{d^2}{16\mu} \qquad (3.31)$$

Hence:

$$\frac{u_x}{u_{CL}} = 1 - \frac{s^2}{r^2} \qquad (3.32)$$

or:

$$= 1 - \frac{4s^2}{d^2} \qquad (3.33)$$

The velocity is thus seen to vary in a parabolic manner over the cross-section of the pipe. This agrees well with experimental measurements.

Volumetric rate of flow and average velocity

If the velocity is taken as constant over an annulus of radii s and $(s + \mathrm{d}s)$, the volumetric rate of flow $\mathrm{d}Q$ through the annulus is given by:

$$\mathrm{d}Q = 2\pi s\,\mathrm{d}s\,u_x$$

$$= 2\pi u_{CL}s\left(1 - \frac{s^2}{r^2}\right)\mathrm{d}s \qquad (3.34)$$

Then the total flow over the cross-section is given by integrating equation 3.34:

$$Q = 2\pi u_{CL} \int_0^r s \left(1 - \frac{s^2}{r^2}\right) ds$$

$$= 2\pi u_{CL} \left[\frac{s^2}{2} - \frac{s^4}{4r^2}\right]_0^r$$

$$= \frac{\pi}{2} r^2 u_{CL} = \frac{\pi}{8} d^2 u_{CL} \tag{3.35}$$

Thus the average velocity u is given by:

$$u = \frac{Q}{(\pi d^2/4)}$$

On substitution from equation 3.35 into equation 3.31:

$$u = \left(-\frac{dP}{dl}\right) \frac{r^2}{8\mu} = \left(-\frac{dP}{dl}\right) \frac{d^2}{32\mu} = \frac{u_{CL}}{2} = \frac{u_{max}}{2} \tag{3.36}$$

This relation was derived by HAGEN[8] in 1839 and independently by POISEUILLE[9] in 1840.

Now: $\quad 32\mu u \dfrac{dl}{d^2} = -dP = 4\dfrac{R}{\rho u^2} \dfrac{dl}{d}(\rho u^2) \quad$ (from equations 3.15 and 3.36)

so that: $\qquad \dfrac{R}{\rho u^2} = \dfrac{8\mu}{u d \rho} = 8Re^{-1} \tag{3.37}$

as shown earlier in this chapter (Figure 3.7).

From equations 3.32 and 3.35:

$$\frac{u_x}{u} = \frac{2(d^2 - 4s^2)}{d^2} = 2\left(1 - \frac{s^2}{r^2}\right) \tag{3.38}$$

Equation 3.38 is plotted in Figure 3.11 which shows the shape of the velocity profile for streamline flow.

The velocity gradient at any point is obtained by differentiating equation 3.38 with respect to s:

$$\frac{du_x}{dy} = -\frac{du_x}{ds} = \frac{4su}{r^2}$$

At the wall, $s = r$ and $\quad \left(\dfrac{du_x}{dy}\right)_{y=0} = \dfrac{4u}{r} = \dfrac{8u}{d} \tag{3.39}$

Thus, the shear stress at the wall $R = \dfrac{8\mu u}{d} \tag{3.40}$

Kinetic energy of fluid

In order to obtain the kinetic energy term for insertion in the energy balance equation, it is necessary to obtain the average kinetic energy per unit mass in terms of the mean velocity.

Figure 3.11.   Shape of velocity profiles for streamline and turbulent flow

The kinetic energy of the fluid flowing per unit time in the annulus between s and $(s + ds)$

$$= (\rho 2\pi s \, ds \, u_x)\frac{u_x^2}{2}$$

$$= \pi \rho s u_x^3 ds$$

$$= \pi \rho u^3 8s \left[1 - \frac{s^2}{r^2}\right]^3 ds \quad \text{(from equation 3.38)}$$

The total kinetic energy per unit time of the fluid flowing in the pipe

$$= -4\pi \rho u^3 r^2 \int_{s=0}^{s=r} \left(1 - \frac{s^2}{r^2}\right)^3 d\left(1 - \frac{s^2}{r^2}\right)$$

Integrating gives: $$= -\frac{4\pi \rho u^3 r^2}{4}\left[\left(1 - \frac{s^2}{r^2}\right)^4\right]_{s=0}^{s=r}$$

$$= \rho \pi r^2 u^3$$

Thus the kinetic energy per unit mass:

$$\frac{\rho \pi r^2 u^3}{\rho \pi r^2 u} = u^2 \tag{3.41}$$

In the energy balance equation, the kinetic energy per unit mass was expressed as $u^2/2\alpha$ and thus $\alpha = 0.5$ for the streamline flow of a fluid in a round pipe.

Flow between two parallel plates

Considering the flow of fluid between two plates of unit width, a distance d_a apart (Figure 3.12), then with respect to the equilibrium of an element $ABCD$, a force balance can be set up in a similar manner to that used for flow through pipes.

$$P\,2s - \left(P + \frac{dP}{dl}\delta l\right) 2s + 2\delta l \mu \frac{du_x}{ds} = 0$$

i.e.:
$$-\frac{dP}{dl}s + \mu \frac{du_x}{ds} = 0 \tag{3.42}$$

and:
$$u_x = \frac{1}{\mu}\frac{dP}{dl}\frac{s^2}{2} + \text{ constant}$$

When $s = d_a/2$, $u_x = 0$.

$\therefore$
$$\text{constant} = -\frac{d_a^2}{8}\frac{1}{\mu}\frac{dP}{dl}$$

i.e.:
$$u_x = \frac{1}{2\mu}\left(-\frac{dP}{dl}\right)\left(\frac{d_a^2}{4} - s^2\right) \tag{3.43}$$

The total rate of flow of fluid between the plates is obtained by calculating the flow through two laminae of thickness ds and situated at a distance s from the centre plane and then integrating. Flow through laminae:

$$dQ' = \frac{1}{2\mu}\left(-\frac{dP}{dl}\right)\left(\frac{d_a^2}{4} - s^2\right)2\,ds$$

Then, on integrating between the limits of s from 0 to $d_a/2$:

Total rate of flow:
$$Q' = \frac{1}{\mu}\left(-\frac{dP}{dl}\right)\left(\frac{d_a^3}{8} - \frac{d_a^3}{24}\right)$$

$$= \left(-\frac{dP}{dl}\right)\frac{d_a^3}{12\mu} \tag{3.44}$$

Figure 3.12.   Streamline flow between parallel plates

The average velocity of the fluid: $u = \dfrac{Q'}{d_a}$

$$= \left(-\frac{dP}{dl}\right)\frac{d_a^2}{12\mu} \tag{3.45}$$

The maximum velocity occurs at the centre plane and its value is obtained by putting $s = 0$ in equation 3.43:

$$\text{Maximum velocity} = u_{max} = \left(-\frac{dP}{dl}\right)\frac{d_a^2}{8\mu} = 1.5u \tag{3.46}$$

It has been assumed that the width of the plates is large compared with the distance between them so that the flow can be considered as unidirectional.

Flow through an annulus

The velocity distribution and the mean velocity of a fluid flowing through an annulus of outer radius r and inner radius r_i is more complex (Figure 3.13). If the pressure changes by an amount dP as a result of friction in a length dl of annulus, the resulting force can be equated to the shearing force acting on the fluid. Consider the flow of the fluid situated at a distance not greater than s from the centre line of the pipes. The shear force acting on this fluid consists of two parts; one is the drag on its outer surface; this can be expressed in terms of the viscosity of the fluid and the velocity gradient at that radius: the other is the drag occurring at the inner boundary of the annulus; this cannot be estimated at present and will be denoted by the symbol λ for unit length of pipe.

Then:

$$\left(\frac{dP}{dl}\right)dl\,\pi(s^2 - r_i^2) = \mu 2\pi s\,dl\left(\frac{du_x}{ds}\right) + \lambda\,dl \tag{3.47}$$

where u_x is the velocity of the fluid at radius s.

$$\therefore \qquad du_x = \frac{s^2 - r_i^2}{2\mu s}\frac{dP}{dl}ds - \lambda\frac{ds}{2\pi\mu s} \tag{3.48}$$

Integrating:

$$u_x = \frac{1}{2\mu}\frac{dP}{dl}\left(\frac{s^2}{2} - r_i^2\ln s\right) - \frac{\lambda}{2\pi\mu}\ln s + u_c \tag{3.49}$$

where u_c is an integration constant with dimensions of velocity.

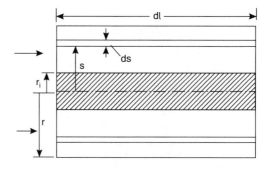

Figure 3.13.   Streamline flow through annulus

Substituting the boundary conditions $s = r_i$, $u_x = 0$, and $s = r$, $u_x = 0$ in equation 3.49 and solving for λ and u_c:

$$\lambda = \pi \frac{dP}{dl} \left(\frac{r^2 - r_i^2}{2 \ln(r/r_i)} - r_i^2 \right) \tag{3.50}$$

and:

$$u_c = \frac{1}{2\mu} \left(-\frac{dP}{dl} \right) \left(\frac{r^2}{2} - \frac{r^2 - r_i^2}{2 \ln(r/r_i)} \ln r \right) \tag{3.51}$$

Now substituting these values of λ and u_c in equation 3.49, and simplifying:

$$u_x = \frac{1}{4\mu} \left(-\frac{dP}{dl} \right) \left(r^2 - s^2 + \frac{r^2 - r_i^2}{\ln(r/r_i)} \ln \frac{s}{r} \right) \tag{3.52}$$

The rate of flow of fluid through a small annulus of inner radius s and outer radius $(s + ds)$, is given by:

$$dQ = 2\pi s \, ds \, u_x$$

$$= \frac{\pi}{2\mu} \left(-\frac{dP}{dl} \right) \left(r^2 s - s^3 + \frac{r^2 - r_i^2}{\ln(r/r_i)} s \ln \frac{s}{r} \right) ds \tag{3.53}$$

Integrating between the limits $s = r_i$ and $s = r$:

$$Q = \frac{\pi}{8\mu} \left(-\frac{dP}{dl} \right) \left(r^2 + r_i^2 - \frac{r^2 - r_i^2}{\ln(r/r_i)} \right) (r^2 - r_i^2) \tag{3.54}$$

The average velocity: $\quad u = \dfrac{Q}{\pi(r^2 - r_i^2)}$

$$= \frac{1}{8\mu} \left(-\frac{dP}{dl} \right) \left(r^2 + r_i^2 - \frac{r^2 - r_i^2}{\ln(r/r_i)} \right) \tag{3.55}$$

3.5.3 The transition from laminar to turbulent flow in a pipe

Laminar flow ceases to be stable when a small perturbation or disturbance in the flow tends to increase in magnitude rather than decay. For flow in a pipe of circular cross-section, the critical condition occurs at a Reynolds number of about 2100. Thus although laminar flow can take place at much higher values of Reynolds number, that flow is no longer stable and a small disturbance to the flow will lead to the growth of the disturbance and the onset of turbulence. Similarly, if turbulence is artificially promoted at a Reynolds number of less than 2100 the flow will ultimately revert to a laminar condition in the absence of any further disturbance.

RYAN and JOHNSON[10] have proposed a *stability parameter* Z. If the critical value Z_c of that parameter is exceeded at any point on the cross-section of the pipe, then turbulence will ensue. Based on a concept of a balance between energy supply to a perturbation and energy dissipation, they proposed that Z could be defined as:

$$Z = \frac{r\rho u_x}{R} \frac{\partial u_x}{\partial y} \tag{3.56}$$

Z will be zero at the pipe wall ($u_x = 0$) and at the axis ($\partial u_x / \partial y = 0$), and will reach a maximum at some intermediate position in the cross-section. From equation 3.6, the wall shear stress R for laminar flow can be expressed in terms of the pressure gradient along the pipe ($-\Delta P/l$):

$$R = \frac{-\Delta Pd}{4l} = \frac{-\Delta Pr}{2l} \qquad \text{(equation 3.6)}$$

Thus:
$$Z = \frac{2l\rho u_x}{-\Delta P}\frac{\partial u_x}{\partial y}$$

The velocity distribution over the pipe cross-section is given by:

$$u_x = \frac{-\Delta P}{4\mu l}(r^2 - s^2)$$

and:
$$\frac{\partial u_x}{\partial y} = -\frac{\partial u_x}{\partial s} = \frac{-\Delta Ps}{2\mu l}$$

Thus:
$$Z = \frac{2l\rho}{-\Delta P}\left[\frac{-\Delta P}{4\mu l}(r^2 - s^2)\right]\frac{-\Delta Px}{2\mu l}$$

$$= \frac{-\Delta P\rho}{4\mu^2 l}(r^2 s - s^3)$$

The value of s at which Z is a maximum is found by differentiating Z with respect to s and equating the derivative to zero.

$$\frac{\partial Z}{\partial s} = \frac{-\Delta P\rho}{4\mu^2 l}(r^2 - 3s^2)$$

and, when $\dfrac{\partial Z}{\partial s} = 0$:
$$s = \frac{r}{\sqrt{3}}$$

Thus, the position in the cross-section where laminar flow will first break down is where $s/r = 1/\sqrt{3}$

and:
$$Z_{\max} = \frac{-\Delta P}{6\sqrt{3}}\frac{\rho r^3}{\mu^2 l} \qquad (3.57)$$

Substituting for ($-\Delta P/l$) from equation 3.36:

$$u = \frac{-\Delta Pr^2}{8\mu l} \qquad \text{(equation 3.36)}$$

$$Z_{\max} = \frac{4}{3\sqrt{3}}\frac{\rho r}{\mu}$$

$$= \frac{2}{3\sqrt{3}}\frac{ud\rho}{\mu} = 0.384\,Re_{\text{crit}} \qquad (3.58)$$

Taking $Re_{\text{crit}} = 2100$:

$$Z_{\max} = 808$$

3.5.4. Velocity distribution, turbulent flow

No exact mathematical analysis of the conditions within a turbulent fluid has yet been developed, though a number of semi-theoretical expressions for the shear stress at the walls of a pipe of circular cross-section have been suggested.[6]

The shear stresses within the fluid are responsible for the frictional force at the walls and the velocity distribution over the cross-section. A given assumption for the shear stress at the walls therefore implies some particular velocity distribution. It will be shown in Chapter 11 that the velocity at any point in the cross-section will be proportional to the one-seventh power of the distance from the walls if the shear stress is given by the Blasius equation (equation 3.10). This may be expressed as follows:

$$\frac{u_x}{u_{CL}} = \left(\frac{y}{r}\right)^{1/7} \tag{3.59}$$

where u_x is the velocity at a distance y from the walls, u_{CL} the velocity at the axis of the pipe, and r the radius of the pipe.

This equation is sometimes referred to as the *Prandtl one-seventh power law*.

Mean velocity

In a thin annulus of inner radius s and outer radius $s + ds$, the velocity u_x may be taken as constant (see Figure 3.10).

$$\therefore \qquad dQ = 2\pi s \, ds u_x$$

$$= -2\pi(r - y)dy \, u_x \text{(since } s + y = r) \tag{3.60}$$

$$= -2\pi(r - y)dy \, u_{CL} \left(\frac{y}{r}\right)^{1/7} \tag{3.61}$$

$$\therefore \qquad Q = \int_{y=r}^{y=0} -2\pi r^2 u_{CL} \left(1 - \frac{y}{r}\right)\left(\frac{y}{r}\right)^{1/7} d\left(\frac{y}{r}\right)$$

$$= -2\pi r^2 u_{CL} \left[\frac{7}{8}\left(\frac{y}{r}\right)^{8/7} - \frac{7}{15}\left(\frac{y}{r}\right)^{15/7}\right]_{y=r}^{y=0}$$

$$= 2\pi r^2 u_{CL} \left[\frac{7}{8} - \frac{7}{15}\right]$$

$$= \frac{49}{60}\pi r^2 u_{CL} = 0.817\pi r^2 u_{CL} \tag{3.62}$$

The mean velocity of flow: $\quad u = \dfrac{Q}{\pi r^2} = \dfrac{49}{60}u_{CL} = 0.817u_{CL} \tag{3.63}$

This relation holds provided that the one-seventh power law can be assumed to apply over the whole of the cross-section of the pipe. This is strictly so only at high Reynolds numbers when the thickness of the laminar sub-layer is small. By combining equations 3.59 and 3.63 it is seen that the velocity profile is given by:

$$\frac{u_x}{u} = \frac{1}{0.817}\left(\frac{y}{r}\right)^{1/7}$$

$$= 1.22\left(1 - \frac{s}{r}\right)^{1/7} \tag{3.64}$$

Equation 3.64 is plotted in Figure 3.11, and it will be noted that the velocity profile is very much flatter than for streamline flow.

The variation of (u/u_{max}) with the Reynolds number is shown in Figure 3.14. The sharp change at a Reynolds number between 2000 and 3000 will be noted.

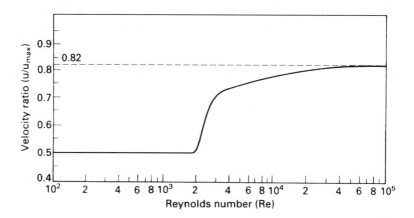

Figure 3.14.   Variation of (u/u_{max}) with Reynolds number in a pipe

Kinetic energy

Since: $$dQ = -2\pi(r - y)dy\, u_x \qquad \text{(equation 3.60)}$$

the kinetic energy per unit time of the fluid passing through the annulus

$$= -2\pi(r - y)\rho\, dy\, u_x \frac{u_x^2}{2}$$

$$= -\pi u_{CL}^3 \rho r^2 \left(1 - \frac{y}{r}\right)\left(\frac{y}{r}\right)^{3/7} d\left(\frac{y}{r}\right) \qquad (3.65)$$

Integration of equation 3.65 gives the total kinetic energy per unit time as:

$$= \pi u_{CL}^3 \rho r^2 \left[\tfrac{7}{10} - \tfrac{7}{17}\right]$$

$$= \tfrac{49}{170}\pi r^2 \rho u_{CL}^3 \qquad (3.66)$$

The mean kinetic energy per unit mass of fluid from equation 3.63 is:

$$= \frac{\left(\tfrac{49}{170}\pi r^2 \rho u_{CL}^3\right)}{\left(\tfrac{49}{60}\pi r^2 \rho u_{CL}\right)}$$

$$= \tfrac{6}{17}u_{CL}^2$$

$$= \tfrac{6}{17}\left(\tfrac{60}{49}u\right)^2 \quad \text{(from equation 3.60)}$$

$$= 0.53u^2 \approx \frac{u^2}{2}$$

$$= \frac{u^2}{2\alpha} \quad \text{(by definition of } \alpha) \qquad (3.67)$$

Thus for turbulent flow at high Reynolds numbers, where the thickness of the laminar sub-layer can be neglected, $\alpha \approx 1$.

When the thickness of the laminar sub-layer cannot be neglected, α will be slightly less than 1.

Flow in non-circular ducts

For turbulent flow in a duct of non-circular cross-section, the hydraulic mean diameter may be used in place of the pipe diameter and the formulae for circular pipes can then be applied without introducing a large error. This method of approach is entirely empirical.

The *hydraulic mean diameter* d_m is defined as four times the cross-sectional area divided by the wetted perimeter: some examples are given. For a circular pipe:

$$d_m = \frac{4(\pi/4)d^2}{\pi d} = d \tag{3.68}$$

For an annulus of outer radius r and inner radius r_i:

$$d_m = \frac{4\pi(r^2 - r_i^2)}{2\pi(r + r_i)} = 2(r - r_i) = d - d_i \tag{3.69}$$

For a duct of rectangular cross-section d_a by d_b:

$$d_m = \frac{4d_a d_b}{2(d_a + d_b)}$$

$$= \frac{2d_a d_b}{d_a + d_b} \tag{3.70}$$

For streamline flow this method is not applicable, and exact expressions relating the pressure drop to the velocity can be obtained for ducts of certain shapes only. Two examples of this have already been given.

Flow through curved pipes

If the pipe is not straight, the velocity distribution over the section is altered and the direction of flow of the fluid is continuously changing. The frictional losses are therefore somewhat greater than for a straight pipe of the same length. If the radius of the pipe divided by the radius of the bend is less than about 0.002 however, the effects of the curvature are negligible.

It has been found[11] that stable streamline flow persists at higher values of the Reynolds number in coiled pipes. Thus, for instance, when the ratio of the diameter of the pipe to the diameter of the coil is 1 to 15, the transition occurs at a Reynolds number of about 8000.

3.5.5. Miscellaneous friction losses

The friction losses occurring as a result of a sudden enlargement or contraction in the cross-section of the pipe, and the resistance of various standard pipe fittings, will now be considered.

Sudden enlargement

If the diameter of the pipe suddenly increases, as shown in Figure 3.15, the effective area available for flow will gradually increase from that of the smaller pipe to that of the larger one and the velocity of flow will progressively decrease. Thus fluid with a relatively high velocity will be injected into relatively slow moving fluid; turbulence will be set up and much of the excess kinetic energy will be converted into heat and therefore wasted. If the change of cross-section is gradual, the kinetic energy can be recovered as pressure energy.

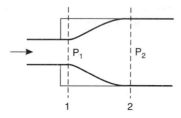

Figure 3.15.   Sudden enlargement

For the fluid flowing as shown in Figure 3.15, from section 1 (the pressure just inside the enlargement is found to be equal to that at the end of the smaller pipe) to section 2, the net force = the rate of change of momentum, i.e.

$$P_1 A_2 - P_2 A_2 = \rho_2 A_2 u_2 (u_2 - u_1)$$

i.e.:

$$(P_1 - P_2)v_2 = u_2^2 - u_1 u_2 \tag{3.71}$$

For an incompressible fluid:
$$(P_1 - P_2)v_2 = -\int_{P_1}^{P_2} v \, dP \tag{3.72}$$

Applying the energy equation between the two sections:

$$\frac{u_1^2}{2\alpha_1} = \frac{u_2^2}{2\alpha_2} + \int_1^2 v \, dP + F \tag{3.73}$$

i.e.:

$$F = \frac{u_1^2}{2\alpha_1} - \frac{u_2^2}{2\alpha_2} + u_2^2 - u_1 u_2 \tag{3.74}$$

For fully turbulent flow: $\alpha_1 = \alpha_2 = 1$ and

$$F = \frac{(u_1 - u_2)^2}{2} \tag{3.75}$$

The change in pressure $-\Delta P_f$ is therefore given by:

$$-\Delta P_f = \rho \frac{(u_1 - u_2)^2}{2} \tag{3.76}$$

The loss of head h_f is given by:

$$h_f = \frac{(u_1 - u_2)^2}{2g} \tag{3.77}$$

Substituting into equation 3.76, from the relation $u_1A_1 = u_2A_2$:

$$-\Delta P_f = \frac{\rho u_1^2}{2} \left[1 - \left(\frac{A_1}{A_2} \right) \right]^2 \tag{3.78}$$

The loss can be substantially reduced if a tapering enlarging section is used. For a circular pipe, the optimum angle is about $7°$ and for a rectangular duct is about $11°$.

If A_2 is very large compared with A_1, as for instance at a pipe exit:

$$-\Delta P_f = \frac{\rho u_1^2}{2} \tag{3.79}$$

The use of a small angle enlarging section will be seen in the venturi meter (Chapter 6).

Example 3.5

Water flows at 7.2 m³/h through a sudden enlargement from a 40 mm to a 50 mm diameter pipe. What is the loss in head?

Solution

$$\text{Velocity in 50 mm pipe} = \frac{7.2/3600}{(\pi/4)(50 \times 10^{-3})^2} = 1.02 \text{ m/s}$$

$$\text{Velocity in 40 mm pipe} = \frac{7.2/3600}{(\pi/4)(40 \times 10^{-3})^2} = 1.59 \text{ m/s}$$

Head lost is given by equation 3.77 as:

$$h_f = \frac{(u_1 - u_2)^2}{2g}$$

$$= \frac{0.57^2}{2 \times 9.81} = 0.0165 \text{ m water}$$

or:

$$\underline{\underline{16.5 \text{ mm water}}}$$

Sudden contraction

The effective area for flow gradually decreases as the sudden contraction is approached and then continues to decrease, for a short distance, to what is known as the *vena contracta*. After the vena contracta the flow area gradually approaches that of the smaller pipe. As the fluid moves towards the vena contracta it is accelerated and pressure energy is converted into kinetic energy; this process does not give rise to eddy formation and losses are very small. However, beyond the vena contracta the velocity falls as the flow area increases and conditions are equivalent to those for a sudden enlargement. The expression for the loss at a sudden enlargement can, therefore, be applied for the fluid flowing from the vena contracta to some section a small distance away, where the whole of the cross-section of the pipe is available for flow.

Applying equation 3.75 between sections C and 2, as shown in Figure 3.16 the frictional loss per unit mass of fluid is given by:

Figure 3.16.   Sudden contraction

$$F = \frac{(u_c - u_2)^2}{2}$$

$$= \frac{u_2^2}{2}\left[\frac{u_c}{u_2} - 1\right]^2 \tag{3.80}$$

Denoting the ratio of the area at section C to that at section 2 by a coefficient of contraction C_c:

$$F = \frac{u_2^2}{2}\left[\frac{1}{C_c} - 1\right]^2 \tag{3.81}$$

Thus the change in pressure ΔP_f is $-(\rho u^2/2)[(1/C_c) - 1]^2$ and the head lost is:

$$\frac{u_2^2}{2g}\left[\frac{1}{C_c} - 1\right]^2$$

C_c varies from about 0.6 to 1.0 as the ratio of the pipe diameters varies from 0 to 1. For a common value of C_c of 0.67:

$$F = \frac{u_2^2}{8} \tag{3.82}$$

It should be noted that the maximum possible frictional loss which can occur at a change in cross-section is the whole of the kinetic energy of the fluid.

Pipe fittings

Most pipes are made from steel with or without small alloying ingredients, and they are made by welding or drawing to give a seamless pipe. Tubes from 6 to 50 mm are frequently made from non-ferrous metals such as copper, brass, or aluminium, and these are very widely used in heat exchangers. For special purposes there is a very wide range of materials including industrial glass, many varieties of plastics, rubber, stoneware, and ceramic materials. The normal metal piping is supplied in standard lengths of about 6 m and these are joined to give longer lengths as required. Such jointing is by screw flanging or welding, and the small diameter copper or brass tubes are often brazed or soldered or jointed by compression fittings.

A very large range of pipe fittings is available to enable branching and changes in size to be arranged for industrial pipe layouts. For the control of the flow of a fluid, valves of various kinds are used. The most important of these are known as gate, globe, and needle valves. In addition, check valves are supplied for relieving the pressure in pipelines, and reducing valves are available for controlling the pressure on the downstream side of the valve. Gate and globe valves are supplied in all sizes and may be controlled by

Table 3.2.   Friction losses in pipe fittings

	Number of pipe diameters	Number of velocity heads ($u^2/2g$)
45° elbows (a)*	15	0.3
90° elbows (standard radius) (b)	30–40	0.6–0.8
90° square elbows (c)	60	1.2
Entry from leg of T-piece (d)	60	1.2
Entry into leg of T-piece (d)	90	1.8
Unions and couplings (e)	Very small	Very small
Globe valves fully open	60–300	1.2–6.0
Gate valves: fully open	7	0.15
$\frac{3}{4}$ open	40	1
$\frac{1}{2}$ open	200	4
$\frac{1}{4}$ open	800	16

*See Figure 3.17.

(a)   (b)   (c)   (d)   (e)

Figure 3.17.   Standard pipe fittings

Globe valve   Gate valve

Figure 3.18.   Standard valves

motor units operated from an automatic control system. Hand wheels should be fitted for emergency use.

On the whole, gate valves give coarse control, globe valves finer and needle valves the finest control of the rate of flow. Diaphragm valves are also very much in use for the handling of corrosive fluids since the diaphragm can be made of corrosion resistant materials.

Some average figures are given in Table 3.2 for the friction losses in various pipe fittings for the turbulent flow of fluid. They are expressed in terms of the equivalent length of straight pipe with the same resistance, and as the number of velocity heads $(u^2/2g)$ lost. Considerable variation will occur according to the exact construction of the fittings.

Typical pipe-fittings are shown in Figures 3.17 and 3.18. For illustrations of other types of valves, see Volume 6.

Example 3.6

2.27 m^3/h water at 320 K is pumped in a 40 mm i.d. pipe through a length of 150 m in a horizontal direction and up through a vertical height of 10 m. In the pipe there is a control valve which may be taken as equivalent to 200 pipe diameters and also other pipe fittings equivalent to 60 pipe diameters. Also in the line is a heat exchanger across which there is a loss in head of 1.5 m of water. If the main pipe has a roughness of 0.2 mm, what power must be supplied to the pump if it is 60 per cent efficient?

Solution

Relative roughness:
$$\frac{e}{d} = \frac{0.2}{40} = 0.005$$

Viscosity at 320 K:
$$\mu = 0.65 \text{ mN s/m}^2 \text{ or } 0.65 \times 10^{-3} \text{ N s/m}^2$$

$$\text{Flowrate} = 2.27 \text{ m}^3/\text{h} = 6.3 \times 10^{-4} \text{ m}^3/\text{s}$$

$$\text{Area for flow} = \frac{\pi}{4}(40 \times 10^{-3})^2 = 1.26 \times 10^{-3} \text{ m}^2$$

$$\text{Velocity} = \frac{6.3 \times 10^{-4}}{1.26 \times 10^{-3}} = 0.50 \text{ m/s}$$

$$\therefore \quad Re = \frac{40 \times 10^{-3} \times 0.50 \times 1000}{0.65 \times 10^{-3}} = 30,770$$

$$\therefore \quad \frac{R}{\rho u^2} = 0.004$$

$$\text{Equivalent length of pipe} = 150 + 10 + (260 \times 40 \times 10^{-3}) = 170.4 \text{ m}$$

$$h_f = 4\frac{R}{\rho u^2}\frac{l}{d}\frac{u^2}{g}$$

$$= 4 \times 0.004\frac{170.4}{40 \times 10^{-3}}\frac{0.5^2}{9.81}$$

$$= 1.74 \text{ m}$$

$$\text{Total head to be developed} = 1.74 + 1.5 + 10 = 13.24 \text{ m}$$

$$\text{Mass throughput} = 6.3 \times 10^{-4} \times 1000 = 0.63 \text{ kg/s}$$

$$\text{Power required} = 0.63 \times 13.24 \times 9.81 = 81.8 \text{ W}$$

Since the pump efficiency is 60 per cent, power taken $= \dfrac{81.8}{0.60} = 136.4$ W or $\underline{\underline{0.136 \text{ kW}}}$

The kinetic energy head, $u^2/2g$ has been omitted as this amounts to $0.5^2/(2 \times 9.81) = 0.013$ m, which may be neglected.

Example 3.7

Water in a tank flows through an outlet 25 m below the water level into a 0.15 m diameter horizontal pipe 30 m long, with a $90°$ elbow at the end leading to a vertical pipe of the same diameter 15 m long. This is connected to a second $90°$ elbow which leads to a horizontal pipe of the same diameter, 60 m long, containing a fully open globe valve and discharging to atmosphere 10 m below the level of the water in the tank. Taking $e/d = 0.01$ and the viscosity of water as 1 mN s/m^2, what is the initial rate of discharge?

Solution

From equation 3.17, the head lost due to friction:

$$h_f = 4 \frac{R}{\rho u^2} \frac{l}{d} \frac{u^2}{g} \quad \text{m water}$$

Now the total head loss:

$$h = \frac{u^2}{2g} + h_f + \text{losses in fittings}$$

Losses in fittings (from Table 3.2)

$$= \frac{2 \times 0.8u^2}{2g} \text{ (for the elbows)} + \frac{5.0u^2}{2g} \text{ (for the valve)}$$

$$= \frac{6.6u^2}{2g} \text{ m water}$$

Taking $\dfrac{R}{\rho u^2} = 0.0045$, then:

$$10 = \frac{(6.6 + 1)u^2}{2g} + 4 \times 0.0045 \left[\frac{30 + 15 + 60}{0.15} \right] \frac{u^2}{2g}$$

$$= \frac{(7.6 + 12.6)u^2}{2g}$$

$$u^2 = 9.71$$

and:

$$u = 3.12 \text{ m/s}$$

The assumed value of $R/\rho u^2$ must now be checked.

$$Re = \frac{du\rho}{\mu} = \frac{0.15 \times 3.12 \times 1000}{1 \times 10^{-3}} = 4.68 \times 10^5$$

For $Re = 4.68 \times 10^5$ and $e/d = 0.01$, from Figure 3.7, $R/\rho u^2 = 0.0046$ which agrees with the assumed value.

$\therefore$ rate of discharge $= 3.12 \times (\pi/4)0.15^2 = 0.055$ m^3/s or $(0.055 \times 1000) = \underline{\underline{55 \text{ kg/s}}}$.

Flow over banks of tubes

The frictional loss for a fluid flowing parallel to the axes of the tubes can be calculated in the normal manner by considering the hydraulic mean diameter of the system. This is applicable only to turbulent flow.

For flow at right angles to the axes of the tubes, the cross-sectional area is continually changing, and the problem can be treated as one involving a series of sudden enlargements and sudden contractions. Thus the friction loss would be expected to be directly proportional to the number of banks of pipes j in the direction of flow and to the kinetic energy of the fluid. Thus the pressure drop $-\Delta P_f$ can be written:

$$-\Delta P_f = \frac{C_f j \rho u_t^2}{6} \qquad (3.83)$$

where C_f is a coefficient dependent on the arrangement of the tubes and the Reynolds number. Values of C_f are given in Chapter 9 (Tables 9.3 and 9.4). u_t is the velocity of flow at the narrowest cross-section.

3.6. FLOW WITH A FREE SURFACE

If a liquid is flowing with a free surface exposed to the atmosphere, the pressure at the liquid surface will everywhere be constant and equal to atmospheric pressure. Flow will only take place therefore as the result of the action of the gravitational force and the surface level will necessarily fall in the direction of flow.

Two cases will now be considered. The first, which is the laminar flow of a thin film down an inclined surface, is important in the heat transfer from a condensing vapour where the main resistance to transfer lies in the condensate film, as discussed in Chapter 9 (Section 9.6.1). The second is the flow in open channels which are frequently used for transporting liquids downhill on an industrial site.

3.6.1. Laminar flow down an inclined surface

In any liquid flowing down a surface, a velocity profile will be established with the velocity increasing from zero at the surface itself to a maximum where it is in contact with the ambient atmosphere. The velocity distribution can be obtained in a manner similar to that used earlier in connection with pipe flow, bearing in mind that the driving force is that due to gravity rather than a pressure gradient.

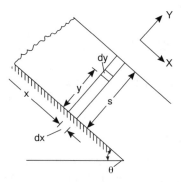

Figure 3.19.   Flow of liquid over a surface

For flow of liquid of depth s down a plane surface of width w inclined at an angle θ to the horizontal (as shown in Figure 3.19), a force balance in the X-direction (parallel to the surface) may be written. In an element of length dx the gravitational force acting on that part of the liquid which is at a distance greater than y from the surface is given by:

$$(s - y)w \, dx \, \rho g \sin \theta$$

If the drag force of the atmosphere is negligible, the retarding force for laminar flow will be attributable to the viscous drag in the liquid at the distance y from the surface and this will be given by:

$$\mu \frac{du_x}{dy} w \, dx$$

where u_x is the velocity of the fluid at that position.

At equilibrium therefore:

$$(s - y)w \, dx \, \rho g \sin \theta = \mu \frac{du_x}{dy} w \, dx \qquad (3.84)$$

Thus, since there will be no slip between the liquid and the surface, $u_x = 0$ when $y = 0$:

$$\int_0^{u_x} du_x = \frac{\rho g \sin \theta}{\mu} \int_0^y (s - y) \, dy$$

and:

$$u_x = \frac{\rho g \sin \theta}{\mu} (sy - \tfrac{1}{2} y^2) \qquad (3.85)$$

The mass rate of flow G of liquid down the surface can now be calculated.

$$G = \int_0^s \left(\frac{\rho g \sin \theta}{\mu} w (sy - \tfrac{1}{2} y^2) \right) \rho \, dy$$

$$= \frac{\rho^2 g \sin \theta}{\mu} w \left(\frac{s^3}{2} - \frac{s^3}{6} \right)$$

$$= \frac{\rho^2 g \sin \theta \, w s^3}{3\mu} \qquad (3.86)$$

Then the mean velocity of the fluid:

$$u = \frac{\rho g \sin \theta \, s^2}{3\mu} \qquad (3.87)$$

For a vertical surface: $\sin \theta = 1$ and $u = \dfrac{\rho g s^2}{3\mu}$

The maximum velocity, which occurs at the free surface:

$$u_s = \frac{\rho g \sin \theta \, s^2}{2\mu} \qquad (3.88)$$

and is 1.5 times the mean velocity of the liquid.

3.6.2. Flow in open channels

For flow in an open channel, consideration will be given only to turbulent flow because streamline flow occurs in practice only when the liquid is flowing as a thin layer, as discussed in the previous section. The transition from streamline to turbulent flow occurs over the range of Reynolds numbers, $u\rho d_m/\mu = 4000 - 11{,}000$, where d_m is the hydraulic mean diameter discussed earlier under *Flow in non-circular ducts*.

Three different types of turbulent flow can be obtained in open channels. They are *tranquil flow*, *rapid flow*, and *critical flow*. In tranquil flow the velocity is less than that at which some disturbance, such as a surge wave, will be transmitted, and the flow is influenced by conditions at both the upstream and the downstream end of the channel. In rapid flow, the velocity of the fluid is greater than the velocity of a surge wave and the conditions at the downstream end do not influence the flow. Critical flow occurs when the velocity is exactly equal to the velocity of a surge wave.

Uniform flow

For a channel of constant cross-section and slope, the flow is said to be uniform when the depth of the liquid D is constant throughout the length of the channel. For these conditions, as indicated in Figure 3.20, for a length l of channel:

The accelerating force acting on liquid $= lA\rho g \sin\theta$.

The force resisting the motion $= R_m pl$

where R_m is the mean value of the shear stress at the solid surface of the channel, p the wetted perimeter and A the cross-sectional area of the flowing liquid.

Figure 3.20.   Uniform flow in open channel

For uniform motion:

$$lA\rho g \sin\theta = R_m pl$$

$$R_m = \frac{A}{p}\rho g \sin\theta$$

$$= \tfrac{1}{4}d_m\rho g \sin\theta \tag{3.89}$$

where d_m is the hydraulic mean diameter $= 4\,A/p$. (see equations 3.68–3.70).

Dividing both sides of the equation by ρu^2, where u is the mean velocity in the channel:

$$\frac{R_m}{\rho u^2} = \frac{d_m}{4}\frac{g\sin\theta}{u^2} \tag{3.90}$$

i.e.:

$$u^2 = \frac{d_m}{4(R_m/\rho u^2)}g\sin\theta \tag{3.91}$$

For turbulent flow, $R_m/\rho u^2$ is almost independent of velocity but is a function of the surface roughness of the channel. Thus the resistance force is proportional to the square of the velocity. $R_m/\rho u^2$ is found experimentally to be proportional to the one-third power of the relative roughness of the channel surface and can conveniently be written:

$$\frac{R_m}{\rho u^2} = \frac{1}{16}\left(\frac{e}{d_m}\right)^{1/3} \tag{3.92}$$

for the values of the roughness e given in Table 3.3.

$$\therefore \qquad u^2 = 4d_m\left(\frac{d_m}{e}\right)^{1/3}g\sin\theta \tag{3.93}$$

$$= 4d_m^{4/3}e^{-1/3}g\sin\theta \tag{3.94}$$

Table 3.3. Values of the roughness e, for use in equation 3.92

	ft	mm
Planed wood or finished concrete	0.00015	0.046
Unplaned wood	0.00024	0.073
Unfinished concrete	0.00037	0.11
Cast iron	0.00056	0.17
Brick	0.00082	0.25
Riveted steel	0.0017	0.51
Corrugated metal	0.0055	1.68
Rubble	0.012	3.66

The volumetric rate of flow Q is then given by:

$$Q = uA$$

$$= 2Ad_m^{2/3}e^{-1/6}\sqrt{g\sin\theta} \tag{3.95}$$

The loss of energy due to friction for unit mass of fluid flowing isothermally through a length l of channel is equal to its loss of potential energy because the other forms of energy remain unchanged:

i.e.:

$$F = gl\sin\theta = 4\frac{R_m}{\rho u^2}\frac{l}{d_m}u^2 \tag{3.96}$$

An empirical equation for the calculation of the velocity of flow in an open channel is the Chezy equation, which can be expressed as follows:

$$u = C\sqrt{\tfrac{1}{4}d_m\sin\theta} \tag{3.97}$$

where the value of the coefficient C is a function of the units of the other quantities in the equation. This expression takes no account of the effect of surface roughness on the velocity of flow.

The velocity of the liquid varies over the cross-section and is usually a maximum at a depth of between $0.05\,D$ and $0.25\,D$ below the surface, at the centre line of the channel. The velocity distribution can be measured by means of a pitot tube as described in Chapter 6.

The shape and the proportions of the channel can be chosen for any given flow rate so that the perimeter, and hence the cost of the channel, is a minimum.

Now:
$$Q = 2Ad_m^{2/3}e^{-1/6}\sqrt{g\sin\theta} \qquad \text{(equation 3.95)}$$

Assuming that slope and the roughness of the channel are fixed then, for a given flowrate Q, from equation 3.95:

$$Ad_m^{2/3} = \text{constant}$$

$\therefore$
$$A^{5/3}p^{-2/3} = \text{constant} \qquad (3.98)$$

The perimeter is therefore a minimum when the cross-section for flow is a minimum.

For a rectangular channel, of depth D and width B:

$$A = DB = D(p - 2D)$$

$\therefore$
$$D^{5/3}(p - 2D)^{5/3}p^{-2/3} = \text{constant}$$

$\therefore$
$$Dp^{3/5} - 2D^2p^{-2/5} = \text{constant}$$

p is a minimum when $\mathrm{d}p/\mathrm{d}D = 0$. Differentiating with respect to D and putting $\mathrm{d}p/\mathrm{d}D$ equal to 0:

$$p^{3/5} - p^{-2/5}4D = 0$$

$\therefore$
$$4D = p = 2D + B$$

i.e.:
$$B = 2D \qquad (3.99)$$

Thus the most economical rectangular section is such that the width is equal to twice the depth of liquid flowing. The most economical proportions for other shapes can be determined in a similar manner.

Specific energy of liquid

Considering a liquid flowing in a channel inclined at an angle θ to the horizontal (Figure 3.21), the various energies possessed by unit mass of fluid at a depth h below the liquid surface (measured at right angles to the bottom of the channel) are as follows:

Internal energy: U

Pressure energy: $(P_a + h\rho g\sec\theta)v = P_a v + hg\sec\theta$

where P_a is atmospheric pressure.

Potential energy: $zg + (D - h)g\sec\theta$

Figure 3.21. Energy of fluid in open channel

where z is the height of the bottom of the channel above the datum level at which the potential energy is reckoned as zero.

Kinetic energy:
$$\frac{u^2}{2}.$$

The total energy per unit mass is, therefore:

$$U + P_a v + zg + Dg \sec\theta + \frac{u^2}{2} \qquad (3.100)$$

It will be seen that at any cross-section the total energy is independent of the depth h below the liquid surface. As the depth is increased, the pressure energy increases at the same rate as the potential energy decreases. If the fluid flows through a length dl of channel, the net change in energy per unit mass is given by:

$$\delta q - \delta W_s = \mathrm{d}U + g\,\mathrm{d}z + g\sec\theta\,\mathrm{d}D + u\,\mathrm{d}u \qquad (3.101)$$

For an irreversible process, from Chapter 2:

$$\mathrm{d}U = T\,\mathrm{d}S - P\,\mathrm{d}v \qquad \text{(equation 2.5)}$$

$$= \delta q + \delta F \qquad \text{(equation 2.9)}$$

assuming the fluid is incompressible, and d$v = 0$.

If no work is done on the surroundings, $\delta W_s = 0$ and:

$$g\,\mathrm{d}z + g\sec\theta\,\mathrm{d}D + u\,\mathrm{d}u + \delta F = 0 \qquad (3.102)$$

For a fluid at a constant temperature, the first three terms in equation 3.100 are independent of the flow conditions within the channel. On the other hand, the last two terms are functions of the velocity and the depth of liquid. The *specific energy* of the fluid is defined by the relation:

$$J = Dg\sec\theta + \frac{u^2}{2} \qquad (3.103)$$

For a horizontal channel, rectangular in section:

$$J = Dg + \frac{u^2}{2} = Dg + \frac{Q^2}{2B^2 D^2}$$

The specific energy will vary with the velocity of the liquid and will be a minimum for some critical value of D; for a given rate of flow Q the minimum will occur when $\mathrm{d}J/\mathrm{d}D = 0$.

Then:
$$g + \frac{(-2Q^2)}{2B^2D^3} = 0$$

$$\frac{u^2}{D} = g$$

and:
$$u = \sqrt{gD} \tag{3.104}$$

This value of u is known as the *critical velocity*.

The corresponding values of D the *critical depth* and J are given by:

$$D = \left(\frac{Q^2}{B^2g}\right)^{1/3} \tag{3.105}$$

and:
$$J = Dg + \frac{Dg}{2} = \frac{3Dg}{2} \tag{3.106}$$

Similarly it may be shown that, at the critical conditions, the flowrate is a maximum for a given value of the specific energy J. At the critical velocity, (u^2/gD) is equal to unity. This dimensionless group is known as the *Froude number Fr*. For velocities greater than the critical velocity it is greater than unity, and vice versa. It will now be shown that the velocity with which a small disturbance is transmitted through a liquid in an open channel is equal to the critical velocity. Thus the Froude number is the criterion by which the type of flow, tranquil or rapid, is determined. Tranquil flow occurs when the Froude number is less than unity and rapid flow when it is greater than unity.

Velocity of transmission of a wave

Consider a liquid which is flowing with a velocity u in a rectangular channel of width B. Initially, the depth of liquid is D_1 but, as a result of a change in conditions at the downstream end of the channel, the level there is suddenly increased to some value D_2. A wave therefore tends to move upstream against the motion of the oncoming fluid. Consider two sections, 1 and 2, one on each side of the wave at any instant (Figure 3.22).

The rate of accumulation of fluid between the two sections is:

$$u_1D_1B - u_2D_2B$$

Figure 3.22.   Transmission of a wave

This accumulation of fluid results from the propagation of the wave and is therefore equal to $u_wB(D_2 - D_1)$.

Thus:
$$u_1D_1 - u_2D_2 = u_w(D_2 - D_1)$$

$$\therefore \qquad u_2 = \frac{u_w(D_1 - D_2) + u_1 D_1}{D_2} \qquad (3.107)$$

The velocity of the fluid is changed from u_1 to u_2 by the passage of the wave. The rate of travel of the wave relative to the upstream liquid is $(u_1 + u_w)$ and therefore the mass of fluid whose velocity is changed in unit time

$$= (u_1 + u_w)BD_1\rho \qquad (3.108)$$

The force acting on the fluid at any section where the liquid depth is D is equal to:

$$\int_0^D (h\rho g)B\,dh = \tfrac{1}{2}B\rho g D^2 \qquad (3.109)$$

where h is any depth below the liquid surface.

The net force acting on the fluid between sections 1 and 2 is:

$$\tfrac{1}{2}B\rho g(D_1^2 - D_2^2) \quad \text{in the direction of flow}$$

Then, neglecting the frictional drag of the walls of the channel between sections 1 and 2, the net force can be equated to the rate of increase of momentum:

i.e.: $$\qquad \tfrac{1}{2}B\rho g(D_1^2 - D_2^2) = (u_1 + u_w)BD_1\rho(u_2 - u_1)$$

$$\therefore \qquad (u_1 + u_w)D_1\left\{\frac{1}{D_2}[u_w(D_1 - D_2) + u_1 D_1] - u_1\right\} = \frac{1}{2}g(D_1^2 - D_2^2)$$

$$\therefore \qquad (u_1 + u_w)^2 = \frac{D_2}{D_1}(D_1 + D_2)\frac{g}{2}$$

$$= \frac{gD_2}{2}\left(1 + \frac{D_2}{D_1}\right) \qquad (3.110)$$

where $u_1 + u_w$ is the velocity of the wave relative to the oncoming fluid. For a very small wave, $D_1 \to D_2$ and:

$$u_1 + u_w = \sqrt{(gD_2)} \qquad (3.111)$$

It is thus seen that the velocity of an elementary wave is equal to the critical velocity, at which the specific energy of the fluid is a minimum for a given flowrate. The criterion for critical conditions is therefore that the Froude number, (u^2/gD), is equal to unity.

Hydraulic jump

If a liquid enters a channel under a gate, it will flow at a high velocity through and just beyond the gate and the depth will be correspondingly low. This is an unstable condition, and at some point the depth of the liquid may suddenly increase and the velocity fall. This change is known as the *hydraulic jump*, and is accompanied by a reduction of the specific energy of the liquid as the flow changes from rapid to tranquil; the excess energy is dissipated as a result of turbulence.

Consider a liquid flowing in a rectangular channel in which a hydraulic jump occurs between sections 1 and 2 (Figure 3.23). Then the conditions after the jump can be determined by equating the net force acting on the liquid between the sections to the rate of change of momentum; the frictional forces at the walls of the channel are neglected.

Figure 3.23.   Hydraulic jump

Net force acting on the fluid

$$= \tfrac{1}{2} B \rho g (D_1^2 - D_2^2)$$

Rate of change of momentum of fluid

$$= u_1 B D_1 \rho (u_2 - u_1)$$

$$\therefore \qquad \tfrac{1}{2} g (D_1^2 - D_2^2) = u_1 D_1 (u_2 - u_1)$$

The volumetric rate of flow of the fluid is the same at sections 1 and 2.

$$\therefore \qquad Q = u_1 B D_1 = u_2 B D_2$$

$$\therefore \qquad \frac{1}{2} g (D_1^2 - D_2^2) = u_1^2 D_1 \left[\left(\frac{D_1}{D_2} \right) - 1 \right]$$

If $D_1 \neq D_2$,

$$\frac{1}{2} g (D_1 + D_2) = \frac{u_1^2 D_1}{D_2}$$

$$\therefore \qquad D_2^2 + D_1 D_2 - \frac{2 u_1^2 D_1}{g} = 0 \qquad\qquad (3.112)$$

$$\therefore \qquad D_2 = \frac{1}{2} \left(-D_1 \,{}^{+}_{[\neq]} \sqrt{D_1^2 + \frac{8 u_1^2 D_1}{g}} \right) \qquad\qquad (3.113)$$

This expression gives D_2 as a function of the conditions at the upstream side of the hydraulic jump. The corresponding velocity u_2 is obtained by substituting in the equation:

$$u_1 D_1 = u_2 D_2$$

Corresponding values of D_1 and D_2 are referred to as *conjugate depths*.

The minimum depth at which a hydraulic jump can occur is found by putting $D_1 = D_2 = D$ in equation 3.112:

i.e.:

$$2 D^2 = 2 D \frac{u_1^2}{g}$$

$$\therefore \qquad D = \frac{u_1^2}{g} \qquad\qquad (3.114)$$

This value of D corresponds to the *critical depth* for flow in a channel.

Thus a hydraulic jump can occur provided that the depth of the liquid is less than the critical depth. After the jump it will be greater than the critical depth, the flow having changed from rapid to tranquil.

The energy loss associated with the hydraulic jump will now be calculated. For a small change in the flow of a fluid in an open channel:

$$u \, du + g \, dz + g \sec \theta \, dD + \delta F = 0 \qquad \text{(equation 3.102)}$$

For a horizontal channel, therefore:

$$F = g(D_1 - D_2) + \frac{(u_1^2 - u_2^2)}{2}$$

From equation 3.112:
$$u_1^2 = \frac{gD_2(D_1 + D_2)}{2D_1}$$

Similarly:
$$u_2^2 = \frac{gD_1(D_1 + D_2)}{2D_2}$$

$$\therefore \qquad F = g(D_1 - D_2) + \frac{1}{4}g(D_1 + D_2)\left(\frac{D_2}{D_1} - \frac{D_1}{D_2}\right)$$

$$= \frac{1}{4}g(D_1 - D_2)\left[4 + (D_1 + D_2)\frac{-(D_1 + D_2)}{D_1 D_2}\right]$$

$$= \frac{(D_2 - D_1)^3 g}{4D_1 D_2} \qquad (3.115)$$

The hydraulic jump may be compared with the shock wave for the flow of a compressible fluid which will be discussed in Chapter 4.

3.7. NON-NEWTONIAN BEHAVIOUR

In the previous sections of this chapter, the calculation of frictional losses associated with the flow of simple *Newtonian* fluids has been discussed. A Newtonian fluid at a given temperature and pressure has a constant viscosity μ which does not depend on the shear rate and, for streamline (laminar) flow, is equal to the ratio of the shear stress (R_y) to the shear rate (du_x/dy) as shown earlier in equation 3.4:

i.e.:
$$\mu = \frac{|R_y|}{|du_x/dy|} \qquad (3.116)$$

The modulus sign is used because shear stresses within a fluid act in both the positive and negative senses. Gases and simple low molecular weight liquids are all Newtonian, and viscosity can be treated as constant in any flow problem unless there are significant variations of temperature or pressure.

Many fluids, including some that are encountered in everyday life, exhibit non-Newtonian behaviour and their apparent viscosities may depend on the rate at which they are sheared and on their previous shear history. At any position and time in the fluid, the apparent viscosity μ_a is defined as the ratio of the shear stress to the shear rate at that point:

i.e.:
$$\mu_a = \frac{|R_y|}{|du_x/dy|} \qquad (3.117)$$

When the apparent viscosity is a function of the shear rate, the behaviour is said to be *shear-dependent*; when it is a function of the duration of shearing at a particular rate, it is referred to as *time-dependent*. Any shear-dependent fluid must to some extent be time-dependent because, if the shear rate is suddenly changed, the apparent viscosity does not alter instantaneously, but gradually moves towards its new value. In many cases, however, the time-scale for the flow process may be sufficiently long for the effects of time-dependence to be negligible.

The apparent viscosity of a fluid may either decrease or increase as the shear rate is raised. The more common effect is for the apparent viscosity to fall as the shear rate is raised; such behaviour is referred to as *shear-thinning*. Most paints are shear-thinning: in the can and when loaded on to the brush, the paint is subject only to low rates of shear and has a high apparent viscosity. When applied to the surface the paint is sheared by the brush, its apparent viscosity becomes less, and it flows readily to give an even film. However, when the brushing ceases, it recovers its high apparent viscosity and does not drain from the surface. This non-Newtonian behaviour is an important characteristic of a good paint. It is frequently necessary to build-in non-Newtonian characteristics to give a product the desired properties. Paints will usually exhibit appreciable time-dependent behaviour. Thus, as the paint is stirred its apparent viscosity will decrease progressively until it reaches an asymptotic equilibrium value characteristic of that particular rate of shear.

Some materials have the characteristics of both solids and liquids. For instance, tooth paste behaves as a solid in the tube, but when the tube is squeezed the paste flows as a plug. The essential characteristic of such a material is that it will not flow until a certain critical shear stress, known as the *yield stress* is exceeded. Thus, it behaves as a solid at low shear stresses and as a fluid at high shear stress. It is a further example of a shear-thinning fluid, with an infinite apparent viscosity at stress values below the yield value, and a falling finite value as the stress is progressively increased beyond this point.

A further important property which may be shown by a non-Newtonian fluid is *elasticity* — which causes the fluid to try to regain its former condition as soon as the stress is removed. Again, the material is showing some of the characteristics of both a solid and a liquid. An ideal (Newtonian) liquid is one in which the stress is proportional to the *rate* of shear (or rate of strain). On the other hand, for an ideal solid (obeying Hooke's Law) the stress is proportional to the strain. A fluid showing elastic behaviour is termed *viscoelastic* or *elastoviscous*.

The branch of science which is concerned with the flow of both simple (Newtonian) and complex (non-Newtonian) fluids is known as *rheology*. The flow characteristics are represented by a *rheogram*, which is a plot of shear stress against rate of shear, and normally consists of a collection of experimentally determined points through which a curve may be drawn. If an equation can be fitted to the curve, it facilitates calculation of the behaviour of the fluid. It must be borne in mind, however, that such equations are approximations to the actual behaviour of the fluid and should not be used outside the range of conditions (particularly shear rates) for which they were determined.

An understanding of non-Newtonian behaviour is important to the chemical engineer from two points of view. Frequently, non-Newtonian properties are desirable in that they can confer desirable properties on the material which are essential if it is to fulfil the purpose for which it is required. The example of paint has already been given. Toothpaste should not flow out of the tube until it is squeezed and should stay in place on the brush until it is applied to the teeth. The texture of foodstuffs is largely attributable to rheology.

Secondly, it is necessary to take account of non-Newtonian behaviour in the design of process plant and pipelines. Heat and mass transfer coefficients are considerably affected by the behaviour of the fluid, and special attention must be devoted to the selection of appropriate mixing equipment and pumps.

In this section, some of the important aspects of non-Newtonian behaviour will be quantified, and some of the simpler approximate equations of state will be discussed. An attempt has been made to standardise nomenclature in BS 5168[12].

Shear stress is denoted by R in order to be consistent with other parts of the book; τ is frequently used elsewhere to denote shear stress. R without suffix denotes the shear stress acting on a surface in the direction of flow and $R_0(=-R)$ denotes the shear stress exerted by the surface on the fluid. R_s denotes the positive value in the fluid at a radius s and R_y the positive value at a distance y from a surface. Strain is defined as the ratio dx/dy, where dx is the shear displacement of two elements a distance dy apart and is often denoted by γ. The rate of strain or rate of shear is $(dx/dt)/dy$ or du_x/dy and is denoted by $\dot{\gamma}$.

Thus equation 3.117 can be written:

$$\mu_a = \frac{\tau}{\dot{\gamma}} \tag{3.118}$$

3.7.1. Steady-state shear-dependent behaviour

In this section, consideration will be given to the equilibrium relationships between shear stress and shear rate for fluids exhibiting non-Newtonian behaviour. Whenever the shear stress or the shear rate is altered, the fluid will gradually move towards its new equilibrium state and, for the present, the period of "adjustment" between the two equilibrium states will be ignored.

The fluid may be either shear-thinning or, less often, shear-thickening, and in either case the shear stress and the *apparent viscosity* μ_a are functions of shear rate:

i.e.:
$$|R_y| = f_1\left(\left|\frac{du_x}{dy}\right|\right) \tag{3.119}$$

and:
$$\mu_a = |R_y| \bigg/ \left|\frac{du_x}{dy}\right| = f_2\left(\left|\frac{du_x}{dy}\right|\right) \tag{3.120}$$

Here y is distance measured from a boundary surface.

Typical forms of curve of shear stress versus shear rate are shown in Figure 3.24 for a shear-thinning (or *pseudoplastic*) fluid, a *Bingham plastic*, a Newtonian fluid and a shear-thickening (or *dilatant*) fluid. For the particular case chosen the apparent viscosity is the same for all four at the shear rate where they intersect. At lower shear rates the shear-thinning fluid is more viscous than the Newtonian fluid, and at higher shear rates it is less viscous. For the shear-thickening fluid, the situation is reversed. The corresponding curves showing the variation of apparent viscosity are given in Figure 3.25. Because the rates of shear of interest can cover several orders of magnitude, it is convenient to replot the curves using log-log coordinates (Figures 3.26 and 3.27).

Figure 3.24.   Shear stress-shear rate behaviour of Newtonian and non-Newtonian fluids plotted using linear coordinates

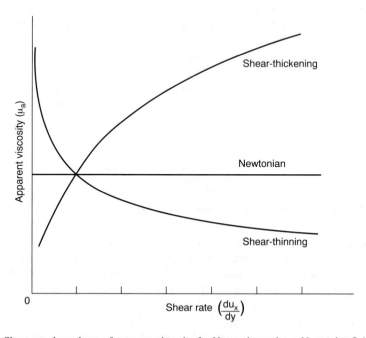

Figure 3.25.   Shear rate dependence of apparent viscosity for Newtonian and non-Newtonian fluids plotted on linear co-ordinates

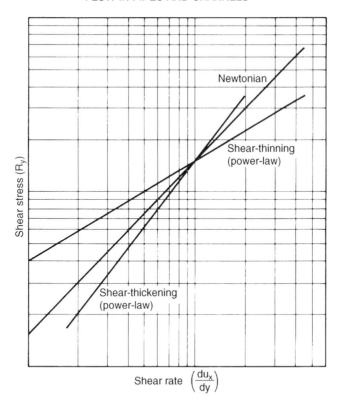

Figure 3.26.   Log-log relation between shear-stress and shear-rate

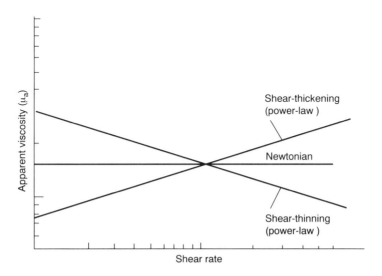

Figure 3.27.   Log-log plot of apparent viscosity-shear rate relation

The relation between shear stress and shear rate for the Newtonian fluid is defined by a single parameter μ, the viscosity of the fluid. No single parameter model will describe non-Newtonian behaviour and models involving two or even more parameters only approximate to the characteristics of real fluids, and can be used only over a limited range of shear rates.

A useful two-parameter model is the *power law* model, or Ostwald–de Waele law to identify its first proponents. The relation between shear stress and shear rate is given by:

$$|R_y| = k \left(\left| \frac{\mathrm{d}u_x}{\mathrm{d}y} \right| \right)^n \tag{3.121}$$

where n is known as the *power law index* and k is known as the *consistency*. This equation can be written:

$$|R_y| = k \left(\left| \frac{\mathrm{d}u_x}{\mathrm{d}y} \right| \right)^{n-1} \left| \frac{\mathrm{d}u_x}{\mathrm{d}y} \right| \tag{3.122}$$

It is therefore seen that the apparent viscosity μ_a is given by:

$$\mu_a = k \left(\left| \frac{\mathrm{d}u_x}{\mathrm{d}y} \right| \right)^{n-1} \tag{3.123}$$

From equation 3.123: when $n > 1$, μ_a increases with increase of shear rate, and shear-thickening behaviour is described;

when $n = 1$, μ_a is constant and equal to the Newtonian viscosity μ of the fluid;

when $n < 1$, μ_a decreases with increase of shear rate, and the behaviour is that of a shear-thinning fluid.

Thus, by selecting an appropriate value of n, both shear-thinning and shear-thickening behaviour can be represented, with $n = 1$ representing Newtonian behaviour which essentially marks the transition from shear-thinning to shear-thickening characteristics.

It will be noted that the dimensions of k are $\mathbf{ML^{-1}T^{n-2}}$, i.e. they are dependent on the value of n. Values of k for fluids with different n values cannot therefore be compared. *Numerically*, k is the value of the apparent viscosity (or shear stress) at unit shear rate and this numerical value will depend on the units used, e.g. the value of k at a shear rate of $1~\mathrm{s}^{-1}$ will be different from that at a shear rate of $1~\mathrm{h}^{-1}$.

In Figure 3.28, the shear stress is shown as a function of shear rate for a typical shear-thinning fluid, using log-log coordinates. Over the shear rate range (*ca.* 10^0 to $10^3~\mathrm{s}^{-1}$), the fluid behaviour is described by the power-law equation with an index n of 0.6, i.e. the line CD has a slope of 0.6. If the power-law were followed at all shear rates, the extrapolated line $C'CDD'$ would be applicable. Figure 3.29 shows the corresponding values of apparent viscosity and the line $C'CDD'$ has a slope of $n - 1 = -0.4$; it is seen that it extrapolates to $\mu_a = \infty$ at zero shear rate and to $\mu_a = 0$ at infinite shear rate.

In practice, the viscosities of most real shear-thinning fluids approach constant values both at low shear rates and at high shear rates; that is they tend to show Newtonian properties at the extremes of shear rates. The limiting viscosity at low shear rates μ_0 is referred to as the lower Newtonian (or zero shear) viscosity (see lines AB in Figures 3.28 and 3.29) and at high shear rates μ_∞ is the upper Newtonian (or infinite shear) viscosity (see lines EF in Figures 3.28 and 3.29).

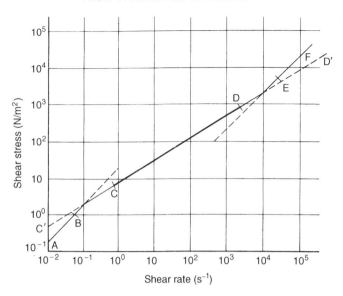

Figure 3.28.   Behaviour of a typical shear-thinning fluid plotted logarithmically for several orders of shear rate

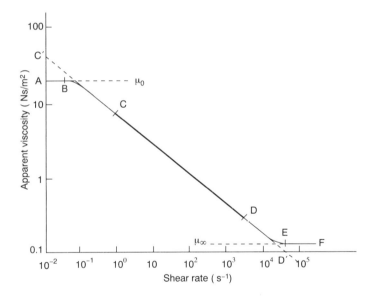

Figure 3.29.   Apparent viscosity corresponding to the data of Figure 3.28

By reference to Figure 3.29, it is seen that the power law model can, at best therefore, represent the behaviour of a real fluid over only a limited range of shear rates. Effectively, it represents what is happening in the region where apparent viscosity is changing most rapidly with shear rate. However, over any limited range of shear rates, it is possible to approximate to the curve in Figure 3.29 by a straight line, whose slope $(n - 1)$ will determine the "best" value of the power law index n which can be used over that range.

A convenient form of 3-parameter equation which extrapolates to a constant limiting apparent viscosity (μ_0 or μ_∞) as the shear rate approaches both zero and infinity has been proposed by CROSS[13]:

$$\frac{\mu_0 - \mu_\infty}{\mu_a - \mu_\infty} = 1 + \beta\dot{\gamma}^{2/3} \tag{3.124}$$

where $\dot{\gamma}$ is the shear rate and μ_0, μ_∞ and β must be determined for each fluid. For some materials a slight modification of the exponent of $\dot{\gamma}(2/3)$ leads to improved agreement between predicted and experimental results.

This equation is based on the assumption that pseudoplastic (shear-thinning) behaviour is associated with the formation and rupture of structural linkages. It is based on an experimental study of a wide range of fluids — including aqueous suspensions of flocculated inorganic particles, aqueous polymer solutions and non-aqueous suspensions and solutions — over a wide range of shear rates ($\dot{\gamma}$). (~ 10 to 10^4 s^{-1}).

For a shear-thickening fluid the same arguments can be applied, with the apparent viscosity rising from zero at zero shear rate to infinity at infinite shear rate, on application of the power law model. However, shear-thickening is generally observed over very much narrower ranges of shear rate and it is difficult to generalise on the type of curve which will be obtained in practice.

In order to get over the shortcomings of the power-law model, several alternative forms of equation between shear rate and shear stress have been proposed. These are all more complex involving three or more parameters. Reference should be made to specialist works on non-Newtonian flow[14-16] for details of these *Constitutive Equations*.

Some fluids exhibit a *yield stress*. When subjected to stresses below the yield stress they do not flow and effectively can be regarded as fluids of infinite viscosities, or alternatively as solids. When the yield stress is exceeded they flow as fluids. Such behaviour cannot be described by a power-law model.

The simplest type of behaviour for a fluid exhibiting a yield stress is known as *Bingham plastic*. The shear rate is directly proportional to the amount by which the stress exceeds the yield stress.

Thus:
$$|R_y| - R_Y = \mu_p \left|\frac{du_x}{dy}\right| \qquad (|R_y| \geqslant R_Y) \tag{3.125}$$

$$\frac{du_x}{dy} = 0 \qquad (|R_y| \leqslant R_Y)$$

μ_p is known as the *plastic viscosity*.

The apparent viscosity μ_a is given, by definition, as:

$$\mu_a = |R_y| \left/ \left|\frac{du_x}{dy}\right|\right. \tag{equation 3.117}$$

$$= \mu_p + R_Y \left/ \left|\frac{du_x}{dy}\right|\right. \qquad (|R_y| \geqslant R_Y) \tag{3.126}$$

Thus, the apparent viscosity falls from infinity at zero shear rate ($|R_y| \leqslant R_Y$) to μ_p at infinite shear rate, i.e. the fluid shows shear-thinning characteristics.

Because it is very difficult to measure the flow characteristics of a material at very low shear rates, behaviour at zero shear rate can often only be assessed by extrapolation of experimental data obtained over a limited range of shear rates. This extrapolation can be difficult, if not impossible. From Example 3.8, later in the chapter, it can be seen that it is sometimes possible to approximate the behaviour of a fluid *over the range of shear rates for which experimental results are available*, either by a power law or by a Bingham plastic equation.

Some materials give more complex behaviour and the plot of shear stress against shear rate approximates to a curve, rather than to a straight line with an intercept R_Y on the shear stress axis. An equation of the following form may then be used:

$$|R_y| - R_Y = \mu'_p \left(\left| \frac{du_x}{dy} \right| \right)^m \tag{3.127}$$

i.e., equation 3.127, which includes three parameters, is effectively a combination of equations 3.121 and 3.125. It is sometimes called the *generalised Bingham equation* or *Herschel—Bulkley equation*, and the fluids are sometimes referred to as *false body*. Figures 3.30 and 3.31 show shear stress and apparent viscosity, respectively, for Bingham plastic and false body fluids, using linear coordinates.

In many flow geometries, the shear stress will vary over the cross-section and there will be regions where the shear stress exceeds the yield stress and the fluid will be sheared. In other regions, the shear stress will be less than the yield value and the fluid will flow as an unsheared plug. Care must be taken therefore in the application of equations 3.125 and 3.127. Thus, for the case of flow in a circular tube which will be considered later, the shear stress varies linearly from a maximum value at the wall, to zero at the centre-line (see equation 3.9). If the shear stress at the wall is less than the yield stress, no flow will occur. If it is greater, shear will take place in the region between the wall and the point

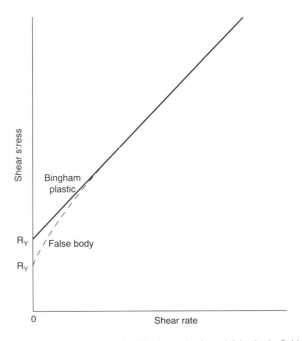

Figure 3.30.   Shear stress-shear rate data for Bingham plastic and false body fluids (linear scales)

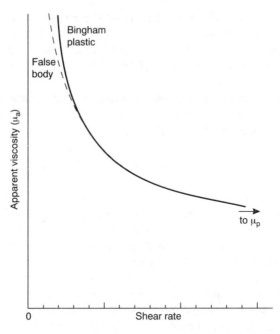

Figure 3.31.   Apparent viscosity for Bingham plastic and false body fluids (linear scales)

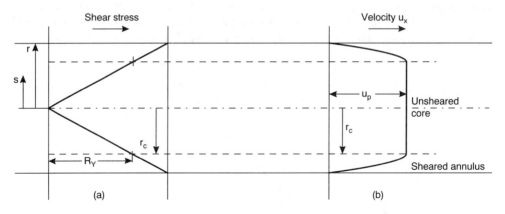

Figure 3.32.   (a) Shear stress distribution in pipe (b) Velocity profile for Bingham plastic fluid in pipe

where the shear stress equals the yield stress (Figure 3.32). Inside this region the fluid will flow as an unsheared plug. As the pressure difference over the tube is increased, the wall shear stress will increase in direct proportion and the unsheared plug will be of smaller radius. For any fluid subjected to a finite pressure gradient there must always be some region, however small, near the pipe axis in which shear does not take place.

3.7.2 Time-dependent behaviour

In a *time-dependent fluid*, the shear rate depends upon the time for which it has been subjected to a given shear stress. Conversely, if the shear rate is kept constant, the shear stress will change with time. However, with all *time-dependent* fluids an equilibrium condition is reached if the imposed condition (e.g. shear rate or shear stress) is maintained constant. Some fluids respond so quickly to changes that the effect of time dependence can be neglected. Others may have a much longer constant, and in changing flow situations will never be in the equilibrium state.

In general, for shear-thinning pseudoplastic fluids the apparent viscosity will gradually decrease with time if there is a step increase in its rate of shear. This phenomenon is known as *thixotropy*. Similarly, with a shear-thickening fluid the apparent viscosity increases under these circumstances and the fluid exhibits *rheopexy* or *negative thixotropy*.

The effect of increasing and then decreasing the rate of shear of a thixotropic fluid is shown in Figure 3.33 in which the shear stress is plotted against shear rate in an experiment; the shear rate is steadily increased from zero and subsequently decreased again. It is seen that a hysteresis loop is formed, with the shear stress always lagging behind its equilibrium value. If the shear rate is changed rapidly the hysteresis loop will have a large area. At low rates of change the area will be small and will eventually become zero as the two curves coincide when sufficient time is allowed for equilibrium to be reached at each point on the curve. In Figure 3.34, the effect on the apparent viscosity is seen of a step increase in the shear rate; it gradually decreases from the initial to the final equilibrium value. This is a picture of what happens to a material which does not suffer any irreversible changes as a result of shearing. Some materials, particular gels, suffer structural breakdown when subjected to a shear field, but this effect is not considered here.

The behaviour of a *rheopectic* fluid is the reverse of that of a thixotropic fluid and is illustrated by the broken lines in Figures 3.33 and 3.34.

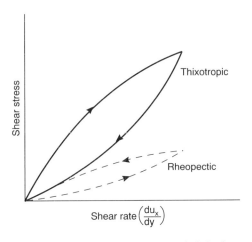

Figure 3.33.   Thixotropic and rheopectic behaviour

3.7.3. Viscoelastic behaviour

A true fluid flows when it is subjected to a shear field and motion ceases as soon as the stress is removed. In contrast, an ideal solid subjected to a stress recovers its original

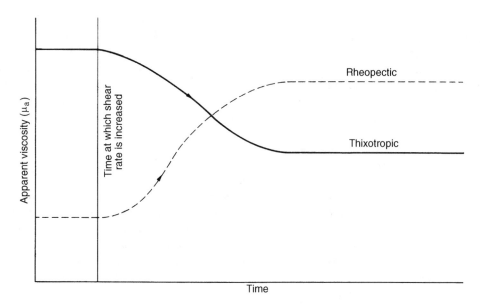

Figure 3.34.   Effect of sudden change of shear rate on apparent viscosity of time-dependent fluid

state as soon as the stress is removed. At the two extremes therefore there are:

1. The ideal Newtonian fluid in which:

$$|R_y| = \mu \left| \frac{du_x}{dy} \right|$$ (equation 3.4)

$$(\text{or } \tau = \mu \dot{\gamma})$$

   where du_x/dy represents the rate of shear.

2. The ideal elastic solid which obeys Hooke's law in which the relation between distortion and stress is:

$$|R_y| = \mathbf{G} \left| \frac{dx}{dy} \right|$$ (3.128)

$$(\text{or } \tau = \mathbf{G}\gamma)$$

   where $\mathbf{G}$ is Young's modulus, and $dx/dy(= \gamma)$ is the ratio of the shear displacement of two elements to their distance apart.

Some materials exhibit *viscoelastic* characteristics having some of the properties of both a solid and a liquid. Thus a solid may be subject to "creep" and a fluid to "elastic" properties. Two examples of viscoelastic behaviour in a liquid are as follows:

1. The liquid in a cylindrical vessel is given a swirling motion by means of a stirrer. When the stirring is stopped, the fluid gradually comes to rest and, if viscoelastic, may then start to rotate in the opposite direction, i.e. to unwind.
2. A viscoelastic fluid, on emerging from a tube or from a die, may form a jet which is of larger diameter than the aperture. The phenomenon is referred to as "die-swell"

and results from the sudden removal of a constraining force on the fluid. Viscoelastic fluids are thus capable of exerting *normal stresses.*

Viscoelastic effects are of particular importance in unsteady state flow and where there are rapid changes in the pressure to which the fluid is subjected. They can give rise to very complex behaviour and mechanical analogues have sometimes been found useful for calculating behaviour. The fluid properties are represented by a "dashpot", a piston in a cylinder with a small outlet in which the flowrate is linearly related to the pressure difference across it. The elastic properties are represented by a spring. By combination of such elements in a variety of ways it is possible to simulate the behaviour of very complex materials. A single dashpot and spring in series is known as the Maxwell model; and the dashpot and spring in parallel is the Voigt model (Figure 3.35).

For further consideration of viscoelastic behaviour, reference should be made to specialist works.[14–16]

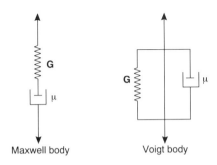

Figure 3.35.   Mechanical analogues for viscoelastic fluids

3.7.4. Characterisation of non-Newtonian fluids

The characterisation of non-Newtonian fluids is a major area of science in itself and the equipment required for measurement of all the relevant properties, including the normal stresses of viscoelastic fluids, is extremely complex. The range of shear rates of interest is very great covering many orders of magnitude, and it is frequently better to use a number of instruments each covering a relatively narrow range, rather than to try to carry out all the measurements with a single instrument. Some instruments are designed to operate at a series of constant shear rates and the resulting stresses are measured. Others, constant stress instruments, are more suitable for measurements at low shear rate conditions, particularly with materials that have yield stresses.

Even the measurement of the steady-state characteristics of shear-dependent fluids is more complex than the determination of viscosities for Newtonian fluids. In simple geometries, such as capillary tubes, the shear stress and shear rate vary over the cross-section and consequently, at a given operating condition, the apparent viscosity will vary with location. Rheological measurements are therefore usually made with instruments in which the sample to be sheared is subjected to the same rate of shear throughout its whole mass. This condition is achieved in concentric cylinder geometry (Figure 3.36) where the fluid is sheared in the annular space between a fixed and a rotating cylinder; if the gap is small compared with the diameters of the cylinders, the shear rate is approximately

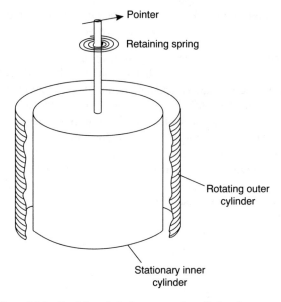

Figure 3.36. Partial section of a concentric-cylinder viscometer

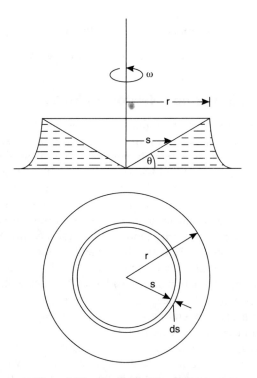

Figure 3.37. Cone and plate viscometer

constant. Alternatively, a cone and plate geometry (Figure 3.37) gives a constant shear rate, provided that the angle θ between the cone and the plate is sufficiently small that $\sin \theta \approx \theta$.

If it is known that a particular form of relation, such as the power-law model is applicable, it is not necessary to maintain a constant shear rate. Thus, for instance, a capillary tube viscometer can be used for determination of the values of the two parameters in the model. Details of the method are given by FAROOQI and RICHARDSON.[17]

3.7.5. Relation between rheology and structure of material

An understanding of the contribution of the relevant physical and chemical properties of the system to rheological behaviour is an area which has made little progress until recent years, but a clearer picture is now emerging of how the various types of rheological behaviour arise.

Most non-Newtonian fluids are either two-phase systems, or single phase systems in which large molecules are in solution in a liquid which itself may be Newtonian or non-Newtonian, or are in the form of a melt.

Flocculated suspensions of fine particles (coal, china clay, pigments, etc.) are usually shear-thinning. The flocs are weak mechanically and tend to break up in a shear field and the individual particles are capable of regrouping to give structures offering a lower resistance to shear. When the shear field is removed, the original structure is regained. The particles are generally of the order of 1 μm in size and the flocs (loose agglomerates occluding liquid) may be 10-100 μm in size. The specific surface of the particles is very high, and surface chemistry plays an important role in determining the structure of the flocs. The rheology of many of these suspensions can be approximated over a reasonable range of flowrates by a power-law equation $(0.1 < n < 0.6)$. There is evidence that some have a finite yield stress and conform to the Bingham-plastic model. However, frequently the rheology can equally well be described by a power-law or Bingham-plastic form of equation. (See Example 3.8.)

Non-flocculated suspensions can exist at very much higher concentrations and, at all but the highest volumetric concentrations, are often Newtonian. When such suspensions are sheared, some *dilation* occurs as a result of particles trying to "climb over each other". If the amount of liquid present is then insufficient fully to fill the void spaces, particle-particle solid friction can come into play and the resistance to shear increases. This is just one way in which shear-thickening can occur.

Many polymers form shear-thinning solutions in water. The molecules are generally long and tend to be aligned and to straighten out in a shear field, and thus to offer less resistance to flow. Such solutions are sometimes viscoelastic and this effect may be attributable to a tendency of the molecules to recover their previous configuration once the stress is removed. Molten polymers are usually viscoelastic.

3.7.6. Flow in pipes and channels of regular geometry

As in the case of Newtonian fluids, one of the most important practical problems in the flow of non-Newtonian fluids is the calculation of the pressure drop for the flow in pipelines. Flow is much more likely to be streamline or laminar because non-Newtonian fluids usually have very much higher apparent viscosities than most simple Newtonian fluids. Furthermore, the difference in behaviour is much greater for laminar flow where viscosity plays such an important role. Attention will be focused on laminar-flow with particular reference to power-law and Bingham plastic fluids.

In order to predict the transition point from stable streamline to stable turbulent flow, it is necessary to define a modified Reynolds number, though it is not clear that the same sharp transition in flow regime always occurs. Particular reference will be made to flow in pipes of circular cross-section, but the methods are applicable to other geometries (annuli, between flat plates, etc.) as was so in the case of Newtonian fluids, and the methods described earlier for flow between plates, through an annulus or down a surface can be adapted to take account of non-Newtonian characteristics of the fluid.

Streamline flow in a pipe

(a) Power-law fluid

The distribution of shear stress over the cross-section of a pipe is determined by a force balance and is independent of the nature of the fluid or the type of flow.

From equation 3.8 and Figure 3.32a it is seen that the shear stress $|R_s|$ at a radius s in a pipe of radius r is given by:

$$\frac{|R_s|}{R} = \frac{s}{r} \qquad (3.129)$$

that is, the shear stress varies linearly from the centre of the pipe ($s = 0$) to the wall ($s = r$).

When the fluid behaviour can be described by a power-law, the apparent viscosity for a shear-thinning fluid will be a minimum at the wall where the shear stress is a maximum, and will rise to a theoretical value of infinity at the pipe axis where the shear stress is zero. On the other hand, for a shear-thickening fluid the apparent viscosity will fall off to zero at the pipe axis. It is apparent, therefore, that there will be some error in applying the power law near the pipe axis since all real fluids have a limiting viscosity at zero shear stress. However, the errors are not large. The method of procedure is exactly analogous to that used for the Newtonian fluid, except that the power law relation is used to relate shear stress to shear rate, as opposed to the simple Newtonian equation.

For a power-law fluid, equation 3.28 becomes:

$$\left(-\frac{dP}{dl}\right)s - 2k\left(-\frac{du_x}{ds}\right)^n = 0$$

$$\therefore \qquad du_x = -\left(-\frac{dP}{dl}\frac{1}{2k}\right)^{1/n} s^{1/n}\,ds$$

and: $\qquad u_x = -\left(-\frac{dP}{dl}\frac{1}{2k}\right)^{1/n}\frac{n}{n+1}s^{(n+1)/n} + \text{constant}$

At the pipe wall, $s = r$ and, for the no-slip condition, $u_x = 0$.

So: $\qquad \text{constant} = \left(-\frac{dP}{dl}\frac{1}{2k}\right)^{1/n}\frac{n}{n+1}r^{(n+1)/n}$

$$\therefore \qquad u_x = \left(-\frac{\mathrm{d}P}{\mathrm{d}l}\frac{1}{2k}\right)^{1/n} \frac{n}{n+1}(r^{(n+1)/n} - s^{(n+1)/n}) \qquad (3.130)$$

The velocity at the centre line, u_{CL}, is then given by putting $s = 0$:

$$u_{CL} = \left(-\frac{\mathrm{d}P}{\mathrm{d}l}\frac{1}{2k}\right)^{1/n} \frac{n}{n+1}r^{(n+1)/n}. \qquad (3.131)$$

Dividing equation 3.130 by equation 3.131:

$$\frac{u_x}{u_{CL}} = 1 - \left(\frac{s}{r}\right)^{(n+1)/n} \qquad (3.132)$$

The mean velocity of flow u is given by:

$$u = \frac{1}{\pi r^2} \int_0^r u_x 2\pi s \, \mathrm{d}s$$

$$= \frac{2}{r^2} u_{CL} r^2 \int_0^1 \left(1 - \frac{s}{r}\right)^{(n+1)/n} \frac{s}{r} \, \mathrm{d}\left(\frac{s}{r}\right)$$

$$= 2u_{CL} \left[\frac{1}{2}\left(\frac{s}{r}\right)^2 - \frac{n}{3n+1}\left(\frac{s}{r}\right)^{(3n+1)/n}\right]_0^1$$

$$= 2u_{CL} \left(\frac{1}{2} - \frac{n}{3n+1}\right)$$

i.e.: $\qquad\qquad \dfrac{u}{u_{CL}} = \dfrac{n+1}{3n+1} \qquad (3.133)$

Combining equations 3.132 and 3.133, gives the velocity profile in terms of the mean velocity u in place of the centre-line velocity u_{CL}:

$$\frac{u_x}{u} = \frac{3n+1}{n+1}\left[1 - \left(\frac{s}{r}\right)^{(n+1)/n}\right] \qquad (3.134)$$

Substituting into equation 3.131:

$$u = \left(-\frac{\mathrm{d}P}{\mathrm{d}l}\frac{1}{2k}\right)^{1/n} \frac{n}{3n+1}r^{(n+1)/n} \qquad (3.135)$$

Working in terms of pipe diameter:

$$u = \left(-\frac{\mathrm{d}P}{\mathrm{d}l}\frac{1}{4k}\right)^{1/n} \frac{n}{6n+2}d^{(n+1)/n} \qquad (3.136)$$

For a Newtonian fluid $n = 1$,

and: $\qquad\qquad u = -\dfrac{\mathrm{d}P}{\mathrm{d}l}\dfrac{d^2}{32k} \qquad\qquad$ (equation 3.35)

This is identical to equation 3.35, bearing in mind that $k = \mu$ for a Newtonian fluid.

The shear rate (velocity gradient) at the tube wall is obtained by differentiating equation 3.134 with respect to s, and then putting $s = r$.

$$\frac{1}{u}\frac{du_x}{ds} = \frac{3n+1}{n+1}\left[-\frac{n+1}{n}\left(\frac{s}{r}\right)^{(n+1)/n}\frac{1}{s}\right]$$

If y is distance from the wall, $y + s = r$ and:

$$\left(-\frac{du_x}{ds}\right)_{s=r} = \left(\frac{du_x}{dy}\right)_{y=0} = \frac{3n+1}{n}\frac{u}{r} = \frac{6n+2}{n}\frac{u}{d} \tag{3.137}$$

For a Newtonian fluid, this gives a wall shear rate of $8u/d$ (equation 3.39) and a shear stress of $8\mu u/d$ (equation 3.40).

For a Newtonian fluid, the data for pressure drop are represented on a pipe friction chart as a friction factor $\phi = (R/\rho u^2)$ as a function of Reynolds number $Re = (ud\rho/\mu)$. The friction factor is independent of the rheological properties of the fluid, but the Reynolds number involves the viscosity which is dependent on shear rate, and there is consequently a difficulty in defining an appropriate Reynolds number. METZNER and REED[18] defined a Reynolds number Re_{MR} for a power-law fluid in such a way that it is related to the friction factor for streamline flow in the same way as for a Newtonian fluid.

From equation 3.39 therefore:

$$\frac{R}{\rho u^2} = 8Re_{MR}^{-1} \tag{3.138}$$

For the flow of a power law fluid in a pipe of length l, the pressure drop $-\Delta P$ is obtained from equation 3.136:

$$-\Delta P = \left(\frac{6n+2}{n}\right)^n 4klu^n d^{-(n+1)}$$

Thus:

$$\frac{R}{\rho u^2} = \frac{-\Delta Pd}{4l\rho u^2} \qquad\qquad \text{(from equation 3.18)}$$

$$= \left(\frac{6n+2}{n}\right)^n \frac{ku^{n-2}d^{-n}}{\rho} \tag{3.139}$$

Substituting into equation 3.137:

$$Re_{MR} = 8\left(\frac{n}{6n+2}\right)^n \frac{\rho u^{2-n}d^n}{k} \tag{3.140}$$

This definition of Re_{MR} can be used to determine the limit of stable streamline flow. The transition value $(Re_{MR})_c$ is approximately the same as for a Newtonian fluid, but there is some evidence that streamline flow may persist to somewhat higher values. Putting $n = 1$ in equation 3.140 leads to the standard definition of the Reynolds number.

The effect of power-law index on the velocity profile is seen by plotting equation 3.134 for various values of n, as shown in Figure 3.38.

Compared with the parabolic profile for a Newtonian fluid ($n = 1$), the profile is flatter for a shear-thinning fluid ($n < 1$) and sharper for a shear-thickening fluid ($n > 1$). The

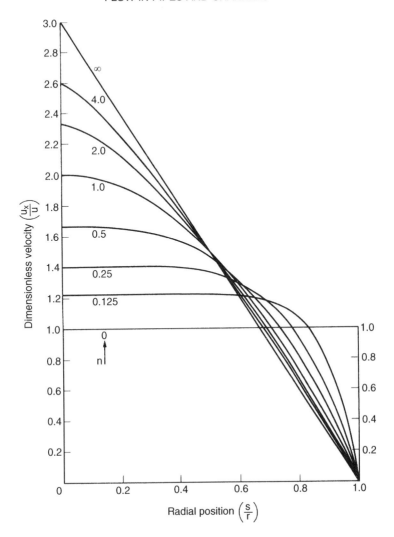

Figure 3.38.   Fully-developed laminar velocity profiles for power-law fluids in a pipe (from equation 3.134)

ratio of the centre line (u_{CL}) to mean (u) velocity, calculated from equation 3.133 is as follows:

n	2.0	1.5	1.0	0.8	0.6	0.4	0.2	0.1
u_{CL}/u	2.33	2.2	2	1.89	1.75	1.57	1.33	1.18

(b) Bingham-plastic fluid

For the flow of a Bingham-plastic fluid, the cross-section can be considered in two parts (Figure 3.38):

(1) A central unsheared core in which the fluid is all travelling at the centre-line velocity.
(2) An annular region separating the core from the pipe wall, over which the whole of the velocity profile is concentrated.

The boundary between the two regions and the radius r_c of the core is determined by the position in the cross-section at which the shear stress is exactly equal to the yield stress R_Y of the fluid. Since the shear stress is linearly related to the radial position:

$$\frac{R_Y}{R} = \frac{r_c}{r}$$ (equation 3.8)

where R is the shear stress at the wall (radius r).

Now, in a given pipe, R is determined solely by the pressure drop $-\Delta P$ since, from equation 3.6:

$$R = -\Delta P \frac{r}{2l}$$ (3.141)

Thus:

$$\frac{r_c}{2l} = \frac{R_Y}{-\Delta P}$$ (3.142)

Thus, for a given fluid (R_Y constant), the critical radius r_c is determined entirely by the pressure drop and becomes progressively larger as the pressure drop is reduced. Flow ceases when the shear stress at the wall R falls to a value equal to the yield stress R_Y.

Flow in the annular region (s > r$_c$)

In this region, the relation between the shear stress R_s and the velocity gradient du_x/ds is given by:

$$|R_s| - R_Y = \mu_p \left(-\frac{du_x}{ds} \right)$$

Writing: $R_s = R(s/r)$ and rearranging:

$$du_x = \frac{1}{\mu_p} (R_Y - \frac{s}{r} R) \, ds$$

Integrating:

$$u_x = \frac{1}{\mu_p} \left\{ R_Y s - R \frac{s^2}{2r} \right\} + \text{constant.}$$

For the no-slip condition at the wall, $u_x = 0$ at $s = r$.

$$\therefore \qquad \text{constant} = -\frac{1}{\mu_p} \left\{ R_Y r - R \frac{r^2}{2r} \right\}$$

Thus:

$$u_x = \frac{1}{\mu_p} \left\{ \frac{R}{2r} (r^2 - s^2) - R_Y (r - s) \right\}$$ (3.143)

This is the equation of the velocity profile, which reduces to the parabolic form for a Newtonian fluid when $R_Y = 0$. It applies in the region $r > s > r_c$.

The volumetric flowrate Q_A in the annular region is given by:

$$Q_A = \int_{s=2l(R_Y/-\Delta P)}^{s=r} \left\{ \frac{1}{\mu_p} \left[\frac{R}{2r}(r^2 - s^2) - R_Y(r-s) \right] \right\} 2\pi s \, ds$$

This reduces to:

$$Q_A = \frac{-\Delta P \pi r^4}{8\mu_p l} \left\{ 1 - \frac{4}{3}X - 2X^2 + 4X^3 - \frac{5}{3}X^4 \right\} \tag{3.144}$$

where $X = \dfrac{2R_Y l}{-\Delta P r} = \dfrac{R_Y}{R} = \dfrac{\text{Yield stress}}{\text{Wall shear stress}}$.

Flow in centre plug

The velocity u_p of the centre plug is obtained by putting:

$$s = r_c = 2l\frac{R_Y}{-\Delta P}$$

in equation 3.143.

Thus: $u_p = \dfrac{1}{\mu_p} \left\{ \dfrac{-\Delta P r}{2l} \dfrac{1}{2r} \left[r^2 - 4l^2 \left(\dfrac{R_Y}{-\Delta P} \right)^2 \right] - R_Y \left(r - 2l\dfrac{R_Y}{-\Delta P} \right) \right\}$

This simplifies to:

$$u_p = \frac{-\Delta P r^2}{8\mu_p l}\{2 - X + 2X^2\} \tag{3.145}$$

The volumetric flowrate in the plug Q_p is given by:

$$Q_p = u_p \pi r_c^2$$

$$= \frac{-\Delta P r^2}{8\mu_p l}(2 - X + 2X^2)\pi \frac{4l^2 R_Y^2}{(-\Delta P)^2}$$

$$= \frac{-\Delta P \pi r^4}{8\mu_p l}\{2X^2 - 4X^3 + 2X^4\} \tag{3.146}$$

Total flow through pipe

The total flowrate Q through the pipe is given by:

$$Q = Q_A + Q_p$$

$$= \frac{-\Delta P \pi r^4}{8\mu_p l}\left\{ 1 - \frac{4}{3}X + \frac{1}{3}X^4 \right\} \tag{3.147}$$

Thus the mean velocity of flow u is given by:

$$u = \frac{-\Delta P r^2}{8\mu_p l}\left\{ 1 - \frac{4}{3}X + \frac{1}{3}X^4 \right\}$$

$$= \frac{-\Delta P d^2}{32\mu_p l}\left\{ 1 - \frac{4}{3}X + \frac{1}{3}X^4 \right\} \tag{3.148}$$

For a Newtonian fluid both R_Y and X are zero and the mean velocity becomes:

$$u = \frac{-\Delta P d^2}{32\mu_p l}$$
(cf. equation 3.35)

Equation 3.148 is sometimes known as Buckingham's equation.

Example 3.8

The rheological properties of a particular suspension can be approximated reasonably well by either a "power law" or a "Bingham plastic" model over the shear rate range of 10 to 50 s^{-1}. If the consistency k is 10 N s^n/m^2 and the flow behaviour index n is 0.2 in the power law model, what will be the approximate values of the yield stress and of the plastic viscosity in the Bingham plastic model?

What will be the pressure drop, when the suspension is flowing under laminar conditions in a pipe 200 m long and 40 mm diameter, when the centre line velocity is 1 m/s, according to the power law model? Calculate the centre line velocity for this pressure drop for the Bingham plastic model.

Solution

Using *power law model* (equation 3.121):

$$|R_y| = k \left(\left| \frac{du_x}{dy} \right| \right)^n = 10 \left(\left| \frac{du_x}{dy} \right| \right)^{0.2}$$

When:
$$\left| \frac{du_x}{dy} \right| = 10 \text{ s}^{-1} : |R_y| = 10 \times 10^{0.2} = 15.85 \text{ N/m}^2$$

$$\left| \frac{du_x}{dy} \right| = 50 \text{ s}^{-1} : |R_y| = 10 \times 50^{0.2} = 21.87 \text{ N/m}^2$$

Using *Bingham plastic model* (equation 3.125):

$$|R_y| = R_Y + \mu_p \left| \frac{du_x}{dy} \right|$$

When:
$$\left| \frac{du_x}{dy} \right| = 10 \text{ s}^{-1} : \quad 15.85 = R_Y + 10\mu_p$$

$$\left| \frac{du_x}{dy} \right| = 50 \text{ s}^{-1} : \quad 21.87 = R_Y + 50\mu_p.$$

Subtracting:
$$6.02 = 40\mu_p.$$

Thus:
$$\mu_p = 0.150 \text{ N s/m}^2$$

and:
$$R_Y = 14.35 \text{ N/m}^2$$

Thus, Bingham plastic equation is:

$$|R_y| = 14.35 + 0.150 \left| \frac{du_x}{dy} \right|$$

For power-law fluid:

Equation 3.131 gives:
$$u_{CL} = \left(\frac{-\Delta P}{l} \frac{1}{2k} \right)^{1/n} \frac{n}{n+1} r^{(n+1)/n}$$

Re-arrangement gives:
$$-\Delta P = 2kl u_{CL}^n \left(\frac{n+1}{n} \right)^n r^{-(n+1)}.$$

Numerical values in SI units are:

$$u_{CL} = 1, \quad l = 200, \quad r = 0.02, \quad k = 10, \quad n = 0.2$$

and:

$$-\Delta P = \underline{\underline{626,000 \ \text{N/m}^2}}$$

For Bingham plastic fluid:

Centre line velocity is given by equation 3.145:

$$u_p = \frac{-\Lambda P r^2}{4 \mu_p l} (1 - 4X + 4X^2)$$

where:

$$X = \frac{l}{r} \frac{R_Y}{-\Delta P}$$

$$= \frac{200}{0.02} \times \frac{12}{626,000} = 0.229$$

$$1 - 4X + 4X^2 = 0.293$$

$$u_p = \frac{626,000 \times (0.02)^2}{4 \times 0.150 \times 200} \times 0.293$$

$$= \underline{\underline{0.61 \ \text{m/s}}}$$

Example 3.9

A Newtonian liquid of viscosity 0.1 N s/m^2 is flowing through a pipe of 25 mm diameter and 20 m in length, and the pressure drop is 10^5 N/m^2. As a result of a process change a small quantity of polymer is added to the liquid and this causes the liquid to exhibit non-Newtonian characteristics; its rheology is described adequately by the "Power Law" model and the "flow index" is 0.33. The apparent viscosity of the modified fluid is equal to the viscosity of the original liquid at a shear rate of 1000 s^{-1}.

If the pressure difference over the pipe is unaltered, what will be the ratio of the volumetric flowrates of the two liquids?

Solution

For power law fluid:
$$|R_y| = k \left(\left| \frac{du_x}{dy} \right| \right)^n = \left[k \left(\left| \frac{du_x}{dy} \right| \right)^{n-1} \right] \frac{du_x}{dy}$$

Apparent viscosity
$$\mu_a = k \left(\left| \frac{du_x}{dy} \right| \right)^{-0.67}$$

When:
$$\left| \frac{du_x}{dy} \right| = 1000 \ \text{s}^{-1} \mu_a = k(1000)^{-0.67} = \frac{k}{100} = 0.1 \ \text{kg/m s}$$

$$\therefore \qquad k = 10$$

Rheological equation is
$$|R_y| = 10 \left(\left| \frac{du_x}{dy} \right| \right)^{0.33}$$

From equation 3.136:
For power law fluid:

$$u = \left(\frac{-\Delta P}{4kL} \right)^{1/n} \frac{n}{2(3n+1)} d^{(n+1)/n}.$$

For polymer solution ($n = 0.33$):

$$u_2 = \left(\frac{10^5}{4 \times 10 \times 20}\right)^3 \frac{1}{12} \cdot 0.025^4$$

$$= 0.0636 \text{ m/s}$$

For original Newtonian fluid ($n = 1$):

$$u_1 = \frac{10^5}{4 \times 0.1 \times 20} \cdot \frac{1}{8} \cdot 0.025^2$$

$$= 0.977 \text{ m/s}$$

$$\frac{u_2}{u_1} = \underline{\underline{0.065}}$$

3.7.7. General equations for pipeline flow

Fluids whose behaviour can be approximated by the power-law or Bingham plastic equation are essentially special cases, and frequently the rheology may be very complex and it may not be possible to fit simple algebraic equations to the flow curves. It is therefore desirable to adopt a more general approach for time-independent fluids in fully-developed flow which will be introduced here. For a more detailed treatment and for examples of its application, reference should be made to more specialist sources.[14−16]

If the shear stress is a function of the shear rate, it is possible to express the relation the other way round to give shear rate, $-du_x/ds$, as a function of the shear stress (the negative sign is included because velocity decreases from the pipe centre outwards).

Thus:
$$\dot{\gamma} = \frac{-du_x}{ds} = f(R_s) = f\left(R\frac{s}{r}\right) \tag{3.149}$$

where R_s is the positive value of the shear stress at radius s.

On integration, and noting that $u_x = 0$ at $s = r$:

$$u_x = \int_r^s f\left(R\frac{s}{r}\right) ds \tag{3.150}$$

The total volumetric flowrate Q through the pipe is found by integrating over the whole cross-section:

$$Q = \int_0^r 2\pi s u_x \, ds$$

$$= \pi \int_0^{r^2} u_x \, d(s^2)$$

$$= \pi \left[s^2 u_x - \int s^2 \, du_x \right]_{s=0}^{s=r}$$

$$= \pi \int_0^r s^2 f\left(R\frac{s}{r}\right) ds \quad \text{(since } u_x = 0 \text{ at } s = r) \tag{3.151}$$

Substituting $s = r(R_s/R)$:

$$Q = \pi \int_0^R \left(r\frac{R_s}{R}\right)^2 f(R_s)\frac{r}{R} \, dR_s$$

or:
$$\frac{Q}{\pi r^3} = \frac{1}{R^3} \int_0^R R_s^2 f(R_s) \, dR_s.$$ (3.152)

For a *Power-law fluid*: from equation 3.121, $f(R_s) = \dot{\gamma} = \left(\frac{R_s}{k}\right)^{1/n}$ and integration of equation 3.152 yields equation 3.136 for the relation between u and $-\Delta P$.

For a *Bingham plastic fluid*: from equation 3.125, $f(R_s) = \dot{\gamma} = \dfrac{R_s - R_Y}{\mu_p}$ and integration of equation 3.152 yields equation 3.148.

Multiplying equation 3.152 by R^3 and differentiating with respect to R_s, noting that the integral is definite with an upper limit of R:

$$R^3 \frac{d(Q/\pi r^3)}{dR} + 3R^2 \frac{Q}{\pi r^3} = R^2 f(R)$$

Dividing by R^2 and putting $R = -\Delta P \dfrac{r}{2l}$:

$$-\Delta P \frac{r}{2l} \frac{d(Q/\pi r^3)}{d\left(-\Delta P \dfrac{r}{2l}\right)} + \frac{3Q}{\pi r^3} = f(R) = \left(-\frac{du_x}{ds}\right)_{s=r}$$

$$\therefore \qquad \frac{4Q}{\pi r^3} \frac{d[\ln(Q/\pi r^3)]}{4 d \ln\left[-\Delta P \dfrac{r}{2l}\right]} + \frac{3Q}{\pi r^3} = \left(-\frac{du_x}{ds}\right)_{s=r}$$ (3.153)

Writing:
$$n' = \frac{d\left[\ln\left(-\Delta P \dfrac{r}{2l}\right)\right]}{d[\ln(Q/\pi r^3)]}$$ (3.154)

(where n' is the slope of the log/log plot of $-\Delta P(r/2l)$ against $Q/\pi r^3$)

$$\frac{4Q}{\pi r^3}\left(\frac{1}{4n'} + \frac{3}{4}\right) = \frac{4Q}{\pi r^3}\left(\frac{3n'+1}{4n'}\right) = \left(-\frac{du_x}{ds}\right)_{s=r}$$ (3.155)

Now the rheological data for a fluid can be represented as a curve of $R = -\Delta P(r/2l)$ plotted against $4Q/\pi r^3$ using log/log coordinates and the slope n' can be measured over any small range, irrespective of whether an equation can be fitted to the curve. This provides a basis of using the rheological data for *any shear-dependent fluid*.

From the definition of n':

$$R = -\Delta P \frac{r}{2l} = \text{constant}\left(\frac{Q}{\pi r^3}\right)^{n'} = k'\left(\frac{4Q}{\pi r^3}\right)^{n'}$$ (3.156)

Thus n' and k' are parameters which can be measured for any fluid. The application of this method will not be dealt with further, but it is important to have a knowledge of the significance of these two parameters which are widely used in the literature on rheology.

For a *power-law fluid*, the slope n' (equation 3.154) has a constant value n;

$$\therefore \qquad n' = n$$ (3.157)

and using equation 3.135 to substitute for $-\Delta P$ in equation 3.156:

$$k' = k\left(\frac{3n+1}{4n}\right)^n$$ (3.158)

3.7.8. Turbulent flow

It may seem surprising but there is no completely reliable method of predicting pressure drop for turbulent flow of non-Newtonian fluids in pipes. There is strong evidence that fluids showing similar flow characteristics in laminar flow do not necessarily behave in the same way in turbulent flow. Thus, for instance, a flocculated suspension and a polymer solution following the power-law model and having similar values of n and k may give different results in turbulent flow. HEYWOOD and CHENG[19] have illustrated the difficulties in a paper in which they have, unsuccessfully, attempted to calculate pressure drops for the turbulent flow of sewage sludge, using the various equations given in the literature.

As indicated earlier, non-Newtonian characteristics have a much stronger influence on flow in the streamline flow region where viscous effects dominate than in turbulent flow where inertial forces are of prime importance. Furthermore, there is substantial evidence to the effect that, for shear-thinning fluids, the standard friction chart tends to over-predict pressure drop if the Metzner and Reed Reynolds number Re_{MR} is used. Furthermore, laminar flow can persist for slightly higher Reynolds numbers than for Newtonian fluids. Overall, therefore, there is a factor of safety involved in treating the fluid as Newtonian when flow is expected to be turbulent.

HARTNETT and KOSTIC[20] have recently examined the published correlations for turbulent flow of shear-thinning "power-law" fluids in pipes and in non-circular ducts. They have concluded that, for smooth pipes, DODGE and METZNER'S[21] modification of equation 3.11 (to which it reduces for Newtonian fluids) is the most satisfactory.

However, they also report that YOO[22] has proposed a simple modification to the Blasius equation for turbulent flow in a pipe, which gives values of the friction factor accurate to within about ±10 per cent. The friction factor is expressed in terms of the METZNER and REED[18] generalised Reynolds number Re_{MR} and the power-law index n.

$$\frac{f}{2} = \phi = \frac{R}{\rho u^2} = 0.0396 n^{0.675} Re_{MR}^{-0.25} \tag{3.159}$$

Equation 3.159 reduces to the simple Blasius relation (equation 3.10) for a Newtonian fluid ($n = 1$).

Rearranging equation 3.159 gives:

$$\frac{R}{\rho u^2} n^{-0.675} = 0.0396 Re_{MR}^{-0.25} \tag{3.160}$$

Thus, the pipe friction chart for a Newtonian fluid (Figure 3.3) may be used for shear-thinning power-law fluids if Re_{MR} is used in place of Re. In the turbulent region, the ordinate is equal to $(R/\rho u^2) n^{-0.675}$. For the streamline region the ordinate remains simply $R/\rho u^2$, because Re_{MR} has been defined so that it shall be so (see equation 3.140). More recently, IRVINE[23] has proposed an improved form of the modified Blasius equation which predicts the friction factor for inelastic shear-thinning polymer-solutions to within ±7 per cent.

$$\phi = \frac{R}{\rho u^2} = \left[\frac{2^{3-2n}}{7^{7n}} \left(\frac{4}{3n+1} \right)^{n(3n+2)} Re_{MR}^{-1} \right]^{1/(3n+1)} \tag{3.161}$$

Both equations 3.160 and 3.161 are easy to use as they give $R/\rho u^2$ as explicit functions of Re_{MR} and n. Equation 3.159 should be used with caution, particularly if the fluid exhibits any elastic properties, bearing in mind the conclusions of HEYWOOD and CHENG[19].

3.8. FURTHER READING

ASTARITA, G. and MARRUCCI, G.: *Principles of Non-Newtonian Fluid Mechanics* (McGraw-Hill, New York, 1974).

BIRD, R. B., ARMSTRONG, R. C. and HASSAGER, O.: *Dynamics of Polymeric Liquids Vol. 1, Fluid Mechanics* (Wiley, New York 1977).

BRASCH, D. J. and WHYMAN, D.: *Problems in Fluid Flow* (Edward Arnold, London, 1986).

CHURCHILL, S. W.: *Viscous Flows: The Practical Use of Theory* (Butterworth, London, 1988).

CRANE: Technical Paper No. 410. *Flow Through Valves, Fittings and Pipes*. 14th printing (Crane, New York, 1974).

DUNCAN, W. J., THOM, A. S. and YOUNG, A. D.: *Mechanics of Fluids*, 2nd edn (Edward Arnold, London, 1972).

FRANCIS, J. R. D.: *A Textbook of Fluid Mechanics*, 3rd edn (Edward Arnold, London, 1971).

HOLLAND, F. A.: *Fluid Flow* (Edward Arnold, London, 1986).

HUTTON, J. F., PEARSON, J. R. A. and WALTERS, K. (eds): *Theoretical Rheology* (Applied Science Publ., London, 1975).

KAY, J. M. and NEDDERMAN, R. M.: *An Introduction to Fluid Mechanics and Heat Transfer*, 3rd edn (Cambridge Univ. Press, Cambridge, 1974).

MASSEY, B. S.: *Mechanics of Fluids*, 5th edn (Van Nostrand/Reinhold, London, 1987).

PATERSON, A. R.: *A First Course in Fluid Mechanics* (Cambridge Univ. Press, Cambridge 1985).

STREETER, V. L. and WYLIE, E. B.: *Fluid Mechanics*, 8th edn (McGraw-Hill, New York, 1985).

ZAHORSKI, S.: *Mechanics of Viscoelastic Fluids* (PNN, Polish Scientific Publishers and Martinus Nijheff, 1981).

3.9. REFERENCES

1. REYNOLDS, O.: *Papers on Mechanical and Physical Subjects* **2** (1881–1901) 51. An experimental investigation of the circumstances which determine whether the motion of water shall be direct or sinuous and the law of resistance in parallel channels. 535. On the dynamical theory of incompressible viscous fluids and the determination of the criterion.

2. PRANDTL, L.: *Z. Ver. deut. Ing.* **77** (1933) 105. Neuere Ergebnisse der Turbulenzforschung.

3. NIKURADSE, J.: *Forsch. Ver. deut. Ing.* **361** (1933). Strömungsgesetze in rauhen Röhren.

4. STANTON, T. and PANNELL, J.: *Phil. Trans. R. Soc.* **214** (1914) 199. Similarity of motion in relation to the surface friction of fluids.

5. MOODY, L.F.: *Trans. Am. Soc. Mech. Engrs.* **66** (1944) 671. Friction factors for pipe flow.

6. BLASIUS, H.: *Forsch. Ver. deut. Ing.* **131** (1913). Das Ähnlichkeitsgesetz bei Reibungsvorgängen in Flüssigkeiten.

7. LAUFER, J.: United States National Advisory Committee for Aeronautics. Report No. 1174 (1955). The structure of turbulence in fully developed pipe flow.

8. HAGEN, G.: *Ann Phys. (Pogg. Ann.)* **46** (1839) 423. Ueber die Bewegung des Wassers in engen zylindrischen Röhren.

9. POISEUILLE, J.: *Inst. de France Acad. des Sci. Mémoires présentés par divers savantes* **9** (1846) 433. Recherches experimentales sur le mouvement des liquides dans les tubes de très petit diamètre.

10. RYAN, N. W. and JOHNSON, M. A.: *A.I.Ch.E.Jl.* **5** (1959) 433. Transition from laminar to turbulent flow in pipes.

11. WHITE, C. M.: *Proc. Roy. Soc.* A, **123** (1929) 645. Streamline flow through curved pipes.

12. BS 5168: 1975: British Standard 5168 (British Standards Institution, London). Glossary of Rheological Terms.

13. CROSS, M. M.: *J. Colloid Sci.* **20** (1965) 417. Rheology of non-Newtonian fluids: a new flow equation for pseudoplastic systems.

14. ASTARITA, G. and MARRUCCI, G.: *Principles of Non-Newtonian Fluid Mechanics* (McGraw-Hill, New York, 1974).

15. SKELLAND, A.H.P.: *Non-Newtonian Flow and Heat Transfer* (Wiley, 1967).

16. HUTTON, J. F., PEARSON, J. R. A. and WALTERS, K. (eds): *Theoretical Rheology* (Applied Science Publishers, London, 1975).

17. FAROOQI, S. I. and RICHARDSON, J. F.: *Trans. I. Chem. E.* **58** (1980) 116. Rheological behaviour of kaolin suspensions in water and water-glycerol mixtures.
18. METZNER, A. B. and REED, J. C.: *A. I. Ch. E. Jl.* **1** (1955) 434. Flow of non-Newtonian fluids—correlation of the laminar, transition and turbulent flow regions.
19. HEYWOOD, N. I. and CHENG, D. C.-H.: *Trans Inst. Measurement and Control* **6** (1984) 33. Comparison of methods for predicting head loss in turbulent pipe flow of non-Newtonian fluids.
20. HARTNETT, J. P. and KOSTIC, M.: *Int. Comm. Heat & Mass Transfer* **17** (1990) 59. Turbulent friction factor correlations for power law fluids in circular and non-circular channels.
21. DODGE, D. W. and METZNER, A. B.: *A. I. Ch. E. Jl.* **5** (1959) 189. Turbulent flow of non-Newtonian systems.
22. YOO, S. S.: Ph.D. Thesis, University of Illinois, Chicago (1974), Heat transfer and friction factors for non-Newtonian fluids in circular tubes.
23. IRVINE, T. F.: *Chem. Eng. Comm.* **65** (1988) 39. A generalized Blasius equation for power law fluids.

3.10. NOMENCLATURE

		Units in SI system	Dimensions in M, L, T, θ
A	Area perpendicular to direction of flow	m^2	L^2
B	Width of rectangular channel or notch	m	L
C_c	Coefficient of contraction	—	—
C_f	Coefficient for flow over a bank of tubes	—	—
D	Depth of liquid in channel	m	L
d	Diameter of pipe	m	L
d_a	Dimension of rectangular duct, or distance apart of parallel plates	m	L
d_b	Dimension of rectangular duct	m	L
d_m	Hydraulic mean diameter $(= 4\,A/p)$	m	L
E	Eddy kinematic viscosity	m^2/s	L^2T^{-1}
e	Surface roughness	m	L
F	Energy per unit mass degraded because of irreversibility of process	J/kg	L^2T^{-2}
f	Fanning friction factor $(= 2\,R/\rho u^2)$	—	—
f'	Moody friction factor $(8R/\rho u^2)$	—	—
G	Mass rate of flow	kg/s	MT^{-1}
$\mathbf{G}$	Young's modulus	N/m^2	$ML^{-1}T^{-2}$
g	Acceleration due to gravity	m/s^2	LT^{-2}
h	Depth below surface measured perpendicular to bottom of channel or notch	m	L
h_f	Head lost due to friction	m	L
i	Hydraulic gradient (h_f/l)	—	—
J	Specific energy of fluid in open channel	J/kg	L^2T^{-2}
j	Number of banks of pipes in direction of flow	—	—
k	Consistency coefficient in power-law equation	N s^n/m^2	$ML^{-1}T^{n-2}$
k'	Coefficient defined by equation 3.93	N s$^{n'}$/m^2	$ML^{-1}T^{n'-2}$
L	Characteristic linear dimension	m	L
l	Length of pipe or channel	m	L
m	Index in equation 3.64	—	—
n	Index in power-law equation	—	—
n'	Slope defined by equation 3.90	—	—
P	Pressure	N/m^2	$ML^{-1}T^{-2}$
P_f	Pressure due to friction	N/m^2	$ML^{-1}T^{-2}$
ΔP	Pressure difference or change	N/m^2	$ML^{-1}T^{-2}$
p	Wetted perimeter	m	L
Q	Volumetric rate of flow	m^3/s	L^3T^{-1}
Q_A	Volumetric flowrate in sheared annulus	m^3/s	L^3T^{-1}
Q_p	Volumetric flowrate in unsheared plug	m^3/s	L^3T^{-1}
q	Net heat flow into system	J/kg	L^2T^{-2}
R	Shear stress on surface	N/m^2	$ML^{-1}T^{-2}$
R_m	Mean value of shear stress at surface	N/m^2	$ML^{-1}T^{-2}$

		Units in SI system	Dimensions in $\mathbf{M, L, T}, \theta$
R_S	Shear stress at radius s	N/m^2	$\mathbf{ML^{-1}T^{-2}}$
R_Y	Yield stress	N/m^2	$\mathbf{ML^{-1}T^{-2}}$
R_y	Shear stress at some point in fluid distance y from surface	N/m^2	$\mathbf{ML^{-1}T^{-2}}$
R_0	Shear stress $(-R)$ in fluid at boundary surface $(y = 0)$	N/m^2	$\mathbf{ML^{-1}T^{-2}}$
r	Radius of pipe, or outer pipe in case of annulus, or radius of cone	m	$\mathbf{L}$
r_c	Radius of unsheared plug	m	$\mathbf{L}$
r_i	Radius of inner pipe of annulus	m	$\mathbf{L}$
S	Entropy per unit mass	J/kg K	$\mathbf{L^2T^{-2}\theta^{-1}}$
s	Distance from axis of pipe or from centre plane or of rotation or thickness of liquid film	m	$\mathbf{L}$
t	Time	s	$\mathbf{T}$
T	Temperature	K	θ
U	Internal energy per unit mass	J/kg	$\mathbf{L^2T^{-2}}$
u	Mean velocity	m/s	$\mathbf{LT^{-1}}$
u_{CL}	Velocity in pipe at centre line	m/s	$\mathbf{LT^{-1}}$
u_{Ex}	Fluctuating velocity component	m/s	$\mathbf{LT^{-1}}$
u_i	Instantaneous value of velocity	m/s	$\mathbf{LT^{-1}}$
u_p	Velocity of unsheared plug	m/s	$\mathbf{LT^{-1}}$
u_s	Velocity at free surface	m/s	$\mathbf{LT^{-1}}$
$u_{\max}$	Maximum velocity	m/s	$\mathbf{LT^{-1}}$
u_t	Velocity at narrowest cross section of bank of tubes	m/s	$\mathbf{LT^{-1}}$
u_x	Velocity in X-direction at distance y from surface	m/s	$\mathbf{LT^{-1}}$
u_w	Velocity of propagation wave	m/s	$\mathbf{LT^{-1}}$
v	Volume per unit mass of fluid	m^3/kg	$\mathbf{M^{-1}L^3}$
W_s	Shaft work per unit mass	J/kg	$\mathbf{L^2T^{-2}}$
w	Width of surface	m	$\mathbf{L}$
X	$(2l/r)(R_Y/-\Delta P)$	—	—
x	Distance in X-direction or in direction of motion or parallel to surface	m	$\mathbf{L}$
y	Distance in Y-direction or perpendicular distance from surface	m	$\mathbf{L}$
Z	Stability criterion defined in equation 3.56	—	—
z	Distance in vertical direction	m	$\mathbf{L}$
α	Constant in expression for kinetic energy of fluid	—	—
β	Constant in equation 3.124	s$^{2/3}$	$\mathbf{T^{2/3}}$
γ	Strain	—	—
$\dot{\gamma}$	Rate of shear or of strain	s^{-1}	$\mathbf{T^{-1}}$
δ	Boundary layer thickness	m	$\mathbf{L}$
λ	Shear force acting on unit length of inner surface of annulus	N/m	$\mathbf{MT^{-2}}$
μ	Viscosity of fluid	N s/m^2	$\mathbf{ML^{-1}T^{-1}}$
μ_a	Apparent viscosity defined by equation 3.117	N s/m^2	$\mathbf{ML^{-1}T^{-1}}$
μ_p	Plastic viscosity	N s/m^2	$\mathbf{ML^{-1}T^{-1}}$
μ_0	Apparent viscosity $(\dot{\gamma} \to 0)$	N s/m^2	$\mathbf{ML^{-1}T^{-1}}$
μ_∞	Apparent viscosity $(\dot{\gamma} \to \infty)$	N s/m^2	$\mathbf{ML^{-1}T^{-1}}$
ϕ	Friction factor $(= R/\rho u^2)$	—	—
ρ	Density of fluid	kg/m^3	$\mathbf{ML^{-3}}$
θ	Angle between cone and plate in viscometers or angle	—	—
τ	Shear stress $(= R)$	N/m^2	$\mathbf{ML^{-1}T^{-2}}$
ω	Angular speed of rotation	s^{-1}	$\mathbf{T^{-1}}$
Re	Reynolds number with respect to pipe diameter	—	—
Re_{MR}	Generalised (Metzner and Reed) Reynolds number	—	—
$Re_{(MR)c}$	Critical value of Re_{MR} at laminar-turbulent transition	—	—
Re_x	Reynolds number with respect to distance x	—	—

CHAPTER 4

Flow of Compressible Fluids

4.1. INTRODUCTION

Although all fluids are to some degree compressible, compressibility is sufficiently great to affect flow under normal conditions only for a gas. Furthermore, if the pressure of the gas does not change by more than about 20 per cent, it is usually satisfactory to treat the gas as an incompressible fluid with a density equal to that at the mean pressure.

When compressibility is taken into account, the equations of flow become very much more complex than they are for an incompressible fluid, even if the simplest possible equation of state (Ideal Gas Law) is used to describe their behaviour. In this chapter, attention will be confined to consideration of the flow of ideal gases. The physical property of a gas which varies, but which is constant for an incompressible fluid, is the density ρ or specific volume $v(= 1/\rho)$. Density is a function of both temperature and pressure and it is necessary therefore to take account of the effects of both of these variables. The relation between the pressure and the density will be affected by the heat transfer to the gas as it passes through the system. Isothermal conditions can be maintained only if there is very good heat transfer to the surroundings and normally exist only at low flowrates in small equipment. At the opposite extreme, in large installations with high flowrates, conditions are much more nearly adiabatic. It should be noted that, except for isothermal flow, the relation between pressure and density is influenced by the way in which the change is caused (e.g. the degree of reversibility).

In this chapter consideration will be given to the flow of gases through orifices and nozzles, and to flow in pipelines. It will be found that, in all these cases, the flow may reach a limiting maximum value which is independent of the downstream pressure; this is a phenomenon which does not arise with incompressible fluids.

4.2. FLOW OF GAS THROUGH A NOZZLE OR ORIFICE

This is one of the simplest applications of the flow of a compressible fluid and it can be used to illustrate many of the features of the process. In practical terms, it is highly relevant to the design of relief valves or bursting discs which are often incorporated into pressurised systems in order to protect the equipment and personnel from dangers which may arise if the equipment is subjected to pressures in excess of design values. In many cases it is necessary to vent gases which are involved in a chemical reaction.

For this purpose, the gas flowrate at an aperture through which it discharges from a vessel maintained at a constant pressure P_1 to surroundings at a pressure P_2 (Figure 4.1) will be considered.

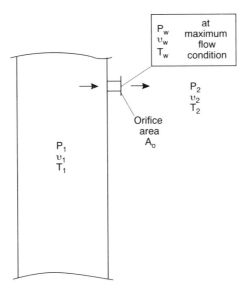

Figure 4.1.   Discharge through an orifice

The energy balance equation (2.54) is as follows:

$$\frac{1}{\alpha} u \, du + g \, dz + v \, dP + \delta W_s + \delta F = 0 \tag{4.1}$$

If the flow is horizontal ($dz = 0$), the velocity distribution is assumed to be flat ($\alpha = 1$) and friction neglected ($\delta F = 0$) then, as no external work is performed:

$$u \, du + v \, dP = 0 \tag{4.2}$$

If the velocity in the vessel at which the gas approaches the outlet is negligible ($u_1 = 0$), integration of equation 4.2 gives:

$$\frac{u_2^2}{2} = -\int_{P_1}^{P_2} v \, dP \tag{4.3}$$

The value of the integral depends on the pressure–volume relation.

4.2.1. Isothermal flow

For the isothermal flow of an ideal gas:

$$-\int_{P_1}^{P_2} v \, dP = -P_1 v_1 \ln \frac{P_2}{P_1} = P_1 v_1 \ln \frac{P_1}{P_2} \tag{4.4}$$

(from equation 2.69)

Substituting into equation 4.3, gives the discharge velocity u_2:

$$\frac{u_2^2}{2} = P_1 v_1 \ln \frac{P_1}{P_2}$$

i.e.:
$$u_2 = \sqrt{2P_1v_1 \ln \frac{P_1}{P_2}}$$
(4.5)

The mass rate of flow is given by:

$$G = u_2 A_2 \rho_2 = \frac{u_2 A_2}{v_2}$$

$$= \frac{A_2}{v_2} \sqrt{2P_1v_1 \ln \frac{P_1}{P_2}}$$
(4.6)

It is shown in Chapter 6 that the minimum flow area A_2 tends to be somewhat smaller than the area A_0 of the aperture because the gas leaves with a small radial inwards velocity component. Furthermore, there will be some reduction of discharge rate because of the frictional effects which have been neglected. Taking these factors together and by means of a coefficient of discharge $C_D (C_D < 1)$:

$$G = \frac{C_D A_0}{v_2} \sqrt{2P_1v_1 \ln \frac{P_1}{P_2}}$$
(4.7)

Substituting for isothermal flow:

$$v_2 = v_1 \frac{P_1}{P_2}$$

$$G = C_D A_0 \frac{P_2}{P_1 v_1} \sqrt{2P_1 v_1 \ln \frac{P_1}{P_2}}$$

$$= C_D A_0 P_2 \sqrt{\frac{2}{P_1 v_1} \ln \frac{P_1}{P_2}} = C_D A_0 P_2 \sqrt{\frac{2M}{\mathbf{R}T} \ln \frac{P_1}{P_2}}$$
(4.8)

where $\mathbf{R}$ is the universal gas constant, and
M is the molecular weight of the gas.

Maximum flow conditions

Equation 4.8 gives $G = 0$ for $P_2 = P_1$, and for $P_2 = 0$. As it is a continuous function, it must yield a maximum value of G at some intermediate pressure $P_w (0 < P_w < P_1)$.

Differentiating both sides with respect to P_2:

$$\frac{dG}{dP_2} = C_D A_0 \sqrt{\frac{2}{P_1 v_1}} \frac{d}{dP_2} \left\{ P_2 \left[\ln \left(\frac{P_1}{P_2} \right) \right]^{1/2} \right\}$$

$$= C_D A_0 \sqrt{\frac{2}{P_1 v_1}} \left\{ \left[\ln \left(\frac{P_1}{P_2} \right) \right]^{1/2} + P_2 \frac{1}{2} \left[\ln \left(\frac{P_1}{P_2} \right) \right]^{-1/2} \left(\frac{P_2}{P_1} \right) \left(\frac{P_1}{-P_2^2} \right) \right\}$$

For a maximum value of G, put $dG/dP_2 = 0$:

Then:
$$\ln \frac{P_1}{P_2} = \frac{1}{2}$$
(4.9)

i.e.:
$$\frac{P_1}{P_2} = 1.65 \tag{4.10}$$

and the critical pressure ratio
$$\frac{P_2}{P_1} = w_c = 0.607 \tag{4.11}$$

Substituting into equation 4.8 to give the maximum value of $G(G_{max})$:

$$G_{max} = C_D A_0 P_2 \sqrt{\frac{2}{P_1 v_1} \cdot \frac{1}{2}}$$

$$= C_D A_0 \sqrt{\frac{P_2}{v_2}}$$

$$= C_D A_0 \rho_2 \sqrt{P_2 v_2} \quad \left(\rho_2 = \frac{1}{v_2}\right) \tag{4.12}$$

Thus the velocity
$$u_w = \sqrt{P_2 v_2} = \sqrt{P_0 v_0} = \sqrt{\frac{RT}{M}} \tag{4.13}$$

It will be shown later (Section 4.3) that this corresponds to the velocity at which a small pressure wave will be propagated under isothermal conditions, sometimes referred to as the "isothermal sonic velocity", though the heat transfer rate will not generally be sufficient to maintain truly isothermal conditions.

Writing equation 4.12 in terms of the upstream conditions (P_1, v_1):

$$G_{max} = C_D A_0 \frac{1}{v_1} \frac{P_2}{P_1} \sqrt{P_1 v_1}$$

or, on substituting from equation 4.11:

$$= 0.607 C_D A_0 \sqrt{\frac{P_1}{v_1}} \tag{4.14}$$

At any given temperature, $P_1 v_1 = P_0 v_0 =$ constant, where v_0 is the value of v at some reference pressure P_0.

Then:
$$G_{max} = 0.607 C_D A_0 P_1 \sqrt{\frac{1}{P_0 v_0}} = 0.607 C_D A_0 P_1 \sqrt{\frac{M}{RT}} \tag{4.15}$$

Thus, G_{max} is linearly related to P_1.

It will be seen when the pressure ratio P_2/P_1 is less than the critical value ($w_c = 0.607$) the flow rate becomes independent of the downstream pressure P_2. The fluid at the orifice is then flowing at the velocity of a small pressure wave; therefore the velocity of the pressure wave *relative to the orifice* is zero; that is the upstream fluid cannot be influenced by the pressure in the downstream reservoir. Thus, the pressure falls to the critical value at the orifice, and further expansion to the downstream pressure takes place in the reservoir with the generation of a *shock wave* (Section 4.6).

4.2.2. Non-isothermal flow

If the flow is non-isothermal, it may be possible to represent the relationship between pressure and volume by an equation of the form:

$$Pv^k = \text{constant} \tag{4.16}$$

Then from equation 2.73:

$$-\int_{P_1}^{P_2} v \, dP = \frac{k}{k-1} P_1 v_1 \left[1 - \left(\frac{P_2}{P_1} \right)^{(k-1)/k} \right] \tag{4.17}$$

The discharge velocity u_2 is then given by:

$$\frac{u_2^2}{2} = \frac{k}{k-1} P_1 v_1 \left[1 - \left(\frac{P_2}{P_1} \right)^{(k-1)/k} \right]$$

i.e.:

$$u_2 = \sqrt{ \frac{2k}{k-1} P_1 v_1 \left[1 - \left(\frac{P_2}{P_1} \right)^{(k-1)/k} \right] } \tag{4.18}$$

As previously, allowing for a discharge coefficient, the mass rate of flow is given by:

$$G = \frac{C_D A_0}{v_2} \sqrt{ \frac{2k}{k-1} P_1 v_1 \left[1 - \left(\frac{P_2}{P_1} \right) \right]^{(k-1)/k} } \tag{4.19}$$

$$= \frac{C_D A_0}{v_1} \left(\frac{P_2}{P_1} \right)^{1/k} \sqrt{ \frac{2k}{k-1} P_1 v_1 \left[1 - \left(\frac{P_2}{P_1} \right)^{(k-1)/k} \right] } \tag{4.20}$$

Maximum flow conditions

This equation again gives $G = 0$ for $P_2 = 0$ and for $P_2 = P_1$.

Differentiating both sides of equation 4.20 with respect to P_2/P_1:

$$\frac{dG}{d(P_2/P_1)} = \frac{C_D A_0}{v_1} \sqrt{ \frac{2k}{k-1} P_1 v_1 } \left\{ \left(\frac{P_2}{P_1} \right)^{1/k} \frac{1}{2} \left[1 - \left(\frac{P_2}{P_1} \right)^{(k-1)/k} \right]^{-1/2} \right.$$

$$\left. \times \left(-\frac{k-1}{k} \right) \left(\frac{P_2}{P_1} \right)^{-1/k} + \left[1 - \left(\frac{P_2}{P_1} \right)^{(k-1)/k} \right]^{1/2} \frac{1}{k} \left(\frac{P_2}{P_1} \right)^{(1/k)-1} \right\}$$

Putting $dG/d(P_2/P_1) = 0$ for the maximum value of $G(G_{max})$:

$$1 - \left(\frac{P_2}{P_1} \right)^{(k-1)/k} = \frac{1}{2}(k-1) \left(\frac{P_2}{P_1} \right) \left(\frac{P_2}{P_1} \right)^{-1/k} \tag{4.21}$$

Substituting into equation 4.19:

$$G_{max} = \frac{C_D A_0}{v_2} \sqrt{ \frac{2k}{k-1} P_1 v_1 \frac{1}{2}(k-1) \frac{P_2}{P_1} \left(\frac{P_2}{P_1} \right)^{-1/k} }$$

$$= C_D A_0 \rho_2 \sqrt{k P_2 v_2} \tag{4.22}$$

Hence: $\qquad u_w = \sqrt{kP_2v_2}$ (4.23)

The velocity $u_w = \sqrt{kP_2v_2}$ will be shown to be the velocity of a small pressure wave if the pressure-volume relation is given by $Pv^k = $ constant. If the expansion approximates to a reversible adiabatic (isentropic) process $k \approx \gamma$ (the ratio of the specific heats of the gases) (see equation 2.30).

Equation 4.22 then becomes:

$$G_{max} = C_DA_0\rho_2\sqrt{\gamma P_2v_2}$$ (4.24)

and: $\qquad u_w = \sqrt{\gamma P_2v_2} = \sqrt{\dfrac{\gamma RT_2}{M}}$ (4.25)

$u_w = \sqrt{\gamma P_2v_2}$ is the velocity of propagation of a small pressure wave under isentropic conditions.

As for isothermal conditions, when maximum flow is occurring, the velocity of a small pressure wave, *relative to the orifice*, is zero, and the fluid at the orifice is not influenced by the pressure further downstream.

Substituting the critical pressure ratio $w_c = P_2/P_1$ in equation 4.21:

$$1 + w_c^{(k-1)/k} = \frac{k-1}{2}w_c^{(k-1)/k}$$

giving: $\qquad w_c = \left[\dfrac{2}{k+1}\right]^{k/(k-1)}$ (4.26)

For isentropic conditions:

$$w_c = \left[\frac{2}{\gamma+1}\right]^{\gamma/(\gamma-1)}$$ (4.26a)

For a diatomic gas at approximately atmospheric pressure, $\gamma = 1.4$, and $w_c = 0.53$. From equation 4.22:

$$G_{max} = C_DA_0\sqrt{\frac{kP_2}{v_2}}$$

$$= C_DA_0\sqrt{\frac{k}{v_1}\left(\frac{P_2}{P_1}\right)^{(k+1)/k}P_2}$$

$$= C_DA_0\sqrt{\frac{kP_1}{v_1}\left(\frac{P_2}{P_1}\right)^{(k+1)/k}}$$ (4.27)

Substituting from equation 4.26:

$$G_{max} = C_DA_0\sqrt{\frac{kP_1}{v_1}\left[\frac{2}{k+1}\right]^{(k+1)/(k-1)}}$$ (4.28)

For isentropic conditions, $k = \gamma$, and:

$$G_{max} = C_DA_0\sqrt{\frac{\gamma P_1}{v_1}\left[\frac{2}{\gamma+1}\right]^{(\gamma+1)/(\gamma-1)}}$$ (4.28a)

For a given upstream temperature T_1, $P_1 v_1 = P_0 v_0$ = constant where v_0 is the value of v at some reference pressure P_0 and temperature T_1.

Thus:

$$G_{max} = C_D A_0 P_1 \sqrt{\frac{k}{P_0 v_0} \left[\frac{2}{k+1}\right]^{(k+1)/(k-1)}} = C_D A_0 P_1 \sqrt{\frac{M}{RT_1} k \left[\frac{2}{k+1}\right]^{(k+1)/(k-1)}}$$

(4.29)

For isentropic conditions $k = \gamma$,

and:

$$G_{max} = C_D A_0 P_1 \sqrt{\frac{\gamma}{P_0 v_0} \left[\frac{2}{\gamma+1}\right]^{(\gamma+1)/(\gamma-1)}} = C_D A_0 P_1 \sqrt{\frac{M}{RT_1} \gamma \left[\frac{2}{\gamma+1}\right]^{(\gamma+1)/(\gamma-1)}}$$

(4.30)

For a diatomic gas, $\gamma \approx 1.4$ for pressures near atmospheric

and:
$$G_{max} = 0.685 C_D A_0 P_1 \sqrt{\frac{1}{P_0 v_0}} = 0.685 C_D A_0 P_1 \sqrt{\frac{M}{RT_1}}$$

(4.31)

It will be noted that equations 4.29, 4.30 and 4.31 give a linear relation between G_{max} and P_1. By comparison with equation 4.15, it is seen that the maximum flowrate G_{max} is $0.685/0.607 = 1.13$ times greater than for isothermal flow for a diatomic gas.

In Figure 4.2, the mass flowrate is plotted as a function of cylinder pressure for discharge through an orifice to an atmosphere at a constant downstream pressure P_2 — for

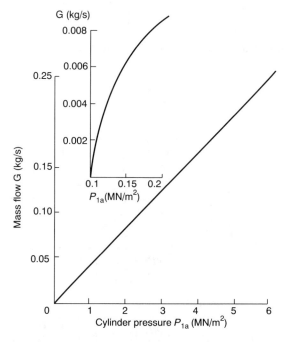

Figure 4.2.   Mass rate of discharge of gas as function of cylinder pressure for constant downstream pressure

Figure 4.3. Mass rate of discharge of gas as function of downstream pressure for constant cylinder pressure

the conditions given in Example 4.1. In Figure 4.3, the cylinder pressure P_1 is maintained constant and the effect of the downstream pressure P_2 on the flowrate is shown.

Example 4.1

A cylinder contains air at a pressure of 6.0 MN/m^2 and discharges to atmosphere through a valve which may be taken as equivalent to a sharp-edged orifice of 6 mm diameter (coefficient of discharge = 0.6). (i) Plot the rate of discharge of air from the cylinder against cylinder pressure. Assume that the expansion at the valve is approximately isentropic and that the temperature of the air in the cylinder remains unchanged at 273 K. (ii) For a constant pressure of 5 MN/m^2 in the cylinder, plot the discharge rate as a function of the pressure P_2 of the surroundings.

Solution

Critical pressure ratio for discharge through the valve

$$= \left[\frac{2}{\gamma + 1} \right]^{[\gamma/(\gamma-1)]} \qquad \text{(equation 4.26)}$$

γ varies from 1.40 to 1.54 over the pressure range 0.1 to 6.0 MN/m^2. Critical conditions occur at a relatively low cylinder pressure. Taking $\gamma = 1.40$:

$$\text{critical ratio} = \left(\frac{2}{2.40} \right)^{1.4/0.4} = 0.53$$

(i) Sonic velocity will occur until the pressure in the cylinder falls to 101.3/0.53 = 191.1 kN/m^2.

Rate of discharge for cylinder pressures greater than 191.1 kN/m² is given by equation 4.30:

$$G_{\max} = C_D A_0 P_1 \sqrt{\frac{\gamma M}{RT_1} \left(\frac{2}{\gamma+1}\right)^{(\gamma+1)/(\gamma-1)}}$$

Take a mean value for γ of 1.47.

Then:
$$\gamma \left(\frac{2}{\gamma+1}\right)^{(\gamma+1)/(\gamma-1)} = 0.485.$$

$$A_0 = \frac{\pi}{4} \times (0.006)^2 = 2.83 \times 10^{-5} \text{ m}^2$$

$$\frac{1}{P_1 v_1} = \frac{M}{RT_1} = \frac{29}{8314 \times 273} = 1.28 \times 10^{-5} \text{ s}^2/\text{m}^2$$

$$P_1 v_1 = \frac{RT_1}{M} = 78{,}100 \text{ s}^2/\text{m}^2$$

$$G_{\max} = 0.6 \times (2.83 \times 10^{-5}) \times P_1 \sqrt{(1.28 \times 10^{-5}) \times 0.485}$$

$$= 4.23 \times 10^{-8} P_1 \text{ kg/s} \quad (P_1 \text{ in N/m}^2)$$

$$= 4.23 \times 10^{-2} P_{1a} \text{ kg/s} \quad (P_{1a} \text{ in MN/m}^2)$$

For cylinder pressures below 191.1 kN/m², the mass flowrate is given by equation 4.20. Putting $k = \gamma$ for isentropic conditions:

$$G = \frac{C_D A_0}{v_1} \left(\frac{P_2}{P_1}\right)^{1/\gamma} \sqrt{\frac{2\gamma}{\gamma-1} P_1 v_1 \left[1 - \left(\frac{P_2}{P_1}\right)^{(\gamma-1)/\gamma}\right]}$$

Using a value of 1.40 for γ in this low pressure region, and writing:

$$\frac{1}{v_1} = \frac{1}{P_1 v_1} \cdot P_1$$

$$G = \frac{0.6 \times (2.83 \times 10^{-5})}{78{,}100} P_1 \left(\frac{P_2}{P_1}\right)^{0.714} \sqrt{7 \times 78{,}100 \left[1 - \left(\frac{P_2}{P_1}\right)^{0.286}\right]}$$

$$= 1.608 \times 10^{-7} P_1 \left(\frac{P_2}{P_1}\right)^{0.714} \sqrt{\left[1 - \left(\frac{P_2}{P_1}\right)^{0.286}\right]}$$

For discharge to atmospheric pressure, $P_2 = 101{,}300$ N/m²:

$$G = 6.030 \times 10^{-4} P_1^{0.286} \sqrt{1 - 27 P_1^{-0.286}} \text{ kg/s} \quad (P_1 \text{ in N/m}^2)$$

Putting pressure P_{1a} in MN/m²:

$$G = 0.0314 P_{1a}^{0.286} \sqrt{1 - 0.519 P_{1a}^{-0.286}} \text{ kg/s} \quad (P_{1a} \text{ in MN/m}^2)$$

The discharge rate is plotted in Figure 4.2.

(ii)

G vs P_{1a} data

(i) Above $P_{1a} = 0.19$ MN/m^2		(ii) Below $P_{1a} = 0.19$ MN/m^2	
P_{1a} (MN/m^2)	G (kg/s)	P_{1a} (MN/m^2)	G (kg/s)
0.19	0.0080		
0.2	0.0084	0.10	0
0.5	0.021	0.125	0.0042
1.0	0.042	0.15	0.0060
2.0	0.084	0.17	0.0070
3.0	0.126	0.19	0.0079
4.0	0.168		
5.0	0.210		
6.0	0.253		

Note: The slight mismatch in the two columns for $P_{1a} = 0.19$ MN/m^2 is attributable to the use of an average value of 1.47 for γ in column 2 and 1.40 for γ in column 4.

From this table, for a constant upstream pressure of 5 MN/m^2, the mass flowrate G remains at 0.210 kg/s for all pressures P_2 below $5 \times 0.53 = 2.65$ MN/m^2.

For higher values of P_2, the flowrate is obtained by substitution of the constant value of P_1 (5 MN/m^2) in equation 4.20.

i.e.:
$$G = 1.608 \times 10^{-7} \times (82.40) \times P_2^{0.714} \sqrt{1 - 0.01214 P_2^{0.286}} \quad (P_2 \text{ in N/m}^2)$$

$$= 0.2548 P_{2a}^{0.714} \sqrt{1 - 0.631 P_{2a}^{0.286}} \text{ kg/s} \quad (P_2 \text{ in MN/m}^2)$$

This relationship is plotted in Figure 4.3 and values of G as a function of P_{2a} are tabulated below:

P_{2a} MN/m^2	G kg/s
<2.65	0.210
3.0	0.206
3.5	0.194
4.0	0.171
4.5	0.123
4.9	0.061
4.95	0.044
5	0

4.3. VELOCITY OF PROPAGATION OF A PRESSURE WAVE

When the pressure at some point in a fluid is changed, the new condition takes a finite time to be transmitted to some other point in the fluid because the state of each intervening element of fluid has to be changed. The velocity of propagation is a function of the bulk modulus of elasticity ε, where ε is defined by the relation:

$$\varepsilon = \frac{\text{increase of stress within the fluid}}{\text{resulting volumetric strain}} = \frac{dP}{-(dv/v)} = -v\frac{dP}{dv} \quad (4.32)$$

Suppose a pressure wave is transmitted at a velocity u_w over a distance dl in a fluid of cross-sectional area A, from section A to section B (Figure 4.4).

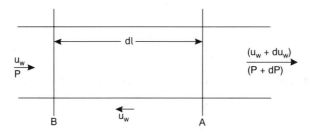

Figure 4.4.   Propagation of pressure wave

Now imagine the pressure wave to be brought to rest by causing the fluid to flow at a velocity u_w in the opposite direction. Let the pressure and specific volume at B be P and v, and at A be $(P + dP)$ and $(v + dv)$, respectively. As a result of the change in pressure, the velocity of the fluid changes from u_w at B to $(u_w + du_w)$ at A.

The mass rate of flow of fluid:

$$G = \frac{u_w A}{v} = \frac{(u_w + du_w)A}{v + dv}$$

The net force acting on the fluid between sections A and B is equal to the rate of change of momentum of the fluid:

i.e.: $$PA - (P + dP)A = G\, du_w$$

i.e.: $$-A\, dP = G\frac{G}{A}\, dv$$

$$-\frac{dP}{dv} = \frac{G^2}{A^2}$$

$$= \frac{\varepsilon}{v} \quad \text{(from equation 4.32)}$$

∴ $$u_w = \sqrt{\varepsilon v} \tag{4.33}$$

For an ideal gas, ε can be calculated from the equation of state. Under isothermal conditions:

$$Pv = \text{constant.}$$

∴ $$-\frac{dP}{dv} = \frac{P}{v}$$

∴ $$\varepsilon = P \tag{4.34}$$

and: $$u_w = \sqrt{Pv} = \sqrt{\frac{RT}{M}} \tag{4.35}$$

Under isentropic conditions:

$$Pv^\gamma = \text{constant}$$

$$-\frac{dP}{dv} = \frac{\gamma P}{v}$$

∴ $$\varepsilon = \gamma P \tag{4.36}$$

and:
$$u_w = \sqrt{\gamma P v} = \sqrt{\gamma \frac{RT}{M}} \qquad (4.37)$$

This value of u_w is found to correspond closely to the velocity of sound in the fluid. That is, for normal conditions of transmission of a small pressure wave, the process is almost isentropic. When the relation between pressure and volume is $Pv^k = \text{constant}$:

$$u_w = \sqrt{kPv} = \sqrt{k \frac{RT}{M}} \qquad (4.38)$$

4.4. CONVERGING–DIVERGING NOZZLES FOR GAS FLOW

Converging-diverging nozzles (Figure 4.5), sometimes known as Laval nozzles, are used for the expansion of gases where the pressure drop is large. If the nozzle is carefully designed so that the contours closely follow the lines of flow, the resulting expansion of the gas is almost reversible. Because the flow rate is large for high-pressure differentials, there is little time for heat transfer to take place between the gas and surroundings and the expansion is effectively isentropic. In the analysis of the nozzle, the change in flow will be examined for various pressure differentials across the nozzle.

The specific volume v_2 at a downstream pressure P_2, is given by:

$$v_2 = v_1 \left(\frac{P_1}{P_2}\right)^{1/\gamma} = v_1 \left(\frac{P_2}{P_1}\right)^{-1/\gamma} \qquad (4.39)$$

If gas flows under turbulent conditions from a reservoir at a pressure P_1, through a horizontal nozzle, the velocity of flow u_2, at the pressure P_2 is given by:

$$\frac{u_2^2}{2} + \int_1^2 v \, dP = 0 \quad \text{(from equation 2.42)}$$

Thus:
$$u_2^2 = \frac{2\gamma}{\gamma - 1} P_1 v_1 \left[1 - \left(\frac{P_2}{P_1}\right)^{(\gamma-1)/\gamma}\right] \qquad (4.40)$$

Since:
$$A_2 = \frac{Gv_2}{u_2} \quad \text{(from equation 2.36)} \qquad (4.41)$$

the required cross-sectional area for flow when the pressure has fallen to P_2 can be found.

4.4.1. Maximum flow and critical pressure ratio

In the flow of a gas through a nozzle, the pressure falls from its initial value P_1 to a value P_2 at some point along the nozzle; at first the velocity rises more rapidly than the specific volume and therefore the area required for flow decreases. For low values of the pressure ratio P_2/P_1, however, the velocity changes much less rapidly than the specific volume so that the area for flow must increase again. The effective area for flow presented by the nozzle must therefore pass through a minimum. It will be shown that this occurs if the pressure ratio P_2/P_1 is less than the critical pressure ratio (usually approximately 0.5) and that the velocity at the throat is then equal to the velocity of sound. For expansion

(a) Throat

Flow area taken as
infinity with $u_1 = 0$

(b)

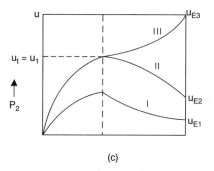

(c)

Figure 4.5.   Flow through converging-diverging nozzles

to pressures below the critical value the flowing stream must be diverging. Thus in a
converging nozzle the velocity of the gas stream will never exceed the sonic velocity
though in a converging-diverging nozzle supersonic velocities may be obtained in the
diverging section. The flow changes which occur as the back pressure P_B is steadily
reduced are shown in Figure 4.5. The pressure at the exit of the nozzle is denoted by
P_E, which is not necessarily the same as the back pressure P_B. In the three cases given
below, the exit and back pressures correspond and the flow is approximately isentropic.
The effect of a mismatch in the two pressures will be considered later.

Case I.   Back-pressure P_B quite high. Curves I show how pressure and velocity change
along the nozzle. The pressure falls to a minimum at the throat and then rises
to a value $P_{E1} = P_B$. The velocity increases to maximum at the throat (less
than sonic velocity) and then decreases to a value of u_{E1} at the exit of the

nozzle. This situation corresponds to conditions in a venturi operating entirely at subsonic velocities.

Case II. Back-pressure reduced (curves II). The pressure falls to the critical value at the throat where the velocity is sonic. The pressure then rises to $P_{E2} = P_B$ at the exit. The velocity rises to the sonic value at the throat and then falls to u_{E2} at the outlet.

Case III. Back-pressure low, with pressure less than critical value at the exit. The pressure falls to the critical value at the throat and continues to fall to give an exit pressure $P_{E3} = P_B$. The velocity increases to sonic at the throat and continues to increase to supersonic in the diverging cone to a value u_{E3}.

With a converging-diverging nozzle, the velocity increases beyond the sonic velocity only if the velocity at the throat is sonic and the pressure at the outlet is lower than the throat pressure. For a converging nozzle the rate of flow is independent of the downstream pressure, provided the critical pressure ratio is reached and the throat velocity is sonic.

4.4.2. The pressure and area for flow

As indicated in the previous section, the area required at any point will depend upon the ratio of the downstream to the upstream pressure P_2/P_1 and it is helpful to establish the minimum value of $A_2 \cdot A_2$ may be expressed in terms of P_2 and $w[= (P_2/P_1)]$ using equations 4.39, 4.40 and 4.41.

Thus:
$$A_2^2 = G^2 \frac{\gamma - 1}{2\gamma} \frac{v_1^2 (P_2/P_1)^{-2/\gamma}}{P_1 v_1 [1 - (P_2/P_1)^{(\gamma-1)/\gamma}]}$$

$$= \frac{G^2 v_1 (\gamma - 1)}{2 P_1 \gamma} \frac{w^{-2/\gamma}}{1 - w^{(\gamma-1)/\gamma}} \tag{4.42}$$

For a given rate of flow G, A_2 decreases from an effectively infinite value at pressure P_1 at the inlet to a minimum value given by:

$$\frac{dA_2^2}{dw} = 0$$

i.e., when:
$$(1 - w^{(\gamma-1)/\gamma}) \frac{-2}{\gamma} w^{-1-2/\gamma} - w^{-2/\gamma} \frac{1 - \gamma}{\gamma} w^{-1/\gamma} = 0$$

i.e., when:
$$w = \left(\frac{2}{\gamma + 1} \right)^{\gamma/(\gamma-1)} \tag{4.43}$$

The value of w given by equation 4.43 is the critical pressure ratio w_c (equation 4.26a). Thus the velocity at the throat is equal to the sonic velocity. Alternatively, equation 4.42 can be put in terms of the flowrate (G/A_2) as:

$$\left(\frac{G}{A_2} \right)^2 = \frac{2\gamma}{\gamma - 1} \left(\frac{P_2}{P_1} \right)^{2/\gamma} \frac{P_1}{v_1} \left[1 - \left(\frac{P_2}{P_1} \right)^{(\gamma-1)/\gamma} \right] \tag{4.44}$$

and the flowrate G/A_2 can then be shown to have a maximum value of $G_{max}/A_2 = \sqrt{\gamma P_2/v_2}$.

The nozzle is correctly designed for any outlet pressure between P_1 and P_{E1} in Figure 4.5. Under these conditions the velocity will not exceed the sonic velocity at any point, and the flowrate will be independent of the exit pressure $P_E = P_B$. It is also correctly designed for supersonic flow in the diverging cone for an exit pressure of P_{E3}.

It has been shown above that when the pressure in the diverging section is greater than the throat pressure, subsonic flow occurs. Conversely, if the pressure in the diverging section is less than the throat pressure the flow will be supersonic beyond the throat. Thus at a given point in the diverging cone where the area is equal to A_2 the pressure may have one of two values for isentropic flow.

As an example, when γ is 1.4:

$$v_2 = v_1 w^{-0.71} \tag{4.45}$$

and:

$$u_2^2 = 7P_1v_1(1 - w^{0.29}) \tag{4.46}$$

Thus:

$$A_2^2 = \frac{v_1 w^{-1.42}}{7P_1(1 - w^{0.29})}G^2 \tag{4.47}$$

In Figure 4.6 values of v_2/v_1, $u_2/\sqrt{P_1/v_1}$, and $(A_2/G)\sqrt{P_1/v_1}$ which are proportional to v_2, u_2, and A_2 respectively are plotted as abscissae against P_2/P_1. It is seen that the area A_2 decreases to a minimum and then increases again. At the minimum cross-section the velocity is equal to the sonic velocity and P_2/P_1 is the critical ratio. $u_w/\sqrt{P_1v_1}$ is also plotted and, by comparing with the curve for $u_2/\sqrt{P_1v_1}$, it is easy to compare the corresponding values of u_2 and u_w. The flow is seen to be sub-sonic for $P_2/P_1 > 0.53$ and supersonic for $P_2/P_1 < 0.53$.

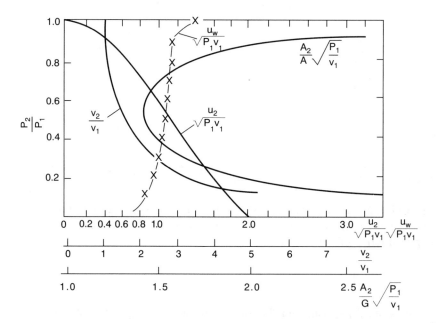

Figure 4.6.   Specific volume, velocity and nozzle area as a function of pressure

4.4.3. Effect of back pressure P_B on flow in nozzle

It is of interest to study how the flow in the nozzle will vary with the back-pressure P_B.

(1) $P_B > P_{E2}$. The flow is subsonic throughout the nozzle and the rate of flow G is determined by the value of the back pressure P_B. Under these conditions $P_E = P_B$. It is shown as P_{E1} in Figure 4.4.

(2) $P_B = P_{E2}$. The flow is subsonic throughout the nozzle, except at the throat where the velocity is sonic. Again $P_E = P_B$.

(3) $P_{E3} < P_B < P_{E2}$. The flow in the converging section and at the throat is exactly as for Case 2. For this range of exit pressures the flow is not isentropic through the whole nozzle and the area at the exit section in greater than the design value for a back pressure P_B. One of two things may occur. Either the flow will be supersonic but not occupy the whole area of the nozzle, or a shock wave will occur at some intermediate point in the diverging cone, giving rise to an increase in pressure and a change from supersonic to subsonic flow.

(4) $P_B = P_{E3}$. The flow in the converging section and at the throat is again the same as for Cases 2, 3, and 4. The flow in the diverging section is supersonic throughout and the pressure of P_B is reached smoothly. P_{E3} is therefore the design pressure for supersonic flow at the exit.

(5) $P_B < P_{E3}$. In this case the flow throughout the nozzle is exactly as for Case 4. The pressure at the exit will again be P_{E3} and the pressure will fall beyond the end of the nozzle from P_{E3} to P_B by continued expansion of the gas.

It should be noted that the flowrate through the nozzle is a function of back pressure only for Case 1.

A practical example of flow through a converging-diverging nozzle is given at the end of the chapter (Example 4.4) after the discussion of the generation of shock waves.

4.5. FLOW IN A PIPE

Compressibility of a gas flowing in a pipe can have significant effect on the relation between flowrate and the pressures at the two ends. Changes in fluid density can arise as a result of changes in either temperature or pressure, or in both, and the flow will be affected by the rate of heat transfer between the pipe and the surroundings. Two limiting cases of particular interest are for isothermal and adiabatic conditions.

Unlike the orifice or nozzle, the pipeline maintains the area of flow constant and equal to its cross-sectional area. There is no possibility therefore of the gas expanding laterally. Supersonic flow conditions can be reached in pipeline installations in a manner similar to that encountered in flow through a nozzle, but *not within the pipe itself* unless the gas enters the pipe at a supersonic velocity. If a pipe connects two reservoirs and the upstream reservoir is maintained at constant pressure P_1, the following pattern will occur as the pressure P_2 in the downstream reservoir is reduced.

(1) Starting with $P_2 = P_1$ there is, of course, no flow and $G = 0$.

(2) Reduction of P_2 initially results in an increase in G. G increases until the gas velocity at the outlet of the pipe just reaches the velocity of propagation of a pressure wave ("sonic velocity"). This value of P_2 will be denoted by P_w.

(3) Further reduction of P_2 has no effect on the flow in the pipeline. The pressure distribution along the length of the pipe remains unaltered and the pressure at its outlet remains at P_w. The gas, on leaving the pipe, expands laterally and its pressure falls to the reservoir pressure P_2.

In considering the flow in a pipe, the differential form of the general energy balance equation 2.54 will be used, and the friction term δF will be written in terms of the energy dissipated per unit mass of fluid for flow through a length dl of pipe. In the first instance, isothermal flow of an ideal gas will be considered and the flowrate will be expressed as a function of upstream and downstream pressures. Non-isothermal and adiabatic flow will be discussed later.

4.5.1. Energy balance for flow of ideal gas

In Chapter 2, the general energy equation for the flow of a fluid through a pipe has been expressed in the form:

$$\frac{u\,du}{\alpha} + g\,dz + v\,dP + \delta W_s + \delta F = 0 \quad \text{(equation 2.54)}$$

For a fluid flowing through a length dl of pipe of constant cross-sectional area A:

$$W_s = 0$$

$$\delta F = 4\left(\frac{R}{\rho u^2}\right)u^2\frac{dl}{d} \qquad \text{(from equation 3.19)}$$

$$\therefore \qquad \frac{u\,du}{\alpha} + g\,dz + v\,dP + 4\frac{R}{\rho u^2}u^2\frac{dl}{d} = 0 \qquad (4.48)$$

This equation cannot be integrated directly because the velocity u increases as the pressure falls and is, therefore, a function of l (Figure 4.7). It is, therefore, convenient to work in terms of the mass flow G which remains constant throughout the length of pipe.

Now: $\qquad\qquad\qquad\qquad u = \dfrac{Gv}{A} \qquad\qquad\qquad\qquad$ (from equation 2.36)

Thus: $\qquad \dfrac{1}{\alpha}\left(\dfrac{G}{A}\right)^2 v\,dv + g\,dz + v\,dP + 4\left(\dfrac{R}{\rho u^2}\right)\left(\dfrac{G}{A}\right)^2 v^2\dfrac{dl}{d} = 0 \qquad (4.49)$

For turbulent flow, which is usual for a gas, $\alpha = 1$.

Figure 4.7.   Flow of compressible fluid through pipe

Then, for flow in a horizontal pipe ($dz = 0$):

$$\left(\frac{G}{A}\right)^2 v\,dv + v\,dP + 4\left(\frac{R}{\rho u^2}\right)\left(\frac{G}{A}\right)^2 v^2 \frac{dl}{d} = 0 \tag{4.50}$$

Dividing by v^2:

$$\left(\frac{G}{A}\right)^2 \frac{dv}{v} + \frac{dP}{v} + 4\left(\frac{R}{\rho u^2}\right)\left(\frac{G}{A}\right)^2 \frac{dl}{d} = 0 \tag{4.51}$$

Now the friction factor $R/\rho u^2$ is a function of the Reynolds number Re and the relative roughness e/d of the pipe surface which will normally be constant along a given pipe:
The Reynolds number is given by:

$$Re = \frac{u\,d\rho}{\mu} = \frac{Gd}{A\mu} = \frac{4G}{\pi d\mu} \tag{4.52}$$

Since G is constant over the length of the pipe, Re will vary only as a result of changes in the viscosity μ. Although μ is a function of temperature, and to some extent of pressure, it is not likely to vary widely over the length of the pipe. Furthermore, the friction factor $R/\rho u^2$ is only a weak function of Reynolds number when Re is high, and little error will therefore arise from regarding it as constant.
    Thus, integrating equation 4.51 over a length l of pipe:

$$\left(\frac{G}{A}\right)^2 \ln\frac{v_2}{v_1} + \int_{P_1}^{P_2} \frac{dP}{v} + 4\left(\frac{R}{\rho u^2}\right)\frac{l}{d}\left(\frac{G}{A}\right)^2 = 0 \tag{4.53}$$

The integral will depend on the P-v relationship during the expansion of the gas in the pipe, and several cases will now be considered.

4.5.2. Isothermal flow of an ideal gas in a horizontal pipe

For isothermal changes in an ideal gas:

$$\int_{P_1}^{P_2} \frac{dP}{v} = \frac{1}{P_1 v_1}\int_{P_1}^{P_2} P\,dP = \frac{P_2^2 - P_1^2}{2P_1 v_1} \tag{4.54}$$

and, therefore, substituting in equation 4.53:

$$\left(\frac{G}{A}\right)^2 \ln\frac{P_1}{P_2} + \frac{P_2^2 - P_1^2}{2P_1 v_1} + 4\left(\frac{R}{\rho u^2}\right)\frac{l}{d}\left(\frac{G}{A}\right)^2 = 0 \tag{4.55}$$

Since v_m, the specific volume at the mean pressure in the pipe, is given by:

$$\frac{P_1 + P_2}{2} v_m = P_1 v_1$$

$$\left(\frac{G}{A}\right)^2 \ln\frac{P_1}{P_2} + (P_2 - P_1)\frac{1}{v_m} + 4\left(\frac{R}{\rho u^2}\right)\frac{l}{d}\left(\frac{G}{A}\right)^2 = 0 \tag{4.56}$$

If the pressure drop in the pipe is a small proportion of the inlet pressure, the first term is negligible and the fluid can be treated as an incompressible fluid at the mean pressure in the pipe.

It is sometimes convenient to substitute RT/M in place of $P_1 v_1$ in equation 4.55 to give:

$$\left(\frac{G}{A}\right)^2 \ln \frac{P_1}{P_2} + \frac{P_2^2 - P_1^2}{2RT/M} + 4\left(\frac{R}{\rho u^2}\right)\frac{l}{d}\left(\frac{G}{A}\right)^2 = 0 \qquad (4.57)$$

Equations 4.55 and 4.57 are the most convenient for calculation of gas flowrate as a function of P_1 and P_2 under isothermal conditions. Some additional refinement can be added if a compressibility factor is introduced as defined by the relation $Pv = ZRT/M$, for conditions where there are significant deviations from the Ideal Gas Law (equation 2.15).

Maximum flow conditions

Equation 4.55 expresses G as a continuous function of P_2, the pressure at the downstream end of the pipe for a given upstream pressure P_1. If there is no pressure change over the pipe $P_2 = P_1$, substitution of P_1 for P_2 in equation 4.55 gives $G = 0$, as would be expected. Furthermore, substituting $P_2 = 0$ also gives $G = 0$. Thus, for some intermediate value of P_2 ($= P_w$, say), where $0 < P_w < P_1$, the flowrate G must be a maximum.

Multiplying equation 4.55 by $(A/G)^2$:

$$-\ln\left(\frac{P_2}{P_1}\right) + \left(\frac{A}{G}\right)^2 \frac{(P_2^2 - P_1^2)}{2P_1 v_1} + 4\left(\frac{R}{\rho u^2}\right)\frac{l}{d} = 0 \qquad (4.58)$$

Differentiating with respect to P_2, for a constant value of P_1:

$$-\frac{P_1}{P_2}\frac{1}{P_1} + \left(\frac{A}{G}\right)^2 \frac{2P_2}{2P_1 v_1} + \frac{A^2}{2P_1 v_1}(P_2^2 - P_1^2)\frac{-2}{G^3}\frac{dG}{dP_2} = 0$$

The rate of flow is a maximum when $dG/dP_2 = 0$. Denoting conditions at the downstream end of the pipe by suffix w, when the flow is a maximum:

$$\frac{1}{P_w} = \left(\frac{A}{G}\right)^2 \frac{P_w}{P_1 v_1}$$

i.e.:
$$\left(\frac{G}{A}\right)^2 = \frac{P_w^2}{P_1 v_1} = \frac{P_w^2}{P_w v_w} = \frac{P_w}{v_w} \qquad (4.59)$$

i.e.:
$$u_w = \sqrt{P_w v_w} \qquad (4.60)$$

It has been shown already (equation 4.35) that this velocity v_w is equal to the velocity of transmission of a small pressure wave in the fluid at the pressure P_w if heat could be transferred sufficiently rapidly to maintain isothermal conditions. If the pressure at the downstream end of the pipe were P_w, the fluid there would then be moving with the velocity of a pressure wave, and therefore a wave could not be transmitted through the fluid in the opposite direction because its velocity relative to the pipe would be zero. If, at the downstream end, the pipe were connected to a reservoir in which the pressure was reduced below P_w, the flow conditions within the pipe would be unaffected and the pressure at the exit of the pipe would remain at the value P_w (Figure 4.8). The drop in pressure from P_w to P_2 would then take place by virtue of lateral expansion of the gas beyond the end of the pipe. If the pressure P_2 in the reservoir at the downstream end were gradually reduced from P_1, the rate of flow would increase until the pressure reached P_w; it would then remain constant at this maximum value as the pressure was further reduced.

Figure 4.8.   Maximum flow conditions

Thus, with compressible flow there is a maximum mass flowrate G_w which can be attained by the gas for a given upstream pressure P_1, and further reduction in pressure in the downstream reservoir below P_w will not give any further increase.

The maximum flowrate G_w is given by equation 4.59:

$$G_w = A\sqrt{\frac{P_w}{v_w}}$$

Substituting $v_w = v_1(P_1/P_w)$:

$$G_w = AP_w\sqrt{\frac{1}{P_1v_1}} \tag{4.61}$$

P_w is given by substitution in equation 4.58,

i.e.:
$$\ln\left(\frac{P_1}{P_w}\right) + \frac{v_w}{P_w}\frac{1}{P_wv_w}\frac{P_w^2 - P_1^2}{2} + 4\left(\frac{R}{\rho u^2}\right)\frac{l}{d} = 0$$

or:
$$\ln\left(\frac{P_1}{P_w}\right)^2 + 1 - \left(\frac{P_1}{P_w}\right)^2 + 8\left(\frac{R}{\rho u^2}\right)\frac{l}{d} - 0 \tag{4.62}$$

or:
$$8\left(\frac{R}{\rho u^2}\right)\frac{l}{d} = \left(\frac{1}{w_c}\right)^2 - \ln\left(\frac{1}{w_c}\right)^2 - 1 \tag{4.63}$$

where $w_c = P_w/P_1$ (the critical value of the pressure ratio $w = (P_2/P_1)$.

Equations 4.62 and 4.63 are very important in that they give:

(a) The maximum value of the pressure ratio $P_1/P_2(= P_1/P_w)$ for which the whole of the expansion of the gas can take place in the pipe.

If $P_1/P_2 > P_1/P_w$, the gas expands:

from P_1 to P_w in the pipe, and
from P_w to P_2 in the downstream reservoir.

Table 4.1. Limiting pressure rates for pipe flow

$\dfrac{P_1}{P_w}$	$\dfrac{P_w}{P_1}$	$\left(\dfrac{1}{w_c}\right)^2$	$\ln\left(\dfrac{1}{w_c}\right)^2$	$8\dfrac{R}{\rho u^2}\dfrac{l}{d}$	For $\dfrac{R}{\rho u^2}=0.0015$ $\dfrac{l}{d}$	For $d=25$ mm $l(m)$	For $P_1=100$ bar P_1-P_w (bar)	Average $\dfrac{P_1-P_w}{l}$ (bar/m)
1	1	1	0	0	0	0	0	—
1.1	0.9091	1.21	0.1906	0.0194	1.62	0.0404	9.1	225
1.78	0.562	3.16	1.151	1.01	84.2	2.11	43.8	20.8
3.16	0.316	10	2.302	6.70	558	14.0	68.4	4.89
5.62	0.178	31.6	3.453	27.1	2258	56.5	82.2	1.45
10	0.1	100	4.605	94.4	7867	197	90	0.457
17.8	0.0562	316	5.756	309.2	25,770	644	94.4	0.146
31.6	0.0316	1000	6.901	992	82,700	2066	96.8	0.046
56.2	0.0178	3160	8.058	3151	263,000	6575	98.2	0.015
100	0.01	10,000	9.210	9990	832,500	20,800	99	0.005

Equations 4.55, 4.56 and 4.57 can be used to calculate what is happening in the pipe only provided $P_1/P_2 \not> P_1/P_w$ as given by equation 4.62.

(b) The minimum value of $8(R/\rho u^2)(l/d)$ for which, for any pressure ratio P_1/P_2, the fall in gas pressure will take place entirely within the pipe.

In Table 4.1, the relation between P_1/P_w and $8(R/\rho u^2)(l/d)$ is seen for a series of numerical values of w_c.

It is seen that P_1/P_w increases with $8(R/\rho u^2)(l/d)$; in other words, in general, the longer the pipe (assuming $R/\rho u^2 \approx$ constant) the greater is the ratio of upstream to downstream pressure P_1/P_2 which can be accommodated in the pipe itself.

For an assumed average pipe friction factor $(R/\rho u^2)$ of 0.0015, this has been expressed as an l/d ratio. For a 25-mm diameter pipeline, the corresponding pipelengths are given in metres. In addition, for an upstream pressure P_1 of 100 bar, the *average* pressure gradient in the 25-mm pipe is given in the last column. At first sight it might have seemed strange that limiting conditions should be reached in such very short pipes for values of $P_1/P_w \approx 1$, but it will be seen that the *average pressure gradients* then become very high.

In Figure 4.9 values of P_1/P_w and w_c are plotted against $8(R/\rho u^2)(l/d)$. This curve gives the limiting value of P_1/P_2 for which the whole of the expansion of the gas can take place within the pipe.

By reference to Table 4.1, it is seen, for instance, that for a 25-mm pipeline and an assumed value of 0.0015 for $R/\rho u^2$ that for a pipe of length 14 m, the ratio of the pressure at the pipe inlet to that at the outlet cannot exceed 3.16 ($w_c = 0.316$). If $P_1/P_2 > 3.16$, the gas expands to the pressure $P_w(= 0.316P_1)$ inside the pipe and then it expands down to the pressure P_2 *within the downstream reservoir*.

Flow with fixed upstream pressure and variable downstream pressure

As an example, the flow of air at 293 K in a pipe of 25 mm diameter and length 14 m will be considered, using the value of 0.0015 for $R/\rho u^2$ employed in the calculation of the figures in Table 4.1; $R/\rho u^2$ will, of course, show some variation with Reynolds number, but this effect will be neglected in the following calculation. The variation in flowrate G will be examined, for a given upstream pressure of 10 MN/m², as a function of downstream pressure P_2. As the critical value of P_1/P_2 for this case is 3.16 (see Table 4.1), the maximum flowrate will occur at all values of P_2 less than $10/3.16 - 3.16$ MN/m². For values of P_2 greater than 3.16 MN/m², equation 4.57 applies:

$$\left(\frac{G}{A}\right)^2 \ln \frac{P_1}{P_2} + \frac{P_2^2 - P_1^2}{2RT/M} + 4\left(\frac{R}{\rho u^2}\right)\frac{l}{d}\left(\frac{G}{A}\right)^2 = 0$$

Multiplying through by $(A/G)^2$ and inserting the numerical values with pressures P_1 and P_2 expressed in MN/m² (P_{1a} and P_{2a}):

$$\ln \frac{P_{1a}}{P_{2a}} - 5.95 \times 10^6(P_{1a}^2 - P_{2a}^2)\left(\frac{A}{G}\right)^2 + 3.35 = 0$$

For $P_{1a} = 10$ MN/m²:

$$2.306 - \ln P_{2a} - 5.95 \times 10^6(100 - P_{2a}^2)\left(\frac{A}{G}\right)^2 + 3.35 = 0$$

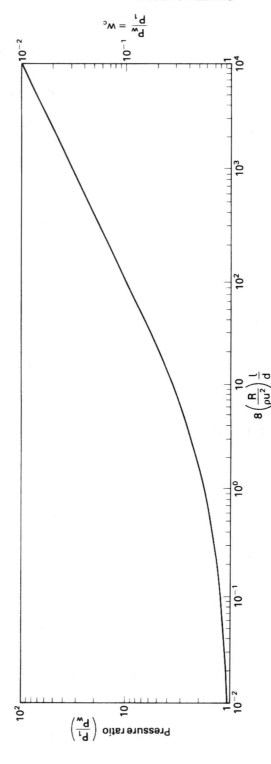

Figure 4.9. Critical pressure ratio P_1/P_w (maximum value of P_1/P_2 which can occur in pipe) as function of $8(R/\rho u^2)(l/d)$

giving:
$$\left(\frac{G}{A}\right)^2 = \frac{5.95 \times 10^6(100 - P_{2a}^2)}{5.656 - \ln P_{2a}} \quad (\text{kg}^2/\text{m}^4 \text{ s}^2)$$

Values of G calculated from this relation are given in Table 4.2 and plotted in Figure 4.10.

In the last column of Table 4.2, values are given for G^*/A calculated by ignoring the effect of changes in the kinetic energy of the gas, i.e. by neglecting the log term in equation 4.57. It will be noted that the effects are small ranging from about 2 per cent for $P_{2a} = 9$ MN/m² to about 13 per cent for $P_{2a} = 4$ MN/m².

Table 4.2. Mass flowrate as function of downstream pressure P_{2a} for constant upstream pressure of 10 MN/m²

Pressure P_{2a} MN/m²	$\dfrac{G}{A} \times 10^{-4}$ kg/m² s	G kg/s	$\dfrac{G^*}{A} \times 10^{-4}$ kg/m² s
10	0	0	0
9.5	0.41	2.0	0.42
9	0.57	2.8	0.58
8	0.78	3.8	0.80
7	0.90	4.4	0.95
6	0.99	4.9	1.07
5	1.05	5.2	1.15
4	1.08	5.3	1.22
<3.16	1.09	5.4	1.26

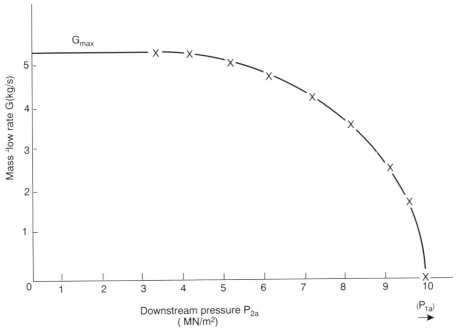

Figure 4.10. Mass flowrate as function of downstream pressure for isothermal flow of air in pipe ($l = 14$ m, $d = 25$ mm)

Values of G_{max} for maximum flow conditions can also be calculated from equation 4.61:

$$G_{max} = AP_w \sqrt{\frac{1}{P_1 v_1}}$$

Substitution of the numerical values gives:

$$G_{max} = 5.4 \text{ kg/s} \quad \text{in agreement with the value in Table 4.2.}$$

For $G/A = 10^4$ kg/m² s, Reynolds number $Re \approx \dfrac{10^4 \times 25 \times 10^{-3}}{10^{-5}} \approx 2.5 \times 10^7$

The chosen value of 0.0015 for $R/\rho u^2$ is reasonable for a smooth pipe over most of the range of Reynolds numbers encountered in this exercise.

Example 4.2

Over a 30 m length of a 150 mm vacuum line carrying air at 295 K, the pressure falls from 0.4 kN/m² to 0.13 kN/m². If the relative roughness e/d is 0.003 what is the approximate flowrate? (The flow of gases in vacuum systems is discussed fully by GRIFFITHS[1].)

Solution

$$\left(\frac{G}{A}\right)^2 \ln \frac{P_1}{P_2} + \frac{P_2^2 - P_1^2}{2 P_1 v_1} + 4 \left(\frac{R}{\rho u^2}\right) \frac{l}{d} \left(\frac{G}{A}\right)^2 = 0 \qquad \text{(equation 4.55)}$$

$$\left(\frac{G}{A}\right)^2 \ln \frac{0.4}{0.13} + \frac{(0.13^2 - 0.40^2) \times 10^6}{2 \times 8.46 \times 10^4} + 4 \times 0.004 \times \frac{30}{0.150} \times \left(\frac{G}{A}\right)^2 = 0$$

$$1.124 \left(\frac{G}{A}\right)^2 - 0.846 + 3.2 \left(\frac{G}{A}\right)^2 = 0$$

$$\frac{G}{A} = 0.44 \text{ kg/m}^2 \text{ s}$$

Viscosity of air at 295 K $= 1.8 \times 10^{-5}$ N s/m²

$$\text{Reynold number, } Re = \frac{0.44 \times 0.15}{1.8 \times 10^{-5}} = 3670 \text{ (from equation 4.52)}$$

$$\text{Again, from chart, } \frac{R}{\rho u^2} = 0.005$$

Inserting new values:

$$1.124 \left(\frac{G}{A}\right)^2 - 0.846 + 4.0 \left(\frac{G}{A}\right)^2 = 0$$

$$\frac{G}{A} = 0.41 \text{ kg/m}^2 \text{ s}$$

Check that this flowrate is possible.

$$8 \frac{R}{\rho u^2} \frac{l}{d} = 8.0$$

From Figure 4.8:

$$\frac{P_1}{P_w} = 3.5$$

Actual value:
$$\frac{P_1}{P_2} = \frac{0.4}{0.13} = 3.1$$

Thus, flow condition just falls within the possible range.

Example 4.3

A flow of 50 m³/s methane, measured at 288 K and 101.3 kN/m², has to be delivered along a 0.6 m diameter line, 3.0 km long and relative roughness 0.0001, linking a compressor and a processing unit. The methane is to be discharged at the plant at 288 K and 170 kN/m² and it leaves the compressor at 297 K. What pressure must be developed at the compressor in order to achieve this flowrate?

Solution

Taking mean temperature of 293 K,

$$P_1 v_1 = \frac{RT}{M} = \frac{8314 \times 293}{16} = 1.5225 \times 10^5 \text{ m}^2/\text{s}^2$$

At 288 K and 101.3 kN/m², $v = \dfrac{1.5225 \times 10^5}{1.013 \times 10^5} \times \dfrac{288}{293} = 1.477 \text{ m}^3/\text{kg}$

Mass flowrate of methane, $G = \dfrac{50}{1.497} = 33.85 \text{ kg/s}$

Cross-sectional area of pipe $A = \dfrac{\pi}{4} \times (0.6)^2 = 0.283 \text{ m}^2$

$$\frac{G}{A} = \frac{33.85}{0.283} = 119.6 \text{ kg/m}^2 \text{ s}$$

$$\left(\frac{G}{A}\right)^2 = 1.431 \times 10^4 (\text{kg/m}^2 \text{ s})^2.$$

Viscosity of methane at 293 K $\approx 0.01 \times 10^{-3}$ N s/m²

$$\text{Reynolds number } Re = \frac{119.6 \times 0.6}{10^{-5}} = 7.18 \times 10^6$$

For $\dfrac{e}{d} = 0.0001$, $\dfrac{R}{\rho u^2} = 0.0015$ (from chart)

The upstream pressure is calculated using equation 4.55:

$$\left(\frac{G}{A}\right)^2 \ln \frac{P_1}{P_2} + \frac{P_2^2 - P_1^2}{2P_1 v_1} + 4\left(\frac{R}{\rho u^2}\right) \frac{l}{d} \left(\frac{G}{A}\right)^2 = 0$$

Substituting: $1.431 \times 10^4 \ln \dfrac{P_1}{1.7 \times 10^5} - \dfrac{P_1^2 - (1.7 \times 10^5)^2}{2 \times 1.5225 \times 10^5} + 4 \times 0.0015 \times \dfrac{3000}{0.6} \times 1.431 \times 10^4 = 0$

Dividing by 1.431×10^4:

$$\ln P_1 - 12.04 - 2.29 \times 10^{-10} P_1^2 + 6.63 + 30.00 = 0$$

$$2.29 \times 10^{-10} P_1^2 - \ln P_1 = 24.59$$

A trial and error solution gives:
$$P_1 = 4.05 \times 10^5 \text{ N/m}^2$$

Check that this degree of expansion is possible *within* pipe

$$\frac{P_1}{P_2} = 2.38$$

By reference to Figure 4.9:

$$8 \frac{R}{\rho u^2} \frac{l}{d} = 60 \quad \text{and} \quad \frac{P_1}{P_w} = 8.1$$

Thus, pressure ratio is within possible range.

Heat flow required to maintain isothermal conditions

As the pressure in the pipe falls, the kinetic energy of the fluid increases at the expense of the internal energy and the temperature tends to fall. The maintenance of isothermal conditions therefore depends on the transfer of an adequate amount of heat from the surroundings. For a small change in the system, the energy balance is given in Chapter 2:

$$\delta q - \delta W_s = dH + g \, dz + u \, du \quad \text{(from equation 2.51)}$$

For a horizontal pipe, $dz = 0$, and for isothermal expansion of an ideal gas $dH = 0$. Thus if the system does no work on the surroundings:

$$\delta q = u \, du \tag{4.64}$$

and the required transfer of heat (in mechanical energy units) per unit mass is $\Delta u^2/2$. Thus the amount of heat required is equivalent to the increase in the kinetic energy of the fluid. If the mass rate of flow is G, the total heat to be transferred per unit time is $G\Delta u^2/2$.

In cases where the change in the kinetic energy is small, the required flow of heat is correspondingly small, and conditions are almost adiabatic.

4.5.3. Non-isothermal flow of an ideal gas in a horizontal pipe

In general, where an ideal gas expands or is compressed, the relation between the pressure P and the specific volume v can be represented approximately by the expression:

$$Pv^k = \text{a constant} = P_1 v_1^k$$

where k will depend on the heat transfer to the surroundings.

Evaluation of the integral gives:

$$\int_{P_1}^{P_2} \frac{dP}{v} = \frac{k}{k+1} \frac{P_1}{v_1} \left[\left(\frac{P_2}{P_1} \right)^{(k+1)/k} - 1 \right] \tag{4.65}$$

Inserting this value in equation 4.53:

$$\left(\frac{G}{A} \right)^2 \frac{1}{k} \ln \left(\frac{P_1}{P_2} \right) + \frac{k}{k+1} \frac{P_1}{v_1} \left[\left(\frac{P_2}{P_1} \right)^{(k+1)/k} - 1 \right] + 4 \left(\frac{R}{\rho u^2} \right) \frac{l}{d} \left(\frac{G}{A} \right)^2 = 0 \tag{4.66}$$

For a given upstream pressure P_1 the maximum flow rate occurs when $u_2 = \sqrt{kP_2v_2}$, the velocity of transmission of a pressure wave under these conditions (equation 4.38). Flow under adiabatic conditions is considered in detail below, but an approximate result is obtained by putting k equal to γ in equation 4.66; this is only approximate because equating k to γ implies reversibility.

4.5.4. Adiabatic flow of an ideal gas in a horizontal pipe

The conditions existing during the adiabatic flow in a pipe can be calculated using the approximate expression $Pv^k = $ a constant to give the relation between the pressure and the specific volume of the fluid. In general, however, the value of the index k may not be known for an irreversible adiabatic process. An alternative approach to the problem is therefore desirable.[2,3]

For a fluid flowing under turbulent conditions in a pipe, $\delta W_s = 0$ and:

$$\delta q = \mathrm{d}H + g\,\mathrm{d}z + u\,\mathrm{d}u \qquad\qquad \text{(from equation 2.51)}$$

In an adiabatic process, $\delta q = 0$, and the equation can then be written as follows for the flow in a pipe of constant cross-sectional area A:

$$\left(\frac{G}{A}\right)^2 v\,\mathrm{d}v + g\,\mathrm{d}z + \mathrm{d}H = 0 \tag{4.67}$$

Now:
$$\mathrm{d}H = \mathrm{d}U + \mathrm{d}(Pv)$$

$$= C_v\,\mathrm{d}T + \mathrm{d}(Pv) \quad \text{for an ideal gas (from equation 2.24)}$$

Further:
$$C_p\,\mathrm{d}T = C_v\,\mathrm{d}T + \mathrm{d}(Pv) \quad \text{for an ideal gas (from equation 2.26)}$$

$$\therefore \qquad\qquad \mathrm{d}T = \frac{\mathrm{d}(Pv)}{C_p - C_v} \tag{4.68}$$

so that:
$$\mathrm{d}H = \mathrm{d}(Pv)\left(\frac{C_v}{C_p - C_v} + 1\right) \tag{4.69}$$

$$= \frac{\gamma}{\gamma - 1}\,\mathrm{d}(Pv) \tag{4.70}$$

Substituting this value of $\mathrm{d}H$ in equation 4.67 and writing $g\,\mathrm{d}z = 0$ for a horizontal pipe:

$$\left(\frac{G}{A}\right)^2 v\,\mathrm{d}v + \frac{\gamma}{\gamma - 1}\,\mathrm{d}(Pv) = 0 \tag{4.71}$$

Integrating, a relation between P and v for adiabatic flow in a horizontal pipe results:

$$\frac{1}{2}\left(\frac{G}{A}\right)^2 v^2 + \frac{\gamma}{\gamma - 1}Pv = \frac{1}{2}\left(\frac{G}{A}\right)^2 v_1^2 + \frac{\gamma}{\gamma - 1}P_1 v_1 = \text{constant, } K\,(\text{say}) \tag{4.72}$$

From equation 4.72:

$$P = \frac{\gamma - 1}{\gamma}\left[\frac{K}{v} - \frac{1}{2}\left(\frac{G}{A}\right)^2 v\right] \tag{4.73}$$

$$\mathrm{d}P = \frac{\gamma - 1}{\gamma}\left[-\frac{K}{v^2} - \frac{1}{2}\left(\frac{G}{A}\right)^2\right]\mathrm{d}v$$

$$\frac{\mathrm{d}P}{v} = \frac{\gamma - 1}{\gamma}\left[-\frac{K}{v^3} - \frac{1}{2}\left(\frac{G}{A}\right)^2\frac{1}{v}\right]\mathrm{d}v \tag{4.74}$$

$$\int_{P_1}^{P_2} \frac{dP}{v} = \frac{\gamma - 1}{\gamma} \left[\frac{K}{2} \left(\frac{1}{v_2^2} - \frac{1}{v_1^2} \right) - \frac{1}{2} \left(\frac{G}{A} \right)^2 \ln \frac{v_2}{v_1} \right] \tag{4.75}$$

Substituting for K from equation 4.72:

$$\int_{P_1}^{P_2} \frac{dP}{v} = \frac{\gamma - 1}{\gamma} \left[\left(\frac{G}{A} \right)^2 \frac{v_1^2}{4} \left(\frac{1}{v_2^2} - \frac{1}{v_1^2} \right) + \frac{\gamma P_1 v_1}{2(\gamma - 1)} \left(\frac{1}{v_2^2} - \frac{1}{v_1^2} \right) - \frac{1}{2} \left(\frac{G}{A} \right)^2 \ln \frac{v_2}{v_1} \right]$$

$$= \frac{\gamma - 1}{4\gamma} \left(\frac{G}{A} \right)^2 \left(\frac{v_1^2}{v_2^2} - 1 - 2 \ln \frac{v_2}{v_1} \right) + \frac{P_1 v_1}{2} \left(\frac{1}{v_2^2} - \frac{1}{v_1^2} \right) \tag{4.76}$$

Inserting the value of $\int_{P_1}^{P_2} dP/v$ from equation 4.76 into equation 4.53:

$$\left(\frac{G}{A} \right)^2 \ln \frac{v_2}{v_1} + \frac{\gamma - 1}{4\gamma} \left(\frac{G}{A} \right)^2 \left(\frac{v_1^2}{v_2^2} - 1 - 2 \ln \frac{v_2}{v_1} \right)$$

$$+ \frac{P_1 v_1}{2} \left(\frac{1}{v_2^2} - \frac{1}{v_1^2} \right) + 4 \left(\frac{R}{\rho u^2} \right) \frac{l}{d} \left(\frac{G}{A} \right)^2 = 0$$

Simplifying:

$$8 \left(\frac{R}{\rho u^2} \right) \frac{l}{d} = \left[\frac{\gamma - 1}{2\gamma} + \frac{P_1}{v_1} \left(\frac{A}{G} \right)^2 \right] \left[1 - \left(\frac{v_1}{v_2} \right)^2 \right] - \frac{\gamma + 1}{\gamma} \ln \frac{v_2}{v_1} \tag{4.77}$$

This expression enables v_2, the specific volume at the downstream end of the pipe, to be calculated for the fluid flowing at a mass rate G from an upstream pressure P_1.

Alternatively, the mass rate of flow G can be calculated in terms of the specific volume of the fluid at the two pressures P_1 and P_2.

The pressure P_2 at the downstream end of the pipe is obtained by substituting the value of v_2 in equation 4.72.

For constant upstream conditions, the maximum flow through the pipe is found by differentiating with respect to v_2 and putting (dG/dv_2) equal to zero. The maximum flow is thus shown to occur when the velocity at the downstream end of the pipe is the sonic velocity $\sqrt{\gamma P_2 v_2}$ (equation 4.37).

The rate of flow of gas under adiabatic conditions is never more than 20 per cent greater than that obtained for the same pressure difference with isothermal conditions. For pipes of length at least 1000 diameters, the difference does not exceed about 5 per cent. In practice the rate of flow may be limited, not by the conditions in the pipe itself, but by the development of sonic velocity at some valve or other constriction in the pipe. Care should, therefore, be taken in the selection of fittings for pipes conveying gases at high velocities.

Analysis of conditions for maximum flow

It will now be shown from purely thermodynamic considerations that for adiabatic conditions supersonic flow cannot develop in a pipe of constant cross-sectional area because the fluid is in a condition of maximum entropy when flowing at the sonic velocity. The condition of the gas at any point in the pipe where the pressure is P is given by the equations:

$$Pv = \frac{1}{M}RT \qquad \text{(equation 2.16)}$$

and:

$$\frac{\gamma}{\gamma - 1}Pv + \frac{1}{2}\left(\frac{G}{A}\right)^2 v^2 = K \qquad \text{(equation 4.72)}$$

It will be noted that if the changes in the kinetic energy of the fluid are small, the process is almost isothermal.

Eliminating v, an expression for T results:

$$\frac{\gamma}{\gamma - 1}\frac{RT}{M} + \frac{1}{2}\left(\frac{G}{A}\right)^2 \frac{R^2 T^2}{P^2 M^2} = K \qquad (4.78)$$

The corresponding value of the entropy is obtained as follows:

$$dH = T\,dS + v\,dP = C_p\,dT \quad \text{for an ideal gas (from equations 2.28 and 2.26)}$$

$$\therefore \qquad dS = C_p\frac{dT}{T} - \frac{R}{MP}dP$$

$$\therefore \qquad S = C_p\ln\frac{T}{T_0} - \frac{R}{M}\ln\frac{P}{P_0} \quad (\text{if}C_p \text{ is constant}) \qquad (4.79)$$

where T_0, P_0 represents the condition of the gas at which the entropy is arbitrarily taken as zero.

The temperature or enthalpy of the gas can then be plotted to a base of entropy to give a *Fanno line*.[4] This line shows the condition of the fluid as it flows along the pipe. If the velocity at entrance is subsonic (normal condition), then the enthalpy will decrease along the pipe and the velocity will increase until sonic velocity is reached. If the flow is supersonic at the entrance, the velocity will decrease along the duct until it becomes sonic. The entropy has a maximum value corresponding to sonic velocity as shown in Figure 4.11. (Mach number $Ma < 1$ represents sub-sonic conditions; $Ma > 1$ supersonic.)

Fanno lines are also useful in presenting conditions in nozzles, turbines, and other units where supersonic flow arises.[5]

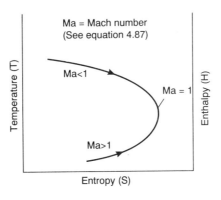

Figure 4.11.   The Fanno line

For small changes in pressure and entropy, the kinetic energy of the gas increases only very slowly, and therefore the temperature remains almost constant. As the pressure is further reduced, the kinetic energy changes become important and the rate of fall of temperature increases and eventually dT/dS becomes infinite. Any further reduction of the pressure would cause a decrease in the entropy of the fluid and is, therefore, impossible.

The condition of maximum entropy occurs when $dS/dT = 0$.

$$\frac{dS}{dT} = \frac{C_p}{T} - \frac{R}{MP}\frac{dP}{dT} \qquad \text{(from equation 4.79)}$$

The entropy is, therefore, a maximum when:

$$\frac{dP}{dT} = \frac{MPC_P}{RT} \qquad (4.80)$$

Now, for an ideal gas:

$$C_p - C_v = \frac{R}{M} \qquad \text{(equation 2.27)}$$

Substituting in equation 4.80:

$$\frac{dP}{dT} = \frac{PC_v}{(C_p - C_v)T} = \frac{P}{T}\frac{\gamma}{\gamma - 1} \qquad (4.81)$$

The general value of dP/dT can be obtained by differentiating equation 4.78 with respect to T.

Then:

$$\frac{R}{M} + \frac{\gamma - 1}{2\gamma}\left(\frac{G}{A}\right)^2\frac{R^2}{M^2}\left(\frac{P^2 2T - T^2 2P(dP/dT)}{P^4}\right) = 0$$

$$\therefore \quad 1 + \frac{\gamma - 1}{\gamma}\left(\frac{G}{A}\right)^2\frac{R}{M}\left(\frac{T}{P^2} - \frac{T^2}{P^3}\frac{dP}{dT}\right) = 0$$

$$\therefore \quad \frac{dP}{dT} = \frac{P}{T} + \frac{\gamma}{\gamma - 1}\left(\frac{A}{G}\right)^2\frac{M}{R}\frac{P^3}{T^2} \qquad (4.82)$$

The maximum value of the entropy occurs when the values of dP/dT given by equations 4.81 and 4.82 are the same:

i.e., when:

$$\frac{\gamma}{\gamma - 1}\frac{P}{T} = \frac{P}{T} - \frac{\gamma}{\gamma - 1}\left(\frac{A}{G}\right)^2\frac{M}{R}\frac{P^3}{T^2}$$

i.e., when:

$$\left(\frac{G}{A}\right)^2 = \frac{\gamma}{\gamma - 1}\frac{M}{R}\frac{P^2}{T}(\gamma - 1)$$

$$= \gamma\frac{P^2}{T}\frac{T}{Pv} - \gamma\frac{P}{v}$$

i.e., when:

$$u = \sqrt{\gamma Pv} = u_w \qquad \text{(from equation 4.37)}$$

which has already been shown to be the velocity of propagation of a pressure wave. This represents the limiting velocity which can be reached in a pipe of constant cross-sectional area.

4.5.5. Flow of non-ideal gases

Methods have been given for the calculation of the pressure drop for the flow of an incompressible fluid and for a compressible fluid which behaves as an ideal gas. If the fluid is compressible and deviations from the ideal gas law are appreciable, one of the approximate equations of state, such as van der Waals' equation, can be used in place of the law $PV = n\mathbf{R}T$ to give the relation between temperature, pressure, and volume. Alternatively, if the enthalpy of the gas is known over a range of temperature and pressure, the energy balance equation, 2.56, which involves a term representing the change in the enthalpy, can be employed:

$$\Delta \frac{u^2}{2\alpha} + g\Delta z + \Delta H = q - W_s \qquad \text{(equation 2.56)}$$

This method of approach is useful in considering the flow of steam at high pressures.

4.6. SHOCK WAVES

It has been seen in deriving equations 4.33 to 4.38 that for a small disturbance the velocity of propagation of the pressure wave is equal to the velocity of sound. If the changes are much larger and the process is not isentropic, the wave developed is known as a shock wave, and the velocity may be much greater than the velocity of sound. Material and momentum balances must be maintained and the appropriate equation of state for the fluid must be followed. Furthermore, any change which takes place must be associated with an increase, never a decrease, in entropy. For an ideal gas in a uniform pipe under adiabatic conditions a material balance gives:

$$\frac{u_1}{v_1} = \frac{u_2}{v_2} \qquad \text{(equation 2.62)}$$

Momentum balance: $\qquad \dfrac{u_1^2}{v_1} + P_1 = \dfrac{u_2^2}{v_2} + P_2 \qquad \text{(equation 2.64)}$

Equation of state: $\qquad \dfrac{u_1^2}{2} + \dfrac{\gamma}{\gamma - 1}P_1 v_1 = \dfrac{u_2^2}{2} + \dfrac{\gamma}{\gamma - 1}P_2 v_2 \qquad \text{(from equation 4.72)}$

Substituting from equation 2.62 into 2.64:

$$v_1 = \frac{u_1^2 - u_1 u_2}{P_2 - P_1} \qquad (4.83)$$

$$v_2 = \frac{u_2^2 - u_1 u_2}{P_1 - P_2} \qquad (4.84)$$

Then, from equation 4.72:

$$\frac{u_1^2 - u_2^2}{2} + \frac{\gamma}{\gamma - 1}\frac{P_1}{P_2 - P_1}u_1(u_1 - u_2) - \frac{\gamma}{\gamma - 1}\frac{P_1}{P_1 - P_2}u_2(u_2 - u_1) = 0$$

i.e. $(u_1 - u_2) = 0$, representing no change in conditions, or:

$$\frac{u_1 + u_2}{2} + \frac{\gamma}{\gamma - 1}\frac{1}{P_2 - P_1}(u_1 P_1 - u_2 P_2) = 0$$

i.e.:
$$\frac{u_2}{u_1} = \frac{(\gamma - 1)(P_2/P_1) + (\gamma + 1)}{(\gamma + 1)(P_2/P_1) + (\gamma - 1)} \tag{4.85}$$

and:
$$\frac{P_2}{P_1} = \frac{(\gamma + 1) - (\gamma - 1)(u_2/u_1)}{(\gamma + 1)(u_2/u_1) - (\gamma - 1)} \tag{4.86}$$

Equation 4.86 gives the pressure changes associated with a sudden change of velocity. In order to understand the nature of a possible velocity change, it is convenient to work in terms of Mach numbers. The Mach number (Ma) is defined as the ratio of the velocity at a point to the corresponding velocity of sound:

i.e.:
$$Ma_1 = \frac{u_1}{\sqrt{\gamma P_1 v_1}} \tag{4.87}$$

$$Ma_2 = \frac{u_2}{\sqrt{\gamma P_2 v_2}} \tag{4.88}$$

From equations 4.83, 4.84, 4.87, and 4.88:

$$v_1 = \frac{u_1^2}{Ma_1^2} \frac{1}{\gamma P_1} = \frac{u_1(u_1 - u_2)}{P_2 - P_1} \tag{4.89}$$

$$v_2 = \frac{u_2^2}{Ma_2^2} \frac{1}{\gamma P_2} = \frac{u_2(u_2 - u_1)}{P_1 - P_2} \tag{4.90}$$

Giving:
$$\frac{Ma_2^2}{Ma_1^2} = \frac{u_2}{u_1} \frac{P_1}{P_2} \tag{4.91}$$

From equation 4.89:

$$\frac{1}{\gamma Ma_1^2} = \frac{1 - (u_2/u_1)}{(P_2/P_1) - 1} \tag{4.92}$$

$$= \frac{2}{(\gamma + 1)(P_2/P_1) + (\gamma - 1)} \quad \text{(from equation 4.85)} \tag{4.93}$$

Thus:
$$\frac{P_2}{P_1} = \frac{2\gamma Ma_1^2 - (\gamma - 1)}{\gamma + 1} \tag{4.94}$$

and:
$$\frac{u_2}{u_1} = \frac{(\gamma - 1)Ma_1^2 + 2}{Ma_1^2(\gamma + 1)} \tag{4.95}$$

From equation 4.91:

$$\frac{Ma_2^2}{Ma_1^2} = \frac{(\gamma - 1)Ma_1^2 + 2}{Ma_1^2(\gamma + 1)} \frac{(\gamma + 1)}{2\gamma Ma_1^2 - (\gamma - 1)}$$

i.e.:
$$Ma_2^2 = \frac{(\gamma - 1)Ma_1^2 + 2}{2\gamma Ma_1^2 - (\gamma - 1)} \tag{4.96}$$

For a sudden change or normal shock wave to occur, the entropy change per unit mass of fluid must be positive.

From equation 4.79, the change in entropy is given by:

$$S_2 - S_1 = C_p \ln \frac{T_2}{T_1} - \frac{R}{M} \ln \frac{P_2}{P_1}$$

$$= C_p \ln \frac{P_2}{P_1} + C_p \ln \frac{v_2}{v_1} - \frac{R}{M} \ln \frac{P_2}{P_1}$$

$$= C_v \ln \frac{P_2}{P_1} + C_p \ln \frac{u_2}{u_1} \quad \text{(from equations 2.62 and 2.27)}$$

$$= C_v \ln \frac{2\gamma Ma_1^2 - (\gamma - 1)}{\gamma + 1} - C_p \ln \frac{Ma_1^2(\gamma + 1)}{(\gamma - 1)Ma_1^2 + 2} \quad (4.97)$$

$S_2 - S_1$ is positive when $Ma_1 > 1$. Thus a normal shock wave can occur only when the flow is supersonic. From equation 4.96, if $Ma_1 > 1$, then $Ma_2 < 1$, and therefore the flow necessarily changes from supersonic to subsonic. If $Ma_1 = 1$, $Ma_2 = 1$ also, from equation 4.96, and no change therefore takes place. It should be noted that there is no change in the energy of the fluid as it passes through a shock wave, but the entropy increases and therefore the change is irreversible.

For flow in a pipe of constant cross-sectional area, therefore, a shock wave can develop only if the gas enters at supersonic velocity. It cannot occur spontaneously.

Example 4.4

A reaction vessel in a building is protected by means of a bursting disc and the gases are vented to atmosphere through a stack pipe having a cross-sectional area of 0.07 m². The ruptured disc has a flow area of 4000 mm² and the gases expand to the full area of the stack pipe in a divergent section. If the gas in the vessel is at a pressure of 10 MN/m² and a temperature of 500 K, calculate: (a) the initial rate of discharge of gas, (b) the pressure and Mach number immediately upstream of the shock wave, and (c) the pressure of the gas immediately downstream of the shock wave.

Assume that isentropic conditions exist on either side of the shock wave and that the gas has a mean molecular weight of 40 kg/kmol, a ratio of specific heats of 1.4, and obeys the ideal gas law.

Solution

The pressure ratio w_c at the throat is given by equation 4.43:

$$w_c = \frac{P_c}{P_1} = \left[\frac{2}{\gamma + 1}\right]^{\gamma/(\gamma-1)} = \left(\frac{2}{2.4}\right)^{1.4/0.4} = 0.53$$

Thus, the throat pressure $= (10 \times 0.53) = 5.3$ MN/m².
Specific volume of gas in reactor:

$$v_1 = \frac{22.4}{40} \frac{500}{273} \frac{101.3}{10,000} = 0.0103 \text{ m}^3/\text{kg}$$

Specific volume of gas at the throat $= (0.0103)(1/0.53)^{1/1.4} = (0.0103 \times 1.575) = 0.0162$ m³/kg

$\therefore$ velocity at the throat $=$ sonic velocity (from equation 4.37)

$$= \sqrt{\gamma P v}$$

$$= \sqrt{1.4 \times 5.3 \times 10^6 \times 0.0162} = 347 \text{ m/s}$$

Initial rate of discharge of gas; $G = uA/v$ (at throat)

$$= \frac{347 \times 4000 \times 10^{-6}}{0.0162}$$

$$= 85.7 \text{ kg/s}$$

The gas continues to expand isentropically and the pressure ratio w is related to the flow area by equation 4.47. If the cross-sectional area of the exit to the divergent section is such that $w^{-1} = (10,000/101.3) = 98.7$, the pressure here will be atmospheric and the expansion will be entirely isentropic. The duct area, however, has nearly twice this value, and the flow is *over-expanded,* atmospheric pressure being reached within the divergent section. In order to satisfy the boundary conditions, a shock wave occurs further along the divergent section across which the pressure increases. The gas then expands isentropically to atmospheric pressure:

Suppose the shock wave occurs when the flow area is A, then the flow conditions at this point can be calculated by solution of the equations for:

(1) *The isentropic expansion from conditions at the vent.* The pressure ratio w (pressure/pressure in the reactor) is given by equation 4.47:

$$w^{-1.42}(1 - w^{0.29}) = \left(\frac{A}{G}\right)^2 \left(\frac{7P_1}{v_1}\right)$$

$$= \left(\frac{A}{85.7}\right)^2 \frac{7 \times 10,000 \times 10^3}{0.0103} = 9.25 \times 10^5 \, A^2 \qquad (1)$$

The pressure at this point is $10 \times 10^6 w \text{ N/m}^2$.

Specific volume of gas at this point is given by equation 4.45:

Thus: $v = v_1 w^{-0.71} = 0.0103 w^{-0.71}$

The velocity is given by equation 4.46.

Thus: $u^2 = 7P_1 v_1(1 - w^{0.29})$

$$= 7 \times 10 \times 10^6 \times 0.0103(1 - w^{0.29})$$

$$\therefore \qquad u = 0.849 \times 10^3(1 - w^{0.29})^{0.5} \text{ m/s}$$

Velocity of sound at a pressure of $10 \times 10^6 w \text{ N/m}^2$

$$= \sqrt{1.4 \times 10 \times 10^6 w \times 0.0103 w^{-0.71}}$$

$$= 380 w^{0.145} \text{ m/s}$$

$$\text{Mach number} = \frac{0.849 \times 10^3(1 - w^{0.29})^{0.5}}{380 w^{0.145}} = 2.23(w^{-0.29} - 1)^{0.5} \qquad (2)$$

(2) *The non-isentropic compression across the shock wave.* The velocity downstream from the shock wave (suffix s) is given by equation 4.95:

$$u_s = u_1 \frac{(\gamma - 1)Ma_1^2 + 2}{Ma_1^2(\gamma + 1)}$$

$$= \frac{0.849 \times 10^3(1 - w^{0.29})^{0.5}[0.4 \times 4.97(w^{-0.29} - 1) + 2]}{4.97(w^{-0.29} - 1) \times 2.4}$$

$$= 141(1 - w^{0.29})^{-0.5} \text{ m/s} \qquad (3)$$

The pressure downstream from the shock wave P_s is given by equation 4.94:

$$\frac{P_s}{10} \times 10^6 w = \left[2\gamma Ma_1^2 - \frac{\gamma - 1}{\gamma + 1}\right] \qquad (4)$$

Substituting from equation 2:

$$P_s = 56.3w(w^{-0.29} - 1) \times 10^6 \text{ N/m}^2$$

(3) *The isentropic expansion of the gas to atmospheric pressure.* The gas now expands isentropically from P_s to P_a (= 101.3 kN/m^2) and the flow area increases from A to the full bore of 0.07 m^2. Denoting conditions at the outlet by suffix a, from equation 4.46:

$$u_a^2 - u_s^2 = 7P_s v_s \left[1 - \left(\frac{P_a}{P_s} \right)^{0.25} \right] \qquad (5)$$

$$\frac{u_a}{v_a} = \frac{85.7}{0.07} = 1224 \text{ kg/m}^2\text{s} \qquad (6)$$

$$\frac{u_s}{v_s} = \frac{85.7}{A} \text{ kg/m}^2\text{s} \qquad (7)$$

$$\frac{v_a}{v_s} = \left(\frac{P_a}{P_s} \right)^{-0.71} \qquad (8)$$

Equations 1, 3 to 8, involving seven unknowns, can be solved by trial and error methods to give $w = 0.0057$. Thus the pressure upstream from the shock wave is:

$$10 \times 10^6 \times 0.0057 = 0.057 \times 10^6 \text{ N/m}^2$$

or: $$\underline{\underline{57 \text{ kN/m}^2}}$$

Mach Number, from equation (2): $$= \underline{\underline{4.15}}$$

Pressure downstream from shock wave P_s, from equation (4):

$$= \underline{\underline{1165 \text{ kN/m}^2}}$$

4.7. FURTHER READING

BECKER, E. (Trans. E. L. CHU): *Gas Dynamics* (Academic Press, New York and London, 1968).
DODGE, B. F.: *Chemical Engineering Thermodynamics* (McGraw-Hill, New York, 1944).
LEE, J. F. and SEARS, F. W.: *Thermodynamics*, 2nd edn (Addison-Wesley, 1962).
LIEPMANN, H. W. and ROSHKO, A.: *Elements of Gasdynamics* (Wiley, New York, 1957).
MAYHEW, Y. R. and ROGERS, G. F. C.: *Thermodynamics and Transport Properties of Fluids*, 2nd edn (Blackwell, Oxford, 1971).
SHAPIRO, A. H.: *The Dynamics and Thermodynamics of Compressible Fluid Flow*, Vols. I and II (Ronald, New York, 1953 and 1954).

4.8. REFERENCES

1. GRIFFITHS, H.: *Trans. Inst. Chem. Eng.* **23** (1945) 113. Some problems of vacuum technique from a chemical engineering standpoint.
2. LAPPLE, C. E.: *Trans. Am. Inst. Chem. Eng.* **39** (1948) 385. Isothermal and adiabatic flow of compressible fluids.
3. WOOLLATT, E.: *Trans. Inst. Chem. Eng.* **24** (1946) 17. Some aspects of chemical engineering thermodynamics with particular reference to the development and use of the steady flow energy balance equations.
4. STODOLA, A. and LOWENSTEIN, L. C.: *Steam and Gas Turbines* (McGraw-Hill, New York, 1945).
5. LEE, J. F. and SEARS, F. W.: *Thermodynamics*, 2nd edn (Addison-Wesley, Reading, Mass., 1962).

4.9. NOMENCLATURE

		Units in SI system	Dimensions in M, L, T, θ
A	Cross-sectional area of flow	m^2	L^2
A_0	Area of orifice	m^2	L^2
C_D	Coefficient of discharge	—	—
C_p	Specific heat capacity at constant pressure	J/kg K	$L^2T^{-2}\theta^{-1}$
C_r	Specific heat capacity at constant volume	J/kg K	$L^2T^{-2}\theta^{-1}$
d	Pipe diameter	m	L
e	Pipe roughness	m	L
F	Energy dissipated per unit mass	J/kg	L^2T^{-2}
G	Mass flowrate	kg/s	MT^{-1}
G_{max}	Mass flowrate under conditions of maximum flow	kg/s	MT^{-1}
G^*	Value of G calculated ignoring changes in kinetic energy	kg/s	MT^{-1}
g	Acceleration due to gravity	m/s^2	LT^{-2}
H	Enthalpy per unit mass	J/kg	L^2T^{-2}
K	Energy per unit mass	J/kg	L^2T^{-2}
k	Gas expansion index	—	—
l	Pipe length	m	L
M	Molecular weight	kg/kmol	MN^{-1}
n	Number of moles	k mol	N
P	Pressure	N/m^2	$ML^{-1}T^{-2}$
P_B	Back-pressure at nozzle	N/m^2	$ML^{-1}T^{-2}$
P_E	Exit pressure of gas	N/m^2	$ML^{-1}T^{-2}$
P_w	Downstream pressure P_2 at maximum flow condition	N/m^2	$ML^{-1}T^{-2}$
q	Heat added per unit mass	J/kg	L^2T^{-2}
R	Shear stress at pipe wall	N/m^2	$ML^{-1}T^{-2}$
$\mathbf{R}$	Universal gas constant	(8314) J/kmol K	$MN^{-1}L^2T^{-2}\theta^{-1}$
S	Entropy per unit mass	J/kg K	$L^2T^{-2}\theta^{-1}$
T	Temperature (absolute)	K	θ
U	Internal energy per unit mass	J/kg	$L^2T^{-2}\theta^{-1}$
u	Velocity	m/s	LT^{-1}
u_E	Velocity at exit of nozzle	m/s	LT^{-1}
u_w	Velocity of pressure wave	m/s	LT^{-1}
v	Specific volume $(= \rho^{-1})$	m^3/kg	$M^{-1}L^3$
W_s	Shaft work per unit mass	J/kg	L^2T^{-2}
w	Pressure ratio P_2/P_1	—	—
w_c	Pressure ratio P_w/P_1	—	—
z	Vertical height	m	L
α	Kinetic energy correction factor	—	—
ε	Bulk modulus of elasticity	N/m^2	$ML^{-1}T^{-2}$
ρ	Density	kg/m^3	ML^{-3}
μ	Viscosity	kg/m s (Ns/m^2)	$ML^{-1}T^{-1}$
γ	Specific heat ratio (C_p/C_v)	—	—
Ma	Mach number	—	—
Re	Reynolds number	—	—

Suffixes

0	Reference condition
1	Upstream condition
2	Downstream condition
w	Maximum flow condition

Superscript

*	Value obtained neglecting kinetic energy changes

CHAPTER 5

Flow of Multiphase Mixtures

5.1. INTRODUCTION

The flow problems considered in the previous chapters have concerned homogeneous fluids, either single phases or suspensions of fine particles whose settling velocities are sufficiently low for the solids to be completely suspended in the fluid. Consideration will now be given to the far more complex problem of the flow of multiphase systems in which the composition of the mixture may vary over the cross-section of the pipe or channel; furthermore, the components may be moving at different velocities to give rise to the phenomenon of "slip" between the phases.

Multiphase flow is important in many areas of chemical and process engineering and the behaviour of the material will depend on the properties of the components, the flowrates and the geometry of the system. In general, the complexity of the flow is so great that design methods depend very much on an analysis of the behaviour of such systems in practice and, only to a limited extent, on theoretical predictions. Some of the more important systems to be considered are:

Mixtures of liquids with gas or vapour.
Liquids mixed with solid particles ("hydraulic transport").
Gases carrying solid particles wholly or partly in suspension ("pneumatic transport").
Multiphase systems containing solids, liquids and gases.

Mixed materials may be transported horizontally, vertically, or at an inclination to the horizontal in pipes and, in the case of liquid-solid mixtures, in open channels. Although there is some degree of common behaviour between the various cases, the range of physical properties is so great that each different type of system must be considered separately. Liquids may have densities up to three orders of magnitude greater than gases but they do not exhibit any significant compressibility. Liquids themselves can range from simple Newtonian liquids such as water, to non-Newtonian fluids with very high apparent viscosities. These very large variations in density and viscosity are responsible for the big differences in behaviour of solid-gas and solid-liquid mixtures which must, in practice, be considered separately. For, all multiphase flow systems, however, it is important to understand the nature of the interactions between the phases and how these influence the *flow patterns*—the ways in which the phases are distributed over the cross-section of the pipe or duct. In design it is necessary to be in a position to predict *pressure drop* which, usually, will depend not only on the flow pattern, but also on the relative velocity of the phases; this *slip velocity* will influence the *hold-up*, the fraction of the pipe volume which is occupied by a particular phase. It is important to note that, in the flow of a

157

two-component mixture, the hold-up (or *in situ* concentration) of a component will differ from that in the mixture discharged at the end of the pipe because, as a result of *slip* of the phases relative to one another, their residence times in the pipeline will not be the same. Special attention will therefore be focused on three aspects of the flow of these complex mixtures.

(1) The flow patterns.
(2) The hold-up of the individual phases and their relative velocities.
(3) The relationship between pressure gradient in a pipe and the flowrates and physical properties of the phases.

The difference in density between the phases is important in determining flow pattern. In gas-solid and gas-liquid mixtures, the gas will always be the lighter phase, and in liquid-solid systems it will be usual for the liquid to be less dense than the solid. In vertical upward flow, therefore, there will be a tendency for the lighter phase to rise more quickly than the denser phase giving rise to a slip velocity. For a liquid-solid or gas-solid system this *slip* velocity will be close to the terminal falling velocity of the particles. In a liquid-gas system the slip velocity will depend on the flow pattern in a complex way. In all cases, there will be a net upwards force resulting in a transference of energy from the faster to the slower moving phase, and a vertically downwards gravitational force will be balanced by a vertically upwards drag force. There will be axial symmetry of flow.

In horizontal flow, the flow pattern will inevitably be more complex because the gravitational force will act perpendicular to the pipe axis, the direction of flow, and will cause the denser component to flow preferentially nearer the bottom of the pipe. Energy transfer between the phases will again occur as a result of the difference in velocity, but the net force will be horizontal and the suspension mechanism of the particles, or the dispersion of the liquid will be a more complex process. In this case, the flow will not be symmetrical about the pipe axis.

In practice, many other considerations will affect the design of an installation. For example, wherever solid particles are present, there is the possibility of *blockage* of the pipe and it is therefore important to operate under conditions where the probability of this occurring is minimised. Solids may be *abrasive* and cause undue wear if the velocities are too high or changes in direction of flow are too sudden; choice of suitable materials of construction and operating conditions is therefore important. In pneumatic transport, *electrostatic charging* may take place and cause considerable increase in pressure gradient.

5.2. TWO-PHASE GAS (VAPOUR)–LIQUID FLOW

5.2.1. Introduction

Some of the important features of the flow of two-phase mixtures composed of a liquid together with a gas or vapour will be discussed in this section. There are many applications in the chemical and process industries. These range from the flow of mixtures of oil and gas from well heads to flow of vapour-liquid mixtures in boilers and evaporators.

Because of the presence of the two phases there are considerable complications in describing and quantifying the nature of the flow compared with conditions with a single phase. The lack of knowledge of the velocities at a point in the individual phases makes it impossible to give any real picture of the velocity distribution. In most cases the gas phase, which may be flowing with a much greater velocity than the liquid, continuously accelerates the liquid thus involving a transfer of energy. Either phase may be in *streamline* or in *turbulent* flow, but the most important case is that in which both phases are turbulent. The criterion for streamline or turbulent flow of a phase is whether the Reynolds number for its flow at the same rate on its own is less or greater than 1000–2000. This distinction is to some extent arbitrary in that injection of a gas into a liquid initially in streamline flow may result in turbulence developing.

If there is no heat transfer to the flowing mixture, the mass rate of flow of each phase will remain substantially constant, but the volumetric flowrates (and velocities) will increase progressively as the gas expands with falling pressure. In a boiler or evaporator, there will be a progressive vaporisation of the liquid leading to a decreased mass flowrate of liquid and corresponding increase for the vapour, with the total mass rate of flow remaining constant. The volumetric flowrate will increase very rapidly as a result of the combined effects of falling pressure and increasing vapour/liquid ratio.

A gas-liquid mixture will have a lower density than the liquid alone. Therefore, if in a U-tube one limb contains liquid and the other a liquid-gas mixture, the equilibrium height in the second limb will be higher than in the first. If two-phase mixture is discharged at a height less than the equilibrium height, a continuous flow of liquid will take place from the first to the second limb, provided that a continuous feed of liquid and gas is maintained. This principle is used in the design of the air lift pump described in Chapter 8.

Consideration will now be given to the various flow regimes which may exist and how they may be represented on a "Flow Pattern Map"; to the calculation and prediction of hold-up of the two phases during flow; and to the calculation of pressure gradients for gas-liquid flow in pipes. In addition, when gas-liquid mixtures flow at high velocities serious erosion problems can arise and it is necessary for the designer to restrict flow velocities to avoid serious damage to equipment.

A more detailed treatment of the subject is given by GOVIER and AZIZ[1], by CHISHOLM[2] and by HEWITT[3].

5.2.2. Flow regimes and flow patterns

Horizontal flow

The flow pattern is complex and is influenced by the diameter of the pipe, the physical properties of the fluids and their flowrates. In general, as the velocities are increased and as the gas-liquid ratio increases, changes will take place from "bubble flow" through to "mist flow" as shown in Figure 5.1[1-7]; the principal characteristics are described in Table 5.1. At high liquid-gas ratios the liquid forms the continuous phase and at low values it forms the disperse phase. In the intervening region, there is generally some instability; and sometimes several flow regimes are lumped together. In plug flow and slug flow, the gas is flowing faster than the liquid and liquid from a slug tends to become detached, to move as a relatively slow moving film along the surface of the pipe and then to be reaccelerated when the next liquid slug catches it up. This process can account for a significant proportion of the total energy losses. Particularly in short pipelines, the flow develops an oscillating pattern arising largely from discontinuities associated with the expulsion of successive liquid slugs.

Figure 5.1.   Flow patterns in two-phase flow

Table 5.1.   Flow regimes in horizontal two-phase flow

Regime	Description	Typical velocities (m/s)	
		Liquid	Vapour
1. Bubble flow[a]	Bubbles of gas dispersed throughout the liquid	1.5–5	0.3–3
2. Plug flow[a]	Plugs of gas in liquid phase	0.6	< 1.0
3. Stratified flow	Layer of liquid with a layer of gas above	< 0.15	0.6–3
4. Wavy flow	As stratified but with a wavy interface due to higher velocities	< 0.3	> 5
5. Slug flow[a]	Slug of gas in liquid phase	Occurs over a wide range of velocities	
6. Annular flow[b]	Liquid film on inside walls with gas in centre		> 6
7. Mist flow[b]	Liquid droplets dispersed in gas		> 60

[a] Frequently lumped together as *intermittent* flow
[b] Sometimes merged as *annular/mist* flow

The regions over which the different types of flow can occur are conveniently shown on a "Flow Pattern Map" in which a function of the gas flowrate is plotted against a function of the liquid flowrate and boundary lines are drawn to delineate the various regions. It should be borne in mind that the distinction between any two flow patterns is not clear-cut and that these divisions are only approximate as each flow regime tends to merge in with its neighbours; in any case, the whole classification is based on highly subjective observations. Several workers have produced their own maps[4–8].

Most of the data used for compiling such maps have been obtained for the flow of water and air at near atmospheric temperature and pressure, and scaling factors have been introduced to extend their applicability to other systems. However, bearing in mind the diffuse nature of the boundaries between the regimes and the relatively minor effect of changes in physical properties, such a refinement does not appear to be justified. The flow pattern map for horizontal flow illustrated in Figure 5.2 which has been prepared by CHHABRA and RICHARDSON[9] is based on those previously presented by MANDHANE et al.[8] and WEISMAN et al.[7]. The axes of this diagram are superficial liquid velocity u_L and superficial gas velocity u_G (in each case the volumetric flowrate of the phase divided by the total cross-sectional area of the pipe).

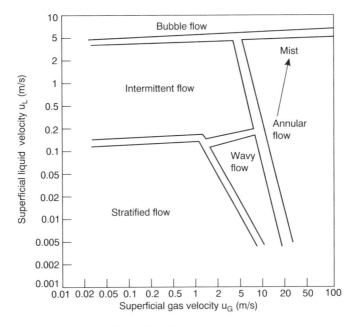

Figure 5.2.   Flow pattern map

Slug flow should be avoided when it is necessary to obviate unsteady conditions, and it is desirable to design so that annular flow still persists at loadings down to 50 per cent of the normal flow rates. Even though in many applications both phases should be turbulent, excessive gas velocity will lead to a high pressure drop, particularly in small pipes.

Although most of the data relate to flow in pipes of small diameters (< 42 mm), results of experiments carried out in a 205 mm pipe fit well on the diagram. The flow pattern map (Figure 5.2) also gives a good representation of results obtained for the flow of mixtures of gas and shear-thinning non-Newtonian liquids, including very highly shear-thinning suspensions (Power Law Index $n \approx 0.1$) and viscoelastic polymer solutions.

Vertical flow

In vertical flow, axial symmetry exists and flow patterns tend to be somewhat more stable. However, with slug flow in particular, oscillations in the flow can occur as a result of sudden changes in pressure as liquid slugs are discharged from the end of the pipe.

The principal flow patterns are shown in Figure 5.1. In general, the flow pattern map (Figure 5.2) is also applicable to vertical flow.

5.2.3. Hold-up

Because the gas always flows at a velocity greater than that of the liquid, the *in situ* volumetric fraction of liquid at any point in a pipeline will be greater than the input volume fraction of liquid; furthermore it will progressively change along the length of the pipe as a result of expansion of the gas.

There have been several experimental studies of two-phase flow in which the hold-up has been measured, either directly or indirectly. The direct method of measurement involves suddenly isolating a section of the pipe by means of quick-acting valves and then determining the quantity of liquid trapped.[10,11] Such methods are cumbersome and are subject to errors arising from the fact that the valves cannot operate instantaneously. Typical of the indirect methods is that in which the pipe cross-section is scanned by γ-rays and the hold-up is determined from the extent of their attenuation.[12,13,14]

LOCKHART and MARTINELLI[15] expressed hold-up in terms of a parameter X, characteristic of the relative flowrates of liquid and gas, defined as follows:

$$X = \sqrt{\frac{-\Delta P_L}{-\Delta P_G}} \tag{5.1}$$

where $-\Delta P_L$ and $-\Delta P_G$ are the frictional pressure drops which would arise from the flow of the respective phases on their own at the same rates. Their correlation is reproduced in Figure 4.9. As a result of more recent work it is now generally accepted that the correlation overpredicts values of liquid hold-up. Thus FAROOQI and RICHARDSON[16], the results of whose work are also shown in Figure 5.3, have given the following expression for liquid hold-up h_L for co-current flow of air and Newtonian liquids in horizontal pipes:

$$\left.\begin{aligned}
h_L &= 0.186 + 0.0191X & 1 < X < 5 \\
h_L &= 0.143X^{0.42} & 5 < X < 50 \\
h_L &= \frac{1}{0.97 + 19/X} & 50 < X < 500
\end{aligned}\right\} \tag{5.2}$$

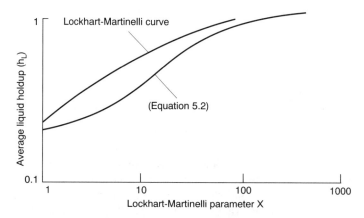

Figure 5.3.   Correlation for average liquid hold-up, h_L

It should be noted that, for turbulent flow of each phase, pressure drop is approximately proportional to the square of velocity and X is then equal to the ratio of the superficial velocities of the liquid and gas.

Equation 5.2 is found to hold well for non-Newtonian shear-thinning suspensions as well, provided that the liquid flow is turbulent. However, for laminar flow of the liquid, equation 5.2 considerably overpredicts the liquid hold-up h_L. The extent of overprediction increases as the degree of shear-thinning increases and as the liquid Reynolds number becomes progressively less. A modified parameter X' has therefore been defined[16,17] in such a way that it reduces to X both at the superficial velocity u_L equal to the transitional velocity $(u_L)_c$ from streamline to turbulent flow and when the liquid exhibits Newtonian properties.

$$X' = X \left(\frac{u_L}{(u_L)_c} \right)^{1-n} \tag{5.3}$$

On the assumption that the rheology of the shear-thinning fluid can be approximated by the Power Law Model (Chapter 3), n is the power law index. It will be seen that the correction factor becomes progressively more important as n deviates from unity and as the velocity deviates from the critical velocity. Equation 5.2 can then be applied provided X' is used in place of X in the equation.

Thus, in summary, liquid hold-up can be calculated using equation 5.2 for:

Newtonian fluids in laminar or turbulent flow.
Non-Newtonian fluids in *turbulent flow only*.

Equation 5.2, with the modified parameter X' used in place of X, can be used for laminar flow of shear-thinning fluids whose behaviour can be described by the Power-Law model.

A knowledge of hold-up is particularly important for vertical flow since the hydrostatic pressure gradient, which is frequently the major component of the total pressure gradient, is directly proportional to liquid hold-up. However, in slug flow, the situation is complicated by the fact that any liquid which is in the form of an annular film surrounding the gas slug does not contribute to the hydrostatic pressure[14].

5.2.4. Pressure, momentum, and energy relations

Methods for determining the drop in pressure start with a physical model of the two-phase system, and the analysis is developed as an extension of that used for single-phase flow. In the *separated flow* model the phases are first considered to flow separately; and their combined effect is then examined.

The total pressure gradient in a horizontal pipe, $(-dP_{TPF}/dl)$, consists of two components which represent the frictional and the acceleration pressure gradients respectively,

i.e.:
$$\frac{-dP_{TPF}}{dl} = \frac{-dP_f}{dl} + \frac{-dP_a}{dl} \tag{5.4}$$

A momentum balance for the flow of a two-phase fluid through a horizontal pipe and an energy balance may be written in an expanded form of that applicable to single-phase fluid flow. These equations for two-phase flow cannot be used in practice since the individual phase velocities and local densities are not known. Some simplification is possible if it is assumed that the two phases flow separately in the channel occupying fixed fractions of the total area, but even with this assumption of separated flow regimes, progress is difficult. It is important to note that, as in the case of single-phase flow of a compressible fluid, it is no longer possible to relate the shear stress to the pressure drop in a simple form since the pressure drop now covers both frictional and acceleration losses. The shear at the wall is proportional to the total rate of momentum transfer (arising from friction and acceleration), so that the total drop in pressure $-\Delta P_{TPF}$ is given by:

$$-\Delta P_{TPF} = (-\Delta P_f) + (-\Delta P_a) \tag{5.5}$$

The pressure drop due to acceleration is important in two-phase flow because the gas is normally flowing much faster than the liquid, and therefore as it expands the liquid phase will accelerate with consequent transfer of energy. For flow in a vertical direction, an additional term $-\Delta P_{\text{gravity}}$ must be added to the right hand side of equation 5.5 to account for the hydrostatic pressure attributable to the liquid in the pipe, and this may be calculated approximately provided that the liquid hold-up is known.

Analytical solutions for the equations of motion are not possible because of the difficulty of specifying the flow pattern and of defining the precise nature of the interaction between the phases. Rapid fluctuations in flow frequently occur and these cannot readily be taken into account. For these reasons, it is necessary for design purposes to use correlations which have been obtained using experimental data. Great care should be taken, however, if these are used outside the limits used in the experimental work.

Practical methods for evaluating pressure drop

Probably the most widely used method for estimating the drop in pressure due to friction is that proposed by LOCKHART and MARTINELLI[15] and later modified by CHISHOLM[18]. This is based on the physical model of separated flow in which each phase is considered separately and then a combined effect formulated. The two-phase pressure drop due to friction $-\Delta P_{TPF}$ is taken as the pressure drop $-\Delta P_L$ or $-\Delta P_G$ that would arise for either phase flowing alone in the pipe at the stated rate, multiplied by some factor Φ_G^2 or Φ_L^2. This factor is presented as a function of the ratio of the individual single-phase pressure drops.

$$\frac{-\Delta P_{TPF}}{-\Delta P_G} = \Phi_G^2 \tag{5.6}$$

$$\frac{-\Delta P_{TPF}}{-\Delta P_L} = \Phi_L^2 \tag{5.7}$$

The relation between Φ_G and Φ_L and X (defined by equation 5.1) is shown in Figure 5.4, where it is seen that separate curves are given according to the nature of the flow of the two phases. This relation was developed from studies on the flow in small tubes of up to 25 mm diameter with water, oils, and hydrocarbons using air at a pressure of up to 400 kN/m². For mass flowrates per unit area of L' and G' for the liquid and gas, respectively, Reynolds numbers $Re_L(L'd/\mu_L)$ and $Re_G(G'd/\mu_G)$ may be used as criteria for defining the flow regime; values less than 1000 to 2000, however, do not necessarily imply that the fluid is in truly laminar flow. Later experimental work showed that the total pressure has an influence and data presented by GRIFFITH[19] should be consulted where

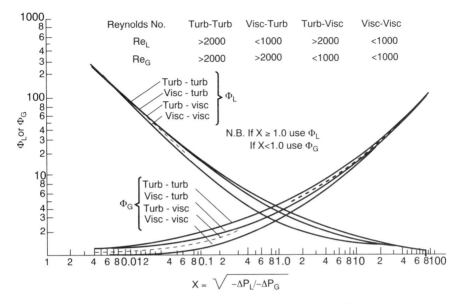

Figure 5.4.   Relation between Φ and X for two-phase flow

pressures are in excess of 3 MN/m². CHISHOLM[18] has developed a relation between Φ_L and X which he puts in the form:

$$\Phi_L^2 = 1 + \frac{c}{X} + \frac{1}{X^2} \tag{5.8}$$

where c has a value of 20 for turbulent/turbulent flow, 10 for turbulent liquid/streamline gas, 12 for streamline liquid/turbulent gas, and 5 for streamline/streamline flow. If the densities of the fluids are significantly different from those of water and air at atmospheric temperature and pressure, the values of c are somewhat modified.

CHENOWETH and MARTIN[20,21] have presented an alternative method for calculating the drop in pressure. Their method is empirical and based on experiments with pipes of 75 mm and pressures up to 0.7 MN/m². They have plotted the volume fraction of the inlet stream that is liquid as abscissa against the ratio of the two-phase pressure drop to that for liquid flowing at the same volumetric rate as the mixture. An alternative technique has been described by BAROCZY[22]. If heat transfer gives rise to evaporation then reference should be made to work by DUKLER et al[23].

An illustration of the method of calculation of two-phase pressure drop is included here as Example 5.1.

Critical flow

For the flow of a compressible fluid, conditions of sonic velocity may be reached, thus limiting the maximum flowrate for a given upstream pressure. This situation can also occur with two-phase flow, and such critical velocities may sometimes be reached with a drop in pressure of only 30 per cent of the inlet pressure.

Example 5.1

Steam and water flow through a 75 mm i.d. pipe at flowrates of 0.05 and 1.5 kg/s respectively. If the mean temperature and pressure are 330 K and 120 kN/m^2, what is the pressure drop per unit length of pipe assuming adiabatic conditions?

Solution

$$\text{Cross-sectional area for flow} = \frac{\pi(0.075)^2}{4} = 0.00442 \text{ m}^2$$

$$\text{Flow of water} = \frac{1.5}{1000} = 0.0015 \text{ m}^3/\text{s}$$

$$\text{Water velocity} = \frac{0.0015}{0.00442} = 0.339 \text{ m/s}$$

Density of steam at 330 K and 120 kN/m^2

$$= \frac{18}{22.4} \frac{273}{330} \frac{120}{101.3} = 0.788 \text{ kg/m}^3$$

$$\text{Flow of steam} = \frac{0.05}{0.788} = 0.0635 \text{ m}^3/\text{s}$$

$$\text{Steam velocity} = \frac{0.0635}{0.00442} = 14.37 \text{ m/s}$$

Viscosities at 330 K and 120 kN/m^2:

$$\text{steam} = 0.0113 \times 10^{-3} \text{ N s/m}^2; \quad \text{water} = 0.52 \times 10^{-3} \text{ N s/m}^2$$

Therefore:
$$Re_L = \frac{0.075 \times 0.339 \times 1000}{0.52 \times 10^{-3}} = 4.89 \times 10^4$$

$$Re_G = \frac{0.075 \times 14.37 \times 0.788}{0.0113 \times 10^{-3}} = 7.52 \times 10^4$$

That is, both the gas and liquid are in turbulent flow.
From the friction chart (Figure 3.7), assuming $e/d = 0.00015$:

$$\left(\frac{R}{\rho u^2}\right)_L = 0.0025 \quad \text{and} \quad \left(\frac{R}{\rho u^2}\right)_G = 0.0022$$

∴ From equation 3.18:

$$-\Delta P_L = 4\left(\frac{R}{\rho u^2}\right)_L \frac{l}{d}\rho u^2 = 4 \times 0.0025 \left(\frac{1}{0.075}\right)(1000 \times 0.339^2) = 15.32 \text{ (N/m}^2)/\text{m}$$

$$-\Delta P_G = 4 \times 0.0022 \left(\frac{1}{0.075}\right)(0.778 \times 14.37^2) = 18.85 \text{ (N/m}^2)/\text{m}$$

∴
$$\frac{-\Delta P_L}{-\Delta P_G} = \frac{15.32}{18.85} = 0.812$$

that is:
$$X^2 = 0.812 \quad \text{and} \quad X = 0.901$$

From Figure 5.4, for turbulent-turbulent flow,

$$\Phi_L = 4.35 \quad \text{and} \quad \Phi_G = 3.95$$

Therefore:
$$\frac{-\Delta P_{TPF}}{-\Delta P_G} = 3.95^2 = 15.60$$

and:
$$-\Delta P_{TPF} = 15.60 \times 18.85 = 294 \text{ (N/m}^2)/\text{m}$$

or:
$$\underline{\underline{-\Delta P_{TPF} = 0.29 \text{ (kN/m}^2)/\text{m}}}$$

Non-Newtonian flow

When a liquid exhibits non-Newtonian characteristics, the above procedures for Newtonian fluids are valid provided that the liquid flow is turbulent.

For streamline flow of non-Newtonian liquids the situation is completely different and the behaviour of two-phase mixtures in which the liquid is a shear-thinning fluid will now be examined.

The injection of air into a shear-thinning liquid in laminar flow can result in a substantial reduction in the pressure drop[24] and values of the drag ratio ($\Phi_L^2 = -\Delta P_{TPF}/-\Delta P_L$) may be substantially below unity. For a constant flowrate of liquid, the drag ratio gradually falls from unity as the gas rate is increased. At a critical air flowrate, the drag ratio passes

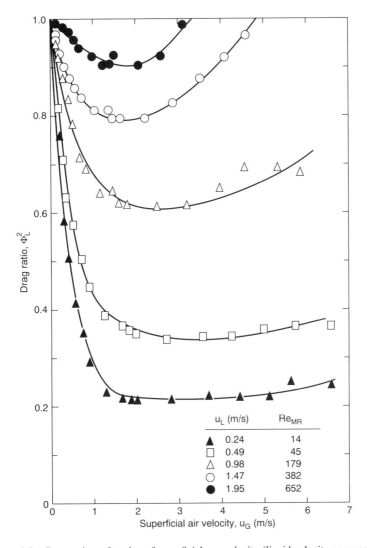

	u_L (m/s)	Re_{MR}
▲	0.24	14
□	0.49	45
△	0.98	179
○	1.47	382
●	1.95	652

Figure 5.5.   Drag ratio as function of superficial gas velocity (liquid velocity as parameter)

through a minimum $(\Phi_L^2)_{min}$ and then increases, reaching values in excess of unity at high gas flowrates. This effect has been observed with shear-thinning solutions of polymers and with flocculated suspensions of fine kaolin and anthracite coal, and is confined to conditions where the liquid flow would be laminar in the absence of air.

A typical graph of drag ratio as a function of superficial air velocity is shown in Figure 5.5 in which each curve refers to a constant superficial liquid velocity. The liquids in question exhibited power law rheology and the corresponding values of the Metzner and Reed Reynolds numbers Re_{MR} based on the superficial liquid velocity u_L (see Chapter 3) are given. The following characteristics of the curves may be noted:

(1) For a given liquid, the value of the minimum drag ratio $(\Phi_L^2)_{min}$ decreases as the superficial liquid velocity is decreased.
(2) The superficial air velocity required to give the minimum drag ratio increases as the liquid velocity decreases.

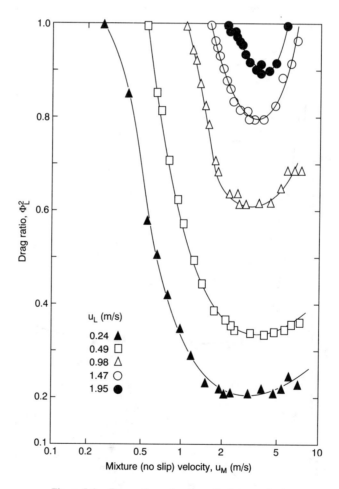

Figure 5.6.   Drag ratio as function of mixture velocity

For a more highly shear-thinning liquid, the minimum drag ratio becomes even smaller but more air must be added to achieve the condition.

If, for a given liquid, drag ratio is plotted (Figure 5.6) against total mixture velocity (superficial air velocity + superficial liquid velocity) as opposed to superficial gas velocity, it is found that the minima all occur at the same mixture velocity, irrespective of the liquid flowrate. For liquids of different rheological properties the minimum occurs at the same Reynolds number Re_{MR} (based on mixture as opposed to superficial liquid velocity)—about 2000 which corresponds to the limit of laminar flow. This suggests that the extent of drag reduction increases progressively until the liquid flow ceases to be laminar. Thus, at low flowrates more air can be injected before the liquid flow becomes turbulent, as indicated above.

At first sight, it seems anomalous that, on increasing the total volumetric flowrate by injection of air, the pressure drop can actually be reduced. Furthermore, the magnitude of the effect can be very large with values of drag ratio as low as 0.2, i.e. the pressure drop can be reduced by a *factor of 5* by air injection. How this can happen can be illustrated by means of a highly simplified model. Suppose that the air and liquid flow as a series of separate plugs, as shown in Figure 5.7. The total pressure drop can then be taken as the sum of the pressure drops across the liquid and gas slugs; the pressure drop across the gas slugs however will be negligible compared with that contributed by the liquid.

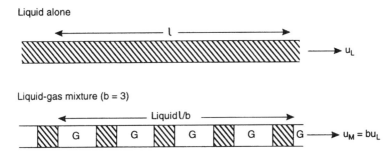

Figure 5.7.   Air-liquid flow in form of separate discrete slugs

For a "power law" fluid in laminar flow at a velocity u_L in a pipe of length l, the pressure drop $-\Delta P_L$ will be given by:

$$-\Delta P_L = K u_L^n l \tag{5.9}$$

If air is injected so that the mixture velocity is increased to bu_L, then the total length of liquid slugs in the pipe will be $(1/b)l$. Then, neglecting the pressure drop across the air slugs, the two-phase pressure drop $-\Delta P_{TP}$ will be given by:

$$-\Delta P_{TPF} = K(bu_L)^n \left(\frac{1}{b}l\right)$$

$$= K b^{n-1} u^n l \tag{5.10}$$

Thus, dividing equation 5.10 by equation 5.9:

$$\Phi_L^2 = \frac{-\Delta P_{TPF}}{-\Delta P_L} = b^{n-1} \tag{5.11}$$

Because $n < 1$ for a shear thinning fluid Φ_L^2 will be less than unity and a reduction in pressure drop occurs. The lower the value of n and the larger the value of b, the greater the effect will be. It will be noted that the effects of expansion of the air as the pressure falls have not been taken into account.

It has been found experimentally that equation 5.11 does apply for small rates of injection of air; at higher injection rates the model is not realistic and the pressure drop is reduced by a smaller amount than predicted by the equation.

It is seen that for a fluid with a power law index of 0.2 (a typical value for a flocculated kaolin suspension) and a flowrate of air equal to that of liquid ($b = 2$), $\Phi_L^2 = 0.57$, i.e. there is a pressure drop reduction of 43 per cent. For $n = 0.1$ and $b = 5$, $\Phi_L^2 = 0.23$, corresponding to a 77 per cent reduction.

It will be noted that for a Newtonian fluid ($n = 1$) equation 5.11 gives $\Phi_L^2 = 1$ for all values of b. In other words, the pressure drop will be unaffected by air injection provided that the liquid flow remains laminar. In practice, because of losses not taken into account in the simplified model, the pressure drop for a Newtonian fluid always increases with air injection.

Furthermore, it can be seen that for turbulent flow when $n \to 2$, air injection would result in substantial increases in pressure drop.

Air injection can be used, in practice, in two ways:

(1) To reduce the pressure drop, and hence the upstream pressure in a pipeline, for a given flowrate of shear-thinning liquid.
(2) To increase the flowrate of liquid for any given pipeline pressure drop.

It should be noted that energy will be required for compressing the air to the injection pressure which must exceed the upstream pressure in the pipeline. The conditions under which power-saving is achieved have been examined by DZIUBINSKI[25], who has shown that the relative efficiency of the liquid pump and the air compressor are critically important factors.

The pressure drop for a fluid exhibiting a yield stress (e.g. a *Bingham plastic* material) can be similarly reduced by air injection.

In a practical situation, air injection can be beneficial in that, when a pipeline is shut down, it may be easier to start up again if the line is not completely full of a slurry. On the other hand, if the pipeline follows an undulating topography difficulties can arise if air collects at the high spots.

Air injection may sometimes be an alternative to deflocculation. In general, deflocculated suspensions flow more readily but they tend to give much more highly consolidated sediments which can be difficult to resuspend on starting up following a shutdown. Furthermore, deflocculants are expensive and may adversely affect the suitability of the solids for subsequent use.

5.2.5. Erosion

The flow of two-phase systems often causes erosion, and many empirical relationships have been suggested to define when the effects are likely to be serious. Since high velocities may be desirable to avoid the instability associated with slug flow, there is a danger that any increase in throughput above the normal operating condition will lead to a situation where erosion may become a serious possibility.

An indication of the velocity at which erosion becomes significant may be obtained from:

$$\rho_M u_M^2 = 15{,}000$$

where ρ_M is the mean density of the two-phase mixture (kg/m^3) and u_M the mean velocity of the two-phase mixture (m/s),

$$\rho = [L' + G'] \bigg/ \left[\frac{L'}{\rho_L} + \frac{G'}{\rho_G} \right]$$

and:

$$u_M = u_L + u_G$$

where u_L and u_G are the superficial velocities of the liquid and gas respectively.

It is apparent that some compromise may be essential between the avoidance of both a slug-flow condition and the velocities which are likely to cause erosion.

5.3. HYDRAULIC TRANSPORT

5.3.1. Introduction

Hydraulic transport is the general name frequently given to the transportation of solid particles in liquids — the term *hydraulic* relates to the fact that most of the earlier applications of the technique involved water as the carrier fluid. Today there are many industrial plants, particularly in the mining industries, where particles are transported in a variety of liquids. Transport may be in vertical or horizontal pipes and, as is usually the case in long pipelines, it may follow the undulations of the land over which the pipeline is constructed. The diameter and length of the pipeline and its inclination, the properties of the solids and of the liquid, and the flowrates all influence the nature of the flow and the pressure gradient. Design methods are, in general, not very reliable, particularly for the transportation of coarse particles in horizontal pipelines and calculated values of pressure drop and energy requirements should be treated with caution. In practice, it may be more important to ensure that the system operates reliably, and without the risk of blockage and without excessive erosion, than to achieve optimum conditions in relation to power requirements.

The most important variables which must be considered in estimating power consumption and pressure drop are as follows:

(1) The pipeline   — length, diameter, inclination to the horizontal, necessity for bends, valves, etc.
(2) The liquid   — its physical properties, including density, viscosity and rheology, and its corrosive nature, if any.
(3) The solids   — their particle size and size distribution, shape and density — all factors which affect their behaviour in a fluid.
(4) The concentration of particles and the flowrates of both solids and liquid.

Suspensions are conveniently divided into two broad classes — fine *suspensions* in which the particles are reasonably uniformly distributed in the liquid; and *coarse suspensions* in which particles tend to travel predominantly in the bottom part of a horizontal pipe (at a lower velocity than the liquid) and to have a significantly lower velocity than the liquid in a vertical pipe. This is obviously not a sharp classification and, furthermore, is influenced by the flowrate and concentration of solids. However, it provides a useful initial basis on which to consider the behaviour of solid-liquid mixtures.

5.3.2. Homogeneous non-settling suspensions

Fine suspensions are reasonably homogeneous and segregation of solid and liquid phases does not occur to any significant extent during flow. The settling velocities of the particles are low in comparison with the liquid velocity and the turbulent eddies within the fluid are responsible for the suspension of the particles. In practice, turbulent flow will always be used, except when the liquid has a very high viscosity or exhibits non-Newtonian characteristics. The particles may be individually dispersed in the liquid or they may be present as flocs.

In disperse, or deflocculated, suspensions, the particles generally all have like charges and therefore repel each other. Such conditions exist when the pH range and the concentration of ions in the liquid is appropriate; this subject is discussed in Volume 2. Disperse suspensions tend to exhibit Newtonian behaviour. High volumetric concentrations (0.4–0.5) are achievable and pressure drops for pipeline flow are comparatively low and correspond closely with those calculated for homogeneous fluids of the same density as the suspension. The particles do not have a very large effect on the viscosity of the liquid except at very high concentrations, when non-Newtonian shear-thickening characteristics may be encountered as a result of the build up of a "structure". There are two possible reasons for not transporting particles in the deflocculated state. First, they tend to form dense coherent sediments and, if the flow in the pipe is interrupted, there may be difficulties in restarting operation. Secondly, the cost of the chemicals which need to be added to maintain the particles in the dispersed state may be considerable.

In most large-scale pipelines the suspensions of fine particles are usually transported in the flocculated state. Flocs consist of large numbers of particles bound loosely together with liquid occluded in the free space between them. They therefore tend to behave as relatively large particles of density intermediate between that of the liquid and the solid. Because some of the liquid is immobilised within the flocs, the maximum solids concentration obtainable is less than for deflocculated suspensions. The flocs are fragile and can break down and deform in the shear fields which exist within a flowing fluid. They therefore tend to exhibit non-Newtonian behaviour. At all but the highest concentrations they are *shear-thinning* and frequently exhibit a *yield stress* (see Chapter 3). Their behaviour, over limited ranges of shear rates, can usually be reasonably well-described by the *power law* or *Bingham plastic* models (see equations 3.120 and 3.122). Thixotropic and viscoelastic effects are usually negligible under the conditions existing in hydraulic transport lines.

Because concentrated flocculated suspensions generally have high apparent viscosities at the shear rates existing in pipelines, they are frequently transported under laminar flow conditions. Pressure drops are then readily calculated from their rheology, as described in Chapter 3. When the flow is turbulent, the pressure drop is difficult to predict accurately and will generally be somewhat *less* than that calculated assuming Newtonian behaviour. As the Reynolds number becomes greater, the effects of non-Newtonian behaviour become progressively less. There is thus a safety margin for design purposes if the suspensions are treated as Newtonian fluids when in turbulent flow.

Since, by definition, the settling velocity of the particles is low in a fine suspension, its behaviour is not dependent on its direction of flow and, if allowance is made for the hydrostatic pressure, pressure gradients are similar in horizontal and vertical pipelines.

In a series of experiments on the flow of flocculated kaolin suspensions in laboratory and industrial scale pipelines[26,27,28], measurements of pressure drop were made as a function of flowrate. Results were obtained using a laboratory capillary-tube viscometer, and pipelines of 42 mm and 205 mm diameter arranged in a recirculating loop. The rheology of all of the suspensions was described by the *power-law* model with a power law index less than unity (i.e. they were all shear-thinning); the behaviour in the laminar region can be described by the equation:

$$|R_y| = k \left| \frac{du_x}{dy} \right|^n \quad \text{(see equation 3.119)} \tag{5.12}$$

where R_y is the shear stress at a distance y from the wall,

u_x is the velocity at that position,

n is the *flow index*, and

k is the *consistency*.

Values of n and k for the suspensions used are given in Table 5.2. Experimental results are shown in Figure 5.8 as wall shear stress R as a function of wall shear rate $(du_x/dy)_{y=0}$ using logarithmic coordinates.

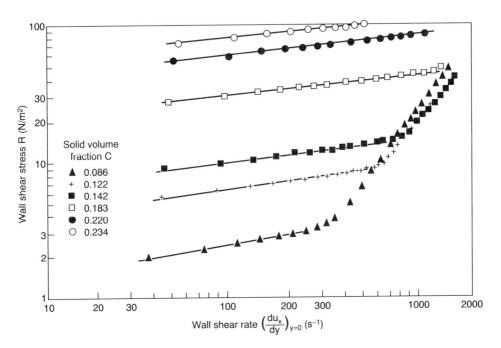

Figure 5.8. Rheograms for flocculated kaolin suspensions

Table 5.2.   Power law parameters for flocculated kaolin
suspensions

Solids volume fraction C	Flow index n	Consistency k (Ns^n/m^2)
0.086	0.23	0.89
0.122	0.18	2.83
0.142	0.16	4.83
0.183	0.15	15.3
0.220	0.14	32.4
0.234	0.13	45.3

It was shown in Chapter 3 (equation 3.134) that:

$$\left[\frac{du_x}{dy}\right]_{y=0} = \frac{6n+2}{n}\frac{u}{d} \tag{5.13}$$

Figure 5.8 shows clearly the transition point from laminar to turbulent flow for each of the suspensions when flowing in the 42 mm diameter pipe.

5.3.3. Coarse solids

The flow behaviour of suspensions of coarse particles is completely different in horizontal and vertical pipes. In horizontal flow, the concentration of particles increases towards the bottom of the pipe, the degree of non-uniformity increasing as the velocity of flow is decreased. In vertical transport, however, axial symmetry is maintained with the solids evenly distributed over the cross-section. The two cases will therefore be considered separately.

5.3.4. Coarse solids in horizontal flow

Only with fine solids are the particles uniformly distributed over the cross-section of a horizontal pipe. For coarse particles, the following principal types of flow are observed as the velocity is decreased:

(a) Heterogeneous suspension with all the particles suspended but with a significant concentration gradient vertically.
(b) Heterogeneous suspension in the upper part of the bed but a sliding bed moving along the bottom of the pipe.
(c) A similar pattern to (b), but with the bed composed of moving layers at the top and a stationary deposit at the bottom.
(d) Transport as a bed with the lower layers stationary and a few particles moving over the surface of the bed in intermittent suspension.

In addition, it is possible to obtain what is known as *dense phase flow*[29] with the particles filling the whole bore of the pipe and sliding with little relative movement between the particles.

In all cases where the two phases are moving with different velocities, it is important to differentiate between the concentration of particles in the pipe (their holdup h_s) and the volume fraction of particles (C) in the discharge. The implications of this will now be considered, together with possible means of experimentally determining the holdup.

Hold-up and slip velocity

In any two-phase flow system in which the two phases are flowing at different velocities, the in-line concentration of a component will differ from that in the stream which leaves the end of the pipe. The in-line concentration of the component with the *lower* velocity, will be greater than its concentration in the exit stream because it will have a longer residence time. It is important to understand what is happening within the pipe because the relative velocity between the phases results in energy transfer from the faster to the slower moving component. Thus, in hydraulic transport the liquid will transfer energy to the solid particles at a rate which is a function of the *slip velocity*. The solid particles will, in turn, be losing energy as a result of impact with the walls and frictional effects. In the steady state, the rate of gain and of loss of energy by the particles will be equal.

It is not possible to calculate the in-line concentrations and slip velocity from purely *external* measurements on the pipe, i.e. a knowledge of the rates at which the two components are delivered from the end of the pipe provides no evidence for what is happening within the pipe. It is thus necessary to measure *one* or more of the following variables:

The absolute linear velocity of the particles u'_S
The absolute linear velocity of the liquid u'_L
The slip velocity $u_R = u'_L - u'_S$
The hold-up of the solids h_S
The hold-up of the liquid $h_L = 1 - h_S$

In a recent study of the transport of solids by liquid in a 38 mm diameter pipe,[30] the following variables were measured:

u mixture velocity (by electromagnetic flowmeter)
u'_L linear velocity of liquid (by salt injection method)
h_S hold-up of solid particles in the pipe (by γ-ray absorption method).

In the salt injection method[31] a pulse of salt solution is injected into the line and the time is measured for it to travel between two electrode pairs situated a known distance apart, downstream from the injection point.

The γ-ray absorption method of determining in-line concentration (hold-up) of particles depends on the different degree to which the solid and the liquid attenuate γ-rays; details of the method are given in the literature[13,14].

All the other important parameters of the systems can be determined from a series of material balances as follows:

The superficial velocity of the liquid: $u_L = u'_L(1 - h_S)$ (5.14)

The superficial velocity of the solids: $u_S = u - u_L$

$$= u - u'_L(1 - h_S) \qquad (5.15)$$

Note: Superficial velocity of a component is defined as the velocity it would have at the same volumetric flowrate if it occupied the total cross-section of the pipe.

The absolute velocity of the solids: $\quad u_S' = \dfrac{u_S}{h_S}$

$$= u_L' - \frac{1}{h_S}(u_L' - u) \tag{5.16}$$

The slip, or relative, velocity: $\qquad u_R = u_L' - u_S'$

$$= \frac{1}{h_S}(u_L' - u) \tag{5.17}$$

The fractional volumetric concentration of solids C in the mixture issuing from the end of the pipe can then be obtained simply as the ratio of the superficial velocity of the solids (u_S) to the mixture flowrate.

$$u = (u_S + u_L)$$

i.e.: $\qquad C = \dfrac{u_S}{u} = \dfrac{u_S' h_S}{u} = 1 - \dfrac{u_L'}{u}(1 - h_S)$

or: $\qquad \dfrac{u_R}{u} = \dfrac{h_S - C}{(1 - h_S)h_S} \tag{5.18}$

The consistency of the data can be checked by comparing values calculated using equation 5.18 with measured values for samples collected at the outlet of the pipe.

When an industrial pipeline is to be designed, there will be no *a priori* way of knowing what the in-line concentration of solids or the slip velocity will be. In general, the rate at which solids are to be transported will be specified and it will be necessary to predict the pressure gradient as a function of the properties of the solid particles, the pipe dimensions and the flow velocity. The main considerations will be to select a pipeline diameter, such that the liquid velocity and concentrations of solids in the discharged mixture will give acceptable pressure drops and power requirements and will not lead to conditions where the pipeline is likely to block.

It is found that the major factor which determines the behaviour of the solid particles is their terminal falling velocity in the liquid. This property gives a convenient way of taking account of particle size, shape and density.

In the experimental study which has just been referred to, it was found that the slip velocity was of the same order as the terminal falling velocity of the particles. Although there is no theoretical basis, its assumption does provide a useful working guide and enables all the internal parameters, including holdup, to be calculated for a given mixture velocity u and delivered concentration C using equations 5.14–5.18. It is of interest to note that, in pneumatic conveying (discussed in Section 5.3), slip velocity is again found to approximate to terminal falling velocity.

Some of the experimental results of different workers and methods of correlating results will now be described.

Predictive methods for pressure drop

A typical curve for the conveying of solids in water is shown in Figure 5.9 which refers to the transport of *ca.* 200 μm sand in a small pipeline of 25 mm diameter.[32] It shows

Figure 5.9. Hydraulic gradient-velocity curves for 200 μm sand in 25 mm diameter hydraulic conveying line

hydraulic gradient i (head lost per unit length of pipe) as a function of mean velocity, with delivered concentration (C) as parameter. Each curve shows a minimum, which corresponds approximately to the transition between flow with a bed and suspended flow. The economic operating condition is frequently close to the critical velocity. It will be noted that as concentration is increased, the minimum occurs at a progressively high velocity. The contribution of the solids to hydraulic gradient is seen to be greatest at low velocities. Great care should be exercised in operating at velocities below the critical values as the system is then unstable and blockage can easily occur; it is a region in which the pressure drop increases as the velocity is reduced. The prediction of conditions under which blockage is liable to occur is a complex area and the subject has been discussed in depth by HISAMITSU, ISE and TAKEISHI[33].

There were several studies of hydraulic transport in the 1950s, sparked off particularly by an interest in the economic possibilities of transportation of coal and other minerals over long distances. NEWITT et al.,[32] working with solids of a range of particle sizes (up to 5 μm) and densities (1180–4600 kg/m^3) in a 25 mm diameter pipe, suggested separate correlations for flow with a bed deposit and for conditions where the particles were predominantly in heterogeneous suspension.

For flow where a bed deposit tends to form:

$$\frac{i - i_w}{Ci_w} = 66\frac{gd}{u^2}(s - 1) = 66\left[\frac{u^2}{gd(s - 1)}\right]^{-1} \tag{5.19}$$

and for heterogeneous suspensions of particles of terminal falling velocity u_0:

$$\frac{i - i_w}{Ci_w} = 1100\frac{gd}{u^2}(s - 1)\frac{u_0}{u} = 1100\frac{u_0}{u}\left[\frac{u^2}{gd(s - 1)}\right]^{-1} \tag{5.20}$$

The following features of equations 5.19 and 5.20 should be noted.

(a) They both represent pressure gradient in the form of the quotient of a dimensionless excess hydraulic gradient $(i - i_w)/i_w$ and the delivered concentration (C). This implies that the excess pressure gradient is linearly related to concentration. More recent work casts doubt on the validity of this assumption, particularly for flow in suspension.[34]

(b) The excess pressure gradient is seen to be inversely proportional to a modified Froude number $u^2/[gd(s - 1)]$ in which s is the ratio of the densities of the solids and the liquid. The pipe diameter d has been included to make the right hand side dimensionless but its effect was not studied.

(c) Particle characteristics do not feature in the correlation for flow with a bed. This is because the additional hydraulic gradient is calculated using a force balance in which the contribution of the particles to pressure drop is attributed to solid-solid friction at the walls of the pipe. This equation does not include the coefficient of friction, the importance of which will be referred to in a later section.

In a comprehensive study carried out at roughly the same time by DURAND[35,36,37] the effect of pipe diameter was examined using pipes of large diameter (40–560 mm) and a range of particle sizes d_p. His correlation took the form:

$$\frac{i - i_w}{Ci_w} = 121 \left\{ \frac{gd}{u^2}(s - 1)\frac{u_0}{[gd_p(s - 1)]^{1/2}} \right\}^{1.5} \tag{5.21}$$

Reference to Volume 2 (Chapter 3) shows that at low particle Reynolds numbers (Stokes' Law region), $u_0 \propto d_p^2$ and that at high Reynolds number $u_0 \propto d_p^{1/2}$ — at intermediate Reynolds numbers the relation between u_0 and d_p is complex, but over a limited range $u_0 \propto d_p^m$ where $\frac{1}{2} < m < 2$. It will be seen, therefore, that the influence of particle diameter is greatest for small particles (low Reynolds numbers) and becomes progressively less as particle size increases, becoming independent of size at high Reynolds numbers — as in equation 5.19 which refers to flow with a moving bed. DURAND and CONDOLIOS[38] found in their experiments using large diameter pipes that particle size when in excess of 20 mm did not affect the pressure difference. JAMES and BROAD[39] also used pipelines ranging from 102 to 207 mm diameter for the transportation of coarse particles under conditions which tended to give rise to the formation of a bed deposit. Their results suggest that the coefficient in equation 5.19 is dependent on pipe diameter and that the constant value of 66 should be replaced by $(60 + 0.24d)$ where d is in mm. Most of their results related to conditions of heterogeneous suspension.

Equations 5.21 and 5.20 give results which are reasonably consistent; they both give $(i - i_w)/Ci_w$ proportional to u^{-3}.

The term $u_0/[gd_p(s - 1)]^{1/2}$ is shown in Volume 2 (Chapter 3) to be proportional to the reciprocal square root of the *drag coefficient* (C_D) for a particle settling at its terminal falling velocity.

Substituting $\frac{4}{3}(s - 1)(gd_p/u_0^2) = C_D$ into equation 5.21 gives:

$$\frac{i - i_w}{Ci_w} = 150 \left\{ \frac{gd}{u^2}(s - 1)\frac{1}{\sqrt{C_D}} \right\}^{1.5} \tag{5.22}$$

If the concentration term C is transferred to the right-hand side of equation 5.22 (taking account of the fact that $(i - i_w)$ is not necessarily linearly related to C), it may be written in the form:

$$\frac{i - i_w}{i_w} = f\left[\frac{u^2 \sqrt{C_D}}{gd(s - 1)C}\right] \tag{5.23}$$

[Note that in Volume 2, the drag coefficient $C_D'(= C_D/2)$ is used in the calculation of the behaviour of single particles. However, C_D is used in this Chapter to facilitate comparison with the results of other workers in the field of Hydraulic Transport].

In Figure 5.10,[40] results of a number of workers[33,34,39,40,41,42,43,44] covering a wide range of experimental conditions, are plotted as:

$$\frac{i - i_w}{i_w} \text{ vs } \frac{u^2 \sqrt{C_D}}{gd(s - 1)C}$$

The scatter of the results is considerable but this arrangement of the groups seems to be the most satisfactory. The best line through all the points has the equation:

$$\frac{i - i_w}{i_w} = 30\left[\frac{u^2 \sqrt{C_D}}{gd(s - 1)C}\right]^{-1} \tag{5.24}$$

ZANDI and GOVATOS[45] suggest that the transition from flow with bed formation to flow as a heterogeneous suspension occurs at the condition where:

$$x = \frac{u^2 \sqrt{C_D}}{Cgd(s - 1)} = 40 \tag{5.25}$$

For $x < 40$, corresponding to flow with a bed, the best value for the slope of the line in Figure 5.10 is about -1. For higher values of x, corresponding to heterogeneous suspension, there is some evidence for a slope nearer -1.5, in line with equations 5.21 and 5.22 of Durand. However, in this region values of $(i - i_w)/i_w$ are so low that the experimental errors will be very great, for example at $x = 300$, $(i - i_w)/i_w \approx 0.1$ and friction losses differ from those for water alone by only about 10 per cent. What the figure does show is that the excess pressure gradient due to the solids becomes very large only under conditions where bed formation tends to occur ($x < 40$). As this is also the region of greatest practical interest, further consideration will be confined to this region.

It will be noted that in this region ($x < 40$) the experimental data show an approximately fourfold spread of ordinate at any given value of x.

Even taking into account the different experimental conditions of the various workers and errors in their measurements, and the generally unstable nature of solid-liquid flow, equation 5.24 is completely inadequate as a design equation.

An alternative approach to the representation of results for solid-liquid flow is to use the *two-layer model* which will be described in the following section. It will be seen that the coefficient of friction between the particles and the wall of the pipe is an important parameter in the model. It is suggested that its complete absence in equation 5.24 may be an important reason for the extent of the scatter. Unfortunately, it is a quantity which has been measured in only a very few investigations. It is interesting to note that the form of equation 5.19 was obtained by NEWITT et al.[32] using a force balance similar to that

Figure 5.10.   Excess hydraulic gradient due to solids as function of modified Froude number. Comparison of results of different workers

employed in the two-layer model and it is implicit in their analysis that the coefficient (66 in equation 5.19) will be a function of the coefficient of friction between the particles at the wall of the pipe.

ROCO and SHOOK[46,47] do not accept that the suspended solids and the bed constitute two identifiable regions in the pipe and have developed a model for the prediction of concentrations and velocities as continuous functions of position in the cross-section. The prediction of pressure drop from their theory involves complex numerical calculations and there is some difficulty in assessing the range of conditions over which it can be used because of the paucity of experimental data available. Some of the earlier experiments were carried out by SHOOK and DANIEL[48] who used a γ-ray absorption system for measuring in line concentration profiles. An extensive programme of experimental work is continuing at the Saskatchewan Research Council with a view to providing much needed data on large diameter pipelines (250 mm).

A simplified form of the two layer model will now be described[30].

The two-layer model

From the preceding section, it is seen that the models put forward do not adequately explain the behaviour of hydraulic conveying systems and they poorly correlate the results of different workers.

A two-layer model was first proposed by WILSON[49] in 1976 to describe the flow of solid-liquid mixtures in pipes under conditions where part of the solids are present in a moving bed and part are in suspended flow in the upper part of the pipe. He carried out a steady-state force balance on the two layers in order to calculate the pressure drop in the pipeline. The original model has been modified considerably by Wilson himself and by others[50,51]. However, it has not been used widely for design purposes, partly because of its complexity and the difficulty of obtaining convergent solutions to the iterative calculations which are necessary. Furthermore, it is necessary to be able to predict both the proportion of the solids which are present in each layer and the interfacial shear at the upper surface of the bed. The former is dependent on the experimental measurement of solids distribution and the generation of a reliable method of predicting it in terms of system properties and flowrates. The latter, the interfacial shear stress, cannot readily be measured and is calculated on the basis that the upper liquid layer is moving over the surface of a bed whose effective roughness is a function of the size of the particles.

The force balance is as follows (see Figure 5.11):

For the bed: $$F_B = -\Delta PA_B = (\mu_F \Sigma F_N + R_B S_B - R_i S_i)l \qquad (5.26)$$

For the upper layer: $$F_L = -\Delta PA_L = (R_i S_i + R_L S_L)l \qquad (5.27)$$

where $-\Delta P$ is the pressure drop over the pipe of length l,

  A_B is the cross-section occupied by bed,

  A_L is the cross-section occupied by the upper layer,

  R_B is the shear stress at the wall in the liquid in the bed,

  R_i is the interfacial shear stress,

  R_L is the shear stress at the wall in the liquid above the bed,

  S_B is the perimeter of the cross-section in contact with the bed,

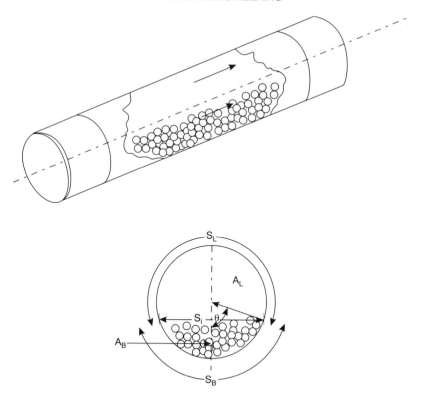

Figure 5.11.   Solids distribution over pipe section

S_i is the length of the chord between the two layers, and

S_L is the perimeter of the cross-section in contact with the upper layer.

The term $\mu_F \Sigma F_N$ is the product of the coefficient of friction μ_F and the sum ΣF_N of the normal forces per unit length between the particles and the wall.

Adding equations 5.26 and 5.27 and dividing by A, the total cross-section area of the pipe:

$$\frac{-\Delta P}{l} = \frac{R_L S_L + \mu_F \Sigma F_N + R_B S_B}{A} \tag{5.28}$$

It will be noted that the term $R_i S_i$ cancels out, the interfacial drag forces on the two layers being equal and opposite.

The two perimeters S_L and S_B are a function of the depth of bed and the angle subtended by the bed surface to the centre of the pipe. In order to calculate their values, it is necessary to know the in-line concentration of solids in the pipe and the voidage of the bed.

The shear stress R_L at the pipe wall in the upper portion of the pipe may be calculated on the assumption that the liquid above the bed is flowing through a non-circular duct, bounded at the top by the wall of the pipe and at the bottom by the upper surface of the bed. The hydraulic mean diameter may then be used in the calculation of wall shear stress. However, this does not take account of the fact that the bottom boundary, the top surface of the bed, is not stationary, and will have a greater effective roughness than the pipe wall. The effects of these two assumptions do, however, operate in opposite directions and therefore partially cancel each other out. The shear stress R_B due to the flow of liquid in the bed, relative to the pipe wall, is calculated on the assumption that it is unaltered by the presence of the particles, since their size is likely to be considerably greater than the thickness of the laminar sub-layer for turbulent flow in a pipe.

Before the pressure drop can be calculated from equation 5.28, it is still necessary to know the in-line concentration of solids in the pipe for a given flowrate and discharge concentration. Furthermore, it is necessary to know how the liquid and solid are distributed between the two layers.

In a recent study of the transport of coarse solids in a horizontal pipeline of 38 mm diameter,[30] measurements were made of pressure drop, as a function not only of mixture velocity (determined by an electromagnetic flowmeter) but also of in-line concentration of solids and liquid velocity. The solids concentration was determined using a *γ-ray absorption* technique, which depends on the difference in the attenuation of γ-rays by solid and liquid. The liquid velocity was determined by a *salt injection* method,[31] in which a pulse of salt solution was injected into the flowing mixture, and the time taken for the pulse to travel between two electrode pairs a fixed distance apart was measured. It was then possible, using equation 5.17, to calculate the relative velocity of the liquid to the solids. This relative velocity was found to increase with particle size and to be of the same order as the terminal falling velocity of the particles in the liquid.

The values of R_L, S_L, R_B and S_B were then calculated on the following assumptions:

(1) That the relative velocity u_R was equal to the terminal falling velocity of the particles. This assumption does not necessarily apply to large pipes.
(2) That the particles all travelled in a bed of porosity equal to 0.5. It had previously been shown that, as soon as bed formation occurred, the major part of the contribution of the solids was attributable to those present within the bed.
(3) That the particles and liquid in *the bed* were travelling at the same velocity, i.e. there was no slip within the bed.

In addition, off-line experiments were carried out to measure the coefficient of solid friction μ_F between the particles and the surface of the pipe. The term $\mu_F \Sigma F_N$ was then calculated on the assumption that ΣF_N was equal to the net weight (actual minus buoyancy) of the particles in the pipeline. This assumption was reasonable provided that the bed did not occupy more than about 30 per cent of the pipe cross-section.

A sensitivity analysis showed that the calculated value of pressure gradient was very sensitive to the value of the coefficient of friction μ_F, but was relatively insensitive to the slip velocity u_R and to the bed voidage.

Some examples of calculated pressure gradients are shown in Figure 5.12 in which solids concentration is a parameter. Experimental points are given, with separate designations for each concentration band. The difference between experimental and predicted value does not generally exceed about 15 per cent.

The sensitivity of the pressure drop to the coefficient of solids-surface friction μ_F may well account for the wide scatter in the results shown earlier in Figure 5.10. Unfortunately this quantity has been measured by only very few investigators. It must be emphasised that in the design of any hydraulic transport system it is extremely important to have a knowledge of the coefficient of friction.

Figure 5.12. Comparison of predicted and experimental results for simplified "two-layer" model (3.5 mm gravel in 38 mm pipe[30])

Transport of coarse particles in heavy and shear-thinning media

There have been several studies involving the use of media consisting of fine dense particles suspended in water for transporting coarse particles. The fine suspension behaves as a homogeneous fluid of increased density, but its viscosity is not sufficiently altered to have a significant effect on the pressure drop during turbulent flow, the normal condition for hydraulic transport. The cost of the dense particles may, however, be appreciable and their complete separation from the coarse particles may be difficult.

More recently attention has been focused on the use of suspensions of fines at very high concentrations; the desired increase in density can then be achieved using lighter particles. Furthermore, if the particles are flocculated, the suspension will generally exhibit shear-thinning rheology conforming approximately to the power-law or the Bingham-Plastic model (see Chapter 3). Contamination of the product can frequently be eliminated by using fines of the same material as the coarse particles. For example, coarse particles of coal can be transported in suspensions of fine coal with beneficial results.

Concentrated (20–30 per cent by volume) flocculated suspensions of coal and of china clay are highly shear-thinning (power-law index $0.1 < n < 0.2$) and, at the shear rates encountered in pipe line flow, have high apparent viscosities, as a result of which the flow is frequently laminar. In these circumstances, the suspension has a very high apparent viscosity in the low shear region in the core of the flow, and therefore suspension of coarse particles is facilitated. At the same time, the apparent viscosity in the high-shear region near the pipe wall is much lower and pressure drops are not excessive. Thus, in effect, a shear-thinning suspension is a tailor-made fluid for hydraulic transport giving a highly desirable variation of apparent viscosity over the cross-section of the pipe.

Solutions of some polymeric materials also have similar rheological properties and behave in the same way as suspensions in pipe flow. However, they do in time break down and they do not give any appreciable increase in buoyancy as their densities differ little from that of water.

One of the earlier studies was carried out in 1971 by CHARLES and CHARLES[52] who investigated the feasibility of transporting coarse materials in heavy media (sand in floccu-lated suspensions of clay in water). They concluded that the power requirement for the transport of one million tonnes/year of solids in a 200 mm diameter pipe could be reduced by a factor of about 6 by using heavy media in place of water. KENCHINGTON[53], who also studied the transport of coarse particles in china clay suspensions, concluded that a significant proportion of the coarse particles may be suspended in the dense medium without a noticeable increase in pressure drop. DUCKWORTH et al.[54−56] showed that it is possible to transport coal with a maximum particle size of 19 mm in a 250 mm diameter pipeline under laminar conditions using a fine coal slurry as the carrier fluid. Work has also been carried out on characterising suspensions of coarse particles in fines by means of a form of rotary viscometer, a "shearometer", and designing pipelines on the basis of the resulting measurements[57].

CHHABRA and RICHARDSON[58] transported coarse particles in a 42 mm pipe in a wide range of fluids, including Newtonian liquids of high viscosity, shear-thinning polymer solutions and shear-thinning suspensions of fine flocculated particles. They correlated their own and Kenchington's experimental results by means of linear regression to give the following expression:

$$\frac{i - i_w}{i_w}\frac{\phi}{C} = 0.20 \left[\frac{u^2}{gd(s-1)}\right]^{-1} \tag{5.29}$$

A somewhat improved correlation is obtained by non-linear regression:

$$\frac{i - i_w}{i_w}\frac{\phi}{C} = 0.28 \left[\frac{u^2}{gd(s-1)}\right]^{-1.25} \tag{5.30}$$

ϕ is the friction factor, $R/\rho u^2$, for the flow of the liquid alone at the same superficial velocity u. Because most of the work was carried out in the laminar-flow region (high apparent viscosity of the medium), ϕ is very sensitive to the flowrate. This contrasts with the work described in the previous section using water. The flow was then turbulent and the friction factor did not vary significantly over the range of velocities studied; in effect, the value of ϕ was implicitly incorporated in the coefficient in the equations. It is of interest to note that equation 5.29 is consistent with equation 5.19 for heterogeneous suspension in water for a value of ϕ equal to 0.003, corresponding to a Reynolds number of about 22,000 for flow in a smooth pipe.

5.3.5. Coarse solids in vertical flow

DURAND[59] has also studied vertical transport of sand and gravel of particle size ranging between 0.18 mm and 4.57 mm in a 150 mm diameter pipe, and WORSTER and DENNY[60] conveyed coal and gravel in vertical pipes of diameters 75, 100, and 150 mm. They concluded that the pressure drop for the slurry was the same as for the water alone, if due allowance was made for the static head attributable to the solids in the pipe.

NEWITT et al.[61] conveyed particles of specific gravities ranging from 1.19 to 4.56, of sizes 0.10 to 3.8 mm, in a 25 mm diameter pipe 12.8 m tall and in a 50 mm pipe 6.7 m tall. The particles used had a thirtyfold variation in terminal falling velocity. It was found that the larger particles had little effect on the frictional losses, provided the static head due to the solids was calculated on the assumption that the particles had a velocity relative to the liquid equal to their terminal falling velocities. Furthermore, it was shown photographically that at high velocities these particles travel in a central core, and thus the frictional forces at the wall will be unaffected by their presence. Very fine particles of sand give suspensions which behave as homogeneous fluids, and the hydraulic gradient due to friction is the same as for horizontal flow. When the settling velocity could not be neglected in comparison with the liquid velocity, the hydraulic gradient was found to be given by:

$$\frac{i - i_w}{C i_w} = 0.0037 \left(\frac{gd}{u^2}\right)^{1/2} \frac{d}{d_p} u_0^2 \tag{5.31}$$

For the transport of coarse particles, the relative velocity between the liquid and solids is an important factor determining the hold-up, and hence the in-line concentration of solids. CLOETE et al.[62] who conveyed sand and glass ballotini particles through vertical pipes of 12.5 and 19 mm diameter at velocities up to 3 m/s, obtained values of the in-line concentration by means of a γ-ray technique similar to that discussed earlier. They measured the vertical pressure gradient in the pipe and subtracted the component due to the hydrostatic pressure of the mixture in order to obtain the frictional component. They found that the frictional component was similar to that for water alone for velocities up to 0.7 m/s, but tended to increase at higher velocities.

In the absence of a direct measurement of the in-line concentration of solids h_s, it is necessary to make an estimate of its value in order that the hydrostatic pressure gradient in the pipe may be calculated. This can be done for a given mixture velocity u and delivered concentration C, provided that the velocity of the particles relative to the liquid u_R is known.

Following the same argument as used for horizontal flow (equations 5.14–5.18):

For unit area of cross-section:

$$\text{Mixture flowrate} = \text{Solids flowrate} + \text{Liquid flowrate}$$

$$u = u_s' h_s + u_L' (l - h_s) \tag{5.32}$$

where u_s' and u_L' are the linear velocities of solid and liquid, respectively

and

$$u_R = u_L' - u_s'. \tag{5.33}$$

Considering the flow of solids:

$$uC = u_s' h_s \tag{5.34}$$

From equations 5.32, 5.33 and 5.34, the relation between h_s and C is given by:

$$C = h_s \left[1 - \frac{u_R}{u}(1 - h_s) \right] \tag{5.35}$$

For the transport of a dilute suspension of solids, u_R will approximate to the free-falling velocity u_0 of the particles in the liquid. For concentrated suspensions, a correction must be applied to take account of the effect of neighbouring particles. This subject is considered in detail in Volume 2 (Chapter 5) from which it will be seen that the simplest form of correction takes the form:

$$\frac{u_R}{u_0} = (1 - h_s)^{n-1} \tag{5.36}$$

where n ranges from about 4.8 for fine particles to 2.4 for coarse particles. Equations 5.35 and 5.36 are then solved simultaneously to give the value of h_s, from which the contribution of the solids to the hydrostatic pressure may be calculated.

KOPKO et al.[63] made measurements of the pressure drop for the flow of suspensions of iron shot (0.0734–0.131 mm diameter) in a vertical pipe 61 mm diameter and 4.3 m tall.

They obtained the frictional pressure drop by subtracting the hydrostatic component from the total measured pressure drop and found that it constituted only a very small proportion of the total measured. More recently, AL-SALIHI[64] has confirmed that, if allowance is made for the hydrostatic component using the procedure outlined above, the frictional pressure drop is largely unaltered by the presence of the solids.

In summary, therefore, it is recommended that the frictional pressure drop for vertical flow be calculated as follows:

For *"non-settling" suspensions*: The standard equation for a single phase fluid is used with the physical properties of the suspension in place of those of the liquid.

For *suspensions of coarse particles*: The value calculated for the carrier fluid flowing alone at the mixture velocity is used.

5.4. PNEUMATIC CONVEYING

5.4.1. General considerations

Pneumatic conveying involves the transport of particulate materials by air or other gases. It is generally suitable for the transport of particles in the size range 20 μm to 50 mm. Finer particles cause problems arising from their tendency to adhere together and to the walls of the pipe and ancillary equipment. Sticky and moist powders are the worst of all. Large particles may need excessively high velocities in order to maintain them in suspension or to lift them from the bottom of the pipe in horizontal systems. Pneumatic conveying lines may be horizontal, vertical or inclined and may incorporate bends, valves and other fittings, all of which may exert a considerable influence on the behaviour of the whole installation. Whereas it is possible to make reasonable predictions of pressure gradients in long straight pipes, the effects of such fittings can be ascertained reliably only from the results of practical studies under conditions close to those that will be experienced in the final installation. The successful operation of a pneumatic conveyor may well depend much more on the need to achieve reliable operation, by removing the risks of blockage and of damage by erosion, than on achieving conditions which optimise the performance of the straight sections of the pipeline. It is important to keep changes in direction of flow as gradual as possible, to use suitable materials of construction (polyurethane lining is frequently employed) and to use velocities of flow sufficiently high to keep the particles moving, but not so high as to cause serious erosion. Whenever possible, it is desirable to carry out pilot scale tests with a sample of the solids.

Two characteristics of the conveying fluid result in considerable differences in the behaviour of pneumatic and hydraulic conveying systems: fluid density and compressibility. In hydraulic conveying the densities of the solids and fluid are of the same order of magnitude, with the solids usually having a somewhat higher density than the liquid. Practical flow velocities are commonly in the range of 1 to 5 m/s. In pneumatic transport, the solids may have a density two to three orders of magnitude greater than the gas and velocities will be considerably greater — up to 20–30 m/s. In a mixture with a volume fraction of solids, of say 0.05, the mass ratio of solids to fluid will be about 0.15 to 0.20 for hydraulic conveying compared with about 50 for pneumatic conveying. Compressibility of the gas is important because it expands as it flows down the pipe and if, for example, the ratio of the pressures at the upstream and downstream ends is 4, the velocity will increase by this factor and it would be necessary to double the diameter of the pipe to maintain the same linear gas velocity as at the beginning of the pipe. In practice, "stepped" pipelines are commonly used with progressively larger pipes used towards the downstream end. Horizontal pipelines up to 500 m long are in common use and a few 2000 m lines now exist; vertical lifts usually do not exceed about 50 m.

There are three basic modes of transport which are employed. The first, and most common, is termed *dilute phase* or *lean phase* transport in which the volume fraction of solids in this suspension does not exceed about 0.05 and a high proportion of the particles spend most of their time in suspension. The second is transport which takes place largely in the form of a *moving bed* in which the solids volume fraction may be as high as 0.6; this is relevant only for horizontal or slightly inclined pipelines. The third form is *dense phase* transport in which fairly close packed "slugs" of particles, with volume fractions of up to 0.5, alternate with slugs of gas and are propelled along the pipe. Because the velocities (~ 3 m/s) are very much lower than in dilute phase transport, there is less attrition of particles and less pipe wear. The gas usage is reduced but the risk of blockage is serious in horizontal pipelines.

Considerably more work has been carried out on horizontal as opposed to vertical pneumatic conveying. A useful review of relevant work and of correlations for the calculation of pressure drops has been given by KLINZING et al.[65]. Some consideration will now be given to horizontal conveying, with particular reference to dilute phase flow, and this is followed by a brief analysis of vertical flow.

5.4.2. Horizontal transport

Flow patterns

In a horizontal pipeline the distribution of the solids over the cross-section becomes progressively less uniform as the velocity is reduced. The following flow patterns which are commonly encountered in sequence at decreasing gas velocities have been observed in pipelines of small diameter.

1. *Uniform suspended flow*

The particles are evenly distributed over the cross-section over the whole length of pipe.

2. *Non-uniform suspended flow*

The flow is similar to that described above but there is a tendency for particles to flow preferentially in the lower portion of the pipe. If there is an appreciable size distribution, the larger particles are found predominantly at the bottom.

3. *Slug flow*

As the particles enter the conveying line, they tend to settle out before they are fully accelerated. They form dunes which are then swept bodily downstream giving an uneven longitudinal distribution of particles along the pipeline.

4. *Dune flow*

The particles settle out as in slug flow but the dunes remain stationary with particles being conveyed above the dunes and also being swept from one dune to the next.

5. *Moving bed*

Particles settle out near the feed point and form a continuous bed on the bottom of the pipe. The bed develops gradually throughout the length of the pipe and moves slowly forward. There is a velocity gradient in the vertical direction in the bed, and conveying continues in suspended form above the bed.

6. *Stationary bed*

The behaviour is similar to that of a moving bed, except that there is virtually no movement of the bed particles. The bed can build up until it occupies about three-quarters of the cross-section. Further reduction in velocity quickly gives rise to a complete blockage.

7. *Plug flow*

Following slug flow, the particles, instead of forming stationary dunes, gradually build up over the cross-section until they eventually cause a blockage. This type of flow is less common than dune flow.

Suspension mechanisms

The mechanism of suspension is related to the type of flow pattern obtained. Suspended types of flow are usually attributable to dispersion of the particles by the action of the turbulent eddies in the fluid. In turbulent flow, the vertical component of the eddy velocity will lie between one-seventh and one-fifth of the forward velocity of the fluid and, if this is more than the terminal falling velocity of the particles, they will tend to be supported in the fluid. In practice it is found that this mechanism is not as effective as might be thought because there is a tendency for the particles to damp out the eddy currents.

If the particles tend to form a bed, they will be affected by the lateral dispersive forces described by BAGNOLD[66,67]. A fluid in passing through a loose bed of particles exerts a dilating action on the system. This gives rise to a dispersion of the particles in a direction at right angles to the flow of fluid.

If a particle presents a face inclined at an angle to the direction of motion of the fluid, it may be subjected to an upward lift due to the *aerofoil* effect. In Figure 5.13 a flat plate

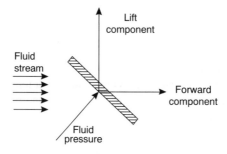

Figure 5.13.   Aerofoil effect

is shown at an angle to a stream of fluid flowing horizontally. The fluid pressure acts normally at the surface and thus produces forces with vertical and horizontal components as shown.

If the particle rotates in the fluid, it will be subjected to a lift on the *Magnus Principle*. Figure 5.14 shows a section through a cylinder rotating in a fluid stream. At its upper edge the cylinder and the fluid are both moving in the same direction, but at its lower edge they are moving in opposite directions. The fluid above the cylinder is therefore accelerated, and that below the cylinder is retarded. Thus the pressure is greater below the cylinder and an upward force is exerted.

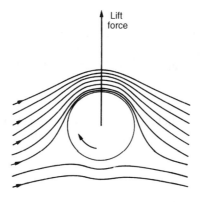

Figure 5.14.   Magnus effect

The above processes, other than the action of the dispersive forces, can result in an upward or downward displacement of an individual particle, there being approximately equal chances of the net force acting upwards or downwards. However, because the gravitational force gives rise to a tendency for the concentration of particles to be greater at the bottom of the pipe, the overall effect is to lift the particles.

The principal characteristic of a particle which determines the dominant suspension mechanism is its terminal falling velocity. Particles with low falling velocities will be readily suspended by the action of the eddies, whereas the dispersive forces will be most important with particles of high falling velocities. In a particular case, of course, the fluid velocity will also be an important factor, full suspension of a given particle occurring more readily at high velocities.

Energy requirements for dilute phase conveying

The energy required for conveying can conveniently be considered in two parts: that required for the flow of the air alone, and the additional energy necessitated by the presence of the particles. It should be noted, however, that the fluid friction will itself be somewhat modified for the following reasons: the total cross-sectional area will not be available for the flow of fluid; the pattern of turbulence will be affected by the solids; and the pressure distribution through the pipeline will be different, and hence the gas density at a given point will be affected by the solids.

The presence of the solids is responsible for an increased pressure gradient for a number of reasons. If the particles are introduced from a hopper, they will have a lower forward velocity than the fluid and therefore have to be accelerated. Because the relative velocity is greatest near the feed point and progressively falls as the particles are accelerated, their velocity will initially increase rapidly and, as the particles approach their limiting velocities, the acceleration will become very small. The pressure drop due to acceleration is therefore greatest near the feed point. Similarly, when solids are transported round a bend, they are retarded and the pressure gradient in the line following the bend is increased as a result of the need to accelerate the particles again. In pneumatic conveying, the air is expanding continuously along the line and therefore the solid velocity is also increasing. Secondly, work must be done against the action of the earth's gravitational field because the particles must be lifted from the bottom of the pipe each time they drop. Finally, particles will collide with one another and with the walls of the pipe, and therefore their velocities will fall and they will need to be accelerated again. Collisions between particles will be less frequent and result in less energy loss than impacts with the wall, because the relative velocity is much lower in the former case.

The transference of energy from the gas to the particles arises from the existence of a relative velocity. The particles will always be travelling at a lower velocity than the gas. The loss of energy by a particle will generally occur on collision and thus be a discontinuous process. The acceleration of the particle will be a gradual process occurring after each collision, the rate of transfer of energy falling off as the particle approaches the gas velocity.

The accelerating force exerted by the fluid on the particle will be a function of the properties of the gas, the shape and size of the particle, and the relative velocity. It will also depend on the dispersion of the particles over the cross-section and the shielding of individual particles. The process is complex and therefore it is not possible to develop a precise analytical treatment, but it is obviously important to know the velocity of the particles.

Determination of solid velocities

The determination of the velocity of individual particles can be carried out in a number of ways. First, the particles in a given section of pipe can be isolated using two rapidly acting shutters or valves, and the quantity of particles trapped measured by removing the intervening section of pipe. This method was used by SEGLER[68], CLARK et al.[69], and MITLIN[70]. It is, however, cumbersome, very time-consuming, and dependent upon extremely good synchronisation of the shutters. Another method is to take two photographs of the particles in the pipeline at short time-intervals and to measure the distance the particles have travelled in the time. This method is restricted to very low concentrations of relatively large particles, and does not permit the measurement of more than a few particles.[71]

A third method consists of measuring the time taken for a "tagged" particle (e.g. radio-active or magnetic) to travel between two points[72]. The method gives results applicable only to an isolated particle which may not be representative of the bulk of the particles. These techniques can readily be used in experimental equipment but are not practicable for industrial plant.

Cross-correlation methods

Because of the general oscillations in the flow condition in pneumatic conveying, it is possible to use sensing probes to measure the high frequency fluctuations at two separate locations a specified distance apart and then to determine the mean time interval between a given pattern appearing at the two locations. KLINZING and MATHUR[73] used a dielectric measuring device for this purpose. Such methods are readily applied to both laboratory and large scale plant.

MCLEMAN[74] used a method which was really a pre-cursor of the cross-correlation technique. He injected a pulse of air into the conveying line as a result of which there was a very short period during which the walls at any particular point were not subject to bombardment by particles, and the noise level was substantially reduced. By placing two transducers in contact with the wall of the pipe at a known distance apart and connecting each to a thyratron, it was possible to arrange for the first to start a frequency counter and for the second to stop the counter. A very accurate method of timing the air pulse was thus provided. It was found that the pulse retained its identity over a long distance, and this suggested that the velocities of all the particles tended to be the same. The method enabled extremely rapid and accurate measurements of the solids velocity to be obtained. Over a 16 m distance the error was less than 1 per cent.

The importance of obtaining accurate measurements of solid velocity is associated with the fact that the drag exerted by the fluid on the particle is approximately proportional to the square of the relative velocity. As the solid velocity frequently approaches the air velocity, the necessity for very accurate values is apparent.

Pressure drops and solid velocities for dilute phase flow

When solid particles are introduced into an air stream a large amount of energy is required to accelerate the particles, and the acceleration period occupies a considerable length of pipe. In order to obtain values of pressure gradients and solid velocities under conditions approaching equilibrium, measurements must be made at a considerable distance from the feed point. Much of the earlier experimental work suffered from the facts that conveying lines were much too short and that the pressure gradients were appreciably influenced by the acceleration of the particles. A typical curve obtained by CLARK et al.[69] for the pressure distribution along a 25 mm diameter conveying line, is shown in Figure 5.15. It will be noted that the pressure gradient gradually diminishes from a very high value in the neighbourhood of the feed point to an approximately constant value at distances greater than about 2 m. It has been checked experimentally that the solid velocity is increasing in the region of decreasing pressure gradient and that the length of the acceleration

Figure 5.15. Pressure in horizontal 25 mm conveying line for transport of cress seed

period increases with the mass of the particles, as might be expected. When the pressure gradient is approximately constant, so is the solid velocity. A further factor which makes it necessary to obtain measurements over a long section of pipe is that the pressure gradient does in many cases exhibit a wave form.[74] This appears to be associated with the tendency for dune formation to occur within the pipe, and thus the measured value of the pressure gradient may be influenced by the exact location of the pressure tappings. It is therefore concluded that measurements should be made over a length of at least 15 m of pipe, and remote from the solids feed point.

If the pressure drop is plotted against air flow rate as in Figure 5.16, it is seen that at a given feed rate the curve always passes through a minimum. At air velocities above the minimum of the curve, the solids are in suspended flow, but at lower velocities particles are deposited on the bottom of the pipe and there is a serious risk of blockage occurring.

Accurate measurements of solid velocities and pressure gradients were made by MCLEMAN[74] using a continuously operating system in which the solids were separated from the discharged air in a cyclone separator and introduced again to the feed hopper at the high pressure end of the system by means of a specially constructed rotary valve. A 25 mm diameter pipeline was used with two straight lengths of about 35 m joined by a semicircular bend. Experiments were carried out with air velocities up to 35 m/s. The solids used and their properties are listed in Table 5.3.

The velocities of solid particles u_s (m/s) could be represented in terms of the air velocity u_G (m/s), the free-falling velocity of the particles u_0 (m/s), and the density of the solid particles ρ_S (kg/m^3) by the following equation:

$$u_G - u_S = \frac{u_0}{0.468 + 7.25\sqrt{u_0/\rho_S}} \qquad (5.37)$$

Over the range studied, the slip velocity u_R ($= u_G - u_S$) is close to the terminal falling velocity of the particles in air ($0.67 < u_R/u_0 < 1.03$).

Figure 5.16. Effect of air rate on pressure drop for transport of cress seed in 4.8 m horizontal length of 25 mm pipe

Table 5.3. Physical properties of the solids used for pneumatic conveying[74]

Material	Shape	Particle size (mm)		Density (kg/m³)	Free-falling velocity u_0 (m/s)
		Range	Mean		
Coal A	Rounded	1.5–Dust	0.75	1,400	2.80
Coal B	Rounded	1.3–Dust	0.63		2.44
Coal C	Rounded	1.0–Dust	0.50		2.13
Coal D	Rounded	2.0–Dust	1.0		3.26
Coal E	Rounded	4.0–Dust	2.0		3.72
Perspex A	Angular	2.0–1.0	1.5	1,185	3.73
Perspex B	Angular	5.0–2.5	3.8		5.00
Perspex C	Spherical	1.0–0.5	0.75		2.35
Polystyrene	Spherical	0.4–0.3	0.36	1,080	1.62
Lead	Spherical	1.0–0.15	0.30	11,080	8.17
Brass	Porous, feathery filings	0.6–0.2	0.40	8,440	4.08
Aluminium	Rounded	0.4–0.1	0.23	2,835	3.02
Rape seed	Spherical	2.0–1.8	1.91	1,080	5.91
Radish seed	Spherical	2.8–2.3	2.5	1,065	6.48
Sand	Nearly spherical	1.5–1.0	1.3	2,610	4.66
Managanese dioxide	Rounded	1.0–0.25	0.75	4,000	5.27

Deviations from equation 5.37 were noted only at high loadings with fine solids of wide size distribution. Experimental results are plotted in Figure 5.17.

Figure 5.17.   Slip velocities $u_G - u_S$ for various materials

The additional pressure drop due to the presence of solids in the pipeline $(-\Delta P_x)$ could be expressed in terms of the solid velocity, the terminal falling velocity of the particles and the feed rate of solids F (kg/s). Equation 5.38 represents the experimental results for a 25 mm pipe to within $\pm$ 10 per cent:

$$\frac{-\Delta P_x}{-\Delta P_{air}} \frac{u_S^2}{F} = \frac{2805}{u_0} \tag{5.38}$$

In Figure 5.18 $(-\Delta P_x / - \Delta P_{air})u_S^2/F$ is plotted against $1/u_0$ for a 21 m length of pipe. $-\Delta P_{air}$ is the pressure drop for the flow of air alone at the pressure existing within the pipe.

The effect of pipe diameter on the pressure drop in a conveying system is seen by examining the results of SEGLER[68] who conveyed wheat grain in pipes up to 400 mm diameter. Taking a value of the solids velocity for wheat from the work of GASTERSTÄDT[75], $[(-\Delta P_x / - \Delta P_{air})u_S^2 u_0]/F$ is plotted against pipe diameter in

Figure 5.18.   Pressure drop for flow of solids in pneumatic conveying

Figure 5.19. The constant in equation 5.38 decreases with pipe diameter and is proportional to $d^{-0.7}$. Equation 5.38 can thus be written (with d in m):

$$\frac{-\Delta P_x}{-\Delta P_{air}} \frac{u_S^2}{F} = \frac{210}{u_0 d^{0.7}} \qquad (5.39)$$

The above correlations all apply in straight lengths of pipe with the solids suspended in the gas stream. Under conditions of slug or dune flow the pressure gradients will be considerably greater. It is generally found that the economic velocity corresponds approximately with the minimum of the curves in Figure 5.16. If there are bends in the conveying line, the energy requirement for conveying will be increased because the particles are retarded and must be re-accelerated following the bend. Bends should be eliminated wherever possible because, in addition, the wear tends to be very high.

Electrostatic charging

Equations 5.37, 5.38 and 5.39 for solid velocity and pressure drop are applicable only in the absence of electrostatic charging of the particles. Many materials, including sand,

Figure 5.19.   Effect of pipe diameter on pressure drop for transport of wheat

become charged during transport and cause the deposition of a charged layer on the surface of the pipe. The charge remains on the earthed pipeline for long periods but can be removed by conveying certain materials, including coal and perspex, through the pipe. The charging is thought to be associated with attrition of the particles and therefore to be a relatively slow process, while the discharging is entirely an electrical phenomenon and therefore more rapid. The effect of electrostatic charging is to increase the frequency of collisions between the particles and the wall. As a result, the velocity of the solids is substantially reduced and the excess pressure difference $(-\Delta P_x)$ may be increased by as much as tenfold. The results included in Figure 5.18 are those obtained before an appreciable charge has built up. Increasing the humidity of the air will substantially reduce the electrostatic effects.

Dense phase conveying

There are several attractive features of dense phase conveying which warrant its use in some circumstances. In general, pipe wear is less and separation of the product is easier. KONRAD[76] has carried out a comprehensive review of the current state of the art. He points out that the technique is not yet well established and that there are considerable design difficulties, not least a satisfactory procedure for predicting pressure drop. The risk of blocking the pipeline if the operating conditions are not correctly chosen is a deterrent to the widescale application of dense phase conveying.

5.4.3. Vertical transport

Vertical pneumatic conveying has received considerably less study than horizontal conveying and interest is usually confined to transport over much smaller distances. It is used for transferring particulate materials between plant units at different levels and for unloading into storage vessels. Another important application is in the recycling of solids in fluidised bed reactor systems (See Volume 2, Chapter 6). As in hydraulic transport, vertical upwards flows are inherently simpler than horizontal flows in that gravitational forces are directly balanced by forces due to the pressure gradient in the pipeline and the drag forces of the fluid on the particles. Furthermore, the flow is axisymmetric and there is no tendency for particles to settle out on to the pipe walls.

Detailed consideration of the interaction between particles and fluids is given in Volume 2 to which reference should be made. Briefly, however, if a particle is introduced into a fluid stream flowing vertically upwards it will be transported by the fluid provided that the fluid velocity exceeds the terminal falling velocity u_0 of the particle; the relative or *slip* velocity will be approximately u_0. As the concentration of particles increases this slip velocity will become progressively less and, for a *slug* of fairly close packed particles, will approximate to the *minimum fluidising velocity* of the particles. (See Volume 2, Chapter 6.)

The evidence from practical studies of vertical transport[72] is that the contribution of the solids to the pressure gradient in the pipeline is attributable predominantly to the weight of the particles. Because the density of the particles ρ_s is much greater than that of the gas, this additional pressure gradient can be written:

$$-\frac{dP}{dh} = (1-e)\rho_s g \qquad (5.40)$$

where e is the voidage of the mixture in the pipeline.

The conveying of fine particles in vertical pipes of diameters 25 mm, 50 mm, and 75 mm has been studied by BOOTHROYD[72]. He measured the pressure gradient in the pipeline, and found that the frictional pressure drop was less than that for air alone in the 25 mm pipe, but was greater in the larger pipes. This effect was attributed to the fact that the extent to which the fluid turbulence was affected by the presence of the particles was markedly influenced by pipe size.

KLINZING *et al.*[65] report the results of experiments on dense-phase plug-flow of a cohesive coal. Pressure drops were measured over plugs up to 0.6 m in length and were found to be linearly related to plug length. There was little wall friction and pressure drops were almost entirely attributable to the gravitational contribution, given by equation 5.40.

LI and KWAUK[77] made a theoretical and practical study of dense phase vertical flow in a tapered rectangular transport line designed to maintain an approximately constant gas velocity. The voidage of the bed was measured at three points along its height, employing a capacitance void detector, and the normal wall shear stress arising from interparticle contact pressure was measured with a probe incorporating a diaphragm and strain gauge. Typically, pressure gradient along the line was about 1.5 times the gravitational component calculated using equation 5.40.

SINCLAIR and JACKSON[78] have presented a theoretical relation between pressure gradient and the flowrates of gas and solids over the whole range of possible conditions for both cocurrent and countercurrent flow. It predicts marked segregation of gas and particles in the radial direction.

MILLER and GIDASPOW[79] have studied transport of 75 μm catalyst particles in a 75 mm diameter vertical pipe. The flow was characterised by a dilute rising core and a dense annular region at the walls which tended to move downwards. The fractional volumetric concentration of the solids was from 0.007 to 0.04 in the core and up to 0.25 in the annular region.

5.4.4. Practical applications

Although pressure drop and energy requirements are important considerations in the design of pneumatic conveying installations, the over-riding factor must be that the plant will operate safely and without giving rise to serious operating and maintenance problems. There is always a considerable reluctance to operate at velocities below that corresponding to the minimum pressure drop shown in Figure 5.16 for horizontal transport because this constitutes an unstable operating region, with the pressure drop increasing as the velocity falls, and the risk of blockage of the line is considerable. As a rule of thumb, it is undesirable to use gas velocities less than about twice the terminal falling velocity of the particles — a figure that is equally valid for vertical transport. Excessive velocities should be avoided, however, not only because of the consequent high costs of power, but also because of the increased effects of abrasion from the solids when moving at high velocity. The effects of abrasion are particularly marked at bends and fittings and changes of flow direction should be as gradual as possible. With combustible solids there are risks of dust explosions and it may be necessary to convey in an inert gas, particularly where there is likely to be a buildup of electrostatic charges.

The solids: gas ratio should be kept as high as possible to avoid the costs of pumping unnecessarily large amounts of gas and of separating the excess gas from the solids at the pipe outlet. However, because fluctuating flow conditions frequently occur it is desirable to avoid high concentrations because of the risk of blockage.

At the downstream end of the pipeline it is necessary to disengage the solids from the gas and this is most usually carried out in cyclone separators (See Volume 2, Chapter 8). However, there may be a carry-over of fine particles which must be eliminated before the gas is vented, and gas filters or electrostatic precipitators may be used for this purpose (See again Volume 2, Chapter 8). At the upstream end, the particles must be introduced using some form of positive feeder, such as a rotary valve or a blow tank.

The design of pneumatic conveying installations presents a number of problems associated with the behaviour of the particles to be transported — particularly their tendency to agglomerate and to electrostatic charging. These factors are very difficult to define, depending as they do on the exact form of solids and the humidity of the air used for conveying. Reliable design methods have therefore not been developed for any but the simplest of problems. It should be borne in mind that variations in excess pressure drops up to 2–10-fold may arise from some of these ill-definable effects. This has led WYPICH and ARNOLD[80] to suggest that it is desirable wherever possible to use a test rig to determine the conveying characteristics of the particles, using standardised procedures, before finalising the design.

5.5. FURTHER READING

BERGLES, A. E., COLLIER, J. G., DELHAYE, J. M., HEWITT, G. F. and MAYINGER, F.: *Two-Phase Flow and Heat Transfer in the Power and Process Industries* (Hemisphere Publishing Corporation and McGraw-Hill, New York, 1981).

BROWN, N. P. and HEYWOOD, N. I. (eds): *Slurry Handling. Design of Solid-Liquid Systems* (Elsevier, Amsterdam, 1991).

CHHABRA, R. P.: In *Civil Engineering Practice*, Volume 2, Cheremisinoff, P. N., Cheremisinoff, N. P. and Cheng, S. L. eds (Technomic Pub. Press PA, 1988). Hydraulic transport of solids in horizontal pipes.

CHHABRA, R. P. and RICHARDSON, J. F.: In *Encyclopedia of Fluid Mechanics*, Volume 3, Gas-Liquid Flow Cheremisinoff, N. P. eds (Gulf Publishing Co. 1986). Co-current horizontal and vertical upwards flow of gas and non-Newtonian liquid.

GOVIER, G. W. and AZIZ, K.: *The Flow of Complex Mixtures in Pipes* (Krieger, Florida, 1982).

HEWITT, G. F.: *Int. J. Multiphase Flow*, **9** (1983) 715–749. Two-phase flow studies in the United Kingdom.

KLINZING, G. E.: *Gas-Solid Transport* (McGraw-Hill, New York, 1981).

MILLS, D.: *Pneumatic Conveying Design Guide* (Butterworth, London, 1990).

SHOOK, C. A.: In *Handbook of Fluids in Motion* by Cheremisinoff, N. P. and Gupta, R. (eds) (Ann Arbor, New York, 1983), pp. 929–943: Pipeline flow of coarse particle slurries.

SHOOK, C. A. and ROCO, M. C.: *Slurry Flow. Principles and Practice* (Butterworth-Heinemann, London, 1991).

WILLIAMS, O. A.: *Pneumatic and Hydraulic Conveying of Solids* (Dekker, New York, 1983).

ZANDI, I. (ed): In *Advances in Solid-Liquid Flow in Pipes and its Applications*. Hydraulic Transport of Bulky Materials (Pergamon, Oxford, 1971).

ZENZ, F. A. and OTHMER, D. F.: *Fluidization and Fluid-Particle Systems* (Reinhold, New York, 1960).

5.6. REFERENCES

1. GOVIER, G. W. and AZIZ, K.: *The Flow of Complex Mixtures in Pipes* (Krieger, Florida, 1982).
2. CHISHOLM, D.: *Two-phase Flow in Pipelines and Heat Exchangers* (George Godwin, London, 1983).
3. HEWITT, G. F.: In *Handbook of Multiphase Systems*, G. Hetsroni ed. (McGraw-Hill, New York, 1982), 2–25.
4. BAKER, O.: *Oil and Gas Jl.* **53** (26 July 1954) 185. Simultaneous flow of gas and oil; **56** (10 Nov. 1958) 156. Multiphase flow in pipelines.
5. HOOGENDOORN, C. J.: *Chem. Eng. Sci.* **9** (1959) 205. Gas-liquid flow in horizontal pipes.
6. GRIFFITH, P. and WALLIS, G. B.: *Trans. Am. Soc. Mech. Eng., J. Heat Transfer* **83** (1961) 307. Two-phase slug flow.
7. WEISMAN, J., DUNCAN, D., GIBSON, J. and CRAWFORD, T.: *Intl. Jl. Multiphase Flow* **5** (1979) 437–462. Effects of fluid properties and pipe diameter on two phase flow patterns in horizontal lines.
8. MANDHANE, J. M., GREGORY, G. A. and AZIZ, K.: *Intl. Jl. Multiphase Flow* **1** (1974) 537–553. A flow pattern map for gas-liquid flow in horizontal pipes.
9. CHHABRA, R. P. and RICHARDSON, J. F.: *Can. J. Chem. Eng.* **62** (1984) 449. Prediction of flow patterns for the cocurrent flow of gas and non-Newtonian liquid in horizontal pipes.
10. HEWITT, G. F., KING, I. and LOVEGROVE, P. C.: *Brit. Chem. Eng.* **8** (1963) 311–318. Holdup and pressure drop measurements in the two phase annular flow of air-water mixtures.
11. CHEN, J. J. J. and SPEDDING, P. L.: *Intl. Jl. Multiphase Flow* **9** (1983) 147–159. An analysis of holdup in horizontal two phase gas-liquid flow.
12. PETRICK, P. and SWANSON, B. S.: *Rev. Sci. Instru.* **29** (1958) 1079–1085. Radiation attenuation method of measuring density of a two phase fluid.
13. HEYWOOD, N. I. and RICHARDSON, J. F.: *Chem. Eng. Sci.* **34** (1979) 17–30. Slug flow of air-water mixtures in a horizontal pipe: determination of liquid holdup by γ-ray absorption.
14. KHATIB, Z. and RICHARDSON, J. F.: *Chem. Eng. Res. Des.* **62** (1984) 139. Vertical co-current flow of air and shear thinning suspensions of kaolin.
15. LOCKHART, R. W. and MARTINELLI, R. C.: *Chem. Eng. Prog.* **45** (1949) 39. Proposed correlation of data for isothermal two-phase, two-component flow in pipes.
16. FAROOQI, S. I. and RICHARDSON, J. F.: *Trans. Inst. Chem. Eng.* **60** (1982) 292–305. Horizontal flow of air and liquid (Newtonian and non-Newtonian) in a smooth pipe: Part I: Correlation for average liquid holdup.
17. CHHABRA, R. P., FAROOQI, S. I. and RICHARDSON, J. F.: *Chem. Eng. Res. Des.* **62** (1984) 22–32. Isothermal two phase flow of air and aqueous polymer solutions in a smooth horizontal pipe.
18. CHISHOLM, D.: *Intl. Jl. Heat and Mass Transfer* **10** (1967) 1767. A theoretical basis for the Lockhart-Martinelli correlation for two-phase flow.

19. GRIFFITH, P.: In *Handbook of Heat Transfer*, edited by W. M. Rohsenow and J. P. Hartnett (McGraw-Hill, New York, 1973) Section 14. Two-phase flow.

20. CHENOWETH, J. M. and MARTIN, M. W.: *Pet. Ref.* **34** (No. 10) (1955) 151. Turbulent two-phase flow.

21. CHENOWETH, J. M. and MARTIN, M. W.: *Trans. Am. Soc. Mech. Eng.*, Paper 55 PET-9 (1955). A pressure drop correlation for turbulent two-phase flow of gas-liquid mixtures in horizontal pipes.

22. BAROCZY, C. J.: *Chem. Eng. Prog. Symp. Ser.* No. 64, **62** (1966) 232. A systematic correlation for two-phase pressure drop.

23. DUKLER, A. E. MOYE WICKS III, and CLEVELAND, R. G.: *A. I. Ch. E. Jl.* **10** (1964) 38. Frictional pressure drop in two-phase flow. A comparison of existing correlations for pressure loss and holdup.

24. FAROOQI, S. I. and RICHARDSON, J. F.: *Trans. Inst. Chem. Eng.* **60** (1982) 323. Horizontal flow of air and liquid (Newtonian and non-Newtonian) in a smooth pipe: Part II: Average pressure drop.

25. DZIUBINSKI, M. and RICHARDSON, J. F.: *J. Pipelines* **5** (1985) 107. Two-phase flow of gas and non-Newtonian liquids in horizontal pipes—superficial velocity for maximum power saving.

26. HEYWOOD, N. I. and RICHARDSON, J. F.: *J. Rheology* **22** (1978) 599. Rheological behaviour of flocculated and dispersed aqueous kaolin suspensions in pipe flow.

27. FAROOQI, S. I. and RICHARDSON, J. F.: *Trans. Inst. Chem. Eng.* **58** (1980) 116. Rheological behaviour of kaolin suspensions in water and water-glycerol mixtures.

28. CHHABRA, R. P., FAROOQI, S. I., RICHARDSON, J. F. and WARDLE, A. P.: *Chem. Eng. Res. Des.* **61** (1983) 56. Cocurrent flow of air and china clay suspensions in large diameter pipes.

29. STREAT, M.: *Hydrotransport* 10 (BHRA, Fluid Engineering, Innsbruck, Austria), (Oct. 1986). Paper B1. Dense phase flow of solid-water mixtures in pipelines: A state of the art review.

30. PIRIE, R. L.: Ph.D. Thesis, University of Wales (1990). Transport of coarse particles in water and shear-thinning suspensions in horizontal pipes.

31. PIRIE, R. L., DAVIES, T., KHAN, A. R. and RICHARDSON, J. F.: 2nd International Conf. on Flow Measurement—BHRA, London (1988) Paper F3, 187. Measurement of liquid velocity in multiphase flow by salt injection method.

32. NEWITT, D. M., RICHARDSON, J. F., ABBOTT, M., and TURTLE, R. B.: *Trans. Inst. Chem. Eng.* **33** (1955) 93. Hydraulic conveying of solids in horizontal pipes.

33. HISAMITSU, N., ISE, T. and TAKEISHI, Y.: *Hydrotransport*-7 (BHRA Fluid Engineering, Sendai, Japan) (Nov. 1980) B4 71. Blockage of slurry pipeline.

34. CHHABRA, R. P. and RICHARDSON, J. F.: *Chem. Eng. Res. Des.* **61** (1983) 313. Hydraulic transport of coarse gravel particles in a smooth horizontal pipe.

35. DURAND, R.: *Houille Blanche* **6** (1951) 384. Transport hydraulique des materiaux solides en conduite.

36. DURAND, R.: *Houille Blanche* **6** (1951) 609. Transport hydraulique de graviers et galets en conduite.

37. DURAND, R.: Proc. of the Minnesota International Hydraulic Convention (1953) 89. Basic relationships of transportation of solids in pipes—experimental research.

38. DURAND, R. and CONDOLIOS, E.: Proceedings of a Colloquium on the Hydraulic Transport of Coal, National Coal Board, London (1952), Paper IV. The hydraulic transportation of coal and solid materials in pipes.

39. JAMES, J. G. and BROAD, B. A.: Transport and Road Research Laboratory, TRRL Supplementary Report 635 (1980): Conveyance of coarse particle solids by hydraulic pipeline: Trials with limestone aggregates in 102, 156 and 207 mm diameter pipes.

40. KHAN, A. R., PIRIE, R. L. and RICHARDSON, J. F.: *Chem. Eng. Sci.* **42** (1987) 767. Hydraulic transport of solids in horizontal pipelines—predictive methods for pressure gradient.

41. ABBOTT, M.: Ph.D. Thesis, University of London (1955). The Hydraulic Conveying of Solids in Pipe Lines.

42. TURTLE, R. B.: Ph.D. Thesis, University of London (1952). The Hydraulic conveying of Granular Material.

43. HAAS, D. B., GILLIES, R., SMALL, M. and HUSBAND, W. H. W.: Saskatchewan Research Council publication No. E-835-1-C80 (March 1980). Study of the hydraulic properties of coarse particles of metallurgical coal when transported in slurry form through pipelines of various diameters.

44. GAESSLER, H.: Experimentelle und theoretische Untersuchungen über die Strömungsvorgänge beim Transport von Feststoffen in Flüssigkeiten durch horizontale Rohrleitungen, Dissertation Technische Hochschule, Karlsruhe, Germany. (1966).

45. ZANDI, I. and GOVATOS, G.: *J. Hydr. Div. Amer. Soc. Civ. Engrs.*, **93** (1967) 145. Heterogeneous flow of solids in pipelines.

46. ROCO, M. and SHOOK, C. A.: *Can. J. Chem. Eng.* **61** (1983) 494–503. Modelling of slurry flow: The effect of particle size.

47. ROCO, M. and SHOOK, C. A.: *J. Pipelines*, **4** (1984) 3–13. A model for turbulent slurry flow.

48. SHOOK, C. A. and DANIEL, S. M.: *Can. J. Chem. Eng.* **43** (1965) 56. Flow of suspensions of solids in pipelines. Part I. Flow with a stable stationary deposit.

49. WILSON, K. C.: *Hydrotransport* 4 (BHRA Fluid Engineering, Banff, Alberta, Canada) (May 1976) A1.1. A unified physically-based analysis of solid-liquid pipeline flow.

50. WILSON, K. C. and BROWN, N. P.: *Can. J. Chem. Eng.* **60** (1982) 83. Analysis of fluid friction in dense-phase pipeline flow.

51. WILSON, K. C. and PUGH, F. J.: *Can. Jl. Chem. Eng.* **66** (1988) 721. Dispersive-force modelling of turbulent suspensions in heterogeneous slurry flow.
52. CHARLES, M. E. and CHARLES, R. A.: Paper A12, *Advances in Solid-Liquid Flow in Pipes and its Application*, Zandi, I. (ed.). Pergamon Press (1971). The use of heavy media in the pipeline transport of particulate solids.
53. KENCHINGTON, J. M.: Hydrotransport 5, (BHRA Fluid Engineering, Hanover, Germany) (May 1978) Paper D7. Prediction of pressure gradient in dense phase conveying.
54. DUCKWORTH, R. A., PULLUM, L. and LOCKYEAR, C. F.: *J. Pipelines* **3** (1983) 251. The hydraulic transport of coarse coal at high concentration.
55. DUCKWORTH, R. A., PULLUM, L., ADDIE, G. R. and LOCKYEAR, C. F.: Hydrotransport 10 (BHRA Fluid Engineering, Innsbruck, Austria) (October 1986). Paper C2. The pipeline transport of coarse materials in a non-Newtonian carrier fluid.
56. DUCKWORTH, R. A., ADDIE, G. R. and MAFFETT, J.: 11th Int. Conf. on Slurry Technology — The Second Decade, organised by STA Washington DC, Hilton Head SC (March 1986) 187. Minewaste disposal by pipeline using a fine slurry carrier.
57. TATSIS, A., JACOBS, B. E. A., OSBORNE, B. and ASTLE, R. D., 13th Int. Conf. on Slurry Technology, organised by STA Washington DC, Hilton Head SC (March 1988). A comparative study of pipe flow prediction for high concentration slurries containing coarse particles.
58. CHHABRA, R. P. and RICHARDSON, J. F.: *Chem. Eng. Res. Des.* **63** (1985) 390. Hydraulic transport of coarse particles in viscous Newtonian and non-Newtonian media in a horizontal pipe.
59. DURAND, R.: *Houille Blanche* **8** (1953) 124. Écoulements de mixture en conduites verticales — influence de la densité des matériaux sur les caractéristiques de refoulement en conduite horizontale.
60. WORSTER, R. C. and DENNY, D. F.: *Proc. Inst. Mech. Eng.* **169** (1955) 563. The hydraulic transport of solid material in pipes.
61. NEWITT, D. M., RICHARDSON, J. F., and GLIDDON, B. J.: *Trans. Inst. Chem. Eng.* **39** (1961) 93. Hydraulic conveying of solids in vertical pipes.
62. CLOETE, F. L. D., MILLER, A. I., and STREAT, M.: *Trans. Inst. Chem. Eng.* **45** (1967) T392. Dense phase flow of solid-water mixtures through vertical pipes.
63. KOPKO, R. J., BARTON, P. and MCCORMICK, R. H.: *Ind. Eng. Chem. (Proc. Des. and Dev.)* **14** (1975) 264. Hydrodynamics of vertical liquid-solids transport.
64. AL-SALIHI, L.: Ph.D. Thesis, University of Wales (1989). Hydraulic transport of coarse particles in vertical pipelines.
65. KLINZING, G. E., ROHATGI, N. D., ZALTASH, A. and MYLER, C. A.: *Powder Technol.* **51** (1987) 135. Pneumatic transport — a review (Generalized phase diagram approach to pneumatic transport).
66. BAGNOLD, R. A.: *Proc. Inst. Civ. Eng.* (iii) **4** (1955) 174. Some flume experiments on large grains but little denser than the transporting fluid, and their implications.
67. BAGNOLD, R. A.: *Phil. Trans.* **249** (1957) 235. The flow of cohesionless grains in fluids.
68. SEGLER, G.: *Z. Ver. deut. Ing.* **79** (1935) 558. Untersuchungen an Körnergebläsen und Grundlagen für ihre Berechnung. (Pneumatic Grain Conveying (1951), National Institute of Agricultural Engineering.)
69. CLARK, R. H., CHARLES, D. E., RICHARDSON, J. F., and NEWITT, D. M.: *Trans. Inst. Chem. Eng.* **30** (1952) 209. Pneumatic conveying. Part I. The pressure drop during horizontal conveyance.
70. MITLIN, L.: University of London, Ph.D. thesis (1954). A study of pneumatic conveying with special reference to solid velocity and pressure drop during transport.
71. JONES, C. and HERMGES, G.: *Brit. J. Appl. Phys.* **3** (1952) 283. The measurement of velocities for solid-fluid flow in a pipe.
72. BOOTHROYD, R. G.: *Trans. Inst. Chem. Eng.* **44** (1966) T306. Pressure drop in duct flow of gaseous suspensions of fine particles.
73. KLINZING, G. E. and MATHUR, M. P.: *Can. J. Chem. Eng.* **59** (1981) 590. The dense and extrusion flow regime in gas-solid transport.
74. RICHARDSON, J. F. and MCLEMAN, M.: *Trans. Inst. Chem. Eng.* **38** (1960) 257. Pneumatic conveying. Part II. Solids velocities and pressure gradients in a one-inch horizontal pipe.
75. GASTERSTÄDT, J.: *Forsch. Arb. Geb. Ing. Wes.* No. 265 (1924) 1–76. Die experimentelle Untersuchung des pneumatischen Fördervorganges.
76. KONRAD, K.: *Powder Technol.* **49** (1986) 1. Dense-phase pneumatic conveying: a review.
77. LI, H. and KWAUK, M.: *Chem. Eng. Sci.* **44** (1989) 249, 261. Vertical pneumatic moving bed transport. I. Analysis of flow dynamics. II Experimental findings.
78. SINCLAIR, J. L. and JACKSON, R.: *A. I. Ch. E. Jl.* **35** (1989) 1473. Gas-particle flow in a vertical pipe with particle — particle interactions.
79. MILLER, A. and GIDASPOW, D.: *A. I. Ch. E. Jl.* **38** (1992) 1801. Dense, vertical gas-solid flow in a pipe.
80. WYPYCH, P. W. and ARNOLD, P. C.: *Powder Technol.* **50** (1987) 281. On improving scale-up procedures for pneumatic conveying design.

5.7. NOMENCLATURE

		Units in SI system	Dimensions in **M**, **L**, **T**
A	Total cross-sectional area of pipe	m^2	L^2
A_B	Cross-sectional area of bed deposit in pipe	m^2	L^2
A_L	Cross-sectional area of pipe over which suspended flow occurs	m^2	L^2
b	Ratio of volumetric flowrates of mixture and liquid	—	—
C	Fractional volumetric concentration of solids in *flowing* mixture	—	—
C_D	Drag coefficient for particle settling at its terminal falling velocity $$\left[= \frac{4}{3}(s'-1)\frac{d_p g}{u_0^2} \right]$$	—	—
c	Constant in equation 5.8	—	—
d	Pipe diameter	m	L
d_p	Particle diameter	m	L
e	Roughness of pipe wall	m	L
	or bed voidage	—	—
F_B	Force on bed layer in length l of pipe	N	MLT^{-2}
F_L	Force on material in suspended flow in length l of pipe	N	MLT^{-2}
ΣF_N	Sum of normal forces between particles and wall in length l of pipe	N	MLT^{-2}
G'	Mass flowrate per unit area for gas	$kg/m^2 s$	$ML^{-2}T^{-1}$
g	Acceleration due to gravity	m/s^2	LT^{-2}
h_G	Average hold-up for gas in two-phase flow	—	—
h_L	Average hold-up for liquid in two-phase flow	—	—
h_s	Average hold-up for solids in two-phase flow	—	—
i	Hydraulic gradient for two-phase flow	—	—
i_w	Hydraulic gradient for liquid (water) in two-phase flow	—	—
K	Proportionality constant in equation 5.9	—	—
k	Consistency coefficient for power-law fluid	$N\ s^n/m^2$	$ML^{-1}T^{n-2}$
L'	Mass flowrate per unit area for liquid	$kg/m^2 s$	$ML^{-2}T^{-1}$
l	Pipe length	m	L
M	Mass rate of feed of solids	kg/s	MT^{-1}
n	Flow index for power-law fluid	—	—
P	Pressure	N/m^2	$ML^{-1}T^{-2}$
$-\Delta P$	Pressure drop	N/m^2	$ML^{-1}T^{-2}$
$-\Delta P_a$	Pressure drop associated with acceleration of fluid	N/m^2	$ML^{-1}T^{-2}$
$-\Delta P_{air}$	Pressure drop for flow of air alone in pipe	N/m^2	$ML^{-1}T^{-2}$
$-\Delta P_f$	Pressure drop associated with friction	N/m^2	$ML^{-1}T^{-2}$
$-\Delta P_G$	Pressure drop for gas flowing *alone* at same superficial velocity as in two-phase flow	N/m^2	$ML^{-1}T^{-2}$
$-\Delta P_{gravity}$	Pressure drop associated with hydrostatic effects	N/m^2	$ML^{-1}T^{-2}$
$-\Delta P_L$	Pressure drop for liquid flowing *alone* at same superficial velocity as in two-phase flow	N/m^2	$ML^{-1}T^{-2}$
$-\Delta P_{TPF}$	Pressure drop for two-phase flow over length l of pipe	N/m^2	$ML^{-1}T^{-2}$
$-\Delta P_x$	Excess pressure drop due to solids in pneumatic conveying	N/m^2	$ML^{-1}T^{-2}$
R	Shear stress at pipe wall	N/m^2	$ML^{-1}T^{-2}$
R_B	Shear stress at pipe wall in fluid within bed	N/m^2	$ML^{-1}T^{-2}$
R_i	Interfacial shear stress at surface of bed	N/m^2	$ML^{-1}T^{-2}$
R_L	Shear stress at pipe wall in fluid above bed	N/m^2	$ML^{-1}T^{-2}$
R_y	Shear stress at distance y from wall	N/m^2	$ML^{-1}T^{-2}$
S_B	Perimeter of pipe in contact with bed	m	L
S_i	Length of chord at top surface of bed	m	L
S_L	Perimeter of pipe in contact with suspended flow	m	L
s	Ratio of density of solid to density of liquid (ρ_s/ρ)	—	—
u	Mean velocity in pipe or mixture velocity	m/s	LT^{-1}
u_G	Superficial velocity of gas	m/s	LT^{-1}
u_L	Superficial velocity of liquid	m/s	LT^{-1}

		Units in SI system	Dimensions in **M, L, T**
$(u_L)_c$	Critical value of u_L for laminar-turbulent transition	m/s	$\mathbf{LT}^{-1}$
u_R	Relative velocity of phases; slip velocity	m/s	$\mathbf{LT}^{-1}$
u_S	Superficial velocity of solids	m/s	$\mathbf{LT}^{-1}$
u_X	Velocity in x-direction at distance y from surface	m/s	$\mathbf{LT}^{-1}$
u_0	Terminal falling velocity of particle in fluid	m/s	$\mathbf{LT}^{-1}$
u'_L	Actual linear velocity of liquid	m/s	$\mathbf{LT}^{-1}$
u'_S	Actual linear velocity of solids	m/s	$\mathbf{LT}^{-1}$
X	Lockhart and Martinelli parameter $\sqrt{-\Delta P_L / - \Delta P_G}$	—	—
X'	Modified parameter X, defined by equation 5.3	—	—
y	Distance perpendicular to surface	m	$\mathbf{L}$
z	Vertical height	m	$\mathbf{L}$
μ_F	Coefficient of friction between solids and pipe wall	—	—
μ_G	Viscosity of gas	Ns/m^2	$\mathbf{ML}^{-1}\mathbf{T}^{-1}$
μ_L	Viscosity of liquid	Ns/m^2	$\mathbf{ML}^{-1}\mathbf{T}^{-1}$
ρ	Density	kg/m^3	$\mathbf{ML}^{-3}$
ρ_G	Density of gas	kg/m^3	$\mathbf{ML}^{-3}$
ρ_L	Density of liquid	kg/m^3	$\mathbf{ML}^{-3}$
ρ_M	Density of mixture	kg/m^3	$\mathbf{ML}^{-3}$
ρ_S	Density of solid	kg/m^3	$\mathbf{ML}^{-3}$
Φ_G^2	Drag ratio for two-phase flow $(-\Delta P_{TPF} / - \Delta P_G)$	—	—
Φ_L^2	Drag ratio for two-phase flow $(-\Delta P_{TPF} / - \Delta P_L)$	—	—
Re	Reynolds number	—	—
Re_G	Reynolds number for gas flowing alone	—	—
Re_L	Reynolds number for liquid flowing alone	—	—
Re_{MR}	Metzner and Reed Reynolds number for power-law fluid (see Chapter 3)	—	—

CHAPTER 6

Flow and Pressure Measurement

6.1. INTRODUCTION

The most important parameters measured to provide information on the operating conditions in a plant are flowrates, pressures, and temperatures. The instruments used may give either an instantaneous reading or, in the case of flow, may be arranged to give a cumulative flow over any given period. In either case, the instrument may be required to give a signal to some control unit which will then govern one or more parameters on the plant. It should be noted that on industrial plants it is usually more important to have information on the change in the value of a given parameter than to use meters that give particular absolute accuracy. To maintain the value of a parameter at a desired value a control loop is used as outlined in Figure 6.1.

A simple control system, or loop, is illustrated in Figure 6.1. The temperature T_0 of the water at Y is measured by means of a thermocouple, the output of which is fed to a controller mechanism. The latter can be divided into two sections (normally housed in the same unit). In the first (the *comparator*), the *measured value* (T_0) is compared with the desired value (T_d) to produce an *error (e)*, where:

$$e = T_d - T_0 \qquad (6.1)$$

The second section of the mechanism (the *controller*) produces an output which is a function of the magnitude of e. This is fed to a control valve in the steam line, so that the valve closes when T_0 increases and vice versa. The system as shown may be used to counteract fluctuations in temperature due to extraneous causes such as variations in water flowrate or upstream temperature — termed *load* changes. It may also be employed to change the water temperature at Y to a new value by adjustment of the desired value.

It is very important to note that in this loop system the parameter T_0, which must be kept constant, is measured, though all subsequent action is concerned with the magnitude of the error and not with the actual value of T_0. This simple loop will frequently be complicated by there being several parameters to control, which may necessitate considerable instrumental analysis and the control action will involve operation of several control valves.

This represents a simple form of control for a single variable, though in a modern plant many parameters are controlled at the same time from various measuring instruments, and the variables on a plant such as a distillation unit are frequently linked together, thus increasing the complexity of control that is required.

On industrial plants, the instruments are therefore required not only to act as indicators but also to provide some link which can be used to help in the control of the plant. In this chapter, pressure measurement is briefly described and methods of measurement of flowrate are largely confined to those which depend on the application of the energy

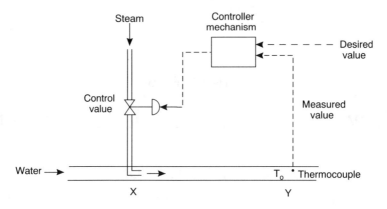

Figure 6.1. Simple feedback control system

balance equation (2.55). For further detail, reference should be made to Volume 3, Chapter 6 (Measurement) and Chapter 7 (Process Control) and to WIGHTMAN[1].

6.2. FLUID PRESSURE

In a stationary fluid the pressure is exerted equally in all directions and is referred to as the static pressure. In a moving fluid, the static pressure is exerted on any plane parallel to the direction of motion. The pressure exerted on a plane at right angles to the direction of flow is greater than the static pressure because the surface has, in addition, to exert sufficient force to bring the fluid to rest. This additional pressure is proportional to the kinetic energy of the fluid; it cannot be measured independently of the static pressure.

6.2.1. Static pressure

The energy balance equation can be applied between any two sections in a continuous fluid. If the fluid is not moving, the kinetic energy and the frictional loss are both zero, and therefore:

$$v \, dP + g \, dz = 0 \quad \text{(from equation 2.57)}$$

For an incompressible fluid:

$$v(P_2 - P_1) + g(z_2 - z_1) = 0$$

or: $$(P_2 - P_1) = -\rho g(z_2 - z_1) \tag{6.2}$$

Thus the pressure difference can be expressed in terms of the height of a vertical column of fluid.

If the fluid is incompressible and behaves as an ideal gas, for isothermal conditions:

$$P_1 v_1 \ln \frac{P_2}{P_1} + g(z_2 - z_1) = 0 \quad \text{(from equation 2.69)}$$

$$\therefore \qquad \frac{P_2}{P_1} = \exp \frac{-gM}{RT}(z_2 - z_1) \quad \text{(from equation 2.16)} \tag{6.3}$$

This expression enables the pressure distribution within an ideal gas to be calculated for isothermal conditions.

When the static pressure in a moving fluid is to be determined, the measuring surface must be parallel to the direction of flow so that no kinetic energy is converted into pressure energy at the surface. If the fluid is flowing in a circular pipe the measuring surface must be perpendicular to the radial direction at any point. The pressure connection, which is known as a *piezometer tube*, should terminate flush with the wall of the pipe so that the flow is not disturbed: the pressure is then measured near the walls where the velocity is a minimum and the reading would be subject only to a small error if the surface were not quite parallel to the direction of flow. A piezometer tube of narrow diameter is used for accurate measurements.

The static pressure should always be measured at a distance of not less than 50 diameters from bends or other obstructions, so that the flow lines are almost parallel to the walls of the tube. If there are likely to be large cross-currents or eddies, a piezometer ring should be used. This consists of four pressure tappings equally spaced at $90°$ intervals round the circumference of the tube; they are joined by a circular tube which is connected to the pressure measuring device. By this means, false readings due to irregular flow are avoided. If the pressure on one side of the tube is relatively high, the pressure on the opposite side is generally correspondingly low; with the piezometer ring a mean value is obtained. The cross-section of the piezometer tubes and ring should be small to prevent any appreciable circulation of the fluid.

6.2.2. Pressure measuring devices

(a) *The simple manometer*, shown in Figure 6.2a, consists of a transparent U-tube containing the fluid A of density ρ whose pressure is to be measured and an immiscible fluid B of higher density ρ_m. The limbs are connected to the two points between which the pressure difference $(P_2 - P_1)$ is required; the connecting leads should be completely full of fluid A. If P_2 is greater than P_1, the interface between the two liquids in limb 2 will be depressed a distance h_m (say) below that in limb 1. The pressure at the level $a - a$ must be the same in each of the limbs and, therefore:

$$P_2 + z_m\rho g = P_1 + (z_m - h_m)\rho g + h_m\rho_m g$$

$$\therefore \qquad \Delta P = P_2 - P_1 = h_m(\rho_m - \rho)g \tag{6.4}$$

If fluid A is a gas, the density ρ will normally be small compared with the density of the manometer fluid ρ_m so that:

$$\Delta P = h_m\rho_m g \tag{6.5}$$

(b) In order to avoid the inconvenience of having to read two limbs, the *well-type manometer* shown in Figure 6.2b can be used. If A_w and A_c are the cross-sectional areas of the well and the column and h_m is the increase in the level of the column and h_w the decrease in the level of the well, then:

$$P_2 = P_1 + \rho g(h_m + h_w)$$

or: $$P_2 - P_1 = \rho g(h_m + h_w)$$

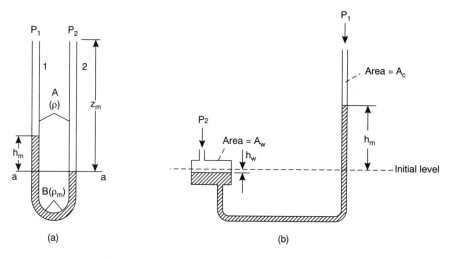

Figure 6.2. (a) The simple manometer (b) The well-type manometer

The quantity of liquid expelled from the well is equal to the quantity pushed into the column so:

$$A_w h_w = A_c h_m$$

and:

$$h_w = \frac{A_c}{A_w} h_m$$

Substituting:

$$P_2 - P_1 = \rho g h_m \left(1 + \frac{A_c}{A_w}\right) \qquad (6.6)$$

If the well is large in comparison to the column then:

$$P_2 - P_1 = \rho g h_m \qquad (6.7)$$

(c) The *inclined manometer* shown in Figure 6.3 enables the sensitivity of the manometers described previously to be increased by measuring the length of the column of liquid.

Figure 6.3.   An inclined manometer

If θ is the angle of inclination of the manometer (typically about 10–20°) and L is the movement of the column of liquid along the limb, then:

$$L = \frac{h_m}{\sin \theta} \qquad (6.8)$$

and if $\theta = 10°$, the manometer reading L is increased by about 5.7 times compared with the reading h_m which would have been obtained from a simple manometer.

(d) The *inverted manometer* (Figure 6.4) is used for measuring pressure differences in liquids. The space above the liquid in the manometer is filled with air which can be admitted or expelled through the tap A in order to adjust the level of the liquid in the manometer.

Figure 6.4.   Inverted manometer

(e) *The two-liquid manometer*. Small differences in pressure in gases are often measured with a manometer of the form shown in Figure 6.5. The reservoir at the top of each limb is of a sufficiently large cross-section for the liquid level to remain approximately the

Figure 6.5.   Two-liquid manometer

same on each side of the manometer. The difference in pressure is then given by:

$$\Delta P = (P_2 - P_1) = h_m(\rho_{m1} - \rho_{m2})g \qquad (6.9)$$

where ρ_{m1} and ρ_{m2} are the densities of the two manometer liquids. The sensitivity of the instrument is very high if the densities of the two liquids are nearly the same. To obtain accurate readings it is necessary to choose liquids which give sharp interfaces: paraffin oil and industrial alcohol are commonly used. According to OWER and PANKHURST[2], benzyl

Figure 6.6.　Bourdon gauge (*a*) Standard mechanism (*b*) With multiple coil

alcohol (specific gravity 1.048) and calcium chloride solutions give the most satisfactory results. The difference in density can be varied by altering the concentration of the calcium chloride solution.

(f) *The Bourdon gauge* (Figure 6.6). The pressure to be measured is applied to a curved tube, oval in cross-section, and the deflection of the end of the tube is communicated through a system of levers to a recording needle. This gauge is widely used for steam and compressed gases, and frequently forms the indicating element on flow controllers. The simple form of the gauge is illustrated in Figure 6.6*a* and Figure 6.6*b* shows a Bourdon type gauge with the sensing element in the form of a helix. This instrument has a very much greater sensitivity and is suitable for very high pressures.

Note: The pressure measuring devices (a) to (e) described above all measure a pressure *difference* $\Delta P(= P_2 - P_1)$. In the case of the Bourdon gauge (f), the pressure indicated is the difference between that communicated by the system to the tube and the external (ambient) pressure, and this is usually referred to as the *gauge* pressure. It is then necessary to add on the ambient pressure in order to obtain the (absolute) pressure. Even the mercury barometer measures, not atmospheric pressure, but the difference between atmospheric pressure and the vapour pressure of mercury which, of course, is negligible. Gauge pressures are not, however, used in the SI System of units.

6.2.3. Pressure signal transmission — the differential pressure cell

The meters described so far provide a measurement usually in the form of a pressure differential, though in most modern plants these readings must be transmitted to a central control facility where they form the basis for either recording or automatic control. The device which converts the quantity being measured into a signal is known as a transducer. Most transducers consist of a sensing element and a conversion or control element. The sensor may respond to movement, heat, light, magnetic or chemical effects and these physical quantities may be converted to change in an electrical parameter such as voltage, resistance, capacitance or inductance. The transmission is most conveniently effected by pneumatic or electrical methods, and one typical arrangement known as a differential pressure (d.p.) cell operating on the force balance system is illustrated in Figures 6.7 and 6.8.

In Figure 6.8 increase in the pressure on the "in" bellows causes the force-bar to turn clockwise, thus reducing the separation in the flapper-nozzle. This nozzle is fed with air, and the increase in back pressure from the nozzle in turn increases the force developed by the negative feedback bellows producing a counterclockwise movement which restores the flapper-nozzle separation. Balance is achieved when the feedback pressure is proportional to the applied pressure.

In this way, the signal pressure is made to respond to the input pressure, and by adjusting the distance of the bellows from the fulcrum, any range of force may be made to give an output pressure in the range $120-200$ kN/m^2. It is important to note that since the flapper-nozzle unit has a high gain (large pressure change for small change in separation), the actual movements are very small, usually within the range $0.004-0.025$ mm.

An instrument in which this feedback arrangement is used is shown in Figure 6.7. The differential pressure from the meter is applied to the two sides of the diaphragm A giving

Figure 6.7.   Differential pressure cell

Figure 6.8. Force balance system

a force on the bar B. The closure E in the force bar B acts as the fulcrum, so that the flapper-nozzle separation at H responds to the initial difference in pressure. This gives the change in air pressure or signal which can be transmitted through a considerable distance. A unit of this kind is thus an essential addition to all pressure meters from which a signal in a central control room is required.

6.2.4. Impact pressure

The pressure exerted on a plane at right angles to the direction of flow of the fluid consists of two components:

(a) static pressure;
(b) the additional pressure required to bring the fluid to rest at the point.

Consider a fluid flowing between two sections, 1 and 2 (Figure 6.9), which are sufficiently close for friction losses to be negligible between the two sections; they are a sufficient distance apart, however, for the presence of a small surface at right angles to the direction of flow at section 2 to have negligible effect on the pressure at section 1. These conditions are normally met if the distance between the sections is one pipe diameter.

Consider a small filament of liquid which is brought to rest at section 2. Applying the energy balance equation between the two sections, since $g\Delta z$, W_s, and F are all zero:

Figure 6.9. Impact pressure

$$\frac{\dot{u}_1^2}{2} = \frac{\dot{u}_2^2}{2} + \int_{P_1}^{P_2} v \, dP \qquad\qquad\text{(from equation 2.55)}$$

where $\dot{u}_1$ and $\dot{u}_2$ are velocities at 1 and 2.

If the fluid is incompressible or if the change in the density of the fluid between the sections is negligible, then (since $\dot{u}_2 = 0$):

$$\dot{u}_1^2 = 2v(P_2 - P_1) = 2h_i g$$

or: $\qquad\qquad\qquad \dot{u}_1 = \sqrt{2v(P_2 - P_1)} = \sqrt{2gh_i} \qquad\qquad\qquad (6.10)$

where h_i is the difference between the impact pressure head at section 2 and the static pressure head at section 1.

Little error is introduced if this expression is applied to the flow of a compressible fluid provided that the velocity is not greater than about 60 m/s. When the velocity is high, the equation of state must be used to give the relation between the pressure and the volume of the gas. For non-isothermal flow, $Pv^k =$ a constant,

and: $\qquad \int_{P_1}^{P_2} v \, dP = \frac{k}{k-1} P_1 v_1 \left[\left(\frac{P_2}{P_1} \right)^{(k-1)/k} - 1 \right] \qquad\qquad \text{(equation 2.73)}$

so that: $\qquad \frac{\dot{u}_1^2}{2} = \frac{k}{k-1} P_1 v_1 \left[\left(\frac{P_2}{P_1} \right)^{(k-1)/k} - 1 \right]$

and: $\qquad\qquad P_2 = P_1 \left(1 + \frac{\dot{u}_1^2}{2} \frac{k-1}{k} \frac{1}{P_1 v_1} \right)^{k/(k-1)} \qquad\qquad (6.11)$

For isothermal flow: $\qquad\qquad \frac{\dot{u}_1^2}{2} = P_1 v_1 \ln \frac{P_2}{P_1} \qquad\qquad \text{(from equation 2.69)}$

$$\therefore \qquad\qquad P_2 = P_1 \exp \frac{\dot{u}_1^2}{2 P_1 v_1}$$

or: $\qquad\qquad\qquad P_2 = P_1 \exp \frac{\dot{u}_1^2 M}{2 \mathbf{R} T} \qquad\qquad\qquad (6.12)$

Equations 6.11 and 6.12 can be used for the calculation of the fluid velocity and the impact pressure in terms of the static pressure a short distance upstream. The two sections are chosen so that they are sufficiently close together for frictional losses to be negligible. Thus P_1 will be approximately equal to the static pressure at both sections and the equations give the relation between the static and impact pressure—and the velocity—at any point in the fluid.

6.3. MEASUREMENT OF FLUID FLOW

The most important class of flowmeter is that in which the fluid is either accelerated or retarded at the measuring section by reducing the flow area, and the change in the kinetic energy is measured by recording the pressure difference produced.

This class includes:

The pitot tube, in which a small element of fluid is brought to rest at an orifice situated at right angles to the direction of flow. The flowrate is then obtained from the difference between the impact and the static pressure. With this instrument the velocity measured is that of a small filament of fluid.

The orifice meter, in which the fluid is accelerated at a sudden constriction (the orifice) and the pressure developed is then measured. This is a relatively cheap and reliable instrument though the overall pressure drop is high because most of the kinetic energy of the fluid at the orifice is wasted.

The venturi meter, in which the fluid is gradually accelerated to a throat and gradually retarded as the flow channel is expanded to the pipe size. A high proportion of the kinetic energy is thus recovered but the instrument is expensive and bulky.

The nozzle, in which the fluid is gradually accelerated up to the throat of the instrument but expansion to pipe diameter is sudden as with an orifice. This instrument is again expensive because of the accuracy required over the inlet section.

The notch or weir, in which the fluid flows over the weir so that its kinetic energy is measured by determining the head of the fluid flowing above the weir. This instrument is used in open-channel flow and extensively in tray towers[3] where the height of the weir is adjusted to provide the necessary liquid depth for a given flow.

Each of these devices will now be considered in more detail together with some less common and special purpose meters.

6.3.1. The pitot tube

The pitot tube is used to measure the difference between the impact and static pressures in a fluid. It normally consists of two concentric tubes arranged parallel to the direction of flow; the impact pressure is measured on the open end of the inner tube. The end of the outer concentric tube is sealed and a series of orifices on the curved surface give an accurate indication of the static pressure. The position of these orifices must be carefully chosen because there are two disturbances which may cause an incorrect reading of the static pressure. These are due to:

(1) the head of the instrument;
(2) the portion of the stem which is at right angles to the direction of flow of the fluid.

These two disturbances cause errors in opposite directions, and the static pressure should therefore be measured at the point where the effects are equal and opposite.

If the head and stem are situated at a distance of 14 diameters from each other as on the standard instrument,[4] the two disturbances are equal and opposite at a section 6 diameters from the head and 8 from the stem. This is, therefore, the position at which the static pressure orifices should be located. If the distance between the head and the stem is too great, the instrument will be unwieldy; if it is too short, the magnitude of each of the disturbances will be relatively great, and a small error in the location of the static pressure orifices will appreciably affect the reading.

The two standard instruments are shown in Figure 6.10; the one with the rounded nose is preferred, since this is less subject to damage.

Figure 6.10.   Pitot tubes

For Reynolds numbers of 500–300,000, based on the external diameter of the pitot tube, an error of not more than 1 per cent is obtained with this instrument. A Reynolds number of 500 with the standard 7.94 mm pitot tube corresponds to a water velocity of 0.070 m/s or an air velocity of 0.91 m/s. Sinusoidal fluctuations in the flowrate up to 20 per cent do not affect the accuracy by more than 1 per cent, and calibration of the instrument is not necessary.

A very small pressure difference is obtained for low rates of flow of gases, and the lower limit of velocity that can be measured is usually set by the minimum difference in pressure that can be measured. This limitation is serious, and various methods have been adopted for increasing the reading of the instrument although they involve the need for calibration. Correct alignment of the instrument with respect to the direction of flow is important; this is attained when the differential reading is a maximum.

For the flow not to be appreciably disturbed, the diameter of the instrument must not exceed about one-fiftieth of the diameter of the pipe; the standard instrument (diameter 7.94 mm) should therefore not be used in pipes of less than 0.4 m diameter. An accurate measurement of the impact pressure can be obtained using a tube of very small diameter with its open end at right angles to the direction of flow; hypodermic tubing is convenient for this purpose. The static pressure is measured using a single piezometer tube or a piezometer ring upstream at a distance equal approximately to the diameter of the pipe: measurement should be made at least 50 diameters from any bend or obstruction.

The pitot tube measures the velocity of only a filament of fluid, and hence it can be used for exploring the velocity distribution across the pipe section. If, however, it is desired to measure the total flow of fluid through the pipe, the velocity must be measured at various distances from the walls and the results integrated. The total flowrate can be calculated from a single reading only if the velocity distribution across the section is already known.

6.3.2. Measurement by flow through a constriction

In measuring devices where the fluid is accelerated by causing it to flow through a constriction, the kinetic energy is thereby increased and the pressure energy therefore decreases. The flowrate is obtained by measuring the pressure difference between the inlet of the meter and a point of reduced pressure, as shown in Figure 6.11 where the

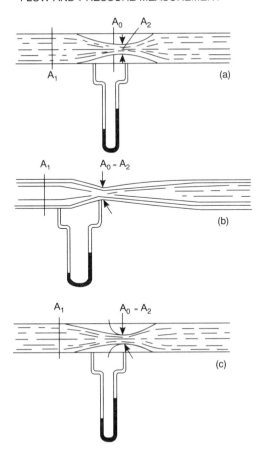

Figure 6.11. (a) Orifice meter (b) Venturi meter (c) Nozzle

orifice meter, the nozzle and the venturi meter are illustrated. If the pressure is measured a short distance upstream where the flow is undisturbed (section 1) and at the position where the area of flow is a minimum (section 2), application of the energy and material balance equations gives:

$$\frac{u_2^2}{2\alpha_2} - \frac{u_1^2}{2\alpha_1} + g(z_2 - z_1) + \int_{P_1}^{P_2} v \, dP + W_s + F = 0 \quad \text{(from equation 2.55)}$$

and the mass flow $\quad G = \dfrac{u_1 A_1}{v_1} = \dfrac{u_2 A_2}{v_2} \quad$ (from equation 2.37)

If the frictional losses are neglected, and the fluid does no work on the surroundings, i.e. putting W_s and F equal to zero:

$$\frac{u_2^2}{2\alpha_2} - \frac{u_1^2}{2\alpha_1} = g(z_1 - z_2) - \int_{P_1}^{P_2} v \, dP \tag{6.13}$$

Inserting the value of u_1 in terms of u_2 in equation 6.13 enables u_2 and G to be obtained.

For an incompressible fluid:

$$\int_{P_1}^{P_2} v\,dP = v(P_2 - P_1) \qquad\qquad \text{(equation 2.65)}$$

and:

$$u_1 = u_2 \frac{A_2}{A_1}$$

Substituting these values in equation 6.13:

$$\frac{u_2^2}{2\alpha_2}\left(1 - \frac{\alpha_2 A_2^2}{\alpha_1 A_1^2}\right) = g(z_1 - z_2) + v(P_1 - P_2)$$

Thus:

$$u_2^2 = \frac{2\alpha_2[g(z_1 - z_2) + v(P_1 - P_2)]}{1 - \frac{\alpha_2}{\alpha_1}\left(\frac{A_2}{A_1}\right)^2} \qquad\qquad (6.14)$$

For a horizontal meter $z = 0$:

$$\therefore \qquad u_2 = \sqrt{\frac{2\alpha_2 v(P_1 - P_2)}{1 - \frac{\alpha_2}{\alpha_1}\left(\frac{A_2}{A_1}\right)^2}}$$

and:

$$G = \frac{u_2 A_2}{v_2} = \frac{A_2}{v}\sqrt{\frac{2\alpha_2 v(P_1 - P_2)}{1 - \frac{\alpha_2}{\alpha_1}\left(\frac{A_2}{A_1}\right)^2}} \qquad\qquad (6.15)$$

For an ideal gas in isothermal flow:

$$\int_{P_1}^{P_2} v\,dP = P_1 v_1 \ln \frac{P_2}{P_1}. \qquad\qquad \text{(equation 2.69)}$$

and:

$$u_1 = u_2 \frac{A_2 v_1}{A_1 v_2}$$

And again neglecting terms in z, from equation 6.13:

$$\frac{u_2^2}{2\alpha_2}\left[1 - \frac{\alpha_2}{\alpha_1}\left(\frac{v_1 A_2}{v_2 A_1}\right)^2\right] = P_1 v_1 \ln \frac{P_1}{P_2}$$

i.e.:

$$u_2^2 = \frac{2\alpha_2 P_1 v_1 \ln \frac{P_1}{P_2}}{1 - \frac{\alpha_2}{\alpha_1}\left(\frac{v_1 A_2}{v_2 A_1}\right)^2} \qquad\qquad (6.16)$$

and the mass flow G is again $u_2 A_2 / v_2$.

For an ideal gas in non-isothermal flow. If the pressure and volume are related by $Pv^k = $ constant, then a similar analysis gives:

$$\int_{P_1}^{P_2} v\,dP = P_1 v_1 \frac{k}{k-1}\left[\left(\frac{P_2}{P_1}\right)^{(k-1)/k} - 1\right] \qquad\qquad \text{(equation 2.73)}$$

and, hence:
$$\frac{u_2^2}{2\alpha_2}\left[1 - \frac{\alpha_2}{\alpha_1}\left(\frac{v_1 A_2}{v_2 A_1}\right)^2\right] = -P_1 v_1 \frac{k}{k-1}\left[\left(\frac{P_2}{P_1}\right)^{(k-1)/k} - 1\right]$$

i.e.:
$$u_2^2 = \frac{2\alpha_2 P_1 v_1 \dfrac{k}{k-1}\left[1 - \left(\dfrac{P_2}{P_1}\right)^{(k-1)/k}\right]}{1 - \dfrac{\alpha_2}{\alpha_1}\left(\dfrac{v_1 A_2}{v_2 A_1}\right)^2} \qquad (6.17)$$

and the mass flow G is again $u_2 A_2 / v_2$.

It should be noted that equations 6.16 and 6.17 apply provided that P_2/P_1 is greater than the critical pressure ratio w_c. This subject was discussed in Chapter 4, where it was shown that when $P_2/P_1 < w_c$, the flowrate becomes independent of the downstream pressure P_2 and conditions of *maximum flow* occur.

6.3.3. The orifice meter (Figure 6.12)

The most important factors influencing the reading of an orifice meter are the size of the orifice and the diameter of the pipe in which it is fitted, though a number of other factors do affect the reading to some extent. Thus the exact position and the method of fixing the pressure tappings are important because the area of flow, and hence the velocity, gradually changes in the region of the orifice. The meter should be located not less than 50 pipe diameters from any pipe fittings. Details of the exact shape of the orifice, the orifice thickness, and other details are given in BS1042[4] and the details must be followed if a standard orifice is to be used without calibration — otherwise the meter must be calibrated. It should be noted that the standard applies only for pipes of at least 150 mm diameter.

A simple instrument can be made by inserting a drilled orifice plate between two pipe flanges and arranging suitable pressure connections. The orifice must be drilled with sharp edges and is best made from stainless steel which resists corrosion and abrasion. The size of the orifice should be chosen to give a convenient pressure drop. Although the flowrate is proportional to the square root of the pressure drop, it is difficult to cover a wide range in flow with any given size of orifice. Unlike the pitot tube, the orifice meter gives the average flowrate from a single reading.

The most serious disadvantage of the meter is that most of the pressure drop is not recoverable, that is it is inefficient. The velocity of the fluid is increased at the throat without much loss of energy. The fluid is subsequently retarded as it mixes with the relatively slow-moving fluid downstream from the orifice. A high degree of turbulence is set up and most of the excess kinetic energy is dissipated as heat. Usually only about 5 or 10 per cent of the excess kinetic energy can be recovered as pressure energy. The pressure drop over the orifice meter is therefore high, and this may preclude it from being used in a particular instance.

The area of flow decreases from A_1 at section 1 to A_0 at the orifice and then to A_2 at the *vena contracta* (Figure 6.11). The area at the vena contracta can be conveniently related to the area of the orifice by the coefficient of contraction C_c, defined by the relation:

$$C_c = \frac{A_2}{A_0}$$

Figure 6.12. (a) General arrangement (b) Orifice plate with corner tappings. Upper half shows construction with piezometer ring. Lower half shows construction with tappings into pipe flange (c) Orifice plate with d and $d/2$ tappings. Nipples must finish flush with wall of pipe without burrs

Inserting the value $A_2 = C_cA_0$ in equation 6.15, then for an *incompressible fluid* in a horizontal meter:

$$G = \frac{C_cA_0}{v} \sqrt{\frac{2\alpha_2 v(P_1 - P_2)}{1 - \frac{\alpha_2}{\alpha_1}\left(C_c\frac{A_0}{A_1}\right)^2}} \qquad (6.18)$$

Using a coefficient of discharge C_D to take account of the frictional losses in the meter and of the parameters C_c, α_1, and α_2:

$$G = \frac{C_D A_0}{v} \sqrt{\frac{2v(P_1 - P_2)}{1 - \left(\dfrac{A_0}{A_1}\right)^2}} \tag{6.19}$$

For a meter in which the area of the orifice is small compared with that of the pipe:

$$\sqrt{1 - \left(\frac{A_0}{A_1}\right)^2} \rightarrow 1$$

and:

$$G = \frac{C_D A_0}{v} \sqrt{2v(P_1 - P_2)}$$

$$= C_D A_0 \sqrt{2\rho(P_1 - P_2)} \tag{6.20}$$

$$= C_D A_0 \rho \sqrt{2gh_0} \tag{6.21}$$

where h_0 is the difference in head across the orifice expressed in terms of the fluid in question.

This gives a simple working equation for evaluating G though the coefficient C_D is not a simple function and depends on the values of the Reynolds number in the orifice and the form of the pressure tappings. A value of 0.61 may be taken for the standard meter for Reynolds numbers in excess of 10^4, though the value changes noticeably at lower values of Reynolds number as shown in Figure 6.13.

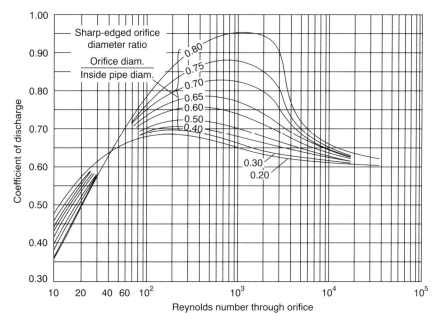

Figure 6.13.   Coefficient for orifice meter

For the isothermal flow of an ideal gas, from equation 6.16 and using C_D as above:

$$G = \frac{C_D A_0}{v_2} \sqrt{\frac{2 P_1 v_1 \ln\left(\frac{P_1}{P_2}\right)}{1 - \left(\frac{v_1 A_0}{v_2 A_1}\right)^2}} \tag{6.22}$$

For a meter in which the area of the orifice is small compared with that of the pipe:

$$G = \frac{C_D A_0}{v_2} \sqrt{2 P_1 v_1 \ln\left(\frac{P_1}{P_2}\right)} \tag{6.23}$$

$$= C_D A_0 \sqrt{2\frac{P_2}{v_2} \ln \frac{P_1}{P_2}}$$

$$= C_D A_0 \sqrt{2\frac{P_2}{v_2} \ln\left(1 - \frac{\Delta P}{P_2}\right)} \quad (\text{where}\, \Delta P = P_2 - P_1)$$

$$= C_D A_0 \sqrt{2\left(\frac{-\Delta P}{v_2}\right)} \quad \text{if } \Delta P \text{ is small compared with } P_2 \tag{6.24}$$

For non-isothermal flow of an ideal gas, from equation 6.17:

$$G = \frac{C_D A_0}{v_2} \sqrt{\frac{2 P_1 v_1 \left[\dfrac{k}{k-1}\right]\left[1 - \left(\dfrac{P_2}{P_1}\right)^{(k-1)/k}\right]}{1 - \left(\dfrac{v_1 A_0}{v_2 A_1}\right)^2}} \tag{6.25}$$

For a horizontal orifice in which (A_0/A_1) is small:

$$G = \frac{C_D A_0}{v_2} \sqrt{2 P_1 v_1 \frac{k}{k-1}\left[1 - \left(\frac{P_2}{P_1}\right)^{(k-1)/k}\right]} \tag{6.26}$$

$$= \frac{C_D A_0}{v_2} \sqrt{2 P_1 v_1 \frac{k}{k-1}\left[1 - \left(1 + \frac{\Delta P}{P_1}\right)^{(k-1)/k}\right]}$$

$$= C_D A_0 \sqrt{2\left(\frac{-\Delta P}{v_2}\right)} \quad \text{if } \Delta P \text{ is small compared with } P_2 \tag{6.27}$$

Equations 6.22, 6.23, 6.25 and 6.26 hold provided that P_2/P_1 is greater than the critical pressure ratio w_c.

For $P_2/P_1 < w_c$, the flowrate through the orifice is maintained at the maximum rate. From equation 4.15, the flowrate is given by equation 6.28 and *not equation* 6.23 for isothermal conditions:

$$G_{max} = 0.607 C_D A_0 \sqrt{\frac{P_1}{v_1}} \tag{6.28}$$

Similarly, when the pressure-volume relation is $Pv^k = $ constant, equation 4.30 replaces equation 6.26:

$$G_{max} = C_D A_0 \sqrt{\frac{kP_1}{v_1} \left(\frac{2}{k+1}\right)^{(k+1)/(k-1)}} \tag{6.29}$$

For isentropic conditions $k = \gamma$ and:

$$G_{max} = C_D A_0 \sqrt{\frac{\gamma P_1}{v_1} \left(\frac{2}{\gamma+1}\right)^{(\gamma+1)/(\gamma-1)}} \tag{6.30}$$

For a diatomic gas, $\gamma = 1.4$ and:

$$G_{max} = 0.685 C_D A_0 \sqrt{\frac{P_1}{v_1}} \tag{6.31}$$

For the flow of steam, a highly non-ideal gas, it is necessary to apply a correction to the calculated flowrate, the magnitude of which depends on whether the steam is saturated, wet or superheated. Correction charts are given by LYLE[5] who also quotes a useful approximation[6] — that a steam meter registers 1 per cent low for every 2 per cent of liquid water in the steam, and 1 per cent high for every 8 per cent of superheat.

Example 6.1

Water flows through an orifice of 25 mm diameter situated in a 75 mm diameter pipe, at a rate of 300 cm³/s. What will be the difference in level on a water manometer connected across the meter? The viscosity of water is 1 mN s/m².

Solution

$$\text{Area of orifice} = \frac{\pi}{4} \times 25^2 = 491 \text{ mm}^2 \text{ or } 4.91 \times 10^{-4} \text{ m}^2$$

$$\text{Flow of water} = 300 \text{ cm}^3/\text{s or } 3.0 \times 10^{-4} \text{ m}^3/\text{s}$$

$$\therefore \text{ Velocity of water through the orifice} = \frac{3.0 \times 10^{-4}}{4.91 \times 10^{-4}} = 0.61 \text{ m/s}$$

$$Re \text{ at the orifice} = \frac{25 \times 10^{-3} \times 0.61 \times 1000}{1 \times 10^{-3}} = 15250$$

From Figure 6.13, the corresponding value of $C_D = 0.61$ (diameter ratio $= 0.33$):

$$\left[1 - \left(\frac{A_0}{A_1}\right)^2\right]^{0.5} = \left[1 - \left(\frac{25^2}{75^2}\right)^2\right]^{0.5} = 0.994 \approx 1$$

Equation 6.21 may therefore be applied:

$$G = 3.0 \times 10^{-4} \times 10^3 = 0.30 \text{ kg/s}.$$

$$\therefore \qquad 0.30 = 0.61 \times 4.91 \times 10^{-4} \times 10^3 \sqrt{(2 \times 9.81 \times h_0)}.$$

Hence: $\sqrt{h_0} = 0.226$

and: $h_0 = 0.051$ m of water

                          $= \underline{\underline{51 \text{ mm of water}}}$

6.3.4. The nozzle

This is similar to the orifice meter but has a converging tube in place of the orifice plate, as shown in Figures 6.11c and 6.14. The velocity of the fluid is gradually increased and the contours are so designed that almost frictionless flow takes place in the converging portion; the outlet corresponds to the vena contracta on the orifice meter. The nozzle has a constant high coefficient of discharge (*ca.* 0.99) over a wide range of conditions because the coefficient of contraction is unity, though because the simple nozzle is not fitted with a diverging cone, the head lost is very nearly the same as with an orifice. Although much more costly than the orifice meter, it is extensively used for metering steam. When the ratio of the pressure at the nozzle exit to the upstream pressure is less than the critical pressure ratio w_c, the flowrate is independent of the downstream pressure and can be calculated from the upstream pressure alone.

Figure 6.14.   (*a*) General arrangement (*b*) Standard nozzle (A_0/A_1) is less than 0.45. Left half shows construction for corner tappings. Right half shows construction for piezometer ring (*c*) Standard nozzle where A_0/A_1 is greater than 0.45

6.3.5. **The venturi meter** (Figures 6.11*b* and 6.15)

In this meter the fluid is accelerated by its passage through a converging cone of angle 15–20°. The pressure difference between the upstream end of the cone (section 1) and the throat (section 2) is measured and provides the signal for the rate of flow. The fluid is then retarded in a cone of smaller angle (5–7°) in which a large proportion of the kinetic energy is converted back to pressure energy. Because of the gradual reduction in the area of flow there is no vena contracta and the flow area is a minimum at the throat so that the coefficient of contraction is unity. The attraction of this meter lies in its high energy recovery so that it may be used where only a small pressure head is available, though its construction[4] is expensive. The flow relationship is given by a similar equation to that for the orifice.

Figure 6.15.   The venturi meter

For an incompressible fluid in horizontal meter:

$$G = \frac{C_D A_2}{v} \sqrt{\frac{2v(P_1 - P_2)}{1 - (A_2/A_1)^2}} \qquad \text{(equation 6.19)}$$

$$= C_D \rho \frac{A_1 A_2}{\sqrt{A_1^2 - A_2^2}} \sqrt{2v(P_1 - P_2)} \qquad (6.32)$$

$$= C_D \rho C' \sqrt{2gh_v} \qquad (6.33)$$

where C' is a constant for the meter and h_v is the loss in head over the converging cone expressed as height of fluid. The coefficient C_D is high, varying from 0.98 to 0.99. The meter is equally suitable for compressible and incompressible fluids.

Example 6.2

The rate of flow of water in a 150 mm diameter pipe is measured with a venturi meter with a 50 mm diameter throat. When the pressure drop over the converging section is 121 mm of water, the flowrate is 2.91 kg/s. What is the coefficient for the converging cone of the meter at this flowrate?

Solution

From equation 6.32, mass rate of flow:

$$G = \frac{C_D \rho A_1 A_2 \sqrt{(2gh_v)}}{\sqrt{(A_1^2 - A_2^2)}}$$

The coefficient for the meter is therefore given by:

$$2.91 = C_D \times 1000 \left(\frac{\pi}{4}\right)^2 (150 \times 10^{-3})^2 (50 \times 10^{-3})^2$$

$$\times \frac{\sqrt{(2 \times 9.81 \times 121 \times 10^{-3})}}{(\pi/4)\sqrt{[(150 \times 10^{-3})^4 - (50 \times 10^{-3})^4]}}$$

$\therefore$ $\qquad\qquad C_D = \underline{\underline{0.985}}$

6.3.6. Pressure recovery in orifice-type meters

The orifice meter has a low discharge coefficient of about 0.6 which means that there is a substantial loss in pressure over the instrument. This is seen in Figure 6.16 where the pressure is shown in relation to the position along the tube. There is a steep drop in pressure of about 10 per cent as the vena contracta is reached and the subsequent pressure recovery is poor, amounting to about 50 per cent of the loss. The *Dall Tube* (Figure 6.17) has been developed to provide a low head loss over the meter while still giving a reasonably high value of pressure head over the orifice and thus offering great advantages. A short length of parallel lead-in tube is followed by the converging upstream cone and then a diverging downstream cone. This recovery cone is formed by a liner which fits into the meter, the *throat* being formed by a circular slot located between the two cones. One pressure connection is taken to the throat through the annular chamber shown, and the second tapping is on the upstream side. The flow leaves the throat as a diverging jet which follows the walls of the downstream cone so that eddy losses are almost eliminated. These instruments are made for pipe sizes greater than 150 mm and the pressure loss is only 2–8 per cent of the differential head. They are cheaper than a venturi, much shorter in length, and correspondingly lighter.

It should be noted that all these restricted flow type meters are primarily intended for pipe sizes greater than 50 mm, and when used on smaller tubes they must be individually calibrated.

Figure 6.16 shows the pressure changes over the four instruments considered. If the upstream pressure is P_1, the throat pressure P_2, and the final recovery pressure P_3, then

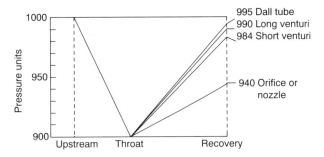

Figure 6.16. Pressure distribution using orifice plate, venturi meter, and Dall tube. Pressure falls by 10% from upper pressure tapping to throat in each case

the pressure recovery is conveniently expressed as $(P_1 - P_3)/(P_1 - P_2)$, and for these meters the values are given in Table 6.1.

Table 6.1. Typical meter pressure recovery

Meter type	Value of C_D	Value of $(P_1 - P_3)/(P_1 - P_2)$ (%)
Orifice	0.6	40–50
Nozzle	0.7	40–50
Venturi	0.9	80–90
Dall	0.9	92–98

Because of its simplicity the orifice meter is commonly used for process measurements, and this instrument is suitable for providing a signal of the pressure to some comparator as indicated in Figure 6.1.

6.3.7. Variable area meters — rotameters

In the meters so far described the area of the constriction or orifice is constant and the drop in pressure is dependent on the rate of flow. In the variable area meter, the drop in pressure is constant and the flowrate is a function of the area of the constriction.

Figure 6.17.   Dall tube

A typical meter of this kind, which is commonly known as a *rotameter* (Figure 6.18), consists of a tapered tube with the smallest diameter at the bottom. The tube contains a freely moving float which rests on a stop at the base of the tube. When the fluid is flowing the float rises until its weight is balanced by the upthrust of the fluid, its position then indicating the rate of flow. The pressure difference across the float is equal to its weight divided by its maximum cross-sectional area in a horizontal plane. The area for flow is the annulus formed between the float and the wall of the tube.

This meter may thus be considered as an orifice meter with a variable aperture, and the formulae already derived are therefore applicable with only minor changes. Both in the orifice-type meter and in the rotameter the pressure drop arises from the conversion of pressure energy to kinetic energy and from frictional losses which are accounted for in the coefficient of discharge. The pressure difference over the float $-\Delta P$, is given by:

$$-\Delta P = \frac{V_f(\rho_f - \rho)g}{A_f} \tag{6.34}$$

where V_f is the volume of the float, ρ_f the density of the material of the float, and A_1 is the maximum cross-sectional area of the float in a horizontal plane.

If the area of the annulus between the float and tube is A_2 and the cross-sectional area of the tube is A_1, then from equation 6.19:

$$G = C_D A_2 \sqrt{\frac{2\rho(-\Delta P)}{1 - (A_2/A_1)^2}} \tag{6.35}$$

Figure 6.18.   Variable area flowmeters

Substituting for $-\Delta P$ from equation 6.34:

$$G = C_D A_2 \sqrt{\frac{2gV_f(\rho_f - \rho)\rho}{A_f[1 - (A_2/A_1)^2]}} \qquad (6.36)$$

The coefficient C_D depends on the shape of the float and the Reynolds number (based on the velocity in the annulus and the mean hydraulic diameter of the annulus) for the flow through the annular space of area A_2. In general, floats which give the most nearly constant coefficient are of such a shape that they set up eddy currents and give low values of C_D. The variation in C_D largely arises from differences in viscous drag of fluid on the float, and if turbulence is artificially increased, the drag force rises quickly to a limiting but high value. As seen in Figure 6.19, float A does not promote turbulence and the coefficient rises slowly to a high value of 0.98. Float C promotes turbulence and C_D rises quickly but only to a low value of 0.60.

The constant coefficient for float C arises from turbulence promotion, and for this reason the coefficient is also substantially independent of the fluid viscosity. The meter

Figure 6.19.   Coefficients for rotameters

can be made relatively insensitive to changes in the density of the fluid by selection of the density of the float, ρ_f. Thus the flowrate for a given meter will be independent of ρ when $dG/d\rho = 0$.

From equation 6.36:

$$\frac{dG}{d\rho} = \frac{C_D A_2}{2} \sqrt{\frac{2gV_f}{A_f[1 - (A_2/A_1)^2]}} \left\{ (\rho_f - \rho)^{1/2} \tfrac{1}{2} \rho^{-1/2} - \rho^{1/2} \tfrac{1}{2} (\rho_f - \rho)^{-1/2} \right\} \quad (6.37)$$

When:
$$\frac{dG}{d\rho} = 0,$$

$$\rho_f = 2\rho \qquad\qquad\qquad (6.38)$$

Thus if the density of the float is twice that of the fluid, then the position of the float for a given flow is independent of the fluid density.

The range of a meter can be increased by the use of floats of different densities, a given float covering a flowrate range of about 10:1. For high pressure work the glass tube is replaced by a metal tube. When a metal tube is used or when the liquid is very dark or dirty an external indicator is required.

Example 6.3

A rotameter tube is 0.3 m long with an internal diameter of 25 mm at the top and 20 mm at the bottom. The diameter of the float is 20 mm, its density is 4800 kg/m³ and its volume is 6.0 cm³. If the coefficient of discharge is 0.7, what is the flowrate of water (density 1000 kg/m³) when the float is halfway up the tube?

Solution

From equation 6.36:

$$G = C_D A_2 \sqrt{\frac{2gV_f(\rho_f - \rho)\rho}{A_f[1 - (A_2/A_1)^2]}}$$

Cross-sectional area at top of tube $= \dfrac{\pi}{4} 25^2 = 491$ mm^2 or 4.91×10^{-4} m^2

Cross-sectional area at bottom of tube $= \dfrac{\pi}{4} 20^2 = 314$ mm^2 or 3.14×10^{-4} m^2

Area of float: $\qquad A_f = 3.14 \times 10^{-4}$ m^2

Volume of float: $\qquad V_f = 6.0 \times 10^{-6}$ m^3

When the float is halfway up the tube, the area at the height of the float A_1 is given by:

$$A_1 = \frac{\pi}{4} 22.5^2 \times 10^{-4}$$

or:
$$A_1 = 3.98 \times 10^{-4} \text{ m}^2$$

The area of the annulus A_2 is given by:

$$A_2 = A_1 - A_f = 0.84 \times 10^{-4} \text{ m}^2$$

$$A_2/A_1 = 0.211 \quad \text{and} \quad [1 - (A_2/A_1)^2] = 0.955$$

Substituting into equation 6.36:

$$G = 0.70 \times 0.84 \times 10^{-4} \sqrt{\left\{ \frac{[2 \times 9.81 \times 6.0 \times 10^{-6}(4800 - 1000)1000]}{(3.14 \times 10^{-4} \times 0.955)} \right\}}$$

$$= 0.072 \text{ kg/s}$$

6.3.8. The notch or weir

The flow of a liquid presenting a free surface can be measured by means of a weir. The pressure energy of the liquid is converted into kinetic energy as it flows over the weir which may or may not cover the full width of the stream, and a calming screen may be fitted before the weir. Then the height of the weir crest gives a measure of the rate of flow. The velocity with which the liquid leaves depends on its initial depth below the surface. For unit mass of liquid, initially at a depth h below the free surface, discharging through a notch, the energy balance (equation 2.67) gives:

$$\Delta \frac{u^2}{2} + v\Delta P = 0 \tag{6.39}$$

for a frictionless flow under turbulent conditions. If the velocity upstream from the notch is small and the flow upstream is assumed to be unaffected by the change of area at the notch:

$$\frac{u_2^2}{2} = -v\Delta P = gh$$

i.e.:
$$u_2 = \sqrt{2gh} \tag{6.40}$$

Rectangular notch

For a rectangular notch (Figure 6.20) the rate of discharge of fluid at a depth h through an element of cross-section of depth dh will be given by:

$$dQ = C_D B \, dh \sqrt{2gh} \tag{6.41}$$

where C_D is the coefficient of discharge (usually about 0.6) and B is the width of the notch.

Figure 6.20.   Rectangular notch

Thus the total rate of flow:

$$Q = C_D B \sqrt{2g} (\tfrac{2}{3}) D^{1.5} \qquad (6.42)$$

where D is the total depth of liquid above the bottom of the notch.

The empirical Francis formula:

$$Q = 1.84(B - 0.1nD)D^{1.5} \qquad (6.43)$$

gives the flowrate in m^3/s if the dimensions of the notch are expressed in metres with $C_D = 0.62$; $n = 0$ if the notch is the full width of the channel; $n = 1$ if the notch is narrower than the channel but is arranged with one edge coincident with the edge of the channel; $n = 2$ if the notch is narrower than the channel and is situated symmetrically (see Figure 6.20b): n is known as the number of end contractions.

Example 6.4

Water flows in an open channel across a weir which occupies the full width of the channel. The length of the weir is 0.5 m and the height of water over the weir is 100 mm. What is the volumetric flowrate of water?

Solution

Use is made of the Francis formula (equation 6.43):

$$Q = 1.84(L - 0.1nD)D^{1.5} \text{ (m}^3/\text{s)} \qquad (6.43)$$

where L is the length of the weir (m), n the number of end contractions (in this case $n = 0$), and D the height of liquid above the weir (m).

Thus: $$Q = 1.84(0.5)0.100^{1.5}$$

$$= 0.030 \text{ m}^3/\text{s}$$

Example 6.5

An organic liquid flows across a distillation tray and over a weir at the rate of 15 kg/s. The weir is 2 m long and the liquid density is 650 kg/m^3. What is the height of liquid flowing over the weir?

Solution

Use is made of the Francis formula (equation 6.43), where, as in the previous example, $n = 0$. In the context of this example the height of liquid flowing over the weir is usually designated h_{ow} and the volumetric liquid flow by Q. Rearrangement of equation 6.43 gives:

$$h_{ow} = 0.666 \left(\frac{Q}{L_w}\right)^{0.67} \quad \text{(m)}$$

where Q is the liquid flowrate (m³/s) and L_w the weir length (m).
In this case:

$$Q = \frac{15}{650} = 0.0230 \text{ m}^3/\text{s} \quad \text{and} \quad L_w = 2.0 \text{ m}$$

$$\therefore \quad h_{ow} = 0.666 \left(\frac{0.0230}{2}\right)^{0.67} = 0.033 \text{ m}$$

$$= \underline{\underline{33.0 \text{ mm}}}$$

Triangular notch

For a triangular notch of angle 2θ, the flow dQ through the thin element of cross-section (Figure 6.21) is given by:

$$dQ = C_D(D - h)2 \tan\theta \, dh \sqrt{2gh} \tag{6.44}$$

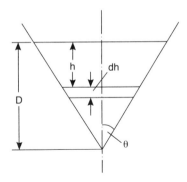

Figure 6.21.   Triangular notch

The total rate of flow:

$$Q = 2C_D \tan\theta \sqrt{2g} \int_0^D (Dh^{1/2} - h^{3/2})dh$$

$$= 2C_D \tan\theta \sqrt{2g}(\tfrac{2}{3}D^{5/2} - \tfrac{2}{5}D^{5/2})$$

$$= \tfrac{8}{15}C_D \tan\theta \sqrt{2g}D^{5/2} \tag{6.45}$$

For a 90° notch for which $C_D = 0.6$, and using SI units:

$$Q = 1.42D^{2.5} \tag{6.46}$$

Thus for a rectangular notch the rate of discharge is proportional to the liquid depth raised to a power of 1.5, and for a triangular notch to a power of 2.5. A triangular notch will, therefore, handle a wider range of flowrates. In general, the index of D is a function of the contours of the walls of the notch, and any desired relation between flowrate and depth can be obtained by a suitable choice of contour. It is sometimes convenient to employ a notch in which the rate of discharge is directly proportional to the depth of liquid above the bottom of the notch. It can be shown that the notch must have curved walls giving a large width to the bottom of the notch and a comparatively small width towards the top.

While open channels are not used frequently for the flow of material other than cooling water between plant units, a weir is frequently installed for controlling the flow within the unit itself, for instance in a distillation column or reactor.

6.3.9. Other methods of measuring flowrates

The meters which have been described so far depend for their operation on the conversion of some of the kinetic energy of the fluid into pressure energy or vice versa. They form by far the largest class of flowmeters. Other meters are used for special purposes, and brief reference will now be made to a few of these.

Hot-wire anemometer

If a heated wire is immersed in a fluid, the rate of loss of heat will be a function of the flowrate. In the hot-wire anemometer a fine wire whose electrical resistance has a high temperature coefficient is heated electrically. Under equilibrium conditions the rate of loss of heat is then proportional to $I^2\Omega$, where Ω is the resistance of the wire and I is the current flowing.

Either the current or the resistance (and hence the temperature) of the wire is maintained constant. The following is an example of a method in which the resistance is maintained constant. The wire is incorporated as one of the resistances of a Wheatstone network (Figure 6.22) in which the other three resistances have low temperature coefficients. The circuit is balanced when the wire is immersed in the stationary fluid but, when the fluid is set in motion, the rate of loss of heat increases and the temperature of the wire falls. Its resistance therefore changes and the bridge is thrown out of balance. The balance can

Figure 6.22.   Typical circuit for a hot-wire anemometer

be restored by increasing the current so that the temperature and resistance of the wire are brought back to their original values; the other three resistances will be unaffected by the change in current because of their low temperature coefficients. The current flowing in the wire is then measured using either an ammeter or a voltmeter. The rate of loss of heat is found to be approximately proportional to $\sqrt{u\rho + b'}$, so that

$$a'\sqrt{u\rho + b'} = I^2\Omega \quad \text{under equilibrium conditions} \tag{6.47}$$

where u is the velocity of the fluid (m/s), ρ the density (kg/m^3), and a' and b' are constants for a given meter.

Since the resistance of the wire is maintained constant:

$$u\rho = \frac{I^4\Omega^2}{a'^2} - b' = a''I^4 - b' \tag{6.48}$$

where $a'' = \Omega^2/a'^2$ remains constant, i.e. the mass rate of flow per unit area is a function of the fourth power of the current, which can be accurately measured.

The hot-wire anemometer is very accurate even for very low rates of flow. It is one of the most convenient instruments for the measurement of the flow of gases at low velocities; accurate readings are obtained for velocities down to about 0.03 m/s. If the ammeter has a high natural frequency, pulsating flows can be measured. Platinum wire is commonly used.

The magnetic meter

The magnetic flowmeter is a volumetric metering device based on a discovery by Faraday that electrical currents are induced in a conductor that is moved through a magnetic field. Faraday's Law states that the voltage induced in a conductor is proportional to the width of the conductor, the intensity of the magnetic field, and the velocity of the conductor. Thus the output of the flowmeter, a millivoltage, is linearly related to the velocity of the conductor, which is the flowing process fluid. The principle of operation and meter itself is shown in Figure 6.23.

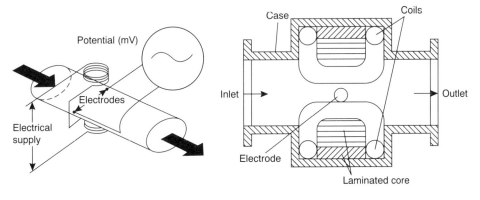

Figure 6.23.   The magnetic meter

The induced voltage is in a plane that is mutually perpendicular to both the motion of the conductor and the magnetic field. The process fluid passes through the magnetic field induced by electromagnetic coils or permanent magnets built around the short length of pipe called the metering section. Because of the magnetic field used for the induction, the metering section must be of nonmagnetic material. Further, if it is metal, it must have a lining that serves as an electrical insulator for the flowing liquid.

These meters have been very useful for handling liquid metal coolants where the conductivity is high though they have now been developed for liquids such as tap water which have a poor conductivity. Designs are now available for all sizes of tubes, and the instrument is particularly suitable for pharmaceutical and biological purposes where contamination of the fluid must be avoided.

This meter has a number of special features which should be noted. Thus it gives rise to a negligible drop in pressure for the fluid being metered; fluids containing a high percentage of solids can be metered and the flow can be in either direction. The liquid must have at least a small degree of electrical conductivity.

Quantity meters

The meters which have been described so far give an indication of the rate of flow of fluid; the total amount passing in a given time must be obtained by integration. Orifice meters are frequently fitted with integrating devices. A number of instruments is available, however, for measuring directly the total quantity of fluid which has passed. An average rate of flow can then be obtained by dividing the quantity by the time of passage.

Figure 6.24.   Vane anemometer

Gas meters

A simple quantity meter which is used for the measurement of the flow of gas in an accessible duct is the anemometer (Figure 6.24). A shaft carrying radial vanes or cups supported on low friction bearings is caused to rotate by the passage of the gas; the relative velocity between the gas stream and the surface of the vanes is low because the frictional resistance of the shaft is small. The number of revolutions of the spindle is counted automatically, using a gear train connected to a series of dials. The meter must be calibrated and should be checked at frequent intervals because the friction of the bearings will not necessarily remain constant. The anemometer is useful for gas velocities above about 0.15 m/s.

Quantity meters, suitable for the measurement of the flow of gas through a pipe, include the standard wet and dry meters. In the wet meter, the gas fills a rotating segment and an equal volume of gas is expelled from another segment (Figure 6.25). The dry gas meter employs a pair of bellows. Gas enters one of the bellows and automatically expels gas from the other; the number of cycles is counted and recorded on a series of dials. Both of these are positive displacement meters and therefore do not need frequent calibration. Gas meters usually appear very bulky for the quantities they are measuring. This is because the linear velocity of a gas in a pipe is normally very high compared with that of a liquid, and the large volume is needed so that the speed of the moving parts can be reduced and wear minimised.

Figure 6.25.   Wet gas meter

Liquid meters

In one form of meter shown in Figure 6.26, which is capable of handling flows between 5 cm³/s and 0.015 m³/s, the flow of the liquid results in the positive displacement of a rotating element, the cumulative flow being obtained by gearing to a counter.

The body, in the form of a cylindrical chamber, is fitted with a radial partition and a central hub. The circular piston has a slot gap in the circumference to fit over the partition and a peg on its upper face to control movement around the central hub. A rolling seal is thus formed between the piston and hub and between the piston and main chamber.

Figure 6.26.   Water meter

The chamber is thus split into four spaces as shown in Figure 6.27. The fluid enters the bottom of the chamber through one port and leaves by the other port, the piston forming a movable division between the inlet and outlet.

(1)

Spaces 1 and 3 are receiving fluid from the inlet port A, and spaces 2 and 4 are discharging through the outlet port B

(2)

The piston has advanced and space 1, in connection with the inlet port, has enlarged, and space 2, in connection with the outlet port, has decreased while spaces 3 and 4, which have combined, are about to move into position to discharge through the outlet port

(3)

Space 1 is still admitting fluid from the inlet port and space 3 is just opening up again to the inlet port, while spaces 2 and 4 are discharging through the outlet port

(4)

Fluid is being received into space 3 and discharged from space 4, while spaces 1 and 2 have combined and are about to begin discharging as the piston moves forward again to occupy position as shown in (1)

Figure 6.27. The oscillating piston meter

In a cycle of operation the liquid enters port *A* and fills the spaces 1 and 3, thus forcing the piston to oscillate counterclockwise opening spaces 2 and 4 to port *B*. Because of the partition, the piston moves downwards so that space 3 is cut off from port *A* and becomes space 4. Further movement allows the exit port to be uncovered, and the measured volume between hub and piston is then discharged. The outer space 1 increases until the piston moves upwards over the partition and space 1 becomes space 2 when a second metered volume is discharged by the filling of the inner space 3.

Turbine flow meters are composed of some form of rotary device such as a helical rotor, Pelton wheel or a vane mounted in the flow stream. The fluid passing the rotor causes the rotor to turn at an angular velocity which is proportional to the flow velocity and hence the volumetric flowrate through the meter. The rotary motion of the rotor is sensed by some form of pick-up device that produces an electrical pulse output. The frequency of this signal is proportional to the flowrate and the total count of pulses is proportional to the total volume of liquid passed through the meter.

Turbine flow meters range in size from 5 to 600 mm in diameter, are suitable for temperatures of between 20 and 750 K at pressures of up to 300 bar. A normal range of flows falls between 20 cm^3/s and 2 m^3/s and a diagram showing a section of an axial turbine flow meter is shown in Figure 6.28.

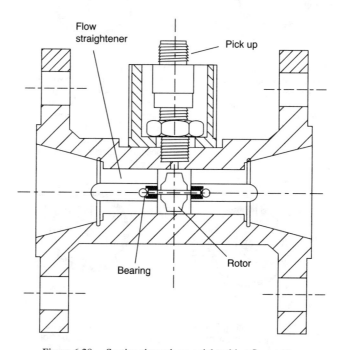

Figure 6.28.   Section through an axial turbine flow meter

Further details of these and other measuring equipment are given in OWER and PANKHURST[2] and LINFORD[7] in the *Process Industries and Controls Handbook*[8] and in the *Instrument Manual*[9].

6.4. FURTHER READING

CHEREMISINOFF, N. P. and CHEREMISINOFF, P. N.: *Flow Measurement for Engineers and Scientists* (Marcel Dekker, New York, 1988).
MILLER, J. T.: *The Revised Course on Industrial Instrument Technology* (United Trade Press, London, 1973).
MILLER, J. T. (ed.): *Instrument Manual*, 4th edn (United Trade Press, London, 1971).
MILLER, R. N.: *Flow Measurement Engineering Handbook* (McGraw-Hill, New York, 1983).
MOORE, R. L.: *Basic Instrumentation Lecture Notes and Study Guide*, Vol. 1 — Measurement Fundamentals, Vol. 2 — Process Analyzers and Recorders. Instrument Society of America (Prentice Hall, Englewood Cliffs, 1982).
SCOTT, R. W. W. (ed.): *Developments in Flow Measurement* (Applied Science, London, 1982).

SYDENHAM, P. H.: *Transducers in Measurement and Control* (Adam Hilger, London, 1984).
WIGHTMAN, E. J.: *Instrumentation in Process Control* (Butterworths, London, 1972).

6.5. REFERENCES

1. WIGHTMAN, E. J.: *Instrumentation in Process Control* (Butterworths, London 1972).
2. OWER, E. and PANKHURST, R. C.: *Measurement of Air Flow*, 4th edn (Pergamon, Oxford 1966).
3. SMITH, B. D.: *Design of Equilibrium Stage Processes* (McGraw-Hill, New York, 1963).
4. BS 1042: 1943: British Standard 1042 (British Standards Institution, London): Code for flow measurement.
5. LYLE, O.: *The Efficient Use of Fuel*, 2nd edn, 2nd impression (HMSO, London, 1964). 237–241.
6. LYLE, O.: *The Efficient Use of Steam* (HMSO, London, 1963) §231.
7. LINFORD, A.: *Flow Measurement and Meters* (Spon, London, 1949).
8. CONSIDINE, D. M.: *Process Industries and Controls Handbook* (McGraw-Hill, New York, 1957).
9. MILLER, J. T. (ed.): *Instrument Manual*, 4th edn (United Trade Press, London, 1971).

6.6. NOMENCLATURE

		Units in SI system	Dimensions in M, N, L, T, θ, I
A	Area perpendicular to direction of flow	m^2	L^2
A_c	Cross-sectional area of column	m^2	L^2
A_f	Area of rotameter float	m^2	L^2
A_o	Area of orifice	m^2	L^2
A_w	Cross-sectional area of well	m^2	L^2
B	Width of rectangular channel or notch	m	L
C_c	Coefficient of contraction	—	—
C_D	Coefficient of discharge	—	—
C'	Constant for venturi meter	m^2	L^2
D	Depth of liquid or above bottom of notch	m	L
d	Diameter of pipe or pitot tube	m	L
d_o	Diameter of orifice	m	L
e	Error		
F	Energy dissipated per unit mass of fluid	J/kg	L^2T^{-2}
G	Mass rate of flow	kg/s	MT^{-1}
g	Acceleration due to gravity	m/s^2	LT^{-2}
h_i	Difference between impact and static heads on pitot tube	m	L
h_m	Reading on manometer	m	L
h_0	Fall in head over orifice meter	m	L
h_v	Fall in head over converging cone of venturi meter	m	L
h_w	Decrease in level in well	m	L
I	Electric current	A	I
k	Numerical constant used as index for compression	—	—
L	Length of weir or of inclined tube of manometer	m	L
M	Molecular weight	$kg/kmol$	MN^{-1}
n	Number of end contractions	—	—
P	Pressure	N/m^2	$ML^{-1}T^{-2}$
ΔP	Pressure difference	N/m^2	$ML^{-1}T^{-2}$
Q	Volumetric rate of flow	m^3/s	L^3T^{-1}
$\mathbf{R}$	Universal gas constant	$(8314)\ J/kmol\ K$	$MN^{-1}L^2T^{-2}\theta^{-1}$
T	Absolute temperature	K	θ
u	Mean velocity	m/s	LT^{-1}
V_f	Volume of rotameter float	m^3	L^3
v	Volume per unit mass of fluid	m^3/kg	$M^{-1}L^3$
W_s	Shaft work per unit mass	J/kg	L^2T^{-2}
w	Pressure ratio P_2/P_1	—	—
w_c	Critical pressure ratio	—	—

		Units in SI system	Dimensions in M, N, L, T, θ, I
z	Distance in vertical direction	m	L
z_m	Vertical distance between level of manometer liquid and axis of venturi meter	m	L
α	Constant in expression for kinetic energy of fluid	—	—
γ	Ratio of specific heats at constant pressure and volume	—	—
ρ	Density of fluid	kg/m^3	ML^{-3}
ρ_m	Density of manometer fluid	kg/m^3	ML^{-3}
ρ_f	Density of rotameter float	kg/m^3	ML^{-3}
θ	Half angle of triangular notch or angle made by inclined manometer tube with horizontal	—	—
Ω	Electrical resistance	ohm	ML^{2}T^{-3}I^{-2}

CHAPTER 7

Liquid Mixing

7.1. INTRODUCTION—TYPES OF MIXING

Mixing is one of the most common operations carried out in the chemical, processing and allied industries. The term "mixing" is applied to the processes used to reduce the degree of non-uniformity, or gradient of a property in a system such as concentration, viscosity, temperature and so on. Mixing is achieved by moving material from one region to another. It may be of interest simply as a means of achieving a desired degree of homogeneity but it may also be used to promote heat and mass transfer, often where a system is undergoing a chemical reaction.

At the outset it is useful to consider some common examples of problems encountered in industrial mixing operations, since this will not only reveal the ubiquitous nature of the process, but will also provide an appreciation of some of the associated difficulties. Several attempts have been made to classify mixing problems and, for example, REAVELL[1] used as a criterion for mixing of powders, the flowability of the final product. HARNBY et al.[2] base their classification on the phases present, that is liquid-liquid, liquid-solid and so on. This is probably the most useful description of mixing as it allows the adoption of a unified approach to the problems encountered in a range of industries. This approach is now followed here.

7.1.1. Single-phase liquid mixing

In many instances, two or more miscible liquids must be mixed to give a product of a desired specification, such as, for example, in the blending of petroleum products of different viscosities. This is the simplest type of mixing as it involves neither heat nor mass transfer, nor indeed a chemical reaction. Even such simple operations can however pose problems when the two liquids have vastly different viscosities. Another example is the use of mechanical agitation to enhance the rates of heat and mass transfer between the wall of a vessel, or a coil, and the liquid. Additional complications arise in the case of highly viscous Newtonian and non-Newtonian liquids.

7.1.2. Mixing of immiscible liquids

When two immiscible liquids are stirred together, one phase becomes dispersed as tiny droplets in the second liquid which forms a continuous phase. Liquid-liquid extraction, a process using successive mixing and settling stages (see Volume 2, Chapter 13) is one important example of this type of mixing. The liquids are brought into contact with a solvent that will selectively dissolve one of the components present in the mixture. Vigorous agitation causes one phase to disperse in the other and, if the droplet size is small, a high interfacial area is created for interphase mass transfer. When the agitation is stopped, phase separation takes place, but care must be taken to ensure that the droplets are not so small that a diffuse layer appears in the region of the interface; this can remain in a semi-stable state over a long period of time and prevent effective separation from occurring. Another important example of dispersion of two immiscible liquids is the production of stable emulsions, such as those encountered in food, brewing and pharmaceutical applications. Because the droplets are very small, the resulting emulsion is usually stable over considerable lengths of time.

7.1.3. Gas–liquid mixing

Numerous processing operations involving chemical reactions, such as aerobic fermentation, wastewater treatment, oxidation of hydrocarbons, and so on, require good contacting between a gas and a liquid. The purpose of mixing here is to produce a high interfacial area by dispersing the gas phase in the form of bubbles into the liquid. Generally, gas-liquid mixtures or dispersions are unstable and separate rapidly if agitation is stopped, provided that a foam is not formed. In some cases a stable foam is needed, and this can be formed by injecting gas into a liquid which is rapidly agitated, often in the presence of a surface-active agent.

7.1.4. Liquid–solid mixing

Mechanical agitation may be used to suspend particles in a liquid in order to promote mass transfer or a chemical reaction. The liquids involved in such applications are usually of low viscosity, and the particles will settle out when agitation ceases.

At the other extreme, in the formation of composite materials, especially filled polymers, fine particles must be dispersed into a highly viscous Newtonian or non-Newtonian liquid. The incorporation of carbon black powder into rubber is one such operation. Because of the large surface areas involved, surface phenomena play an important role in such applications.

7.1.5. Gas–liquid–solid mixing

In some applications such as catalytic hydrogenation of vegetable oils, slurry reactors, froth flotation, evaporative crystallisation, and so on, the success and efficiency of the process is directly influenced by the extent of mixing between the three phases. Despite its great industrial importance, this topic has received only scant attention.

7.1.6. Solid–solid mixing

Mixing together of particulate solids, sometimes referred to as blending, is a very complex process in that it is very dependent, not only on the character of the particles — density, size, size distribution, shape and surface properties — but also on the differences of these properties in the components. Mixing of sand, cement and aggregate to form concrete and of the ingredients in gunpowder preparation are longstanding examples of the mixing of solids.

Other industrial sectors employing solids mixing include food, drugs, and the glass industries, for example. All these applications involve only physical contacting, although in recent years, there has been a recognition of the industrial importance of solid-solid reactions, and solid-solid heat exchangers[3]. Unlike liquid mixing, research on solids mixing has not only been limited but also has been carried out only relatively recently. The problems involved in the blending of solids are discussed in Volume 2, Chapter 1.

7.1.7. Miscellaneous mixing applications

Mixing equipment may be designed not only to achieve a predetermined level of homogeneity, but also to improve heat transfer. For example, the rotational speed of an impeller in a mixing vessel is selected so as to achieve a required rate of heat transfer, and the agitation may then be more than sufficient for the mixing duty. Excessive or overmixing should be avoided as it is not only wasteful of energy but may be detrimental to product quality. For example, in biological applications, excessively high impeller speeds or energy input may give rise to shear rates which damage the micro-organisms present. In a similar way, where the desirable rheological properties of some polymer solutions may be attributable to structured long-chain molecules, excessive impeller speeds or agitation over prolonged periods, may damage the structure of the polymer molecules thereby altering their properties. It is therefore important to appreciate that overmixing may often be undesirable because it may result in both excessive energy consumption and impaired product quality. From the examples given of its application, it is abundantly clear that mixing cuts across the boundaries between industries, and indeed it may be required to mix virtually anything with anything else — be it a gas or a liquid or a solid; it is clearly not possible to consider the whole range of mixing problems here. Instead attention will be given primarily to *batch liquid mixing*, and reference will be made to the literature, where appropriate, in this field. This choice has been made largely because it is liquid mixing which is most commonly encountered in the processing industries. In addition, an extensive literature, mainly dealing with experimental studies, now exists which provides the basis for the design and selection of mixing equipment. It also affords some insight into the nature of the mixing process itself.

In mixing, there are two types of problems to be considered — how to design and select mixing equipment for a given duty, and how to assess whether a mixer is suitable for a particular application. In both cases, the following aspects of the mixing process must be understood:

  (i) Mechanisms of mixing.
 (ii) Scale-up or similarity criteria.
(iii) Power consumption.
 (iv) Flow patterns.
  (v) Rate of mixing and mixing time.
 (vi) The range of mixing equipment available and its selection.

Each of these factors is now considered.

7.2. MIXING MECHANISMS

If mixing is to be carried out in order to produce a uniform mixture, it is necessary to understand how liquids move and approach this condition. In liquid mixing devices, it is necessary that two requirements are fulfilled. Firstly, there must be bulk or convective flow so that there are no dead (stagnant) zones. Secondly, there must be a zone of intensive or high-shear mixing in which the inhomogeneities are broken down. Both these processes are energy-consuming and ultimately the mechanical energy is dissipated as heat; the proportion of energy attributable to each varies from one application to another. Depending upon the fluid properties, primarily viscosity, the flow in mixing vessels may be laminar or turbulent, with a substantial transition zone in between the two, and frequently both flow types will occur simultaneously in different parts of the vessel. Laminar and turbulent flow arise from different mechanisms, and it is convenient to consider them separately.

7.2.1. Laminar mixing

Laminar flow is usually associated with high viscosity liquids (in excess of 10 N s/m^2) which may be either Newtonian or non-Newtonian. The inertial forces therefore tend to die out quickly, and the impeller of the mixer must cover a significant proportion of the cross-section of the vessel to impart sufficient bulk motion. Because the velocity gradients close to the rotating impeller are high, the fluid elements in that region deform and stretch. They repeatedly elongate and become thinner each time the fluid elements pass through the high shear zone. Figure 7.1 shows the sequence for a fluid element undergoing such a process.

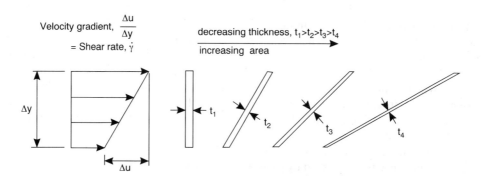

Figure 7.1.   Schematic representation of the thinning of fluid elements due to laminar shear flow

In addition, extensional or elongational flow usually occurs simultaneously. As shown in Figure 7.2, this is a result of the convergence of the streamlines and consequential increased velocity in the direction of flow. As the volume remains constant, there must, be a thinning or flattening of the fluid elements, as shown in Figure 7.2. Both of these mechanisms (shear and elongation), give rise to stresses in the liquid which then effect a reduction in droplet size and an increase in interfacial area, by which means the desired degree of homogeneity is obtained.

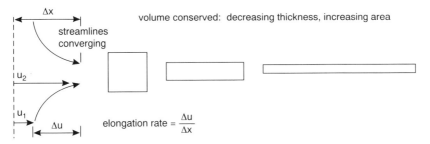

Figure 7.2.   Schematic representation of the thinning of fluid elements due to extensional flow

In addition, molecular diffusion is always tending to reduce inhomogeneities but its effect is not significant until the fluid elements have been reduced in size sufficiently for their specific areas to become large. It must be recognized, however, that the ultimate homogenisation of miscible liquids, can be only brought about by molecular diffusion. In the case of liquids of high viscosity, this is a slow process.

In laminar flow, a similar mixing process occurs when the liquid is sheared between two rotating cylinders. During each revolution, the thickness of the fluid element is reduced, and molecular diffusion takes over when the elements are sufficiently thin. This type of mixing is shown schematically in Figure 7.3 in which the tracer is pictured as being introduced perpendicular to the direction of motion.

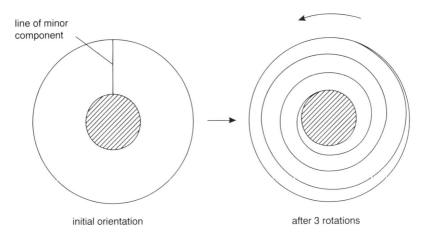

Figure 7.3.   Laminar shear mixing in a coaxial cylinder arrangement

Finally, mixing can be induced by physically splicing the fluid into smaller units and re-distributing them. In-line mixers rely primarily on this mechanism, which is shown in Figure 7.4.

Thus, mixing in liquids is achieved by several mechanisms which gradually reduce the size or scale of the fluid elements and then redistribute them in the bulk. If there are initially differences in concentration of a soluble material, uniformity is gradually achieved, and molecular diffusion becomes progressively more important as the element

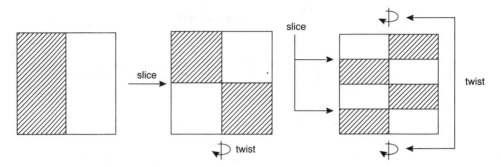

Figure 7.4.   Schematic representation of mixing by cutting and folding of fluid elements

size is reduced. OTTINO[4] has recently illustrated the various stages by means of a series
of coloured photographs.

7.2.2. Turbulent mixing

For low viscosity liquids (less than 10 mN s/m^2), the bulk flow pattern in mixing vessels
with rotating impellers is turbulent. The inertia imparted to the liquid by the rotating
impeller is sufficient to cause the liquid to circulate throughout the vessel and return
to the impeller. Turbulent eddy diffusion takes place throughout the vessel but is a
maximum in the vicinity of the impeller. Eddy diffusion is inherently much faster than
molecular diffusion and, consequently, turbulent mixing occurs much more rapidly than
laminar mixing. Ultimately homogenisation at the molecular level depends on molecular
diffusion, which takes place more rapidly in low viscosity liquids. Mixing is most rapid
in the region of the impeller because of the high shear rates due to the presence of
trailing vortices, generated by disc-turbine impellers, and associated Reynolds stresses
(Chapter 12); furthermore, a high proportion of the energy is dissipated here.

Turbulent flow is inherently complex, and calculation of the flow fields prevailing in a
mixing vessel is not amenable to rigorous theoretical treatment. If the Reynolds numbers
of the main flow is sufficiently high, some insight into the mixing process can be gained by
using the theory of local isotropic turbulence[5]. As is discussed in Chapter 12, turbulent
flow may be considered as a spectrum of velocity fluctuations and eddies of different sizes
superimposed on an overall time-averaged mean flow. In a mixing vessel, it is reasonable
to suppose that the large primary eddies, of a size corresponding approximately to the
impeller diameter, would give rise to large velocity fluctuations but would have a low
frequency. Such eddies are anisotropic and account for the bulk of the kinetic energy
present in the system. Interaction between these primary eddies and slow moving streams
produces smaller eddies of higher frequency which undergo further disintegration until
finally, dissipate their energy as heat.

During the course of this disintegration process, kinetic energy is transferred from
fast moving large eddies to the smaller eddies. The description given here is a gross
oversimplification, but it does give a qualitative representation of the salient features of
turbulent mixing. This whole process is similar to that for the turbulent flow of a fluid
close to a boundary surface described in Chapter 11. Although some quantitative results

for the scale size of eddies, have been obtained[5] and some workers[6,7] have reported experimental measurements on the structure of turbulence in mixing vessels, it is not at all clear how these data can be used for design purposes.

7.3. SCALE-UP OF STIRRED VESSELS

One of the problems confronting the designers of mixing equipment is that of deducing the most satisfactory arrangement for a large unit from experiments with small units. In order to achieve the same kind of flow pattern in two units, geometrical, kinematic, and dynamic similarity and identical boundary conditions must be maintained. This problem of scale-up has been discussed by a number of workers including HARNBY *et al.*[2], RUSHTON *et al.*[8], KRAMERS *et al.*[9], SKELLAND[10] and OLDSHUE[11]. It has been found convenient to relate the power used by the agitator to the geometrical and mechanical arrangement of the mixer, and thus to obtain a direct indication of the change in power arising from alteration of any of the factors relating to the mixer. A typical mixer arrangement is shown in Figure 7.5.

D   Diameter of agitator
D_T Diameter of tank
Z_A Height of agitator from base of tank
H   Depth of liquid
R   No. of baffles
W_B Width of baffles
N   Speed of agitator
P   Pitch of agitator
W   Width of blades of agitator or paddle

Figure 7.5.   Typical configuration and dimensions of an agitated vessel

For similarity in two mixing systems, it is important to achieve geometric kinematic and dynamic similarity.

Geometric similarity prevails between two systems of different sizes if all counterpart length dimensions have a constant ratio. Thus the following ratios must be the same in two systems:

$$\frac{D_T}{D}; \frac{Z_A}{D}; \frac{W_B}{D}; \frac{W}{D}; \frac{H}{D} \text{ and so on.}$$

Kinematic similarity exists in two geometrically similar units when the velocities at corresponding points have a constant ratio. Also, the paths of fluid motion (flow patterns) must be alike.

Dynamic similarity occurs in two geometrically similar units of different sizes if all corresponding forces at counterpart locations have a constant ratio. It is necessary here to distinguish between the various types of force: inertial, gravitational, viscous, surface tension and other forms, such as normal stresses in the case of viscoelastic non-Newtonian liquids. Some or all of these forms may be significant in a mixing vessel. Considering corresponding positions in systems 1 and 2 which refer to the laboratory and large scale, respectively, when the different types of force occurring are F_a, F_b, F_c and so on dynamic similarity requires that:

$$\frac{F_{a1}}{F_{a2}} = \frac{F_{b1}}{F_{b2}} = \frac{F_{c1}}{F_{c2}} = \cdots \cdots \text{ a constant} \tag{7.1}$$

or:

$$\frac{F_{a1}}{F_{b1}} = \frac{F_{a2}}{F_{b2}}; \frac{F_{a1}}{F_{c1}} = \frac{F_{a2}}{F_{c2}} \text{ etc.} \tag{7.2}$$

Kinematic and dynamic similarities both require geometrical similarity, so the corresponding positions "1" and "2" can be identified in the two systems. Some of the various types of forces that may arise during mixing or agitation will now be formulated.

Inertial force is associated with the reluctance of a body to change its state of rest or motion. Considering a mass m of liquid flowing with a linear velocity u through an area A during the time interval dt; then $dm = \rho uA \, dt$ where ρ is the fluid density. The inertial force $F_i = (\text{mass} \times \text{acceleration})$ or:

$$dF_i = (\rho uA \, dt)\left(\frac{du}{dt}\right) = \rho uA \, du$$

$$\therefore \qquad F_i = \int_0^{F_i} dF_i = \int_0^u \rho uA \, du = \rho A \frac{u^2}{2} \tag{7.3}$$

The area for flow is however, $A = (\text{constant}) L^2$, where L is the characteristic linear dimension of the system. In mixing applications, L is usually chosen as the impeller diameter D, and, likewise, the representative velocity u is taken to be the velocity at the tip of impeller (πDN), where N is revolutions per unit time. Therefore, the expression for inertial force may be written as:

$$F_i \propto \rho D^4 N^2 \tag{7.4}$$

where the constants have been omitted.

The rate change in u due to F_i, du/dt, may be countered by the rate of change in u due to *viscous forces* F_v, which for a Newtonian fluid is given by:

$$F_v = \mu A'\left(\frac{du}{dy}\right) \tag{7.5}$$

where du/dy, the velocity gradient, may be taken to be proportional to u/L, and A' is proportional to L^2. The viscous force is then given by:

$$F_v \propto \mu uL \tag{7.6}$$

which, for an agitated system, becomes $\mu D^2 N$. In a similar way, it is relatively simple to show that the *force due to gravity*, F_g, is given by:

$$F_g = (\text{mass of fluid} \times \text{acceleration due to gravity})$$

or:

$$F_g \propto \rho D^3 g \tag{7.7}$$

and finally the *surface tension* force, $F_s \propto \sigma D$.

Taking F_a, F_b, and so on, in equation 7.1 to represent F_i, F_v, F_g, F_s respectively for systems 1 and 2, dynamic similarity of the two systems requires that:

$$\left(\frac{F_i}{F_v}\right)_1 = \left(\frac{F_i}{F_v}\right)_2 \tag{7.8}$$

which upon substitution of the respective expressions for F_i and F_v leads to:

$$\left(\frac{\rho D^2 N}{\mu}\right)_1 = \left(\frac{\rho D^2 N}{\mu}\right)_2 \tag{7.9}$$

This is, of course, the *Reynolds number*, which determines the nature of the flow, as shown in Chapter 3. In similar fashion, the constancy of the ratios between other forces results in Froude and Weber numbers as follows:

$$\left(\frac{F_i}{F_g}\right)_1 = \left(\frac{F_i}{F_g}\right)_2 \rightarrow \left(\frac{DN^2}{g}\right)_1 = \left(\frac{DN^2}{g}\right)_2 \qquad (\text{Froude number}) \tag{7.10}$$

$$\left(\frac{F_i}{F_s}\right)_1 = \left(\frac{F_i}{F_s}\right)_2 \rightarrow \left(\frac{D^3 N^2 \rho}{\sigma}\right)_1 = \left(\frac{D^3 N^2 \rho}{\sigma}\right)_2 \qquad (\text{Weber number}) \tag{7.11}$$

Thus, the ratios of the various forces occurring in mixing vessels can be expressed as the above dimensionless groups which, in turn, serve as similarity parameters for scale-up of mixing equipment. It can be shown that the existence of geometric and dynamic similarities also ensures kinematic similarity.

In the case of non-viscoelastic, time-independent, non-Newtonian fluids, the fluid viscosity μ, occurring in Reynolds numbers, must be replaced by an apparent viscosity μ_a evaluated at an appropriate value of shear rate. METZNER and co-workers[12,13] have developed a procedure for calculating an average value of shear rate in a mixing vessel as a function of impeller/vessel geometry and speed of rotation; this, in turn, allows the value of the apparent viscosity to be obtained either directly from the rheological data, or by the use of a rheological model. Though this scheme has proved satisfactory for time-independent non-Newtonian fluids, further complications arise in the scale-up of mixing equipment for viscoelastic liquids (ULBRECHT[14], ASTARITA[15] and SKELLAND[16]).

7.4. POWER CONSUMPTION IN STIRRED VESSELS

From a practical point of view, power consumption is perhaps the most important parameter in the design of stirred vessels. Because of the very different flow patterns and mixing mechanisms involved, it is convenient to consider power consumption in low and high viscosity systems separately.

7.4.1. Low viscosity systems

Typical equipment for low viscosity liquids consists of a vertical cylindrical tank, with a height to diameter ratio of 1.5 to 2, fitted with an agitator. For low viscosity liquids, high-speed propellers of diameter about one-third that of the vessel are suitable, running at 10-25 Hz. Although work on single-phase mixing of low viscosity liquids is of limited value in industrial applications it does, however, serve as a useful starting point for the subsequent treatment of high viscosity liquids.

Considering a stirred vessel in which a Newtonian liquid of viscosity μ, and density ρ is agitated by an impeller of diameter D rotating at a speed N; the tank diameter is D_T, and the other dimensions are as shown in Figure 7.5, then, the functional dependence of the power input to the liquid $\mathbf{P}$ on the independent variables (μ, ρ, N, D, D_T, g, other geometric dimensions) may be expressed as:

$$\mathbf{P} = \mathrm{f}(\mu, \rho, N, g, D, D_T, \text{ other dimensions}) \tag{7.12}$$

In equation 7.12, $\mathbf{P}$ is the impeller power, that is, the energy per unit time dissipated within the liquid. Clearly, the electrical power required to drive the motor will be greater than $\mathbf{P}$ on account of transmission losses in the gear box, motor, bearings, and so on.

It is readily acknowledged that the functional relationship in equation 7.12 cannot be established from first principles. However, by using dimensional analysis, the number of variables can be reduced to give:

$$\frac{\mathbf{P}}{\rho N^3 D^5} = \mathrm{f}\left(\frac{\rho N D^2}{\mu}, \frac{N^2 D}{g}, \frac{D_T}{D}, \frac{W}{D}, \frac{H}{D}, \cdots\right) \tag{7.13}$$

where the dimensionless group on the left-hand side is called the *Power number* (N_p); ($\rho N D^2/\mu$) is the *Reynolds number (Re)* and ($N^2 D/g$) is the *Froude number (Fr)*. Other dimensionless length ratios, such as (D_T/D), (W/D) and so on, relate to the specific impeller/vessel arrangement. For geometrically similar systems, these ratios must be equal, and the functional relationship between the Power number and the other dimensionless groups reduces to:

$$N_p = \mathrm{f}(Re, Fr) \tag{7.14}$$

The simplest form of the function in equation 7.14 is a power law, giving:

$$N_p = K' Re^b Fr^c \tag{7.15}$$

where the values of K', b and c must be determined from experimental measurements, and are dependent upon impeller/vessel configuration and on the flow regime, that is laminar, transition or turbulent, prevailing in the mixing vessel. There are several ways of measuring the power input to the impeller, including the mounting of Prony brakes, using a dynamometer, or a simple calculation from the electrical measurements. Detailed descriptions, along with the advantages and disadvantages of different experimental techniques, have been given by OLDSHUE[11], and HOLLAND and CHAPMAN[17]. It is appropriate to note that the uncertainty regarding the actual losses in gear box, bearings, and so on makes the estimation of $\mathbf{P}$ from electrical measurements rather less accurate than would be desirable.

In equation 7.15, the Froude number is usually important only in situations where gross vortexing occurs, and can be neglected if the value of Reynolds number is less than about 300. Thus, in a plot of Power number (N_p) against Reynolds number (Re) with Froude number as parameter, all data fall on a single line for values of $Re < 300$ confirming that in this region Fr has no significant effect on N_p. This behaviour is clearly seen in Figure 7.6 where the data for a propeller, as reported by RUSHTON et al.[8], are plotted. Such a plot is known as a *power curve*.

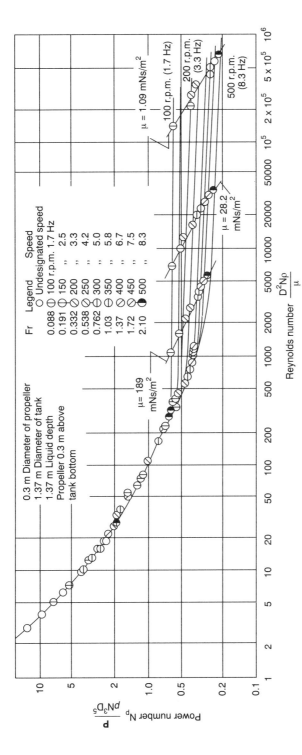

Figure 7.6. Power number as a function of Reynolds number for a propeller mixer

Thus at $Re < 300$:

$$N_p = K'Re^b \tag{7.16}$$

For values of $Re < 10$, b is found to be -1 (see Figure 7.6).
Therefore, power **P** is then given by:

$$\mathbf{P} = K'\mu N^2 D^3 \tag{7.17}$$

The value of K' depends on the type of impeller/vessel arrangement, and whether the tank is fitted with baffles. For marine-type three-bladed propellers with pitch equal to diameter, K' has been found to have a value of about 41.

For higher values of Re, the Froude number seems to exert some influence on the value of N_p, and separate lines are drawn for various speeds in Figure 7.6. It will be noted that, on this graph, lines of constant speed of rotation relate to constant values of the Froude number because the diameter of the impeller is constant and equal to 0.3 m. Thus, $Fr = 0.0305N^2$. The Reynolds number was varied by using liquids of different viscosities, as well as different rotational speeds, and the slanting lines in Figure 7.6 represent conditions of constant viscosity. In this region, the effect of Froude number may be minimised, or indeed eliminated, by the use of baffles or installing the impeller off-centre. This point is discussed in more detail in Section 7.5.

Example 7.1

On the assumption that the power required for mixing in a stirred tank is a function of the variables given in equation 7.12, obtain the dimensionless groups which are important in calculating power requirements for geometrically similar arrangements.

Solution

The variables in this problem, together with their dimensions, are as follows:

P	$\mathbf{ML^2T^{-3}}$
μ	$\mathbf{ML^{-1}T^{-1}}$
ρ	$\mathbf{ML^{-3}}$
N	$\mathbf{T^{-1}}$
g	$\mathbf{LT^{-2}}$
D	$\mathbf{L}$
D_T	$\mathbf{L}$

This list includes seven variables and there are three fundamentals (**M, L, T**). By Buckingham's Π theorem, there will be $7 - 3 = 4$ dimensionless groups.

Choosing as the recurring set ρ, N and D, then these three variables themselves cannot be grouped together to give a dimensionless number. **M, L, T** can now be expressed in terms of combinations of ρ, N, D.

$$\mathbf{L} \equiv D$$

$$\mathbf{T} \equiv N^{-1}$$

$$\mathbf{M} \equiv \rho D^3.$$

Dimensionless group 1 : $\mathbf{PM^{-1}L^{-2}T^3} = \mathbf{P}(\rho D^3)^{-1}(D)^{-2}(N^{-1})^3 = \mathbf{P}\rho^{-1}D^{-5}N^{-3}$

Dimensionless group 2 : $\mu\mathbf{M^{-1}LT}$ $= \mu(\rho D^3)^{-1}DN^{-1} = \mu\rho^{-1}D^{-2}N^{-1}$

Dimensionless group 3 : $g\mathbf{L^{-1}T^2}$ $= gD^{-1}(N^{-1})^2 = gD^{-1}N^{-2}$

Dimensionless group 4 : $D_T \mathbf{L}^{-1}$ $= D_T(D^{-1}) = D_T D^{-1}$

Thus, $\mathbf{P}\rho^{-1}D^{-5}N^{-3}$ $= \mathrm{f}(\mu\rho^{-1}D^{-2}N^{-1}, gD^{-1}N^2, D_T D^{-1})$

Re-arranging:

$$\frac{\mathbf{P}}{\rho N^3 D^5} = \mathrm{f}\left(\frac{\rho N D^2}{\mu}, \frac{N^2 D}{g}, \frac{D_T}{D} \ldots\right)$$

which corresponds with equation 7.13.

Example 7.2

A solution of sodium hydroxide of density 1650 kg/m³ and viscosity 50 mN s/m² is agitated by a propeller mixer of 0.5 m diameter in a tank of 2.28 m diameter, and the liquid depth is 2.28 m. The propeller is situated 0.5 m above the bottom of the tank. What is the power which the propeller must impart to the liquids for a rotational speed of 2 rev/s?

Solution

In this problem the geometrical arrangement corresponds with the configuration for which the curves in Figure 7.6 are applicable.

$$\frac{D_T}{D} = \frac{2.28}{0.5} = 4.56; \quad \frac{H}{D} = \frac{2.28}{0.5} = 4.56; \quad \frac{Z_A}{D} = \frac{0.5}{0.5} = 1$$

$$Re = \frac{D^2 N \rho}{\mu} = \frac{0.5^2 \times 2 \times 1650}{50 \times 10^{-3}} = 16,500$$

$$Fr = \frac{N^2 D}{g} = \frac{2^2 \times 0.5}{9.81} = 0.20$$

From Figure 7.6:

$$N_p = \frac{\mathbf{P}}{\rho N^3 D^5} = 0.5$$

and $\quad \mathbf{P} = 0.5 \times 1650 \times (2)^3 \times (0.5)^5$

$\qquad = \underline{\underline{206 \text{ W}}}$

Figure 7.7, also taken from the work of RUSHTON et al.[8], shows N_p vs Re data for a 150 mm diameter turbine with six flat blades. In addition, this figure also shows the effect of introducing baffles in the tank. BISSELL et al.[18] have studied the effect of different types of baffles and their configuration on power consumption.

Evidently, for most conditions of practical interest, the Froude number is not a significant variable and the Power number is a unique function of Reynolds number for a fixed impeller/vessel configuration and surroundings. The vast amount of work[2] reported on the mixing of low viscosity liquids suggests that, for a given geometrical design and configuration of impeller and vessel, all single phase experimental data fall on a single power curve. Typically there are three discernible zones in a power curve.

At low values of the Reynolds number, less than about 10, a laminar or viscous zone exists and the slope of the power curve on logarithmic coordinates is −1, which is typical of most viscous flows. This region, which is characterised by slow mixing at both macro- and micro-levels, is where the majority of the highly viscous (Newtonian as well as non-Newtonian) liquids are processed.

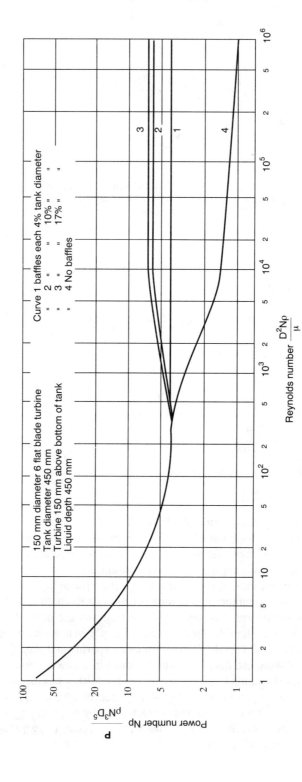

Figure 7.7.   Power number as a function of Reynolds number for a turbine mixer

At very high values of Reynolds number, greater than about 10^4, the flow is fully turbulent and inertia dominated, resulting in rapid mixing. In this region the Power number is virtually constant and independent of Reynolds number, as shown in Figures 7.6 and 7.7, but depends upon the impeller/vessel configuration. Often gas-liquid, solid-liquid and liquid-liquid contacting operations are carried out in this region. Though the mixing itself is quite rapid, the limiting factor controlling the process may be mass transfer.

Between the laminar and turbulent zones, there exists a transition region in which the viscous and inertial forces are of comparable magnitudes. No simple mathematical relationship exists between N_p and Re in this flow region and, at a given value of Re, N_p must be obtained from the appropriate power curve.

For purposes of scale-up, it is generally most satisfactory in the laminar region to maintain a constant speed for the tip of the impeller, and mixing time will generally increase with scale. The most satisfactory basis for scale-up in the turbulent region is to maintain a constant power input per unit volume.

Power curves for many different impeller geometries, baffle arrangements, and so on are to be found in the literature[10,11,17,19−21], but it must always be remembered that though the power curve is applicable to any single phase Newtonian liquid, at any impeller speed, the curve will be valid for only *one system geometry*.

Sufficient information is available on low viscosity systems for the estimation of the power requirements for a given duty under most conditions of practical interest.

7.4.2. High viscosity systems

As noted previously, mixing in highly viscous liquids is slow both at the molecular scale, on account of the low values of diffusivity, as well as at the macroscopic scale, due to poor bulk flow. Whereas in low viscosity liquids momentum can be transferred from a rotating impeller through a relatively large body of fluid, in highly viscous liquids only the fluid in the immediate vicinity of the impeller is influenced by the agitator and the flow is normally laminar.

For the mixing of highly viscous and non-Newtonian fluids it is usually necessary to use specially designed impellers involving close clearances with the vessel walls; these are discussed in a later section. High-speed stirring with small impellers merely wastefully dissipates energy at the central portion of the vessel. The power curve approach is usually applicable and the proportionality constant K' in equation 7.17 is a function of the type of rotating member, and the geometrical configuration of the system. Most of the highly viscous fluids of interest in the processing industries exhibit non-Newtonian behaviour, though highly viscous Newtonian fluids include glycerol and many lubricating oils.

A simple relationship has been shown to exist, however, between much of the data on power consumption with time-independent non-Newtonian liquids and Newtonian liquids in the laminar region. This link, which was first established by METZNER and OTTO[12] for pseudoplastic liquids, depends on the fact that there appears to be an average angular shear rate $\dot{\gamma}_{ang}$ for a mixer which characterises power consumption, and which is directly proportional to the rotational speed of impeller:

$$\dot{\gamma}_{ang} = k_s N \qquad (7.18)$$

Table 7.1. Values of k_s for various types of impellers and key to Figure 7.8. (From ref. 16)

Curve	Impeller	Baffles	D (m)	D_T/D	N(Hz)	k_s (n < 1)
A-A	Single turbine with 6 flat blades	4, $W_B/D_T = 0.1$	0.051–0.20	1.3–5.5	0.05–1.5	11.5 ± 1.5
A-A$_1$	Single turbine with 6 flat blades	None.	0.051–0.20	1.3–5.5	0.18–0.54	11.5 ± 1.4
B-B	Two turbines, each with 6 flat blades and $D_T/2$ apart	4, $W_B/D_T = 0.1$	—	3.5	0.14–0.72	11.5 ± 1.4
B-B$_1$	Two turbines, each with 6 flat blades and $D_T/2$ apart	4, $W_B/D_T = 0.1$ or none	—	1.023–1.18	0.14–0.72	11.5 ± 1.4
C-C	Fan turbine with 6 blades at 45°	4, $W_B/D_T = 0.1$ or none	0.10–0.20	1.33–3.0	0.21–0.26	13 ± 2
C-C$_1$	Fan turbine with 6 blades at 45°	4, $W_B/D_T = 0.1$ or none	0.10–0.30	1.33–3.0	1.0–1.42	13 ± 2
D-D	Square-pitch marine propellers with 3 blades (downthrusting)	None, (i) shaft vertical at vessel axis, (ii) shaft 10° from vertical, displaced r/3 from centre	0.13	2.2–4.8	0.16–0.40	10 ± 0.9
D-D$_1$	Same as for D-D but upthrusting	None, (i) shaft vertical at vessel axis, (ii) shaft 10° from vertical, displaced r/3 from centre	0.13	2.2–4.8	0.16–0.40	10 ± 0.9
D-D$_2$	Same as for D-D	None, position (ii)	0.30	1.9–2.0	0.16–0.40	10 ± 0.9
D-D$_3$	Same as for D-D	None, position (i)	0.30	1.9–2.0	0.16–0.40	10 ± 0.9
E-E	Square-pitch marine propeller with 3 blades	4, $W_B/D_T = 0.1$	0.15	1.67	0.16–0.60	10
F-F	Double-pitch marine propeller with 3 blades (downthrusting)	None, position (ii)	—	1.4–3.0	0.16–0.40	10 ± 0.09
F-F$_1$	Double-pitch marine propeller with 3 blades (downthrusting)	None, position (i)	—	1.4–3.0	0.16–0.40	10 ± 0.9
G-G	Square-pitch marine propeller with 4 blades	4, $W_B/D_T = 0.1$	0.12	2.13	0.05–0.61	10
G-G$_1$	Square-pitch marine propeller with 4 blades	4, $W_B/D_T = 0.1$	0.12	2.13	1.28–1.68	—
H-H	2-bladed paddle	4, $W_B/D_T = 0.1$	0.09–0.13	2–3	0.16–1.68	10
—	Anchor	None	0.28	1.02	0.34–1.0	11 ± 5
—	Cone impellers	0 or 4, $W_B/D_T = 0.08$	0.10–0.15	1.92–2.88	0.34–1.0	11 ± 5

where k_s is a function of the type of impeller and the vessel configuration. If the apparent viscosity corresponding to the average shear rate defined above is used in the equation for a Newtonian liquid, the power consumption for laminar conditions is satisfactorily predicted for the non-Newtonian fluid.

The validity of the linear relationship given in equation 7.18 was subsequently confirmed by METZNER and TAYLOR[13]. The experimental evaluation of k_s for a given geometry is as follows:

(i) The Power number (N_p) is determined for a particular value of N.
(ii) The corresponding value of Re is obtained from the appropriate power curve as if the liquid were Newtonian.
(iii) The equivalent viscosity is computed from the value of Re.
(iv) The value of the corresponding shear rate is obtained, either directly from a flow curve obtained by independent experiment, or by use of an appropriate fluid model such as the power-law model.
(v) The value of k_s is calculated for a particular impeller configuration.

This procedure can be repeated for different values of N. A compilation of the experimental values of k_s for a variety of impellers, turbine, propeller, paddle, anchor, and so on, has been given by SKELLAND[16], and an examination of Table 7.1 suggests that for pseudo-plastic liquids, k_s lies approximately in the range of 10–13 for most configurations of practical interest.[22,23] SKELLAND[16] has also correlated much of the data on the agitation of purely viscous non-Newtonian fluids, and this is shown in Figure 7.8.

The prediction of power consumption for agitation of a given non-Newtonian fluid in a particular mixer, at a desired impeller speed, may be evaluated by the following procedure.

(i) Estimate the average shear rate from equation 7.18.
(ii) Evaluate the corresponding apparent viscosity, either from a flow curve, or by means of the appropriate flow model.
(iii) Estimate the value of the Reynolds number as ($\rho N D^2 / \mu_a$) and then the value of the Power number, and hence **P**, from the appropriate curve in Figure 7.8.

Although this approach of METZNER and OTTO[12], has gained wide acceptance, it has come under some criticism. For instance, SKELLAND[10] and MITSUISHI and HIRAI[24] have argued that this approach does not always yield a unique power curve for a wide range of the flow behaviour index, n. Despite this, it is safe to conclude that this method predicts power consumption with an accuracy of within about 25–30 per cent. Furthermore, GODFREY[25] has asserted that the constant k_s is independent of equipment size, and thus there are few scale-up problems. It is not yet established, however, how strongly the value of k_s depends upon the rheology. For example, CALDERBANK and MOO-YOUNG[26] and BECKNER and SMITH[27] have related k_s to the impeller/vessel configuration and rheological properties (n in the case of power law liquids); the dependence on n, however, is quite weak.

Data for power consumption of Bingham plastic fluids have been reported and correlated by NAGATA et al.[28] and of dilatant fluids by NAGATA et al.[28] and METZNER et al.[29]. EDWARDS et al.[30] have dealt with the mixing of time-dependent thixotropic materials.

Figure 7.8.   Power curve for pseudoplastic fluids agitated by different types of impeller

Very little is known about the effect of the viscoelasticity of a fluid on power consumption, but early work[31,32] seems to suggest that it is negligible under creeping flow conditions. More recent work by OLIVER *et al.*[33] and by PRUD'HOMME and SHAQFEH[34] has indicated that the power consumption for Rushton-type turbine impellers, illustrated in Figure 7.20, may increase or decrease depending upon the value of Reynolds number and the magnitude of elastic effects or the Deborah number. At higher Reynolds numbers, it appears that the elasticity suppresses secondary flows and this results in a reduction in power consumption in comparison with a purely viscous liquid.[35] Many useful review articles have been published on this subject[2,14,16,36], and theoretical developments relating to the mixing of high viscosity materials have been dealt with by IRVING and SAXTON[37].

Finally, it should be noted that the calculation of the power requirement requires a knowledge of the impeller speed which is necessary to blend the contents of a tank in a given time, or of the impeller speed required to achieve a given mass transfer rate in a gas-liquid system. A full understanding of the mass transfer/mixing mechanism is not yet available, and therefore the selection of the optimum operating speed remains primarily a matter of experience. Before concluding this section, it is appropriate to indicate typical power consumptions in kW/m^3 of liquid for various duties, and these are shown in Table 7.2.

Table 7.2. Typical power consumptions

Duty	Power (kW/m^3)
Low power	
Suspending light solids, blending of low viscosity liquids	0.2
Moderate power	
Gas dispersion, liquid-liquid contacting, some heat transfer, etc.	0.6
High power	
Suspending heavy solids, emulsification, gas dispersion, etc.	2
Very high power	
Blending pastes, doughs	4

7.5. FLOW PATTERNS IN STIRRED TANKS

A qualitative picture of the flow field created by an impeller in a mixing vessel in a single-phase liquid is useful in establishing whether there are stagnant or dead regions in the vessel, and whether or not particles are likely to be suspended. In addition, the efficiency of mixing equipment, as well as product quality, are influenced by the flow patterns prevailing in the vessel.

The flow patterns for single phase, Newtonian and non-Newtonian liquids in tanks agitated by various types of impeller have been reported in the literature.[13,27,38,39] The experimental techniques which have been employed include the introduction of tracer liquids, neutrally buoyant particles or hydrogen bubbles, and measurement of local velocities by means of Pitot tubes, laser-doppler anemometers, and so on. The salient features of the flow patterns encountered with propellers and disc turbines are shown in Figures 7.9 and 7.10.

Figure 7.9.   Flow pattern from propeller mixer

Figure 7.10.   Radial flow pattern from disc turbine

Basically, the propeller creates an axial flow through the impeller, which may be upwards or downwards depending upon the direction of rotation. The velocities at any point are three-dimensional, and unsteady, but circulation patterns such as those shown in Figure 7.9 are useful in avoiding dead zones.

If a propeller is mounted centrally, and there are no baffles in the tank, there is a tendency for the lighter fluid to be drawn in to form a vortex, and for the degree of agitation to be reduced. To improve the rate of mixing, and to minimise vortex formation, baffles are usually fitted in the tank, and the resulting flow pattern is as shown in Figure 7.11. The power requirements are, however, considerably increased by the incorporation of baffles. Another way of minimising vortex formation is to mount the agitator, off-centre; where the resulting flow pattern is depicted in Figure 7.12.

The flat-bladed turbine impeller produces a strong radial flow outwards from the impeller (Figure 7.10), thereby creating circulating zones in the top and bottom of the tank. The flow pattern can be altered by changing the impeller geometry and, for example, if the turbine blades are angled to the vertical, a stronger axial flow component is produced. This can be useful in applications where it is necessary to suspend solids. A flat paddle produces

Side view

Bottom view

Figure 7.11.   Flow pattern in vessel with vertical baffles

Propeller turning
counterclockwise
looking down on
shaft

Off-centre top-entering
propeller position

Figure 7.12.   Flow pattern with agitator offset from centre

a flow field with large tangential components of velocity, and this does not promote good mixing. Propellers, turbines and paddles are the principal types of impellers used for low viscosity systems operating in the transition and turbulent regimes.

A brief mention will be made of flow visualisation studies for some of the agitators used for high viscosity liquids; these include anchors, helical ribbons and screws. Detailed flow pattern studies[40] suggest that both gate and anchor agitators promote fluid motion close to the vessel wall but leave the region in the vicinity of the shaft relatively stagnant. In addition, there is modest top to bottom turnover and thus vertical concentration gradients may exist, but these may be minimised by means of a helical ribbon or a screw added to the shaft. Such combined impellers would have a ribbon pumping upwards near the wall with the screw twisted in the opposite sense, pumping the fluid downwards near the shaft region. Typical flow patterns for a gate-type anchor are shown in Figure 7.13. In these systems the flow patterns change[27] as the stirring speed is increased and the average shear rate cannot be completely described by a linear relationship. Furthermore, any rotational motion induced within a vessel will tend to produce a secondary flow in the vertical direction; the fluid in contact with the bottom of the tank is stationary, while that at higher levels is rotating and will experience centrifugal forces. Consequently, there exist unbalanced pressure forces within the fluid which lead to the formation of a toroidal vortex. This secondary system may be single-celled or double-celled, as shown in Figures 7.14 and 7.15, depending upon the viscosity and type of fluid.

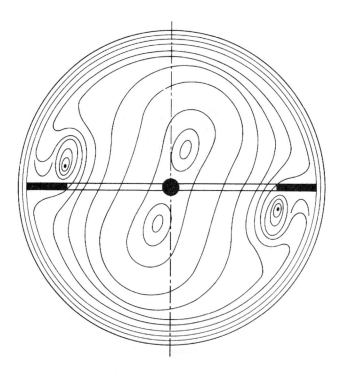

Figure 7.13.   Streamlines in a tank with a gate agitator, drawn relative to the arm of the stirrer

Clearly, the flow pattern established in a mixing vessel depends critically upon the vessel/impeller configuration and on the physical properties of the liquid (particularly viscosity). In selecting the appropriate combination of equipment, it must be ensured that the resulting flow pattern is suitable for the required application.

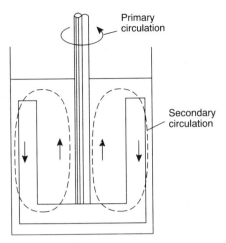

Figure 7.14.   Single-celled secondary flow with an anchor agitator

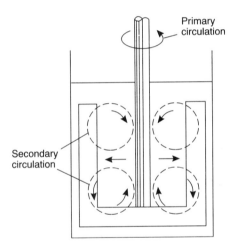

Figure 7.15.   Double-celled secondary flow

7.6. RATE AND TIME FOR MIXING

Before considering rate of mixing and mixing time, it is necessary to have some means of assessing the quality of a mixture, which is the product of a mixing operation. Because of the wide scope and range of mixing problems, several criteria have been developed to assess mixture quality, none of which is universally applicable. One intuitive and convenient, but perhaps unscientific, criterion is whether the product or mixture meets the required specification. Other ways of judging the quality of mixing have been described by HARNBY *et al.*[2]. Whatever the criteria used, mixing time is defined as the time required to produce a mixture or a product of predetermined quality, and the rate of mixing is the rate at which the mixing progresses towards the final state. For a single-phase liquid in a stirred tank to which a volume of tracer material is added, the mixing time is measured from the instant the tracer is added to the time when the contents of the vessel have reached the required degree of uniformity or mixedness. If the tracer is completely miscible and has the same viscosity and density as the liquid in the tank, the tracer concentration may be measured as a function of time at any point in the vessel by means of a suitable detector, such as a colour meter, or by electrical conductivity. For a given amount of tracer, the equilibrium concentration C_∞ may be calculated; this value will be approached asymptotically at any point as shown in Figure 7.16.

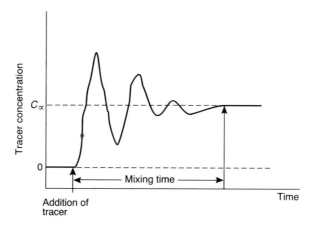

Figure 7.16.   Mixing-time measurement curve

The mixing time will be that required for the mixture composition to come within a specified deviation from the equilibrium value and this will be dependent upon the way in which the tracer is added and the location of the detector. It may therefore be desirable to record the tracer concentration at several locations, and to define the variance of concentration σ^2 about the equilibrium value as:

$$\sigma^2 = \frac{1}{n-1} \sum_{i=0}^{i=n} (C_i - C_\infty)^2 \qquad (7.19)$$

where C_i is the tracer concentration at time t recorded by the ith detector. A typical variance curve is shown in Figure 7.17.

Several experimental techniques may be used, such as acid/base titration, electrical conductivity measurement, temperature measurement, or measurement of optical properties such as refractive index, light absorption, and so on. In each case, it is necessary to specify the manner of tracer addition, the position and number of recording stations, the sample volume of the detection system, and the criteria used in locating the end-point. Each of these factors will influence the measured value of mixing time, and therefore care must be exercised in comparing results from different investigations.

For a given experiment and configuration, the mixing time will depend upon the process and operating variables as follows:

$$t_m = \text{f}(\rho, \mu, N, D, g, \text{geometrical dimensions of the system}) \qquad (7.20)$$

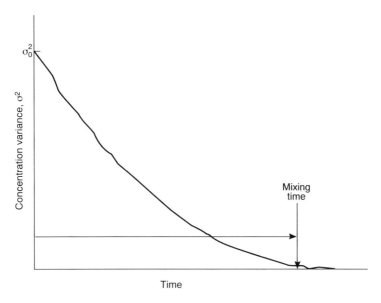

Figure 7.17.   Reduction in variance of concentration with time

Using dimensional analysis, the functional relationship may be rearranged as:

$$Nt_m = \theta_m = f\left(\frac{\rho ND^2}{\mu}, \frac{DN^2}{g}, \text{geometrical dimensions as ratios}\right) \qquad (7.21)$$

For geometrically similar systems, and assuming that the Froude number DN^2/g is not important:

$$\theta_m = f\left(\frac{\rho ND^2}{\mu}\right) = f(Re) \qquad (7.22)$$

The available experimental data[2] seem to suggest that the dimensionless mixing time (θ_m) is independent of Reynolds number for both laminar and turbulent zones, in each of which it attains a constant value, and changes from the one value to the other in the transition region. $\theta_m - Re$ behaviour is schematically shown in Figure 7.18, and some typical mixing data obtained by NORWOOD and METZNER[41] for turbine impellers in baffled vessels using an acid/base titration technique are shown in Figure 7.19. It is widely reported that θ_m is quite sensitive to impeller/vessel geometry for low viscosity liquids. BOURNE and BUTLER[42] have made some interesting observations which appear to suggest that the rates of mixing and, hence mixing time, are not very sensitive to the fluid properties for both Newtonian and non-Newtonian materials. On the other hand, GODLESKI and SMITH[43] have reported mixing times for pseudoplastic non-Newtonian fluids fifty times greater than those expected from the corresponding Newtonian behaviour, and this emphasises the care which must be exercised in applying any generalised conclusions to a particular system.

Although very little is known about the influence of fluid elasticity on rates of mixing and mixing times, the limited work appears to suggest that the role of elasticity strongly

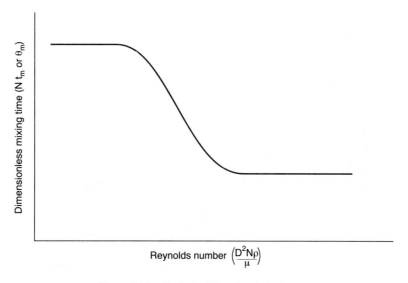

Figure 7.18.   Typical mixing-time behaviour

Figure 7.19.   Typical mixing time data for turbines

depends on the impeller geometry. For instance, CHAVAN *et al.*[44] reported a decrease in mixing rate with increasing levels of viscoelasticity for helical ribbons, whereas HALL and GODFREY[45] found the mixing rates to increase with elasticity for sigma blades. Similar conflicting conclusions regarding the influence of viscoelasticity can be drawn from the work of CARREAU *et al.*[39] who found that the direction of impeller rotation had an appreciable effect on mixing rates and times.

Considerable confusion thus seems to exist in the literature regarding the influence of rheological properties on mixing rates and times. Exhaustive reviews are available on this subject[16,36].

7.7. MIXING EQUIPMENT

The wide range of mixing equipment available commercially reflects the enormous variety of mixing duties encountered in the processing industries. It is reasonable to expect therefore that no single item of mixing equipment will be able to carry out such a range of duties effectively. This has led to the development of a number of distinct types of mixer over the years. Very little has been done, however, by way of standardisation of equipment, and no design codes are available. The choice of a mixer type and its design is therefore primarily governed by experience. In the following sections, the main mechanical features of commonly used types of equipment together with their range of applications are described qualitatively. Detailed description of design and selection of various types of mixers have been presented by OLDSHUE[11].

7.7.1. Mechanical agitation

This is perhaps the most commonly used method of mixing liquids, and essentially there are three elements in such devices.

Vessels

These are often vertically mounted cylindrical tanks, up to 10 m in diameter, which typically are filled to a depth equal to about one diameter, although in some gas-liquid contacting systems tall vessels are used and the liquid depth is up to about three tank diameters; multiple impellers fitted on a single shaft are then frequently used. The base of the tanks may be flat, dished, or conical, or specially contoured, depending upon factors such as ease of emptying, or the need to suspend solids, etc., and so on.

For the batch mixing of viscous pastes and doughs using ribbon impellers and Z-blade mixers, the tanks may be mounted horizontally. In such units, the working volume of pastes and doughs is often relatively small, and the mixing blades are massive in construction.

Baffles

To prevent gross vortexing, which is detrimental to mixing, particularly in low viscosity systems, baffles are often fitted to the walls of the vessel. These take the form of thin strips about one-tenth of the tank diameter in width, and typically four equi-spaced baffles may be used. In some cases, the baffles are mounted flush with the wall, although occasionally a small clearance is left between the wall and the baffle to facilitate fluid motion in the wall region. Baffles are, however, generally not required for high viscosity liquids because the viscous shear is then sufficiently great to damp out the rotary motion. Sometimes, the problem of vortexing is circumvented by mounting impellers off-centre.

Impellers

Figure 7.20 shows some of the impellers which are frequently used. Propellers, turbines, paddles, anchors, helical ribbons and screws are usually mounted on a central vertical shaft in a cylindrical tank, and they are selected for a particular duty largely on the basis of liquid viscosity. By and large, it is necessary to move from a propeller to a turbine and then, in order, to a paddle, to an anchor and then to a helical ribbon and finally to a screw as the viscosity of the fluids to be mixed increases. In so doing the speed of agitation or rotation decreases.

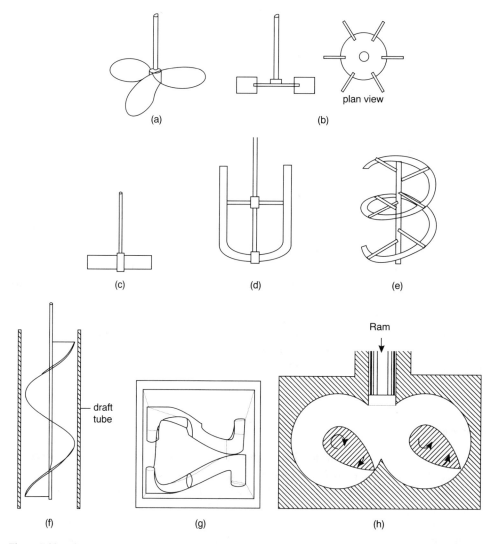

Figure 7.20. Commonly used impellers (*a*) Three-bladed propeller (*b*) Six-bladed disc turbine (Rushton turbine) (*c*) Simple paddle (*d*) Anchor impeller (*e*) Helical ribbon (*f*) Helical screw with draft tube (*g*) Z-blade mixer (*h*) Banbury mixer

Propellers, turbines and paddles are generally used with relatively low viscosity systems and operate at high rotational speeds. A typical velocity for the tip of the blades of a turbine is of the order of 3 m/s, with a propeller being a little faster and the paddle a little slower. These are classed as remote-clearance impellers, having diameters in the range $(0.13-0.67) \times$ (tank diameter). Furthermore, minor variations within each type are possible. For instance, Figure 7.20*b* shows a six-flat bladed Rushton turbine, whereas possible variations are shown in Figure 7.21. Hence it is possible to have retreating-blade turbines, angled-blade turbines, four- to twenty-bladed turbines, and so on. For dispersion of gases in liquid, turbines are usually employed.

Figure 7.21.   Variation in turbine impeller designs (*a*) Flat blade (*b*) Disc flat blade (*c*) Pitched vane (*d*) Curved blade (*e*) Tilted blade (*f*) Shrouded (*g*) Pitched blade (*h*) Pitched curved blade (*i*) Arrowhead

Propellers are frequently of the three-bladed marine type and are used for in-tank blending operations with low viscosity liquids, and may be arranged as angled side-entry units, as shown in Figure 7.22. REAVELL[1] has shown that the fitting of a cruciform baffle at the bottom of the vessel enables much better dispersion to be obtained, as shown in Figure 7.23. For large vessels, and when the liquid depth is large compared with the tank diameter, it is a common practice to mount more than one impeller on the same shaft. With this arrangement the unsupported length of the propeller shaft should not exceed about 2 m. In the case of large vessels, there is some advantage to be gained by using side- or bottom-entry impellers to avoid the large length of unsupported shaft,

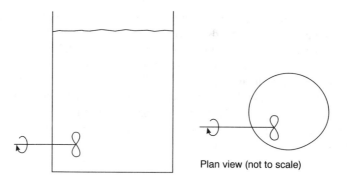

Figure 7.22.   Side entering propeller

Figure 7.23.   Flow pattern in vessel with cruciform baffle

though a good gland or mechanical seal is needed for such installations or alternatively, a foot bearing is employed. Despite a considerable amount of practical experience, foot bearings can be troublesome owing to the difficulties of lubrication, especially when handling corrosive liquids.

In comparing propellers and turbines, the following features may be noted:

Propellers

(a) are self-cleaning in operation,
(b) can be used at a wide range of speeds,
(c) give an excellent shearing effect at high speeds,
(d) do not damage dispersed particles at low speeds,
(e) are reasonably economical in power, provided the pitch is adjusted according to the speed,
(f) by offset mounting, vortex formation is avoided,

(g) if horizontally mounted, a stuffing box is required in the liquid, and they are not effective in viscous liquids.

Shrouded turbines

(a) are excellent for providing circulation,
(b) are normally mounted on a vertical shaft with the stuffing box above the liquid,
(c) are effective in fluids of high viscosity,
(d) are easily fouled or plugged by solid particles,
(e) are expensive to fabricate,
(f) are restricted to a narrow range of speeds, and
(g) do not damage dispersed particles at economical speeds.

Open impellers

(a) are less easily plugged than the shrouded type,
(b) are less expensive, and
(c) give a less well-controlled flow pattern.

Anchors, helical ribbons and screws, are generally used for high viscosity liquids. The anchor and ribbon are arranged with a close clearance at the vessel wall, whereas the helical screw has a smaller diameter and is often used inside a draft tube to promote fluid motion throughout the vessel. Helical ribbons or interrupted ribbons are often used in horizontally mounted cylindrical vessels.

Kneaders, Z- and sigma-blade, and Banbury mixers as shown in Figure 7.20, are generally used for the mixing of high-viscosity liquids, pastes, rubbers, doughs, and so on. The tanks are usually mounted horizontally and two impellers are often used. The impellers are massive and the clearances between blades, as well as between the wall and blade, are very small thereby ensuring that the entire mass of liquid is sheared.

7.7.2. Portable mixers

For a wide range of applications, a portable mixer which can be clamped on the top or side of the vessel is often used. This is commonly fitted with two propeller blades so that the bottom rotor forces the liquid upwards and the top rotor forces the liquid downwards. The power supplied is up to about 2 kW, though the size of the motor becomes too great at higher powers. To avoid excessive strain on the armature, some form of flexible coupling should be fitted between the motor and the main propeller shaft. Units of this kind are usually driven at a fairly high rate (15 Hz), and a reduction gear can be fitted to the unit fairly easily for low-speed operations although this increases the mass of the unit.

7.7.3. Extruders

Mixing duties in the plastics industry are often carried out in either single or twin screw extruders. The feed to such units usually contains the base polymer in either granular or powder form, together with additives such as stabilisers, pigments, plasticisers, and so on. During processing in the extruder the polymer is melted and the additives mixed. The extrudate is delivered at high pressure and at a controlled rate from the extruder for shaping by means of either a die or a mould. Considerable progress has been made in the design of extruders in recent years, particularly by the application of finite element methods. One of the problems is that a considerable amount of heat is generated and the fluid properties may change by several orders of magnitude as a result of temperature changes. It is therefore always essential to solve the coupled equations of flow and heat transfer.

In the typical single-screw shown in Figure 7.24, the shearing which occurs in the helical channel between the barrel and the screw is not intense, and therefore this device does not give good mixing. Twin screw extruders, as shown in Figure 7.25, may be co- or counter-rotatory, and here there are regions where the rotors are in close proximity thereby generating extremely high shear stresses. Clearly, twin-screw units can yield a product of better mixture quality than a single-screw machine. Detailed accounts of the design and performance of extruders are available in literature[46−48].

Figure 7.24.   Single-screw extruder

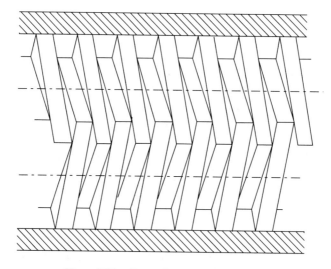

Figure 7.25.   Co-rotating twin-screw extruder

7.7.4. Static mixers

All the mixers so far described have been of the dynamic type in the sense that moving blades are used to impart motion to the fluid and produce the mixing effect. In static mixers,[2] sometimes called "in-line" or "motionless" mixers, the fluids to be mixed are pumped through a pipe containing a series of specially shaped stationary blades. Static mixers can be used with liquids of a wide range of viscosities in either the laminar or turbulent regimes, but their special features are perhaps best appreciated in relation to laminar flow mixing. The flow patterns within the mixer are complex, but a numerical simulation of the flow has been carried out by LANG et al.[49]

Figure 7.26 shows a particular type of static mixer in which a series of stationary helical blades mounted in a circular pipe is used to divide and twist the flowing streams. In laminar flow, (Figure 7.27) the material divides at the leading edge of each of these elements and follows the channels created by the element shape. At each succeeding element the two channels are further divided, and mixing proceeds by a distributive process similar to the cutting and folding mechanism shown in Figure 7.4. In principle,

Figure 7.26.   Twisted-blade type of static mixer elements

Figure 7.27.   Twisted-blade type of static mixer operating in the laminar flow regime (*a*) Distributive mixing mechanism showing, in principle, the reduction in striation thickness produced (*b*) Radial mixing contribution from laminar shear mechanism

if each element divided the streams neatly into two, feeding two dissimilar streams to the mixer would give a striated material in which the thickness of each striation would be of the order $D_T/2^n$ where D_T is the diameter of the tube and **n** is the number of elements. However, the helical elements shown in Figure 7.26 also induce further mixing by a laminar shear mechanism (see Figures 7.1 and 7.3). This, combined with the twisting action of the element, helps to promote radial mixing which is important in eliminating any radial gradients of composition, velocity and possibly temperature that might exist in the material. Figure 7.28 shows how these mixing mechanisms together produce after only 10–12 elements, a well-blended material. Figures 7.26 to 7.28 all refer to static mixers made by CHEMINEER Ltd, Derby, U.K.

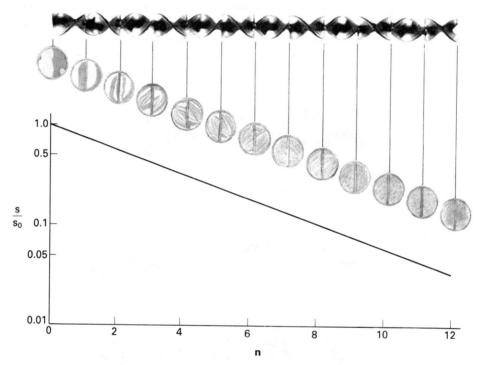

Figure 7.28.   Static mixer in laminar flow: reduction in relative standard deviation of samples indicating improvement in mixture quality with increasing number **n** of elements traversed

Figure 7.29 shows a Sulzer type SMX static mixer where the mixing element consists of a lattice of intermeshing and interconnecting bars contained in a pipe 80 mm diameter. It is recommended for viscous materials in laminar flow. The mixer shown is used in food processing, for example mixing fresh cheese with whipped cream.

Quantitatively, a variety of methods[50] has been proposed to describe the degree or quality of mixing produced in a static mixer. One of these measures of mixing quality is the relative standard deviation s/s_0, where s is the standard deviation in composition of a set of samples taken at a certain stage of the mixing operation, and s_0 is the standard deviation for a similar set of samples taken at the mixer inlet. Figure 7.28 shows schematically how the relative standard deviation falls as the number **n** of elements through which the

Figure 7.29.   Static mixer for viscous materials

material has passed increases, a perfectly mixed product having a zero relative standard deviation. One of the problems in using relative standard deviation or a similar index as a measure of mixing is that this depends on sample size which therefore needs to be taken into account in any assessment.

One of the most important considerations in choosing or comparing static mixers is the power consumed by the mixer to achieve a desired mixture quality. The pressure drop characteristics of a mixer are most conveniently described as the ratio of mixer pressure drop to empty pipe pressure drop for the same flowrate and diameter. Different static mixer designs can be compared[50] on a basis of mixing quality, pressure-drop ratio, initial cost and convenience of installation.

Static mixers for viscous fluids are widely used in processes producing polymers, fibres and plastics of all kinds where the materials tend to be viscous, hot and often at high pressures. However, static mixers have also achieved widespread use for low viscosity fluid mixing for blending, liquid-liquid, and even gas-liquid dispersions. In some cases the designs used for high viscosity liquids have also proved effective in the turbulent mixing regime for low viscosity fluids. In other cases manufacturers have developed special designs for turbulent flow mixing, and a wide variety of static mixer devices is now available[2].

7.7.5. Other types of mixer

Only a selection of commercially available mixing equipment has been described here. Indeed, the devices described all exist in a variety of configurations. In addition, there are many items of equipment based on altogether different principles; typical examples include jet mixers, in-line dynamic mixers, mills, valve homogenisers, ultrasonic homogenisers, etc. These, as well as many other types, have been discussed by HARNBY et al.[2], OLDSHUE[11], and NAGATA[19].

Frequently, it is convenient to mix the contents of a tank without the necessity for introducing an agitator and this may present difficulties in the construction of the vessel. It may then be possible to use an external circulating pump (usually centrifugal). If it is desirable not to make special connections to the vessel, it may be possible to connect the suction side of the pump to the bottom outlet of the tank by means of a T-piece and to discharge the liquid into the tank through its open top or through an existing entry point. In such a system, dispersion is effected in the high-shear region in the pump, and the liquid in the tank is maintained in a state of continuous circulation.

Such an arrangement may well be suitable when it is necessary to prevent fine particles from settling out at the bottom of the tank.

7.8. MIXING IN CONTINUOUS SYSTEMS

The mixing problems considered so far have related to batch systems in which two materials are mixed together and uniformity is maintained by continued operation of the agitator.

Frequently, stirred tanks are used with a continuous flow of material in on one side of the tank and with a continuous outflow from the other. A particular application is the use of the tank as a continuous-stirred-tank-reactor (CSTR). Inevitably, there will be a very wide range of residence times for elements of fluid in the tank. Even if the mixing is so rapid that the contents of the tank are always virtually uniform in composition, some elements of fluid will almost immediately flow to the outlet point and others will continue circulating in the tank for a very long period before leaving. The *mean residence time* of fluid in the tank is given by:

$$t_r = \frac{V}{Q} \tag{7.23}$$

where V is the volume of the contents of the tank (assumed constant), and Q is the volumetric throughput.

In a completely mixed system, the composition of the outlet stream will be equal to the composition in the tank.

The variation of time for which fluid elements remain with the tank is expressed as a *residence time distribution* and this can be calculated from a simple material balance if mixing is complete. For incomplete mixing, the calculation presents difficulties.

The problem is of great significance in the design of reactors because a varying residence time will, in general, lead to different degrees of chemical conversion of various fluid elements, and this is discussed in some detail in Volume 3, Chapter 1.

7.9. FURTHER READING

HARNBY, N., EDWARDS, M. F., and NIENOW, A. W. (eds): *Mixing in the Process Industries* 2nd edn (Butterworth-Heinemann, Oxford, 1992).

HOLLAND, F. A. and CHAPMAN, F. S.: *Liquid Mixing and Processing in Stirred Tanks* (Reinhold, New York, 1966).

METZNER, A. B.: In *Handbook of Fluid Dynamics*, section 7, V. L. Streeter (ed.) (McGraw-Hill, New York, 1961).

NAGATA, S.: *Mixing: Principles and Applications* (Wiley, London, 1975).

OLDSHUE, J. Y.: *Fluid Mixing Technology* (McGraw-Hill, New York, 1983).

TATTERSON, G. B.: *Fluid Mixing and Gas Dispersion in Agitated Tanks* (McGraw-Hill, New York, 1991).

UHL, V. W. and GRAY, J. B. (eds): *Mixing: Theory and Practice,* Vol. 1 (Academic Press, London, 1966); Vol. 2 (1967); Vol. 3 (1986).

7.10. REFERENCES

1. REAVELL, B. N.: *Trans. Inst. Chem. Eng.* **29** (1951) 301. Practical aspects of liquid mixing and agitation.
2. HARNBY, N., EDWARDS, M. F., and NIENOW, A. W. (eds): *Mixing in the Process Industries*, 2nd edn (Butterworth-Heinemann, London, 1992).
3. LEVENSPIEL, O.: *Engineering Flow and Heat Exchange* (Plenum, London, 1985).
4. OTTINO, J. M.: *Sci. Amer.* **260** (1989) 56. The mixing of fluids.
5. LEVICH, V. G.: *Physico-chemical Hydrodynamics* (Prentice-Hall, London, 1962).
6. MOLEN, VAN DER, K. and MAANEN, VAN, H. R. E.: *Chem. Eng. Sci.* **33** (1978) 1161. Laser-Doppler measurements of the turbulent flow in stirred vessels to establish scaling rules.
7. RAO, M. A. and BRODKEY, R. S.: *Chem. Eng. Sci.* **27** (1972) 137. Continuous flow stirred tank turbulence parameters in the impeller stream.
8. RUSHTON, J. H., COSTICH, E. W., and EVERETT, H. J.: *Chem. Eng. Prog.* **46** (1950) 395, 467. Power characteristics of mixing impellers. Parts I and II.
9. KRAMERS, H., BAARS, G. M., and KNOLL, W. H.: *Chem. Eng. Sci.* **2** (1953) 35. A comparative study on the rate of mixing in stirred tanks.
10. SKELLAND, A. H. P.: *Non-Newtonian Flow and Heat Transfer* (Wiley, New York, 1967).
11. OLDSHUE, J. Y.: *Fluid Mixing Technology* (McGraw-Hill, New York, 1983).
12. METZNER, A. B. and OTTO, R. E.: *A.I.Ch.E. Jl.* **3** (1957) 3. Agitation of non-Newtonian fluids.
13. METZNER, A. B. and TAYLOR, J. S.: *A.I.Ch.E. Jl.* **6** (1960) 109. Flow patterns in agitated vessels.
14. ULBRECHT, J.: *The Chemical Engineer* (London) No. 286 (1974) 347. Mixing of viscoelastic fluids by mechanical agitation.
15. ASTARITA, G.: *J. Non-Newt. Fluid Mech.* **4** (1979) 285. Scaleup problems arising with non-Newtonian fluids.
16. SKELLAND, A. H. P.: Mixing and agitation of non-Newtonian fluids. In *Handbook of Fluids in Motion*, Cheremisinoff, N. P. and Gupta, R. (eds) (Ann Arbor, New York, 1983).
17. HOLLAND, F. A. and CHAPMAN, F. S.: *Liquid Mixing and Processing in Stirred Tanks* (Reinhold, New York, 1966).
18. BISSELL, E. S., HESSE, H. C., EVERETT, H. J., and RUSHTON, J. H.: *Chem. Eng. Prog.* **43** (1947) 649. Design and utilization of internal fittings for mixing vessels.
19. NAGATA, S.: *Mixing: Principles and Applications* (Wiley, London, 1975).
20. UHL, V. W. and GRAY, J. B. (eds): *Mixing: Theory and Practice*, Vol. 1 (Academic Press, London, 1966).
21. ŠTĚRBÁČEK Z. and TAUSK, P.: *Mixing in the Chemical Industry*. Translated by Mayer, K. and Bourne, J. R. (Pergamon, Oxford, 1965).
22. HOOGENDOORN, C. J. and DEN HARTOG, A. P.: *Chem. Eng. Sci.* **22** (1967) 1689. Model studies on mixers in the viscous flow region.
23. ULLRICH, H. and SCHREIBER, H.: *Chem. Ing. Tech.* **39** (1967) 218. Rühren in zähen Flüssigkeiten.
24. MITSUISHI, N. and HIRAI, N. J.: *J. Chem. Eng. Japan* **2** (1969) 217. Power requirements in the agitation of non-Newtonian fluids.
25. GODFREY, J. C.: Mixing of high viscosity liquids. In: *Mixing in the Processing Industries*: Harnby, N., Edwards, M. F., and Nienow, A. W. (eds) (Butterworths, London, 1985) pp. 185–201.
26. CALDERBANK, P. H. and MOO-YOUNG, M. B.: *Trans. Inst. Chem. Eng.* **37** (1959) 26. The prediction of power consumption in the agitation of non-Newtonian fluids.
27. BECKNER, J. L. and SMITH, J. M.: *Trans. Inst. Chem. Eng.* **44** (1966) T224. Anchor-agitated systems: Power input with Newtonian and pseudo-plastic fluids.
28. NAGATA, S., NISHIKAWA, M., TADA, H., HIRABAYASHI, H., and GOTOH, S.: *J. Chem. Eng. Japan* **3** (1970) 237. Power consumption of mixing impellers in Bingham plastic fluids.
29. METZNER, A. B., FEEHS, R. H., RAMOS, H. L., OTTO, R. E., and TUTHILL, J. P.: *A.I.Ch.E. Jl.* **7** (1961) 3. Agitation of viscous Newtonian and non-Newtonian fluids.
30. EDWARDS, M. F., GODFREY, J. C., and KASHANI, M. M.: *J. Non-Newt. Fluid Mech.* **1** (1976) 309. Power requirements for the mixing of thixotropic fluids.
31. KELKAR, J. V., MASHELKAR, R. A., and ULBRECHT, J.: *Trans. Inst. Chem. Eng.* **50** (1972) 343. On the rotational viscoelastic flows around simple bodies and agitators.
32. RIEGER, F. and NOVAK, V.: *Trans. Inst. Chem. Eng.* **52** (1974) 285. Power consumption for agitating viscoelastic liquids in the viscous regime.
33. OLIVER, D. R., NIENOW, A. W., MITSON, R. J., and TERRY, K.: *Chem. Eng. Res. Des.* **62** (1984) 123. Power consumption in the mixing of Boger fluids.
34. PRUD'HOMME, R. K. and SHAQFEH, E.: *A.I.Ch.E. Jl.* **30** (1984) 485. Effect of elasticity on mixing torque requirements for Rushton turbine impellers.
35. KALE, D. D., MASHELKAR, R. A., and ULBRECHT, J.: *Chem. Ing. Tech.* **46** (1974) 69. High speed agitation of non-Newtonian fluids: Influence of elasticity and fluid inertia.

36. CHAVAN, V. V. and MASHELKAR, R. A.: *Adv. Transport Proc.* **1** (1980) 210. Mixing of viscous Newtonian and non-Newtonian fluids.
37. IRVING, H. F. and SAXTON, R. L.: In *Mixing Theory and Practice*, Uhl, V.W. and Gray, J. B. (eds) Vol. 2, pp. 169–224, (Academic Press, 1967).
38. GRAY, J. B.: In *Mixing Theory and Practice*, Uhl, V.W. and Gray, J. B. (eds) Vol. 1, pp. 179–278. (Academic Press, 1966).
39. CARREAU, P. J., PATTERSON, I., and YAP, C. Y.: *Can J. Chem. Eng.* **54** (1976) 135. Mixing of viscoelastic fluids with helical ribbon agitators I. Mixing time and flow pattern.
40. PETERS, D. C. and SMITH, J. M.: *Trans. Inst. Chem. Eng.* **45** (1967) 360. Fluid flow in the region of anchor agitator blades.
41. NORWOOD, K. W. and METZNER, A. B.: *A.I.Ch.E. Jl.* **6** (1960) 432. Flow patterns and mixing rates in agitated vessels.
42. BOURNE, J. R. and BUTLER, H.: *Trans. Inst. Chem. Eng.* **47** (1969) T11. On analysis of the flow produced by helical ribbon impellers.
43. GODLESKI, E. S. and SMITH, J. C.: *A.I.Ch.E. Jl.* **8** (1962) 617. Power requirements and blend times in the agitation of pseudoplastic fluids.
44. CHAVAN, V. V., ARUMUGAM, M. and ULBRECHT, J.: *A.I.Ch.E. Jl.* **21** (1975) 613. On the influence of liquid elasticity on mixing in a vessel agitated by a combined ribbon-screw impeller.
45. HALL, K. R. and GODFREY, J. C.: *Trans. Inst. Chem. Eng.* **46** (1968) 205. The mixing rates of highly viscous Newtonian and non-Newtonian fluids in a laboratory sigma blade mixer.
46. McKELVEY, J. M.: *Polymer Processing* (Wiley, New York, 1962).
47. JANSSEN, L. P. B. M.: *Twin Screw Extrusion* (Elsevier, Amsterdam, 1978).
48. SCHENKEL, G.: *Plastics Extrusion Technology* (Cliffe Books, London, 1966).
49. LANG, E., DRTINA, P., STREIFF, F. and FLEISCHLI, M.: *Chem. Eng. Sci.* **38** (1995) 2239. Numerical simulation of the fluid flow and mixing process in a static mixer.
50. HEYWOOD, N. I., VINEY, L. J., and STEWART, I. W.: *Inst. Chem. Eng.* Symposium Series No. 89, Fluid Mixing II (1984) 147. Mixing efficiencies and energy requirements of various motionless mixer designs for laminar mixing applications.

7.11. NOMENCLATURE

		Units in SI system	Dimensions in M, L, T
A	Area of flow	m^2	L^2
A'	Area for shear	m^2	L^2
C	Concentration	—	—
D	Impeller diameter	m	L
D_T	Tank on tube diameter	m	L
F	Force	N	MLT^{-2}
Fr	Froude number (N^2D/g)	—	—
g	Acceleration due to gravity	m/s^2	LT^{-2}
H	Depth of liquid	m	L
k_s	Function of impeller speed (equation 7.18)	—	—
L	Length	m	L
m	Mass of liquid	kg	M
N	Speed of rotation (revs/unit time)	Hz	T^{-1}
N_p	Power number ($P/\rho N^3 D^5$)	—	—
n	Power law index	—	—
n	Number of elements	—	—
P	Power	W	ML^2T^{-3}
P	Pitch of agitator	m	L
Q	Volumetric throughput	m^3/s	L^3T^{-1}
R	Number of baffles	—	—
r	Radius	m	L
s, s_0	Standard deviations	—	—
t	Time	s	T
t_m	Mixing time	s	T
t_r	Residence time	s	T

		Units in SI system	Dimensions in M, L, T
u	Velocity	m/s	$\mathbf{LT^{-1}}$
V	Volume of tank	m³	$\mathbf{L^3}$
W	Blade width	m	$\mathbf{L}$
W_B	Width of baffles	m	$\mathbf{L}$
y	Distance	m	$\mathbf{L}$
Z_A	Height of agitator from base of tank	m	$\mathbf{L}$
$\dot{\gamma}_{\mathrm{ang}}$	Shear rate (angular)	Hz	$\mathbf{T^{-1}}$
μ	Viscosity	N s/m²	$\mathbf{ML^{-1}T^{-1}}$
μ_a	Apparent viscosity	N s/m²	$\mathbf{ML^{-1}T^{-1}}$
ρ	Density	kg/m³	$\mathbf{ML^{-3}}$
σ	Surface tension	N/m	$\mathbf{MT^{-2}}$
σ^2	Variance	—	—
θ_m	Dimensionless mixing time Nt_m	—	—
Re	Reynolds number $(\rho ND^2/\mu)$	—	—

CHAPTER 8

Pumping of Fluids

8.1. INTRODUCTION

For the pumping of liquids or gases from one vessel to another or through long pipes, some form of mechanical pump is usually employed. The energy required by the pump will depend on the height through which the fluid is raised, the pressure required on delivery, the length and diameter of the pipe, the rate of flow, together with the physical properties of the fluid, particularly its viscosity and density. The pumping of liquids such as sulphuric acid or petroleum products like benzene or naphtha from bulk store to process buildings, or the pumping of fluids round reaction units and through heat exchangers, are typical illustrations of the use of pumps. The pumping of crude oil or natural gas over very long distances, or of fluids from one section of a plant to another, are further examples. On the one hand, it may be necessary to inject reactants or catalyst into a reactor at a low, but accurately controlled rate, and on the other to pump cooling water to a power station or refinery at a very high rate. The fluid may be a gas or liquid of low viscosity, or it may be a highly viscous liquid, possibly with non-Newtonian characteristics. It may be clear or it may contain suspended particles and be very corrosive. All these factors influence the choice of pump.

Because of the wide variety of requirements, many different types are in use including centrifugal, piston, gear, screw, and peristaltic pumps, though in the chemical and petroleum industries the centrifugal type is much the most important and is discussed in some detail later in the chapter.

Understanding the criteria for pump selection, determination of size and power requirements, assessing the positioning of pumps in relation to pipe systems form the main features of this chapter, though for greater detail the reader is referred to specialist publications included in section 8.7.

Pump design and construction is a specialist field, and manufacturers should always be consulted. In general, pumps used for circulating gases work at higher speeds than those used for liquids, and lighter valves are used. Moreover, the clearances between moving parts are smaller on gas pumps because of the much lower viscosity of gases, giving rise to an increased tendency for leakage to occur. When a pump is used to provide a vacuum, it is even more important to guard against leakage.

The work done by the pump is found by setting up an energy balance equation. If W_s is the shaft work done by unit mass of fluid on the surroundings, then $-W_s$ is the shaft work done on the fluid by the pump. From equations developed in Chapter 2:

from equation 2.55: $$-W_s = \Delta \frac{u^2}{2\alpha} + g\Delta z + \int_{P_1}^{P_2} v\, \mathrm{d}P + F \qquad (8.1)$$

and from equation 2.56: $-W_s = \Delta\dfrac{u^2}{2\alpha} + g\Delta z + \Delta H - q$ (8.2)

In any practical system, the pump will not be 100 per cent efficient, and more energy must be supplied by the motor driving the pump than is given by $-W_s$. If liquids are considered to be incompressible, there is no change in specific volume from the inlet to the delivery side of the pump. The physical properties of gases are, however, considerably influenced by the pressure, and the work done in raising the pressure of a gas is influenced by the rate of heat flow between the gas and the surroundings. Thus, if the process is carried out adiabatically, all the energy added to the system appears in the gas and its temperature rises. If an ideal gas is compressed and then cooled to its initial temperature, its enthalpy will be unchanged and the whole of the energy supplied by the compressor is dissipated to the surroundings. If, however, the compressed gas is allowed to expand it will absorb heat and is therefore capable of doing work at the expense of heat energy from the surroundings.

8.2. PUMPING EQUIPMENT FOR LIQUIDS

As already indicated, the liquids used in the chemical industries differ considerably in physical and chemical properties, and it has been necessary to develop a wide variety of pumping equipment. The two main forms are the positive displacement type and centrifugal pumps. In the former, the volume of liquid delivered is directly related to the displacement of the piston and therefore increases directly with speed and is not appreciably influenced by the pressure. In this group are the reciprocating piston pump and the rotary gear pump, both of which are commonly used for delivery against high pressures and where nearly constant delivery rates are required. The centrifugal type depends on giving the liquid a high kinetic energy which is then converted as efficiently as possible into pressure energy. For some applications, such as the handling of liquids which are particularly corrosive or contain abrasive solids in suspension, compressed air is used as the motive force instead of a mechanical pump. An illustration of the use of this form of equipment is the transfer of the contents of a reaction mixture from one vessel to another.

The following factors influence the choice of pump for a particular operation.

(1) The quantity of liquid to be handled. This primarily affects the size of the pump and determines whether it is desirable to use a number of pumps in parallel.
(2) The head against which the liquid is to be pumped. This will be determined by the difference in pressure, the vertical height of the downstream and upstream reservoirs and by the frictional losses which occur in the delivery line. The suitability of a centrifugal pump and the number of stages required will largely be determined by this factor.
(3) The nature of the liquid to be pumped. For a given throughput, the viscosity largely determines the friction losses and hence the power required. The corrosive nature will determine the material of construction both for the pump and the packing. With suspensions, the clearances in the pump must be large compared with the size of the particles.

(4) The nature of the power supply. If the pump is to be driven by an electric motor or internal combustion engine, a high-speed centrifugal or rotary pump will be preferred as it can be coupled directly to the motor. Simple reciprocating pumps can be connected to steam or gas engines.

(5) If the pump is used only intermittently, corrosion troubles are more likely than with continuous working.

The cost and mechanical efficiency of the pump must always be considered, and it may be advantageous to select a cheap pump and pay higher replacement or maintenance costs rather than to install a very expensive pump of high efficiency.

8.2.1. Reciprocating pump

The piston pump

The piston pump consists of a cylinder with a reciprocating piston connected to a rod which passes through a gland at the end of the cylinder as indicated in Figure 8.1. The liquid enters from the suction line through a suction valve and is discharged through a delivery valve. These pumps may be single-acting, with the liquid admitted only to the portion of the cylinder in front of the piston or double-acting, in which case the feed is admitted to both sides of the piston. The majority of pumps are of the single-acting type typically giving a low flowrate of say 0.02 m³/s at a high pressure of up to 100 MN/m².[1]

Figure 8.1.   A typical steam-driven piston pump

The velocity of the piston varies in an approximately sinusoidal manner and the volumetric rate of discharge of the liquid shows corresponding fluctuations. In a single-cylinder pump the delivery will rise from zero as the piston begins to move forward to a maximum when the piston is fully accelerated at approximately the mid point of its stroke; the delivery will then gradually fall off to zero. If the pump is single-acting there will be an interval during the return stroke when the cylinder will fill with liquid and the delivery will remain zero. On the other hand, in a double-acting pump the delivery will be similar in the forward and return strokes. In many cases, however, the cross-sectional area of the piston rod may be significant compared with that of the piston and the volume delivered during the return stroke will therefore be less than that during the forward stroke. A more even delivery is obtained if several cylinders are suitably compounded. If two double-acting cylinders are used there will be a lag between the deliveries of the two cylinders, and the total delivery will then be the sum of the deliveries from the individual cylinders. Typical curves of delivery rate for a single-cylinder (Simplex) pump are shown in Figure 8.2a. The delivery from a two-cylinder (Duplex) pump in which both the cylinders are double-acting is shown in Figure 8.2b; the broken lines indicate the deliveries from the individual cylinders and the unbroken line indicates the total delivery. It will be seen that the delivery is much smoother than that obtained with the simplex pump, the minimum delivery being equal to the maximum obtained from a single cylinder.

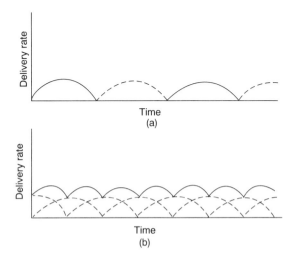

Figure 8.2.   Delivery from (a) simplex and (b) duplex pumps

The theoretical delivery of a piston pump is equal to the total swept volume of the cylinders. The actual delivery may be less than the theoretical value because of leakage past the piston and the valves or because of inertia of the valves. In some cases, however, the actual discharge is greater than theoretical value because the momentum of the liquid in the delivery line and sluggishness in the operation of the delivery valve may result in continued delivery during a portion of the suction stroke. The volumetric efficiency, which is defined as the ratio of the actual discharge to the swept volume, is normally greater than 90 per cent.

The size of the suction and delivery valves is determined by the throughput of the pump. Where the rate of flow is high, two or more valves may be used in parallel.

The piston pump can be directly driven by steam, in which case the piston rod is common to both the pump and the steam engine. Alternatively, an electric motor or an internal combustion engine may supply the motive power through a crankshaft; because the load is very uneven, a heavy flywheel should then be fitted and a regulator in the steam supply may often provide a convenient form of speed control.

The pressure at the delivery of the pump is made up of the following components:

(1) The static pressure at the delivery point.

(2) Pressure required to overcome the frictional losses in the delivery pipe.

(3) The pressure for the acceleration of the fluid at the commencement of the delivery stroke.

The liquid in the delivery line is accelerated and retarded in phase with the motion of the piston and therefore the whole of the liquid must be accelerated at the commencement of the delivery stroke and retarded at the end of it. Every time the fluid is accelerated, work has to be done on it and therefore in a long delivery line the expenditure of energy is very large since the excess kinetic energy of the fluid is not usefully recovered during the suction stroke. Due to the momentum of the fluid, the pressure at the pump may fall sufficiently low for separation to occur. The pump is then said to *knock*. The flow in the delivery line can be evened out and the energy at the beginning of each stroke reduced, by the introduction of an air vessel at the pump discharge. This consists of a sealed vessel which contains air at the top and liquid at the bottom. When the delivery stroke commences, liquid is pumped into the air vessel and the air is compressed. When the discharge from the pump decreases towards the end of the stroke, the pressure in the air vessel is sufficiently high for some of the liquid to be expelled into the delivery line. If the air vessel is sufficiently large and is fitted close to the pump, the velocity of the liquid in the delivery line can be maintained approximately constant. The frictional losses are also reduced by the incorporation of an air vessel because the friction loss under turbulent conditions is approximately proportional to the linear velocity in the pipe raised to the power 1.8; i.e. the reduced friction losses during the period of minimum discharge do not compensate for the greatly increased losses when the pump is delivering at maximum rate (see Section 8.6). Further, the maximum stresses set up in the pump are reduced by the use of an air vessel.

Air vessels are also incorporated in the suction line for a similar reason. Here they may be of even greater importance because the pressure drop along the suction line is necessarily limited to rather less than one atmosphere if the suction tank is at atmospheric pressure. The flowrate may be limited if part of the pressure drop available must be utilised in accelerating the fluid in the suction line; the air vessel should therefore be sufficiently large for the flowrate to be maintained approximately constant.

The plunger or ram pump

This pump is the same in principle as the piston type but differs in that the gland is at one end of the cylinder making its replacement easier than with the standard piston type. The sealing of piston and ram pumps has been much improved but, because of the nature of the fluids frequently used, care in selecting and maintaining the seal is very important. The piston or ram pump may be used for injections of small quantities of inhibitors to polymerisation units or of corrosion inhibitors to high pressure systems, and also for boiler feed water applications.

The diaphragm pump

The diaphragm pump has been developed for handling corrosive liquids and those containing suspensions of abrasive solids. It is in two sections separated by a diaphragm of rubber, leather, or plastics material. In one section a plunger or piston operates in a cylinder in which a non-corrosive fluid is displaced. The movement of the fluid is transmitted by means of the flexible diaphragm to the liquid to be pumped. The only moving parts of the pump that are in contact with the liquid are the valves, and these can be specially designed to handle the material. In some cases the movement of the diaphragm is produced by direct mechanical action, or the diaphragm may be air actuated as shown in Figure 8.3, in which case a particularly simple and inexpensive pump results, capable of operating up to 0.2 MN/m^2.

Figure 8.3.   Diaphragm pump

When pumping non-Newtonian fluids, difficulties are sometimes experienced in getting pseudoplastic materials to flow initially. Positive displacement pumps can overcome the problem and the diaphragm pump in particular is useful in dealing with agglomerates in suspension as the liquid is subjected to a minimum of shearing. Care must always to be taken that the safe working pressure for the pump is not exceeded. This can be achieved conveniently by using a hydraulic drive for a diaphragm pump, equipped with a pressure limiting relief valve which ensures that no damage is done to the system as shown in Figure 8.4.

Figure 8.4.   A hydraulic drive to protect a positive displacement pump

By virtue of their construction, diaphragm pumps cannot be used for high pressure applications. In the Mars pump, there is no need for a diaphragm as the working fluid (oil), of lower density than the liquid to be pumped, forms an interface with it in a vertical chamber. The pump, which is used extensively for concentrated slurries, is really a development of the old Ferrari's acid pump which was designed for corrosive liquids.

The metering pump

Metering pumps are positive displacement pumps driven by constant speed electric motors. They are used where a constant and accurately controlled rate of delivery of a liquid is required, and they will maintain this constant rate irrespective of changes in the pressure against which they operate. The pumps are usually of the plunger type for low throughput and high-pressure applications; for large volumes and lower pressures a diaphragm is used. In either case, the rate of delivery is controlled by adjusting the stroke of the piston element, and this can be done whilst the pump is in operation. A single-motor driver may operate several individual pumps and in this way give control of the actual flows and of the flow ratio of several streams at the same time. The output may be controlled from zero to maximum flowrate, either manually on the pump or remotely. These pumps may be used for the dosing of works effluents and water supplies, and the feeding of reactants, catalysts, or inhibitors to reactors at controlled rates, and although a simple method for controlling flowrate is provided, high precision standards of construction are required.

Example 8.1

A single-acting reciprocating pump has a cylinder diameter of 110 mm and a stroke of 230 mm. The suction line is 6 m long and 50 mm in diameter and the level of the water in the suction tank is 3 m below the cylinder of the pump. What is the maximum speed at which the pump can run without an air vessel if separation is not to occur in the suction line? The piston undergoes approximately simple harmonic motion. Atmospheric pressure is equivalent to a head of 10.36 m of water and separation occurs at an absolute pressure corresponding to a head of 1.20 m of water.

Solution

The tendency for separation to occur will be greatest at:

(a) the inlet to the cylinder because here the static pressure is a minimum and the head required to accelerate the fluid in the suction line is a maximum;
(b) the commencement of the suction stroke because the acceleration of the piston is then a maximum.

If the maximum permissible speed of the pump is N Hz:
Angular velocity of the driving mechanism $= 2\pi N$ radians/s
Acceleration of piston $= 0.5 \times 0.230(2\pi N)^2 \cos(2\pi Nt)$ m/s²
Maximum acceleration (i.e. when $t = 0$) $= 4.54N^2$ m/s²
Maximum acceleration of the liquid in the suction pipe

$$= \left(\frac{0.110}{0.05}\right)^2 \times 4.54N^2 = 21.97N^2 \text{ m/s}^2$$

Accelerating force acting on the liquid

$$= 21.97N^2 \frac{\pi}{4}(0.050)^2 \times 6 \times 1000 \text{ N}$$

Pressure drop in suction line due to acceleration $= 21.97N^2 \times 6 \times 1000$ N/m²

$$= 1.32 \times 10^5 N^2 \text{ N/m}^2$$

or:
$$\frac{1.32 \times 10^5 N^2}{1000 \times 9.81} = 13.44N^2 \text{ m water}$$

Pressure head at cylinder when separation is about to occur,

$$1.20 = (10.36 - 3.0 - 13.44N^2) \text{ m water}$$

∴
$$N = 0.675 \text{ Hz}$$

8.2.2. Positive displacement rotary pumps

The gear pump

The gear pump (Figure 8.5) is the most widely used of the positive action rotary pumps. Two gear wheels operate inside a casing; one of the gear wheels is driven and the other rotates in mesh with it. The liquid is carried round in the space between consecutive teeth and the casing and is then ejected as the teeth come into mesh. The pump has no valves and depends for its seal on the small clearance between the gear wheels and the case. It is a positive displacement pump and will deliver against high pressures of the order of 35 MN/m². The delivery is almost independent of pressure and priming is unnecessary.

HARVEST[2] has discussed recent developments and applications of gear pumps but the main advantage of the gear pump over the reciprocating pump is that it gives an even

Figure 8.5.   Gear pump

Figure 8.6.   Flow inducer

delivery and can be directly connected to an electric motor drive. Gear pumps will handle liquids of very high viscosities and are extensively used in the oil industry for pumping viscous residual oils from distillation plant. Because the spaces between the gear teeth are comparatively small, the pump cannot be used for suspensions.

The flow inducer or peristaltic pump

This is a special form of pump in which a length of silicone rubber or other elastic tubing, typically of 3 to 25 mm diameter, is compressed in stages by means of a rotor as shown in Figure 8.6. The tubing is fitted to a curved track mounted concentrically with a rotor carrying three rollers. As the rollers rotate, they flatten the tube against the track at the points of contact. These 'flats' move the fluid by positive displacement, and the flow can be precisely controlled by the speed of the motor.

These pumps have been particularly useful for biological fluids where all forms of contact must be avoided. They are being increasingly used and are suitable for pumping emulsions, creams, and similar fluids in laboratories and small plants where the freedom from glands, avoidance of aeration, and corrosion resistance are valuable, if not essential. Recent developments[1] have produced thick-wall, reinforced moulded tubes which give a pumping performance of up to 0.02 m^3/s at 1 MN/m^2 and these pumps have been further discussed by GADSDEN[3]. The control is such that these pumps may conveniently be used as metering pumps for dosage processes.

The Mono pump

Another example of a positive acting rotary pump is the Mono pump, illustrated in Figure 8.7, in which a specially shaped helical metal rotor revolves eccentrically within a double-helix, resilient rubber stator of twice the pitch length of the metal rotor. A continuous forming cavity is created as the rotor turns — the cavity progressing towards the discharge, advancing in front of a continuously forming seal line and thus carrying the pumped material with it as shown in Figure 8.8.

The Mono pump gives a uniform flow and is quiet in operation. It will pump against high pressures; the higher the required pressure, the longer are the stator and the rotor

Figure 8.7. Mono pump

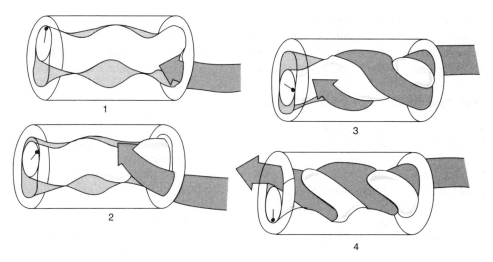

Figure 8.8. The mode of operation of a Mono pump

and the greater the number of turns. The pump can handle corrosive and gritty liquids and is extensively used for feeding slurries to filter presses. It must never be run dry.

Screw pumps

A most important class of pump for dealing with highly viscous material is represented by the screw extruder used in the polymer industry. Extruders find their main application in the manufacture of simple and complex sections (rods, tubes, beadings, curtain rails, rainwater gutterings and a multitude of other shapes). However, the shape of section produced in a given material is dependent only on the profile of the hole through which the fluid is pushed just before it cools and solidifies. The screw pump is of more general application and will be considered first.

The principle is shown in Figure 8.9. The fluid is sheared in the channel between the screw and the wall of the barrel. The mechanism that generates the pressure can be

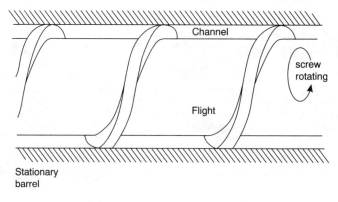

Figure 8.9. Section of a screw pump

visualised in terms of a model consisting of an open channel covered by a moving plane surface (Figure 8.10). This representation of a screw pump takes as the frame of reference a stationary screw with rotating barrel. The planar simplification is not unreasonable, provided that the depth of the screw channel is small with respect to the barrel diameter. It should also be recognised that the distribution of centrifugal forces would be different according to whether the rotating member is the wall or the screw: this distinction would have to be drawn if a detailed force balance were to be attempted, but in any event the centrifugal (inertial) forces are generally far smaller than the viscous forces.

Figure 8.10. Planar model of part of a screw pump

If the upper plate moved along in the direction of the channel, then a velocity profile would be established that would approximate to the linear velocity gradient that exists between planar walls moving parallel to each other. If it moved at right angles to the channel axis, however, a circulation would be developed in the gap, as drawn in Figure 8.11. In fact, the relative movement of the barrel wall is somewhere in between, and is determined by the pitch of the screw. The fluid path in a screw pump is therefore of a complicated helical form within the channel section. The nature of the velocity components along the channel depends on the pressure generated and the amount of resistance at the discharge end. If there is no resistance, the velocity distribution in the channel direction will be the Couette simple shear profile shown in Figure 8.12a. With a totally closed discharge end the net flow would be zero, but the velocity components at the walls would not be affected. As a result, the flow field necessarily would be of the form shown in Figure 8.12b.

Figure 8.11. Fluid displacement resulting from movement of plane surface

Viscous forces within the fluid will always prevent a completely unhindered discharge, but in extrusion practice an additional die head resistance is used to generate backflow and mixing, so that a more uniform product is obtained. The flow profile along the channel is then of some intermediate form, such as that shown in Figure 8.12c.

It must be emphasised that flow in a screw pump is produced as a result of viscous forces. Pressures achieved with low viscosity materials are negligible. The screw pump is

Figure 8.12. Velocity profile produced between screw pump surfaces (*a*) with no restriction on fluid flow
(*b*) with no net flow (total restriction) (*c*) with a partially restricted discharge

not therefore a modification of the Archimedes screw used in antiquity to raise water — that
was essentially a positive displacement device using a deep cut helix mounted at an angle
to the horizontal, and not running full. If a detailed analysis of the flow in a screw pump is
to be carried out, then it is also necessary to consider the small but finite leakage flow that
can occur between the flight and the wall. With the large pressure generation in a polymer
extruder, commonly 100 bar (10^7 N/m^2), the flow through this gap, which is typically
about 2 per cent of the barrel internal diameter, can be significant. The pressure drop over
a single pitch length may be of the order of 10 bar (10^6 N/m^2), and this will force fluid
through the gap. Once in this region the viscous fluid is subject to a high rate of shear
(the rotation speed of the screw is often about 2 Hz), and an appreciable part of the total
viscous heat generation occurs in this region of an extruder.

8.2.3. The centrifugal pump

The centrifugal pump is by far the most widely used type in the chemical and petroleum
industries. It will pump liquids with very wide ranging properties and suspensions with
a high solids content including, for example, cement slurries, and may be constructed
from a very wide range of corrosion resistant materials. The whole pump casing may
be constructed from plastics such as polypropylene or it may be fitted with a corrosion-
resistant lining. Because it operates at high speed, it may be directly coupled to an electric
motor and it will give a high flowrate for its size.

In this type of pump, the fluid is fed to the centre of a rotating impeller and is thrown
outward by centrifugal action. As a result of the high speed of rotation the liquid acquires
a high kinetic energy and the pressure difference between the suction and delivery sides
arises from the conversion of kinetic energy into pressure energy.

The impeller (Figure 8.13) consists of a series of curved vanes so shaped that the flow
within the pump is as smooth as possible. The greater the number of vanes on the impeller,
the greater is the control over the direction of motion of the liquid and hence the smaller
are the losses due to turbulence and circulation between the vanes. In the open impeller,
the vanes are fixed to a central hub, whereas in the closed type the vanes are held between
two supporting plates and leakage across the impeller is reduced. As will be seen later,
the angle of the tips of the blades very largely determines the operating characteristics of
the pump.

The liquid enters the casing of the pump, normally in an axial direction, and is picked up
by the vanes of the impeller. In the simple type of centrifugal pump, the liquid discharges
into a volute, a chamber of gradually increasing cross-section with a tangential outlet.
A volute type of pump is shown in Figure 8.14*a*. In the turbine pump (Figure 8.14*b*)
the liquid flows from the moving vanes of the impeller through a series of fixed vanes

Figure 8.13.   Types of impeller (*a*) fully-shrouded radial-flow impeller (*b*) fully-shrouded mixed-flow impeller
(*c*) semi-open mixed-flow impeller (*d*) solid-passing impeller

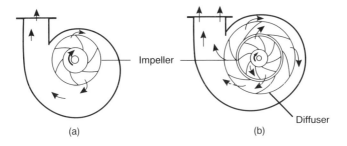

Figure 8.14.   Radial flow pumps (*a*) with volute (*b*) with diffuser vanes

forming a diffusion ring. This gives a more gradual change in direction to the fluid and
more efficient conversion of kinetic energy into pressure energy than is obtained with the
volute type. The angle of the leading edge of the fixed vanes should be such that the fluid
is received without shock. The liquid flows along the surface of the impeller vane with a
certain velocity whilst the tip of the vane is moving relative to the casing of the pump.
The direction of motion of the liquid relative to the pump casing — and the required angle

of the fixed vanes — is found by compounding these two velocities. In Figure 8.15, u_v is the velocity of the liquid relative to the vane and u_t is the tangential velocity of the tip of the vane; compounding these two velocities gives the resultant velocity u_2 of the liquid. It is apparent, therefore, that the required vane angle in the diffuser is dependent on the throughput, the speed of rotation, and the angle of the impeller blades. The pump will therefore operate at maximum efficiency only over a narrow range of conditions.

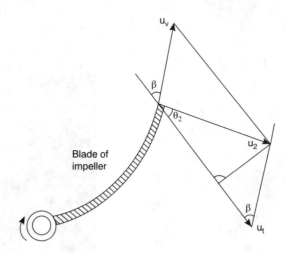

Figure 8.15.   Velocity diagram

Virtual head of a centrifugal pump

The maximum pressure is developed when the whole of the excess kinetic energy of the fluid is converted into pressure energy. As indicated below, the head is proportional to the square of the radius and to the speed, and is of the order of 60 m for a single-stage centrifugal pump; for higher pressures, multistage pumps must be used. Consider the liquid which is rotating at a distance of between r and $r + dr$ from the centre of the pump (Figure 8.16). The mass of this element of fluid dM is given by $2\pi r\,drb\rho$, where ρ is the density of the fluid and b is the width of the element of fluid.

If the fluid is travelling with a velocity u and at an angle θ to the tangential direction, the angular momentum of this mass of fluid

$$= dM\,(ur\cos\theta)$$

The torque acting on the fluid $d\tau$ is equal to the rate of change of angular momentum with time, as it goes through the pump:

$$\therefore \qquad d\tau = dM\,\frac{\partial}{\partial t}(ur\cos\theta)$$

$$= 2\pi rb\rho\,dr\frac{\partial}{\partial t}(ur\cos\theta) \qquad\qquad (8.3)$$

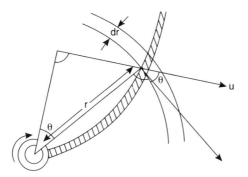

Figure 8.16.   Virtual head

The volumetric rate of flow of liquid through the pump:

$$Q = 2\pi rb\frac{\partial r}{\partial t} \tag{8.4}$$

$$\therefore \qquad d\tau = Q\rho\, d(ur\cos\theta) \tag{8.5}$$

The total torque acting on the liquid in the pump is therefore obtained by integrating $d\tau$ between the limits denoted by suffix 1 and suffix 2, where suffix 1 refers to the conditions at the inlet to the pump and suffix 2 refers to the conditions at the discharge.

Thus: $\qquad\qquad \tau = Q\rho(u_2 r_2 \cos\theta_2 - u_1 r_1 \cos\theta_1) \tag{8.6}$

The power **P** developed by the pump is equal to the product of the torque and the angular velocity ω:

$$\therefore \qquad \mathbf{P} = Q\rho\omega(u_2 r_2 \cos\theta_2 - u_1 r_1 \cos\theta_1) \tag{8.7}$$

The power can also be expressed as the product Ghg, where G is the mass rate of flow of liquid through the pump, g is the acceleration due to gravity, and h is termed the virtual head developed by the pump.

Thus: $\qquad\qquad Ghg = Q\rho\omega(u_2 r_2 \cos\theta_2 - u_1 r_1 \cos\theta_1)$

i.e.: $\qquad\qquad h = \dfrac{\omega(u_2 r_2 \cos\theta_2 - u_1 r_1 \cos\theta_1)}{g} \tag{8.8}$

Since u_1 will be approximately zero, the virtual head:

$$h = \frac{\omega u_2 r_2 \cos\theta_2}{g} \tag{8.9}$$

where g, ω and r_2 are known in any given instance, and u_2 and θ_2 are to be expressed in terms of known quantities.

From the geometry of Figure 8.15:

$$u_v \sin\beta = u_2 \sin\theta_2 \tag{8.10}$$

and: $\qquad\qquad u_t = u_v \cos\beta + u_2 \cos\theta_2 \tag{8.11}$

(where β is the angle between the tip of the blade of the impeller and the tangent to the direction of its motion. If the blade curves backwards, β lies between 0 and $\pi/2$ and if it curves forwards, β lies between $\pi/2$ and π).

The volumetric rate of flow through the pump Q is equal to the product of the area available for flow at the outlet of the impeller and the radial component of the velocity,

i.e.:
$$Q = 2\pi r_2 b u_2 \sin \theta_2$$

$$= 2\pi r_2 b u_v \sin \beta \quad \text{(from equation 8.10)} \tag{8.12}$$

$\therefore$
$$u_v = \frac{Q}{2\pi r_2 b \sin \beta} \tag{8.13}$$

Thus:
$$h = \frac{\omega r_2 (u_t - u_v \cos \beta)}{g} \quad \text{(from equations 8.9 and 8.11)}$$

$$= \frac{\omega}{g} r_2 \left(r_2 \omega - \frac{Q}{2\pi r_2 b \tan \beta} \right) \quad \text{(since } u_t = r_2 \omega\text{)}$$

$$= \frac{r_2^2 \omega^2}{g} - \frac{Q\omega}{2\pi b g \tan \beta} \tag{8.14}$$

The virtual head developed by the pump is therefore independent of the density of the fluid, and the pressure will thus be directly proportional to the density. For this reason, a centrifugal pump needs priming. If the pump is initially full of air, the pressure developed is reduced by a factor equal to the ratio of the density of air to that of the liquid, and is insufficient to drive the liquid through the delivery pipe.

For a given speed of rotation, there is a linear relation between the head developed and the rate of flow. If the tips of the blades of the impeller are inclined backwards, β is less than $\pi/2$, $\tan \beta$ is positive, and therefore the head decreases as the throughput increases. If β is greater than $\pi/2$ (i.e. the tips of the blades are inclined forwards), the head increases as the delivery increases. The angle of the blade tips therefore profoundly affects the performance and characteristics of the pump. For radial blades the head should be independent of the throughput.

Specific speed

If θ_2 remains approximately constant, u_v, u_t, and u_2 will all be directly proportional to one another, and since $u_t = r_2 \omega$, these velocities are proportional to r_2; thus u_v will vary as $r_2 \omega$.

The output from a pump is a function of its linear dimensions, the shape, number, and arrangement of the impellers, the speed of rotation, and the head against which it is operating. From equation 8.12, for a radial pump with $\beta = \pi/2$ and $\sin \beta = 1$:

$$Q \propto 2\pi r_2 b u_v$$

$$Q \propto 2\pi r_2 b r_2 \omega$$

$$Q \propto r_2^2 b \omega \tag{8.15}$$

When $\tan \beta = \tan \pi/2 = \infty$, then from equation 8.14:

$$h = \frac{r_2^2 \omega^2}{g} \tag{8.16}$$

$$gh = r_2^2 \omega^2$$

or:

$$(gh)^{3/4} = r_2^{3/2} \omega^{3/2} \tag{8.17}$$

For a series of geometrically similar pumps, b is proportional to the radius r_2, and thus, from equation 8.15:

$$Q \propto r_2^3 \omega \tag{8.18}$$

or:

$$Q^{1/2} \propto r_2^{3/2} \omega^{1/2} \tag{8.19}$$

Eliminating r_2 between equations 8.17 and 8.19:

$$\frac{Q^{1/2}}{(gh)^{3/4}} = \frac{\omega^{1/2}}{\omega^{3/2}}$$

or:

$$\frac{\omega Q^{1/2}}{(gh)^{3/4}} = \text{constant} = N_s \text{ for geometrically similar pumps} \tag{8.20}$$

Criteria for similarity

The dimensionless quantity $\omega Q^{1/2}/(gh)^{3/4}$ is a characteristic for a particular type of centrifugal pump, and, noting that the angular velocity is proportional to the speed N, this group may be rewritten as:

$$N_s = \frac{N Q^{1/2}}{(gh)^{3/4}} \tag{8.21}$$

and is constant for geometrically similar pumps. N_s is defined as the specific speed and is frequently used to classify types of centrifugal pumps. Specific speed may be defined as the speed of the pump which will produce unit flow Q against unit head h under conditions of maximum efficiency.

Equation 8.21 is dimensionless but specific speed is frequently quoted in the form:

$$N_s = \frac{N Q^{1/2}}{h^{3/4}} \tag{8.22}$$

where the impeller speed N is in rpm, the volumetric flowrate Q in US gpm and the total head developed is in ft. In this form, specific speed has dimensions of $(\mathbf{LT}^{-2})^{3/4}$ and for centrifugal pumps has values between 400 and 10,000 depending on the type of impeller. The use of specific speed in pump selection is discussed more fully in Volume 6.

Operating characteristics

The operating characteristics of a pump are conveniently shown by plotting the head h, power $\mathbf{P}$, and efficiency η against the flow Q as shown in Figure 8.17. It is important to note that the efficiency reaches a maximum and then falls, whilst the head at first falls

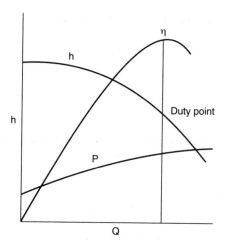

Figure 8.17.   Radial flow pump characteristics

slowly with Q but eventually falls off rapidly. The optimum conditions for operation are
shown as the duty point, i.e. the point where the head curve cuts the ordinate through the
point of maximum efficiency.

A set of curves for h, η, and Q for various speeds are shown in Figure 8.18 from
which it is seen that the efficiency remains reasonably constant over a range of speeds.
A more general indication of the variation of efficiency with specific speed is shown in

Figure 8.18.   Characteristic curves for centrifugal pump

Figure 8.19 for different types of centrifugal pumps. The power developed by a pump is proportional to $Qgh\rho$:

i.e.:
$$\mathbf{P} \propto r_2^2 \omega r_2^2 \omega^2 \rho$$

or:
$$\mathbf{P} \propto r_2^4 b \omega^3 \rho \qquad\qquad (8.23)$$

so that $Q \propto \omega$; $h \propto \omega^2$; $\mathbf{P} \propto \omega^3$, from equations 8.15, 8.16 and 8.23.

Figure 8.19.   Specific speed and efficiency

Cavitation

In designing any installation in which a centrifugal pump is used, careful attention must be paid to check the minimum pressure which will arise at any point. If this pressure is less than the vapour pressure at the pumping temperature, vaporisation will occur and the pump may not be capable of developing the required suction head. Moreover, if the liquid contains gases, these may come out of solution giving rise to pockets of gas. This phenomenon is known as *cavitation* and may result in mechanical damage to the pump as the bubbles collapse. The tendency for cavitation to occur is accentuated by any sudden changes in the magnitude or direction of the velocity of the liquid in the pump. The onset of cavitation is accompanied by a marked increase in noise and vibration as the vapour bubbles collapse, and also a loss of head.

Suction head

Pumps may be arranged so that the inlet is under a suction head or the pump may be fed from a tank. These two systems alter the duty point curves as shown in Figure 8.20. In developing such curves, the normal range of liquid velocities is 1.5 to 3 m/s, but lower values are used for pump suction lines. With the arrangement shown in Figure 8.20a, there can be problems in priming the pump and it may be necessary to use a self-priming centrifugal pump.

For any pump, the manufacturers specify the minimum value of the *net positive suction head* (NPSH) which must exist at the suction point of the pump. The NPSH is the amount by which the pressure at the suction point of the pump, expressed as a head of the liquid to be pumped, must exceed the vapour pressure of the liquid. For any installation this must be calculated, taking into account the absolute pressure of the liquid, the level of the pump, and the velocity and friction heads in the suction line. The NPSH must allow for the fall in pressure occasioned by the further acceleration of the liquid as it flows on to the impeller and for irregularities in the flow pattern in the pump. If the required value

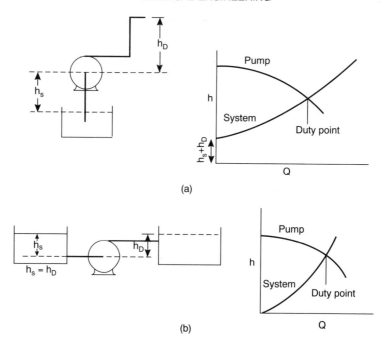

(a)

(b)

Figure 8.20.   Effect of suction head: (a) systems with suction lift and friction; (b) systems with
friction losses only

of NPSH is not obtained, partial vaporisation is liable to occur, with the result that both
suction head and delivery head may be reduced. The loss of suction head is the more
important because it may cause the pump to be starved of liquid.

Consider the system shown in Figure 8.21, in which the pump is taking liquid from a
reservoir at an absolute pressure P_0 in which the liquid level is at a height h_0 above the
suction point of the pump. Then if the liquid in the reservoir can be regarded as at rest,
the *absolute* pressure head h_i at the suction point of the pump is obtained by applying

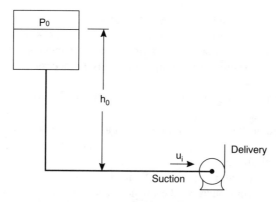

Figure 8.21.   Suction system of centrifugal pump

the energy or momentum balance:

$$h_i = \frac{P_0}{\rho g} + h_0 - \frac{u_i^2}{2g} - h_f \qquad (8.24)$$

where h_f is the head lost in friction, and u_i is the velocity at the inlet of the pump. If the vapour pressure of the liquid is P_v, the NPSH is given by the difference between the *total* head at the suction inlet and the head corresponding to the vapour pressure P_v of the liquid at the pump inlet.

$$\text{NPSH} = \left(h_i + \frac{u_i^2}{2g} \right) - \frac{P_v}{\rho g} \qquad (8.25)$$

$$= \frac{P_0}{\rho g} - \frac{P_v}{\rho g} + h_0 - h_f \qquad (8.26)$$

In equation 8.26, it is implicitly assumed that the kinetic head of the inlet liquid is available for conversion into pressure head. If this is not so, $u_i^2/2g$ must be deducted from the NPSH.

If cavitation and loss of suction head does occur, it can sometimes be cured by increasing the pressure in the system, either by alteration of the layout to provide a greater hydrostatic pressure or a reduced pressure drop in the suction line. Sometimes, slightly closing the valve on the pump delivery or reducing the pump speed by a small amount may be effective. Generally a small fast-running pump will require a larger NPSH than a larger slow-running pump.

The efficiency of a centrifugal pump and the head which it is capable of developing are dependent upon achieving a good seal between the rotating shaft and the casing of the pump and inefficient operation is frequently due to a problem with that seal or with the gland. When the pump is fitted with the usual type of packed gland, maintenance costs are often very high, especially when organic liquids of low viscosity are being pumped. A considerable reduction in expenditure on maintenance can be effected at the price of a small increase in initial cost by fitting the pump with a mechanical seal, in which the sealing action is achieved as a result of contact between two opposing faces, one stationary and the other rotating. In Figure 8.22, a mechanical seal is shown in position in a centrifugal pump. The stationary seat *A* is held in position by means of the clamping plate *D*. The seal is made with the rotating face on a carbon ring *B*. The seal between the rotating face and the shaft is made by means of the wedge ring *C*, usually made of polytetrafluoroethylene (PTFE). The drive is through the retainer *E* secured to the shaft, usually by Allen screws. Compression between the fixed and rotating faces is provided by the spiral springs *F*.

It is advantageous to ensure that the seal is fed with liquid which removes any heat generated at the face. In the illustration this is provided by the connection *G*.

Centrifugal pumps must be fitted with good bearings since there is a tendency for an axial thrust to be produced if the suction is taken only on one side of the impeller. This thrust can be balanced by feeding back a small quantity of the high-pressure liquid to a specially designed thrust bearing. By this method, the risk of air leaking into the pump at the gland — and reducing the pressure developed — is minimised. The glandless centrifugal pump, which is used extensively for corrosive liquids, works on a similar principle, and the use of pumps without glands has been increasing both in the nuclear power industry and in the chemical process industry. Such units are totally enclosed and

Figure 8.22.   Mechanical seal for centrifugal pump

lubrication is provided by the fluid handled. Figure 8.23 gives the performance characteristics at a particular speed of one group of pumps of this type and similar data are
available for pump selection of other centrifugal types.

Centrifugal pumps are made in a wide range of materials, and in many cases the
impeller and the casing are covered with resistant material. Thus stainless steel, nickel,
rubber, polypropylene, stoneware, and carbon are all used. When the pump is used with
suspensions, the ports and spaces between the vanes must be made sufficiently large to
eliminate the risk of blockage. This does mean, however, that the efficiency of the pump
is reduced. The *Vacseal pump*, developed by the International Combustion Company for
pumping slurries, will handle suspensions containing up to 50 per cent by volume of
solids. The whole impeller may be rubber covered and has three small vanes, as shown in
Figure 8.24. The back plate of the impeller has a second set of vanes of larger diameter.
The pressure at the gland is thereby reduced below atmospheric pressure and below the
pressure in the suction line; there is, therefore, no risk of the gritty suspension leaking
into the gland and bearings. If leakage does occur, air will enter the pump. As mentioned
previously, this may reduce the pressure which the pump can deliver, but this is preferable

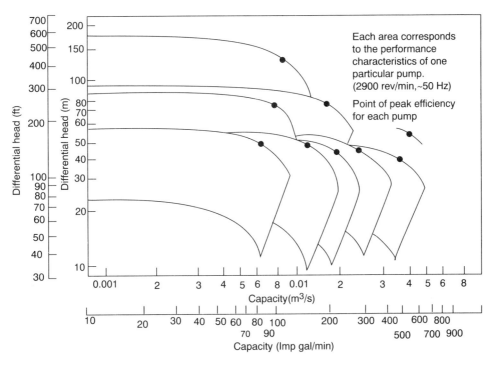

Figure 8.23.   Performance characteristics of a family of glandless centrifugal pumps

Figure 8.24.   Sectioned arrangement of a 38 mm moulded rubber V-type Vacseal pump

to damaging the bearings by allowing them to become contaminated with grit. This is another example of the necessity for tolerating rather low efficiencies in pumps that handle difficult materials. The pumping of slurries has been considered by STEELE and ODROWAZ-PIENIAZEK[4] who also discuss the many aspects of selection and efficient operation of centrifugal pumps in these demanding circumstances.

The advantages and disadvantages of the centrifugal pump

The main advantages are:

(1) It is simple in construction and can, therefore, be made in a wide range of materials.
(2) There is a complete absence of valves.
(3) It operates at high speed (up to 100 Hz) and, therefore, can be coupled directly to an electric motor. In general, the higher the speed the smaller the pump and motor for a given duty.
(4) It gives a steady delivery.
(5) Maintenance costs are lower than for any other type of pump.
(6) No damage is done to the pump if the delivery line becomes blocked, provided it is not run in this condition for a prolonged period.
(7) It is much smaller than other pumps of equal capacity. It can, therefore, be made into a sealed unit with the driving motor and immersed in the suction tank.
(8) Liquids containing high proportions of suspended solids are readily handled.

The main disadvantages are:

(1) The single-stage pump will not develop a high pressure. Multistage pumps will develop greater heads but they are very much more expensive and cannot readily be made in corrosion-resistant material because of their greater complexity. It is generally better to use very high speeds in order to reduce the number of stages required.
(2) It operates at a high efficiency over only a limited range of conditions: this applies especially to turbine pumps.
(3) It is not usually self-priming.
(4) If a non-return valve is not incorporated in the delivery or suction line, the liquid will run back into the suction tank as soon as the pump stops.
(5) Very viscous liquids cannot be handled efficiently.

Pumping of non-Newtonian fluids

As stated in the introduction to centrifugal pumps, the principle of operation is the conversion of kinetic energy into a static pressure head which produces the flow. For a pump of this type the distribution of shear within the pump will vary with throughput. Considering Figure 8.25, when the discharge is completely closed off the highest degree of shearing is in the gap between the rotor and the shell, i.e. at B. Within the vanes of the rotor (region A) there will be some circulation as sketched in Figure 8.26, but in the discharge line C the fluid will be essentially static. If fluid is flowing through the pump, the differences between these shear rates will still be present but not as extreme. If the fluid has pseudoplastic properties, then the effective viscosity will vary in these different

Figure 8.25.   Zones of differing shear in a centrifugal pump

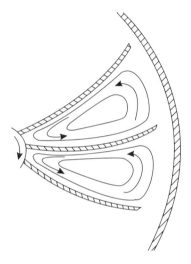

Figure 8.26.   Circulation within a centrifugal pump impeller

regions, being less at *B* than at *A* and *C*. Under steady conditions the pressure developed in the rotor produces a uniform flow through the pump. However, there may be problems on starting, when the very high apparent viscosities in the fluid as the system starts from rest might lead to overloading of the pump motor. At this time too the apparent viscosity of the liquid in the delivery line is at its maximum value, and the pump may take an inordinately long time to establish the required flow. Many pseudoplastic materials are damaged and degraded by prolonged shearing, and such a pump would be unsuitable. The whole question of pump selection for all types of fluids has been given a systems approach by DAVIDSON[5].

8.3. PUMPING EQUIPMENT FOR GASES

Essentially the same basic types of mechanical equipment are used for handling gases and liquids, though the construction may be very different in the two cases. Under the normal range of operating pressures, the density of a gas is considerably less than that of a liquid so that higher speeds of operation can be employed and lighter valves fitted to the delivery and suction lines. Because of the lower viscosity of a gas there is a greater tendency for leakage to occur, and therefore gas compressors are designed with smaller clearances between the moving parts. Further differences in construction are necessitated by the decrease in volume of gas as it is compressed, and this must be allowed for in the design. Since a large proportion of the energy of compression appears as heat in the gas, there will normally be a considerable increase in temperature which may limit the operation of the compressor unless suitable cooling can be effected. For this reason, gas compression is often carried out in a number of stages and the gas is cooled between each stage. Any gas which is not expelled from the cylinder at the end of compression (the clearance volume) must be expanded again to the inlet pressure before a fresh charge can be admitted. This continual compression and expansion of the residual gas results in loss of efficiency because neither the compression nor the expansion can be carried out completely reversibly. With liquids this factor has no effect on the efficiency because the residual fluid is not compressed.

The principal types of compressors for gases are described in the following pages.

8.3.1. The reciprocating piston compressor

This type of compressor which may consist of from one to twelve stages is the only one capable of developing very high pressures, such as for example over 350 MN/m^2

Figure 8.27.   Reciprocating compressor

as required for polythene manufacture. A single-stage, two-cylinder unit, shown in Figure 8.27, operates at relatively slow speed and is being replaced wherever possible by the centrifugal type which can now develop all but the very highest pressures.

8.3.2. Rotary blowers and compressors

These can be divided into two classes — those which develop a high compression ratio and those which have very low ratios. The former include the *sliding-vane type* in which the compression ratio is achieved by eccentric mounting of the rotor (Figure 8.28) and the *Nash Hytor pump* (Figure 8.29) in which the compression ratio is achieved by means of a specially shaped casing and a liquid seal which rotates with the impeller. In the sliding-vane unit the rotor is slotted to take the sliding blades which subdivide the crescent-shaped space between the rotor and cylinder. On rotation the blades move out trapping the gas which is compressed during rotation and discharged at the delivery port as shown.

Figure 8.28.   Vane-type blower

Figure 8.29.   Nash Hytor pump

The Nash Hytor (Figure 8.29) type, which is also known as the *liquid-ring pump*, is a positive displacement type with a specially shaped casing and a liquid seal which rotates with the impeller. The shaft and impeller are the only moving parts and there is no sliding contact, so that no lubricants are required and the gas under compression is not contaminated. With this arrangement the liquid leaves and re-enters the impeller cells in the manner of a piston. The service liquid is supplied at a pressure equal to the discharge pressure of the gas and is drawn in automatically to compensate for that discharged from the ports. During compression the energy is converted to heat so that this is a nearly isothermal process. Downstream the liquid is separated from the gas and recirculated with any necessary make up.

A different kind of unit is represented by the *cycloidal* or *Rootes blower*. In this type, compression is achieved by the rotation of two elements as shown in Figure 8.30. These rotors turn in opposite directions so that, as each passes the inlet, it takes in air which is compressed between impeller and the wall and is then expelled outwards.

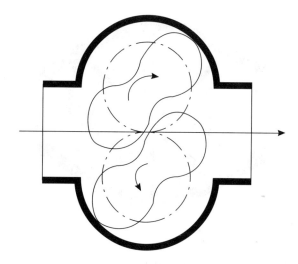

Figure 8.30.   Rootes blower

8.3.3. Centrifugal blowers and compressors, including turbocompressors

These depend on the conversion of kinetic energy into pressure energy. Fans are used for low pressures, and can be made to handle very large quantities of gases. For the higher pressure ratios now in demand, multistage centrifugal compressors are mainly used, particularly for the requirements of high capacity chemical plants. Thus in catalytic reforming, petrochemical separation plants (ethylene manufacture), ammonia plants with a production rate of 12 kg/s, and for the very large capacity needed for natural gas fields, this type of compressor is now supreme. These units now give flowrates up to 140 m^3/s and pressures up to 5.6 MN/m^2 with the newest range going to 40 MN/m^2. It is important to accept that the very large units offer considerable savings over multiple units and that their reliability is remarkably high. The power required is also very high; thus a typical compressor operating on the process gas stream in the catalytic production of ethylene from naphtha will take 10 MW for a 6.5 kg/s plant. A centrifugal compressor is illustrated in Figure 8.31.

Figure 8.31.   A turbocompressor

8.3.4. Power required for the compression of gases

Considering the compression of unit mass of gas, if the volume of the gas changes by an amount dv at a pressure P, the net work done on the gas, $-\delta W$, for a reversible change is given by:

$$-\delta W = -P\,dv$$
$$= v\,dP - d(Pv) \tag{8.27}$$

and for an irreversible change by:

$$-\delta W = -P\,dv + \delta F \qquad \text{(from equation 2.6)}$$
$$= v\,dP - d(Pv) + \delta F \tag{8.28}$$

The work done in a reversible compression will be considered first because this refers to the ideal condition for which the work of compression is a minimum; a reversible compression would have to be carried out at an infinitesimal rate and therefore is not relevant in practice. The actual work done will be greater than that calculated, not only because of irreversibility, but also because of frictional loss and leakage in the compressor. These two factors are difficult to separate and will therefore be allowed for in the overall efficiency of the machine.

The total work of compression from a pressure P_1 to a pressure P_2 is found by integrating equation 8.27. For an ideal gas undergoing an isothermal compression:

$$-\int_{v_1}^{v_2} P \, dv = -W = P_1 v_1 \ln \frac{P_2}{P_1} \qquad \text{(from equation 2.69)}$$

For the isentropic compression of an ideal gas (since $-P \, dv = v \, dP - d(Pv)$):

$$-\int_{v_1}^{v_2} P \, dv = -W = \left[\frac{\gamma}{\gamma - 1}(P_2 v_2 - P_1 v_1) \right] - (P_2 v_2 - P_1 v_1) \qquad \text{(from equation 2.71)}$$

$$= \frac{1}{\gamma - 1}(P_2 v_2 - P_1 v_1) \qquad (8.29)$$

Thus: $\qquad \int_{P_1}^{P_2} v \, dP = -\gamma \int_{v_1}^{v_2} P \, dv \quad$ (comparing equations 2.71 and 8.29)

for isentropic conditions and then:

$$-W = \frac{1}{\gamma - 1} P_1 v_1 \left[\left(\frac{P_2}{P_1} \right)^{(\gamma-1)/\gamma} - 1 \right] \qquad (8.30)$$

Under these conditions the whole of the energy of compression appears as heat in the gas. For unit mass of an ideal gas undergoing an isentropic compression:

$$P_1 v_1^{\gamma} = P_2 v_2^{\gamma} \qquad \text{(from equation 2.30)}$$

and: $$\frac{P_1 v_1}{T_1} = \frac{P_2 v_2}{T_2} \qquad \text{(from equation 2.16)}$$

Eliminating v_1 and v_2 gives the ratio of the outlet temperature T_2 to the inlet temperatures T_1:

$$\frac{T_2}{T_1} = \left(\frac{P_2}{P_1} \right)^{(\gamma-1)/\gamma} \qquad (8.31)$$

In practice, there will be irreversibilities (inefficiencies) associated with the compression and the additional energy needed will appear as heat, giving rise to an outlet temperature higher than T_2 as given by equation 8.31.

For the isentropic compression of a mass m of gas:

$$-Wm = \frac{1}{\gamma - 1} P_1 V_1 \left[\left(\frac{P_2}{P_1} \right)^{(\gamma-1)/\gamma} - 1 \right] \qquad (8.32)$$

where V_1 is the volume of a mass m of gas at a pressure P_1.

If the conditions are intermediate between isothermal and isentropic, k must be used in place of γ, where $\gamma > k > 1$.

If the gas deviates appreciably from the ideal gas laws over the range of conditions considered, the work of compression is most conveniently calculated from the change in the thermodynamic properties of the gas.

Thus: $$dU = T\,dS - P\,dv \qquad \text{(from equation 2.5)}$$

$\therefore$ $$-\delta W = -P\,dv = dU - T\,dS \qquad \text{(for a reversible process)}$$

Under isothermal conditions: $-W = \Delta U - T\Delta S$ \hfill (8.33)

Under isentropic conditions: $-W = \Delta U$ \hfill (8.34)

These equations give the work done during a simple compression of gas in a cylinder but do not take account of the work done either during the admission of the gas prior to compression or during the expulsion of the compressed gas.

Suppose that, after the compression of a volume V_1 of gas at a pressure P_1 to a pressure P_2, the whole of the gas is expelled at constant pressure P_2 and a fresh charge of gas is admitted at a pressure P_1, then the cycle can be followed in Figure 8.32, where the pressure P is plotted as ordinate against the volume V as abscissa.

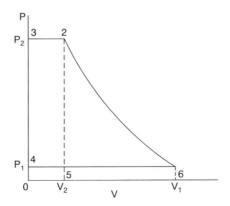

Figure 8.32.   Single-stage compression cycle — no clearance

Point 1 represents the initial condition of the gas (P_1 and V_1).

Line 1–2 represents the compression of gas to pressure P_2, volume V_2.

Line 2–3 represents the expulsion of the gas at a constant pressure P_2.

Line 3–4 represents a sudden reduction in the pressure in the cylinder from P_2 to P_1. As the whole of the gas has been expelled, this can be regarded as taking place instantaneously.

Line 4–1 represents the suction stroke of the piston, during which a volume V_1 of gas is admitted at constant pressure, P_1.

It will be noted that the mass of gas in the cylinder varies during the cycle. The work done by the compressor during each phase of the cycle is as follows:

$$\text{Compression} \qquad -\int_{V_1}^{V_2} P\,\mathrm{d}V \qquad \text{(Area 1-2-5-6)}$$

$$\text{Expulsion} \qquad P_2 V_2 \qquad \text{(Area 2-3-0-5)}$$

$$\text{Suction} \qquad -P_1 V_1 \qquad -\text{(Area 4-0-6-1)}$$

The total work done per cycle

$$= -\int_{V_1}^{V_2} P\,\mathrm{d}V + P_2 V_2 - P_1 V_1 \quad \text{(Area 1-2-3-4)}$$

$$= \int_{P_1}^{P_2} V\,\mathrm{d}P \tag{8.35}$$

The work of compression for an ideal gas per cycle under isothermal conditions:

$$= P_1 V_1 \ln \frac{P_2}{P_1} \tag{8.36}$$

Under isentropic conditions, the work of compression

$$= P_1 V_1 \frac{\gamma}{\gamma - 1} \left[\left(\frac{P_2}{P_1} \right)^{(\gamma-1)/\gamma} - 1 \right] \tag{8.37}$$

Again, working in terms of the thermodynamic properties of the gas:

$$v\,\mathrm{d}P = \mathrm{d}(Pv) - P\,\mathrm{d}v$$

$$= \mathrm{d}(Pv) + \mathrm{d}U - T\,\mathrm{d}S \qquad \text{(from equation 2.5)}$$

$$= \mathrm{d}H - T\,\mathrm{d}S \tag{8.38}$$

For an isothermal process: $-mW = m(\Delta H - T\Delta S)$ \hfill (8.39)

For an isentropic process: $-mW = m\Delta H$ \hfill (8.40)

where m is the mass of gas compressed per cycle.

Clearance volume

In practice, it is not possible to expel the whole of the gas from the cylinder at the end of the compression; the volume remaining in the cylinder after the forward stroke of the piston is termed the *clearance volume*. The volume displaced by the piston is termed the *swept volume*, and therefore the total volume of the cylinder is made up of the clearance volume plus the swept volume. The *clearance c* is defined as the ratio of the clearance volume to the swept volume.

A typical cycle for a compressor with a finite clearance volume can be followed by reference to Figure 8.33. A volume V_1 of gas at a pressure P_1 is admitted to the cylinder; its condition is represented by point 1.

Line 1-2 represents the compression of the gas to a pressure P_2 and volume V_2.

Line 2-3 represents the expulsion of gas at constant pressure P_2, so that the volume remaining in the cylinder is V_3.

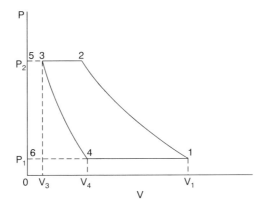

Figure 8.33.   Single-stage compression cycle—with clearance

Line 3–4 represents an expansion of this residual gas to the lower pressure P_1 and volume V_4 during the return stroke.

Line 4–1 represents the introduction of fresh gas into the cylinder at constant pressure P_1. The work done on the gas during each stage of the cycle is as follows.

Compression $\qquad\qquad\qquad -\displaystyle\int_{V_1}^{V_2} P \, dV$

Expulsion $\qquad\qquad\qquad P_2(V_2 - V_3)$

Expansion $\qquad\qquad\qquad -\displaystyle\int_{V_3}^{V_4} P \, dV$

Suction $\qquad\qquad\qquad -P_1(V_1 - V_4)$

The total work done during the cycle is equal to the sum of these four components. It is represented by the area 1–2–3–4, which is equal to area 1–2–5–6 less area 4–3–5–6. If the compression and expansion are taken as isentropic, the work done per cycle therefore:

$$= P_1 V_1 \frac{\gamma}{\gamma - 1}\left[\left(\frac{P_2}{P_1}\right)^{(\gamma-1)/\gamma} - 1\right] - P_1 V_4 \frac{\gamma}{\gamma - 1}\left[\left(\frac{P_2}{P_1}\right)^{(\gamma-1)/\gamma} - 1\right]$$

$$= P_1(V_1 - V_4)\frac{\gamma}{\gamma - 1}\left[\left(\frac{P_2}{P_1}\right)^{(\gamma-1)/\gamma} - 1\right] \qquad\qquad (8.41)$$

Thus, theoretically, the clearance volume does not affect the work done per unit mass of gas, since $V_1 - V_4$ is the volume admitted per cycle. It does, however, influence the quantity of gas admitted and therefore the work done per cycle. In practice, however, compression and expansion are not reversible, and losses arise from the compression and expansion of the clearance gases. This effect is particularly serious at high compression ratios.

The value of V_4 is not known explicitly, but can be calculated in terms of V_3, the clearance volume.

For isentropic conditions:

$$V_4 = V_3 \left(\frac{P_2}{P_1}\right)^{1/\gamma}$$

and:
$$V_1 - V_4 = (V_1 - V_3) + V_3 - V_3 \left(\frac{P_2}{P_1}\right)^{1/\gamma}$$

$$= (V_1 - V_3)\left[1 + \frac{V_3}{V_1 - V_3} - \frac{V_3}{V_1 - V_3}\left(\frac{P_2}{P_1}\right)^{1/\gamma}\right]$$

Now $(V_1 - V_3)$ is the swept volume, V_s, say; and $V_3/(V_1 - V_3)$ is the clearance c.

$$\therefore \qquad V_1 - V_4 = V_s\left[1 + c - c\left(\frac{P_2}{P_1}\right)^{1/\gamma}\right] \tag{8.42}$$

The total work done on the fluid per cycle is therefore:

$$P_1 V_s \frac{\gamma}{\gamma - 1}\left[\left(\frac{P_2}{P_1}\right)^{(\gamma-1)/\gamma} - 1\right]\left[1 + c - c\left(\frac{P_2}{P_1}\right)^{1/\gamma}\right] \tag{8.43}$$

The factor $[1 + c - c(P_2/P_1)^{1/\gamma}]$ is called the theoretical volumetric efficiency and is a measure of the effect of the clearance on an isentropic compression. The actual volumetric efficiency will be affected, in addition, by the inertia of the valves and leakage past the piston.

The gas is frequently cooled during compression so that the work done per cycle is less than that given by equation 8.43, and γ is replaced by some smaller quantity k. The greater the rate of heat removal, the less is the work done. The isothermal compression is usually taken as the condition for least work of compression, but clearly the energy consumption can be reduced below this value if the gas is artificially cooled below its initial temperature as it is compressed. This is not a practicable possibility because of the large amount of energy required to refrigerate the cooling fluid. It can be seen that the theoretical volumetric efficiency decreases as the rate of heat removal is increased since γ is replaced by the smaller quantity k.

In practice the cylinders are usually water-cooled. The work of compression is thereby reduced though the effect is usually small. The reduction in temperature does, however, improve the mechanical operation of the compressor and makes lubrication easier.

Multistage compressors

If the required pressure ratio P_2/P_1 is large, it is not practicable to carry out the whole of the compression in a single cylinder because of the high temperatures which would be set up and the adverse effects of clearance volume on the efficiency. Further, lubrication would be difficult due to carbonisation of the oil, and there would be a risk of causing oil mist explosions in the cylinders when gases containing oxygen were being compressed. The mechanical construction also would be difficult because the single cylinder would have to be strong enough to withstand the final pressure and yet large enough to hold the gas at the initial pressure P_1. In the multistage compressor, the gas passes through a number of cylinders of gradually decreasing volume and can be cooled between the stages. The maximum pressure ratio normally obtained in a single cylinder is 10 but values above 6 are unusual.

The operation of the multistage compressor can conveniently be followed again on a pressure-volume diagram (Figure 8.34). The effect of clearance volume will be neglected at first. The area 1–2–3–4 represents the work done in compressing isentropically from P_1 to P_2 in a single stage. The area 1–2–5–4 represents the necessary work for an isothermal compression. Now consider a multistage isentropic compression in which the intermediate pressures are P_{i1}, P_{i2}, etc. The gas will be assumed to be cooled to its initial temperature in an interstage cooler before it enters each cylinder.

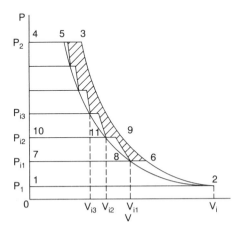

Figure 8.34.   Multistage compression cycle with interstage cooling

Line 1–2 represents the suction stroke of the first stage where a volume V_1 of gas is admitted at a pressure P_1.

Line 2–6 represents an isentropic compression to a pressure P_{i1}.

Line 6–7 represents the delivery of the gas from the first stage at a constant pressure P_{i1}.

Line 7–8 represents the suction stroke of the second stage. The volume of the gas has been reduced in the interstage cooler to V_{i1}; that which would have been obtained as a result of an isothermal compression to P_{i1}.

Line 8–9 represents an isentropic compression in the second stage from a pressure P_{i1} to a pressure P_{i2}.

Line 9–10 represents the delivery stroke of the second stage.

Line 10–11 represents the suction stroke of the third stage. Point 11 again lies on the line 2–5, representing an isothermal compression.

It is seen that the overall work done on the gas is intermediate between that for a single stage isothermal compression and that for an isentropic compression. The net saving in energy is shown as the shaded area in Figure 8.34.

The total work done per cycle W'

$$= P_1 V_1 \frac{\gamma}{\gamma - 1} \left[\left(\frac{P_{i1}}{P_1} \right)^{(\gamma-1)/\gamma} - 1 \right] + P_{i1} V_{i1} \frac{\gamma}{\gamma - 1} \left[\left(\frac{P_{i2}}{P_{i1}} \right)^{(\gamma-1)/\gamma} - 1 \right] + \cdots$$

for an isentropic compression.

For perfect interstage cooling:

$$P_1 V_1 = P_{i1} V_{i1} = P_{i2} V_{i2} = \cdots$$

$$\therefore \qquad W' = P_1 V_1 \frac{\gamma}{\gamma - 1} \left[\left(\frac{P_{i1}}{P_1} \right)^{(\gamma - 1)/\gamma} + \left(\frac{P_{i2}}{P_{i1}} \right)^{(\gamma - 1)/\gamma} + \cdots - n \right]$$

where n is the number of stages.

It is now required to find how the total work per cycle W' is affected by the choice of the intermediate pressures P_{i1}, P_{i2}, etc. The work will be a minimum when

$$\frac{\partial W'}{\partial P_{i1}} = \frac{\partial W'}{\partial P_{i2}} = \frac{\partial W'}{\partial P_{i3}} = \cdots = 0.$$

When:

$$\frac{\partial W'}{\partial P_{i1}} = 0$$

$$P_1 V_1 \frac{\gamma}{\gamma - 1} \left[\frac{\gamma - 1}{\gamma} \left(\frac{P_{i1}}{P_1} \right)^{(\gamma - 1)/\gamma} P_{i1}^{-1} + \frac{1 - \gamma}{\gamma} \left(\frac{P_{i2}}{P_{i1}} \right)^{(\gamma - 1)/\gamma} P_{i1}^{-1} \right] = 0$$

i.e.:

$$\frac{P_{i1}}{P_1} = \frac{P_{i2}}{P_{i1}} \tag{8.44}$$

The same procedure is then adopted for obtaining the optimum value of P_{i2} and hence:

$$\frac{P_{i2}}{P_{i1}} = \frac{P_{i3}}{P_{i2}} \tag{8.45}$$

Thus the intermediate pressures should be arranged so that the compression ratio is the same in each cylinder; and equal work is then done in each cylinder.

The minimum work of compression in a compressor of n stages is therefore:

$$P_1 V_1 \frac{\gamma}{\gamma - 1} \left[n \left(\frac{P_2}{P_1} \right)^{(\gamma - 1)/n\gamma} - n \right] = n P_1 V_1 \frac{\gamma}{\gamma - 1} \left[\left(\frac{P_2}{P_1} \right)^{(\gamma - 1)/n\gamma} - 1 \right] \tag{8.46}$$

The effect of clearance volume can now be taken into account. If the clearances in the successive cylinders are $c_1, c_2, c_3, \ldots$, the theoretical volumetric efficiency of the first cylinder

$$= 1 + c_1 - c_1 \left(\frac{P_{i1}}{P_1} \right)^{1/\gamma} \qquad \text{(from equation 8.42)}$$

Assuming that the same compression ratio is used in each cylinder, then the theoretical volumetric efficiency of the first stage is:

$$1 + c_1 - c_1 \left(\frac{P_2}{P_1} \right)^{1/n\gamma}$$

If the swept volumes of the cylinders are $V_{s1}, V_{s2}, \ldots$, the volume of gas admitted to the first cylinder

$$= V_{s1} \left[1 + c_1 - c_1 \left(\frac{P_2}{P_1} \right)^{1/n\gamma} \right] \tag{8.47}$$

The same mass of gas passes through each of the cylinders and, therefore, if the inter-stage coolers are assumed perfectly efficient, the ratio of the volumes of gas admitted to successive cylinders is $(P_1/P_2)^{1/n}$. The volume of gas admitted to the second cylinder is then:

$$V_{s2}\left[1+c_2-c_2\left(\frac{P_2}{P_1}\right)^{1/n\gamma}\right]=V_{s1}\left[1+c_1-c_1\left(\frac{P_2}{P_1}\right)^{1/n\gamma}\right]\left(\frac{P_1}{P_2}\right)^{1/n}$$

$$\therefore \quad \frac{V_{s1}}{V_{s2}}=\frac{1+c_2-c_2(P_2/P_1)^{1/n\gamma}}{1+c_1-c_1(P_2/P_1)^{1/n\gamma}}\left(\frac{P_2}{P_1}\right)^{1/n} \qquad\qquad (8.48)$$

In this manner the swept volume of each cylinder can be calculated in terms of V_{s1} and $c_1, c_2, \ldots$, and the cylinder dimensions determined.

When the gas does not behave as an ideal gas, the change in its condition can be followed on a temperature-entropy or an enthalpy-entropy diagram. The intermediate pressures $P_{i1}, P_{i2}, \ldots$, are then selected so that the enthalpy change (ΔH) is the same in each cylinder.

Several opposing factors will influence the number of stages selected for a given compression. The larger the number of cylinders the greater is the mechanical complexity. Against this must be balanced the higher theoretical efficiency, the smaller mechanical strains set up in the cylinders and the moving parts, and the greater ease of lubrication at the lower temperatures that are experienced. Compressors with as many as nine stages are used for very high pressures.

Compressor efficiencies

The efficiency quoted for a compressor is usually either an isothermal efficiency or an isentropic efficiency. The isothermal efficiency is the ratio of the work required for an ideal isothermal compression to the energy actually expended in the compressor. The isentropic efficiency is defined in a corresponding manner on the assumption that the whole compression is carried out in a single cylinder. Since the energy expended in an isentropic compression is greater than that for an isothermal compression, the isentropic efficiency is always the greater of the two. Clearly the efficiencies will depend on the heat transfer between the gas undergoing compression and the surroundings and on how closely the process approaches a reversible compression.

The efficiency of the compression will also be affected by a number of other factors which are all connected with the mechanical construction of the compressor. Thus the efficiency will be reduced as a result of leakage past the piston and the valves and because of throttling of the gas at the valves. Further, the mechanical friction of the machine will lower the efficiency and the overall efficiency will be affected by the efficiency of the driving motor and transmission.

Example 8.2

A single-acting air compressor supplies 0.1 m³/s of air (at STP) compressed to 380 kN/m² from 101.3 kN/m². If the suction temperature is 289 K, the stroke is 0.25 m, and the speed is 4.0 Hz, what is the cylinder diameter? Assuming the cylinder clearance is 4 per cent and compression and re-expansion are isentropic ($\gamma = 1.4$), what are the theoretical power requirements for the compression?

Solution

Volume per stroke $= \left(\dfrac{0.1}{4.0}\right)\left(\dfrac{289}{273}\right) = 0.0264 \text{ m}^3$

Compression ratio $= \dfrac{380}{101.3} = 3.75$

The swept volume is given by equation 8.42:

$$0.0264 = V_s[1 + 0.04 - 0.04(3.75)^{1/1.4}]$$

$$\therefore \qquad V_s = \dfrac{0.0264}{1.04 - 0.04 \times 2.7} = 0.0283 \text{ m}^3$$

Thus cross-sectional area of cylinder $= \dfrac{0.0283}{0.25} = 0.113 \text{ m}^2$

$$\text{and diameter} = \left[\dfrac{0.113}{(\pi/4)}\right]^{0.5} = \underline{0.38 \text{ m}}$$

From equation 8.41, work of compression per cycle

$$= 101,300 \times 0.0264\left[\dfrac{1.4}{1.4 - 1.0}\right][(3.75)^{0.4/1.4} - 1]$$

$$= 9360(1.457 - 1) = 4278 \text{ J}$$

Theoretical power requirements

$$= (4278 \times 4) = 17,110 \text{ W} \quad \text{or} \quad \underline{17.1 \text{ kW}}$$

Example 8.3

Air at 290 K is compressed from 101.3 kN/m^2 to 2065 kN/m^2 in a two-stage compressor operating with a mechanical efficiency of 85 per cent. The relation between pressure and volume during the compression stroke and expansion of the clearance gas is $PV^{1.25} = $ constant. The compression ratio in each of the two cylinders is the same, and the interstage cooler may be assumed 100 per cent efficient. If the clearances in the two cylinders are 4 per cent and 5 per cent respectively, calculate:

(a) the work of compression per kg of air compressed;
(b) the isothermal efficiency;
(c) the isentropic efficiency ($\gamma = 1.4$), and
(d) the ratio of the swept volumes in the two cylinders.

Solution

Overall compression ratio $= \dfrac{2065}{101.3} = 20.4$

Specific volume of air at 290 K $= \left(\dfrac{22.4}{28.8}\right)\left(\dfrac{290}{273}\right) = 0.826 \text{ m}^3/\text{kg}$

From equation 8.46, work of compression

$$= 101.3 \times 0.826 \times 2\left[\dfrac{1.25}{1.25 - 1}\right][(20.4)^{0.25/2.5} - 1]$$

$$= 836.7(1.351 - 1)$$

$$= 293.7 \text{ kJ/kg}$$

Energy supplied to the compressor, i.e. work of compression

$$= \frac{293.7}{0.85} = \underline{\underline{345.5 \text{ kJ/kg}}}$$

From equation 8.36, work done in isothermal compression of 1 kg of gas

$$= 101.3 \times 0.826 \ln 20.4$$

$$= 83.7 \times 3.015 = 252.3 \text{ kJ/kg}$$

Isothermal efficiency $\qquad = 100 \times \frac{252.3}{345.5} = \underline{\underline{73\%}}$

From equation 8.37, work done in isentropic compression of 1 kg of gas

$$= 101.3 \times 0.826 \frac{1.4}{0.4} [(20.4)^{0.4/1.4} - 1]$$

$$= 292.9(2.36 - 1) = 398.3 \text{ kJ/kg}$$

Isentropic efficiency $\qquad = 100 \times \frac{398.3}{345.5} = \underline{\underline{115\%}}$

From equation 8.47, volume swept out in first cylinder in compression of 1 kg of gas is given by:

$$0.826 = V_{s_1}[1 + 0.04 - 0.04(20.4)^{1/(2 \times 1.25)}]$$

$$= V_{s_1} \times 0.906$$

$$\therefore \qquad V_{s_1} = 0.912 \text{ m}^3/\text{kg}$$

Similarly, the swept volume of the second cylinder is given by:

$$0.826 \left(\frac{1}{20.4}\right)^{0.5} = V_{s_2}[1 + 0.05 - 0.05(20.4)^{1/(2 \times 1.25)}]$$

$$0.183 = 0.883 V_{s_2}$$

$$\therefore \qquad V_{s_2} = 0.207 \text{ m}^3/\text{kg}$$

$$\therefore \qquad \frac{V_{s_1}}{V_{s_2}} = \underline{\underline{4.41}}$$

8.4. THE USE OF COMPRESSED AIR FOR PUMPING

Compressed gas is sometimes used for transferring liquid from one position to another in a chemical plant, but more particularly for emptying vessels. It is frequently more convenient to apply pressure by means of compressed gas rather than to install a pump, particularly when the liquid is corrosive or contains solids in suspension. Furthermore, to an increasing extent chemicals are being delivered in tankers and are discharged by the application of gas under pressure. For instance, phthalic anhydride is now distributed in heated tankers which are discharged into storage vessels by connecting them to a supply of compressed nitrogen or carbon dioxide.

Several devices have been developed to eliminate the necessity for manual operation of valves, and the automatic acid elevator is an example of equipment incorporating such a device. However, such equipment is becoming less important now that it is possible to construct centrifugal pumps in a wide range of corrosion-resistant materials. The *air-lift pump* makes more efficient use of the compressed air and is used for pumping corrosive liquids. Although it is not used extensively in the chemical industry, it is used for pumping oil from wells, and the principle is incorporated into a number of items of chemical engineering equipment, including the climbing film evaporator.

8.4.1. The air-lift pump

In the air-life pump (Figure 8.35) a high efficiency is obtained by using the air expansively so that it expands to atmospheric pressure in contact with the liquid. It can be regarded simply as a U-tube in a state of dynamic equilibrium. One limb containing only liquid is relatively short and is connected to the feed tank, whilst air is injected near the bottom of the longer limb which therefore contains a mixture of lower density (liquid and air). If the air is introduced sufficiently rapidly, liquid will flow from the short to the long limb and be discharged into the delivery tank. The rate of flow will depend on the difference in density and will, therefore, rise as the air rate is increased, but will reach a maximum because the frictional resistance increases with the volumetric rate of flow.

Figure 8.35.   Air-lift pump

The liquid-feed limb is known as the *submergence line* and the line carrying the aerated mixture is the *rising main*. The ratio of the submergence (h_s) to the total height of rising main above the air injection point ($h_r + h_s$) is known as the *submergence ratio*, $[1 + (h_r/h_s)]$.

If a mass M of liquid is raised through a net height h_r by a mass m of air, the net work done on the liquid is Mgh_r. If the pressure of the entering air is P, the work done by the air in expanding isothermally to atmospheric pressure P_a is given by:

$$W = P_a v_a m \ln \frac{P}{P_a} \qquad (8.49)$$

where v_a is the specific volume of air at atmospheric pressure. The expansion will be almost exactly isothermal because of the intimate contact between the liquid and the air.

The efficiency of the pump η is therefore given by:

$$\eta = \frac{Mgh_r}{mP_a u_a \ln(P/P_a)} \qquad (8.50)$$

The mass of air required to pump unit mass of liquid is, therefore, given by:

$$\frac{m}{M} = \frac{gh_r}{\eta P_a v_a \ln(P/P_a)} \qquad (8.51)$$

If all losses in the operation of the pump were neglected, the pressure at the point of introduction of the compressed air would be equal to atmospheric pressure together with the pressure due to the column of liquid of height h_s, the vertical distance between the liquid level in the suction tank, and the air inlet point. Therefore:

$$P_a = h_a \rho g \quad \text{(say)} \qquad (8.52)$$

and:
$$P = (h_a + h_s)\rho g \qquad (8.53)$$

where ρ is the density of the liquid.

Thus from equation 8.51 the mass of air required to pump unit mass of liquid (m/M) would be equal to:

$$\frac{h_r g}{P_a v_a \ln[(h_s + h_a)/h_a]} \qquad (8.54)$$

This is the minimum air requirement for the pump if all losses are neglected. It will be seen that (m/M) decreases as h_s increases; if h_s is zero (m/M) is infinite and therefore the pump will not work. A high submergence h_s is therefore desirable.

There are a number of important applications of the air-lift pump in the process industries due to its simplicity. It is particularly useful for handling radioactive materials as there are no mechanical parts in contact with the fluid, and the pump will operate virtually indefinitely without the need for maintenance which can prove very difficult when handling radioactive liquids.

Example 8.4

An air-lift pump is used for raising 750 cm³/s of a liquid of specific gravity 1.2 to a height of 20 m. Air is available at 450 kN/m². Assuming isentropic compression of the air, what is the power requirement of the pump, if the efficiency is 30 per cent? ($\gamma = 1.4$.)

Solution

$$\text{Work done by pump} = (750 \times 10^{-6})(1.2 \times 1000) \times 20 \times 9.81$$

$$= 176.6 \text{ W}$$

$$\text{Actual work of expansion of air} = \frac{176.6}{0.30} = 588.6 \text{ W}$$

Flow of air required at 1 bar (101.3 kN/m²) and 273 K is given by:

$$588.6 = 101{,}300 V \ln \frac{450}{101.3}$$

$$\therefore \qquad Q = 0.0039 \text{ m}^3/\text{s}$$

Work done in the isentropic compression of 0.0039 m³ air (from equation 8.37)

$$= 101{,}300 \times 0.0039 \frac{1.4}{0.4} \left[\left(\frac{450}{101.3} \right)^{0.4/1.4} - 1 \right]$$

$$= 1383(1.528 - 1) = 730 \text{ J}$$

$$\therefore \qquad \text{Power required} = 730 \text{ J/s} \quad \text{or} \quad \underline{\underline{0.730 \text{ kW}}}$$

8.5. VACUUM-PRODUCING EQUIPMENT

A great deal of chemical plant, particularly in the area of distillation, operates under low pressures, and pumping equipment is required to create the vacuum and to maintain the low pressure. Vacuum pumps must take in gas at low pressure where the volume is very large, and discharge at atmospheric pressure, and a high compression ratio is necessary. A rotary pump is preferred to a piston type because of capacity and ease of providing a suitable drive. An alternative method very commonly used with chemical plant is the use of a *steam-jet ejector* in which the low-pressure gas is entrained with the high-pressure steam. This unit is described later.

The *sliding-vane* and the *liquid-ring* types are typical of rotary vacuum pumps. The sliding-vane pump illustrated in Figure 8.36 consists of an eccentrically mounted rotor slotted to take the sliding vanes as shown. On rotation the vanes move outwards and the vapour is sucked in at the position of maximum area between rotor and casing and discharged through the delivery port at conditions of minimum area. Such a unit will give a pressure down to about 1.3 N/m².

Liquid-ring pumps (Figure 8.37) are rotary displacement units in which the sealing liquid, which is usually water, is made to act as the piston. The impeller is mounted eccentrically and the liquid circulates within the case and part leaves with the exit gas

Figure 8.36.   The sliding vane rotary compressor and vacuum pump

stream. During rotation, the liquid enters and leaves the individual sections acting as a piston. The sealing liquid is fed in at a pressure equal to the discharge gas stream pressure and is separated in an external tank and recirculated. Here the energy of the compression operation is taken up by the gas, and the process is practically isothermal. Care must be taken to see that cavitation does not occur, and this can arise if the vapour fraction in the cells is greater than 50 per cent. The vapour pressure of the water restricts the lower limit of pressure in a straight unit to about 3.3 kN/m^2 and a liquid with a lower boiling point must be used if a lower pressure is required. This is usually achieved by operating the liquid ring pump in series with an additional device such as an ejector.

Figure 8.37.   Single-stage, single-action liquid ring gas pump: 1, shaft 2, impeller 3, casing 4, guide plate 5, suction port 6, discharge port 7, liquid ring

The steam-jet ejector, illustrated in Figure 8.38, is very commonly used in the process industries since it has no moving parts and will handle large volumes of vapour at low pressures. The operation of the unit is shown in Figure 8.39. The steam is fed at constant pressure P_1 and expands nearly isentropically along AC; mixing of the steam and vapour which is drawn in is represented by CE and vapour compression continues to the throat F and to the exit at G. The steam required increases with the compression ratio so that a single-stage unit will operate down to about 17 kN/m^2 corresponding to a compression ratio of 6:1. For lower final pressures multistage units are used, and Figure 8.40 shows the relationship between the number of stages, steam pressure, and vacuum required. If cooling water is applied between the stages an improved performance will be obtained.

As a guide, a single-stage unit gives a vacuum to 13.5 kN/m^2, a double stage from 3.4 to 13.5 kN/m^2, and a three-stage unit from 0.67 to 2.7 kN/m^2. For very low pressures, a diffusion pump is used with a rotary pump as the first stage unit. The principle of operation is that the gas diffuses into a stream of oil or mercury and is driven out of the pump by molecular bombardment. These are very specialised units, and information and advice on their application and operation should be sought direct from the manufacturers.

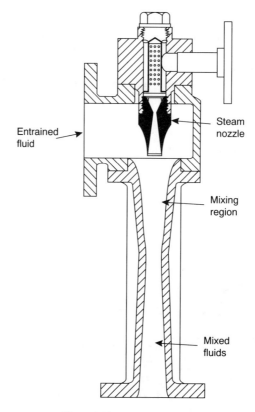

Figure 8.38.   Steam jet ejector

8.6. POWER REQUIREMENTS FOR PUMPING THROUGH PIPELINES

A fluid will flow of its own accord so long as the energy per unit mass of fluid decreases in the direction of flow. It can be made to flow in the opposite direction only by the action of some external agent, such as a pump which supplies energy, and increases the pressure at the upstream end of the system.

The energy balance from equation 2.55 is:

$$\Delta \frac{u^2}{2\alpha} + g\Delta z + \int_{P_1}^{P_2} v\, dP + W_s + F = 0$$

The work done on unit mass of fluid is $-W_s$ and the total rate at which energy must be transferred to the fluid is $-GW_s$, when the mass rate of flow is G:

$$\mathbf{P} = -GW_s = G\left(\Delta \frac{u^2}{2\alpha} + g\Delta z + \int_{P_1}^{P_2} v\, dP + F\right) \qquad (8.55)$$

8.6.1. Liquids

F can be calculated directly for the flow of liquid through a uniform pipe. If a liquid is pumped through a height Δz from one open tank to another and none of the kinetic

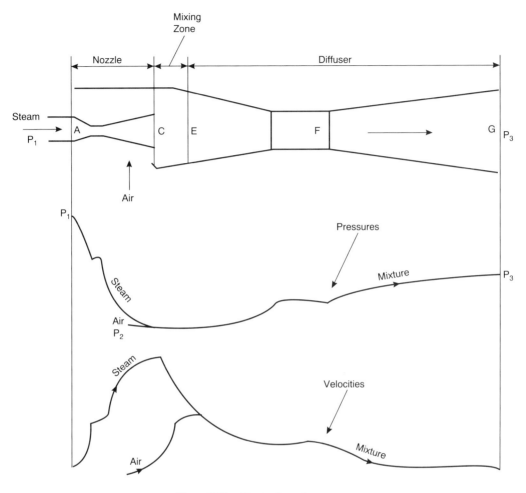

Figure 8.39.   Ejector flow phenomena

energy is recoverable as pressure energy, the fluid pressure is the same at both ends of the system and $\int_{P_1}^{P_2} v\,dP$ is zero. The power requirement, is, therefore:

$$\mathbf{P} = G\left(\frac{u^2}{2\alpha} + g\Delta z + F\right) \tag{8.56}$$

or, in practical terms, taking into account the pump efficiency η the power requirement is given by the product of the mass flowrate, the total head and the acceleration due to gravity as:

$$\mathbf{P} = \frac{1}{\eta}Ghg \tag{8.57}$$

and the use of this equation is best illustrated by the following examples.

Figure 8.40. (a) Diagram showing most suitable number of ejector stages for varying vacua and steam pressures (b) Comparative performance curves of ejectors

Example 8.5

2.16 m³/h (600 × 10⁻⁶ m³/s) water at 320 K is pumped through a 40 mm i.d. pipe, through a length of 150 m in a horizontal direction, and up through a vertical height of 10 m. In the pipe there are a control valve, equivalent to 200 pipe diameters, and other pipe fittings equivalent to 60 pipe diameters. Also in the line is a heat exchanger across which the head lost is 2 m water. Assuming the main pipe has a roughness of 0.0002 m, what power must be supplied to the pump if it is 60 per cent efficient?

Solution

$$\text{Area for flow} = (0.040)^2 \frac{\pi}{4} = 0.0012 \text{ m}^2$$

$$\text{Flow of water} = 600 \times 10^{-6} \text{ m}^3/\text{s}$$

$$\therefore \qquad \text{velocity} = \frac{600 \times 10^{-6}}{0.0012} = 0.50 \text{ m/s}$$

At 320 K, $\mu = 0.65$ mN s/m$^2 = 0.65 \times 10^{-3}$ N s/m^2

$$\rho = 1000 \text{ kg/m}^3$$

$$\therefore \qquad Re = \frac{0.040 \times 0.50 \times 1000}{0.65 \times 10^{-3}} = 30{,}780$$

$$\therefore \qquad \frac{R}{\rho u^2} = 0.004 \text{ for a relative roughness of } \frac{e}{d} = \frac{0.0002}{0.040} = 0.005 \qquad \text{(from Figure 3.7)}$$

Equivalent length of pipe $= 150 + 10 + (260 \times 0.040) = 170.4$ m

$$\therefore \qquad h_f = 4 \frac{R}{\rho u^2} \frac{l}{d} \frac{u^2}{g} \qquad\qquad\qquad \text{(from equation 3.20)}$$

$$= 4 \times 0.004 \left(\frac{170.4}{0.040}\right) \left(\frac{0.50^2}{9.81}\right)$$

$$= 1.74 \text{ m}$$

$$\therefore \qquad \text{Total head to be developed} = 1.74 + 10 + 2 = 13.74 \text{ m}$$

$$\text{Mass flow of water} = 600 \times 10^{-6} \times 1000 = 0.60 \text{ kg/s}$$

$$\therefore \qquad \text{Power required} = 0.60 \times 13.74 \times 9.81 = 80.9 \text{ W}$$

$$\therefore \qquad \text{Power to be supplied} = 80.9 \times \frac{100}{60} = \underline{\underline{135 \text{ W}}}$$

In this solution the kinetic energy head, $u^2/2g$, has been neglected as this represents only some $[0.5^2/(2 \times 9.81)] = 0.013$ m.

Example 8.6

It is required to pump cooling water from a storage pond to a condenser in a process plant situated 10 m above the level of the pond. 200 m of 74.2 mm i.d. pipe is available and the pump has the characteristics given below. The head lost in the condenser is equivalent to 16 velocity heads based on the flow in the 74.2 mm pipe.
    If the friction factor $\phi = 0.003$, estimate the rate of flow and the power to be supplied to the pump assuming an efficiency of 50 per cent.

Discharge (m^3/s)	0.0028	0.0039	0.0050	0.0056	0.0059
Head developed (m)	23.2	21.3	18.9	15.2	11.0

Solution

The head to be developed, $h = 10 + 4\phi \dfrac{l}{d} \dfrac{u^2}{g} + \dfrac{16\, u^2}{2\, g}$ $\qquad\qquad$ (from equation 3.20)

$$= 10 + 4 \times 0.003 \left(\frac{200}{0.0742}\right) \frac{u^2}{g} + \frac{8u^2}{g}$$

$$= 10 + 4.12\, u^2 \text{ m water}$$

Discharge, $Q = (\pi d^2/4)u = 0.00138\pi u$ m^3/s

$$\therefore \qquad u = \frac{Q}{0.00138\pi} = 231.3Q \text{ m/s}$$

$$\therefore \qquad h = 10 + 4.12(231.3Q)^2$$

$$= 10 + 2.205 \times 10^5 Q^2 \text{ m water}$$

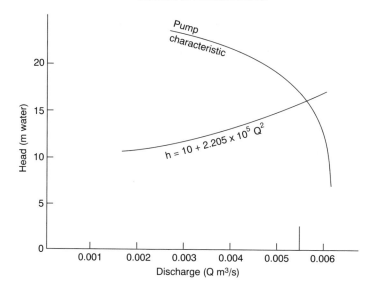

Figure 8.41.   Data for Example 8.6

Values of Q and h are plotted in Figure 8.41 and the discharge at the point of intersection between the pump characteristic equation and the line of the above equation is 0.0054 m³/s.

The head developed is thus, $h = 10 + 2.205 \times 10^5 \times 0.0054^2$

$$= 16.43, \text{ say 16 m water}$$

$$\text{Power required} = \frac{0.0054 \times 1000 \times 16.43 \times 9.81}{0.50}$$

$$= 1741 \text{ W or } \underline{\underline{1.74 \text{ kW}}}$$

If a fluid is to be pumped between two points, the diameter of the pipeline should be chosen so that the overall cost of operation is a minimum. The smaller the diameter, the lower is the initial cost of the line but the greater is the cost of pumping; an economic balance must, therefore, be achieved.

The initial cost of a pipeline and the depreciation and maintenance costs will be approximately proportional to the diameter raised to a power of between 1.0 and 1.5. The power for pumping an incompressible fluid at a given rate G is made up of two parts:

(1) that necessitated by the difference in static pressure and vertical height at the two ends of the system (this is independent of the diameter of the pipe);
(2) that attributable to the kinetic energy of the fluid and the work done against friction. If the kinetic energy is small, this is equal to:

$$G \left[4 \frac{R}{\rho u^2} \frac{l}{d} u^2 \right] \tag{8.58}$$

(from equations 3.19 and 8.56)

which is proportional to $d^{-4.5 \text{ to } -5}$ for turbulent flow, since $u \propto d^{-2}$ and $R/\rho u^2 \propto u^{-0.25 \text{ to } 0}$, according to the roughness of the pipe.

The power requirement can, therefore, be calculated as a function of d and the cost obtained. The total cost per annum is then plotted against the diameter of pipe and the optimum conditions are found either graphically or by differentiation as shown in Example 8.7.

Example 8.7

The relation between cost per unit length C of a pipeline installation and its diameter d is given by:

$$C = a + bd$$

where a and b are independent of pipe size. Annual charges are a fraction β of the capital cost. Obtain an expression for the optimum pipe diameter on a minimum cost basis for a fluid of density ρ and viscosity μ flowing at a mass rate of G. Assume that the fluid is in turbulent flow and that the Blasius equation is applicable, i.e. the friction factor is proportional to the Reynolds number to the power of minus one quarter. Indicate clearly how the optimum diameter depends on flowrate and fluid properties.

Solution

The total annual cost of a pipeline consists of a capital charge plus the running costs. The chief element of the running cost will be the power required to overcome the head loss which is given by equation 3.20:

$$h_f = 8\frac{R}{\rho u^2}\frac{l}{d}\frac{u^2}{2g}$$

If $R/\rho u^2 = 0.04/Re^{0.25}$, the head loss per unit length l is given by equation 3.20:

$$\frac{h_f}{l} = 8\frac{0.04}{Re^{0.25}}\frac{1}{d}\frac{u^2}{2g}$$

$$= 0.16\frac{u^2}{gd}\left(\frac{\mu}{\rho ud}\right)^{0.25}$$

$$= 0.16\frac{u^{1.75}\mu^{0.25}}{\rho^{0.25}gd^{1.25}}$$

Now the velocity $u = G/\rho A = G/\rho(\pi/4)d^2 = 1.27G/\rho d^2$

$$\frac{h_f}{l} = \frac{0.16(1.27G/\rho d^2)^{1.75}\mu^{0.25}}{\rho^{0.25}gd^{1.25}}$$

$$= \frac{0.244G^{1.75}\mu^{0.25}}{\rho^2 gd^{4.75}}$$

The power required for pumping if the pump efficiency is η is:

$$P = Gg\frac{0.244G^{1.75}\mu^{0.25}}{\rho^2 gd^{4.75}}\frac{1}{\eta}$$

If $\eta = 0.5$ (say) $P = 0.488G^{2.75}\mu^{0.25}/(\rho^2 d^{4.75})$ (W)
If $c = $ power cost/W, the cost of pumping is given by:

$$\frac{0.488cG^{2.75}\mu^{0.25}}{\rho^2 d^{4.75}}$$

The total annual cost is then $C = (\beta a + \beta bd) + \dfrac{\gamma G^{2.75}\mu^{0.25}}{\rho^2 d^{4.75}}$

where $\gamma = 0.488c$.

Differentiating the total cost with respect to the diameter gives:

$$\frac{dC}{dd} = \beta b - \frac{4.75\gamma\,G^{2.75}\mu^{0.25}}{\rho^2 d^{5.75}}$$

For minimum cost, $dC/dd = 0$ and:

$$d^{5.75} = \frac{4.75\gamma\,G^{2.75}\mu^{0.25}}{\rho^2\beta b}$$

and:

$$d = \frac{KG^{0.48}\mu^{0.043}}{\rho^{0.35}}$$

where:

$$K = \left(\frac{4.75\gamma}{\beta b}\right)^{0.174} = \left(\frac{2.32c}{\beta b}\right)^{0.174}$$

Effect of fluctuations in flowrate on power for pumping

The importance of maintaining the flowrate in a pipeline constant can be seen by considering the effect of a sinusoidal variation in flowrate. This corresponds approximately to the discharge conditions in a piston pump during the forward movement of the piston. The flowrate $\dot{Q}$ is given as a function of time t by the relation:

$$\dot{Q} = Q_m \sin \omega t \tag{8.59}$$

The discharge rate rises from zero at $t = 0$ to Q_m when $\omega t = \pi/2$, i.e. $t = \pi/2\omega$.

The mean discharge rate Q is then given by:

$$Q = \frac{1}{\pi/2\omega} \int_0^{\pi/2\omega} Q_m \sin \omega t\, dt$$

$$= \frac{Q_m \cdot 2\omega}{\pi}\left[-\frac{1}{\omega}\cos \omega t\right]_0^{\pi/2\omega}$$

$$= \frac{2}{\pi}Q_m \tag{8.60}$$

The power required for pumping will be given by the product of the volumetric flowrate and the pressure difference between the pump outlet and the discharge end of the pipeline. Taking note of the fluctuating nature of the flow, it is necessary to consider the energy transferred to the fluid over a small time interval and to integrate over the cycle to obtain the mean value of the power.

Thus the power requirement $\mathbf{P} = \dfrac{1}{\pi/2\omega} \displaystyle\int_0^{\pi/2\omega} Q_m \sin \omega t(-\Delta P)\, dt \tag{8.61}$

The form of the integral will depend on the relation between $-\Delta P$ and $\dot{Q}$.

Streamline flow

It may be assumed that $-\Delta P = K\dot{Q} = KQ_m \sin \omega t$.

Then:

$$\mathbf{P} = \frac{2\omega}{\pi}KQ_m^2 \int_0^{\pi/2\omega} \sin^2 \omega t\, dt$$

$$= \frac{2\omega}{\pi} K \left(Q^2 \frac{\pi^2}{4} \right) \frac{1}{2} \int_0^{\pi/2\omega} (1 - \cos 2\omega t)\, dt$$

$$= \frac{KQ^2 \pi \omega}{4} \left[t - \frac{1}{2\omega} \sin 2\omega t \right]_0^{\pi/2\omega}$$

$$= \frac{KQ^2 \pi^2}{8}$$

For steady flow:

$$\mathbf{P} = Q(-\Delta P) = KQ^2 \tag{8.62}$$

Thus power is increased by a factor of $\pi^2/8$ (= 1.23) as a result of the sinusoidal fluctuations.

Turbulent flow

As a first approximation it may be assumed that pressure drop is proportional to the square of flowrate

i.e.:
$$-\Delta P = K^1 Q^2 = K^1 Q_m^2 \sin^2 \omega t \tag{8.63}$$

Then:
$$\mathbf{P} = \frac{2\omega}{\pi} K^1 Q_m^3 \int_0^{\pi/2\omega} \sin^3 \omega t\, dt$$

$$= \frac{2\omega}{\pi} K^1 \left(Q^3 \frac{\pi^3}{8} \right) \frac{1}{4} \int_0^{\pi/2\omega} (3 \sin \omega t - \sin 3\omega t)\, dt$$

$$= \frac{K^1 Q^3 \omega \pi^2}{16} \left[-\frac{3}{\omega} \cos \omega t + \frac{1}{3\omega} \cos 3\omega t \right]_0^{\pi/2\omega}$$

$$= \frac{K^1 Q^3 \omega \pi^2}{16} \left(\frac{3}{\omega} - \frac{1}{3\omega} \right)$$

$$= \frac{K^1 Q^3 \pi^2}{6} \tag{8.64}$$

For steady flow:

$$\mathbf{P} = Q(-\Delta P) = K^1 Q^3 \tag{8.65}$$

Thus the power is increased by a factor of $\pi^2/6$ (= 1.64) as a result of fluctuations, compared with 1.23 for streamline flow.

In addition to the increased power to overcome friction, it will be necessary to supply the energy required to accelerate the liquid at the beginning of each cycle.

8.6.2 Gases

If a gas is pumped under turbulent flow conditions from a reservoir at a pressure P_1 to a second reservoir at a higher pressure P_2 through a uniform pipe of cross-sectional area A

by means of a pump situated at the upstream end, the power required is:

$$\mathbf{P} = G\left(\frac{u^2}{2} + \int_{P_1}^{P_2} v\,dP + F\right) \tag{8.66}$$

(from equation 8.55)

neglecting pressure changes arising from change in vertical height.

In order to make the gas flow, the pump must raise the pressure at the upstream end of the pipe to some value P_3, which is greater than P_2. The required value of P_3 will depend somewhat on the conditions of flow in the pipe. Thus for isothermal conditions, P_3 may be calculated from equation 4.55, since the downstream pressure P_2 and the mass rate of flow G are known. For non-isothermal conditions, the appropriate equation, such as 4.66 or 4.77, must be used.

The power requirement is then that for compression of the gas from pressure P_1 to P_3 and for imparting the necessary kinetic energy to it. Under normal conditions, however, the kinetic energy term is negligible. Thus for an isothermal efficiency of compression of η, the power required is:

$$\mathbf{P} = \frac{1}{\eta}GP_1v_1\ln\frac{P_3}{P_1} \tag{8.67}$$

Example 8.8

Hydrogen is pumped from a reservoir at 2 MN/m^2 through a clean horizontal mild steel pipe 50 mm in diameter and 500 m long. The downstream pressure is also 2 MN/m^2 and the pressure of the gas is raised to 2.5 MN/m^2 by a pump at the upstream end of the pipe. The conditions of flow are isothermal and the temperature of the gas is 295 K. What is the flowrate and what is the effective rate of working of the pump if it operates with an efficiency of 60 per cent?

Viscosity of hydrogen $= 0.009$ mN s/m^2 at 295 K.

Solution

Viscosity of hydrogen $= 0.009$ mN s/m^2 or 9×10^{-6} N s/m^2

Density of hydrogen at the mean pressure of 2.25 MN/m^2

$$= \left(\frac{2}{22.4}\right)\left(\frac{2250}{101.3}\right)\left(\frac{273}{295}\right) = 1.83 \text{ kg/m}^3$$

Firstly, an approximate value of G is obtained by neglecting the kinetic energy of the fluid. Taking P_1 and P_2 as the pressures at the upstream and downstream ends of the pipe, 2.5×10^6 and 2.0×10^6 N/m^2, then in equation 4.56:

$$P_1 - P_2 = 4\frac{R}{\rho u^2}\frac{l}{d}\rho_m u_m^2$$

or:

$$0.5 \times 10^6 = 4\frac{R}{\rho u^2}\frac{500}{0.050} \times 1.83u_m^2 \text{ N/m}^2$$

$$\therefore \qquad \frac{R}{\rho u^2}u_m^2 = 6.83$$

$$\therefore \qquad \frac{R}{\rho u^2}Re^2 = \frac{6.83 \times 0.050^2 \times 1.83^2}{(9 \times 10^{-6})^2}$$

$$= 7.02 \times 10^8$$

Taking the roughness of the pipe surface, e as 0.00005 m:

$$\frac{e}{d} = 0.001 \text{ and } Re = 5.7 \times 10^5 \text{ from Figure 3.8}$$

$$\therefore \quad \frac{4G}{\pi \mu d} = 5.7 \times 10^5$$

$$\therefore \quad G = 5.7 \times 10^5 \times \frac{\pi}{4} \times 9 \times 10^{-6} \times 0.050$$

$$= 0.201 \text{ kg/s}$$

From Figure 3.7, $(R/\rho u^2) = 0.0024$

Taking the kinetic energy of the fluid into account, equation 4.56 may be used:

$$\left(\frac{G}{A}\right)^2 \ln \frac{P_1}{P_2} + (P_2 - P_1)\rho_m + 4\frac{R}{\rho u^2}\frac{l}{d}\left(\frac{G}{A}\right)^2 = 0$$

Using the value of $(R/\rho u^2)$ obtained by neglecting the kinetic energy:

$$\left(\frac{G}{A}\right)^2 \ln \frac{2.5}{2.0} - (0.5 \times 10^6 \times 1.83) + \left[4 \times 0.0024 \times \left(\frac{500}{0.05}\right)\right]\left(\frac{G}{A}\right)^2 = 0$$

$$0.223\left(\frac{G}{A}\right)^2 - 915,000 + 96.0\left(\frac{G}{A}\right)^2 = 0$$

$$\therefore \quad \frac{G}{A} = 97.5$$

$$\therefore \quad G = 97.5 \times \frac{\pi}{4} \times 0.050^2$$

$$= 0.200 \text{ kg/s}$$

Thus, as is commonly the case, when the pressure drop is a relatively small proportion of the total pressure, the change in kinetic energy is negligible compared with the frictional losses. This would not be true had the pressure drop been much greater.

The power requirements for the pump can now be calculated from equation 8.67.

$$\text{Power} = \frac{GP_m v_m \ln(P_1/P_2)}{\eta}$$

$$= \frac{0.200 \times 2.25 \times 10^6 \times (1/1.83) \times 0.223}{0.60}$$

$$= 9.14 \times 10^4 \text{ W or } \underline{91.4 \text{ kW}}$$

8.7 FURTHER READING

ANDERSON, J. D. Jr.: *Modern Compressible Flow.* (McGraw-Hill, New York, 1982).

ENGINEERING EQUIPMENT USERS' ASSOCIATION: *Vacuum Producing Equipment,* EEUA Handbook No. 11 (Constable, London, 1961).

ENGINEERING EQUIPMENT USERS' ASSOCIATION: *Electrically Driven Glandless Pumps,* EEUA Handbook No. 26 (Constable, London, 1968).

ENGINEERING EQUIPMENT USERS' ASSOCIATION: *Guide to the Selection of Rotodynamic Pumps,* EEUA Handbook No. 30 (Constable, London, 1972).

GREENE, W. (ed.): *The Chemical Engineering Guide to Compressors* (McGraw-Hill, New York, 1985).

HOLLAND, F. A. and CHAPMAN, F. S.: *Pumping of Liquids* (Reinhold, New York, 1966).

M. W. KELLOGG CO.: *Design of Piping Systems,* 2nd edn (Wiley, Chichester, 1964).

LAZARKIENICZ, S. and TROSKOLANSKI, A. T.: *Impeller Pumps* (Pergamon Press, Oxford, 1965).
McNAUGHTON, K. (ed.): *The Chemical Engineering Guide to Pumps* (McGraw-Hill, New York, 1985).
PERRY, R. H. and GREEN, D. W (eds): *Chemical Engineers Handbook.* 6th edn (McGraw-Hill, New York, 1984).
STREETER, V. L. and WYLIE, E. B.: *Fluid Mechanics* (McGraw-Hill, New York, 1985).
WARRING, R. H.: *Seals and Sealing Handbook* (Trade and Technical Press, London, 1983).

8.8 REFERENCES

1. MARSHALL, P.: *The Chemical Engineer (London)* No. 418 (1985) 52. Positive displacement pumps — a brief survey.
2. HARVEST, J.: *The Chemical Engineer (London)* No. 403 (1984) 28. Recent developments in gear pumps.
3. GADSDEN, C.: *The Chemical Engineer (London)* No. 404 (1984) 42. Squeezing the best from peristaltic pumps.
4. STEELE, K. and ODROWAZ-PIENIAZEK, S.: *The Chem. Engr.* No. 422, (1986) 34 and No. 423, (1986) 30. Advances in slurry pumps — Parts 1 and 2.
5. DAVIDSON, J. (ed.): *Process Pump Selection — a Systems Approach* (I. Mech. E., Bury St Edmunds, Suffolk, 1986).

8.9. NOMENCLATURE

		Units in SI system	Dimensions in M, L, T, θ
A	Cross-sectional area	m^2	L^2
b	Width of pump impeller	m	L
c	Clearance in cylinder, i.e. ratio clearance volume/swept volume	—	—
d	Diameter	m	L
F	Energy degraded due to irreversibility per unit mass of fluid	J/kg	L^2T^{-2}
G	Mass flow rate	kg/s	MT^{-1}
g	Acceleration due to gravity	m/s^2	LT^{-2}
H	Enthalpy per unit mass	J/kg	L^2T^{-2}
h	Head	m	L
h_a	Heat equivalent to atmospheric pressure	m	L
h_f	Friction head	m	L
h_i	Head at pump inlet	m	L
h_0	Height of liquid above pump inlet	m	L
h_r	Height through which liquid is raised	m	L
h_s	Submergence of air lift pump	m	L
K	Coefficient	$(N/m^2)/(m^3/s)$	$ML^{-4}T^{-1}$
K^1	Coefficient	$(N/m^2)/(m^3/s)^2$	ML^{-7}
l	Length	m	L
m	Mass of gas	kg	M
M	Mass of liquid	kg	M
N	Revolutions per unit time	Hz	T^{-1}
N_s	Specific speed of a pump	—	—
n	Number of stages of compression	—	—
P	Pressure	N/m^2	$ML^{-1}T^{-2}$
ΔP	Pressure drop or difference	N/m^2	$ML^{-1}T^{-2}$
P_a	Atmospheric pressure	N/m^2	$ML^{-1}T^{-2}$
P_i	Intermediate pressure	N/m^2	$ML^{-1}T^{-2}$
P_v	Vapour pressure of liquid	N/m^2	$ML^{-1}T^{-2}$
P_0	Pressure at pump suction tank	N/m^2	$ML^{-1}T^{-2}$
$\mathbf{P}$	Power	W	ML^2T^{-3}
Q	Volumetric rate of flow	m^3/s	L^3T^{-1}
Q_m	Maximum rate of flow	m^3/s	L^3T^{-1}
$\dot{Q}$	Rate of flow at time t	m^3/s	L^3T^{-1}

		Units in SI system	Dimensions in M, L, T, θ
q	Heat added from surrounding per unit mass of fluid	J/kg	L^2T^{-2}
R	Shear stress at pipe wall	N/m^2	$ML^{-1}T^{-2}$
r	Radius	m	L
S	Entropy per unit mass	J/kg K	$L^2T^{-2}\theta^{-1}$
T	Temperature	K	θ
t	Time	s	T
U	Internal energy per unit mass	J/kg	L^2T^{-2}
u	Velocity of fluid	m/s	LT^{-1}
u_i	Velocity at pump inlet	m/s	LT^{-1}
u_t	Tangential velocity of tip of vane of compressor or pump	m/s	LT^{-1}
u_v	Velocity of fluid relative to vane of compressor	m/s	LT^{-1}
u_1	Velocity at inlet to centrifugal pump	m/s	LT^{-1}
u_2	Velocity at outlet to centrifugal pump	m/s	LT^{-1}
V	Volume	m^3	L^3
V_s	Swept volume	m^3	L^3
v	Specific volume	m^3/kg	$M^{-1}L^3$
v_a	Specific volume at atmospheric pressure	m^3/kg	$M^{-1}L^3$
W	Net work done per unit mass	J/kg	L^2T^{-2}
W_s	Shaft work per unit mass	J/kg	L^2T^{-2}
W'	Work of compressor per cycle	J	ML^2T^{-2}
z	Height	m	L
α	Correction factor for kinetic energy of fluid	—	—
β	Angle between tangential direction and blade of impeller at its tip	—	—
γ	Ratio of specific heat at constant pressure to specific heat at constant volume	—	—
ρ	Density	kg/m^3	ML^{-3}
θ	Angle between tangential direction and direction of motion of fluid	—	—
τ	Torque	Nm	ML^2T^{-2}
ω	Angular velocity	rad/s	T^{-1}
Δ	Finite difference in quantity	—	—
η	Efficiency	—	—
Π	Dimensionless group	—	—
ϕ	Friction factor	—	—

CHAPTER 9

Heat Transfer

9.1. INTRODUCTION

In the majority of chemical processes heat is either given out or absorbed, and fluids must often be either heated or cooled in a wide range of plant, such as furnaces, evaporators, distillation units, dryers, and reaction vessels where one of the major problems is that of transferring heat at the desired rate. In addition, it may be necessary to prevent the loss of heat from a hot vessel or pipe system. The control of the flow of heat at the desired rate forms one of the most important areas of chemical engineering. Provided that a temperature difference exists between two parts of a system, heat transfer will take place in one or more of three different ways.

Conduction. In a solid, the flow of heat by conduction is the result of the transfer of vibrational energy from one molecule to another, and in fluids it occurs in addition as a result of the transfer of kinetic energy. Heat transfer by conduction may also arise from the movement of free electrons. This process is particularly important with metals and accounts for their high thermal conductivities.

Convection. Heat transfer by convection arises from the mixing of elements of fluid. If this mixing occurs as a result of density differences as, for example, when a pool of liquid is heated from below, the process is known as *natural convection*; if the mixing results from eddy movement in the fluid, for example when a fluid flows through a pipe heated on the outside, it is called *forced convection*. It is important to note that convection requires mixing of fluid elements and is not governed just by temperature difference as is the case in conduction and radiation.

Radiation. All materials radiate thermal energy in the form of electromagnetic waves. When this radiation falls on a second body it may be partially reflected, transmitted, or absorbed. It is only the fraction that is absorbed that appears as heat in the body.

9.2. BASIC CONSIDERATIONS

9.2.1. Individual and overall coefficients of heat transfer

In many of the applications of heat transfer in process plants, one or more of the mechanisms of heat transfer may be involved. In the majority of heat exchangers heat passes through a series of different intervening layers before reaching the second fluid (Figure 9.1). These layers may be of different thicknesses and of different thermal conductivities. The problem of transferring heat to crude oil in the primary furnace before it enters the first distillation column may be considered as an example. The heat from the flames passes by radiation and convection to the pipes in the furnace, by conduction through the

338

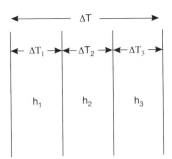

Figure 9.1. Heat transfer through a composite wall

pipe walls, and by forced convection from the inside of the pipe to the oil. Here all three modes of transfer are involved. After prolonged usage, solid deposits may form on both the inner and outer walls of the pipes, and these will then contribute additional resistance to the transfer of heat. The simplest form of equation which represents this heat transfer operation may be written as:

$$Q = UA\Delta T \tag{9.1}$$

where Q is the heat transferred per unit time, A the area available for the flow of heat, ΔT the difference in temperature between the flame and the boiling oil, and U is known as the overall heat transfer coefficient (W/m^2 K in SI units).

At first sight, equation 9.1 implies that the relationship between Q and ΔT is linear. Whereas this is approximately so over limited ranges of temperature difference for which U is nearly constant, in practice U may well be influenced both by the temperature difference and by the absolute value of the temperatures.

If it is required to know the area needed for the transfer of heat at a specified rate, the temperature difference ΔT, and the value of the overall heat-transfer coefficient must be known. Thus the calculation of the value of U is a key requirement in any design problem in which heating or cooling is involved. A large part of the study of heat transfer is therefore devoted to the evaluation of this coefficient.

The value of the coefficient will depend on the mechanism by which heat is transferred, on the fluid dynamics of both the heated and the cooled fluids, on the properties of the materials through which the heat must pass, and on the geometry of the fluid paths. In solids, heat is normally transferred by conduction; some materials such as metals have a high thermal conductivity, whilst others such as ceramics have a low conductivity. Transparent solids like glass also transmit radiant energy particularly in the visible part of the spectrum.

Liquids also transmit heat readily by conduction, though circulating currents are frequently set up and the resulting convective transfer may be considerably greater than the transfer by conduction. Many liquids also transmit radiant energy. Gases are poor conductors of heat and circulating currents are difficult to suppress; convection is therefore much more important than conduction in a gas. Radiant energy is transmitted with only limited absorption in gases and, of course, without any absorption *in vacuo*. Radiation is the only mode of heat transfer which does not require the presence of an intervening medium.

If the heat is being transmitted through a number of media in series, the overall heat transfer coefficient can be broken down into individual coefficients h each relating to a single medium. This is as shown in Figure 9.1. It is assumed that there is good contact between each pair of elements so that the temperature is the same on the two sides of each junction.

Suppose heat is being transferred through three media, each of area A, that the individual coefficients for each of the media are h_1, h_2, and h_3, and the corresponding temperature changes are ΔT_1, ΔT_2, and ΔT_3. If there is no accumulation of heat in the media, the heat transfer rate Q will be the same through each. Three equations, analogous to equation 9.1 can therefore be written:

$$\left.\begin{array}{l} Q = h_1 A \Delta T_1 \\ Q = h_2 A \Delta T_2 \\ Q = h_3 A \Delta T_3 \end{array}\right\} \tag{9.2}$$

Rearranging:
$$\Delta T_1 = \frac{Q}{A} \frac{1}{h_1}$$

$$\Delta T_2 = \frac{Q}{A} \frac{1}{h_2}$$

$$\Delta T_3 = \frac{Q}{A} \frac{1}{h_3}$$

Adding:
$$\Delta T_1 + \Delta T_2 + \Delta T_3 = \frac{Q}{A} \left(\frac{1}{h_1} + \frac{1}{h_2} + \frac{1}{h_3} \right) \tag{9.3}$$

Noting that $(\Delta T_1 + \Delta T_2 + \Delta T_3) =$ total temperature difference ΔT:

then:
$$\Delta T = \frac{Q}{A} \left(\frac{1}{h_1} + \frac{1}{h_2} + \frac{1}{h_3} \right) \tag{9.4}$$

But from equation 9.1:
$$\Delta T = \frac{Q}{A} \frac{1}{U} \tag{9.5}$$

Comparing equations 9.4 and 9.5:

$$\frac{1}{U} = \frac{1}{h_1} + \frac{1}{h_2} + \frac{1}{h_3} \tag{9.6}$$

The reciprocals of the heat transfer coefficients are resistances, and equation 9.6 therefore illustrates that the resistances are additive.

In some cases, particularly for the radial flow of heat through a thick pipe wall or cylinder, the area for heat transfer is a function of position. Thus the area for transfer applicable to each of the three media could differ and may be A_1, A_2 and A_3. Equation 9.3 then becomes:

$$\Delta T_1 + \Delta T_2 + \Delta T_3 = Q \left(\frac{1}{h_1 A_1} + \frac{1}{h_2 A_2} + \frac{1}{h_3 A_3} \right) \tag{9.7}$$

Equation 9.7 must then be written in terms of one of the area terms A_1, A_2, and A_3, or sometimes in terms of a mean area. Since Q and ΔT must be independent of the particular area considered, the value of U will vary according to which area is used as the basis. Thus equation 9.7 can be written, for example:

$$Q = U_1 A_1 \Delta T \quad \text{or} \quad \Delta T = \frac{Q}{U_1 A_1}$$

This will then give U_1 as:

$$\frac{1}{U_1} = \frac{1}{h_1} + \frac{A_1}{A_2}\left(\frac{1}{h_2}\right) + \frac{A_1}{A_3}\left(\frac{1}{h_3}\right) \tag{9.8}$$

In this analysis it is assumed that the heat flowing per unit time through each of the media is the same.

Now that the overall coefficient U has been broken down into its component parts, each of the individual coefficients h_1, h_2, and h_3 must be evaluated. This can be done from a knowledge of the nature of the heat-transfer process in each of the media. A study will therefore be made of how these individual coefficients can be calculated for conduction, convection, and radiation.

9.2.2. Mean temperature difference

Where heat is being transferred from one fluid to a second fluid through the wall of a vessel and the temperature is the same throughout the bulk of each of the fluids, there is no difficulty in specifying the overall temperature difference ΔT. Frequently, however, each fluid is flowing through a heat exchanger such as a pipe or a series of pipes in parallel, and its temperature changes as it flows, and consequently the temperature difference is continuously changing. If the two fluids are flowing in the same direction (*co-current flow*), the temperature of the two streams progressively approach one another (Figure 9.2). In these circumstances the outlet temperature of the heating fluid must always be higher than that of the cooling fluid. If the fluids are flowing in opposite directions (countercurrent flow), the temperature difference will show less variation throughout the heat exchanger (Figure 9.3). In this case it is possible for the cooling liquid to leave at a higher temperature than the heating liquid, and one of the great advantages of *countercurrent flow* is that it is possible to extract a higher proportion of the heat content of the heating fluid. It will

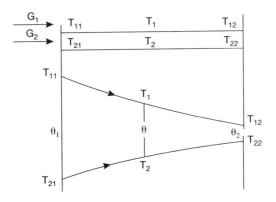

Figure 9.2.   Mean temperature difference for co-current flow

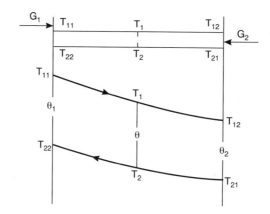

Figure 9.3.   Mean temperature difference for countercurrent flow

now be shown how to calculate the appropriate value of the temperature difference for co-current and for countercurrent flow. It will be assumed that the overall heat transfer coefficient U remains constant throughout in the heat exchanger.

It is necessary to find the average value of the temperature difference θ_m to be used in the general equation:

$$Q = UA\theta_m \qquad\qquad \text{(equation 9.1)}$$

Figure 9.3 shows the temperature conditions for the fluids flowing in opposite directions, a condition known as countercurrent flow.

The outside stream (specific heat C_{p1}) at mass flow rate G_1 falls in temperature from T_{11} to T_{12}.

The inside stream (specific heat C_{p2}) at mass flow rate G_2 rises in temperature from T_{21} to T_{22}.

Over a small element of area dA where the temperatures of the streams are T_1 and T_2. The temperature difference:

$$\theta = T_1 - T_2$$

$$\therefore \qquad d\theta = dT_1 - dT_2$$

Heat given out by hot stream $\qquad = dQ = -G_1 C_{p1}\, dT_1$

Heat taken up by cold stream $\qquad = dQ = G_2 C_{p2}\, dT_2$

$$\therefore \qquad d\theta = -\frac{dQ}{G_1 C_{p1}} - \frac{dQ}{G_2 C_{p2}} = -dQ\left(\frac{G_1 C_{p1} + G_2 C_{p2}}{G_1 C_{p1} \times G_2 C_{p2}}\right) = -\psi\, dQ \quad \text{(say)}$$

$$\therefore \qquad\qquad \theta_1 - \theta_2 = \psi Q$$

Over this element: $\qquad\qquad U\, dA\theta = dQ$

$$\therefore \qquad\qquad U\, dA\theta = -\frac{d\theta}{\psi}$$

If U can be taken as constant:

$$-\psi U \int_0^A \mathrm{d}A = \int_{\theta_1}^{\theta_2} \frac{\mathrm{d}\theta}{\theta}$$

$$\therefore \qquad -\psi UA = -\ln \frac{\theta_1}{\theta_2}$$

But, from the definition of θ_m, $Q = UA\theta_m$.

$$\therefore \qquad \theta_1 - \theta_2 = \psi Q = \psi UA\theta_m = \ln \frac{\theta_1}{\theta_2}(\theta_m)$$

$$\therefore \qquad \theta_m = \frac{\theta_1 - \theta_2}{\ln(\theta_1/\theta_2)} \qquad\qquad (9.9)$$

where θ_m is known as the *logarithmic mean temperature difference*.

UNDERWOOD[1] proposed an approximation for the logarithmic mean temperature difference as follows:

$$(\theta_m)^{1/3} = \tfrac{1}{2}(\theta_1^{1/3} + \theta_2^{1/3}) \qquad\qquad (9.10)$$

and for example, when $\theta_1 = 1$ K and $\theta_2 = 100$ K, θ_m is 22.4 K compared with a logarithmic mean of 21.5 K. When $\theta_1 = 10$ K and $\theta_2 = 100$ K, both the approximation and the logarithmic mean values coincide at 39 K.

If the two fluids flow in the same directions on each side of the tube, co-current flow is taking place and the general shape of the temperature profile along the tube is as shown in Figure 9.2. A similar analysis will show that this gives the same expression for θ_m, the logarithmic mean temperature difference. For the same terminal temperatures it is important to note that the value of θ_m for countercurrent flow is appreciably greater than the value for co-current flow. This is seen from the temperature profiles, for with co-current flow the cold fluid cannot be heated to a higher temperature than the exit temperature of the hot fluid and is illustrated in Example 9.1.

Example 9.1

A heat exchanger is required to cool 20 kg/s of water from 360 K to 340 K by means of 25 kg/s water entering at 300 K. If the overall coefficient of heat transfer is constant at 2 kW/m²K, calculate the surface area required in (a) a countercurrent concentric tube exchanger, and (b) a co-current flow concentric tube exchanger.

Solution

Heat load: $Q = 20 \times 4.18(360 - 340) = 1672$ kW

The cooling water outlet temperature is given by:

$$1672 = 25 \times 4.18(\theta_2 - 300) \quad \text{or} \quad \theta_2 = 316 \text{ K}$$

(a) *Counterflow*

In equation 9.9:
$$\theta_m = \frac{44 - 40}{\ln(44/40)} = 41.9 \text{ K}$$

Heat transfer area:
$$A = \frac{Q}{U\theta_m}$$

$$= \frac{1672}{2 \times 41.9}$$

$$= \underline{\underline{19.95 \ m^2}}$$

(b) *Co-current flow*

In equation 9.9:

$$\theta_m = \frac{60 - 24}{\ln(60/24)} = 39.3 \ K$$

Heat transfer area:

$$A = \frac{1672}{2 \times 39.3}$$

$$= \underline{\underline{21.27 \ m^2}}$$

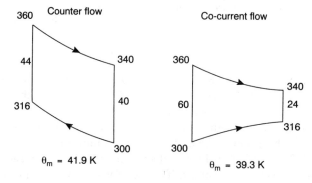

Figure 9.4.   Data for Example 9.1

It may be noted that using Underwood's approximation (equation 9.10), the calculated values for the mean temperature driving forces are 41.9 K and 39.3 K for counter- and co-current flow respectively, which agree exactly with the logarithmic mean values.

9.3. HEAT TRANSFER BY CONDUCTION

9.3.1. Conduction through a plane wall

This important mechanism of heat transfer will now be considered in more detail for the flow of heat through a plane wall of thickness x as shown in Figure 9.5.

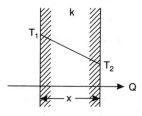

Figure 9.5.   Conduction of heat through a plane wall

The rate of heat flow Q over the area A and a small distance dx may be written as:

$$Q = -kA\left(\frac{dT}{dx}\right) \qquad (9.11)$$

often known as *Fourier's equation*, where the negative sign indicates that the temperature gradient is in the opposite direction to the flow of heat and k is the thermal conductivity of the material. Integrating for a wall of thickness x with boundary temperatures T_1 and T_2 as shown in Figure 9.5,

then: $$Q = \frac{kA(T_1 - T_2)}{x} \qquad (9.12)$$

Thermal conductivity is a function of temperature and experimental data may often be expressed by a linear relationship of the form:

$$k = k_0(1 + k'T) \qquad (9.13)$$

where k is the thermal conductivity at the temperature T and k_0 and k' are constants. Combining equations 9.11 and 9.13:

$$-k\,dT = -k_0(1 + k'T)dT = \frac{Q\,dx}{A}$$

Integrating between the temperature limits T_1 and T_2,

$$-\int_{T_1}^{T_2} k\,dT = (T_1 - T_2)k_0\left\{1 + k'\left(\frac{T_1 + T_2}{2}\right)\right\} = Q\int_{x_1}^{x_2} \frac{dx}{A} \qquad (9.14)$$

Where k is a linear function of T, the following equation may therefore be used:

$$k_a(T_1 - T_2) = Q\int_{x_1}^{x_2} \frac{dx}{A} \qquad (9.15)$$

where k_a is the arithmetic mean of k_1 and k_2 at T_1 and T_2 respectively or the thermal conductivity at the arithmetic mean of T_1 and T_2.

Where k is a non-linear function of T, some mean value, k_m will apply, where:

$$k_m = \frac{1}{T_2 - T_1}\int_{T_1}^{T_2} k\,dT \qquad (9.16)$$

It will be seen in Table 9.1 that metals have very high thermal conductivities, non-metallic solids lower values, non-metallic liquids low values, and gases very low values. It is important to note that amongst metals, stainless steel has a low value, that water has a very high value for liquids (due to partial ionisation), and that hydrogen has a high value for gases (due to the high mobility of the molecules). With gases, k decreases with increase in molecular mass and increases with the temperature. In addition, for gases the dimensionless *Prandtl group* $C_p\mu/k$ is approximately constant (C_p is the specific heat at constant pressure and μ is the viscosity) and can be used to evaluate k at high temperatures where it is difficult to determine experimentally because of the formation of convection currents. k does not vary significantly with pressure until this is reduced to a value so

Table 9.1. Thermal conductivities

	Temp (K)	k (Btu/h ft² °F/ft)	k (W/mK)
Solids—Metals			
Aluminium	573	133	230
Cadmium	291	54	94
Copper	373	218	377
Iron (wrought)	291	35	61
Iron (cast)	326	27.6	48
Lead	373	19	33
Nickel	373	33	57
Silver	373	238	412
Steel 1% C	291	26	45
Tantalum	291	32	55
Admiralty metal	303	65	113
Bronze	—	109	189
Stainless Steel	293	9.2	16
Solids—Non-metals			
Asbestos sheet	323	0.096	0.17
Asbestos	273	0.09	0.16
Asbestos	373	0.11	0.19
Asbestos	473	0.12	0.21
Bricks (alumina)	703	1.8	3.1
Bricks (building)	293	0.4	0.69
Magnesite	473	2.2	3.8
Cotton wool	303	0.029	0.050
Glass	303	0.63	1.09
Mica	323	0.25	0.43
Rubber (hard)	273	0.087	0.15
Sawdust	293	0.03	0.052
Cork	303	0.025	0.043
Glass wool	—	0.024	0.041
85% Magnesia	—	0.04	0.070
Graphite	273	87	151

	Temp (K)	k (Btu/h ft² °F/ft)	k (W/mK)
Liquids			
Acetic acid 50%	293	0.20	0.35
Acetone	303	0.10	0.17
Aniline	273–293	0.1	0.17
Benzene	303	0.09	0.16
Calcium chloride brine 30%	303	0.32	0.55
Ethyl alcohol 80%	293	0.137	0.24
Glycerol 60%	293	0.22	0.38
Glycerol 40%	293	0.26	0.45
n-Heptane	303	0.08	0.14
Mercury	301	4.83	8.36
Sulphuric acid 90%	303	0.21	0.36
Sulphuric acid 60%	303	0.25	0.43
Water	303	0.356	0.62
Water	333	0.381	0.66
Gases			
Hydrogen	273	0.10	0.17
Carbon dioxide	273	0.0085	0.015
Air	273	0.014	0.024
Air	373	0.018	0.031
Methane	273	0.017	0.029
Water vapour	373	0.0145	0.025
Nitrogen	273	0.0138	0.024
Ethylene	273	0.0097	0.017
Oxygen	273	0.0141	0.024
Ethane	273	0.0106	0.018

low that the mean free path of the molecules becomes comparable with the dimensions of the vessel; further reduction of pressure then causes k to decrease.

Typical values for Prandtl numbers are as follows:

Air	0.71	n-Butanol	50
Oxygen	0.63	Light oil	600
Ammonia (gas)	1.38	Glycerol	1000
Water	5 to 10	Polymer melts	10,000
		Mercury	0.02

The low conductivity of heat insulating materials, such as cork, glass wool, and so on, is largely accounted for by their high proportion of air space. The flow of heat through the materials is governed mainly by the resistance of the air spaces, which should be sufficiently small for convection currents to be suppressed.

It is convenient to rearrange equation 9.12 in the form:

$$Q = \frac{(T_1 - T_2)A}{(x/k)} \tag{9.17}$$

where x/k is known as the *thermal resistance* and k/x is the *transfer coefficient*.

Example 9.2.

Find the heat loss per square metre of surface through a brick wall 0.5 m thick when the inner surface is at 400 K and the outside is at 300 K. The thermal conductivity of the brick may be taken as 0.7 W/mK.

Solution

From equation 9.12:

$$Q = \frac{0.7 \times 1 \times (400 - 300)}{0.5}$$

$$= 140 \text{ W/m}^2$$

9.3.2. Thermal resistances in series

It has been seen earlier that thermal resistances may be added together for the case of heat transfer through a complete section formed from different media in series.

Figure 9.6 shows a composite wall made up of three materials with thermal conductivities k_1, k_2, and k_3, with thicknesses as shown and with the temperatures T_1, T_2, T_3, and T_4 at the faces. Applying equation 9.12 to each section in turn and noting that the same quantity of heat Q must pass through each area A:

$$T_1 - T_2 = \frac{x_1}{k_1 A} Q, \quad T_2 - T_3 = \frac{x_2}{k_2 A} Q \quad \text{and} \quad T_3 - T_4 = \frac{x_3}{k_3 A} Q$$

On addition:

$$(T_1 - T_4) = \left(\frac{x_1}{k_1 A} + \frac{x_2}{k_2 A} + \frac{x_3}{k_3 A} \right) Q \tag{9.18}$$

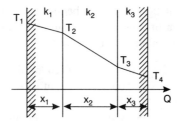

Figure 9.6.   Conduction of heat through a composite wall

or:
$$Q = \frac{T_1 - T_4}{\Sigma(x_1/k_1 A)}$$

$$= \frac{\text{Total driving force}}{\text{Total (thermal resistance/area)}} \qquad (9.19)$$

Example 9.3

A furnace is constructed with 0.20 m of firebrick, 0.10 m of insulating brick, and 0.20 m of building brick. The inside temperature is 1200 K and the outside temperature 330 K. If the thermal conductivities are as shown in Figure 9.7, find the heat loss per unit area and the temperature at the junction of the firebrick and the insulating brick.

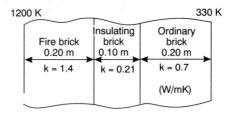

Figure 9.7.   Data for Example 9.3

Solution

From equation 9.19:

$$Q = (1200 - 330) \bigg/ \left[\left(\frac{0.20}{1.4 \times 1} \right) + \left(\frac{0.10}{0.21 \times 1} \right) + \left(\frac{0.20}{0.7 \times 1} \right) \right]$$

$$= \frac{870}{(0.143 + 0.476 + 0.286)} = \frac{870}{0.905}$$

$$= \underline{961 \text{ W/m}^2}$$

(Temperature drop over firebrick)/(Total temperature drop) = (0.143/0.905)

$$\therefore \qquad \text{Temperature drop over firebrick} = \left(\frac{870 \times 0.143}{0.905} \right) = 137 \text{ K}$$

Hence temperature at the firebrick-insulating brick interface = (1200 − 137) = <u>1063 K</u>

9.3.3. Conduction through a thick-walled tube

The conditions for heat flow through a thick-walled tube when the temperatures on the inside and outside are held constant are shown in Figure 9.8. Here the area for heat flow is proportional to the radius so that the temperature gradient is inversely proportional to the radius.

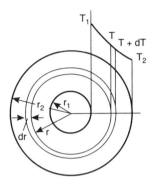

Figure 9.8.   Conduction through thick-walled tube or spherical shell

The heat flow at any radius r is given by:

$$Q = -k2\pi rl \frac{dT}{dr} \qquad (9.20)$$

where l is the length of tube.
    Integrating between limits r_1 and r_2:

$$Q \int_{r_1}^{r_2} \frac{dr}{r} = -2\pi lk \int_{T_1}^{T_2} dT$$

or:
$$Q = \frac{2\pi lk(T_1 - T_2)}{\ln(r_2/r_1)} \qquad (9.21)$$

This equation can be put into the form of equation 9.12:

i.e.:
$$Q = \frac{k(2\pi r_m l)(T_1 - T_2)}{r_2 - r_1} \qquad (9.22)$$

where $r_m = (r_2 - r_1)/\ln(r_2/r_1)$, and is known as the *logarithmic mean radius*. For thin-walled tubes it will be sufficient to use the arithmetic mean radius r_a, giving:

$$Q = \frac{k(2\pi r_a l)(T_1 - T_2)}{r_2 - r_1} \qquad (9.23)$$

9.3.4. Conduction through a spherical shell and to a particle

For heat conduction through a spherical shell, the heat flow at radius r is given by:

$$Q = -k4\pi r^2 \frac{dT}{dr} \qquad (9.24)$$

Integrating between limits r_1 and r_2:

$$Q \int_{r_1}^{r_2} \frac{dr}{r^2} = -4\pi k \int_{T_1}^{T_2} dT$$

$$\therefore \qquad Q = \frac{4\pi k(T_1 - T_2)}{(1/r_1) - (1/r_2)} \qquad\qquad (9.25)$$

An important application of heat transfer to a sphere is that of conduction through a stationary fluid surrounding a spherical particle or droplet of radius r as encountered for example in fluidised beds, rotary kilns, spray dryers and plasma devices. If the temperature difference $T_1 - T_2$ is spread over a very large distance so that $r_2 = \infty$ and T_1 is the temperature of the surface of the drop, then:

$$\frac{Qr}{(4\pi r^2)(T_1 - T_2)k} = 1$$

or: $\qquad\qquad\qquad\qquad \frac{hd}{k} = Nu = 2 \qquad\qquad (9.26)$

where $Q/4\pi r^2(T_1 - T_2) = h$ is the heat transfer coefficient, hd/k is a dimensionless group known as the *Nusselt number* (Nu') discussed in Section 9.4 and d is the diameter of the particle or droplet. This value of 2 for the Nusselt number is the theoretical minimum for heat transfer through a *continuous* medium. It will be greater if the temperature difference is applied over a finite distance, when equation 9.25 must be used. When there is relative motion between the particle and the fluid the heat transfer rate will be further increased, as discussed in Section 9.4.6.

In this approach, heat transfer to a spherical particle by conduction through the surrounding fluid has been the prime consideration. In many practical situations the flow of heat from the surface to the internal parts of the particle is of importance. For example, if the particle is a poor conductor then the rate at which the particulate material reaches some desired average temperature may be limited by conduction inside the particle rather than by conduction to the outside surface of the particle. This problem involves unsteady state transfer of heat which is considered in Section 9.3.5.

Equations may be developed to predict the rate of change of diameter d of evaporating droplets. If the latent heat of vaporisation is provided by heat conducted through a hotter stagnant gas to the droplet surface, and heat transfer is the rate controlling step, it is shown by SPALDING[2] that d^2 decreases linearly with time. A closely related and important practical problem is the prediction of the residence time required in a combustion chamber to ensure virtually complete burning of oil droplets. Complete combustion is desirable to obtain maximum utilisation of energy and to minimise pollution of the atmosphere by partially burned oil droplets. Here a droplet is surrounded by a flame and heat conducted back from the flame to the droplet surface provides the heat to vaporise the oil and sustain the surrounding flame. Again d^2 decreases approximately linearly with time but derivation of the equation is more complex due to mass transfer effects, steep temperature gradients[3] and circulation in the drop[4].

9.3.5. Unsteady state conduction

Basic considerations

In the problems which have been considered so far, it has been assumed that the conditions at any point in the system remain constant with respect to time. The case of heat transfer by conduction in a medium in which the temperature is changing with time, will now be considered. This problem is of importance in the calculation of the temperature distribution in a body which is being heated or cooled. If, in an element of dimensions dx by dy by dz (Figure 9.9), the temperature at the point (x, y, z) is θ and at the point $(x + dx, y + dy, z + dz)$ is $(\theta + d\theta)$, then assuming that the thermal conductivity k is constant and that no heat is generated in the medium, the rate of conduction of heat through the element is:

$$= -k \, dy \, dz \left(\frac{\partial \theta}{\partial x} \right)_{yz} \qquad \text{in the } x\text{-direction}$$

$$= -k \, dz \, dx \left(\frac{\partial \theta}{\partial y} \right)_{zx} \qquad \text{in the } y\text{-direction}$$

$$= -k \, dx \, dy \left(\frac{\partial \theta}{\partial z} \right)_{xy} \qquad \text{in the } z\text{-direction}$$

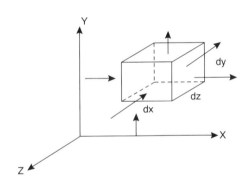

Figure 9.9.   Unsteady state heat conduction to element

The rate of change of heat content of the element will be equal to *minus* the rate of increase of heat flow from (x, y, z) to $(x + dx, y + dy, z + dz)$. Thus the rate of change of the heat content of the element is:

$$= k \, dy \, dz \left(\frac{\partial^2 \theta}{\partial x^2} \right)_{yz} dx + k \, dz \, dx \left(\frac{\partial^2 \theta}{\partial y^2} \right)_{zx} dy + k \, dx \, dy \left(\frac{\partial^2 \theta}{\partial z^2} \right)_{xy} dz$$

$$= k \, dx \, dy \, dz \left[\left(\frac{\partial^2 \theta}{\partial x^2} \right)_{yz} + \left(\frac{\partial^2 \theta}{\partial y^2} \right)_{zx} + \left(\frac{\partial^2 \theta}{\partial z^2} \right)_{zy} \right] \qquad (9.27)$$

The rate of increase of heat content is also equal however to the product of the heat capacity of the element and the rate of rise of temperature.

Thus: $k \, dx \, dy \, dz \left[\left(\dfrac{\partial^2 \theta}{\partial x^2} \right)_{yz} + \left(\dfrac{\partial^2 \theta}{\partial y^2} \right)_{zx} + \left(\dfrac{\partial^2 \theta}{\partial z^2} \right)_{xy} \right] = C_p \rho \, dx \, dy \, dz \dfrac{\partial \theta}{\partial t}$

or: $\dfrac{\partial \theta}{\partial t} = \dfrac{k}{C_p \rho} \left[\left(\dfrac{\partial^2 \theta}{\partial x^2} \right)_{yz} + \left(\dfrac{\partial^2 \theta}{\partial y^2} \right)_{zx} + \left(\dfrac{\partial^2 \theta}{\partial z^2} \right)_{xy} \right]$

$$= D_H \left[\left(\dfrac{\partial^2 \theta}{\partial x^2} \right)_{yz} + \left(\dfrac{\partial^2 \theta}{\partial y^2} \right)_{zx} + \left(\dfrac{\partial^2 \theta}{\partial z^2} \right)_{xy} \right] \qquad (9.28)$$

where $D_H = k/C_p\rho$, is known as the *thermal diffusivity*.

The above partial differential equation is most conveniently solved by the use of the Laplace transform of temperature with respect to time. As an illustration of the method of solution, the problem of the unidirectional flow of heat in a continuous medium will be considered. The basic differential equation for the X-direction is:

$$\dfrac{\partial \theta}{\partial t} = D_H \dfrac{\partial^2 \theta}{\partial x^2} \qquad (9.29)$$

This equation cannot be integrated directly since the temperature θ is expressed as a function of two independent variables, distance x and time t. The method of solution involves transforming the equation so that the Laplace transform of θ with respect to time is used in place of θ. The equation then involves only the Laplace transform $\bar{\theta}$ and the distance x. The Laplace transform of θ is defined by the relation:

$$\bar{\theta} = \int_0^\infty \theta e^{-pt} \, dt \qquad (9.30)$$

where p is a parameter.

Thus $\bar{\theta}$ is obtained by operating on θ with respect to t with x constant.

Then: $\dfrac{\partial^2 \bar{\theta}}{\partial x^2} = \overline{\dfrac{\partial^2 \theta}{\partial x^2}} \qquad (9.31)$

Now: $\overline{\dfrac{\partial \theta}{\partial t}} = \int_0^\infty \dfrac{\partial \theta}{\partial t} e^{-pt} \, dt$

$$= \left[\theta e^{-pt} \right]_0^\infty + p \int_0^\infty e^{-pt} \theta \, dt$$

$$= -\theta_{t=0} + p\bar{\theta} \qquad (9.32)$$

Then, taking the Laplace transforms of each side of equation 9.29:

$$\overline{\dfrac{\partial \theta}{\partial t}} = D_H \overline{\dfrac{\partial^2 \theta}{\partial x^2}}$$

i.e.: $p\bar{\theta} - \theta_{t=0} = D_H \dfrac{\partial^2 \bar{\theta}}{\partial x^2}$ (from equations 9.31 and 9.32)

i.e.: $\dfrac{\partial^2 \bar{\theta}}{\partial x^2} - \dfrac{p}{D_H} \bar{\theta} = -\dfrac{\theta_{t=0}}{D_H}$

If the temperature everywhere is constant initially, $\theta_{t=0}$ is a constant and the equation may be integrated as a normal second-order differential equation since p is not a function of x.

Thus:
$$\bar{\theta} = B_1 e^{\sqrt{(p/D_H)}x} + B_2 e^{-\sqrt{(p/D_H)}x} + \theta_{t=0} p^{-1} \tag{9.33}$$

and therefore:
$$\frac{\partial \bar{\theta}}{\partial x} = B_1 \sqrt{\frac{p}{D_H}} e^{\sqrt{(p/D_H)}x} - B_2 \sqrt{\frac{p}{D_H}} e^{-\sqrt{(p/D_H)}x} \tag{9.34}$$

The temperature θ, corresponding to the above transform $\bar{\theta}$, can now be found by reference to tables of the Laplace transform. It is first necessary, however, to evaluate the constants B_1 and B_2 using the boundary conditions for the particular problem because these constants will in general involve the parameter p which was introduced in the transformation.

Considering the particular problem of the unidirectional flow of heat through a body with plane parallel faces a distance l apart, the heat flow is normal to these faces and the temperature of the body is initially constant throughout. The temperature scale will be so chosen that this uniform initial temperature is zero. At time, $t = 0$, one face (at $x = 0$) will be brought into contact with a source at a constant temperature θ' and the other face (at $x = l$) will be assumed to be perfectly insulated thermally.

The boundary conditions are therefore:

$$t = 0, \qquad \theta = 0$$

$$t > 0, \qquad \theta = \theta' \quad \text{when } x = 0$$

$$t > 0, \qquad \frac{\partial \theta}{\partial x} = 0 \quad \text{when } x = l$$

Thus:
$$\bar{\theta}_{x=0} = \int_0^\infty \theta' e^{-pt}\, dt = \frac{\theta'}{p}$$

and:
$$\left(\frac{\partial \bar{\theta}}{\partial x} \right)_{x=l} = 0$$

Substitution of these boundary conditions in equations 9.33 and 9.34 gives:

$$B_1 + B_2 = \frac{\theta'}{p}$$

and:
$$B_1 e^{\sqrt{(p/D_H)}l} - B_2 e^{-\sqrt{(p/D_H)}l} = 0 \tag{9.35}$$

Hence:
$$B_1 = \frac{(\theta'/p) e^{-\sqrt{(p/D_H)}l}}{e^{\sqrt{(p/D_H)}l} + e^{-\sqrt{(p/D_H)}l}}$$

and:
$$B_2 = \frac{(\theta'/p) e^{\sqrt{(p/D_H)}l}}{e^{\sqrt{(p/D_H)}l} + e^{-\sqrt{(p/D_H)}l}}$$

Then:

$$\bar{\theta} = \frac{e^{(l-x)\sqrt{(p/D_H)}} + e^{-(l-x)\sqrt{(p/D_H)}}}{e^{\sqrt{(p/D_H)}l} + e^{-\sqrt{(p/D_H)}l}} \frac{\theta'}{p}$$

$$= \frac{\theta'}{p}(e^{(l-x)\sqrt{(p/D_H)}} + e^{-(l-x)\sqrt{(p/D_H)}})(1 + e^{-2\sqrt{(p/D_H)}l})^{-1}(e^{-\sqrt{(p/D_H)}l})$$

$$= \frac{\theta'}{p}(e^{-x\sqrt{(p/D_H)}} + e^{-(2l-x)\sqrt{(p/D_H)}})(1 - e^{-2l\sqrt{(p/D_H)}} + \cdots$$

$$+ (-1)^N e^{-2Nl\sqrt{(p/D_H)}} + \cdots)$$

$$= \sum_{N=0}^{N=\infty} \frac{\theta'}{p}(-1)^N(e^{-(2lN+x)\sqrt{(p/D_H)}} + e^{-\{2(N+1)l-x\}\sqrt{(p/D_H)}}) \qquad (9.36)$$

The temperature θ is then obtained from tables of inverse Laplace transforms (included in the Appendix (Table 12, No 83)) and is given by:

$$\theta = \sum_{N=0}^{N=\infty} (-1)^N \theta' \left(\text{erfc} \frac{2lN + x}{2\sqrt{D_H t}} + \text{erfc} \frac{2(N+1)l - x}{2\sqrt{D_H t}} \right) \qquad (9.37)$$

where:
$$\text{erfc } x = \frac{2}{\sqrt{\pi}} \int_x^\infty e^{-\xi^2} \, d\xi$$

Values of erfc x ($= 1 - \text{erf } x$) are to be found in the Appendix (Table 13) and in specialist sources.[5]

Equation 9.37 can be written in the form:

$$\frac{\theta}{\theta'} = \sum_{N=0}^{N=\infty} (-1)^N \left\{ \text{erfc} \left[Fo_l^{-1/2} \left(N + \frac{1}{2}\frac{x}{l} \right) \right] + \text{erfc} \left[Fo_l^{-1/2} \left((N+1) - \frac{1}{2}\frac{x}{l} \right) \right] \right\} \qquad (9.38)$$

where $Fo_l = (D_H t/l^2)$ and is known as the *Fourier number*.

Thus:
$$\frac{\theta}{\theta'} = f\left(Fo_l, \frac{x}{l} \right) \qquad (9.39)$$

The numerical solution to the above problem is then obtained by inserting the appropriate values for the physical properties of the system and using as many terms in the series as are necessary for the degree of accuracy required. In most cases, the above series converge quite rapidly.

This method of solution of problems of unsteady flow is particularly useful because it is applicable when there are discontinuities in the physical properties of the material.[6] The boundary conditions, however, become a little more complicated, but the problem is intrinsically no more difficult.

Example 9.4

Calculate the time taken for the distant face of a brick wall, of thermal diffusivity $D_H = 0.0043$ cm^2/s and thickness $l = 0.45$ m, to rise from 295 to 375 K, if the whole wall is initially at a constant temperature of

295 K and the near face is suddenly raised to 900 K and maintained at this temperature. Assume that all the
flow of heat is perpendicular to the faces of the wall and that the distant face is perfectly insulated.

Solution

The temperature at any distance x from the near face at time t is given by:

$$\theta = \sum_{N=0}^{N=\infty}(-1)^N\theta'\left\{\text{erfc}\left[\frac{2lN+x}{2\sqrt{D_Ht}}\right] + \text{erfc}\left[\frac{2(N+1)l-x}{2\sqrt{D_Ht}}\right]\right\} \qquad\text{(equation 9.37)}$$

The temperature at the distant face is therefore given by:

$$\theta = \sum_{N=0}^{N=\infty}(-1)^N\theta'2\ \text{erfc}\left[\frac{(2N+1)l}{2\sqrt{D_Ht}}\right]$$

Choosing the temperature scale so that the initial temperature is everywhere zero:

$$\therefore\qquad \frac{\theta}{2\theta'} = \frac{375-295}{2(900-295)} = 0.066$$

$$D_H = 4.2\times10^{-7}\ \text{m}^2/\text{s} \quad\therefore\quad \sqrt{D_H} = 6.5\times10^{-4}$$

$$l = 0.45\ \text{m}$$

Thus: $$0.066 = \sum_{N=0}^{N=\infty}(-1)^N\ \text{erfc}\left[\frac{0.45(2N+1)}{2\times6.5\times10^{-4}t^{0.5}}\right]$$

$$= \sum_{N=0}^{N=\infty}(-1)^N\ \text{erfc}\left[\frac{346(2N+1)}{t^{0.5}}\right]$$

$$= \text{erfc}(346t^{-0.5}) - \text{erfc}(1038t^{-0.5}) + \text{erfc}(1730t^{-0.5}) - \cdots$$

An approximate answer is obtained by taking the first term only.

Then: $$346t^{-0.5} = 1.30$$

$$\therefore\qquad t = 70\,840\ \text{s}$$

$$= 70.8\ \text{ks} = 19.7\ \text{h}$$

Use of dimensionless groups

The exact mathematical solution of problems involving unsteady thermal conduction may
be very difficult, and sometimes impossible, especially where bodies of irregular shapes
are concerned, and other methods are therefore required.

When a body of characteristic linear dimension L, initially at a uniform temperature
θ_0, is exposed suddenly to surroundings at a temperature θ', the temperature distribution
at any time t is found by applying the method of dimensional analysis to be:

$$\frac{\theta'-\theta}{\theta'-\theta_0} = f\left(\frac{hL}{k},\quad D_H\frac{t}{L^2},\quad \text{and}\ \frac{x}{L}\right) \qquad (9.40)$$

where x is the distance from the surface and h is the heat transfer coefficient at the
surface. Where the resistance to heat transfer at the surface can be neglected, for example,

where the temperature of the surface of the body is maintained at the temperature θ', the temperature will no longer be a function of h and only three dimensionless groups exist:

i.e.:
$$\frac{\theta' - \theta}{\theta' - \theta_0} = f\left(D_H \frac{t}{L^2}, \frac{x}{L}\right) = f\left(F_{OL}, \frac{x}{L}\right) \tag{9.41}$$

or:
$$\frac{\theta' - \theta}{\theta' - \theta_0} = f\left(D_H \frac{t}{x^2}, \frac{x}{L}\right) = f\left(F_{Ox}, \frac{x}{L}\right) \tag{9.42}$$

Curves connecting these groups have been plotted by a number of workers for bodies of various shapes. The method is limited, however, to those shapes which have been studied experimentally.

In Figure 9.10, the value of $(\theta' - \theta)/(\theta' - \theta_0)$ is plotted to give the temperature θ at the centre of bodies of various shapes, initially at a uniform temperature θ_0, at a time t after the surfaces have been suddenly altered to and maintained at a constant temperature θ'.

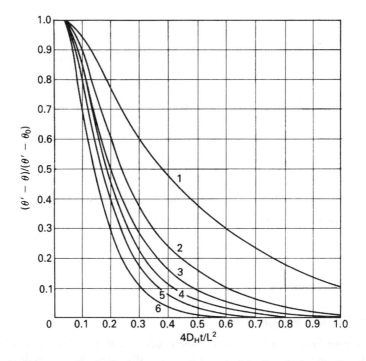

Figure 9.10. Cooling curve for bodies of various shapes: 1, slab ($L =$ thickness); 2, square bar ($L =$ side); 3, long cylinder ($L =$ diameter); 4, cube ($L =$ length $=$ diameter); 5, cylinder ($L =$ length $=$ diameter); 6, sphere ($L =$ diameter)

In this case (x/L) is constant at 0.5 and the results are shown as a function of the particular value $(4D_H t/L^2)$ of the Fourier number F_{Ox}. The results are taken from reference 5.

A general method of estimating the temperature distribution in a body of any shape consists of replacing the heat flow problem by the analogous electrical problem and measuring the electrical potentials at various points. The heat capacity per unit volume $C_p \rho$ is represented by an electrical capacitance, and the thermal conductivity k by an electrical conductivity. The method can be used to take account of variations in the thermal properties over the body.

Schmidt's method

Numerical methods have been developed by replacing the differential equation by a finite difference equation. Thus in a problem of unidirectional flow of heat,

$$\frac{\partial \theta}{\partial t} \approx \frac{\theta_{x(t+\Delta t)} - \theta_{x(t-\Delta t)}}{2\Delta t} \approx \frac{\theta_{x(t+\Delta t)} - \theta_{xt}}{\Delta t}$$

$$\frac{\partial^2 \theta}{\partial x^2} \approx \frac{\left(\dfrac{\theta_{(x+\Delta x)t} - \theta_{xt}}{\Delta x} - \dfrac{\theta_{xt} - \theta_{(x-\Delta x)t}}{\Delta x}\right)}{\Delta x}$$

$$= \frac{\theta_{(x+\Delta x)t} + \theta_{(x-\Delta x)t} - 2\theta_{xt}}{(\Delta x)^2}$$

where θ_{xt} is the value of θ at time t and distance x from the surface, and the other values of θ are at intervals Δx and Δt as shown in Figure 9.11.

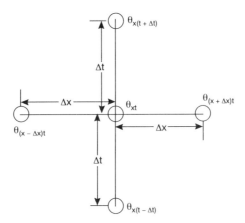

Figure 9.11.   Variation of temperature with time and distance

Substituting these values in equation 9.29:

$$\theta_{x(t+\Delta t)} - \theta_{x(t-\Delta t)} = D_H \frac{2\Delta t}{(\Delta x)^2} (\theta_{(x+\Delta x)t} + \theta_{(x-\Delta x)t} - 2\theta_{xt}) \qquad (9.43)$$

and:

$$\theta_{x(t+\Delta t)} - \theta_{xt} = D_H \frac{\Delta t}{(\Delta x)^2} (\theta_{(x+\Delta x)t} + \theta_{(x-\Delta x)t} - 2\theta_{xt}) \qquad (9.44)$$

Thus, if the temperature distribution at time t, is known, the corresponding distribution at time $t + \Delta t$ can be calculated by the application of equation 9.44 over the whole extent of the body in question. The intervals Δx and Δt are so chosen that the required degree of accuracy is obtained.

A graphical method of procedure is due to SCHMIDT[7]. Suppose that the temperature distribution at time t is represented by the curve shown in Figure 9.12. If the points representing the temperatures at $x - \Delta x$ and $x + \Delta x$ are joined by a straight line, the distance θ_a is given by:

$$\theta_a = \frac{\theta_{(x+\Delta x)t} + \theta_{(x-\Delta x)t}}{2} - \theta_{xt}$$

$$= \frac{(\Delta x)^2}{2D_H \Delta t}(\theta_{x(t+\Delta t)} - \theta_{xt}) \quad \text{(from equation 9.44)} \tag{9.45}$$

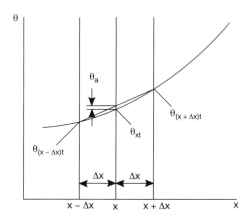

Figure 9.12. Schmidt's method

Thus, θ_a represents the change in θ_{xt} after a time interval Δt, such that:

$$\Delta t = \frac{(\Delta x)^2}{2D_H} \tag{9.46}$$

If this simple construction is carried out over the whole of the body, the temperature distribution after time Δt is obtained. The temperature distribution after an interval $2\Delta t$ is then obtained by a repetition of this procedure.

The most general method of tackling the problem is the use of the *finite-element* technique[8] to determine the temperature distribution at any time by using the finite difference equation in the form of equation 9.43.

Example 9.5

The previous example (9.4) will be solved by Schmidt's method.

Solution

The development of the temperature profile is shown in Figure 9.13. At time $t = 0$ the temperature is constant at 295 K throughout and the temperature of the hot face is raised to 900 K. The problem will be solved by taking relatively large intervals for Δx.

Figure 9.13.   Development of temperature profile

Choosing $\Delta x = 50$ mm, the construction shown in Figure 9.12 is carried out starting at the hot face.

Points corresponding to temperature after a time interval Δt are marked 1, after a time interval $2\Delta t$ by 2, and so on. Because the second face is perfectly insulated, the temperature gradient must be zero at this point. Thus, in obtaining temperatures at $x = 450$ mm it is assumed that the temperature at $x = 500$ mm will be the same as at $x = 400$ mm, i.e. horizontal lines are drawn on the diagram. It is seen that the temperature is less than 375 K after time $23\Delta t$ and greater than 375 K after time $25\Delta t$.

Thus: $t \approx 24\Delta t$

From equation 9.46: $\Delta t = 5.0^2/(2 \times 0.0042) = 2976$ s

Thus time required $= 24 \times 2976 = 71400$ s

or: $\underline{\underline{71.4 \text{ ks} = 19.8 \text{ h}}}$

This value is quite close to that obtained by calculation, especially so in view of the coarse increments in Δx.

Unsteady state heating of particles

When solids or liquid drops are being heated by a surrounding fluid, for example in a rotary kiln, fluidised bed or in a plasma jet, then it is often necessary to predict the time needed for the interior of the particle to reach some specified temperature. If the solids are in the form of small particles and/or the solid is a good thermal conductor, then the rate at which the centre of the particle reaches a specified temperature will be controlled by the rate at which heat arrives at the particle surface. Conversely, with larger particles and solids of low thermal conductivity, internal conduction will be the controlling factor. In some situations internal and external heat transfer may be of similar importance. A useful means of analysing such situations quantitatively is to evaluate the so-called *Biot modulus* or *Biot number*. For a sphere, the Biot modulus:

$$Bi = \frac{hr}{k_p}$$

where h is the external heat transfer coefficient which may include contributions from conduction, convection and radiation, r the radius of sphere and k_p the thermal conductivity of the particle (not that of the surrounding fluid). If the external heat transfer is purely by conduction with the temperature difference spread over a very large distance, then as seen previously:

$$Nu' = \frac{hd}{k} = 2 \qquad \text{(equation 9.26)}$$

Hence:
$$h = \frac{2k}{d} = \frac{k}{r}$$

and:
$$Bi = \frac{k}{k_p} = \frac{\text{thermal conductivity of fluid}}{\text{thermal conductivity of particle}}$$

Thus the higher the Biot number the more likely it is that internal conduction will be the factor controlling the time taken for the centre of a particle to reach some specified temperature.

If both the Fourier number and the Biot modulus are evaluated, then *temperature response* or *Gurney–Lurie charts*[9] can be used to ease the calculation of the time dependent temperature profiles for the surface and inside of the particle. It should be noted that in reference 9 the charts are plotted in terms of the reciprocal of the Biot modulus. Charts are available for other shapes besides spheres, for example, cubes, slabs, and cylinders.

An important problem involving unsteady state conduction of heat to particles is in the melting of powders in plasma spraying[10] where Biot numbers can range from 0.005 to 5. In this case, there is initially a very high relative velocity between the fluid and the powder. The plasmas referred to here are partially ionised gases with temperatures around 10,000 K formed by electric discharges such as arcs. There is an increasing industrial use of the technique of plasma spraying in which powders are injected into a high velocity plasma jet so that they are both melted and projected at velocities of several hundred metres per second on to a surface. The molten particles with diameters typically of the order 10–100 μm impinge to form an integral layer on the surface. Applications include the building up of worn shafts of pumps, for example and the deposition of erosion-resistant ceramic layers on centrifugal pump impellers and other equipment prone to erosion damage. When a powder particle first enters the plasma jet, the relative velocity may be hundreds of metres per second and heat transfer to the particle is enhanced by convection, as discussed in Section 9.4.6. Often, and more particularly for smaller particles, the particle is quickly accelerated to essentially the same velocity as the plasma jet[2] and conduction becomes the main mechanism of heat transfer from plasma to particle. Certainly from a design point of view, neglecting the convective contribution will ease calculations and give a more conservative and safer estimate of the size of the largest particle which can be melted before it strikes the surface. In the absence of complications due to non-continuum conditions discussed elsewhere, the value of $Nu' = hd/k$ is therefore often taken as 2 (see equation 9.26).

One complication which arises in the application of this equation to powder heating in high temperature plasmas lies in the dependence of k, the thermal conductivity of the gas or plasma surrounding the particle, on temperature. For example, the temperature of the particle surface may be 1000 K say, whilst that of the plasma away from the particle may be about 10,000 K or even higher. The thermal conductivity of argon increases by a factor of about 20 over this range of temperature and that of nitrogen gas passes through a pronounced peak at about 7100 K due to dissociation-recombination effects. Thus, the temperature at which the thermal conductivity k is evaluated will have a pronounced effect on the value of the external heat transfer coefficient. A mean value of k would seem appropriate where:

$$(k)_{mean} = \frac{1}{T_2 - T_1} \int_{T_1}^{T_2} k \, dT \qquad \text{(equation 9.16)}$$

Some workers have correlated experimental results in terms of k at the arithmetic mean temperature, and some at the temperature of the bulk plasma. Experimental validation of the true effective thermal conductivity is difficult because of the high temperatures, small particle sizes and variations in velocity and temperature in plasma jets.

In view of the high temperatures involved in plasma devices and the dependence of radiation heat transfer on T^4, as discussed in Section 9.5, it is surprising at first sight that conduction is more significant than radiation in heating particles in plasma spraying. The explanation lies in the small values of d and relatively high values of k for the gas, both of which contribute to high values of h for any given value of Nu'. Also the emissivities of most gases are, as seen later in Section 9.5, rather low.

In situations where the surrounding fluid behaves as a non-continuum fluid, for example, at very high temperatures and/or at low pressures, it is possible to have values of Nu' less than 2. A gas begins to exhibit non-continuum behaviour when the mean free path between collisions of gas molecules or atoms with each other is greater than about 1/100 of the characteristic size of the surface considered. The molecules or atoms are then sufficiently far apart on average for the gas to begin to lose the character of a homogeneous or continuum fluid which is normally assumed in the majority of chemical engineering heat transfer or fluid dynamics problems. For example, with a particle of diameter 25 μm as encountered in, for example, oil-burner sprays, pulverised coal flames, and in plasma spraying in air at room temperature and atmospheric pressure, the mean free path of gas molecules is about 0.06 μm and the air then behaves as a continuum fluid. If, however, the temperature were, say 1800 K as in a flame, then the mean free path would be about 0.33 μm, which is greater than 1/100 of the particle diameter. Non-continuum effects, leading to values of Nu' lower than 2 would then be likely according to theory[11,12]. The exact value of Nu' would depend on the surface accommodation coefficient. This is a difficult parameter to measure for the examples considered here, and hence experimental confirmation of the theory is difficult. At the still higher temperatures that exist in thermal plasma devices, non-continuum effects should be more pronounced and there is limited evidence that values of Nu' below 1 are obtained[10]. In general, non-continuum effects, leading in particular to values of Nu' less than 2, would be more likely at high temperatures, low pressures, and small particle sizes. Thus there is an interest in these effects in the aerospace industry, for example when considering the behaviour of small particles present in rocket engine exhausts.

9.3.6. Conduction with internal heat source

If an electric current flows through a wire then the heat generated internally will result in a temperature distribution between the central axis and the surface of the wire. This type of problem will also arise in chemical or nuclear reactors where heat is generated internally. It is necessary to determine the temperature distribution in such a system and the maximum temperature which will occur.

If the temperature at the surface of the wire is T_0 and the rate of heat generation per unit volume is Q_G; then considering unit length of a cylindrical element of radius r, the heat generated must be transmitted in an outward direction by conduction so that:

$$-k2\pi r \frac{dT}{dr} = \pi r^2 Q_G$$

Hence:

$$\frac{dT}{dr} = -\frac{Q_G r}{2k}$$

Integrating:

$$T = -\frac{Q_G r^2}{4k} + C$$

But $T = T_0$ when $r = r_0$ the radius of wire.

$\therefore$

$$T = T_0 + Q_G \frac{r_0^2 - r^2}{4k}$$

or:

$$T - T_0 = \frac{Q_G r_0^2}{4k} \left(1 - \frac{r^2}{r_0^2} \right) \tag{9.47}$$

This gives a parabolic distribution of temperature and the maximum temperature will occur at the axis of the wire where $(T - T_0) = Q_G r_0^2/4k$. The arithmetic mean temperature difference $(T - T_0)_{av} = Q_G r_0^2/8k$.

Since $Q_G \pi r_0^2$ is the rate of heat release per unit length of the wire then, putting T_1 as the temperature at the centre:

$$T_1 - T_0 = \frac{\text{rate of heat release per unit length}}{4\pi k} \tag{9.48}$$

Example 9.6

A fuel channel in a natural uranium reactor is 5 m long and has a heat release of 0.25 MW. If the thermal conductivity of the uranium is 33 W/mK, what is the temperature difference between the surface and the centre of the uranium element, assuming that the heat release is uniform along the rod?

Solution

Heat release $= 0.25 \times 10^6$ W

$$= \frac{0.25 \times 10^6}{5} = 5 \times 10^4 \text{ W/m}$$

Thus, from equation 9.48: $\qquad\qquad T_1 - T_0 = \dfrac{5 \times 10^4}{4\pi \times 33}$

$$= \underline{\underline{121 \text{ K}}}$$

It should be noted that the temperature difference is independent of the diameter of the fuel rod for a cylindrical geometry and the heat release per unit volume is uniform.

In practice the assumption of the uniform heat release per unit length of the rod is never valid since the neutron flux and hence the heat flux varies sinusoidally along the length of the channel. In the simplest case of an unreflected reactor where the neutron flux may be taken as zero at the ends of the fuel element, the conditions become as shown in Figure 9.14.

Figure 9.14.   Variation of neutron flux along length of fuel rod

In most cases, it is found that the neutron flux may be represented by a function of the form:

$$\phi = \phi_m \sin\left(\frac{\pi x}{L}\right).$$

Since the heat flux is proportional to the neutron flux, the heat dQ developed per unit time in a differential element of the fuel rod dx may be written as:

$$dQ = C \sin\left(\frac{\pi x}{L}\right) dx$$

The total heat generated by the rod Q is then given by:

$$Q = C \int_0^L \sin\left(\frac{\pi x}{L}\right) dx = \frac{2CL}{\pi}$$

Thus, $C = \pi Q/2L$. Then the heat release per unit length at any point is given by:

$$\frac{dQ}{dx} = \frac{\pi Q}{2L} \sin\left(\frac{\pi x}{L}\right)$$

Substituting into equation 9.48 gives:

$$T_1 - T_0 = \frac{\left(\dfrac{\pi Q}{2L}\right) \sin\left(\dfrac{\pi x}{L}\right)}{4\pi k}$$

It should be noted that when $x = 0$ or $x = L$, then $T_1 - T_0$ is zero as would be expected since the neutron flux is also zero under these conditions.

9.4. HEAT TRANSFER BY CONVECTION

9.4.1. Natural and forced convection

Heat transfer by convection occurs as a result of the movement of fluid on a macroscopic scale in the form of eddies or circulating currents. If the currents arise from the heat transfer process itself, *natural convection* occurs. An example of a natural convection process is the heating of a vessel containing liquid by means of a heat source such as a gas flame situated underneath. The liquid at the bottom of the vessel becomes heated and expands and rises because its density has become less than that of the remaining liquid. Cold liquid of higher density takes its place and a circulating current is thus set up.

In *forced convection* the circulating currents are produced by an external agency such as an agitator in a reaction vessel or as a result of turbulent flow in a pipe. In general, the magnitude of the circulation in forced convection is greater, and higher rates of heat transfer are obtained than in natural convection.

In most cases where convective heat transfer is taking place from a surface to a fluid, the circulating currents die out in the immediate vicinity of the surface and a film of fluid, free of turbulence, covers the surface. In this film, heat transfer is by thermal conduction and, as the thermal conductivity of most fluids is low, the main resistance to transfer lies there. Thus an increase in the velocity of the fluid over the surface gives rise to improved heat transfer mainly because the thickness of the film is reduced. As a guide, the film coefficient increases as (fluid velocity)^n, where $0.6 < n < 0.8$, depending upon the geometry.

If the resistance to transfer is regarded as lying within the film covering the surface, the rate of heat transfer Q is given by equation 9.11 as:

$$Q = kA\frac{T_1 - T_2}{x}$$

The effective thickness x is not generally known and therefore the equation is usually rewritten in the form:

$$Q = hA(T_1 - T_2) \tag{9.49}$$

where h is the heat transfer coefficient for the film and $(1/h)$ is the thermal resistance.

9.4.2. Application of dimensional analysis to convection

So many factors influence the value of h that it is almost impossible to determine their individual effects by direct experimental methods. By arranging the variables in a series of dimensionless groups, the problem is made more manageable in that the number of groups is significantly less than the number of parameters. It is found that the heat transfer rate per unit area q is dependent on those physical properties which affect flow pattern — viscosity μ and density ρ —, the thermal properties of the fluid — the specific heat capacity C_p and the thermal conductivity k — a linear dimension of the surface l, the velocity of flow u of the fluid over the surface, the temperature difference ΔT and the factor determining the natural circulation effect caused by the expansion of the fluid on heating — the product of the coefficient of cubical expansion β and the acceleration due to gravity g. Writing this as a functional relationship:

$$q = \phi[u, l, \rho, \mu, C_p, \Delta T, \beta g, k] \tag{9.50}$$

Noting the dimensions of the variables in terms of length **L**, mass **M**, time **T**, temperature θ, heat **H**:

q	Heat transferred/unit area and unit time	$\mathbf{HL^{-2}T^{-1}}$
u	Velocity	$\mathbf{LT^{-1}}$
l	Linear dimension	$\mathbf{L}$
μ	Viscosity	$\mathbf{ML^{-1}T^{-1}}$
ρ	Density	$\mathbf{ML^{-3}}$
k	Thermal conductivity	$\mathbf{HT^{-1}L^{-1}\theta^{-1}}$
C_p	Specific heat capacity at constant pressure	$\mathbf{HM^{-1}\theta^{-1}}$
ΔT	Temperature difference	θ
(βg)	The product of the coefficient of thermal expansion and the acceleration due to gravity	$\mathbf{LT^{-2}\theta^{-1}}$

It will be noted that both temperature and heat are taken as fundamental units as heat is not expressed here in terms of **M, L, T**.

With nine parameters and five dimensions equation 9.50 will be rearranged in four dimensionless groups.

Using the Π-theorem for solution of the equation:

Taking as the recurring set: $\qquad l, \rho, \mu, \Delta T, k$

The non-recurring variables are: $\qquad q, u, (\beta g), C_p$

Then:

$$l \equiv \mathbf{L} \qquad\qquad \mathbf{L} = l$$
$$\rho \equiv \mathbf{ML^{-3}} \qquad\qquad \mathbf{M} = \rho \mathbf{L}^3 = \rho l^3$$
$$\mu \equiv \mathbf{ML^{-1}T^{-1}} \qquad\qquad \mathbf{T} = \mathbf{ML^{-1}}\mu^{-1} = \rho l^3 l^{-1}\mu^{-1} = \rho l^2 \mu^{-1}$$
$$\Delta T \equiv \theta \qquad\qquad \theta = \Delta T$$
$$k \equiv \mathbf{HL^{-1}T^{-1}\theta^{-1}} \qquad\qquad \mathbf{H} = k\mathbf{LT}\theta = kl\rho l^2 \mu^{-1}\Delta T = kl^3 \rho \mu^{-1}\Delta T$$

The Π groups will then be:

$$\Pi_1 = q\mathbf{H^{-1}L^2T} = qk^{-1}l^{-3}\rho^{-1}\mu\Delta T^{-1}l^2 \rho l^2 \mu^{-1} = qk^{-1}l\Delta T^{-1}$$
$$\Pi_2 = u\mathbf{L^{-1}T} = ul^{-1}\rho l^2 \mu^{-1} = u\rho l\mu^{-1}$$

$$\Pi_3 = C_p \mathbf{H}^{-1} \mathbf{M}\theta = C_p k^{-1} l^{-3} \rho^{-1} \mu \Delta T^{-1} \rho l^3 \Delta T = C_p k^{-1} \mu$$

$$\Pi_4 = \beta g \mathbf{L}^{-1} \mathbf{T}^2 \theta = \beta g l^{-1} \rho^2 l^4 \mu^{-2} \Delta T = \beta g \Delta T \rho^2 \mu^{-2} l^3$$

So that the relation in equation 9.50 becomes:

$$\frac{ql}{k\Delta T} = \frac{hl}{k} = \phi \left(\frac{lu\rho}{\mu} \right) \left(\frac{C_p \mu}{k} \right) \left(\frac{\beta g \Delta T l^3 \rho^2}{\mu^2} \right) \tag{9.51}$$

or: $Nu = \phi[Re, Pr, Gr]$

This general equation involves the use of four dimensionless groups but may frequently be simplified for design purposes. In equation 9.51:

hl/k is known as the *Nusselt* group Nu,

$lu\rho/\mu$ the *Reynolds* group Re,

$C_p\mu/k$ the *Prandtl* group Pr, and

$\beta g \Delta T l^3 \rho^2/\mu^2$ the *Grashof* group Gr

It is convenient at this stage to define other dimensionless groups which are also used in the analysis of heat transfer as follows:

$lu\rho C_p/k$ is known as the *Peclet* group $Pe = RePr$,

GC_p/kl the *Graetz* group Gz, and

$h/C_p \rho u$ the *Stanton* group, $St = Nu/(RePr)$

It will be noted that many of these dimensionless groups are ratios. For example, the Nusselt group $h/(k/l)$ is the ratio of the actual heat transfer to that by conduction over a thickness l, whilst the Prandtl group, $(\mu/\rho)/(k/C_p\rho)$ is the ratio of the kinematic viscosity to the thermal diffusivity.

For conditions in which only natural convection occurs the velocity is dependent solely on the buoyancy effects, represented by the Grashof number and the Reynolds group can be omitted. Again, when forced convection occurs the effects of natural convection are usually negligible and the Grashof number may be omitted. Thus:

for natural convection: $Nu = f(Gr, Pr)$ (9.52)

and for forced convection: $Nu = f(Re, Pr)$ (9.53)

For most gases over a wide range of temperature and pressure, the product $C_p\mu/k$ is constant and the Prandtl group may often be omitted, making the design equations for the calculation of film coefficients with gases simpler.

9.4.3 Forced convection in tubes

Turbulent flow

The results of a number of workers who have used a variety of gases such as air, carbon dioxide, and steam and of others who have used liquids such as water, acetone, kerosene, and benzene have been expressed by DITTUS and BOELTER[13] in the general form of equation 9.53 as:

$$Nu = 0.023Re^{0.8}Pr^n \qquad (9.54)$$

where n has a value of 0.4 for heating and 0.3 for cooling. In this equation all of the physical properties are taken at the mean bulk temperature of the fluid $(T_i + T_0)/2$, where T_i and T_0 are the inlet and outlet temperatures. Equation 9.54 is valid for Reynolds numbers greater than 10,000 and has been tested for values of the Prandtl group lying between 0.7 and 160. The difference in the value of the index for heating and cooling occurs because in the former case the film temperature will be greater than the bulk temperature and in the latter case less. Conditions in the film, particularly the viscosity of the fluid, exert an important effect on the heat transfer process.

An alternative equation which is in many ways more convenient has been proposed by COLBURN[14] and includes the Stanton number $(St = h/C_p\rho u)$ instead of the Nusselt number. This equation is:

$$StPr^{0.67} = 0.023Re^{-0.2} \qquad (9.55)$$

It will be noted that:

$$\frac{h}{C_p\rho u} = \frac{hd}{k}\frac{\mu}{ud\rho}\frac{k}{C_p\mu}$$

i.e.:
$$St = NuRe^{-1}Pr^{-1} \qquad (9.56)$$

Thus, multiplying equation 9.55 by $RePr^{0.33}$:

$$Nu = 0.023Re^{0.8}Pr^{0.33} \qquad (9.57)$$

which is a form of equation 9.54.

Again, the physical properties are taken at the bulk temperature except the viscosity in the Reynolds group which is evaluated at the mean film temperature taken as being $(T_{surface} + T_{bulkfluid})/2$. The group on the left-hand side of equation 9.55 is known as the j_h factor for heat transfer and the equation is often written as:

$$j_h = 0.023Re^{-0.2} \qquad (9.58)$$

Writing a heat balance for the flow through a tube of diameter d and length l with a rise in temperature for the fluid from T_i to T_o:

$$h\pi dl\,\Delta T = \frac{\pi d^2}{4}C_p u\rho(T_o - T_i)$$

or:
$$St = \frac{h}{C_p\rho u} = \frac{d(T_o - T_i)}{4l\,\Delta T} \qquad (9.59)$$

where ΔT is the mean temperature difference between the bulk fluid and the walls.

With *very viscous liquids* there will be a marked difference at any cross-section between the viscosity of the fluid adjacent to the surface and the value at the axis or at the bulk temperature of the fluid. SIEDER and TATE[15] have presented a modified equation for

Figure 9.15.   Film coefficients of convection for flow of water through a tube at 289 K

Figure 9.16.   Film coefficients of convection for flow of air through a tube, velocity curves
(289 K, 101.3 kN/m^2)

these conditions including a viscosity correction term:

$$Nu = 0.027Re^{0.8}Pr^{0.33}\left(\frac{\mu}{\mu_s}\right)^{0.14} \tag{9.60}$$

where μ is the viscosity at the bulk temperature and μ_s the viscosity at the wall or surface. This equation may also be written in the form of the Colburn equation (9.55).

When the above equations are applied to *heating or cooling of gases* for which the Prandtl group usually has a value of about 0.74, substitution of $Pr = 0.74$ in equation 9.54 gives the equation:

$$Nu = 0.02Re^{0.8} \tag{9.61}$$

Water is very frequently used as the cooling liquid and, if the variation in value of physical properties with temperature is included, then equation 9.54 becomes:

$$h = 150(1 + 0.011T)\frac{u^{0.8}}{d^{0.2}} \text{ Btu/h ft}^2{}^\circ\text{F} \tag{9.62}$$

where d is in inches, u in ft/s, and T in degrees F. If T is in degrees K, d in m, u in m/s, then:

$$h = 4280(0.00488T - 1)\frac{u^{0.8}}{d^{0.2}} \text{ W/m}^2\text{ K} \tag{9.63}$$

There is a very big difference in the values of h for water and air for the same linear velocity. This is shown in Figures 9.15–9.17 and Table 9.2, all of which are based on the work of FISHENDEN and SAUNDERS[16]. It will be noted that values of h given by equation 9.63 and by Figure 9.15 differ by about 10 per cent; this is within the limits of reproducibility of engineering data of this type.

The effect of length to diameter ratio (l/d) on the value of the heat transfer coefficient can be seen in Figure 9.18. It is important at low Reynolds numbers but ceases to be significant at a Reynolds number of about 10^4.

It is also important to note that the film coefficient varies with the distance from the entrance to the tube. This is especially important at low (l/d) ratios and an average value is given approximately by:

$$\frac{h_{\text{average}}}{h_\infty} = 1 + \left(\frac{d}{l}\right)^{0.7} \tag{9.64}$$

where h_∞ is the limiting value for a very long tube.

The roughness of the surface of the inside of the pipe can have an important bearing on rates of heat transfer to the fluid, although COPE[17], using degrees of artificial roughness ranging from 0.022 to 0.14 of the pipe diameter, found that, although the friction loss was some six times greater than for smooth tubes, the heat transfer was only 100–120 per cent higher. He concluded that, for the same pressure drop, greater heat transfer was obtained from a smooth rather than a rough tube. The effect of a given scale deposit is usually less serious for gases than water because of the higher thermal resistance of the gas film, although layers of dust or of materials which sublime may seriously reduce heat transfer between gas and solid by as much as 40 per cent.

Figure 9.17.   Film coefficients of convection for flow of air through a tube, mass flow curves
(289 K, 101.3 kN/m^2)

Figure 9.18.   Effect of length:diameter ratio on heat transfer coefficient

Streamline flow

Although heat transfer to a fluid in streamline flow takes place solely by conduction, it is convenient to consider it here so that the results may be compared with those for turbulent flow.

In Chapter 3 it has been seen that for streamline flow through a tube the velocity distribution across a diameter is parabolic (Figure 9.19). If a liquid enters a section heated

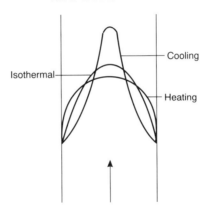

Figure 9.19. Effect of heat transfer on velocity distribution for liquid

Table 9.2. Film coefficients for air and water (289 K and 101.3 kN/m^2)

Inside diameter of tube		Velocity		Mass velocity		Film coefficient of heat transfer h			
(mm)	(in)	(m/s)	(ft/s)	(kg/m^2 s)	(lb/ft^2 h)	(W/m^2 K) [Ref.16]	[Eqn.9.63]	(Btu/h ft^2 °F) [Ref.16]	[Eqn.9.62]
Air									
25	1.0	5	16.4	6.11	4530	31.2	—	5.5	—
		10	32.8	12.2	9050	50.0	—	8.8	—
		20	65.6	24.5	18,100	84.0	—	14.8	—
		40	131	48.9	36,200	146	—	25.7	—
		60	197	73.4	54,300	211	—	37.2	—
50	2.0	5	16.4	6.11	4530	23.8	—	4.2	—
		10	32.8	12.2	9050	44.9	—	7.9	—
		20	65.6	24.5	18,100	77.8	—	13.7	—
		40	131	48.9	36,200	127	—	22.4	—
		60	197	73.4	54,300	181	—	31.9	—
75	3.0	5	16.4	6.11	4530	21.6	—	3.8	—
		10	32.8	12.2	9050	39.7	—	7.0	—
		20	65.6	24.5	18,100	71.0	—	12.5	—
		40	131	48.9	36,200	119	—	21.0	—
		60	197	73.4	54,300	169	—	29.8	—
Water									
25	1.0	0.5	1.64	488	361,000	2160	2080	380	370
		1.0	3.28	975	722,000	3750	3630	660	644
		1.5	4.92	1460	1,080,000	5250	5020	925	891
		2.0	6.55	1950	1,440,000	6520	6320	1150	1120
		2.5	8.18	2440	1,810,000	7780	7560	1370	1340
50	2.0	0.5	1.64	488	361,000	1870	1820	330	357
		1.0	3.28	975	722,000	3270	3160	575	563
		1.5	4.92	1460	1,080,000	4540	4370	800	778
		2.0	6.55	1950	1,440,000	5590	5500	985	978
		2.5	8.18	2440	1,810,000	6700	6580	1180	1170
75	3.0	0.5	1.64	488	361,000	1760	1670	310	298
		1.0	3.28	975	722,000	3070	2910	540	519
		1.5	4.92	1460	1,080,000	4200	4030	740	717
		2.0	6.55	1950	1,440,000	5220	5070	920	902
		2.5	8.18	2440	1,810,000	6220	6070	1100	1080

on the outside, the fluid near the wall will be at a higher temperature than that in the centre and its viscosity will be lower. The velocity of the fluid near the wall will therefore be greater in the heated section, and correspondingly less at the centre. The velocity distribution will therefore be altered, as shown. If the fluid enters a section where it is cooled, the same reasoning will show that the distribution in velocity will be altered to that shown. With a gas the conditions are reversed, because of the increase of viscosity with temperature. The heat transfer problem is therefore complex.

For the common problem of heat transfer between a fluid and a tube wall, the boundary layers are limited in thickness to the radius of the pipe and, furthermore, the effective area for heat flow decreases with distance from the surface. The problem can conveniently be divided into two parts. Firstly, heat transfer in the entry length in which the boundary layers are developing, and, secondly, heat transfer under conditions of fully developed flow. Boundary layer flow is discussed in Chapter 11.

For the region of fully developed flow in a pipe of length L, diameter d and radius r, the rate of flow of heat Q through a cylindrical surface in the fluid at a distance y from the wall is given by:

$$Q = -k2\pi L(r - y)\frac{d\theta}{dy}$$

Close to the wall, the fluid velocity is low and a negligible amount of heat is carried along the pipe by the flowing fluid in this region and Q is independent of y.

Thus:
$$\frac{d\theta}{dy} = -\frac{Q}{k2\pi L}(r - y)^{-1} \quad \text{and} \quad \left(\frac{d\theta}{dy}\right)_{y=0} = -\frac{Q}{2\pi kLr}$$

$$\frac{d^2\theta}{dy^2} = -\frac{Q}{k2\pi L}(r - y)^{-2} \quad \text{and} \quad \left(\frac{d^2\theta}{dy^2}\right)_{y=0} = -\frac{Q}{2\pi kLr^2}$$

Thus:
$$\left(\frac{d^2\theta}{dy^2}\right)_{y=0} = r^{-1}\left(\frac{d\theta}{dy}\right)_{y=0} \qquad (9.65)$$

Assuming that the temperature of the walls remains constant at the datum temperature and that the temperature at any distance y from the walls is given by a polynomial:

$$\theta = a_0 y + b_0 y^2 + c_0 y^3 \qquad (9.66)$$

Thus:
$$\frac{d\theta}{dy} = a_0 + 2b_0 y + 3c_0 y^2 \quad \text{and} \quad \left(\frac{d\theta}{dy}\right)_{y=0} = a_0$$

$$\frac{d^2\theta}{dy^2} = 2b_0 + 6c_0 y \quad \text{and} \quad \left(\frac{d^2\theta}{dy^2}\right)_{y=0} = 2b_0$$

Thus:
$$2b_0 = \frac{a_0}{r} \quad \text{(from equation 9.65)}$$

i.e.:
$$b_0 = \frac{a_0}{2r}$$

Now the temperature of the fluid at the axis of the pipe is θ_s, say, and the temperature gradient at the axis, from symmetry, must be zero.

Thus:
$$0 = a_0 + 2r \left(\frac{a_0}{2r} \right) + 3c_0 r^2$$

$$\therefore \qquad c_0 = -\frac{2a_0}{3r^2}$$

and:
$$\theta_s = a_0 r + r^2 \left(\frac{a_0}{2r} \right) + r^3 \left(\frac{-2a_0}{3r^2} \right)$$

$$= \frac{5}{6} a_0 r$$

$$\therefore \qquad a_0 = \frac{6}{5} \frac{\theta_s}{r}$$

$$b_0 = \frac{3}{5} \frac{\theta_s}{r^2}$$

and:
$$c_0 = -\frac{4}{5} \frac{\theta_s}{r^3}$$

Thus:
$$\frac{\theta}{\theta_s} = \frac{6}{5} \frac{y}{r} + \frac{3}{5} \left(\frac{y}{r} \right)^2 - \frac{4}{5} \left(\frac{y}{r} \right)^3 \qquad (9.67)$$

Thus the rate of heat transfer per unit area at the wall:

$$q = -k \left(\frac{d\theta}{dy} \right)_{y=0}$$

$$= -\frac{6}{5} \frac{k\theta_s}{r} \qquad (9.68)$$

In general, the temperature θ_s at the axis is not known, and the heat transfer coefficient is related to the temperature difference between the walls and the bulk fluid. The bulk temperature of the fluid is defined as the ratio of the heat content to the heat capacity of the fluid flowing at any section. Thus the bulk temperature θ_B is given by:

$$\theta_B = \frac{\displaystyle\int_0^r C_p \rho \theta u_x 2\pi (r - y)\, dy}{\displaystyle\int_0^r C_p \rho u_x 2\pi (r - y)\, dy}$$

$$= \frac{\displaystyle\int_0^r \theta u_x (r - y)\, dy}{\displaystyle\int_0^r u_x (r - y)\, dy} \qquad (9.69)$$

Now from Poiseuille's law (equation 3.30):

$$u_x = \frac{-\Delta P}{4\mu L} [r^2 - (r - y)^2] = \frac{-\Delta P}{4\mu L} (2ry - y^2)$$

Hence:

$$u_s = \frac{-\Delta P}{4\mu L}r^2$$

where u_s is the velocity at the pipe axis, and:

$$\frac{u_x}{u_s} = \frac{2y}{r} - \left(\frac{y}{r}\right)^2$$

Thus:

$$\int_0^r u_x(r-y)\,\mathrm{d}y = r^2 u_s \int_0^1 \left[2\frac{y}{r} - \left(\frac{y}{r}\right)^2\right]\left(1 - \frac{y}{r}\right)\mathrm{d}\left(\frac{y}{r}\right)$$

$$= r^2 u_s \int_0^1 \left[2\left(\frac{y}{r}\right) - 3\left(\frac{y}{r}\right)^2 + \left(\frac{y}{r}\right)^3\right]\mathrm{d}\left(\frac{y}{r}\right)$$

$$= \frac{1}{4}r^2 u_s \tag{9.70}$$

Since:

$$\frac{\theta}{\theta_s} = \frac{6}{5}\frac{y}{r} + \frac{3}{5}\left(\frac{y}{r}\right)^2 - \frac{4}{5}\left(\frac{y}{r}\right)^3 \tag{equation 9.67}$$

$$\int_0^r \theta u_x(r-y)\,\mathrm{d}y = r^2 u_s \theta_s \int_0^1 \left[\frac{6}{5}\frac{y}{r} + \frac{3}{5}\left(\frac{y}{r}\right)^2 - \frac{4}{5}\left(\frac{y}{r}\right)^3\right]\left[2\left(\frac{y}{r}\right) - 3\left(\frac{y}{r}\right)^2 + \left(\frac{y}{r}\right)^3\right]\mathrm{d}\left(\frac{y}{r}\right)$$

$$= r^2 u_s \theta_s \int_0^1 \left[\frac{12}{5}\left(\frac{y}{r}\right)^2 - \frac{12}{5}\left(\frac{y}{r}\right)^3 - \frac{11}{5}\left(\frac{y}{r}\right)^4 + 3\left(\frac{y}{r}\right)^5 - \frac{4}{5}\left(\frac{y}{r}\right)^6\right]\mathrm{d}\left(\frac{y}{r}\right)$$

$$= r^2 u_s \theta_s \left(\frac{4}{5} - \frac{3}{5} - \frac{11}{25} + \frac{1}{2} - \frac{4}{35}\right)$$

$$= \tfrac{51}{350}r^2 u_s \theta_s \tag{9.71}$$

Substituting from equations 9.70 and 9.71 in equation 9.69:

$$\theta_B = \frac{\frac{51}{350}r^2 u_s \theta_s}{\frac{1}{4}r^2 u_s}$$

$$= \tfrac{102}{175}\theta_s = 0.583\theta_s \tag{9.72}$$

Then, the heat transfer coefficient h is given by:

$$h = -\frac{q}{\theta_B}$$

where q is the rate of heat transfer per unit area of tube.
Thus, from equations 9.68 and 9.72:

$$h = \frac{6k\theta_s/5r}{0.583\theta_s} = \frac{2.06k}{r} = 4.1\frac{k}{d}$$

$$Nu = \frac{hd}{k} = 4.1 \tag{9.73}$$

$\therefore$

This expression is applicable only to the region of fully developed flow. The heat transfer coefficient for the inlet length can be calculated approximately, using the expressions that are obtained in Chapter 11 for the development of the boundary layers for the flow over a plane surface. It should be borne in mind that it has been assumed throughout that the physical properties of the fluid are not appreciably dependent on temperature and therefore the expressions will not be expected to hold accurately if the temperature differences are large and if the properties vary widely with temperature.

For values of $(RePr\, d/l)$ greater than 12, the following empirical equation is applicable:

$$Nu = 1.62 \left(RePr\frac{d}{l} \right)^{1/3} = 1.75 \left(\frac{GC_p}{kl} \right)^{1/3} \qquad (9.74)$$

where $G = (\pi d^2/4)\rho u$, i.e. the mass rate of flow.

The product $RePr$ is termed the Peclet number Pe.

Thus:
$$Pe = \frac{ud\rho}{\mu}\frac{C_p\mu}{k} = \frac{C_p\rho ud}{k}$$

Thus equation 9.74 can be written:

$$Nu = 1.62 \left(Pe\frac{d}{l} \right)^{1/3} \qquad (9.75)$$

In this equation the temperature difference is taken as the arithmetic mean of the terminal values, that is:

$$\frac{[(T_w - T_1) + (T_w - T_2)]}{2}$$

where T_w is the temperature of the tube wall which is taken as constant.

If the liquid is heated almost to the wall temperature T_w (i.e. when GC_p/kl is very small) then, on equating the heat gained by the liquid to that transferred from the pipe:

$$GC_p(T_2 - T_1) = \pi dlh\frac{T_2 - T_1}{2}$$

or:
$$h = \frac{2}{\pi}\frac{GC_p}{dl} \qquad (9.76)$$

For values of $(RePr\, d/l)$ less than about 17, the Nusselt group becomes approximately constant at 4.1; the value given in equation 9.73.

Experimental values of h for *viscous oils* are greater than those given by equation 9.74 for heating and less for cooling. This is due to the large variation of viscosity with temperature and the correction introduced for turbulent flow may also be used here, giving the equation as:

$$Nu \left(\frac{\mu_s}{\mu} \right)^{0.14} = 1.86 \left(RePr\frac{d}{l} \right)^{1/3} = 2.01 \left(\frac{GC_p}{kl} \right)^{1/3} \qquad (9.77)$$

or:
$$Nu \left(\frac{\mu_s}{\mu} \right)^{0.14} = 1.86 \left(Pe\frac{d}{l} \right)^{1/3} \qquad (9.78)$$

When $(GC_p/kl) < 10$, the outlet temperature closely approaches that of the wall and equation 9.76 applies. These equations have been obtained with tubes from about 10 mm to 40 mm diameter and the length of unheated tube preceding the heated section is important. They are not entirely consistent since for very small values of ΔT the constants in equations 9.74 and 9.77 would be expected to be the same. It is important to note, when using these equations for design purposes, that the error may be as much as ± 25 per cent for turbulent flow and greater for streamline conditions.

With laminar flow there is a marked influence of tube length and the curves shown in Figure 9.18 show the parameter l/d from 50 to 400.

Whenever possible, streamline conditions of flow are avoided in heat-exchange equipment because of the very low heat transfer coefficients which are obtained. With very viscous liquids, however, turbulent conditions can be produced only if a very high pressure drop across the plant is permissible. In the chemical industry, streamline flow in heat exchangers is most commonly experienced with heavy oils and brines at low temperatures. Since the viscosity of these materials is critically dependent on temperature, the equations would not be expected to apply with a high degree of accuracy.

9.4.4. Forced convection outside tubes

Flow across single cylinders

If a fluid passes at right angles across a single tube, the distribution of velocity around the tube will not be uniform. In the same way the rate of heat flow around a hot pipe across which air is passed is not uniform but is a maximum at the front and back and a minimum at the sides, where the rate is only some 40 per cent of the maximum. The general picture is shown in Figure 9.20 but for design purposes reference is made to the average value.

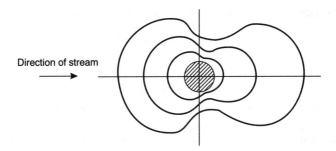

Figure 9.20.   Distribution of film heat transfer coefficient round a cylinder with flow normal to axis for three different values of *Re*

A number of workers (REIHER[18], HILPERT[19], GRIFFITHS and AWBERY[20]) have studied the flow of a hot gas past a single cylinder varying from a thin wire to a tube of 150 mm diameter. They have used temperatures up to 1073 K and air velocities up to 30 m/s with Reynolds numbers $(d_o u \rho / \mu)$ from 1000 to 100,000 (d_o is the cylinder diameter, or the outside tube diameter). Their results may be expressed by the relation:

$$Nu = 0.26 Re^{0.6} Pr^{0.3} \tag{9.79}$$

Taking Pr as 0.74 for gases, this reduces to

$$Nu = 0.24Re^{0.6} \tag{9.80}$$

DAVIS[21] has also worked with water, paraffin, and light oils and obtained similar results. For very low values of Re (from 0.2 to 200) with liquids the data are better represented by the equation:

$$Nu = 0.86Re^{0.43}Pr^{0.3} \tag{9.81}$$

In each case the physical properties of the fluid are measured at the mean film temperature T_f, taken as the average of the surface temperature T_w and the mean fluid temperature T_m; where $T_m = (T_1 + T_2)/2$.

Flow at right angles to tube bundles

One of the great difficulties with this geometry is that the area for flow is continually changing. Moreover the degree of turbulence is considerably less for banks of tubes in line, as at (a) than for staggered tubes, as at (b) in Figure 9.21. With the small bundles which are common in the chemical industry the selection of the true mean area for flow is further complicated by the change in number of tubes in the rows.

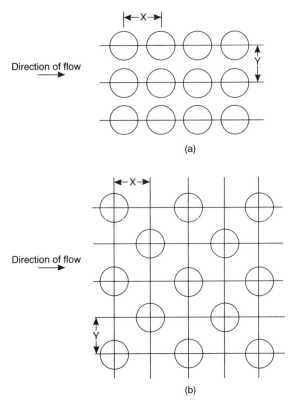

Figure 9.21. Arrangements of tubes in heat exchangers (a) in-line (b) staggered

The results of a number of workers for the flow of gases across tube banks may be expressed in the form:

$$Nu = 0.33C_h Re_{max}^{0.6} Pr^{0.3} \tag{9.82}$$

where C_h depends on the geometrical arrangement of the tubes (Table 9.3). GRIMISON[22] proposed this form of expression to correlate the data of HUGE[23] and PIERSON[24] who worked with small electrically heated tubes in rows of ten deep. Other workers have used similar equations. Some correction factors have been given by PIERSON[24] for bundles with less than ten rows but there are insufficient reported data from commercial exchangers to fix these values with accuracy. Thus for five rows a factor of 0.92 and for eight rows 0.97 is suggested.

Table 9.3.[16]   Values of C_h and C_f

| | $X = 1.25d_0$ | | | | $X = 1.5d_0$ | | | |
| | In-line | | Staggered | | In-line | | Staggered | |
Re_{max}	C_h	C_f	C_h	C_f	C_h	C_f	C_h	C_f
			$Y = 1.25d_0$					
2000	1.06	1.68	1.21	2.52	1.06	1.74	1.16	2.58
20,000	1.00	1.44	1.06	1.56	1.00	1.56	1.05	1.74
40,000	1.00	1.20	1.03	1.26	1.00	1.32	1.02	1.50
			$Y = 1.5d_0$					
2000	0.95	0.79	1.17	1.80	0.95	0.97	1.15	1.80
20,000	0.96	0.84	1.04	1.10	0.96	0.96	1.02	1.16
40,000	0.96	0.74	0.99	0.88	0.96	0.85	0.98	0.96

These equations are based on the maximum velocity through the bundle. Thus for an in-line arrangement as in Figure 9.21, $G'_{max} = G'Y/(Y - d_o)$, where Y is the pitch of the pipes at right angles to direction of flow; it is more convenient here to use the mass flowrate per unit area G' in place of velocity. For staggered arrangements the maximum velocity may be based on the distance between the tubes in a horizontal line or on the diagonal of the tube bundle, whichever is the less.

It has been suggested that for in-line arrangements the constant in equation 9.82 should be reduced to 0.26, but there is insufficient evidence from commercial exchangers to settle this question.

With liquids the same equation may be used, but for Re less than 2000, there is insufficient published work to justify an equation. MCADAMS,[25] however, has given a curve for h for a bundle with staggered tubes ten rows deep.

An alternative approach has been suggested by KERN[26] who works in terms of the hydraulic mean diameter d_e for flow *parallel* to the tubes:

i.e.:
$$d_e = 4 \times \frac{\text{Free area for flow}}{\text{Wetted perimeter}}$$

$$= 4 \left(\frac{Y^2 - (\pi d_o^2/4)}{\pi d_o} \right)$$

for a square pitch as shown in Figure 9.22. The maximum cross-flow area A_s is then given by:

$$A_s = \frac{d_s l_B C'}{Y}$$

where C' is clearance, l_B the baffle spacing, and d_s the internal diameter of the shell.

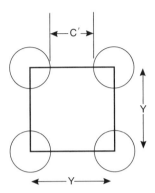

Figure 9.22.   Tube arrangement

The mass rate of flow per unit area G'_s is then given as rate of flow divided by A_s, and the film coefficient is obtained from a Nusselt type expression:

$$\frac{h_o d_e}{k} = 0.36 \left(\frac{d_e G'_s}{\mu}\right)^{0.55} \left(\frac{C_p \mu}{k}\right)^{1/3} \left(\frac{\mu}{\mu_s}\right)^{0.14} \tag{9.83}$$

There are insufficient published data to assess the relative merits of equations 9.82 and 9.83.

For 19 mm tubes on 25 mm square pitch:

$$d_e = 4\frac{[25^2 - (\pi/4)19^2]}{\pi \times 19}$$

$$= 22.8 \text{ mm}$$

$$= 0.023 \text{ m}$$

Example 9.7

14.4 tonne/h (4.0 kg/s) of nitrobenzene is to be cooled from 400 to 315 K by heating a stream of benzene from 305 to 345 K.

Two tubular heat exchangers are available each with a 0.44 m i.d. shell fitted with 166 tubes, 19.0 mm o.d. and 15.0 mm i.d., each 5.0 m long. The tubes are arranged in two passes on 25 mm square pitch with a baffle spacing of 150 mm. There are two passes on the shell side and operation is to be countercurrent. With benzene passing through the tubes, the anticipated film coefficient on the tube side is 1000 W/m²K.

What value of scale resistance could be allowed if these units were used?

For nitrobenzene:

$$C_p = 2380 \text{ J/kg K}$$

$$k = 0.15 \text{ W/m K}$$

$$\mu = 0.70 \text{ mN s/m}^2$$

Solution

(i) Tube side coefficient.

$$h_i = 1000 \text{ W/m}^2 \text{ K based on inside area}$$

or:
$$\frac{1000 \times 15.0}{19.0} = 790 \text{ W/m}^2 \text{ K based on outside area}$$

(ii) Shell side coefficient.

$$\text{Area for flow} = \text{shell diameter} \times \text{baffle spacing} \times \text{clearance/pitch}$$

$$= \frac{0.44 \times 0.150 \times 0.006}{0.025} = 0.0158 \text{ m}^2$$

Hence:
$$G'_s = \frac{4.0}{0.0158} = 253.2 \text{ kg/m}^2\text{s}$$

Taking $\mu/\mu_s = 1$ in equation 9.83:

$$h_0 = \frac{k}{d_e} 0.36 \left(\frac{d_e G'_s}{\mu}\right)^{0.55} \left(\frac{C_p \mu}{k}\right)^{0.33}$$

$$d_e = 4 \left[\left(25^2 - \frac{\pi \times 19.0^2}{4}\right) \Big/ \pi \times 19.0\right] = 22.8 \text{ mm} \quad \text{or} \quad 0.023 \text{ m}$$

$$\therefore \quad h = \frac{0.15}{0.023} 0.36 \left(\frac{0.023 \times 253.2}{0.70 \times 10^{-3}}\right)^{0.55} \left(\frac{2380 \times 0.70 \times 10^{-3}}{0.15}\right)^{0.33}$$

$$= 2.35 \times 143 \times 2.23 = 750 \text{ W/m}^2 \text{ K}$$

(iii) Overall coefficient.
   Log mean temperature difference,

$$\Delta T_m = \frac{(400 - 345) - (315 - 305)}{\ln(400 - 345)/(315 - 305)}$$

$$= 26.4 \text{ K} \quad \text{(see Example 9.1)}$$

$$\text{True mean} = \Delta T_m \times F = 26.4 \times 0.8 = 21.1 \text{ K}$$

(For details of the correction factor for ΔT_m see section on multipass exchangers)

Heat load:
$$Q = 4.0 \times 2380(400 - 315) = 8.09 \times 10^5 \text{ W}$$

Surface area of each tube
$$= 0.0598 \text{ m}^2/\text{m}$$

Thus:
$$U_o = \frac{Q}{A_o \Delta T_m F} = \frac{8.09 \times 10^5}{2 \times 166 \times 5.0 \times 0.0598 \times 21.1}$$

$$= 386.2 \text{ W/m}^2 \text{ K}$$

(iv) Scale resistance.
   If scale resistance is R_d,

$$R_d = \frac{1}{386.2} - \frac{1}{750} - \frac{1}{1000} = 0.00026 \text{ m}^2 \text{ K/W}$$

This is a rather low value, though the heat exchangers would probably be used for this duty.

To increase the velocity over the tubes it is common practice to fit *baffles* across the bundle. The commonest form is shown in Figure 9.23 where it is seen that the cut-away section is about 25 per cent of the total area. With such an arrangement, the flow

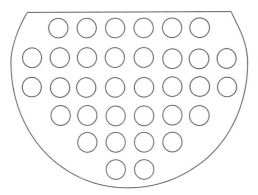

Figure 9.23.   Baffle for heat exchanger

pattern becomes more complex and the extent of leakage between the tubes and the baffle, and between the baffle and the inside of the shell of the exchanger, complicates the problem. For more detail the reader is referred to papers by SHORT[27], DONOHUE[28], and TINKER[29]. The various methods are all concerned with developing a method of calculating the true area of flow and of assessing the probable influence of leaks. When using baffles, the value of h_0, as found from equation 9.81, is commonly multiplied by 0.6 to allow for leakage.

The *drop in pressure* for the flow of a fluid across a tube bundle (Table 9.4) may be important because of the small pressure head available and because by good design it is possible to get a better heat transfer for the same drop in pressure. $-\Delta P_f$ depends on the velocity u_t through the minimum area of flow and in Chapter 3 an equation proposed by GRIMISON[22] was given as:

$$-\Delta P_f = \frac{C_f j \rho u_t^2}{6} \qquad \text{(equation 3.80)}$$

Table 9.4.[16]   Ratio of heat transfer to friction for tube bundles ($Re_{max} = 20,000$)

	X = 1.25d_o			X = 1.5d_o		
	C_h	C_f	C_h/C_f	C_h	C_f	C_h/C_f
In-line						
$Y = 1.25d_o$	1	1.44	0.69	1	1.56	0.64
$Y = 1.5d_o$	0.96	0.84	1.14	0.96	0.96	1.0
Staggered						
$Y = 1.25d_o$	1.06	1.56	0.68	1.05	1.74	0.60
$Y = 1.5d_o$	1.04	1.10	0.95	1.02	1.16	0.88

where C_f depends on the geometry of the tube layout and j is the number of rows of tubes. It is found that the ratio of C_h, the heat transfer factor in equation 9.82, to C_f is greater for the in-line arrangement but that the actual heat transfer is greater for the staggered arrangement.

To calculate the drop in pressure $-\Delta P_f$ over the tube bundles of a heat exchanger the following equation may also be used:

$$-\Delta P_f = \frac{f' G_s'^2 (n+1) d_v}{2\rho d_e} \tag{9.84}$$

where f' is the friction factor given in Figure 9.24, G_s' the mass velocity through bundle, n the number of baffles in the unit, d_v the inside shell diameter, ρ the density of fluid, d_e the equivalent diameter, and $-\Delta P_f$ the drop in pressure.

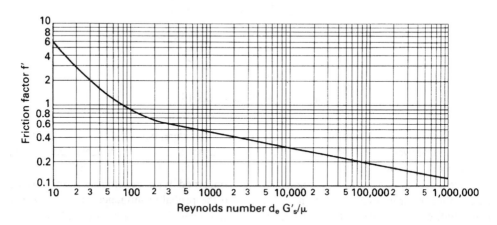

Figure 9.24.   Friction factor for flow over tube bundles

Example 9.8

54 t/h of benzene is cooled by passing the stream through the shell side of a tubular heat exchanger, 1 m i.d., fitted with 5 m tubes, 19 mm o.d. arranged on a 25 mm square pitch with 6 mm clearance. If the baffle spacing is 0.25 m (19 baffles), what will be the pressure drop over the tube bundle? ($\mu = 0.5$ mN s/m^2).

Solution

Cross-flow area: $\qquad A_s = \dfrac{1.0 \times 0.25 \times 0.006}{0.025} = 0.06 \text{ m}^2$

Mass flow: $\qquad G_s' = \dfrac{(54 \times 1000/3600)}{0.06} = 250 \text{ kg/m}^2 \text{ s}$

Equivalent diameter: $\qquad d_e = \dfrac{4[0.025^2 - (\pi/4)0.019^2]}{\pi \times 0.019} = 0.0229 \text{ m}$

Reynolds number through the tube bundle $= \dfrac{250 \times 0.0229}{0.5 \times 10^{-3}} = 11450$

From Figure 9.23: $\qquad f' = 0.280$

Density of benzene $\qquad = 881 \text{ kg/m}^3$

From equation 9.84:

$$-\Delta P_f = \frac{0.280 \times 250^2 \times 20 \times 1.0}{2 \times 881 \times 0.0229} = 8674 \text{ N/m}^2$$

or: $\qquad \dfrac{8674}{881 \times 9.81} = \underline{\underline{1.00 \text{ m of benzene}}}$

9.4.5. Flow in non-circular sections

Rectangular ducts

For the heat transfer for fluids flowing in non-circular ducts, such as rectangular ventilating ducts, the equations developed for turbulent flow inside a circular pipe may be used if some equivalent diameter is used in the place of d.

The use of the hydraulic mean diameter d_e has already been discussed in the previous section.

The data for heating and cooling water in turbulent flow in rectangular ducts are reasonably well expressed by the use of equation 9.54 in the form:

$$\frac{hd_e}{k} = 0.023 \left(\frac{d_e G'}{\mu}\right)^{0.8} \left(\frac{C_p \mu}{k}\right)^{0.4} \tag{9.85}$$

Whilst the experimental data of COPE and BAILEY[30] are somewhat low, data by WASHINGTON and MARKS[31] for heating air in ducts are well represented by this equation.

Annular sections between concentric tubes

Concentric tube heat exchangers are widely used because of their simplicity of construction and the ease with which additions may be made to increase the area. They are also used to give turbulent conditions where the quantity flowing is small.

In presenting equations for the film coefficient in the annulus, one of the difficulties has been to select the best equivalent diameter to use. When considering the film on the outside of the inner tube, DAVIS[32] has proposed the equation:

$$\frac{hd_1}{k} = 0.031 \left(\frac{d_1 G'}{\mu}\right)^{0.8} \left(\frac{C_p \mu}{k}\right)^{0.33} \left(\frac{\mu}{\mu_s}\right)^{0.14} \left(\frac{d_2}{d_1}\right)^{0.15} \tag{9.86}$$

where d_1 and d_2 are the outer diameter of the inner tube, and the inner diameter of the outer tube, respectively.

CARPENTER et al.[33] suggested using the hydraulic mean diameter $d_e = d_2 - d_1$ in the Sieder and Tate equation (9.60) and recommend the expression:

$$\frac{hd_e}{k} \left(\frac{\mu_s}{\mu}\right)^{0.14} = 0.027 \left(\frac{d_e G'}{\mu}\right)^{0.8} \left(\frac{C_p \mu}{k}\right)^{0.33} \tag{9.87}$$

Their results, which were obtained using a small annulus, were somewhat below those given by equation 9.87 for values of $d_e G'/\mu$ less than 10,000, but this may have been because the flow was not fully turbulent: with an index on the Reynolds group of 0.9, the equation fitted the points much better. There is little to choose between these two equations, but they both give rather high values for h.

For the viscous region Carpenter's results are reasonably well covered by the equation:

$$\frac{hd_e}{k} \left(\frac{\mu_s}{\mu}\right)^{0.14} = 2.01 \left(\frac{GC_p}{kl}\right)^{0.33}$$

$$= 1.86 \left[\left(\frac{d_e G'}{\mu}\right) \left(\frac{C_p \mu}{k}\right) \left(\frac{d_1 + d_2}{l}\right)\right]^{1/3}$$

This equation is the same as equation 9.77 with d_e in the place of d.

These results have all been obtained with small units and mainly with water as the fluid in the annulus.

Flow over flat plates

For the turbulent flow of a fluid over a flat plate the Colburn type of equation may be used with a different constant:

$$j_h = 0.037 Re_x^{-0.2} \tag{9.88}$$

where the physical properties are taken as for equation 9.58 and the characteristic dimension in the Reynolds group is the actual distance x along the plate. This equation therefore gives a point value for j_h.

9.4.6. Convection to spherical particles

In Section 9.3.4, consideration is given to the problem of heat transfer by conduction through a surrounding fluid to spherical particles or droplets. Relative motion between the fluid and particle or droplet causes an increase in heat transfer, much of which may be due to convection. Many investigators have correlated their data in the form:

$$Nu' = 2 + \beta'' Re'^n Pr^m \tag{9.89}$$

where values of β'', a numerical constant, and exponents n and m are found by experiment. Here $Nu' = hd/k$ and $Re' = du\rho/\mu$, the Reynolds number for the particle, u is the relative velocity between particle and fluid, and d is the particle diameter. As the relative velocity goes to zero, Re' goes to zero and the equation reduces to $Nu' = 2$ for pure conduction.

ROWE et al.[34], having analysed a large number of previous studies on this subject and provided further experimental data, conclude that for particle Reynolds numbers in the range 20–2000, equation 9.89 can be written as:

$$Nu' = 2.0 + \beta'' Re'^{0.5} Pr^{0.33} \tag{9.90}$$

β'' lies between 0.4 and 0.8 and has a value of 0.69 for air and 0.79 for water. In some practical situations the relative velocity between particle and fluid may change due to particle acceleration or deceleration and the value of Nu' can then be time-dependent.

For mass transfer, which is considered in more detail in Chapter 10, an analogous equation (10.194) applies, with the Sherwood number replacing the Nusselt number and the Schmidt number replacing the Prandtl number.

9.4.7. Natural convection

If a beaker containing water rests on a hot plate, the water at the bottom of the beaker becomes hotter than that at the top. Since the density of the hot water is lower than that of the cold, the water in the bottom rises and heat is transferred by natural convection. In the same way air in contact with a hot plate will be heated by natural convection currents, the air near the surface being hotter and of lower density than that some distance away. In these cases there is no external agency providing forced convection currents, and the transfer of heat occurs at a correspondingly lower rate since the natural convection currents move rather slowly.

For these processes which depend on buoyancy effects, it has been seen from equation 9.52 that the rate of heat transfer would be expected to follow a relation of the form:

$$Nu = f(Gr, Pr) \qquad\qquad\text{(equation 9.52)}$$

Measurements by SCHMIDT[10] of the upward air velocity near a 300 mm vertical plate show that the velocity rises rapidly to a maximum at a distance of about 2 mm from the plate and then falls rapidly. However, the temperature evens out at about 10 mm from the plate. Temperature measurements around horizontal cylinders have been made by RAY[35].

Natural convection from horizontal surfaces to air, nitrogen, hydrogen, and carbon dioxide, and to liquids (water, aniline, carbon tetrachloride, glycerol) has been studied by a number of workers (DAVIS[36], ACKERMANN[37], FISHENDEN and SAUNDERS[16], SAUNDERS[38] and others). Most of the results are for thin wires and tubes up to about 50 mm diameter; the temperature differences used are up to about 1100 K with gases and about 85 K with liquids. The general form of the results is shown in Figure 9.25, where log Nu is plotted against log $(Pr\ Gr)$ for streamline conditions. The curve can be represented by the relation:

$$Nu = C'(Gr\ Pr)^n \qquad\qquad (9.91)$$

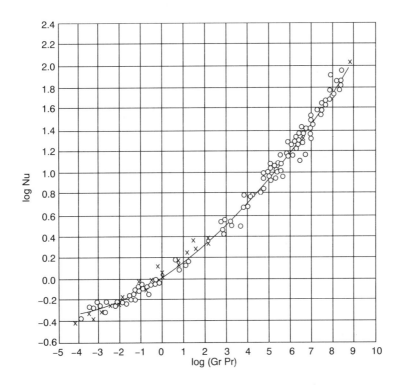

Figure 9.25.   Natural convection from horizontal tubes

Table 9.5.   Values of C', C'' and n for use in equations 9.92 and 9.95[39]

Geometry	GrPr	C'	n	C'' (for air at 294 K)
Vertical surfaces	$< 10^4$	1.36	0.20	
(l = vertical dimension < 1 m)	$10^4 - 10^9$	0.59	0.25	1.37
	$> 10^9$	0.13	0.33	1.24
Horizontal cylinders				
(l = diameter < 0.2 m)	$10^{-5} - 10^{-3}$	0.71	0.04	
	$10^{-3} - 1.0$	1.09	0.10	
	$1.0 - 10^4$	1.09	0.20	
	$10^4 - 10^9$	0.53	0.25	1.32
	$> 10^9$	0.13	0.33	1.24
Horizontal flat surfaces				
(facing upwards)	$10^5 - 2 \times 10^7$	0.54	0.25	1.86
(facing upwards)	$2 \times 10^7 - 3 \times 10^{10}$	0.14	0.33	
(facing downwards)	$3 \times 10^5 - 3 \times 10^{10}$	0.27	0.25	0.88

Numerical values of C' and n, determined experimentally for various geometries, are given in Table 9.5[39]. Values of coefficients may then be predicted using the equation:

$$\frac{hl}{k} = C' \left(\frac{\beta g \Delta T l^3 \rho^2}{\mu^2} \frac{C_p \mu}{k} \right)^n \quad \text{or} \quad h = C' \left(\frac{\Delta T}{l} \right)^n k \left(\frac{\beta g \rho^2 C_p}{\mu k} \right)^n \tag{9.92}$$

where the physical properties are at the mean of the surface and bulk temperatures and, for gases, the coefficient of cubical expansion β is taken as $1/T$, where T is the absolute temperature.

For vertical plates and cylinders, KATO et al.[40] have proposed the following equations for situations where $1 < Pr < 40$:

For $Gr > 10^9$: $Nu = 0.138 Gr^{0.36} (Pr^{0.175} - 0.55)$ (9.93)

and for $Gr < 10^9$: $Nu = 0.683 Gr^{0.25} Pr^{0.25} \left(\dfrac{Pr}{0.861 + Pr} \right)^{0.25}$ (9.94)

Natural convection to air

Simplified dimensional equations have been derived for air (and water and organic liquids) by grouping the fluid properties into a single factor in a rearrangement of equation 9.92 to give the equation:

$$h = C'' (\Delta T)^n l^{3n-1} \quad (\text{W/m}^2\text{K}) \tag{9.95}$$

Values of C'' (in SI units) are also given in Table 9.5 for air at 294 K. Typical values for water and organic liquids are 127 and 59 respectively.

Example 9.9

Fluids between two surfaces
   Estimate the heat transfer coefficient for natural convection from a horizontal pipe 0.15 m diameter, with a surface temperature of 400 K to air at 294 K

Solution

Over a wide range of temperature, $k^4 (\beta g \rho^2 C_p / \mu k) = 36.0$

For air at $0.5(400 + 294) = 347$ K, $k = 0.0310$ W/m K (Table 7, Appendix A2)

$$\therefore \quad \frac{\beta g \rho^2 C_p}{\mu k} = \frac{36.0}{0.0310^4} = 3.9 \times 10^7$$

From Equation 9.92:

$$GrPr = 3.9 \times 10^7 (400 - 294) \times 0.15^3$$

$$= 1.39 \times 10^7$$

From Table 9.5:

$$n = 0.25 \quad \text{and} \quad C'' = 1.32$$

$\therefore$ in Equation 9.94:

$$h = 1.32(400 - 294)^{0.25}(0.15)^{(3 \times 0.25)-1}$$

$$= 1.32 \times 106^{0.25} \times 0.15^{-0.25}$$

$$= \underline{\underline{6.81 \text{ W/m}^2 \text{ K}}}$$

Fluids between two surfaces

For the transfer of heat from a hot surface across a thin layer of fluid to a parallel cold surface:

$$\frac{Q}{Q_k} = \frac{h\Delta T}{(k/x)\Delta T} = \frac{hx}{k} = Nu \tag{9.96}$$

where Q_k is the rate at which heat would be transferred by pure thermal conduction between the layers, a distance x apart, and Q is the actual rate.

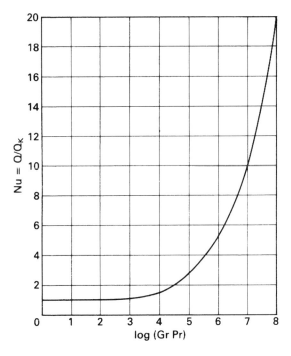

Figure 9.26.   Natural convection between surfaces

For $(Gr\ Pr)$ equal to 10^3, the heat transferred is approximately equal to that due to conduction alone, but for $(Gr\ Pr)$ from 10^4 to 10^6, the heat transferred is given (Figure 9.26) by:

$$\frac{Q}{Q_k} = 0.15(Gr\ Pr)^{0.25} \tag{9.97}$$

In this equation the characteristic dimension to be used for the Grashof group is x, the distance between the planes, and the heat transfer is independent of surface area, provided that the linear dimensions of the surfaces are large compared with x. For higher values of $(Gr\ Pr)$, Q/Q_k is proportional to $(Gr\ Pr)^{1/3}$, showing that the heat transferred is not entirely by convection and is not influenced by the distance x between the surfaces.

A similar form of analysis has been given by KRAUSSOLD[41] for air between two concentric cylinders. It is important to note from this general analysis that a single layer of air will not be a good insulator because convection currents set in before it becomes 25 mm thick. The good insulating properties of porous materials are attributable to the fact that they offer a series of very thin layers of air in which convection currents do not arise.

9.5. HEAT TRANSFER BY RADIATION

9.5.1. Definitions

A heated body emits energy in the form of electromagnetic waves. This energy is radiated in all directions and on falling on a second body is partially absorbed, partially reflected, and partially transmitted as indicated in Figure 9.27. The fraction of the incident radiation absorbed is known as the *absorptivity a*, and the fraction reflected the *reflectivity* of the body. The amount transmitted will therefore depend on these two properties. If the amount transmitted is negligible the material is termed *opaque*.

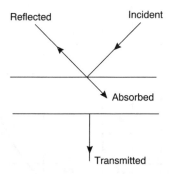

Figure 9.27.   Reflection, absorption, and transmission of radiation

9.5.2. Kirchhoff's Law

Suppose two bodies **A** and **B** of areas A_1 and A_2 to be in a large enclosure from which no energy is lost to the outside. Then the energy absorbed by **A** from the enclosure will be $A_1 a_1 I'$, where I' is the rate at which radiation is falling on **A** per unit area, and a_1 is the absorptivity. The energy given out by **A** will be $E_1 A_1$, where E_1, the *emissive power*, is the energy emitted per unit area per unit time. At equilibrium these quantities will be equal. In the same way the energy emitted by **B** will equal the energy received. Thus:

$$I'A_1 a_1 = A_1 E_1 \quad \text{and} \quad I'A_2 a_2 = A_2 E_2$$

or:

$$\frac{E_1}{a_1} = \frac{E_2}{a_2} = \frac{E}{a} \quad \text{for any other body}$$

Thus, for all bodies: the ratio of the emissive power to the absorptivity will be the same. The maximum possible value for E will occur when a has its maximum value of unity. This condition applies to a *black body* which is defined as one which absorbs all the radiation falling on it, i.e. its absorptivity, $a_b = 1$.

Now the *emissivity* e of a body is defined as the ratio of its emitting power to that of a black body; i.e. $e = E/E_b$. Since E/a is constant for all bodies, $E/a = E_b/a_b$.

Thus:

$$e = \frac{E}{E_b} = \frac{a}{a_b} \tag{9.98}$$

Then, since $a_b = 1$, the emissivity e of any body is equal to its absorptivity a.

9.5.3. Energy emitted by a black body

The energy emitted per unit time by a black body depends only on its temperature. This energy is given out over a range of wavelengths and the general distribution of energy from a black body at various temperatures is shown in Figure 9.28. At any temperature there

Figure 9.28.   Emission from a black body

is a wavelength Z_m at which the maximum energy is emitted. The energy at wavelengths less than this falls off rapidly, but for wavelengths greater than Z_m the drop in energy is much less. It will be seen that the total emission increases rapidly with temperature and that the higher the temperature the shorter is the wavelength at which the maximum occurs and the more pronounced is the peak.

The wavelength of the energy radiated at temperatures below 820 K is too great for it to be visible, but at higher temperatures the wavelengths correspond with those of the visible spectrum, and the colour of a radiating body under these conditions is a good indication of its temperature.

The total energy emitted per unit area per unit time is given by the *Stefan-Boltzmann law*:

$$E = \sigma T^4 \tag{9.99}$$

where T is the absolute temperature and σ is the Stefan-Boltzmann constant which has the value 1.73×10^{-9} Btu/h ft^2 °R^4 or 5.67×10^{-8} W/m^2 K^4. The wavelength Z_m at which the maximum energy is radiated was found by Wien to vary inversely as the absolute temperature. Taking the temperature in degrees K, $TZ_m = 0.00288$, where the wavelength is in metres.

If $E_z' \, dZ$ is the radiation between wavelengths Z and $Z + dZ$, the total radiation is given by:

$$E = \int_0^\infty E_z' \, dZ \tag{9.100}$$

It can be shown that this integral is equal to σT^4, so that:

$$E = \sigma T^4 \tag{equation 9.99}$$

The ratio of the energy emitted by a body to that emitted by a black body at the same temperature has been defined above as the emissivity. Strictly this should be taken for each wavelength, since the ratio e will not remain constant over a wide range of wavelengths. A *grey body* is defined as one which has a constant value of e, so that for any temperature range it radiates the same proportion of the energy radiated by a black body. Similarly it will have a constant absorptivity.

Values of e have been measured for many materials, and it is found that for most industrial non-metallic surfaces and for non-polished metals e may be taken as about 0.9; for highly polished surfaces such as copper or aluminium values of e as low as 0.03 are obtained. A small cavity in a body has an effective emissivity of unity and therefore behaves as a black body.

9.5.4. Heat transferred by radiation

A body of emissivity e at an absolute temperature T_1 emits energy $e\sigma T_1^4$ per unit area. If the surroundings are black, they reflect back none of this radiation, but if they are at an absolute temperature T_2 they will emit radiation σT_2^4. If the body is grey it will absorb a fraction e, so that the net radiation per unit area from the grey body will be:

$$q = e\sigma(T_1^4 - T_2^4) \tag{9.101}$$

This relation will still be true where the grey body is so small that a negligible proportion of its own radiation is reflected back to it from the surroundings, e.g. a body radiating to the atmosphere.

For a material that does not behave as a grey body but as a selective emitter, the absorptivity of the surface at T_1 for radiation from surroundings at T_2 will be a_{T_2}. This will not be equal to its emissivity (e_{T_1}) at T_1 but to its emissivity at T_2, i.e. e_{T_2}. Under these conditions the general equation for the net exchange of heat becomes:

$$q = \sigma(e_{T_1} T_1^4 - e_{T_2} T_2^4) \tag{9.102}$$

The net heat transfer from unit surface of a grey body at a temperature T_1 to a black enclosure at T_2 can be written as:

$$q = h_r(T_1 - T_2)$$

Thus:
$$h_r = \frac{q}{T_1 - T_2} = \frac{\sigma e}{T_1 - T_2}(T_1^4 - T_2^4) \tag{9.103}$$

and h_r may be looked upon as a radiation transfer coefficient. Equation 9.103 is also applicable if the surroundings are not black, provided that the body is small and none of its own radiation is reflected back to it.

Example 9.10

Calculate the total heat loss by radiation and convection from an unlagged horizontal steam pipe, 50 mm o.d. at 377 K to air at 283 K.

Solution

From equation 9.101 assuming an emissivity of 0.9:

loss by radiation:
$$q_r = 5.67 \times 10^{-8} \times 0.9(377^4 - 283^4)$$
$$= 704 \text{ W/m}^2$$

loss by natural convection, assuming streamline flow:

$$= 1.18 \left(\frac{\Delta T}{d_0}\right)^{0.25} \times \Delta T$$

$$= 1.18 \left(\frac{94}{0.050}\right)^{0.25} \times 94$$

$$= 732 \text{ W/m}^2$$

The total loss is therefore:
$$(704 + 732) = \underline{\underline{1436 \text{ W/m}^2}}$$

In practice the heat loss from an unlagged vertical pipe will be rather higher, since here the convection currents are stronger.

If experiments are made on the intensity of radiation emitted by a small plane surface in a given direction it is found that the strongest radiation is in a line normal to the surface. This intensity I normal to the plane is defined as the radiation given out per unit time per unit area of a small surface dA over unit solid angle. Then the intensity of radiation at an angle α to the normal is seen from Figure 9.29 to be $I \cos \alpha$. This form of relation shows

Figure 9.29.   Intensity of radiation in a given direction

the similarity of the transfer process to the radiation of light. It is followed accurately by black bodies but metallic surfaces may show wide divergences.

The radiation intensity normal to surface, I = energy per unit area per unit time per unit solid angle = $dq/d\omega$.

The radiation from small surface dA, through a small solid angle $d\omega$, at an angle α to normal is $I\, dA \cos \alpha \, d\omega$.

$$\therefore \qquad \text{Total energy emitted in forward direction} = \int_{\alpha=0}^{\alpha=\pi/2} I\, dA \cos \alpha \, d\omega$$

But $d\omega = 2\pi r \sin \alpha r\, d\alpha / r^2$, where r is radius of described circle.

$$\therefore \qquad \text{Emissive power } (\sigma T^4 \text{ for black body}) = \int_0^{\pi/2} I \cos \alpha\, 2\pi \sin \alpha \, d\alpha = \pi I \quad (9.104)$$

9.5.5. Geometric or angle factors

Summation rule

In many cases when the *non-black* bodies are situated in fairly close proximity, part of the energy emitted by one body will be reflected back to it by the second body and will then be partly reabsorbed and partly reflected again. Thus the radiation undergoes a series of internal reflections and absorptions.

The simplest case to consider is the heat exchange between two large parallel plates. In this case, all of the heat radiated by the one surface will fall on the other. In many cases, however, the second surface will intercept only part of the radiation. The fraction of the total radiation from surface 1 which is intercepted by surface 2 is referred to as the *geometric* or *angle factor* F_{12} and its value depends on the geometrical arrangement of the two surfaces. For simple configurations the geometric factors can be calculated, but for more complex arrangements they are most readily determined by experimenting with models.

The geometric factor is best understood by considering the projected area which an element of the radiating surface *sees*. Thus, with large parallel plates the field of view of any element of surface 1 remote from the edges consists entirely of surface 2 and therefore the geometric factor F_{12} is unity; and, of course, F_{21} is also unity. Furthermore, since the element does not radiate directly on to any part of its own surface, the geometric factor with respect to itself, F_{11} (and F_{22}) is zero. This latter condition also applies to any convex surface and therefore for all convex surfaces F_{11} is zero.

From the above definition of geometric factor, it follows that the sum of the geometric factors with respect to a surface must be unity if the surfaces $1, 2, \ldots$ form a complete enclosure:

i.e.:
$$F_{11} + F_{12} + \cdots = 1 \tag{9.105}$$

This is known as the *summation rule*.

Two particular cases of radiation between surfaces are of particular practical interest and are capable of analytical solution. The first is the radiation between two long concentric cylinders and the other is the radiation between concentric spheres. In both cases, the inner surface (1) is convex and therefore F_{11} is zero. Since all the radiation from this surface falls on the concave outer surface, F_{12} is unity. However, part of the radiation from surface 2 falls on the inner surface and part falls on itself. The fraction of the field of view of an element of the concave surface which is occupied by the inner convex surface is A_1/A_2 and therefore $F_{21} = A_1/A_2$ and $F_{22} = 1 - (A_1/A_2)$ in both cases.

The net transfer of heat can be considered by summing the individual components. This is done below for two parallel plates, but a second method, the net radiation method, is more general in its application.

For transfer between two large parallel plates, each of which has a grey surface, it will be shown that the heat transfer rate per unit area is given by:

$$q = \frac{e_1 e_2 \sigma}{e_1 + e_2 - e_1 e_2}(T_1^4 - T_2^4) \tag{9.106}$$

For long concentric cylinders or concentric spheres, the heat transfer rate per unit area of surface 1 is given by:

$$q = \frac{e_1 e_2 \sigma}{e_2 + e_1(1 - e_2)A_1/A_2}(T_1^4 - T_2^4) \tag{9.107}$$

Radiation between parallel plates

Considering the radiation of heat between two parallel grey surfaces which are large compared with their distance apart, then only a negligible fraction of the total heat transmitted will be lost to the surroundings at the ends, and all the heat radiated by one surface will be intercepted by the other, i.e. the geometric factors will both be unity.

Then, per unit area and unit time:

$$\text{heat radiated from surface 1} = \sigma e_1 T_1^4 \tag{A}$$

$$\text{of this, amount absorbed by surface 2} = \sigma e_1 T_1^4 e_2 \tag{B}$$

$$\text{and amount reflected by surface 2} = \sigma e_1 T_1^4 (1 - e_2) \tag{C}$$

$$\text{of this, amount re-absorbed by surface 1} = \sigma e_1 T_1^4 (1 - e_2) e_1 \tag{D}$$

$$\text{and amount re-reflected by surface 1} = \sigma e_1 T_1^4 (1 - e_2)(1 - e_1) \tag{E}$$

$$\text{and, of this, amount absorbed by surface 2} = \sigma e_1 T_1^4 (1 - e_2)(1 - e_1) e_2 \tag{F}$$

Thus, as a result of each complete cycle of internal reflections, it is seen by comparing (B) and (F) that the absorption is reduced by a factor $(1 - e_1)(1 - e_2)$.

As the energy suffers an infinite number of reflections:

total transfer of energy from surface 1 to surface 2 per unit area and unit time

$$= \sigma e_1 e_2 T_1^4 [1 + (1 - e_1)(1 - e_2) + (1 - e_1)^2 (1 - e_2)^2 + \cdots \text{ to } \infty]$$

$$= \sigma e_1 e_2 T_1^4 \frac{1}{1 - (1 - e_1)(1 - e_2)}$$

$$= \frac{\sigma e_1 e_2}{e_1 + e_2 - e_1 e_2} T_1^4$$

Since this expression is symmetrical in e_1 and e_2, the total transfer of energy from surface 2 to surface 1 per unit area and unit time

$$= \frac{e_1 e_2 \sigma}{e_1 + e_2 - e_1 e_2} T_2^4$$

Thus net energy transferred per unit area and unit time:

$$q = \frac{e_1 e_2 \sigma}{e_1 + e_2 - e_1 e_2} (T_1^4 - T_2^4) \qquad \text{(equation 9.106)}$$

This approach is known as the *multiple reflection method*. The same result may be obtained by what is known as the *net radiation method* as follows.

If Q_{o1} is the total radiation per unit time leaving surface 1, and Q_{i1} is the total incident radiation, then:

$$Q_{o1} = Q_{i2} \quad \text{and} \quad Q_{o2} = Q_{i1}$$

Q_{o1} consists of emitted radiation Q_{e1} and reflected radiation Q_{r1}.

Thus: $Q_{o1} = Q_{e1} + Q_{r1}$

But: $Q_{r1} = (1 - e_1)Q_{i1}$

Thus: $Q_{o1} = Q_{e1} + (1 - e_1)Q_{i1}$

$$= Q_{e1} + (1 - e_1)Q_{o2} \qquad (9.108)$$

Similarly: $Q_{o2} = Q_{e2} + (1 - e_2)Q_{o1}$

Hence: $Q_{o1} = \dfrac{Q_{e1} + (1 - e_1)Q_{e2}}{1 - (1 - e_1)(1 - e_2)} \quad \text{and} \quad Q_{o2} = \dfrac{Q_{e2} + (1 - e_2)Q_{e1}}{1 - (1 - e_1)(1 - e_2)}$

The net transfer of radiation per unit time:

$$Q = Q_{o1} - Q_{o2} = \frac{e_2 Q_{e1} - e_1 Q_{e2}}{1 - (1 - e_1)(1 - e_2)}$$

But: $Q_{e1} = A\sigma e_1 T_1^4 \quad \text{and} \quad Q_{e2} = A\sigma e_2 T_2^4$

Thus: $q = \dfrac{Q}{A} = \dfrac{e_1 e_2 \sigma}{e_1 + e_2 - e_1 e_2}(T_1^4 - T_2^4) \qquad \text{(equation 9.106)}$

Example 9.11

Heat is transferred by radiation between two large grey surfaces of emissivities e_1 and e_2, respectively. By what faction will the heat transfer rate be reduced by the introduction between the radiating surfaces of a thin metal radiation shield (Figure 9.30), both surfaces of which have an emissivity e_{sh}? Neglect any temperature gradients across the shield.

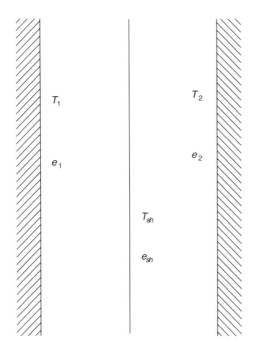

Figure 9.30.   Radiation shield

If the emissivities of all the surfaces are the same ($e_1 = e_2 = e_{sh}$), show that the heat transfer rate will be halved.

Solution

In the absence of the shield:

$$q = \frac{\sigma}{\dfrac{1}{e_1} + \dfrac{1}{e_2} - 1}(T_1^4 - T_2^4) \qquad \text{(from equation 9.103)}$$

If the temperature of the shield is T_{sh} and the new value of q is q_{sh}, considering the transfer between plate 1 and the shield:

$$q_{sh} = \frac{\sigma}{\dfrac{1}{e_1} + \dfrac{1}{e_{sh}} - 1}(T_1^4 - T_{sh}^4)$$

i.e.:

$$T_1^4 - T_{sh}^4 = \frac{q_{sh}}{\sigma}\left(\frac{1}{e_1} + \frac{1}{e_{sh}} - 1\right)$$

Similarly for transfer between the shield and plate 2:

$$T_{sh}^4 - T_2^4 = \frac{q_{sh}}{\sigma}\left(\frac{1}{e_{sh}} + \frac{1}{e_2} - 1\right)$$

Adding:

$$T_1^4 - T_2^4 = \frac{q_{sh}}{\sigma}\left(\frac{1}{e_1} + \frac{2}{e_{sh}} + \frac{1}{e_2} - 2\right)$$

or:

$$q_{sh} = \frac{\sigma}{\dfrac{1}{e_1} + \dfrac{2}{e_{sh}} + \dfrac{1}{e_2} - 2}(T_1^4 - T_2^4)$$

Thus:

$$\frac{q_{sh}}{q} = \frac{\dfrac{1}{e_1} + \dfrac{2}{e_{sh}} + \dfrac{1}{e_2} - 2}{\dfrac{1}{e_1} + \dfrac{1}{e_2} - 1}$$

When $e_1 = e_2 = e_{sh}$:

$$\frac{q_{sh}}{q} = \frac{1}{2}$$

Note: It may be shown in a similar manner that, if there are n shields in series:

$$\frac{q_{sh}}{q} = \frac{1}{n+1}$$

The reciprocal rule

In the *general case*, the radiation incident on a surface will have come partly from itself and partly from the other surface, the proportion being determined by the geometric factors. Thus:

$$Q_{i2} = Q_{o1}F_{12} + Q_{o2}F_{22}$$

$$Q_{i1} = Q_{o2}F_{21} + Q_{o1}F_{11}$$

Again:

$$Q_{o1} = Q_{e1} + (1 - e_1)Q_{i1} \qquad\qquad \text{(equation 9.108)}$$

i.e.:

$$Q_{o1} = Q_{e1} + (1 - e_1)(Q_{o2}F_{21} + Q_{o1}F_{11})$$

Similarly:

$$Q_{o2} = Q_{e2} + (1 - e_2)(Q_{o1}F_{12} + Q_{o2}F_{22})$$

Solving for Q_{o1} and Q_{o2}:

$$Q_{o1} = \frac{Q_{e1}[1 - (1 - e_2)F_{22}] + Q_{e2}(1 - e_1)F_{21}}{[1 - (1 - e_2)F_{22}][1 - (1 - e_1)F_{11}] - (1 - e_1)(1 - e_2)F_{12}F_{21}}$$

and:

$$Q_{o2} = \frac{Q_{e2}[1 - (1 - e_1)F_{11}] + Q_{e1}(1 - e_2)F_{12}}{[1 - (1 - e_2)F_{22}][1 - (1 - e_1)F_{11}] - (1 - e_1)(1 - e_2)F_{12}F_{21}}$$

Now the net rate of transfer of heat is given by:

$$Q = Q_{o1}F_{12} - Q_{o2}F_{21}$$

$$= \frac{Q_{e1}F_{12}[1 - (1 - e_2)(F_{22} + F_{21})] - Q_{e2}F_{21}[1 - (1 - e_1)(F_{11} + F_{12})]}{[1 - (1 - e_2)F_{22}][1 - (1 - e_1)F_{11}] - (1 - e_1)(1 - e_2)F_{12}F_{21}}$$

Substituting $Q_{e1} = A_1 \sigma e_1 T_1^4$ and $Q_{e2} = A_2 \sigma e_2 T_2^4$ and noting from the *summation rule*, (equation 9.105) that $F_{11} + F_{12}$ and $F_{21} + F_{22}$ are both unity:

$$Q = \frac{e_1 e_2 \sigma (A_1 F_{12} T_1^4 - A_2 F_{21} T_2^4)}{[1 - (1 - e_2)F_{22}][1 - (1 - e_1)F_{11}] - (1 - e_1)(1 - e_2)F_{12}F_{21}} \qquad (9.109)$$

Since there can be no net heat transfer between the surfaces when T_1 and T_2 are equal:

$$A_1 F_{12} = A_2 F_{21} \qquad (9.110)$$

This is known as the *reciprocal rule*.

For *long concentric cylinders* and *concentric spheres*:

$$F_{11} = 0 \quad F_{12} = 1 \quad F_{21} = \frac{A_1}{A_2} \quad F_{22} = 1 - \left(\frac{A_1}{A_2}\right)$$

Thus:
$$Q = \frac{A_1 e_1 e_2 \sigma}{e_2 + e_1(1 - e_2)(A_1/A_2)}(T_1^4 - T_2^4)$$

and the heat transfer rate per unit area of surface 1 (Q/A_1) is given by:

$$q = \frac{e_1 e_2 \sigma}{e_2 + e_1(1 - e_2)(A_1/A_2)}(T_1^4 - T_2^4) \qquad \text{(equation 9.107)}$$

9.5.6. Gas radiation

Most of the simple monatomic and diatomic gases such as helium, hydrogen, oxygen, and nitrogen are transparent to thermal radiation, but some polyatomic gases — notably carbon dioxide, water vapour, carbon monoxide, ammonia, and hydrocarbons — absorb a considerable amount of radiation of certain frequencies. These gases, which are industrially very important, radiate appreciably in the same wave bands. In contrast with the behaviour of solids, a considerable thickness of gas is required to absorb a large fraction of the radiation falling on it. Thus if I' is the intensity of the incident radiation, the intensity I after the radiation has passed through a layer of thickness x is given by:

$$I = I'e^{-m'x} \qquad (9.111)$$

where m' will, in general, vary with the wavelength and is approximately proportional to the partial pressure of the gas for any one wavelength, i.e. to the number of molecules per unit volume. The absorption will then be a function of the product $P_g l$, where P_g is the partial pressure of the gas and l is the equivalent thickness of the gas stream. The absorptivity of a gas is conveniently plotted against temperature for a constant value of $P_g l$. The thickness l is assumed the same in all directions but this will only be true for the case of a hemisphere where the gas is radiating to the mid-point of its base. It has been found that for a wide range of conditions l may be taken as 3.4 times the gas volume divided by the area of the retaining walls. A few values on this basis are given in Table 9.6.

Curves showing the emissivity of water vapour at a total pressure of one atmosphere, (Figure 9.31) are taken from the work of HOTTEL and MANGELSDORF[42]. The net radiant

Table 9.6. Equivalent dimensions for hemispherical radiation

Shape	Characteristic dimension L	Equivalent dimension for hemispherical radiation
Sphere	Diameter	$0.57L$
Infinite cylinder radiating to walls	Diameter	$0.86L$
Space between infinite parallel planes	Distance apart	$1.70L$
Space outside bank of tubes with centres on equilateral triangle; clearance = diameter of tube	Clearance	$2.89L$
Ditto, with tube centres on square pitch; tube diameter = clearance	Clearance	$3.49L$

Figure 9.31. Absorptivity and emissivity of water vapour

heat exchange between a gas at T_1 and unit area of enclosure at T_2 acting as a black body will be:

$$q = \sigma(e_g T_1^4 - a_g T_2^4) \qquad (9.112)$$

where e_g and a_g are the emissivity and absorptivity of the gas.

If the enclosure acts as a non-black surface some of the radiation falling on it is reflected back; of this, part is absorbed by the gas and part by the surface. For these general conditions it is usually sufficiently accurate to take the emissivity of the enclosure as e', where $e' = (e_s + 1)/2$. This is permissible because e_s, the emissivity of the surface, will probably lie between 0.7 and 1. The radiation exchange will then be:

$$q = e'\sigma(e_g T_1^4 - a_g T_2^4) \qquad (9.113)$$

Example 9.12

A cylindrical peep-hole in a 0.3 m thick furnace wall is 0.15 m in diameter. The cylindrical surface of the peep-hole behaves as a diffuse reflector and heat losses through it to the material of the furnace wall are negligible. If the inside surfaces of the furnace are at a uniform temperature of 1350 K and the outside surroundings are at 285 K, determine the heat loss by radiation through the peep-hole.

The angle factor F_{12} for radiation between two parallel discs of equal radii in opposite location (i.e. a line joining the centres is normal to the discs) is given by:

$$F_{12} = \frac{[1 + 2x^2 - \sqrt{(1 + 4x^2)}]}{2x^2}$$

where x is the ratio of the disc radius to the distance of separation of the discs.

The peep-hole connects two uniform temperature enclosures, one at 1350 K and one at 285 K; hence the ends of the peep-hole may be treated as plane black body surfaces at these two temperatures. (Stefan-Boltzmann constant = 5.67×10^{-8} W/m^2 K^4.)

Solution

(a) *Determination of angle factors* (Figure 9.32).

Figure 9.32.   Angle factors for peep-hole

The formula given for the angle-factor between surfaces (1) and (2) may be used:

$$x = \frac{r}{z} = \frac{0.075}{0.3} = 0.25$$

$$F_{12} = F_{21} = \frac{[1 + 0.125 - \sqrt{(1 + 0.25)}]}{0.125} = 0.056$$

Since (1) and (2) are plane surfaces,

$$F_{11} = F_{22} = 0$$

∴ By the *summation rule* (equation 9.105):

$$F_{13} = 1 - F_{12} - F_{11} = 1 - 0.056 - 0 = 0.944$$

By the *reciprocal rule* (equation 9.110):

$$A_3 F_{31} = A_1 F_{13}$$

$$\therefore \qquad \frac{F_{13}\pi r^2}{2\pi rz} = F_{31} = \frac{0.944 \times 0.075}{2 \times 0.3} = 0.118$$

By symmetry:
$$F_{23} = F_{13} \quad \text{and} \quad F_{32} = F_{31}$$

By the summation rule:
$$F_{33} = 1 - F_{31} - F_{32} = 1 - 2 \times 0.118$$
$$= 0.764$$

(b) *Heat balances.* Since (1) and (2) are black surfaces, all incident radiation is absorbed and the only outgoing radiation is the emitted radiation.

Using Stefan's law:
$$Q_{o1} = Q_{e1} = \pi r^2 \times 5.67 \times 10^{-8} T_1^4$$
$$= 0.075^2 \pi \times 5.67 \times 10^{-8} (1350)^4 = 3328 \text{ W}$$

Similarly:
$$Q_{o2} = Q_{e2} = 0.075^2 \pi \times 5.67 \times 10^{-8} (285)^4 = 6.6 \text{ W}$$

Since surface (3) is perfectly insulated, the total outgoing radiation (reflected + emitted) is equal to the incident radiation, or:
$$Q_{i3} = Q_{o3} \tag{1}$$

However, from the definition of angle-factors:
$$\text{total incident radiation, } Q_{i3} = Q_{o1}F_{13} + Q_{o2}F_{23} + Q_{o3}F_{33} \tag{2}$$

From equations (1) and (2):
$$Q_{o3} = \frac{Q_{o1}F_{13} + Q_{o2}F_{23}}{1 - F_{33}}$$
$$= \frac{0.944(3328 + 6.6)}{1 - 0.764} = 13,338 \text{ W}$$

Similarly:
$$Q_{i2} = Q_{o1}F_{12} + Q_{o2}F_{22} + Q_{o3}F_{32}$$
$$= 3328 \times 0.056 + 6.6 \times 0 + 13,338 \times 0.118$$
$$= 1760.3 \text{ W}$$

Net radiation from the peep-hole $= Q_{i2} - Q_{o2}$
$$= 1760.3 - 6.6 = \underline{1753.7 \text{ W}} = \underline{1.75 \text{ k W}}$$

Alternatively, the radiation entering by surface (1) is 3328 W and, by definition of F_{12}, the proportion of this leaving by surface (2) is:
$$Q_{e1}F_{12} = 3328 \times 0.056 = 186 \text{ W}$$

The rest, 3142 W, hits surface (3).

Similarly: radiation entering by surface (2) is 6.6 W

proportion leaving by surface (1) is 0.4 W

proportion hitting surface (3) is 6.2 W

$\therefore$ Total radiation incident on surface (3) $= 3142 + 6.2 = 3148.2 \text{ W}$

Since all radiation from surface (3) is emitted or reflected uniformly, by symmetry, half of this incident radiation leaves from each end.

$\therefore \qquad$ Net radiation to atmosphere $= (0.5 \times 3148.2) + 186 - 6.6$
$$= \underline{1753.5 \text{ W}} = \underline{1.75 \text{ k W}}$$

The estimation of the radiation from pulverised-fuel flames, from dust particles in flames and from flames made luminous by the thermal decomposition of hydrocarbons to soot, involves an evaluation of radiation from clouds of particles. In pulverised-fuel flames, these have a mean particle size of typically 0.025 mm and vary in composition from a very high carbon content to virtually pure ash, whereas the suspended matter in luminous flames, resulting from the thermal decomposition of hydrocarbons in the flame due to incomplete mixing with air before being heated, consists of carbon and very heavy hydrocarbons and an initial particle size of some 0.0003 mm. In the main, pulverised-fuel particles are sufficiently large to be substantially opaque to incident radiation whilst the particles of a luminous flame are so small as to act as semi-transparent bodies with respect to thermal or long wave-length radiation.

According to SCHACK[43], a single particle of soot transmits approximately 95 per cent of the incident radiation and only a cloud containing a very large number of particles can result in an appreciable emission. If the concentration of particles is described by a factor K, then the product of K and the thickness of the layer l play the same role as the product $P_g l$ in the radiation of gases. For a known or measured emissivity of the flame, e_F, the heat transfer with a wall is given by:

$$q = e_F e_s \sigma [(T_F)^4 - (T_w)^4] \tag{9.114}$$

where e_s is the effective emissivity of the wall and T_F and T_w the temperatures of the flame and the wall respectively. e_F not only varies from point to point in a flame but depends on the type of fuel, the shape of the burner and combustion chamber and also on the air supply and the preheating of the air and the fuel.

9.6. HEAT TRANSFER IN THE CONDENSATION OF VAPOURS

9.6.1. Film coefficients for vertical and inclined surfaces

When a saturated vapour is brought into contact with a cool surface, heat is transferred from the vapour to the surface and a film of condensate is produced.

In considering the heat that is transferred in such a case, the method which was first put forward by NUSSELT[44] and later modified by subsequent workers will be followed. If the vapour is condensing on a vertical surface, the condensate film flows downwards under the influence of gravity, but is retarded by the viscosity of the liquid. This flow will normally be streamline and the heat will flow through the film by conduction. Nusselt supposed that the temperature of the film at the cool surface was equal to that of the surface, and at the other side was at the temperature of the vapour. Actually there must be some small difference in temperature between the vapour and the film, but this may generally be neglected except where non-condensable gas is present in the vapour.

It has been shown in Chapter 3, that the mean velocity of a fluid flowing down a surface inclined at an angle ϕ to the horizontal is given by:

$$u = \frac{\rho g \sin \phi s^2}{3\mu} \qquad \text{(equation 3.87)}$$

For a vertical surface: $\sin \phi = 1$ and $u = \dfrac{\rho g s^2}{3\mu}$

The maximum velocity u_s occurs at the free surface:

$$u_s = \frac{\rho g \sin \phi \, s^2}{2\mu} \qquad \text{(equation 3.88)}$$

and is 1.5 times the mean velocity of the liquid.

Since the liquid is produced by condensation, the thickness of the film will be zero at the top and will gradually increase towards the bottom. Under stable conditions the difference in the mass rates of flow at distances x and $x + \mathrm{d}x$ from the top of the surface will result from condensation over the small element of the surface of length $\mathrm{d}x$ and width w (Figure 9.33).

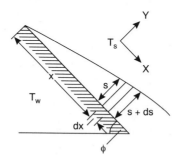

Figure 9.33.   Condensation of an inclined surface

If the thickness of the liquid film increases from s to $s + \mathrm{d}s$ in that distance, the increase in the mass rate of flow of liquid $\mathrm{d}G$

$$= \frac{\mathrm{d}}{\mathrm{d}s}\left(\frac{\rho^2 g \sin \phi \, s^3 w}{3\mu}\right)\mathrm{d}s$$

$$= \frac{\rho^2 g \sin \phi}{\mu} w s^2 \, \mathrm{d}s$$

If the vapour temperature is T_s and the wall temperature is T_w, the heat transferred by thermal conduction to an element of surface of length $\mathrm{d}x$

$$= \frac{k(T_s - T_w)}{s} w \, \mathrm{d}x$$

where k is the thermal conductivity of the condensate.

Thus the mass rate of condensation on this small area of surface

$$= \frac{k(T_s - T_w)}{s\lambda} w \, \mathrm{d}x$$

where λ is the latent heat of vaporisation of the liquid.

Thus: $\qquad \dfrac{k(T_s - T_w)}{s\lambda} w \, \mathrm{d}x = \dfrac{\rho^2 g \sin \phi}{\mu} w s^2 \, \mathrm{d}s$

On integration, this equation gives:

$$\mu k(T_s - T_w)x = \tfrac{1}{4}\rho^2 g \sin \phi \, s^4 \lambda$$

since $s = 0$ when $x = 0$.

Thus:
$$s = \sqrt[4]{\frac{4\mu kx(T_s - T_w)}{g \sin \phi \, \lambda \rho^2}} \qquad (9.115)$$

Now the heat transfer coefficient h at $x = x$, $= k/s$.

Thus:
$$h = \sqrt[4]{\frac{\rho^2 g \sin \phi \, \lambda k^3}{4\mu x(T_s - T_w)}} \qquad (9.116)$$

and:
$$Nu = \frac{hx}{k} = \sqrt[4]{\frac{\rho^2 g \sin \phi \, \lambda x^3}{4\mu k(T_s - T_w)}} \qquad (9.117)$$

These expressions give point values of h and Nu_x at $x = x$. It is seen that the coefficient decreases from a theoretical value of infinity at the top as the condensate film thickens. The mean value of the heat transfer coefficient over the whole surface, between $x = 0$ and $x = x$ is given by:

$$h_m = \frac{1}{x} \int_0^x h \, dx = \frac{1}{x} \int_0^x Kx^{-1/4} \, dx \quad \text{(where K is independent of x)}$$

$$= \frac{1}{x} K \frac{x^{3/4}}{\frac{3}{4}} = \frac{4}{3} Kx^{-1/4} = \frac{4}{3} h$$

$$= 0.943 \sqrt[4]{\frac{\rho^2 g \sin \phi \lambda k^3}{\mu x \Delta T_f}} \qquad (9.118)$$

where ΔT_f is the temperature difference across the condensate film.
For a vertical surface, $\sin \phi = 1$ and:

$$h_m = 0.943 \sqrt[4]{\frac{\rho^2 g \lambda k^3}{\mu x \Delta T_f}} \qquad (9.119)$$

9.6.2. Condensation on vertical and horizontal tubes

The Nusselt equation

If vapour is condensing on the outside of a vertical tube of diameter d_o, then the hydraulic mean diameter for the film

$$= \frac{4 \times \text{flow area}}{\text{wetted perimeter}} = \frac{4S}{b} \quad \text{(say)}$$

If G is the mass rate of flow of condensate, the mass rate of flow per unit area G' is G/S and the Reynolds number for the condensate film is then given by:

$$Re = \frac{(4S/b)(G/S)}{\mu} = \frac{4G}{\mu b} = \frac{4M}{\mu}$$

where M is the mass rate of flow of condensate per unit length of perimeter, i.e.:

$$M = \frac{G}{\pi d_o}$$

For streamline conditions in the film, $4M/\mu \not> 2100$ and:

$$h_m = \frac{Q}{A \Delta T_f} = \frac{G\lambda}{bl \Delta T_f} = \frac{\lambda M}{l \Delta T_f}.$$

From equation 9.119:

$$h_m = 0.943 \left(\frac{k^3 \rho^2 g}{\mu} \frac{\lambda}{l \Delta T_f} \right)^{1/4} = 0.943 \left(\frac{k^3 \rho^2 g}{\mu} \frac{h_m}{M} \right)^{1/4}$$

whence:

$$h_m \left(\frac{\mu^2}{k^3 \rho^2 g} \right)^{1/3} = 1.47 \left(\frac{4M}{\mu} \right)^{-1/3} \tag{9.120}$$

For horizontal tubes, Nusselt gives:

$$h_m = 0.72 \left(\frac{k^3 \rho^2 g \lambda}{d_o \mu \Delta T_f} \right)^{1/4} \tag{9.121}$$

This can be rearranged to give:

$$h_m \left(\frac{\mu^2}{k^3 \rho^2 g} \right)^{1/3} = 1.51 \left(\frac{4M}{\mu} \right)^{-1/3} \tag{9.122}$$

where M is the mass rate of flow per unit length of tube.

This is approximately the same as equation 9.119 for vertical tubes and is a universal equation for condensation if it is remembered that for vertical tubes $M = G/\pi d_o$ and for horizontal tubes $M = G/l$, where l is the length of the tube. Comparison of the two equations shows that, provided the length is more than three times the diameter, the horizontal tube will give a higher transfer coefficient for the same temperature conditions.

For j vertical rows of horizontal tubes, equation 9.121 is modified to give:

$$h_m = 0.72 \left(\frac{k^3 \rho^2 g \lambda}{j d_o \mu \Delta T_f} \right)^{1/4} \tag{9.123}$$

KERN[26] suggests that, based on the performance of commercial exchangers, this equation is too conservative and that the exponent of j should be nearer $-\frac{1}{6}$ than $-\frac{1}{4}$. (See Volume 6, Chapter 12.)

Experimental values

In testing Nusselt's equation it is important to see that the conditions comply with the requirements of the theory. In particular, it is necessary for the condensate to form a uniform film on the tubes, for the drainage of this film to be by gravity, and the flow streamline. Although some of these requirements have probably not been entirely fulfilled, results for pure vapours such as steam, benzene, toluene, diphenyl, ethanol, etc., are sufficiently near to give support to the theory. Some results by HASELDEN and PROSAD[45] for condensing oxygen and nitrogen vapours on a vertical surface, where precautions were taken to see that the conditions were met, are in very good agreement with Nusselt's theory. The results for most of the workers are within 15 per cent for horizontal tubes but tend to be substantially higher than the theoretical for vertical tubes. Some typical values are given in Table 9.7 taken from MCADAMS[25] and elsewhere.

Table 9.7. Average values of film coefficients h_m for condensation of pure saturated vapours on horizontal tubes

Vapour	Value of h_m (Btu/h ft^2 °F)	Value of h_m (W/m^2 K)	Range of ΔT_f (K)
Steam	1700–5000	10,000–28,000	1–11
Steam	3200–6500	18,000–37,000	4–37
Benzene	240–380	1400–2200	23–37
Diphenyl	220–400	1300–2300	4–15
Toluene	190–240	1100–1400	31–40
Methanol	500–600	2800–3400	8–16
Ethanol	320–450	1800–2600	6–22
Propanol	250–300	1400–1700	13–20
Oxygen	570–1400	3300–8000	0.08–2.5
Nitrogen	400–1000	2300–5700	0.15–3.5
Ammonia	1000	6000	—
Freon-12	200–400	1100–2200	—

When considering commercial equipment, there are several factors which prevent the true conditions of Nusselt's theory being met. Thus the temperature of the tube wall will not be constant, and for a vertical condenser with a ratio of ΔT at the bottom to ΔT at the top of five, the film coefficient should be increased by about 15 per cent.

Influence of vapour velocity

A high vapour velocity upwards tends to increase the thickness of the film and thus reduce h though the film may sometimes be disrupted mechanically as a result of the formation of small waves. For the downward flow of vapour, TEN BOSCH[46] has shown that h increases considerably at high vapour velocities and may increase to two or three times the value given by the Nusselt equation. It must be remembered that when a large fraction of the vapour is condensed, there may be a considerable change in velocity over the surface.

Under conditions of high vapour velocity CARPENTER and COLBURN[47] have shown that turbulence may occur with low values of the Reynolds number, e.g. 200–400. When the vapour velocity is high there will be an appreciable drag on the condensate film and the expression obtained for the heat transfer coefficient is difficult to manage.

Carpenter and Colburn have put forward a simple experimental correlation of their results for condensation at varying vapour velocities on the inner surface of a vertical tube:

$$h_m = 0.065 G'_m \sqrt{\frac{C_p \rho k (R'/\rho_v u^2)}{\mu \rho_v}} \qquad (9.124)$$

where:

$$G'_m = \sqrt{\frac{(G_1'^2 + G_1' G_2' + G_2'^2)}{3}}$$

and u is the velocity calculated from G'_m.

In this equation C_p, k, ρ, and μ are properties of the condensate and ρ_v refers to the vapour. G_1' is the mass rate of flow per unit area at the top of the tube and G_2' the corresponding value at the bottom. R' is the shear stress at the free surface of the condensate film.

As pointed out by COLBURN[48], the group $C_p \rho k / \mu \rho_v$ does not vary much for a number of organic vapours so that a plot of h_m and G'_m will provide a simple approximate correlation with separate lines for steam and for organic vapours as shown in Figure 9.34. Whilst this must be regarded as an empirical approximation it is very useful for obtaining a good idea of the effect of vapour velocity.

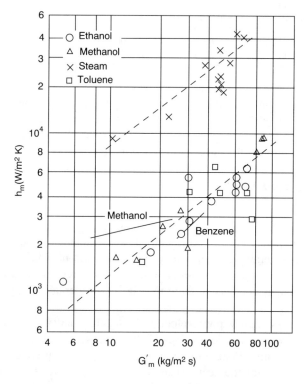

Figure 9.34. Average heat transfer data of CARPENTER and COLBURN[47] (shown as points) compared with those of TEPE and MUELLER[49] (solid lines). Dashed lines represent equation 9.124

Turbulence in the film

If Re is greater than 2100 during condensation on a vertical tube the mean coefficient h_m will increase. Data of KIRKBRIDE[50] and BADGER[51,52] for the condensation of diphenyl vapour and Dowtherm on nickel tubes were expressed in the form:

$$h_m \left(\frac{\mu^2}{k^3 \rho^2 g} \right)^{1/3} = 0.0077 \left(\frac{4M}{\mu} \right)^{0.4} \tag{9.125}$$

Comparing equation 9.120 for streamline flow of condensate and equation 9.122 for turbulent flow, it is seen that, with increasing Reynolds number, h_m decreases with streamline flow but increases with turbulent flow. These results are indicated in Figure 9.35.

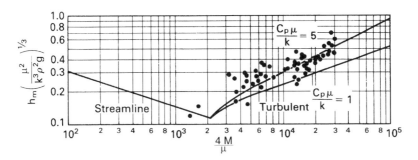

Figure 9.35.   Effects of turbulence in condensate film

In Volume 6, Chapter 12, design equations are given for condensation both inside and outside horizontal and vertical tubes, and the importance of avoiding flooding in vertical tubes is stressed.

9.6.3. Dropwise condensation

In the discussion so far, it has been assumed that the condensing vapour, on coming into contact with the cold surface, wets the tube so that a continuous film of condensate is formed. If the droplets initially formed do not wet the surface, after growing slightly they will fall from the tube exposing fresh condensing surface. This gives what is known as dropwise condensation and, since the heat does not have to flow through a film by conduction, much higher transfer coefficients are obtained. Steam is the only pure vapour for which definite dropwise condensation has been obtained, and values of h from 40 to 114 kW/m² K have been obtained, with much higher values on occasions. This question has been discussed by DREW, NAGLE and SMITH[53] who have shown that there are many materials which make the surface non-wettable; of these, only those which are firmly held to the surface are of any practical use. Mercaptans and oleic acid have been used to promote dropwise condensation, but at present there is little practical application of this technique. Exceptionally high values of h will not give a corresponding increase in the overall coefficient, since for a condenser with steam, a value of about 11 kW/m² K can be obtained with film condensation. On the other hand, it may be helpful in experimental work to reduce the thermal resistance on one side of a surface to negligible value.

9.6.4. Condensation of mixed vapours

In the previous discussion it has been assumed that the vapour is a pure material, such as steam or organic vapour. If it contains a proportion of non-condensable gas and is cooled below its dew point, a layer of condensate is formed on the surface with a mixture of non-condensable gas and vapour above it. The heat flow from the vapour to the surface then takes place by two methods. Firstly, sensible heat is passed to the surface because of the temperature difference. Secondly, since the concentration of vapour in the main stream is greater than that in the gas film at the condensate surface, vapour molecules diffuse to the surface and there condense, giving up their latent heat. The actual rate of condensation is then determined by the combination of these two effects, and its calculation requires a knowledge of mass transfer by diffusion discussed in Chapter 10.

In the design of a cooler-condenser for a vapour and permanent gas, the method of COLBURN and HOUGEN[54] is considered. The method requires the point to point calculation of the condensate-vapour interface conditions T_c and P_s. This calculation is performed by a trial and error solution of the equation:

$$q_v \quad + \quad q_\lambda \quad = \quad q_c \quad = U \Delta T \tag{9.126}$$
$$h_g(T_s - T_c) + k_G \lambda (P_g - P_s) = h_o(T_c - T_{cm}) = U \Delta T \tag{9.127}$$

where the first term q_v represents the sensible heat transferred to the condensing surface, the second term q_λ the latent heat transferred by the diffusing vapour molecules, and the third term q_c the heat transferred from the condensing surface through the pipe wall, dirt and scales, and water film to the cooling medium. h_g is the heat transfer coefficient over the gas film, h_o the conductance of the combined condensate film, tube wall, dirt and scale films, and the cooling medium film and U the overall heat transfer coefficient. T_s is the vapour temperature, T_c the temperature of the condensate, T_{cm} the cooling medium temperature, ΔT the overall temperature difference $= (T_s - T_{cm})$, P_g is the partial pressure of diffusing vapour, P_s the vapour pressure at T_c, λ the latent heat of vaporisation per unit mass, and k_G the mass transfer coefficient in mass per unit time, unit area and unit partial pressure difference.

To evaluate the required condenser area, point values of the group $U \Delta T$ as a function of q_c must be determined by a trial and error solution of equation 9.127. A plot of q_c against $1/U \Delta T$ upon graphical integration will yield the required condenser area. The method takes into account point variations in temperature difference, overall coefficient and mass velocities and consequently produces a reasonably good value for surface area required.

Individual terms in equation 9.127 will now be examined to enable a trial solution to proceed. Values for h_g and k_G are most conveniently obtained from the CHILTON and COLBURN[55] analogy discussed in Chapter 10.

Thus:
$$h_g = \frac{j_h G' C_p}{(C_p \mu / k)^{0.67}} \tag{9.128}$$

$$k_G = \frac{j_d G'}{P_{Bm}(\mu / \rho D)^{0.67}} \tag{9.129}$$

Values of j_h and j_d are obtained from a knowledge of the Reynolds number at a given point in the condenser. The combined conductance h_o is evaluated by determining the condensate film coefficient h_c from the Nusselt equation and combining this with the dirt and tube wall conductances and a cooling medium film conductance predicted from the Sieder-Tate relationships. Generally, h_o may be considered constant throughout the exchanger.

From a knowledge of h_g, k_G, and h_o and for a given T_s and T_{cm} values of the condensate surface temperature T_c can now be estimated until equation 9.127 is satisfied. The calculations are repeated, and in this manner several point values of the group $U \Delta T$ throughout the condenser may be obtained.

The design of a cooler condenser for the case of condensation of two vapours is more complicated than the preceding single vapour-permanent gas case[56]. For an example the reader is referred to the work of JEFFREYS[57].

As an example of the condensation of a vapour in the presence of a non-condensable gas, the following design is considered which is based on an example worked by KERN[26].

Example 9.13

A mixture of 0.57 kg/s of steam and 0.20 kg/s of carbon dioxide at 308 kN/m² and its dew point enters a heat exchanger consisting of 246 tubes, 19 mm o.d., wall thickness 1.65 mm, 3.65 m long, arranged in four passes on 25 mm square pitch in a 0.54 m diameter shell and leaves at 322 K. Condensation is effected by cooling water entering and leaving the unit at 300 and 319 K respectively. If the diffusivity of steam-carbon dioxide mixtures is 0.000011 m²/s and the group $(\mu/\rho D)^{0.67}$ may be taken to be constant at 0.62, estimate the overall coefficient of heat transfer and the dirt factor for the condenser.

Solution

In the steam entering the condenser, there is $\dfrac{0.57}{18} = 0.032$ kmol water

and $\dfrac{0.20}{44} = 0.0045$ kmol CO_2 $\bigg\}$ total $= 0.0365$ kmol.

∴ Partial pressure of water $= (308 \times 0.032/0.0365) = 270$ kN/m² and from Table 11A in the Appendix, the dew point $= 404$ K.

Mean molecular weight of the mixture $= (0.57 + 0.20)/0.0365 = 21.1$ kg/kmol.

At the *inlet*: vapour pressure of water $= 270$ kN/m²

inert pressure $= (308 - 270) = 38$ kN/m² $\bigg\}$ total $= 308$ kN/m²

At the *outlet*: partial pressure of water at 322 K $= 11.7$ kN/m²

inert pressure $= (308 - 11.7) = 296.3$ kN/m² $\bigg\}$ total $= 308$ kN/m²

∴ steam at the outlet $= \dfrac{0.0045 \times 11.7}{296.3} = 0.000178$ kmol

∴ steam condensed $= (0.032 - 0.000178) = 0.03182$ kmol.

The heat load will now be estimated at each interval between the temperatures 404, 401, 397, 380, 339 and 322 K.

For the interval 404 to 401 K

From Table 11A, partial pressure of steam at 401 K $= 252.2$ kN/m² and hence pressure of $CO_2 = (308 - 252.2) = 55.8$ kN/m². Steam remaining $= (0.0045 \times 252.2/55.8) = 0.0203$ kmol.

∴ Steam condensed $= (0.032 - 0.0203) = 0.0117$ kmol

Heat of condensation $= (0.0117 \times 18)(2180 + 1.93(404 - 401)) = 466$ kW

Heat from uncondensed steam $= (0.0203 \times 18 \times 1.93(404 - 401)) = 1.9$ kW

Heat from carbon dioxide $= (0.020 \times 0.92(404 - 401)) = 0.5$ kW

and total for interval $= 468.4$ kW

Heat balance:

Interval (K)	Heat load (kW)
404–401	468.4
401–397	323.5
397–380	343.5
380–339	220.1
339–322	57.9
Total	1407.3

and flow of water $= 1407.3/(4.187(319 - 300)) = 17.7$ kg/s.

With this flow of water and a flow area per pass of 0.0120 m^2, the mass velocity of water is 1425 kg/m^{2}s, equivalent to a velocity of 1.44 m/s at which $h_i = 6.36$ kW/m^2 K. Basing this on the outside area, $h_{i0} = 5.25$ kW/m^2 K.

Shell-side coefficient for entering gas mixture:

$$\text{Mean } C_p = \frac{(0.20 \times 0.92) + (0.57 \times 1.93)}{0.77} = 1.704 \text{ kJ/kg K.}$$

Similarly, the mean $k = 0.025$ kW/m K and mean $\mu = 0.015$ mN s/m^2

The area for flow through the shell $= 0.0411$ m^2 and the mass velocity on the shell side

$$= \frac{0.20 + 0.57}{0.0411} = 18.7 \text{ kg/m}^2\text{s}$$

Taking the equivalent diameter as 0.024 m, $Re = 29,800$

and: $\qquad\qquad\qquad\qquad h_g = 0.107$ kW/m^2 K or 107 W/m^2 K.

Now: $\qquad\qquad \left(\frac{\mu}{\rho D}\right)^{0.67} = 0.62, \quad \left(\frac{C_p \mu}{k}\right)^{0.67} = 1.01$

and: $\qquad\qquad k_G = \dfrac{h_g (C_p \mu / k)^{0.67}}{C_p P_{sF} (\mu / \rho D)^{0.67}} = \dfrac{107 \times 1.01}{1704 P_{sF} \times 0.62}$

$$= \frac{0.102}{P_{sF}}$$

At point 1

Temperature of the gas $T = 404$ K, partial pressure of steam $P_g = 270$ kN/m^2, partial pressure of the inert $P_s = 38$ kN/m^2, water temperature $T_w = 319$ K and $\Delta T = (404 - 319) = 85$ K. A guess is now made for the temperature of the condensate film of $T_c = 391$ K. In this case $P_s = 185.4$ kN/m^2 and $P_s' = (308 - 185.4) = 122.6$ kN/m^2.

$$\therefore \qquad\qquad P_{sF} = \frac{122.6 - 38}{\ln(122.6 - 38)} = 72.4 \text{ kN/m}^2.$$

In equation 9.127:

$$h_g(T_s - T_c) + k_G \lambda (P_g - P_s) = h_{i0}(T_c - T_{cm})$$

$$0.107(404 - 391) + \left(\frac{0.102}{724}\right) 2172(270 - 185.4) = 5.25(391 - 319)$$

$$259 = 378 \quad \text{i.e. no balance.}$$

Try $T_c = 378$ K, $P_s = 118.5$ kN/m^2, $P_g = (308 - 118.5) = 189.5$ kN/m^2

and:

$$P_{sF} = \frac{189.5 - 38}{\ln(189.5 - 38)} = 94.2 \text{ kN/m}^2$$

∴

$$0.107(404 - 378) + \left(\frac{0.102}{94.2}\right) 2172(270 - 118.5) = 5.25(378 - 319)$$

$$310 = 308 \quad \text{which checks out.}$$

∴

$$U\Delta T = 309 \text{ kW/m}^2 \quad \text{and} \quad U = \frac{309}{404 - 319}$$

$$= 3.64 \text{ kW/m}^2 \text{ K.}$$

Repeating this procedure at the various temperature points selected, the area may then be obtained as $A = \Sigma q/\Sigma U \Delta T$ or the area under a plot of Σq and $1/U\Delta T$, although the following tabulation approach is also acceptable.

Point	T_s (K)	T_c (K)	$U\Delta T$ (kW/m^2)	$(U\Delta T)_{ow}$ (kW/m^2)	q (kW)	$A = q/(U\Delta T)_{ow}$ (m^2)	ΔT (K)	ΔT_{ow} (K)	$q/\Delta T_{ow}$ (kW/K)
1	404	378	309	—	—	—	84.4	—	—
2	401	356	228	268.5	468.4	1.75	88.1	86.3	5.42
3	397	336	145	186.5	323.5	1.74	88.6	88.4	3.66
4	380	312	40.6	88.1*	343.5	3.89	76.7	82.7	4.15
5	339	302	5.4	17.5*	220.1	₁2.58	38.1	55.2*	4.00
6	322	300	2.1	3.5*	51.9	14.83	22.2	29.6*	1.75
				Total:	1407.3	34.8			18.98

*based on LMTD.

If no condensation takes place, the logarithmic mean temperature difference is 46.6 K. In practice the value $= (1407.3/18.98) = 74.2$ K.

Assuming no scale resistance, the overall coefficient $= \dfrac{1407.3}{34.8 \times 74.2} = 0.545$ kW/m^2 K.

The available surface area on the outside of the tubes $= 0.060$ m^2/m

or $(246 \times 3.65 \times 0.060) = 53.9$ m^2

The actual coefficient is therefore $\dfrac{1407.3}{53.9 \times 74.2} - 0.352$ kW/m^2 K.

and the dirt factor $= \dfrac{0.545 - 0.352}{0.545 \times 0.352} = \underline{\underline{1.01 \text{ m}^2\text{K/kW.}}}$

As shown in Figure 9.36, the clean coefficient varies from $\underline{\underline{3.64 \text{ kW/m}^2 \text{ K}}}$ at the inlet to $\underline{\underline{0.092 \text{ kW/m}^2 \text{ Ka}}}$ at the outlet.

Condensation of mixed vapours is considered further in Volume 6, Chapter 12, where it is suggested that the local heat transfer coefficient can be expressed in terms of the local gas-film and condensate-film coefficients. For partial condensation where:

(i) *non-condensables < 0.5 per cent*; their effect can be ignored,
(ii) *non-condensables > 70 per cent*; the heat transfer can be taken as being by forced convection alone, and
(iii) *non-condensables 0.5–70 per cent*; both mechanisms are effective.

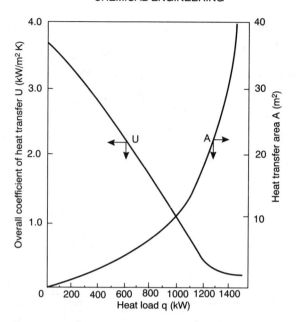

Figure 9.36.   Results for Example 9.13

9.7. BOILING LIQUIDS

9.7.1 Conditions for boiling

In processing plants, liquids are boiled either on submerged surfaces or on the inside of vertical tubes. Mechanical agitation may be applied in the first case and, in the second, the liquid may be driven through the tubes by means of an external pump. The boiling of liquids under either of these conditions normally leads to the formation of vapour first in the form of bubbles and later as a distinct vapour phase above a liquid interface. The conditions for boiling on the submerged surface will be discussed in this chapter and the problems arising with boiling inside tubes are considered in Volume 2, Chapter 14. Much of the fundamental work on the ideas of boiling has been presented by WESTWATER[58] and JAKOB[59], and subsequently by ROHSENOW and CLARK[60] and ROHSENOW[61] and by FORSTER[62]. The boiling of solutions in which a solid phase is separated after evaporation has proceeded to a sufficient extent is considered in Volume 2.

For a bubble to be formed in a liquid, for instance steam in water, it is necessary for a surface of separation to be produced. Kelvin has shown that, as a result of the surface tension between the liquid and vapour, the vapour pressure on the inside of a concave surface will be less than that at a plane surface. As a result, the vapour pressure P_r inside the bubble is less than the saturation vapour pressure P_s at a plane surface. The relation between P_r and P_s is:

$$P_r = P_s - \left(\frac{2\sigma}{r}\right) \tag{9.130}$$

where r is the radius of curvature of the bubble, and σ is the surface tension.

Hence the liquid must be superheated near the surface of the bubble, the extent of the superheat increasing with decrease in the radius of the bubble. On this basis it follows that very small bubbles are difficult to form without excessive superheat. The formation of bubbles is made much easier by the fact that they will form on curved surfaces or on irregularities on the heating surface, so that only a small degree of superheat is normally required.

Nucleation at much lower values of superheat is believed to arise from the presence of existing nuclei such as non-condensing gas bubbles, or from the effect of the shape of the cavities in the surface. Of these, the current discussion on the influence of cavities is the most promising. In many cavities the angle θ will be greater than $90°$ and the effective contact angle, which includes the contact angle of the cavity β, will be considerably greater $[= \theta + (180 - \beta)/2]$, so that a much-reduced superheat is required to give nucleation. Thus the size of the mouth of the cavity and the shape of the cavity plays a significant part in nucleation[63].

It follows that for boiling to occur a small difference in temperature must exist between the liquid and the vapour. JAKOB and FRITZ[64] have measured the temperature distribution for water boiling above an electrically heated hot plate. The temperature dropped very steeply from about 383 K on the actual surface of the plate to 374 K about 0.1 mm from it. Beyond this the temperature was reasonably constant until the water surface was reached. The mean superheat of the water above the temperature in the vapour space was about 0.5 K and this changed very little with the rate of evaporation. At higher pressures this superheating became smaller becoming 0.2 K at 5 MN/m^2 and 0.05 K at 101 MN/m^2. The temperature drop from the heating surface, however, depends very much on the rate of heat transfer and on the nature of the surface. Thus in order to maintain a heat flux of about 25.2 kW/m^2, a temperature difference of only 6 K was required with a rough surface as against 10.6 K with a smooth surface. The heat transfer coefficient on the boiling side is therefore dependent on the nature of the surface and on the difference in temperature available. For water boiling on copper plates JAKOB and FRITZ[64] give the following coefficients for a constant temperature difference of 5.6 K, with different surfaces:

(1) Surface after 8 h (28.8 ks) use and 48 h (172.8 ks)
    immersion in water $h = 8000$ W/m^2 K
(2) Freshly sandblasted $h = 3900$ W/m^2 K
(3) Sandblasted surface after long use $h = 2600$ W/m^2 K
(4) Chromium plated $h = 2000$ W/m^2 K

The initial surface, with freshly cut grooves, gave much higher figures than case (1) above.

The nature of the surface will make a marked difference to the physical form of the bubble and the area actually in contact with the surface, as shown in Figure 9.37.

(a) *Non-wettable surface.* Here the vapour bubbles spread out thus reducing the area available for heat transfer from the hot surface to the liquid.

(b) *Partially wettable surface*—which is the commonest form. Here the bubbles rise from a larger number of sites and the rate of transfer is increased.

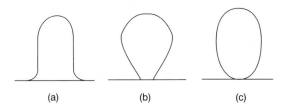

Figure 9.37.   Shapes of bubbles (*a*) screen surface — thin oil layer (*b*) chromium plated and polished surface
(*c*) screen surface — clean

(c) *Entirely wetted surface*, such as that formed by a screen. This gives the minimum area of contact between vapour and surface and the bubbles leave the surface when still very small. It will therefore follow that if the liquid has detergent properties this may give rise to much higher rates of heat transfer.

9.7.2. Types of boiling

Interface evaporation

In boiling liquids on a submerged surface it is found that the heat transfer coefficient depends very much on the temperature difference between the hot surface and the boiling liquid. The general relation between the temperature difference and heat transfer coefficient was first presented by NUKIYAMA[65] who boiled water on an electrically heated wire. His results have been confirmed and extended by others, and Figure 9.38 shows the data of FARBER and SCORAH[66]. This relationship is complex and is best considered in stages.

In interface evaporation, the bubbles of vapour formed on the heated surface move to the vapour-liquid interface by natural convection and exert very little agitation on the liquid. The results are given by:

$$Nu = 0.61(Gr\ Pr)^{1/4} \tag{9.131}$$

which may be compared with the expression for natural convection:

$$Nu = C'(Gr\ Pr)^n \tag{equation 9.91}$$

where $n = 0.25$ for streamline conditions and $n = 0.33$ for turbulent conditions.

Nucleate boiling

At higher values of ΔT the bubbles form more rapidly and form more centres of nucleation. Under these conditions the bubbles exert an appreciable agitation on the liquid and the heat transfer coefficient rises rapidly. This is the most important region for boiling in industrial equipment.

Film boiling

With a sufficiently high value of ΔT, the bubbles are formed so rapidly that they cannot get away from the hot surface, and they therefore form a blanket over the surface. This

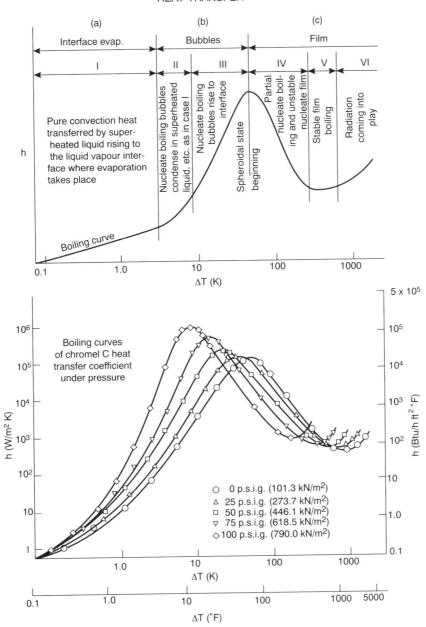

Figure 9.38.   Heat transfer results of FARBER and SCORAH[66]

means that the liquid is prevented from flowing on to the surface by the bubbles of vapour and the coefficient falls. The maximum coefficient occurs during nucleate boiling but in an unstable region for operation. In passing from the nucleate boiling region to the film boiling region, two critical changes occur in the process. The first manifests itself in a decrease in the heat flux, the second is the prelude to stable film boiling. The intermediate region is generally known as the transition region. It should be noted that the first change in the process is an important hydrodynamic phenomenon which is common to other two-phase systems, for example, *flooding* in countercurrent gas-liquid or vapour-liquid systems.

With very high values of ΔT, the heat transfer coefficient rises again because of heat transfer by radiation. These very high values are rarely achieved in practice and the aim will be to operate the plant at a temperature difference a little below the value giving the maximum heat transfer coefficient.

9.7.3. Heat transfer coefficients and heat flux

The values of the heat transfer coefficients for low values of temperature difference have been given by equation 9.131. Figure 9.39 shows the values of h and for q for water boiling on a submerged surface. Whilst the actual values vary somewhat between observers, they all give a maximum for a temperature difference of about 22 K. The maximum value of h is about 50 kW/m^2 K and the flux about 1100 kW/m^2.

Similar results have been obtained by BONILLA and PERRY[67], INSINGER and BLISS[68], and others for a number of organic liquids such as benzene, alcohols, acetone, and carbon

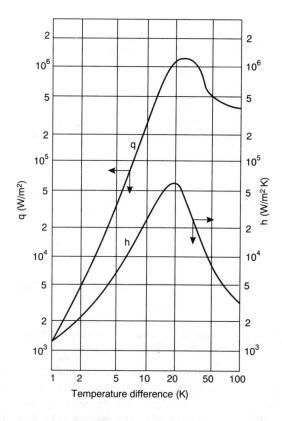

Figure 9.39.   Effect of temperature difference on heat flux and heat transfer coefficient to water boiling at 373 K on a submerged surface

tetrachloride. The data in Table 9.8 for liquids boiling at atmospheric pressure show that the maximum heat flux is much smaller with organic liquids than with water and the temperature difference at this condition is rather higher. In practice the critical value of ΔT may be exceeded. SAUER et al.[69] found that the overall transfer coefficient U for boiling ethyl acetate with steam at 377 kN/m^2 was only 14 per cent of that when the steam pressure was reduced to 115 kN/m^2.

Table 9.8.   Maximum heat flux for various liquids boiling at atmospheric pressure

Liquid	Surface	Critical ΔT (K)	Maximum flux (kW/m^2)
Water	Chromium	25	910
50 mol% ethanol-water	Chromium	29	595
Ethanol	Chromium	33	455
n-Butanol	Chromium	44	455
iso-Butanol	Nickel	44	370
Acetone	Chromium	25	455
iso-Propanol	Chromium	33	340
Carbon tetrachloride	Copper	—	180
Benzene	Copper	—	170–230

In considering the problem of nucleate boiling, the nature of the surface, the pressure, and the temperature difference must be taken into consideration as well as the actual physical properties of the liquid.

Apart from the question of scale, the nature of the clean surface has a pronounced influence on the rate of boiling. Thus BONILLA and PERRY[67] boiled ethanol at atmospheric pressure and a temperature difference of 23 K, and found the heat flux at atmospheric pressure to be 850 kW/m^2 for polished copper, 450 for gold plate, and 370 for fresh chromium plate, and only 140 for old chromium plate. This wide fluctuation means that care must be taken in judging the anticipated heat flux, since the high values that may be obtained initially will not be obtained in practice because of tarnishing of the surface.

Effect of temperature difference

CRYDER and FINALBORGO[70] boiled a number of liquids on a horizontal brass surface, both at atmospheric and at reduced pressure. Some of their results are shown in Figure 9.40, where the coefficient for the boiling liquid h is plotted against the temperature difference between the hot surface and the liquid. The points for the various liquids in Figure 9.40 lie on nearly parallel straight lines, which can be represented by:

$$h = \text{constant} \times \Delta T^{2.5} \qquad (9.132)$$

This value for the index of ΔT has been found by other workers, but JAKOB and LINKE[71] found values as high as 4 for some of their work. It is important to note that this value of 2.5 is true only for temperature differences up to 19 K.

In some ways it is more convenient to show the results in the form of heat flux versus temperature difference. This is shown in Figure 9.41, where some results from a number of workers are given.

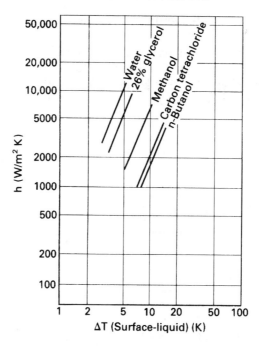

Figure 9.40.   Effect of temperature difference on heat transfer coefficient for boiling liquids
(CRYDER and FINALBORGO[70])

Figure 9.41.   Effect of temperature difference on heat flux to boiling liquids (BONILLA and PERRY[67])

Effect of pressure

CRYDER and FINALBORGO[71] found that h decreased uniformly as the pressure and hence the boiling point was reduced, according to the relation $h = \text{constant} \times B^{T''}$, where T'' is numerically equal to the temperature in K and B is a constant. Combining this with equation 9.132, they gave their results for h in the empirical form:

$$h = \text{constant} \times \Delta T^{2.5} B^{T''}$$

or, using SI units: $\log \left(\dfrac{h}{5.67} \right) = a' + 2.5 \log \Delta T + b'(T'' - 273)$ (9.133)

where $(T'' - 273)$ is now in $°$C.

If a' and b' are given the following values, h is expressed in W/m² K:

	a'	b'		a'	b'
Water	−0.96	0.025	Kerosene	−4.13	0.022
Methanol	−1.11	0.027	10% Na_2SO_4	−1.47	0.029
CCl₄	−1.55	0.022	24% NaCl	−2.43	0.031

The values of a' will apply only to a given apparatus but a value of b' of 0.025 is of more general application. If h_n is the coefficient at some standard boiling point T_n, and h at some other temperature T, equation 9.133 can be rearranged to give:

$$\log \frac{h}{h_n} = 0.025(T'' - T''_n)$$ (9.134)

for a given material and temperature difference.

As the pressure is raised above atmospheric pressure, the film coefficient increases for a constant temperature difference. CICHELLI and BONILLA[72] have examined this problem for pressures up to the critical value for the vapour, and have shown that ΔT for maximum rate of boiling decreases with the pressure. They obtained a single curve (Figure 9.42) by plotting q_{max}/P_c against P_R, where P_c is the critical pressure and P_R the reduced pressure $= P/P_c$. This curve represented the data for water, ethanol, benzene, propane, n-heptane, and several mixtures with water. For water their results cover only a small range of P/P_c because of the high value of P_c. For the organic liquids investigated, they were able to show that the maximum value of heat flux q occurs at a pressure P of about one-third of the critical pressure P_c. The range of physical properties of the organic liquids is not wide (Table 9.9), and further data are required to substantiate the previous relation.

9.7.4. Analysis based on bubble characteristics

Why are such high values of heat flux obtained with the boiling process? It was once thought that the bubbles themselves were carriers of latent heat which was added to the liquid by their movement. It has now been shown, by determining the numbers of bubbles, that this mechanism would result in the transfer of only a moderate part of the heat that is actually transferred. The current views of the mechanism are that the high flux arises

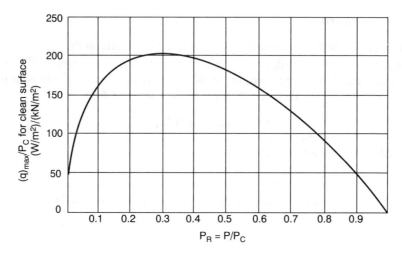

Figure 9.42. Effect of pressure on maximum heat flux in nucleate boiling

Table 9.9. Heat transfer coefficients for boiling liquids

Liquid	Boiling point (K)	ΔT (K)	h (Btu/h ft^2 °F)	(W/m^2 K)
Water	372	4.7	1600	9000
	372	2.9	500	2700
	326	8.8	850	4700
	326	6.1	250	1300
Methanol	337	8.9	850	4800
	337	5.6	250	1500
	306	14.4	500	3000
	306	9.3	150	900
Carbon tetrachloride	349	12.6	600	3500
	349	7.2	200	1100
	315	20.1	400	2000
	315	11.8	100	700

from the agitation produced by the bubbles, and two rather different explanations have been put forward. ROHSENOW and CLARK[60] and ROHSENOW[61] base their argument on the condition of the bubble on leaving the hot surface. Thus by calculating the velocity and size of the bubble he is able to derive an expression for the heat transfer coefficient in the form of a Nusselt type equation, relating the Nusselt group to the Reynolds and Prandtl groups. FORSTER and ZUBER[73,74], however, argue that the important velocity is that of the growing bubble, and this is the term used to express the velocity. In either case the bubble movement is vital for obtaining a high flux. The liquid adjacent to the surface is agitated and exerts a mixing action by pushing hot liquid from the surface to the bulk of the stream.

Considering in more detail the argument proposed by ROHSENHOW and CLARK[60] and ROHSENHOW[61], the size of a bubble at the instant of breakaway from the surface has been determined by FRITZ[75] who showed that d_b was given by the expression:

$$d_b = C_1\phi \left(\frac{2\sigma}{g(\rho_l - \rho_v)} \right)^{1/2} \tag{9.135}$$

where σ is the surface tension, ρ_l and ρ_v the density of the liquid and vapour, ϕ is the contact angle, and C_1 is a constant depending on conditions.

The flowrate of vapour per unit area as bubbles u_b is given by:

$$u_b = \frac{f n \pi d_b^3}{6} \tag{9.136}$$

where f is the frequency of bubble formation at each bubble site and n is the number of sites of nucleation per unit area.

The heat transferred by the bubbles q_b is to a good approximation given by:

$$q_b = \tfrac{1}{6}\pi d_b^3 f n \rho_v \lambda \tag{9.137}$$

where λ is the latent heat of vaporisation.

It has been shown that for heat flux rates up to 3.2 kW/m^2 the product $f d_b$ is constant and that the total heat flow per unit area q is proportional to n. From equation 9.137 it is seen that q_b is proportional to n at a given pressure, so that $q \propto q_b$.

Hence:
$$q = C_2 \frac{\pi}{6} d_b^3 f n \rho_v \lambda \tag{9.138}$$

Substituting from equations 9.136 and 9.138 the mass flow per unit area:

$$\rho_v u_b = f n \frac{\pi}{6} d_b^3 \rho_v = \frac{q}{C_2 \lambda} \tag{9.139}$$

Then a Reynolds number for the bubble flow which represents the term for agitation may be defined by:

$$Re_b = \frac{d_b \rho_v u_b}{\mu_l}$$
$$= C_1\phi \left(\frac{2\sigma}{g(\rho_l - \rho_v)} \right)^{1/2} \left(\frac{q}{C_2\lambda} \right) \frac{1}{\mu_l}$$
$$= C_3\phi \frac{q}{\lambda \mu_l} \left(\frac{\sigma}{g(\rho_l - \rho_v)} \right)^{1/2} \tag{9.140}$$

The Nusselt group for bubble flow $Nu_b = h_b C_1 \dfrac{\phi}{k_l} \left(\dfrac{2\sigma}{g(\rho_l - \rho_v)} \right)^{1/2}$

$$= C_4 h_b \frac{\phi}{k_l} \left(\frac{\sigma}{g(\rho_l - \rho_v)} \right)^{1/2} \tag{9.141}$$

Then a final correlation is obtained in the form:

$$Nu_b = \text{constant } Re_b^n Pr^m \tag{9.142}$$

or:
$$Nu_b = \text{constant} \left[\frac{C_3 \phi q}{\mu_l \lambda} \left(\frac{\sigma}{g(\rho_l - \rho_v)} \right)^{1/2} \right]^n \left(\frac{C_l \mu_l}{k_l} \right)^m \qquad (9.143)$$

where n and m have been found from experimental data to be 0.67 and -0.7 respectively and the constant, which depends on the metal surface, ranges from 67–100 for polished chromium, 77 for platinum wire and 166 for brass.[61]

FORSTER and ZUBER[73,74] employ a similar basic approach but use the radial rate of growth dr/dt for the bubble velocity in the Reynolds group. They are able to show that:

$$\frac{dr}{dt} = \frac{\Delta T C_l \rho_l}{2\lambda \rho_v} \left(\frac{\pi D_{Hl}}{t} \right)^{1/2} \qquad (9.144)$$

where D_{Hl} is the thermal diffusivity $(k_l/C_l \rho_l)$ of the liquid. Using this method they present a final correlation in the form of equation 9.142.

Although these two forms of analysis give rise to somewhat similar expressions, the basic terms are evaluated in quite different ways and the final expressions show many differences. Some data fit the Rohsenow equation reasonably well and other data fit Forster's equation.

These expressions all indicate the importance of the bubbles on the rate of transfer, but as yet they have not been used for design purposes. INSINGER and BLISS[68] made the first approach by dimensional analysis and MCNELLY[76] has subsequently obtained a more satisfactory result. The influence of ΔT is taken into account by using the flux q, and the last term allows for the change in volume when the liquid vaporises. He was able to obtain the following expression, the numerical values of the indices being deduced from existing data:

$$\frac{hd}{k_l} = 0.225 \left(\frac{C_l \mu_l}{k_l} \right)^{0.69} \left(\frac{qd}{\lambda \mu} \right)^{0.69} \left(\frac{Pd}{\sigma} \right)^{0.31} \left(\frac{\rho_l}{\rho_v} - 1 \right)^{0.33} \qquad (9.145)$$

9.7.5. Sub-cooled boiling

If bubbles are formed in a liquid which is much below its boiling point, then the bubbles will collapse in the bulk of the liquid. Thus if a liquid flows over a very hot surface then the bubbles formed are carried away from the surface by the liquid and sub-cooled boiling exists. Under these conditions a very large number of small bubbles are formed and a very high heat flux is obtained. Some results for these conditions are given in Figure 9.43.

If a liquid flows through a tube heated on the outside then the heat flux q will increase with ΔT as shown in Figure 9.43. Beyond a certain value of ΔT the increase in q is very rapid. If the velocity through the tube is increased, then a similar plot is obtained with a higher value of q at low values of ΔT and then the points follow the first line. Over the first section, forced convection boiling exists where an increase in Reynolds number does not bring about a very great increase in q because the bubbles are themselves producing agitation in the boundary layer near the wall. Over the steep section, sub-cooled boiling exists where the velocity is not important provided it is sufficient to remove the bubbles rapidly from the surface. In the same way, mechanical agitation of a liquid boiling on a submerged surface will not markedly increase the heat flux.

Figure 9.43. Heat flux in sub-cooled boiling

9.7.6. Design considerations

In the design of vaporisers and reboilers, two types of boiling are important: that is nucleate boiling in a pool of liquid as in a kettle-type reboiler or a jacketed vessel, and convective boiling which occurs where the vaporising liquid flows over a heated surface and heat transfer is by both forced convection and nucleate boiling as, for example, in forced circulation or thermosyphon reboilers. The following discussion is a summary of that given in Volume 6, Chapter 12, where a worked example is given.

In the absence of experimental data, a correlation given by FORSTER and ZUBER[74] may be used to estimate *pool boiling* coefficients, but the following reduced pressure correlation given by MOSTINSKI[77] is much simpler to use and gives reliable results for h (in W/m^2 K):

$$h = 0.104 P_c^{0.69} q^{0.7} \left[1.8 \left(\frac{P}{P_c} \right)^{0.17} + 4 \left(\frac{P}{P_c} \right)^{1.2} + 10 \left(\frac{P}{P_c} \right)^{10} \right] \qquad (9.146)$$

P_c and P are the critical and operating pressures (bar), respectively, and q is the heat flux (W/m^2). Both equations are for single component fluids, but they can also be used for close-boiling mixtures and for wider boiling ranges with a factor of safety. In reboiler and vaporiser design, it is important that the heat flux is well below the critical value. A correlation is given for the heat transfer coefficient for the case where film-boiling takes place on tubes submerged in the liquid.

Convective boiling, which occurs when the boiling liquid flows through a tube or over a tube bundle, depends on the state of the fluid at any point. The effective heat transfer coefficient can be considered to be made up of the convective and nucleate boiling components. The convective boiling coefficient is estimated using an equation for single-phase forced-convection heat transfer (equation 9.54, for example) modified by a factor to allow for the effects of two-phase flow. Similarly, the nucleate boiling coefficient is obtained from the Forster and Zuber or Mostinski correlation, modified by a factor dependent on the liquid Reynolds number and on the effects of two-phase flow. The estimation of convective boiling coefficients is illustrated by means of an Example in Volume 6, Chapter 12.

One of the most important areas of application of heat transfer to boiling liquids is in the use of evaporators to effect an increase in the concentration of a solution. This topic is considered in Volume 2, Chapter 14.

For vaporising the liquid at the bottom of a distillation column a reboiler is used, as shown in Figure 9.44. The liquid from the still enters the boiler at the base, and, after flowing over the tubes, passes out over a weir. The vapour formed, together with entrained liquid, passes from the top of the unit to the column. The liquid flow may be either by gravity or by forced circulation. In such equipment, provision is made for expansion of the tubes either by having a floating head as shown, or by arranging the tubes in the form of a hairpin bend (Figure 9.45). A vertical reboiler may also be used with steam condensing on the outside of the tube bundle. With all systems it is undesirable to vaporise more than a small percentage of the feed since a good liquid flow over the tubes is necessary to avoid scale building up.

Figure 9.44. Reboiler installed on distillation column

Figure 9.45. Reboiler with hairpin tubes

In the design of *forced convection* reboilers, the normal practice is to calculate the heat transfer coefficient on the assumption that heat is transferred by forced convection only, and this gives safe values. KERN[26] recommends that the heat flux should not exceed 60 kW/m^2 for organics and 90 kW/m^2 for dilute aqueous solutions. In *thermosyphon reboilers*, the fluid circulates at a rate at which the pressure losses in the system are just balanced by the hydrostatic head and the design involves an iterative procedure based on an assumed circulation rate through the exchanger. *Kettle reboilers*, such as that shown in Figure 9.45, are essentially pool boiling devices and their design, based on nucleate boiling data, uses the Zuber equation for single tubes, modified by a tube-density factor. This general approach is developed further in Volume 6, Chapter 12.

9.8. HEAT TRANSFER IN REACTION VESSELS

9.8.1. Helical cooling coils

A simple jacketed pan or kettle is very commonly used in the chemical industries as a reaction vessel. In many cases, such as in nitration or sulphonation reactions, heat has to be removed or added to the mixture in order either to control the rate of reaction or to bring it to completion. The addition or removal of heat is conveniently arranged by passing steam or water through a jacket fitted to the outside of the vessel or through a helical coil fitted to the inside. In either case some form of agitator is used to obtain even distribution in the vessel. This may be of the anchor type for very thick mixes or a propeller or turbine if the contents are not too viscous.

In this case, the thermal resistances to heat transfer arise from the water film on the inside of the coil, the wall of the tube, the film on the outside of the coil, and any scale that may be present on either surface. The overall transfer coefficient may be expressed by the relation:

$$\frac{1}{UA} = \frac{1}{h_i A_i} + \frac{x_w}{k_w A_w} + \frac{1}{h_o A_o} + \frac{R_o}{A_o} + \frac{R_i}{A_i} \qquad (9.147)$$

where R_o and R_i are the scale resistances and the other terms have the usual meaning.

Inside film coefficient

The value of h_i can be found from a form of equation 9.54:

$$\frac{h_i d}{k} = 0.023 \left(\frac{du\rho}{\mu}\right)^{0.8} \left(\frac{C_p\mu}{k}\right)^{0.4} \tag{9.148}$$

if water is used in the coil, and the Sieder and Tate equation 9.60 if a viscous brine is used for cooling.

These equations have been obtained for straight tubes; with a coil somewhat greater transfer is obtained for the same physical conditions. JESCHKE[78] cooled air in a 31 mm steel tube wound in the form of a helix and expressed his results in the form:

$$h_i(\text{coil}) = h_i(\text{straight pipe}) \left(1 + 3.5\frac{d}{d_c}\right) \tag{9.149}$$

where d is the inside diameter of the tube and d_c the diameter of the helix. PRATT[79] has examined this problem in greater detail for liquids and has given almost the same result. Combining equations 9.148 and 9.149, the inside film coefficient h_i for the coil may be calculated.

Outside film coefficient

The value of h_0 is determined by the physical properties of the liquor and by the degree of agitation achieved. This latter quantity is difficult to express in a quantitative manner and the group $L^2N\rho/\mu$ has been used both for this problem and for the allied one of power used in agitation (see Chapter 7). In this group L is the length of the paddle and N the revolutions per unit time. CHILTON, DREW and JEBENS[80], working with a small tank only 0.3 m in diameter d_v, expressed their results in the form:

$$\frac{h_0 d_v}{k} \left(\frac{\mu_s}{\mu}\right)^{0.14} = 0.87 \left(\frac{C_p\mu}{k}\right)^{1/3} \left(\frac{L^2N\rho}{\mu}\right)^{0.62} \tag{9.150}$$

where the factor $(\mu_s/\mu)^{0.14}$ was introduced to cover the difference between the viscosity adjacent to the coil (μ_s) and that in the bulk of the liquor. They obtained a range in physical properties by using water, two oils, and glycerol.

PRATT[79] used both circular and square tanks up to 0.6 m across and a series of different arrangements of a simple paddle (Figure 9.46). He also examined the influence of alteration in the arrangement of the coil and varied the tube diameter d_o, the gap between the turns d_g, the diameter of the helix d_c, the height of the coil d_p, and the width of the stirrer W. His final equations for cylindrical tanks were:

$$\frac{h_0 d_v}{k} = 34 \left(\frac{L^2N\rho}{\mu}\right)^{0.5} \left(\frac{C_p\mu}{k}\right)^{0.3} \left(\frac{d_g}{d_p}\right)^{0.8} \left(\frac{W}{d_c}\right)^{0.25} \left(\frac{L^2 d_v}{d_o^3}\right)^{0.1} \tag{9.151}$$

and for square tanks:

$$\frac{h_0 l_v}{k} = 39 \left(\frac{L^2N\rho}{\mu}\right)^{0.5} \left(\frac{C_p\mu}{k}\right)^{0.3} \left(\frac{d_g}{d_p}\right)^{0.8} \left(\frac{W}{d_c}\right)^{0.25} \left(\frac{L^2 l_v}{d_o^3}\right)^{0.1} \tag{9.152}$$

where l_v is the length of the side of the vessel.

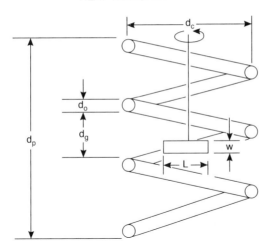

Figure 9.46.   Arrangement of coil in Pratt's work[79]

These give almost the same results as the earlier equations over a wide range of conditions. CUMMINGS and WEST[81] have tested these results with a much larger tank of 0.45 m³ capacity and have given an expression similar to equation 9.150 but with a constant of 1.01 instead of 0.87. They used a retreating blade turbine impeller and in many cases mounted a second impeller above the first; this agitation was probably more intense than that given by the other workers. A constant of 0.9 seems a reasonable average from existing work.

Example 9.14

Toluene is continuously nitrated to mononitro-toluene in a cast-iron vessel, 1 m diameter, fitted with a propeller agitator 0.3 m diameter rotating at 2.5 Hz. The temperature is maintained at 310 K by circulating 0.5 kg/s cooling water through a stainless steel coil 25 mm o.d. and 22 mm i.d. wound in the form of a helix, 0.80 m in diameter. The conditions are such that the reacting material may be considered to have the same physical properties as 75 per cent sulphuric acid. If the mean water temperature is 290 K, what is the overall coefficient of heat transfer?

Solution

The overall coefficient U_o based on the outside area of the coil is obtained from equation 9.147:

$$\frac{1}{U_o} = \frac{1}{h_o} + \frac{x_w d_o}{k_w d_w} + \frac{d_o}{h_i d} + R_o + \frac{R_i d_o}{d}$$

where d_w is the mean diameter of the pipe.

From equations 9.148 and 9.149 the inside film coefficient for the water is given by:

$$h_i = \frac{k}{d}\left(1 + 3.5\frac{d}{d_c}\right) 0.023 \left(\frac{du\rho}{\mu}\right)^{0.8} \left(\frac{C_p\mu}{k}\right)^{0.4}$$

In this equation:

$$u\rho = \frac{0.5}{(\pi/4) \times 0.022^2} = 1315 \text{ kg/m}^2\text{s}$$

$$d = 0.022 \text{ m}$$

$$d_c = 0.80 \text{ m}$$

$$k = 0.59 \text{ W/m K}$$

$$\mu = 1.08 \text{ mN s/m}^2 \quad \text{or} \quad 1.08 \times 10^{-3} \text{ N s/m}^2$$

$$C_p = 4.18 \times 10^3 \text{ J/kg K}$$

$$\therefore \quad h_i = \frac{0.59}{0.022}\left(1 + 3.5 \times \frac{0.022}{0.80}\right) 0.023 \left(\frac{0.022 \times 1315}{1.08 \times 10^{-3}}\right)^{0.8} \left(\frac{4.18 \times 10^3 \times 1.08 \times 10^{-3}}{0.59}\right)^{0.4}$$

$$= 0.680(26,780)^{0.8}(7.65)^{0.4} = 5490 \text{ W/m}^2 \text{ K}$$

The external film coefficient is given by equation 9.150:

$$\frac{h_o d_v}{k}\left(\frac{\mu_s}{\mu}\right)^{0.14} = 0.87 \left(\frac{C_p \mu}{k}\right)^{0.33} \left(\frac{L^2 N \rho}{\mu}\right)^{0.62}$$

For 75 per cent sulphuric acid:

$$k = 0.40 \text{ W/m K}$$

$$\mu_s = 8.6 \times 10^{-3} \text{ N s/m}^2 \text{ at } 300 \text{ K}$$

$$\mu = 6.5 \times 10^{-3} \text{ N s/m}^2 \text{ at } 310 \text{ K}$$

$$C_p = 1.88 \times 10^3 \text{ J/kg K}$$

$$\rho = 1666 \text{ kg/m}^3$$

$$\therefore \quad \frac{h_o \times 1.0}{0.40}\left(\frac{8.6}{6.5}\right)^{0.14} = 0.87 \left(\frac{1.88 \times 10^3 \times 6.5 \times 10^{-3}}{0.40}\right)^{0.33} \left(\frac{0.3^2 \times 2.5 \times 1666}{6.5 \times 10^{-3}}\right)^{0.62}$$

$$2.5 h_o \times 1.04 = 0.87 \times 3.09 \times 900$$

$$\therefore \quad h_o = 930 \text{ W/m}^2 \text{ K}$$

Taking $k_w = 15.9$ W/m K and R_o and R_i as 0.0004 and 0.0002 m^2 K/W:

$$\frac{1}{U_o} = \frac{1}{930} + \frac{0.0015 \times 0.025}{15.9 \times 0.0235} + \frac{0.025}{5490 \times 0.022} + 0.0004 + \frac{0.0002 \times 0.025}{0.022}$$

$$= 0.00107 + 0.00010 + 0.00021 + 0.00040 + 0.00023 = 0.00201$$

$$\therefore \quad U_o = 498 \text{ W/m}^2 \text{ K}$$

In this calculation a mean area of surface might have been used with sufficient accuracy. It is important to note the great importance of the scale terms which together form a major part of the thermal resistance.

9.8.2. Jacketed vessels

For many purposes, heating or cooling of a reaction mixture is most satisfactorily achieved by condensing steam in a jacket or passing water through it; this is commonly done in organic reactions where the mixture is too viscous for the use of coils and a high-speed agitator. CHILTON et al.[80] and CUMMINGS and WEST[81] measured the transfer coefficients for this case by using an arrangement as shown in Figure 9.47, where heat is applied to the jacket and simultaneously removed by passing water through the coil. Chilton measured the temperatures of the inside of the vessel wall, of the bulk liquid, and of the surface of

Figure 9.47. Reaction vessel with jacket and coil

the coil by means of thermocouples and thus obtained the film heat transfer coefficients directly. Cummings and West used an indirect method to give the film coefficient from measurements of the overall coefficients.

CHILTON *et al.*[80] expressed their results in the form:

$$\frac{h_b d_v}{k} \left(\frac{\mu_s}{\mu}\right)^{0.14} = 0.36 \left(\frac{L^2 N \rho}{\mu}\right)^{0.67} \left(\frac{C_p \mu}{k}\right)^{0.33} \tag{9.153}$$

where h_b is the film coefficient for the liquor adjacent to the wall of the vessel. CUMMINGS and WEST[81] used the same equation but gave the value of the constant as 0.40. Considering that Chilton's vessel was only 0.3 m in diameter and was fitted with a single paddle of 150 mm length, and that Cummings and West used a 0.45 m³ vessel with two turbine impellers, agreement between their results is remarkably good. The group $(\mu_s/\mu)^{0.14}$ is again used to cover the difference in the viscosities at the surface and in the bulk of the fluid.

BROWN *et al.*[82] have given results of measurements on 1.5 m diameter sulphonators and nitrators of 3.4 m³ capacity as used in the dyestuffs industry. The sulphonators were of cast iron and had a wall thickness of 25.4 mm; the annular space in the jacket was also 25.4 mm. The agitator of the sulphonator was of the anchor type with a 127 mm clearance at the walls and was driven at 0.67 Hz. The nitrators were fitted with four-bladed propellers of 0.61 m diameter driven at 2 Hz. For cooling, they expressed the film coefficient h_b for the inside of the vessel by the relation:

$$\frac{h_b d_v}{k} \left(\frac{\mu_s}{\mu}\right)^{0.14} = 0.55 \left(\frac{L^2 N \rho}{\mu}\right)^{0.67} \left(\frac{C_p \mu}{k}\right)^{0.25} \tag{9.154}$$

which is very similar to that given by equation 9.153.

They also measured the film coefficient for the water jacket and gave values of 635–1170 W/m² K for water rates of 1440–9230 cm³/s, respectively, in the sulphonator. It should be noted that 7580 cm³/s corresponds to a vertical velocity of only 0.061 m/s and to a Reynolds number in the annulus of 5350. The thermal resistance of the wall of the pan was important, since with the sulphonator it accounted for 13 per cent of the total resistance at 323 K and 31 per cent at 403 K. The change in viscosity with temperature is important when considering these processes, since, for example, the viscosity of the sulphonation liquors ranged from 340 mN s/m^2 at 323 K to 22 mN s/m^2 at 403 K.

In discussing equations 9.153 and 9.154 FLETCHER[83] has summarised correlations obtained for a wide range of impeller and agitator designs in terms of the constant before the Reynolds number and the index on the Reynolds number as shown in Table 9.10.

Table 9.10.   Data on common agitators for use in equations 9.153
and 9.154 $(Re = L^2N\rho/\mu)$

Type of agitator	Constant	Index
Flat blade disc turbine		
unbaffled, or baffled vessel, $Re < 400$	0.54	0.67
baffled, $Re > 400$	0.74	0.67
Retreating-blade turbine with three		
blades, jacketed and baffled vessel,		
$Re = 2 \times 10^4$ to 2×10^6		
glassed steel impeller	0.33	0.67
alloy steel impeller	0.37	0.67
Propeller with three blades		
baffled vessel, $Re = 5500$ to 37,000	0.64	0.67
Flat blade paddle		
baffled or unbaffled vessel, $Re \geq 4000$	0.36	0.67
Anchor		
$Re = 30$ to 300	1.00	0.50
$Re = 300$ to 5000	0.38	0.67

9.8.3. Time of heating or cooling

It is frequently necessary to heat or cool the contents of a large batch reactor or storage tank. In this case the physical constants of the liquor may alter and the overall transfer coefficient change during the process. In practice, however, it is often possible to assume an average value of the transfer coefficient so as to simplify the calculation of the time of heating or cooling. The heating of the contents of a storage tank is commonly effected by condensing steam, either in a coil or in some form of hairpin tube heater.

For the case of a storage tank with liquor of mass m and specific heat C_p, heated by steam condensing in a helical coil, suppose that the overall transfer coefficient U is constant. Let T_s be the temperature of the condensing steam, T_1 and T_2 the initial and final temperatures of the liquor, and A the area of heat transfer surface. If T is the temperature of the liquor at any time t, the rate of transfer of heat is given by:

$$Q = mC_p\frac{dT}{dt} = UA(T_s - T)$$

$$\therefore \qquad \frac{dT}{dt} = \frac{UA}{mC_p}(T_s - T)$$

$$\therefore \qquad \int_{T_1}^{T_2} \frac{dT}{T_s - T} = \frac{UA}{mC_p}\int_0^t dt$$

$$\therefore \qquad \ln\frac{T_s - T_1}{T_s - T_2} = \frac{UA}{mC_p}t \qquad\qquad (9.155)$$

From this equation, the time t of heating from T_1 to T_2, can be found. The same analysis can be used if the steam condenses in a jacket of a reaction vessel.

The analysis does not allow for any heat losses during the heating, or, for that matter, cooling operation. Obviously the higher the temperature of the contents of the vessel, the greater are the heat losses and, in the limit, the heat supplied to the vessel is equal to the heat losses, at which stage no further rise in the temperature of the contents of the vessel is possible. This situation is illustrated in Example 9.15.

The heating-up time could be reduced by improving the rate of heat transfer to the fluid, by agitation of the fluid for example and by reducing heat losses from the vessel by insulation. In the case of a large vessel there is a limit to the degree of agitation and circulation of the fluid through an external heat exchanger is an attractive possibility.

Example 9.15

A vessel contains 1 tonne of a liquid of specific heat capacity 4.0 kJ/kg K. The vessel is heated by steam at 393 K which is fed to a coil immersed in the stirred liquid and heat is lost to the surroundings at 293 K from the outside of the vessel. How long does it take to heat the liquid from 293 to 353 K and what is the maximum temperature to which the liquid can be heated? When the liquid temperature has reached 353 K, the steam supply is turned off for 2 hours (7.2 ks) and the vessel cools. How long will it take to reheat the material to 353 K? The surface area of the coil is 0.5 m^2 and the overall coefficient of heat transfer to the liquid may be taken as 600 W/m^2 deg K. The outside area of the vessel is 6 m^2 and the coefficient of heat transfer to the surroundings may be taken as 10 W/m^2 deg K.

Solution

If T K is the temperature of the liquid at time t s, then a heat balance on the vessel gives:

$$(1000 \times 4000)\frac{dT}{dt} = (600 \times 0.5)(393 - T) - (10 \times 6)(T - 293)$$

$$4,000,000\frac{dT}{dt} = 135,480 - 360T$$

$$11,111\frac{dT}{dt} = 376.3 - T.$$

The equilibrium temperature occurs when $dT/dt = 0$,

that is when:
$$T = 376.3 \text{ K.}$$

In heating from 293 to 353 K, the time taken is:

$$t = 11,111 \int_{293}^{353} \frac{dT}{(376.3 - T)}$$

$$= 11,111 \ln \frac{83.3}{23.3}$$

$$= 14,155 \text{ s} \quad (\text{or } 3.93 \text{ h}).$$

The steam is turned off for 7200 s and during this time a heat balance gives:

$$(1000 \times 4000)\frac{dT}{dt} = -(10 \times 6)(T - 293)$$

$$66,700\frac{dT}{dt} = 293 - T$$

The change in temperature is then given by:

$$\int_{353}^{T} \frac{dT}{(293 - T)} = \frac{1}{66,700} \int_{0}^{7200} dt$$

$$\ln \frac{-60}{293 - T} = \frac{7200}{66,700} = 0.108$$

and: $T = 346.9 \text{ K}.$

The time taken to reheat the liquid to 353 K is then given by:

$$t = 11,111 \int_{346.9}^{353} \frac{dT}{(376.3 - T)}$$

$$= 11,111 \ln \frac{29.4}{23.3}$$

$$= \underline{\underline{2584 \text{ s}}} \quad (0.72\text{h}).$$

9.9. SHELL AND TUBE EXCHANGERS

9.9.1. Description

These units form the most important class of heat transfer equipment because they can be constructed with very large surfaces in a relatively small volume, can be fabricated from alloy steels to resist corrosion, and can be used for heating and condensing all kinds of fluids. Figures 9.48–9.50 show various forms of construction of varying complexity, and Figure 9.51 shows the tube bundle. The simplest type, shown in Figure 9.48, has fixed tube plates at each end and the tubes are expanded into the plates. The tubes are so connected that the water or internal fluid makes several passes up and down the exchanger, thus enabling a high velocity of flow to be obtained for a given heat transfer area and given throughput of liquid. The fluid flowing in the shell is made to flow first in one sense and then in the opposite sense across the tube bundle by fitting a series of baffles along the length. These baffles are frequently of the segmental form, with about 25 per cent cut for the free space, and increase the velocity of flow across the tubes, thus giving better heat transfer. The difficulty with this type of construction is that the tube bundle cannot be removed for cleaning and no provision is made for differential expansion between the tubes and the shell, although an expansion joint is often fitted in the shell.

Figure 9.48.   Heat exchanger with fixed tube plates (four tube pass, one shell pass)

Figure 9.49.   Heat exchanger with floating head (two tube pass, one shell pass)

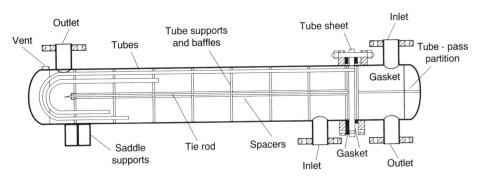

Figure 9.50.   Heat exchanger with hairpin tubes

In order to make removal of the tube bundle possible and to allow for considerable expansion of the tubes, a floating head exchanger is used, as shown in Figure 9.49. In this arrangement one tube plate is fixed as before but the second is bolted to a floating head cover so that the tube bundle can move relative to the shell. This floating tube sheet is clamped between the floating head and a split backing flange in such a way that it is relatively easy to break the flanges at both ends and pull out the tube bundle. It will be seen that the shell cover at the floating head end is larger than that at the other end; this is to enable the tubes to be placed as near as possible to the edge of the fixed tube plate and leave very little unused space between the outer ring of tubes and the shell.

Another arrangement which provides for expansion involves the use of hairpin tubes, as shown in Figure 9.50. This method is very commonly used for the reboilers on large fractionating columns where steam is condensed inside the tubes.

In these designs there is one pass for the fluid on the shell side and a number of passes on the tube side. As will be shown later it is often desirable to have two or more shell side passes, but they considerably increase the difficulty of construction and therefore several smaller exchangers are often connected together to obtain the same effect.

The essential requirements in the design of a heat exchanger are, firstly, the provision of a unit which is reliable and has the desired capacity, and, secondly, the economic

Figure 9.51.   Expanding the ends of the tubes into the tube plate of a heat exchanger bundle

consideration of making an exchanger at the minimum overall cost. This will generally involve using standard parts and fittings and making the design as simple as possible. In most cases it will be necessary to balance the capital cost, as measured by depreciation, against the operating cost. Thus, for instance, in a condenser a high heat transfer coefficient is obtained, and hence a smaller exchanger is required, if a high water velocity is used through the tubes; on the other hand, the cost of pumping increases rapidly with increase in velocity and an economic balance must be struck. A typical graph showing the operating cost, depreciation, and total overall cost plotted against water velocity through the tubes for a condenser is given in Figure 9.52.

The tubes used in heat exchangers are usually 19 mm or 25.4 mm outside diameter d_o, and 14 or 16 BWG (2.1 or 1.65 mm) for admiralty metal or other alloy and 10 or 12 BWG (3.4 or 2.8 mm) for carbon steel. The pitch of the tubes is dictated largely by the necessity for cleaning the outside of the bundle if the fluid is likely to give rise to considerable scale. A square pitch is preferred as this permits cleaning round the tubes;

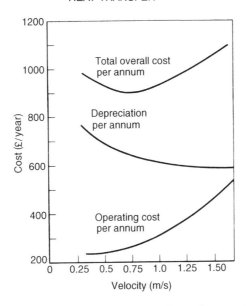

Figure 9.52.   Effect of water velocity on annual operating cost of condenser

the minimum spacing is $1.25d_o$, between centres. If space is at a premium, or if the fluids are very clean, a triangular pitch is used with a 6.35 mm space between the tubes, as this enables considerably more tubes to be put into a given shell.

The shell of a heat exchanger is commonly made of carbon steel, and standard pipes are used for the smaller sizes and rolled welded plate for larger sizes (say 0.4–1 m). The thickness of the shell can be calculated from the formula for thin-walled cylinders, but a minimum of 9.5 mm is used for shells over 0.33 m outside diameter and 11.1 mm for shells over 0.9 m outside diameter. Unless the exchanger works at very high pressure, the calculated thickness will usually be less than these figures, but a corrosion allowance of 3.2 mm is commonly added to all carbon steel parts; the thickness may be determined more by questions of rigidity than of simple internal pressure.

As mentioned earlier, baffles and tube support plates are fitted to increase the rate of flow over the tube bundle. The commonest type is the segmental baffle with about 25 per cent cut, as shown in Figure 9.23. The diameter is governed by the fit of the bundle in the shell, but for good work the clearance between the baffle and the inside of the shell should be not more than 5–10 mm for vapours, and 2.5–5 mm for liquids. Baffles are not normally spaced closer than one-fifth of the shell diameter; very close spacing is avoided because the increase in heat transfer is then small compared with the increase in pressure drop. Support plates to hold the tubes in position are similar to the baffles, but the holes for the tubes are only 0.4 mm greater than the tube diameter, whereas for baffles the clearance may be twice this. These support plates are fitted at least every 1 m for 19 mm tubes and every 60 tube diameters for larger tubes.

In order to keep the tube bundle straight, tie rods are fitted to the fixed tube sheet and to the baffle nearest the floating tube sheet. Usually between four and six rods of thickness 9.5–12.7 mm are necessary. These can be used to hold the baffles and support plates in position if sleeves are fitted over the rods between the baffles.

The thickness of the fixed tube sheet is frequently calculated from the relation:

$$d_t = d_G \sqrt{\frac{0.25P}{f}} \tag{9.156}$$

where d_G is the diameter of the gasket on the tube sheet, P the design pressure, f the allowable working stress, and d_t the thickness of the sheet measured at the bottom of the partition baffle grooves. The floating tube sheet may be made $\sqrt{2}$ times as thick.

Where dished ends are used for the cover plates, these are made to the designs supplied by the maker, but details are given in the *Standards of the Tubular Exchanger Manufactures Association* (TEMA)[84] and in BS 5500[85]. Flanges are also designed from one of these codes. Extensive practical details are given in the TEMA Code which applies particularly to the petroleum industry.

9.9.2. Mean temperature difference in multipass exchangers

In an exchanger with one shell pass and several tube passes, the fluids in the tubes and shell will flow co-currently in some of the passes and countercurrently in the others. The mean temperature difference for countercurrent flow is greater than that for co-current or parallel flow, so that there is no easy way of finding the true temperature difference for the unit. The problem has been examined by UNDERWOOD[86] and by BOWMAN *et al.*[87] who have presented a graphical method for calculating the true mean temperature difference in terms of the value of θ_m for countercurrent flow and a correction factor F. Provided the following conditions are maintained or assumed then F can be found from the curves presented in Figures 9.53–9.57.

(a) The shell fluid temperature is uniform over the cross-section considered as a pass.
(b) There is equal heat transfer surface in each pass.
(c) U is constant over the length.
(d) The specific heats of the two fluids are constant over the temperature range considered.
(e) There is no change in phase of either fluid.
(f) Heat loss from the unit is negligible.

Then: $Q = UAF\theta_m \tag{9.157}$

F is expressed in relation to the two terms:

$$X = \frac{\theta_2 - \theta_1}{T_1 - \theta_1} \quad \text{and} \quad Y = \frac{T_1 - T_2}{\theta_2 - \theta_1} \tag{9.158}$$

Then if one shell side system is used Figure 9.53 applies, for two shell side passes Figure 9.54, for three shell side passes Figure 9.55, and for four shell passes Figure 9.56. For the case of a single shell side pass and two tube passes illustrated in Figure 9.57a and *b* the temperature profile is as shown. Because one of the passes constitutes a parallel flow arrangement the exit temperature of the cold fluid θ_2 cannot be brought up very near to the hot fluid temperature T_1. This is true for the conditions shown in both Figures 9.57a and *b* and UNDERWOOD[86] has shown that F is the same in both cases.

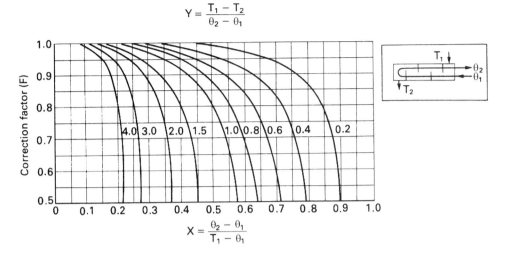

Figure 9.53. Correction for logarithmic mean temperature difference for single shell pass exchanger

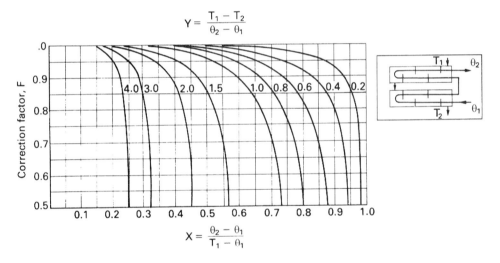

Figure 9.54. Correction for logarithmic mean temperature difference for double shell pass exchanger

Suppose an exchanger is required to operate over the following temperature ranges:

$$T_1 = 455 \text{ K}, \quad T_2 = 372 \text{ K}$$

$$\theta_1 = 283 \text{ K}, \quad \theta_2 = 388 \text{ K}$$

Then:

$$X = \frac{\theta_2 - \theta_1}{T_1 - \theta_1} = \frac{388 - 283}{455 - 283} = 0.6$$

$$Y = \frac{T_1 - T_2}{\theta_2 - \theta_1} = \frac{455 - 372}{388 - 283} = 0.8$$

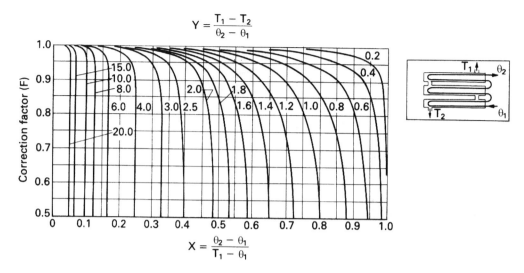

Figure 9.55.   Correction for logarithmic mean temperature difference for three shell pass exchanger

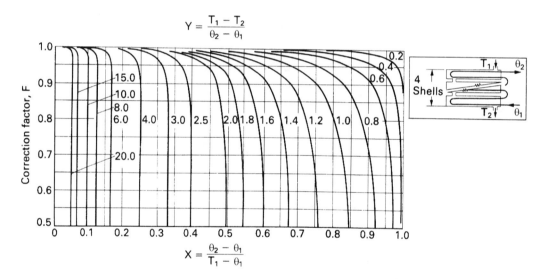

Figure 9.56.   Correction for logarithmic mean temperature difference for four shell pass exchanger

If a single shell side pass is used, then F is found to be 0.65 and with a two shell side unit F has a value of 0.95. For such a requirement the two shell side unit should be used.

To obtain maximum heat recovery from the hot fluid, θ_2 must be as high as possible. $T_2 - \theta_2$ is known as the *approach temperature*, and when $\theta_2 > T_2$ a *temperature cross* is said to occur and the value of F falls rapidly for a single pass on the shell side. This can be seen by taking as an illustration the values opposite where equal ranges of temperature are considered.

	Case 1	Case 2	Case 3
	T_1 613 θ_2 463	T_1 573 θ_2 473	T_1 543 θ_2 463
	T_2 513 θ_1 363	T_2 473 θ_1 373	T_2 443 θ_1 363
Approach temperature $(T_2 - \theta_2)$	50	0	Cross of 20
$X = \dfrac{\theta_2 - \theta_1}{T_1 - \theta_1}$	$\dfrac{100}{250} = 0.4$	$\dfrac{100}{200} = 0.5$	$\dfrac{100}{180} = 0.55$
$Y = \dfrac{T_1 - T_2}{\theta_2 - \theta_1}$	1	1	1
F	0.92	0.80	0.66

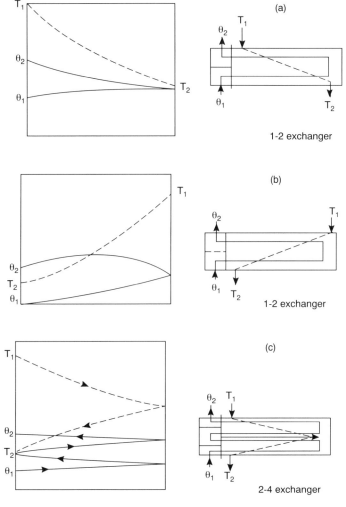

Figure 9.57.   Temperature profiles in single and double shell pass exchangers

Thus, as soon as a temperature cross occurs, a unit with two shell passes should be used. It may be seen in Figure 9.57*b* that there may be some point where the temperature of the cold liquid is greater than θ_2, so that beyond this point the stream will be cooled rather than heated. This is avoided by using the increased number of shell passes. The general form of the temperature profile for a two shell side unit is as shown in Figure 9.57*c*. Longitudinal shell side baffles are rather difficult to fit and there is a serious chance of leakage. For this reason two exchangers arranged one below the other in series are to be preferred. The use of separate exchangers is even more desirable when a three shell side unit is required. On the very largest installations it may be necessary to link up in parallel a number of exchangers arranged as, say, sets of three in series, as indicated in Figure 9.58. This type of arrangement is desirable for any very large unit which is likely to be unwieldy as a single system. When the total surface is much greater than 250 m^2, consideration should be given to using two smaller units even though the initial cost will be rather higher.

Figure 9.58.   Set of three heat exchangers in series

In many chemical engineering operations there may be a large number of process streams, some of which need to be heated and others cooled. An overall heat balance will indicate whether, in total, there is a net surplus or deficit of heat available. It is of great economic importance to achieve the most effective match of the hot and cold streams in the heat exchanger network so as to reduce to a minimum both the heating and cooling duties placed on the works utilities (e.g. supplies of steam and cooling water). This is carried out by making the best use of the temperature driving forces. In considering the overall requirements there will be some point where the temperature difference between the hot and cold streams is a minimum and this is referred to as the *pinch*. The lower the temperature difference at the pinch point, the lower will be the demand on the utilities, but it must be remembered that a greater area, (and hence cost) will be involved and an economic balance must therefore be achieved. Heat exchanger networks are discussed in Volume 6 and, in detail, in the User Guide published by the Institution of Chemical Engineers[88]. Subsequently, LINNHOFF[89] has given an overview of the industrial application of pinch analysis to the design of networks in order to reduce both capital costs and energy requirements.

Example 9.16

Using the data of Example 9.1, calculate the surface area required to effect the given duty using a multipass heat exchanger in which the cold water makes two passes through the tubes and the hot water makes a single pass outside the tubes.

Solution

As in Example 9.1, the heat load $= 1672$ kW

With reference to Figure 9.51: $T_1 = 360$ K, $T_2 = 340$ K

$$X = \frac{\theta_2 - \theta_1}{T_1 - \theta_1} = \frac{316 - 300}{360 - 300} = 0.267$$

$$Y = \frac{T_1 - T_2}{\theta_2 - \theta_1} = \frac{360 - 340}{316 - 300} = 1.25$$

$$F = 0.97$$

$$F\theta_m = (41.9 \times 0.97) = 40.6 \text{ K}$$

and the heat transfer area, $$A = \frac{1672}{2 \times 40.6} = \underline{\underline{20.6 \text{ m}^2}}$$

9.9.3. Values of film coefficients

If in any heat transfer apparatus the overall temperature difference, the area of surface and the heat transferred are measured directly, the overall transfer coefficient U is obtained from the simple relation, $Q = UA\Delta T$. The determination of the individual film coefficients has proved difficult even for the simplest cases, and it is quite common for equipment to be designed on the basis of practical values of U rather than from a series of film coefficients. For the important case of the transfer of heat from one fluid to another across a metal surface, two methods have been developed for measuring the film coefficients. The first requires a knowledge of the temperature difference across each film and therefore involves measuring the temperatures of both fluids and the surface of separation. With a concentric tube system it is very difficult to insert a thermocouple into the thin tube and to prevent the thermocouple wires from interfering with the flow of the fluid. Nevertheless, this method is commonly adopted, particularly when electric heating is used. It must be remembered that when the heat flux is very high, as with boiling liquids, there will be an appreciable temperature drop across the tube wall and the position of the thermocouple is then important. For this reason working with stainless steel, which has a relatively low value of thermal conductivity, is difficult.

The second method uses a technique put forward by WILSON[90]. If steam is condensing on the outside of a horizontal tube through which water is passed at various velocities, then the overall and film transfer coefficients can be related by the equation:

$$\frac{1}{U} = \frac{1}{h_o} + \frac{x_w}{k_w} + R_i + \frac{1}{h_i}$$ (from equation 9.147)

provided that the transfer area on each side of the tube is approximately the same.

For conditions of turbulent flow the transfer coefficient for the water side $h_i = \varepsilon u^{0.8}$, R_i the scale resistance is constant, and h_0 the coefficient for the condensate film is almost independent of the water velocity. Thus the above equation reduces to:

$$\frac{1}{U} = (\text{constant}) + \frac{1}{\varepsilon u^{0.8}}$$ (9.159)

and if $1/U$ is plotted against $1/u^{0.8}$ (a Wilson plot) a straight line is obtained with an intercept equal to the value of the constant. For a clean tube R_i should be nil, and hence h_o

Table 9.11.   Thermal resistance of heat exchanger tubes

Gauge (BWG)	Thickness (in.)	Values of x_w/k_w (ft² h °F/Btu)				
		Copper	Steel	Stainless Steel	Admiralty Metal	Aluminium
18	0.049	0.000018	0.00011	0.00047	0.000065	0.000031
16	0.065	0.000024	0.00014	0.00062	0.000086	0.000042
14	0.083	0.000031	0.00018	0.0008	0.00011	0.000053
12	0.109	0.000041	0.00024	0.001	0.00026	0.000070
	(mm)	Values of x_w/k_w (m² K/kW)				
18	1.24	0.0031	0.019	0.083	0.011	0.0054
16	1.65	0.0042	0.025	0.109	0.015	0.0074
14	2.10	0.0055	0.032	0.141	0.019	0.0093
12	2.77	0.0072	0.042	0.176	0.046	0.0123

Table 9.12.   Thermal resistances of scale deposits from water, etc.

	(ft² h °F/Btu)	(m²K/kW)		(ft² h °F/Btu)	(m²K/kW)
Water (1 m/s velocity, temperatures less than 320K)*			Steam		
			Good quality — oil free	0.0003	0.052
Distilled	0.0005	0.09	Poor quality — oil free	0.0005	0.09
Sea	0.0005	0.09	Exhaust from reciprocating		
Clear river	0.0012	0.21	engines	0.001	0.18
Untreated cooling tower	0.0033	0.58			
Treated cooling tower	0.0015	0.26	Liquids		
Treated boiler feed	0.0015	0.26	Treated brine	0.0015	0.27
Hard well	0.0033	0.58	Organics	0.001	0.18
			Fuel oils	0.006	1.0
			Tars	0.01	2.0
			Gases		
(*3 ft/s velocity, temperatures less than 50°C)					
			Air	0.0015– 0.003	0.25–0.5
			Solvent vapours	0.0008	0.14

Table 9.13. Approximate overall heat transfer coefficients U for shell and tube equipment

Hot side	Cold side	Overall U	
		Btu/h ft^2 °F	W/m^{2}K
Condensers			
Steam (pressure)	Water	350–750	2000–4000
Steam (vacuum)	Water	300–600	1700–3400
Saturated organic solvents (atmospheric)	Water	100–200	600–1200
Saturated organic solvents (vacuum some non-condensable)	Water–brine	50–120	300–700
Organic solvents (atmospheric and high non-condensable)	Water–brine	20–80	100–500
Organic solvents (vacuum and high non-condensable)	Water–brine	10–50	60–300
Low boiling hydrocarbons (atmospheric)	Water	80–200	400–1200
High boiling hydrocarbons (vacuum)	Water	10–30	60–200
Heaters			
Steam	Water	250–750	1500–4000
Steam	Light oils	50–150	300–900
Steam	Heavy oils	10–80	60–400
Steam	Organic solvents	100–200	600–1200
Steam	Gases	5–50	30–300
Dowtherm	Gases	4–40	20–200
Dowtherm	Heavy oils	8–60	50–400
Evaporators			
Steam	Water	350–750	2000–4000
Steam	Organic solvents	100–200	600–1200
Steam	Light oils	80–180	400–1000
Steam	Heavy oils (vacuum)	25–75	150–400
Water	Refrigerants	75–150	400–900
Organic solvents	Refrigerants	30–100	200–600
Heat exchangers (no change of state)			
Water	Water	150–300	900–1700
Organic solvents	Water	50–150	300–900
Gases	Water	3–50	20–300
Light oils	Water	60–160	400–900
Heavy oils	Water	10–50	60–300
Organic solvents	Light oil	20–70	100–400
Water	Brine	100–200	600–1200
Organic solvents	Brine	30–90	200–500
Gases	Brine	3–50	20–300
Organic solvents	Organic solvents	20–60	100–400
Heavy oils	Heavy oils	8–50	50–300

can be found from the value of the intercept. h_i may also be obtained at a given velocity from the difference between $1/U$ at that velocity and the intercept.

This technique has been applied by RHODES and YOUNGER[91] to obtain the values of h_o for condensation of a number of organic vapours, by PRATT[79] to obtain the inside coefficient for a coiled tube, and by COULSON and MEHTA[92] to obtain the coefficient for an annulus. If the results are repeated over a period of time, R_i can also be obtained by this method.

Typical values of thermal resistances and individual and overall heat transfer coefficients are given in Tables 9.11–9.14.

Table 9.14.   Approximate film coefficients (h_i or h_o)

	Approximate film coefficients	
	Btu/hft $°$F	W/m^2 K
No change of state		
Water	300–2000	1700–11,000
Gases	3–50	20–300
Organic solvents	60–500	350–3000
Oils	10–120	60–700
Condensing		
Steam	1000–3000	6000–17,000
Organic solvents	150–500	900–2800
Light oils	200–400	1200–2300
Heavy oils (vacuum)	20–50	120–300
Ammonia	500–1000	3000–6000
Evaporation		
Water	30–200	2000–12,000
Organic solvents	100–300	600–2000
Ammonia	200–400	1100–2300
Light oils	150–300	800–1700
Heavy oils	10–50	60–300

Example 9.17

27 tonne/h of pure isobutane is to be condensed at 332 K in a horizontal tubular exchanger using water entering at 300 K (Figure 9.59). It is proposed to use 19.0 mm o.d. tubes of wall thickness 1.65 mm arranged on a 25 mm triangular pitch. The resistance due to scale may be taken as 0.0005 m^2 K/W. What is the number of tubes and the arrangement of passes to be specified?

Figure 9.59.   Temperatures in heat exchanger

Solution

The latent heat of vaporisation of isobutane is 286 kJ/kg and hence the heat load Q is:

$$Q = \frac{27 \times 1000}{3600} \times 286 = 2145 \text{ kW}$$

The cooling water outlet temperature should not exceed 325 K and a value of 315 K will be used. The water requirement is then $2145/[4.18(315 - 300)] = 34.2$ kg/s.

An approximate size of the unit is now obtained, assuming an overall coefficient, based on the outside area of the tubes, of 510 W/m^2 K.

The various temperatures are:

isobutane (shell):	in 332 K	out 332 K
water (tube side):	in 300 K	out 315 K (mean 307.5 K)
ΔT:	32 K	17 K

$$\therefore \qquad \theta_m = \frac{32 - 17}{\ln(32/17)} = 23.8 \text{ K}$$

The approximate heat transfer area is therefore:

$$A = \frac{2145 \times 10^3}{510 \times 23.8} = 176.7 \text{ m}^2$$

19.0 mm tubes have an outside area of 0.059 m^2/m and hence the total length of tube is $176.7/0.059 = 2995$ m.
  Using the standard length of 4.88 m, number of tubes $= 613$. With this large quantity of water it will be possible to use a four-pass exchanger and still have a reasonable water velocity. In a shell of 0.78 m i.d., 678 tubes can be arranged in four passes using 25 mm triangular pitch and this size will be selected. This result will now be checked by calculating the individual film coefficients and hence the assumed value of the overall film coefficient.

Calculation of inside and outside film coefficients

(a) Inside coefficient, h_i
  Water flow through each tube $= 34.2/(678/4) = 0.202$ kg/s
  For a tube i.d. of $(19.0 - 2 \times 1.65) = 15.7$ mm, cross-sectional area is 193.6 mm^2 or 0.000194 m^2.
  The water velocity is therefore $0.202/(1000 \times 0.000194) = 1.04$ m/s.

From equation 9.63 therefore:

$$h_i = \frac{4280(0.00488 \times 307.5 - 1)1.04^{0.8}}{0.0157^{0.2}}$$

$$= \frac{4280 \times (1.5006 - 1) \times 1.032}{0.4357} = 5075 \text{ W/m}^2 \text{ K}$$

or, based on the outside diameter;

$$h_{io} = \frac{5075 \times 15.7}{19.0} = 4200 \text{ W/m}^2 \text{ K}$$

or: $$4.2 \text{ kW/m}^2 \text{ K}$$

(b) Outside coefficient, h_o
  The temperature drop across the condensate film ΔT_f is given by:

$$\frac{\text{Thermal resistance, water film + scale}}{\text{Total thermal resistance}} = \frac{(1/4.2) + 0.5}{1/0.510} = \frac{\theta_m - \Delta T_f}{\theta_m}$$

$$\therefore \qquad \Delta T_f = 23.8 - \frac{23.8(1/4.2 + 1/2.0)}{1/0.510}$$

$$= 23.8 - \frac{23.8(6.2 \times 0.510)}{4.2 \times 2.0}$$

$$= 14.8 \text{ K}$$

The condensate film is thus at $332 - 14.8 = 317.2$ K

From equation 9.123:

$$h_o = 0.72 \left[\frac{k^3 \rho^2 g \lambda}{j d_o \mu \Delta T_f} \right]^{0.25}$$

Taking $k = 0.13$ W/m K, $\rho = 508$ kg/m^3, $j = \sqrt{678} = 26$, and $\mu = 0.136 \times 10^{-3}$ N s/m^2,

$$h_o = 0.72 \left\{ \frac{0.13^3 \times 508^2 \times 9.81 \times 286 \times 10^3}{26 \times 19.0 \times 10^{-3} \times 0.136 \times 10^{-3} \times 14.8} \right\}^{0.25}$$

$$= 810 \text{ W/m}^2 \text{ K} \quad \text{or} \quad 0.810 \text{ kW/m}^2 \text{ K}$$

Hence the overall coefficient, based on the outside area, is:

$$\frac{1}{U} = \frac{1}{4.2} + \frac{1}{0.810} + 0.50 = 2.0$$

∴ $U = 0.50$ kW/m^2 K   or   500 W/m^2 K

This is sufficiently near the assumed value. The total heat load of the condenser under these conditions will be:

$$(UA\theta_m) = 0.50 \times 4.88 \times 678 \times 0.059 \times 23.8 = 2300 \text{ kW}$$

It should be noted that this is not a complete solution in that the exit water temperature and the optimum velocity should be determined on economic grounds, although it serves as an illustration of the general method of approach.

Example 9.18

135 t/h (37.5 kg/s) of a crude oil is to be heated from 295 to 330 K by heat exchange with the bottom product from a distillation unit, which is to be cooled from 420 to 380 K, flowing at 106 t/h (29.6 kg/s). A tubular heat exchanger is available with a 0.60 m i.d. shell and one shell side pass with two passes on the tube side. The tube bundle consists of 324 19.0 mm o.d. tubes, wall thickness 2.1 mm, each 3.65 m long arranged on 25.4 mm square pitch and supported by baffles with a 25 per cent cut, spaced at 0.23 m intervals.
Is this exchanger suitable?

Solution

Temperatures: bottom product,   in 420 K   out 380 K   mean 400 K

                  crude oil,   out 330 K   in 295 K   mean 313 K

                  ΔT 90 K   85 K   mean 87.5 K

For the crude oil at 313 K: $C_p = 1.986$ kJ/kg K   or   1986 J/kg K

                              $\mu = 2.9$ mN s/m^2 or 2.9×10^{-3} N s/m^2

                              $k = 0.136$ W/m K

                              $\rho = 824$ kg/m^3

For the product at 400 K: $C_p = 2.20$ kJ/kg K   or   2200 J/kg K

Heat load on the tube side: $Q = 37.5 \times 1.986(330 - 295) = 2607$ kW

Heat load on the shell side: $Q = 29.6 \times 2.20(420 - 380) = 2605$ kW

Shell-side coefficient, h_o

Temperature of pipe wall $\approx \dfrac{400 + 313}{2} = 356.5$ K

Film temperature: $T_f = \dfrac{400 + 356.5}{2} = 378$ K

At T_f: $\rho = 867$ kg/m^3

                    $\mu = 5.2$ mN s/m^2 or 5.2×10^{-3} N s/m^2

                    $k = 0.119$ W/m K

Cross area for flow $= \dfrac{\text{shell i.d.} \times \text{clearance} \times \text{baffle spacing}}{\text{pitch}}$

$= \dfrac{0.60 \times 6.4 \times 10^{-3} \times 0.23}{0.025} = 0.0353$ m^2

Hence:
$$G'_{max} = \frac{29.6}{0.0353} = 838.5 \text{ kg/m}^2\text{s}$$

and:
$$Re_{max} = \frac{0.019 \times 838.5}{5.2 \times 10^{-3}} = 3064$$

∴
$$(Re_{max})^{0.6} = 125$$

$$Pr^{0.3} = \left(\frac{2200 \times 5.2 \times 10^{-3}}{0.119}\right)^{0.3} = (96.1)^{0.3} = 3.94$$

Taking $C_h = 1$, in equation 9.82:

$$\frac{h_o \times 0.019}{0.119} = 0.33 \times 125 \times 3.94$$

∴
$$h_o = 1018 \text{ W/m}^2\text{ K} \quad \text{or} \quad 1.02 \text{ kW/m}^2\text{ K}$$

Tube side coefficient, h_{io}

Number of tubes per pass $= 162$

Cross-sectional area of tube $= \frac{\pi}{4}(19.0 - 2 \times 2.1)^2 = 172 \text{ mm}^2 \quad \text{or} \quad 1.72 \times 10^{-4}\text{m}^2$

Mass flow:
$$G' = \frac{37.5}{1.72 \times 10^{-4} \times 162} = 1346 \text{ kg/m}^2\text{s}$$

$$Re = \frac{0.0148 \times 1346}{2.9 \times 10^{-3}} = 6869 \quad (\text{taking } d = 14.8 \text{ mm})$$

∴
$$Re^{0.8} = 1175$$

$$Pr^{0.4} = \left(\frac{1986 \times 2.9 \times 10^{-3}}{0.136}\right)^{0.4} = 4.47$$

From equation 9.54:

$$\frac{h_i \times 0.0148}{0.136} = 0.023 \times 1175 \times 4.47$$

∴
$$h_i = 1110 \text{ W/m}^2\text{ K}$$

Based on outside area: $h_{io} = \dfrac{1110 \times 14.8}{19.0} = 865 \text{ W/m}^2\text{ K} \quad \text{or} \quad 0.865 \text{ kW/m}^2\text{ K}$

Neglecting the thermal resistances of the wall and the scale, the clean overall coefficient based on the outside area is:

$$\frac{1}{U} = \frac{1}{1.02} + \frac{1}{0.865}$$

$$= 2.136 \text{ K m}^2/\text{kW}$$

Outside area available: $A = 324 \times 3.65 \times 0.0598 = 70.7 \text{ m}^2$

(where the outside area $= 0.0598 \text{ m}^2/\text{m}$).

During operation, scale would form and the actual overall coefficient would not have to fall below U_d, given by:

$$\frac{1}{U_d} = \frac{A\Delta T}{Q} = \frac{70.7 \times 87.5}{2607} = 2.37 \text{ K m}^2/\text{kW}$$

The maximum allowable scale resistance is then:

$$R = \frac{1}{U_d} - \frac{1}{U} = 2.37 - 2.136 = 0.23 \text{ K m}^2/\text{kW}$$

This value is very low, as seen from Table 9.12, and, if installed, the exchanger would not give the required temperatures without frequent cleaning. It should be noted that heat losses have been neglected in making this calculation.

Whilst heat transfer coefficients for most fluids may be estimated by the techniques described in this section, special mention must be made of liquid metals, used in recent years in the nuclear industry with the aim of reducing the volume of fluid in the heat transfer circuits. Such fluids have a high thermal conductivity, though in terms of heat capacity per unit volume, liquid sodium, for example, which finds relatively widespread application has a value of $C_p\rho = 1275$ kJ/m^3 K.

Although water has a much greater value it is unsuitable because of its high vapour pressure at the desired temperatures and the corresponding need to use high-pressure piping. Because of their high thermal conductivities, liquid metals have particularly low values of the Prandtl number (of about 0.01) and they behave rather differently from the normal fluids under conditions of forced convection. Some values for typical liquid metals are given in Table 9.15.

Table 9.15.   Prandtl numbers, Pr, of liquid metals

Material	Temp (K)	Pr
Potassium	975	0.003
Sodium	975	0.004
Na/K alloy (56:44)	975	0.06
Mercury	575	0.008
Lithium	475	0.065

The results of work on sodium, lithium, and mercury for forced convection in a pipe have been correlated by the expression:

$$Nu = 0.625(RePr)^{0.4} \qquad (9.160)$$

but the accuracy of the correlation is not very good. With values of Reynolds number of about 18,000 it is quite possible to obtain a value of h of about 11 kW/m^2 K for flow in a pipe system.

9.9.4. Effectiveness of heat exchangers

The effectiveness η of a heat exchanger is defined as the actual rate of heat transfer Q to the maximum rate Q_{max} that is thermodynamically possible.

$$\eta = \frac{Q}{Q_{max}} \qquad (9.161)$$

Q_{max} is the heat transfer rate which would be achieved if it were possible to bring the outlet temperature of the stream with the lower heat capacity to the inlet temperature of the other stream. Using the nomenclature in Figure 9.60 and taking stream 1 as having the lower value of GC_p:

$$Q_{max} = G_1 C_{p_1}(T_{11} - T_{21}) \qquad (9.162)$$

An overall heat balance gives:

$$Q = G_1 C_{p_1}(T_{11} - T_{12}) = G_2 C_{p_2}(T_{22} - T_{21})$$

Thus, based on stream 1:

$$\eta = \frac{G_1 C_{p_1}(T_{11} - T_{12})}{G_1 C_{p_1}(T_{11} - T_{21})} = \frac{T_{11} - T_{12}}{T_{11} - T_{21}} \qquad (9.163)$$

and, based on stream 2:

$$\eta = \frac{G_2 C_{p_2}(T_{22} - T_{21})}{G_1 C_{p_1}(T_{11} - T_{21})} \qquad (9.164)$$

Example 9.19

A flow of 1 kg/s of an organic liquid of specific heat 2.0 kJ/kg K is cooled from 350 to 330 K by a stream of water flowing countercurrently through a double-pipe heat exchanger. Estimate the effectiveness of the unit if the water enters the exchanger at 290 K and leaves at 320 K.

Solution

Heat load, $Q = 1 \times 2.0(350 - 330) = 40$ kW

$$\text{Flow of water, } G_{cool} = \frac{40}{4.187(320 - 290)} = 0.318 \text{ kg/s}$$

For organic: $(GC_p)_{hot} = (1 \times 2.0) = 2.0$ kW/K $(G_2 C_{p_2})$

For water: $(GC_p)_{cold} = (0.318 \times 4.187) = 1.33$ kW/K $= (GC_p)_{min} (G_1 C_{p_1})$

∴. From equation 9.164:

$$\text{effectiveness } \eta = \frac{2.0(350 - 330)}{1.33(350 - 290)}$$

$$= \underline{\underline{0.50}}$$

9.9.5 Transfer units

The concept of a *transfer unit* is useful in the design of heat exchangers, since its magnitude is less dependent on the flowrate of the fluids than the heat transfer coefficient which has been used so far. The number of transfer units **N** is defined by:

$$\mathbf{N} = \frac{UA}{(GC_p)_{min}} \qquad (9.165)$$

where $(GC_p)_{min}$ is the lower of the two values $G_1 C_{p_1}$ and $G_2 C_{p_2}$. It is the ratio of the heat transferred for a unit temperature driving force to the heat absorbed by the fluid stream when its temperature is changed by 1 K. Thus, the number of transfer units gives a measure of the amount of heat which the heat exchanger can transfer. The relation for the effectiveness of the heat exchanger in terms of the heat capacities of the streams will now be given for a number of flow conditions. The relevant nomenclature is given in Figure 9.60a–e.

Transfer units are also used extensively in the calculation of mass transfer rates in countercurrent columns and reference should be made to Section 10.5.7.

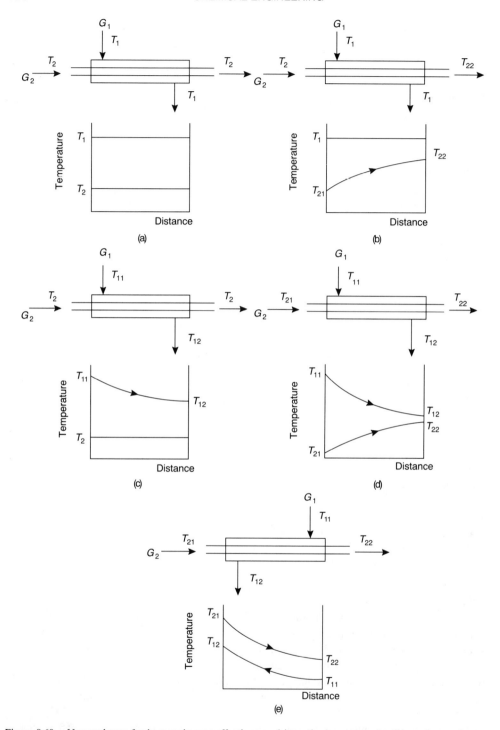

Figure 9.60.   Nomenclature for heat-exchanger effectiveness (*a*) condenser–evaporator (*b*) condenser–heater (*c*) cooler–evaporator (*d*) co-current flow (*e*) countercurrent flow

Co-current flow (Figure 9.60d)

For an elemental area dA of a heat exchanger, the rate of transfer of heat dQ is given by:

$$dQ = U \, dA(T_1 - T_2) = U \, dA \, \theta \tag{9.166}$$

where T_1 and T_2 are the temperatures of the two streams and θ is the point value of the temperature difference between the streams.

Also:
$$dQ = G_2 C_{p_2} \, dT_2 = -G_1 C_{p_1} \, dT_1$$

Thus:
$$dT_2 = \frac{dQ}{G_2 C_{p_2}} \quad \text{and} \quad dT_1 = \frac{-dQ}{G_1 C_{p_1}}$$

Then:
$$dT_1 - dT_2 = d(T_1 - T_2) = d\theta = -dQ \left(\frac{1}{G_1 C_{p_1}} + \frac{1}{G_2 C_{p_2}} \right)$$

Substituting from equation 9.166 for dQ:

$$\frac{d\theta}{\theta} = -U dA \left[\frac{1}{G_1 C_{p_1}} + \frac{1}{G_2 C_{p_2}} \right] \tag{9.167}$$

Integrating:
$$\ln \frac{\theta_2}{\theta_1} = -UA \left[\frac{1}{G_1 C_{p_1}} + \frac{1}{G_2 C_{p_2}} \right]$$

or:
$$\ln \frac{T_{12} - T_{22}}{T_{11} - T_{21}} = \frac{UA}{G_1 C_{p_1}} \left[1 + \frac{G_1 C_{p_1}}{G_2 C_{p_2}} \right] \tag{9.168}$$

If $G_1 C_{p_1} < G_2 C_{p_2}$, $G_1 C_{p_1} = (GC_p)_{\min}$

From equation 9.165:
$$N = \frac{UA}{G_1 C_{p_1}}$$

Thus:
$$\frac{T_{12} - T_{22}}{T_{11} - T_{21}} = \exp \left[-N \left(1 + \frac{G_1 C_{p_1}}{G_2 C_{p_2}} \right) \right] \tag{9.169}$$

From equations 9.163 and 9.164:

$$T_{11} - T_{12} = \eta(T_{11} - T_{21})$$

$$T_{22} - T_{21} = \eta \frac{G_1 C_{p_1}}{G_2 C_{p_2}} (T_{11} - T_{21})$$

Adding:
$$T_{11} - T_{12} + T_{22} - T_{21} = \eta \left(1 + \frac{G_1 C_{p_1}}{G_2 C_{p_2}} \right) (T_{11} - T_{21})$$

$$1 - \frac{T_{12} - T_{22}}{T_{11} - T_{21}} = \eta \left(1 + \frac{G_1 C_{p_1}}{G_2 C_{p_2}} \right)$$

Substituting in equation 9.169:

$$\eta = \frac{1 - \exp \left[-N \left(1 + \dfrac{G_1 C_{p_1}}{G_2 C_{p_2}} \right) \right]}{1 + \dfrac{G_1 C_{p_1}}{G_2 C_{p_2}}} \tag{9.170}$$

For the particular case where $G_1C_{p_1} = G_2C_{p_2}$:

$$\eta = \tfrac{1}{2}[1 - \exp(-2\mathbf{N}]$$

(9.171)

For a very large exchanger ($\mathbf{N} \to \infty$), $\eta \to 0.5$.

Countercurrent flow (Figure 9.60e)

A similar procedure may be followed for countercurrent flow, but it should be noted that, in this case, $\theta_1 = T_{11} - T_{22}$ and $\theta_2 = T_{12} - T_{21}$.

The corresponding equation for the effectiveness factor η is then:

$$\eta = \frac{1 - \exp\left[-\mathbf{N}\left(1 - \dfrac{G_1C_{p_1}}{G_2C_{p_2}}\right)\right]}{1 - \dfrac{G_1C_{p_1}}{G_2C_{p_2}} \exp\left[-\mathbf{N}\left(1 - \dfrac{G_1C_{p_1}}{G_2C_{p_2}}\right)\right]}$$

(9.172)

For the case where $G_1C_{p_1} = G_2C_{p_2}$, it is necessary to expand the exponential terms to give:

$$\eta = \frac{\mathbf{N}}{1 + \mathbf{N}}$$

(9.173)

For a very large exchanger ($\mathbf{N} \to \infty$), $\eta \to 1$ in this case.

Heat transfer with change of phase (Figure 9.60b and c)

If one component is merely undergoing a phase change at constant temperature, $G_1C_{p_1}$ is effectively zero and both equations 9.170 and 9.172 reduce to:

$$\eta = 1 - \exp(-\mathbf{N})$$

(9.174)

Effectiveness factors η are plotted against number of transfer units $\mathbf{N}$ with $(G_1C_{p_1}/G_2C_{p_2})$ as parameter for a number of different configurations by KAYS and LONDON[93]. Examples for countercurrent flow (based on equation 9.164) and an exchanger with one shell pass and one tube pass are plotted in Figures 9.61a and b respectively.

Example 9.20

A process requires a flow of 4 kg/s of purified water at 340 K to be heated from 320 K by 8 kg/s of untreated water which can be available at 380, 370, 360 or 350 K. Estimate the heat transfer surfaces of one shell pass, two tube pass heat exchangers suitable for these duties. In all cases, the mean specific heat of the water streams is 4.18 kJ/kg K and the overall coefficient of heat transfer is 1.5 kW/m² K.

Solution

For the untreated water: $GC_p = (8.0 \times 4.18) = 33.44$ kW/K

For the purified water: $GC_p = (4.0 \times 4.18) = 16.72$ kW/K

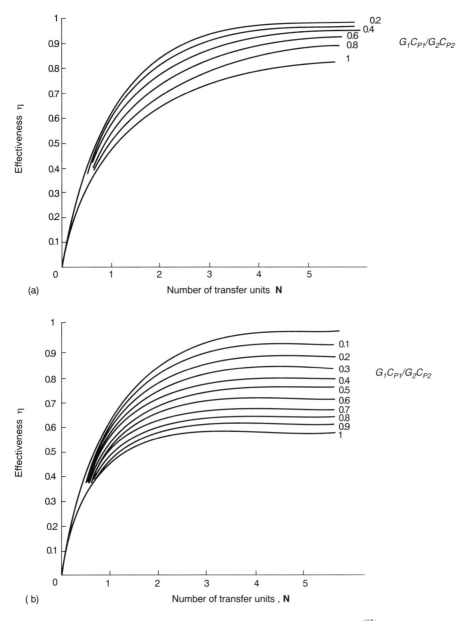

Figure 9.61.   Heat-exchanger effectiveness as a function of number of transfer units[93] (*a*) true countercurrent flow; (*b*) one shell pass, two-tube pass exchanger

Thus:

$$(GC_p)_{\min} = 16.72 \text{ kW/K} = G_1 C_{p_1}$$

$$\frac{G_1 C_{p_1}}{G_2 C_{p_2}} = \frac{16.72}{33.44} = 0.5$$

From equation 9.164:

$$\eta = \frac{4.0 \times 4.18(340 - 320)}{4.0 \times 4.18(T_{11} - 320)}$$

$$= \frac{20}{T_{11} - 320}.$$

Thus η may be calculated from this equation using values of $T_{11} = 380, 370, 360$ or 350 K and then $\mathbf{N}$ obtained from Figure 9.61*b*. The area required is then calculated from:

$$A = \frac{\mathbf{N}(GC_p)_{\min}}{U} \qquad \text{(equation 9.165)}$$

to give the following results:

T_{11} (K)	η (−)	$\mathbf{N}$ (−)	A (m^2)
380	0.33	0.45	5.0
370	0.4	0.6	6.6
360	0.5	0.9	10.0
350	0.67	1.7	18.9

Obviously, the use of a higher untreated water temperature is attractive in minimising the area required, although in practice any advantages would be offset by increased water costs, and an optimisation procedure would be necessary in obtaining the most effective design.

9.10. OTHER FORMS OF EQUIPMENT

9.10.1. Finned tube units

Film coefficients

When viscous liquids are heated in a concentric tube or standard tubular exchanger by condensing steam or hot liquid of low viscosity, the film coefficient for the viscous liquid will be much smaller than that on the hot side and it will therefore control the rate of heat transfer. This condition also arises with air or gas heaters where the coefficient on the gas side will be very low compared with that for the liquid or condensing vapour on the other side. It is often possible to obtain a much better performance by increasing the area of surface on the side with the limiting coefficient. This may be done conveniently by using a finned tube as in Figure 9.62 which shows one typical form of such equipment which may have either longitudinal or transverse fins.

The calculation of the film coefficients on the fin side is complex because each unit of surface on the fin is less effective than a unit of surface on the tube wall. This arises because there will be a temperature gradient along the fin so that the temperature difference between the fin surface and the fluid will vary along the fin. To calculate the film coefficients it is convenient in the first place to consider the extended surface as shown in Figure 9.63. Here a cylindrical rod of length L and cross-sectional area A and perimeter b is heated at one end by a surface at temperature T_1 and cooled throughout its length by a medium at temperature T_G so that the cold end is at a temperature T_2.

(a)

(b)

Figure 9.62.   Finned tubes. (a) Longitudinal fins; (b) Radial fins

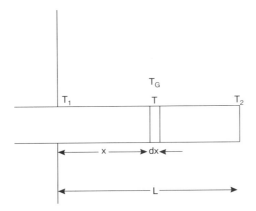

Figure 9.63.   Heat flow in rod with heat loss to surroundings

A heat balance over a length dx at distance x from the hot end gives:

$$\text{heat in} = \text{heat out along rod} + \text{heat lost to surroundings}$$

$$-kA\frac{\mathrm{d}T}{\mathrm{d}x} = \left[-kA\frac{\mathrm{d}T}{\mathrm{d}x} + \frac{\mathrm{d}}{\mathrm{d}x}\left(-kA\frac{\mathrm{d}T}{\mathrm{d}x}\right)\mathrm{d}x\right] + hb\,\mathrm{d}x(T - T_G)$$

where h is the film coefficient from fin to surroundings.
Writing the temperature difference $T - T_G$ equal to θ:

$$kA\frac{\mathrm{d}^2T}{\mathrm{d}x^2}\mathrm{d}x = hb\,\mathrm{d}x\,\theta$$

Since T_G is constant, $d^2T/dx^2 = d^2\theta/dx^2$.

$$\therefore \qquad \frac{d^2\theta}{dx^2} = \frac{hb}{kA}\theta = m^2\theta \quad \left(\text{where } m^2 = \frac{hb}{kA}\right) \qquad (9.175)$$

$$\therefore \qquad \theta = C_1 e^{mx} + C_2 e^{-mx} \qquad (9.176)$$

In solving this equation there are three important cases:

(a) Long rod with temperature falling to that of surroundings, i.e. $\theta = 0$ when $x = \infty$.

Then: $$\theta = \theta_1 e^{-mx} \qquad (9.177)$$

(b) Short rod from which heat loss from end is neglected.

$$\text{Hot end:} \quad x = 0, \quad \theta = \theta_1 = C_1 + C_2$$

$$\text{Cold end:} \quad x = L, \quad \frac{d\theta}{dx} = 0$$

Then: $$0 = C_1 m e^{mL} - C_2 m e^{-mL}$$

and: $$\theta_1 = C_1 + C_1 e^{2mL}$$

$$\therefore \qquad C_1 = \frac{\theta_1}{1 + e^{2mL}}, \quad C_2 = \frac{\theta_1}{1 + e^{-2mL}}$$

Hence: $$\theta = \frac{\theta_1 e^{mx}}{1 + e^{2mL}} + \frac{\theta_1 e^{2mL} e^{-mx}}{1 + e^{2mL}}$$

$$= \frac{\theta_1}{1 + e^{2mL}}[e^{mx} + e^{2mL} e^{-mx}] \qquad (9.178)$$

i.e.: $$\frac{\theta}{\theta_1} = \frac{e^{-mL} e^{mx} + e^{mL} e^{-mx}}{e^{-mL} + e^{mL}}$$

This may be written: $$\frac{\theta}{\theta_1} = \frac{\cosh m(L - x)}{\cosh mL} \qquad (9.179)$$

(c) More accurately, allowing for heat loss from the end:

$$\text{Hot end:} \quad x = 0, \quad \theta = \theta_1 = C_1 + C_2$$

$$\text{Cold end:} \quad x = L, \quad Q = hA\theta_{x=L} = -kA\left(\frac{d\theta}{dx}\right)_{x=L}$$

Determination of C_1 and C_2 in equation 9.176 then gives:

$$\theta = \frac{\theta_1}{1 + Je^{-2mL}}(Je^{-2mL} e^{mx} + e^{-mx}) \qquad (9.180)$$

where:
$$J = \frac{km - h}{km + h}$$

or, again, noting that $\cosh x = \frac{1}{2}(e^x + e^{-x})$ and $\sinh x = \frac{1}{2}(e^x - e^{-x})$:

$$\frac{\theta}{\theta_1} = \frac{\cosh m(L - x) + (h/mk)\sinh m(L - x)}{\cosh mL + (h/mk)\sinh mL} \quad (9.181)$$

The *heat loss* from a finned tube is obtained in the first place by determining the heat flow into the base of the fin from the tube surface. Thus the heat flow to the root of the fin is:

$$Q_f = -kA\left(\frac{dT}{dx}\right)_{x=0} = -kA\left(\frac{d\theta}{dx}\right)_{x=0}$$

For case (a):

$$Q_f = -kA(-m\theta_1) = kA\sqrt{\frac{hb}{kA}}\theta_1 = \sqrt{hbkA}\,\theta_1 \quad (9.182)$$

For case (b):

$$Q_f = -kAm\theta_1\frac{1 - e^{2mL}}{1 + e^{2mL}} = \sqrt{hbkA}\,\theta_1\tanh mL \quad (9.183)$$

For case (c):

$$Q_f = \sqrt{hbkA}\,\theta_1\left(\frac{1 - Je^{-2mL}}{1 + Je^{-2mL}}\right) \quad (9.184)$$

The above expressions are valid provided that the cross-section for heat flow remains constant. When it is not constant, e.g. for a radial or tapered fin, the temperature distribution is in the form of a Bessel function.

If the fin were such that there was no drop in temperature along its length, then the maximum rate of heat loss from the fin would be:

$$Q_{f\,max} = bLh\theta_1$$

The *fin effectiveness* is then given by $Q_f/Q_{f\,max}$ and for case (b) above this becomes:

$$\frac{kAm\theta_1\tanh mL}{bLh\theta_1} = \frac{\tanh ml}{mL} \quad (9.185)$$

Example 9.21

In order to measure the temperature of a gas flowing through a copper pipe, a thermometer pocket is fitted perpendicularly through the pipe wall, the open end making very good contact with the pipe wall. The pocket is made of copper tube, 10 mm o.d. and 0.9 mm wall, and it projects 75 mm into the pipe. A thermocouple is welded to the bottom of the tube and this gives a reading of 475 K when the wall temperature is at 365 K. If the coefficient of heat transfer between the gas and the copper tube is 140 W/m^2 K, calculate the gas temperature. The thermal conductivity of copper may be taken as 350 W/m K (Figure 9.64).

Figure 9.64.   Heat transfer to thermometer pocket

Solution

If θ is the temperature difference $(T - T_G)$:

$$\theta = \theta_1 \frac{\cosh m(L - x)}{\cosh mL}$$

At $x = L$:
$$\theta = \frac{\theta_1}{\cosh mL}$$

$$m^2 = \frac{hb}{kA}$$

where perimeter: $b = \pi \times 0.010$ m,   tube i.d. $= 8.2$ mm   or   0.0082 m

cross-sectional area of metal: $A = \dfrac{\pi}{4}(10.0^2 - 8.2^2) = 8.19\ \pi$ mm^2 or $8.19\ \pi \times 10^{-6}$ m^2

$$\therefore \qquad m^2 = \frac{140 \times 0.010\pi}{350 \times 8.19\pi \times 10^{-6}}$$

$$= 488 \text{ m}^{-2}$$

$$\therefore \qquad m = 22.1 \text{ m}^{-1}$$

$$\theta_1 = T_G - 365, \qquad \theta_2 = T_G - 475$$

$$\frac{\theta_1}{\theta_2} = \cosh mL$$

$$\therefore \qquad \frac{T_G - 365}{T_G - 475} = \cosh(22.1 \times 0.075) = 2.72$$

$$\therefore \qquad T_G = 539 \text{ K}$$

Example 9.22

A steel tube fitted with transverse circular steel fins of constant cross-section has the following specification:

tube o.d.: $d_2 = 54.0$ mm   fin diameter $d_1 = 70.0$ mm

fin thickness: $w = 2.0$ mm   number of fins/metre run $= 230$

Determine the heat loss per metre run of the tube when the surface temperature is 370 K and the temperature of the surroundings 280 K. The heat transfer coefficient between gas and fin is 30 W/m^2 K and the thermal conductivity of steel is 43 W/m K.

Solution

Assume that the height of the fin is small compared with its circumference and that it may be treated as a straight fin of length $(\pi/2)(d_1 + d_2)$.

Perimeter:
$$b = \frac{2\pi(d_1 + d_2)}{2} = \pi(d_1 + d_2)$$

Area:
$$A = \frac{\pi(d_1 + d_2)w}{2}, \quad \text{i.e. the average area at right angles to the heat flow}$$

Then:
$$m = \left(\frac{hb}{kA}\right)^{0.5} = \left\{\frac{h\pi(d_1 + d_2)}{[k\pi(d_1 + d_2)w/2]}\right\}^{0.5}$$

$$= \left(\frac{2h}{kw}\right)^{0.5}$$

$$= \left(\frac{2 \times 30}{43 \times 0.002}\right)^{0.5}$$

$$= 26.42 \text{ m}^{-1}$$

From equation 9.183, the heat flow is given for case (b) as:

$$Q_f = mkA\theta_1 \frac{e^{2mL} - 1}{1 + e^{2mL}}$$

In this equation:
$$A = \frac{\pi(70.0 + 54.0) \times 2.0}{2} = 390 \text{ mm}^2 \quad \text{or} \quad 0.00039 \text{ m}^2$$

$$L = \frac{d_1 - d_2}{2} = 8.0 \text{ mm} \quad \text{or} \quad 0.008 \text{ m}$$

$$mL = 26.42 \times 0.008 = 0.211$$

$$\theta_1 = 370 - 280 = 90 \text{ K}$$

$$\therefore \quad Q_f = \frac{26.42 \times 43 \times 3.9 \times 10^{-4} \times 90(e^{0.422} - 1)}{1 + e^{0.422}}$$

$$= \frac{39.9 \times 0.525}{2.525} = 8.29 \text{ W per fin}$$

$\therefore$ Heat loss per metre run of tube $= 8.29 \times 230 = 1907$ W/m

or:
$$\underline{\underline{1.91 \text{ kW/m}}}$$

In this case the low value of mL indicates a fin efficiency of almost 1.0, though where mL tends to 1.0 the efficiency falls to about 0.8.

Practical data

A neat form of construction has been designed by the Brown Fintube Company of America. On both prongs of a hairpin tube are fitted horizontal fins which fit inside concentric tubes, joined at the base of the hairpin. Units of this form can conveniently be arranged in banks to give large heat transfer surfaces. It is usual for the extended surface to be at least five times greater than the inside surface, so that the low coefficient on the fin side is balanced by the increase in surface. An indication of the surface obtained is given in Table 9.16.

A typical hairpin unit with an effective surface on the fin side of 9.4 m² has an overall length of 6.6 m, height of 0.34 m, and width of 0.2 m. The free area for flow on the fin

Table 9.16.   Data on surface of finned tube units[a]

Pipe size outside diameter (in)	Outside surface of pipe (ft²/ft length)	Number of fins	Surface of finned pipe			
			(ft²/ft)		(m²/m)	
			Height of fin		Height of fin	
			0.5 in	1 in	12.7 mm	25.4 mm
1	0.262	12	1.262	2.262	0.385	0.689
(25.4 mm)	(0.08 m²/m)	16	1.595	2.929	0.486	0.893
		20	1.927	3.595	0.587	1.096
1.9	0.497	20	2.164	3.830	0.660	1.167
(48.3 mm)	(0.15 m²/m)	24	2.497	4.497	0.761	1.371
		28	2.830	5.164	0.863	1.574
		36	3.498	6.497	1.066	1.980

[a] Brown Fintube Company.

side is 2.645 mm² against 1.320 mm² on the inside; the ratio of the transfer surface on the fin side to that inside the tubes is 5.93:1.

The fin side film coefficient h_f has been expressed by plotting:

$$\frac{h_f}{C_p G'} \left(\frac{C_p \mu}{k} \right)^{2/3} \left(\frac{\mu}{\mu_s} \right)^{-0.14} \qquad \text{against} \qquad \frac{d_e G'}{\mu}$$

where h_f is based on the total finside surface area (fin and tube), G' is the mass rate of flow per unit area, and d_e is the equivalent diameter, i.e.:

$$d_e = \frac{4 \times \text{cross-sectional area for flow on fin side}}{\text{total wetted perimeter for flow (fin + outside of tube + inner surface of shell tube)}}$$

Experimental work has been carried out with exchangers in which the inside tube was 48 mm outside diameter and was fitted with 24, 28, or 36 fins (12.5 mm by 0.9 mm) in a 6.1 m length; the finned tubes were inserted inside tubes 90 mm inside diameter. With steam on the tube side, and tube oils and kerosene on the fin side, the experimental data were well correlated by plotting:

$$\frac{h_f}{C_p G'} \left(\frac{C_p \mu}{k} \right)^{2/3} \left(\frac{\mu}{\mu_s} \right)^{-0.14} \qquad \text{against} \qquad \frac{d_e G'}{\mu}$$

Typical values were:

$$\frac{h_f}{C_p G'} \left(\frac{C_p \mu}{k} \right)^{2/3} \left(\frac{\mu}{\mu_s} \right)^{-0.14} = 0.25 \quad 0.055 \quad 0.012 \quad 0.004$$

$$\frac{d_e G'}{\mu} = \quad 1 \quad 10 \quad 100 \quad 1000$$

Some indication of the performance obtained with *transverse finned tubes* is given in Table 9.17. The figures show the heat transferred per unit length of pipe when heating air on the fin side with steam or hot water on the tube side, using a temperature difference of 100 K. The results are given for three different spacings of the fins.

Table 9.17. Data on finned tubes

Inside diam. of tube	$\frac{3}{4}$ in	1 in	$1\frac{1}{2}$ in	2 in	3 in
Outside diam. of fin	$2\frac{1}{2}$ in	$2\frac{3}{4}$ in	$3\frac{7}{8}$ in	$4\frac{5}{16}$ in	$5\frac{1}{2}$ in
No. of fins/ft run	Heat transferred (Btu/h ft)				
20	485	650	1010	1115	1440
24	505	665	1060	1170	1495
30	565	720	1190	1295	1655
Inside diam. of tube	19 mm	25 mm	38 mm	50 mm	75 mm
Outside diam. of fin	64 mm	70 mm	100 mm	110 mm	140 mm
No. of fins/m run	Heat transferred (kW/m)				
65	0.47	0.63	0.37	1.07	1.38
80	0.49	0.64	1.02	1.12	1.44
100	0.54	0.69	1.14	1.24	1.59

(Data taken from catalogue of G. A. Harvey and Co. Ltd. of London.)

9.10.2. Plate type exchangers

A series of plate type heat exchangers which present some special features was first developed by the APV Company. The general construction is shown in Figure 9.65, which shows an Alfa-Laval exchanger and from which it is seen that the equipment consists of a series of parallel plates held firmly together between substantial head frames. The plates are one-piece pressings, frequently of stainless steel, and are spaced by rubber sealing gaskets cemented into a channel around the edge of each plate. Each plate has a number of troughs pressed out at right angles to the direction of flow and arranged so that they interlink with each other to form a channel of constantly changing direction and section. With normal construction the gap between the plates is 1.3–1.5 mm. Each liquid flows in alternate spaces and a large surface can be obtained in a small volume.

Because of the shape of the plates, the developed area of surface is appreciably greater than the projected area; this is shown below for the four common sizes of plate (Table 9.18).

Table 9.18. Plate areas

Plate type	Projected area		Developed area	
	ft^2	m^2	ft^2	m^2
HT	1.00	0.09	1.35	0.13
HX	1.45	0.13	1.81	0.17
HM	2.88	0.27	3.73	0.35
HF	3.85	0.36	4.60	0.43

A high degree of turbulence is obtained even at low flowrates and the high heat transfer coefficients obtained are illustrated by the data in Table 9.19: these refer to the heating of cold water by the equal flow of hot water in an HF type exchanger (aluminium or copper), at an average temperature of 310 K.

Using a stainless steel plate with a flow of 0.00114 m^3/s, the heat transferred is 1760 W/K for each plate.

Figure 9.65.   Alfa-Laval plate heat exchanger

Table 9.19.   Performance of plate-type exchanger type HF

Heat transferred per plate		Water flow		U based on developed area	
Btu/h °F	W/K	gal/h	cm^3/s	Btu/h ft^2 °F	kW/m^2 K
3000	1580	550	700	650	3.70
4000	2110	850	1075	870	4.94
5000	2640	1250	1580	1080	6.13

(Courtesy of the APV Company.)

The high transfer coefficient enables these exchangers to be operated with very small temperature differences, so that a high heat recovery is obtained. These units have been particularly successful in the dairy and brewing industries, where the low liquid capacity and the close control of temperature have been valuable features. A further advantage is that they are easily dismantled for inspection of the whole plate. The necessity for the long gasket is an inherent weakness, but the exchangers have been worked successfully up to 423 K and at pressures of 930 kN/m^2. They are now being used in the processing and gas industries with solvents, sugar, acetic acid, ammoniacal liquor, etc.

9.10.3. Spiral heat exchangers

A spiral plate exchanger is illustrated in Figure 9.66. Here two fluids flow through the channels formed between the spiral plates. With this form of construction the velocity may be as high as 2.1 m/s and overall transfer coefficients of 2.8 kW/m^2 K are frequently obtained. The size can therefore be kept relatively small and the cost becomes comparable or even less than that of tubular units, particularly when they are fabricated from alloy steels.

Figure 9.66.   Spiral plate heat exchanger

A further design of spiral heat exchanger, described by NEIL[94], is essentially a single pass counterflow heat exchanger with fixed tube plates distinguished by the spiral winding of the tubes, each consisting, typically of a 10 mm o.d. tube wound on to a 38 mm o.d. coil such that the inner heat transfer coefficient is 1.92 times greater than for a straight tube. The construction overcomes problems of differential expansion rates of the tubes and the shell and the characteristics of the design enable the unit to perform well on superheated steam where the combination of counterflow, high surface area per unit volume of shell and the high inside coefficient of heat transfer enables the superheat to be removed effectively in a unit of reasonable size and cost.

9.10.4. Compact heat exchangers

Advantages of compact units

In general, heat exchanger equipment accounts for some 10 per cent of the cost of a plant; a level at which there is no great incentive for innovation. However, trends such as the growth in energy conservation schemes, the general move from bulk chemicals to value-added products and plant problems associated with drilling rigs have all prompted the development and increased application of compact heat exchangers. Here compactness is a matter of degree, maximising the heat transfer area per unit volume of exchanger and this leads to narrow channels. Making shell and tube exchangers more compact presents construction problems and a more realistic approach is to use plate or plate and fin heat exchangers. The relation between various types of exchanger, in terms of the heat transfer area per unit volume of exchanger is shown in Figure 9.67, taken from REDMAN[95]. In order to obtain a thermal effectiveness in excess of 90 per cent, countercurrent flow is highly desirable and this is not easily achieved in shell and tube units which often have a number of tube-side passes in order to maintain reasonable velocities. Because of the baffle design on the shell-side, the flow at the best may be described as cross-flow, and the situation is only partly redeemed by having a train of exchangers which is an expensive solution. Again in dealing with high value added products, which could well be heat sensitive, a more controllable heat exchanger than a shell and tube unit, in which not

Figure 9.67.   Surface area as a function of volume of exchanger for different types

all the fluid is heated to the same extent, might be desirable. One important application of compact heat exchangers is the cooling of natural gas on offshore rigs where space (costing as much as £120,000/m² is of paramount importance.

Plate and fin exchangers

Plate and fin heat exchangers, used in the motor, aircraft and transport industries for many years, are finding increased application in the processing industry and in particular in natural gas liquefaction, cryogenic air separation, the production of olefins and in the separation of hydrogen and carbon monoxide. Potential applications include ammonia production, offshore processing, nuclear engineering and syngas production. As described by GREGORY[96], the concept is that of flat plates of metal, usually aluminium, with corrugated metal, which not only holds the plates together but also acts as a secondary surface for heat transfer. Bars at the edges of the plates retain each fluid between adjacent plates; and the space between each pair of plates, apportioned to each fluid according to the heat transfer and pressure drop requirements, is known as a layer. The heights of the bars and corrugations are standardised in the U.K. at 3.8, 6.35, and 8.9 mm and typical designs are shown in Figure 9.68.

Figure 9.68.   Plate and fin exchangers

There are four basic forms of corrugation: *plain* in which a sheet of metal is corrugated in the simplest way with fins at right angles to the plates; *serrated* where each cut is offset in relation to the preceding fin; *herringbone* corrugation made by displacing the fins sideways every 9.5 mm to give a zig-zag path; and *perforated* corrugation, a term used for a plain corrugation made from perforated material. Each stream to be heated or cooled can have different corrugation heights, different corrugation types, different numbers of layers and different entry and exit points including part length entry and exit. Because the surface in the hot fluid (or fluids) can vary widely from that in the cold fluid (or fluids) it is unrealistic to quote surface areas for either side, as in shell and tube units, though the overall surface area available to all the fluids is 1000 m²/m³ of exchanger.

In design, the general approach is to obtain the term (hA) for each stream, sum these for all the cold and all the hot streams and determine an overall value of (hA) given by:

$$\frac{1}{(hA)_{ov}} = \frac{1}{(hA)_{w}} + \frac{1}{(hA)_{c}} \qquad\qquad (9.186)$$

where ov, w and c refer, respectively, to overall, hot-side and cold-side values.

Printed-circuit exchangers

Various devices such as small diameter tubes, fins, tube inserts and porous boiling surfaces may be used to improve the surface density and heat transfer coefficients in shell and tube heat exchangers and yet these are not universally applicable and, in general, such units remain essentially bulky. Plate and fin exchangers have either limited fluid compatibility or limited fin efficiency, problems which are overcome by using printed-circuit exchangers as described by JOHNSTON[97]. These are constructed from flat metal plates which have fluid flow passages chemically milled into them by means of much the same techniques as are used to produce electrical printed circuits. These plates are diffusion-bonded together to form blocks riddled with precisely sized, shaped, routed and positioned passages; and these blocks are in turn welded together to form heat exchange cores of the required capacity. Fluid headers are attached to the core faces and sometimes the assembly is encapsulated. Passages are typically 0.3–1.5 mm deep giving surface areas of 1000–5000 m^2/m^3, an order of magnitude higher than surface densities in shell and tube designs; and in addition the fine passages tend to sustain relatively high heat transfer coefficients, undiminished by fin inefficiencies, and so less surface is required.

In designing a unit, each side of the exchanger is independently tailored to the duty required and the exchanger effectiveness (Section 9.9.4) can range from 2–5 per cent to values in excess of 98 per cent without fundamental design or construction problems arising. Countercurrent, co-current and cross-flow contacting can be employed individually or in combination.

A note of caution on the use of photo-etched channels has been offered by RAMSHAW[98] who points out that the system is attractive in principle provided that severe practical problems such as fouling are not encountered. With laminar flow in matrices with a mean plate spacing of 0.3–1 mm, volumetric heat transfer coefficients of 7 MW/m^3 K have been obtained with modest pressure drops. Such values compare with 0.2 MW/m^3 K for shell and tube exchangers and 1.2 MW/m^3 K for plate heat exchangers.

9.10.5. Scraped-surface heat exchangers

In cases where a process fluid is likely to crystallise on cooling or the degree of fouling is very high or indeed the fluid is of very high viscosity, use is often made of scraped-surface heat exchangers in which a rotating element has spring-loaded scraper blades which wipe the inside surface of a tube which may typically be 0.15 m in diameter. Double-pipe construction is often employed with a jacket, say 0.20 m in diameter, and one common arrangement is to connect several sections in series or to install several pipes within a common shell. Scraped-surface units of this type are used in paraffin-wax plants and for evaporating viscous or heat-sensitive materials under high vacuum; this is an application to which the *thin-film device* is especially suited because of the very short residence times involved. In such a device the clearance between the agitator and the wall may be either fixed or variable since both rigid and hinged blades may be used. The process liquid is continuously spread in a thin layer over the vessel wall and it moves through the device either by the action of gravity or of the agitator (a tapered or helical agitator produces longitudinal forces on the liquid) or of both.

In describing chillers for the production of wax distillates, NELSON[99] points out that the rate of cooling depends very much on the effectiveness of the scrapers, and quotes overall coefficients of heat transfer ranging from 15 W/m^2 K with a poorly fitting scraper to 90 W/m^2 K where close fitting scrapers remove the wax effectively from the chilled surface.

The *Votator* design has two or more floating scraper-agitators which are forced against the cylinder wall by the hydrodynamic action of the fluid on the agitator and by centrifugal action; the blades are loosely attached to a central shaft called the mutator. The votator is used extensively in the food processing industries and also in the manufacture of greases and detergents. As the blades are free to move, the clearance between the blades and the wall varies with operating conditions and a typical installation may be 75–100 mm in diameter and 0.6–1.2 m long. In the *spring-loaded* type of scraped-surface heat exchanger, the scrapers are held against the wall by leaf springs, and again there is a variable clearance between the agitator and the cylinder wall since the spring force is balanced by the radial hydrodynamic force of the liquid on the scraper. Typical applications of this device are the processing of heavy waxes and oils and crystallising solutions; generally the units are 0.15–0.3 m in diameter and up to 12 m long. Some of the more specialised heat exchangers and chemical reactors employ helical ribbons, augers or twisted tapes as agitators and, in general, these are fixed clearance devices used for high viscosity materials. There is no general rule as to maximum or minimum dimensions since each application is a special case.

One of the earliest investigations into the effectiveness of scrapers for improving heat transfer was that of HUGGINS[100] who found that although the improvement with water was negligible, cooling times for more viscous materials could be considerably reduced. This was confirmed by LAUGHLIN[101] who has presented operating data on a system where the process fluid changes from a thin liquid to a paste and finally to a powder. HOULTON[102], in tests on the votator, found that back-mixing was negligible and some useful data on a number of food products have been obtained by BOLANOWSKI and LINEBERRY[103] who, in addition to discussing the operation and uses of the votator, quote overall heat transfer coefficients for each food tested. Using a *liquid-full* system, SKELLAND et al.[104–106] have proposed the following general design correlation for the votator:

$$\frac{hd_v}{k} = c_1 \left(\frac{C_p\mu}{k}\right)^{c_2} \left(\frac{(d_v - d_r)u\rho}{\mu}\right) \left(\frac{d_v N}{u}\right)^{0.82} \left(\frac{d_r}{d_v}\right)^{0.55} (n_B)^{0.53} \qquad (9.187)$$

where for cooling viscous liquids $c_1 = 0.014$ and $c_2 = 0.96$, and for thin mobile liquids $c_1 = 0.039$ and $c_2 = 0.70$. In this correlation d_v is the diameter of the vessel and d_r is the diameter of the rotor and u is the average axial velocity of the liquid. This correlation can only be applied to the range of experimental data upon which it is based, since h will not approach zero as n_B, d_r, $(d_v - d_r)$, N and u approach zero. Reference to the use of the votator for crystallisation is made in Volume 2, Chapter 15.

The majority of work on heat transfer in *thin-film* systems has been directed towards obtaining data on specific systems rather than developing general design methods, although BOTT et al.[107–109] have developed the following correlations for heating without change of phase:

$$\frac{hd_v}{k} = Nu = 0.018 Re''^{0.6} Re^{0.46} Pr^{0.87} \left(\frac{d_v}{l}\right)^{0.48} (n_B)^{0.24} \qquad (9.188)$$

and for evaporation:

$$\frac{hd_v}{k} = Nu = 0.65Re''^{0.43}Re^{0.25}Pr^{0.3}(n_B)^{0.33} \tag{9.189}$$

From both of these equations, it will be noted that the heat transfer coefficient is *not* a function of the temperature difference. Here $Re'' = (d_v^2N\rho/\mu)$ and $Re = (ud_v\rho/\mu)$, where d_v is the tube diameter and u is the average velocity of the liquid in the film in the axial direction.

It is also of significance that the agitation suppresses nucleation in a fluid which might otherwise deposit crystals.

9.11. THERMAL INSULATION

9.11.1. Heat losses through lagging

A hot reaction or storage vessel or a steam pipe will lose heat to the atmosphere by radiation, conduction, and convection. The loss by radiation is a function of the fourth power of the absolute temperatures of the body and surroundings, and will be small for low temperature differences but will increase rapidly as the temperature difference increases. Air is a very poor conductor, and the heat loss by conduction will therefore be small except possibly through the supporting structure. On the other hand, since convection currents form very easily, the heat loss from an unlagged surface is considerable. The conservation of heat, and hence usually of fuel, is an economic necessity, and some form of lagging should normally be applied to hot surfaces. Lagging of plant operating at high temperatures is also necessary in order to achieve acceptable working conditions in the vicinity. In furnaces, as has already been seen, the surface temperature is reduced substantially by using a series of insulating bricks which are poor conductors.

The two main requirements of a good lagging material are that it should have a low thermal conductivity and that it should suppress convection currents. The materials that are frequently used are cork, asbestos, 85 per cent magnesia, glass wool, and vermiculite. Cork is a very good insulator but it becomes charred at moderate temperatures and is used mainly in refrigerating plants. Eighty-five per cent magnesia is widely used for lagging steam pipes and may be applied either as a hot plastic material or in preformed sections. The preformed sections are quickly fitted and can frequently be dismantled and re-used whereas the plastic material must be applied to a hot surface and cannot be re-used. Thin metal sheeting is often used to protect the lagging.

The rate of heat loss per unit area is given by:

$$\frac{\text{total temperature difference}}{\text{total thermal resistance}}$$

For the case of heat loss to the atmosphere from a lagged steam pipe, the thermal resistance is due to that of the condensate film and dirt on the inside of the pipe, that of the pipe wall, that of the lagging, and that of the air film outside the lagging. Thus for unit length of a lagged pipe:

$$\frac{Q}{l} = \Sigma\Delta T \left/ \left[\frac{1}{h_i\pi d} + \frac{x_w}{k_w\pi d_w} + \frac{x_l}{k_r\pi d_m} + \frac{1}{(h_r + h_c)\pi d_s}\right]\right. \tag{9.190}$$

where d is the inside diameter of pipe, d_w the mean diameter of pipe wall, d_m the logarithmic mean diameter of lagging, d_s the outside diameter of lagging, x_w, x_l are the pipe wall and lagging thickness respectively, k_w, k_r the thermal conductivity of the pipe wall and lagging, and h_i, h_r, h_c the inside film, radiation, and convection coefficients.

Example 9.23

A steam pipe (150 mm i.d. and 168 mm o.d.) is carrying steam at 444 K and is lagged with 50 mm of 85 per cent magnesia. What is the heat loss to air at 294 K?

Solution

In this case:

$$d = 150 \text{ mm} \quad \text{or} \quad 0.150 \text{ m}$$

$$d_o = 168 \text{ mm} \quad \text{or} \quad 0.168 \text{ m}$$

$$d_w = 159 \text{ mm} \quad \text{or} \quad 0.159 \text{ m}$$

$$d_s = 268 \text{ mm} \quad \text{or} \quad 0.268 \text{ m}$$

$$d_m, \text{ the log mean of } d_o \text{ and } d_s = 215 \text{ mm} \quad \text{or} \quad 0.215 \text{ m}.$$

The coefficient for condensing steam together with that for any scale will be taken as 8500 W/m² K, k_w as 45 W/m K, and k_l as 0.073 W/m K.

The temperature on the outside of the lagging is estimated at 314 K and $(h_r + h_c)$ will be taken as 10 W/m² K.

The thermal resistances are therefore:

$$\frac{1}{h_i \pi d} = \frac{1}{8500 \times \pi \times 0.150} = 0.00025$$

$$\frac{x_w}{k_w \pi d_w} = \frac{0.009}{45 \times \pi \times 0.159} = 0.00040$$

$$\frac{x_l}{k_l \pi d_m} = \frac{0.050}{0.073 \times \pi \times 0.215} = 1.013$$

$$\frac{1}{(h_r + h_c) \pi d_s} = \frac{1}{10 \times \pi \times 0.268} = 0.119$$

The first two terms may be neglected and hence the total thermal resistance is 1.132 m K/W. The heat loss per metre length $= (444 - 294)/1.132 = \underline{\underline{132.5 \text{ W/m}}}$ (from equation 9.190).

The temperature on the outside of the lagging can now be checked as follows:

$$\frac{\Delta T (\text{lagging})}{\Sigma \Delta T} = \frac{1.013}{1.132} = 0.895$$

$$\Delta T (\text{lagging}) = 0.895(444 - 294) = 134 \text{ K}$$

Thus the temperature on the outside of the lagging is $444 - 134 = 310$ K, which approximates to the assumed value.

Taking an emissivity of 0.9, from equation 9.103:

$$h_r = \frac{5.67 \times 10^{-8} \times 0.9(310^4 - 294^4)}{310 - 294} = 7.40 \text{ W/m}^2 \text{ K}$$

From Table 9.5 for air $(Gr \, Pr = 10^4 - 10^9)$, $n = 0.25$ and $C'' = 1.32$.
Substituting in equation 9.94 (putting $l = $ diameter $= 0.268$ m):

and:
$$h_c = C''(\Delta T)^n l^{3n-1} = 1.32 \left[\frac{310 - 294}{0.268} \right]^{0.25} = 3.67 \text{ W/m}^2 \text{ K}$$

Thus $(h_r + h_c) = 11.1$ W/m^2 K, which is close to the assumed value. In practice it is rare for forced convection currents to be absent, and the heat loss is probably higher than this value.

If the pipe were unlagged, $(h_r + h_c)$ for $\Delta T = 150$ K would be about 20 W/m^2 K and the heat loss would then be:

$$\frac{Q}{l} = (h_r + h_c)\pi d_o \Delta T$$

$$= 20 \times \pi \times 0.168 \times 150 = 1584 \text{ W/m}$$

or: 1.58 kW/m

Under these conditions it is seen that the heat loss has been reduced by more than 90 per cent by the addition of 50 mm of lagging.

9.11.2. Economic thickness of lagging

Increasing the thickness of the lagging will reduce the loss of heat and thus give a saving in the operating costs. The cost of the lagging will increase with thickness and there will be an optimum thickness when further increase does not save sufficient heat to justify the cost. In general the smaller the pipe the smaller the thickness used, but it cannot be too strongly stressed that some lagging everywhere is better than excellent lagging in some places and none in others. For temperatures of 373–423 K, and for pipes up to 150 mm diameter, LYLE[110] recommends a 25 mm thickness of 85 per cent magnesia lagging and 50 mm for pipes over 230 mm diameter. With temperatures of 470–520 K he suggests 38 mm for pipes less than 75 mm diameter and 50 mm for pipes up to 230 mm diameter.

9.11.3. Critical thickness of lagging

As the thickness of the lagging is increased, resistance to heat transfer by thermal conduction increases. However, the outside area from which heat is lost to the surroundings also increases, giving rise to the possibility of increased heat loss. It is perhaps easiest to think of the lagging as acting as a fin of very low thermal conductivity. It will now be shown that, for a cylindrical pipe, there is the possibility of heat losses being increased by the application of lagging, only if $hr/k < 1$, where k is the thermal conductivity of the lagging, h is the outside film coefficient, and r is the outside diameter of the pipe. In practice, this situation is likely to arise only for pipes of small diameters.

The heat loss from a pipe at a temperature T_S to surroundings at temperature T_A will be considered. Heat flows through lagging of thickness x across which the temperature falls from a constant value T_S at its inner radius r, to an outside temperature T_L which is a function of x. (See Figure 9.69.) The rate of heat loss Q from a length l of pipe will be given by equation 9.191, by considering the heat loss from the outside of the lagging; and by equations 9.192 and 9.193, which give the transfer rate by thermal conduction through the lagging of logarithmic mean radius r_m:

$$Q = 2\pi l(r + x)(T_L - T_A)h \qquad (9.191)$$

$$Q = \frac{2\pi l r_m}{x}k(T_S - T_L) \qquad (9.192)$$

$$= \frac{2\pi l k}{\ln \dfrac{r + x}{r}}(T_S - T_L) \qquad (9.193)$$

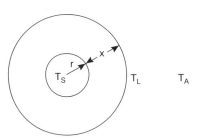

Figure 9.69.   Critical lagging thickness

Equating the values given in equations 9.191 and 9.193:

$$(r + x)h(T_L - T_A) = \frac{k}{\ln \dfrac{r+x}{r}}(T_S - T_L)$$

Then:

$$a = (r + x)\frac{h}{k} \ln \frac{r+x}{r} = \frac{T_S - T_L}{T_L - T_A}$$

$$aT_L - aT_A = T_S - T_L$$

$$T_L = \frac{T_S + aT_A}{a + 1}$$

Substituting in equation 9.191:

$$Q = 2\pi l(r + x)\left\{ \left(\frac{T_S + aT_A}{a + 1} - T_A \right) \right\} h$$

$$= \frac{2\pi l h(r + x)}{a + 1}(T_S - T_A)$$

$$= 2\pi l h(T_S - T_A)\left\{ (r + x)\frac{1}{1 + (r + x)\dfrac{h}{k} \ln \left(\dfrac{r + x}{r} \right)} \right\} \qquad (9.194)$$

Differentiating with respect to x:

$$\frac{1}{2\pi l h(T_S - T_A)}\frac{dQ}{dx} = \frac{\left\{ 1 + (r + x)\dfrac{h}{k} \ln \dfrac{r+x}{r} - (r + x)\left[\dfrac{h}{k} \ln \dfrac{r+x}{r} + (r + x)\dfrac{h}{k}\dfrac{r}{(r+x)}\dfrac{1}{r} \right] \right\}}{\left[1 + (r + x)\dfrac{h}{k} \ln \left(\dfrac{r+x}{r} \right) \right]^2}$$

The maximum value of $Q(Q_{max})$ occurs when $dQ/dx = 0$.

i.e. when:

$$1 - (r + x)\frac{h}{k} = 0$$

or:

$$x = \frac{k}{h} - r \qquad (9.195)$$

When the relation between heat loss and lagging thickness exhibits a maximum for the unlagged pipe ($x = 0$):

$$\frac{hr}{k} = 1 \qquad (9.196)$$

When $hr/k > 1$, the addition of lagging always reduces the heat loss.

When $hr/k < 1$, thin layers of lagging increase the heat loss and it is necessary to exceed the critical thickness given by equation 9.195 before any benefit is obtained from the lagging.

Substituting in equation 9.194 gives the maximum heat loss as:

$$Q_{max} = 2\pi l h(T_S - T_A) \left\{ \frac{k}{h} \frac{1}{1 + \dfrac{kh}{hk} \ln \dfrac{k}{hr}} \right\}$$

$$= 2\pi l (T_S - T_A) k \frac{1}{1 + \ln \dfrac{k}{hr}} \qquad (9.197)$$

For an unlagged pipe, $x = 0$ and $T_L = T_S$. Substitution in equation 9.191 gives the rate of heat loss Q_o:

$$Q_o = 2\pi l r(T_S - T_A)h \qquad (9.197)$$

Thus:

$$\frac{Q_{max}}{Q_o} = \frac{k}{rh} \bigg/ \left(1 + \ln \frac{k}{rh}\right) \qquad (9.198)$$

The ratio Q/Q_o is plotted as a function of thickness of lagging (x) in Figure 9.70.

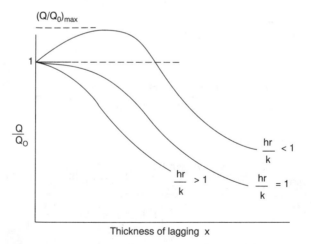

Figure 9.70.   Critical thickness of lagging

Example 9.24

A pipeline of 100 mm outside diameter, carrying steam at 420 K, is to be insulated with a lagging material which costs £10/m^3 and which has a thermal conductivity of 0.1 W/m K. The ambient temperature may be taken as

285 K, and the coefficient of heat transfer from the outside of the lagging to the surroundings as 10 W/m² K. If the value of heat energy is 7.5×10^{-4} £/MJ and the capital cost of the lagging is to be depreciated over 5 years with an effective simple interest rate of 10 per cent per annum based on the initial investment, what is the economic thickness of the lagging?

Is there any possibility that the heat loss could actually be increased by the application of too thin a layer of lagging?

Solution

For a thick-walled cylinder, the rate of conduction of heat through lagging is given by equation 9.21:

$$Q = \frac{2\pi l k(T_i - T_o)}{\ln(d_o/d_i)} \quad \text{W}$$

where d_o and d_i are the external and internal diameters of the lagging and T_o and T_i the corresponding temperatures.

Substituting $k = 0.1$ W/mK, $T_o = 420$K (neglecting temperature drop across pipe wall), and $d_i = 0.1$ m:

$$\frac{Q}{l} = \frac{2\pi \times 0.1(420 - T_o)}{\ln(d_o/0.1)} \quad \text{W/m}$$

The term Q/l must also equal the heat loss from the outside of the lagging,

or:

$$\frac{Q}{l} = h_o(T_o - 285)\pi d_o = 10(T_o - 285)\pi d_o \quad \text{W/m}$$

Thus:

$$T_o = \left\{ \frac{Q}{l} \frac{1}{10\pi d_o} + 285 \right\} \quad \text{K}$$

Substituting:

$$\frac{Q}{l} = \frac{2\pi \times 0.1 \left[135 - \dfrac{Q}{l} \dfrac{1}{10\pi d_o} \right]}{\ln(d_o/0.1)}$$

or:

$$\frac{Q}{l} = \frac{2\pi \times 0.1 \times 135}{\ln(d_o/0.1) + 2\pi \times 0.1 \times \dfrac{1}{10\pi d_o}} = \frac{84.82}{\ln(d_o/0.1) + (0.02/d_o)} \quad \text{W/m}$$

Value of heat lost $= £7.5 \times 10^{-4}$/MJ

or:

$$\frac{84.82}{\ln(d_o/0.1) + (0.02/d_o)} \times 7.5 \times 10^{-4} \times 10^{-6} = \frac{6.36 \times 10^{-8}}{\ln(d_o/2.1) + (0.02/d_o)} \quad £/\text{m s}$$

Volume of lagging per unit pipe length $= \dfrac{\pi}{4}[d_o^2 - (0.1)^2]$ m³/m

Capital cost of lagging $= £10/\text{m}^3$ or $\dfrac{\pi}{4}[d_o^2 - 0.01]10 = 7.85(d_o^2 - 0.01)$ £/m

Then, noting that 1 year $= 31.5$ m s :

Depreciation $= 7.85[d_o^2 - 0.01]/(5 \times 31.5 \times 10^6) = 4.98 \times 10^{-8}(d_o^2 - 0.01)$ £/m s

Interest charges $= (0.1 \times 7.85)(d_o^2 - 0.01)/(31.5 \times 10^6) = 2.49 \times 10^{-8}(d_o^2 - 0.01)$ £/m s

Total capital charges $= 7.47 \times 10^{-8}(d_o^2 - 0.01)$ £/m s

Total cost (capital charges + value of heat lost)

$$C = \left\{ \frac{6.36}{\ln(d_o/0.1) + (0.02/d_o)} + 7.47(d_o^2 - 0.01) \right\} 10^{-8} \quad £/\text{m s}$$

Differentiating with respect to d_o:

$$10^8 \frac{dC}{dd_o} = 6.36 \left[\frac{-1}{[\ln(d_o/0.1) + (0.02/d_o)]^2} \right] \left[\frac{1}{d_o} - \frac{0.02}{d_o^2} \right] + 7.47(2d_o)$$

In order to obtain the minimum value of C, dC/dd_o must be put equal to zero.

Then:
$$\frac{1}{[\ln(d_o/0.1) + (0.02/d_o)]^2} = \frac{7.47 \times 2.}{6.36} \left[\frac{d_o}{(1/d_o) - (0.02/d_o^2)} \right]$$

i.e.:
$$\frac{1}{[\ln(d_o/0.1) + (0.02/d_o)]^2} = 2.35 \frac{d_o^3}{d_o - 0.02}$$

A trial and error solution gives $d_o = 0.426$m or 426 mm
Thus, the economic thickness of lagging $= (426 - 100)/2 = 163$ mm

For this pipeline:
$$\frac{hr}{k} = \frac{10 \times (50 \times 10^{-3})}{0.1} = 5$$

From equation 9.196, the critical value of hr/k, below which the heat loss may be increased by a thin layer of lagging, is 1. For $hr/k > 1$, as in this problem, the situation will not arise.

9.12. FURTHER READING

AZBEL, D.: *Heat Transfer Applications in Process Engineering* (Noyes, New York, 1984).
BACKHURST, J. R. and HARKER, J. H.: *Process Plant Design* (Heinemann, London, 1973).
BIRD, R. B., STEWART, W. E., and LIGHTFOOT, E. N.: *Transport Phenomena* (Wiley, New York, 1960).
CHAPMAN, A. J.: *Heat Transfer*, 2nd edn (Macmillan, New York, 1967).
CHEREMISINOFF, N. P. (ed.): *Handbook of Heat and Mass Transfer*: Vol. 1, Heat Transfer Operations. (Gulf Publications, 1986).
COLLIER, J. G.: *Convective Boiling and Condensation* (McGraw-Hill, New York, 1972).
ECKERT, E. R. G. and DRAKE, R. M., Jr.: *Analysis of Heat and Mass Transfer* (McGraw-Hill, New York, 1972).
GEBHART, B.: *Heat Transfer*, 2nd edn (McGraw-Hill, New York, 1971).
GROBER, H., ERK, E., and GRIGULL, U.: *Fundamentals of Heat Transfer* (McGraw-Hill, New York, 1961).
HALLSTRÖM, B., SKJÖLDEBRAND, C. and TRÄCÄRDH, C.: *Heat Transfer and Food Products* (Elsevier Applied Science, London, 1988).
HEWITT, G. F., SHIRES, G. L., and BOTT, T. R.: *Process Heat Transfer* (CRC Press, 1994).
INCROPERO, F. D. and DE WITT, D. P.: *Fundamentals of Heat and Mass Transfer* (Wiley, New York, 1985).
JAKOB, M.: *Heat Transfer,* Vol. 1 (Wiley, New York, 1949).
JAKOB, M.: *Heat Transfer,* Vol. II (Wiley, New York, 1957).
KAKAC, S., SHAH, R. K. and AUNG, W.: *Handbook of Single-phase Convective Heat Transfer* (Wiley, New York, 1987).
KAYS, W. M.: *Convective Heat and Mass Transfer* (McGraw-Hill, New York, 1966).
KAYS, W. M. and LONDON, A. L.: *Compact Heat Exchangers*, 2nd edn (McGraw-Hill, New York, 1964).
KERN, D. Q.: *Process Heat Transfer* (McGraw-Hill, New York, 1950).
KREITH, F.: *Principles of Heat Transfer* (Feffer & Simons, Scranton, Penn. 1965).
MCADAMS, W. H.: *Heat Transmission*, 3rd edn (McGraw-Hill, New York, 1954).
MINKOWYCZ, W.J.: *Handbook of Numerical Heat Transfer* (Wiley, New York, 1988).
MINTON, P.E.: *Handbook of Evaporation Technology* (Noyes, New York, 1986).
OZISIK, M.N.: *Boundary Value Problems of Heat Conduction* (Int. Textbook Co., Stanton, Penn., 1965).
ROHSENHOW, W.M., HARTNETT, J.P. and GANIC, E.N. (eds): *Handbook of Heat Transfer Fundamentals* (McGraw-Hill, New York, 1985).
SCHACK, A.: *Industrial Heat Transfer* (Chapman & Hall, London, 1965).
SCHNEIDER, P.J.: *Conduction Heat Transfer* (Addison-Wesley, Cambridge, Mass., 1955).
SMITH, R.A.: *Vaporisers: Selection, Design and Operation* (Longman, London, 1987).
TAYLOR, M. (ed.): *Plate-fin Heat Exchangers: Guide to their Specification and Use* (HTFS, Harwell, 1987).

9.13. REFERENCES

1. UNDERWOOD, A.J.V.: *Industrial Chemist*, **9** (1933) 167. Graphical computation of logarithmic mean temperature difference.
2. SPALDING, D.B.: *Combustion and Mass Transfer* (Pergamon Press, Oxford, 1979).
3. LONG, V.D.: *J. Inst. Fuel.* **37** (1964) 522. A simple model of droplet combustion.
4. MONAGHAN, M.T., SIDDAL, R.G., and THRING, M.W.: *Comb and Flame* **17** (1968) 45. The influence of initial diameter on the combustion of single drops of liquid fuel.
5. CARSLAW, H.S. and JAEGER, J.C.: *Conduction of Heat in Solids.* 2nd edn (Oxford University Press, Oxford, 1959).
6. RICHARDSON, J.F.: *Fuel* **28** (1949) 265. Spread of fire by thermal conduction.
7. SCHMIDT, E.: *Föppl Festschrift* (Springer, 1924), and *Z. ges. Kälte-Ind.* **35** (1928) 213. Versuche über den Wärmeübergang in ruhender Luft.
8. GALLAGHER, R.H.: *Finite Element Analysis. Fundamentals* (Prentice Hall, Englewood Cliffs, N.J., 1975).
9. SCHNEIDER, P.J.: *Temperature Response Charts* (Wiley, New York, 1963).
10. WALDIE, B.: *The Chemical Engineer (London)* No. 261 (1972) 188. Review of recent work on the processing of powders in high temperature plasmas: Pt. II. Particle dynamics, heat transfer and mass transfer.
11. TAYLOR, T.D.: *Physics of Fluids* **6** (1963) 987. Heat transfer from single spheres in a low Reynolds number slip flow.
12. FIELD, M.A., GILL, D.W., MORGAN, B.B., and HAWKSLEY, P.G.W.: *Combustion of Pulverised Coal* (BCURA, Leatherhead, 1967).
13. DITTUS, F.W. and BOELTER, L.M.K.: *Univ. of California, Berkeley, Pubns in Engineering* **2** (1930) 443. Heat transfer in automobile radiators of the tubular type.
14. COLBURN, A.P.: *Trans. Am. Inst. Chem. Eng.* **29** (1933) 174. A method of correlating forced convection heat transfer data and a comparison with fluid friction.
15. SIEDER, E.N. and TATE, G.E.: *Ind. Eng. Chem.* **28** (1936) 1429. Heat transfer and pressure drop of liquids in tubes.
16. FISHENDEN, M. and SAUNDERS, O.A.: *An Introduction to Heat Transfer* (Oxford University Press, Oxford, 1950).
17. COPE, W.F.: *Proc. Inst. Mech. Engrs.* **45** (1941) 99. The friction and heat transmission coefficients of rough pipes.
18. REIHER, M.: *Mitt. Forsch.* **269** (1925) 1. Wärmeübergang von strömender Luft an Röhren und Röhrenbündeln im Kreuzstrom.
19. HILPERT, R.: *Forsch. Geb. IngWes.* **4** (1933) 215. Wärmeabgabe von geheizten Drähten und Röhren.
20. GRIFFITHS, E. and AWBERY, J.H.: *Proc. Inst. Mech. Eng.* **125** (1933) 319. Heat transfer between metal pipes and a stream of air.
21. DAVIS, A.H.: *Phil. Mag.* **47** (1924) 1057. Convective cooling of wires in streams of viscous liquids.
22. GRIMISON, E.D.: *Trans. Am. Soc. Mech. Eng.* **59** (1937) 583: and *ibid.* **60** (1938) 381. Correlation and utilization of new data on flow resistance and heat transfer for cross flow of gases over tube banks.
23. HUGE, E.C.: *Trans. Am. Soc. Mech. Eng.* **59** (1937) 573. Experimental investigation of effects of equipment size on convection heat transfer and flow resistance in cross flow of gases over tube banks.
24. PIERSON, O.L.: *Trans. Am. Soc. Mech. Eng.* **59** (1937) 563. Experimental investigation of influence of tube arrangement on convection heat transfer and flow resistance in cross flow of gases over tube banks.
25. McADAMS, W.H.: *Heat Transmission,* 3rd edn (McGraw-Hill, 1954).
26. KERN, D.Q.: *Process Heat Transfer* (McGraw-Hill, 1950).
27. SHORT, B.E.: *Univ. of Texas Pub.* No. 4324 (1943). Heat transfer and pressure drop in heat exchangers.
28. DONOHUE, D.A.: *Ind. Eng. Chem.* **41** (1949) 2499. Heat transfer and pressure drop in heat exchangers.
29. TINKER, T.: *Proceedings of the General Discussion on Heat Transfer, September,* 1951, p. 89. Analysis of the fluid flow pattern in shell and tube exchangers and the effect of flow distribution on the heat exchanger performance, (Inst. of Mech. Eng. and Am. Soc. Mech. Eng.).
30. COPE, W.F. and BAILEY, A.: *Aeronautical Research Comm. (Gt. Britain). Tech. Rept.* **43** (1933) 199. Heat transmission through circular, square, and rectangular tubes.
31. WASHINGTON, L. and MARKS, W.M.: *Ind. Eng. Chem.* **29** (1937) 337. Heat transfer and pressure drop in rectangular air passages.
32. DAVIS, E.S.: *Trans. Am. Soc. Mech. Eng.* **65** (1943) 755. Heat transfer and pressure drop in annuli.
33. CARPENTER, F.G., COLBURN, A.P., SCHOENBORN, E.M., and WURSTER, A.: *Trans. Am. Inst. Chem. Eng.* **42** (1946) 165. Heat transfer and friction of water in an annular space.
34. ROWE, P.N., CLAXTON, K.T., and LEWIS, J.B.: *Trans. Inst. Chem. Eng.* **41** (1965) T14. Heat and mass transfer from a single sphere in an extensive flowing fluid.
35. RAY, B.B.: *Proc. Indian Assoc. Cultivation of Science* **6** (1920) 95. Convection from heated cylinders in air.

36. Davis, A.H.: *Phil. Mag.* **44** (1922) 920. Natural convective cooling in fluids.
37. Ackermann, G.: *Forsch. Geb. IngWes.* **3** (1932) 42. Die Wärmeabgabe einer horizontal geheizten Röhre an kaltes Wasser bei natürlicher Konvektion.
38. Saunders, O.A.: *Proc. Roy. Soc.* **A, 157** (1936) 278. Effect of pressure on natural convection to air.
39. Perry, R. H. and Green D.W. (eds): *Perry's Chemical Engineers' Handbook*, 6th edn (McGraw-Hill, 1984).
40. Kato H., Nishiwaki, N. and Hirata, M.: *Int. Jl. Heat Mass Transfer* **11** (1968) 1117. On the turbulent heat transfer by free convection from a vertical plate.
41. Kraussold, H.: *Forsch. Geb. IngWes.* **5** (1934) 186. Wärmeabgabe von zylindrischen Flüssigkeiten bei natürlicher Konvektion.
42. Hottel, H.C. and Mangelsdorf, H.G.: *Trans. Am. Inst. Chem. Eng.* **31** (1935) 517. Heat transmission by radiation from non-luminous gases. Experimental study of carbon dioxide and water vapour.
43. Schack, A.: *Z. tech. Phys.* **6** (1925) 530. Die Strahlung von leuchtenden Flammen.
44. Nusselt, W.: *Z. Ver. deut. Ing.* **60** (1916) 541 and 569 Die Oberflächenkondensation des Wasserdampfes.
45. Haselden, G.G. and Prosad, S.: *Trans. Inst. Chem. Eng.* **27** (1949) 195. Heat transfer from condensing oxygen and nitrogen vapours.
46. Ten Bosch, M.: *Die Wärmeübertragung* (Springer, 1936).
47. Carpenter, E.F. and Colburn, A.P.: *Proceedings of the General Discussion on Heat Transfer* (*September 1951*) 20. The effect of vapour velocity on condensation inside tubes (*Inst. of Mech. Eng. and Am. Soc. Mech. Eng.*).
48. Colburn, A.P.: *Proceedings of the General Discussion on Heat Transfer, September, 1951*, p. 1. Problems in design and research on condensers of vapours and vapour mixtures. (*Inst. of Mech. Eng. and Am. Soc. Mech. Eng.*).
49. Tepe, J.B. and Mueller, A.C.: *Chem. Eng. Prog.* **43** (1947) 267. Condensation and subcooling inside an inclined tube.
50. Kirkbride, G.C.: *Ind. Eng. Chem.* **26** (1934) 425. Heat Transfer by condensing vapours on vertical tubes.
51. Badger, W.L.: *Ind. Eng. Chem.* **22** (1930) 700. The evaporation of caustic soda to high concentrations by means of diphenyl vapour.
52. Badger, W.L.: *Trans. Am. Inst. Chem. Eng.* **33** (1937) 441. Heat transfer coefficient for condensing Dowtherm films.
53. Drew, T.B., Nagle, W.M., and Smith, W.Q.: *Trans. Am. Inst. Chem. Eng.* **31** (1935) 605. The conditions for drop-wise condensation of steam.
54. Colburn, A.P. and Hougen, O.A.: *Ind. Eng. Chem.* **26** (1934) 1178. Design of cooler condensers for mixtures of vapors with non-condensing gases.
55. Chilton, T.H. and Colburn, A.P.: *Ind. Eng. Chem.* **26** (1934) 1183. Mass transfer (absorption) coefficients.
56. Revilock, J.F., Hurlburt, H.Z., Brake, D.R., Lang, E.G., and Kern, D.Q.: *Chem. Eng. Prog. Symposium Series No. 30*, **56** (1960) 161. Heat and mass transfer analogy: An appraisal using plant scale data.
57. Jeffreys, G.V.: *A Problem in Chemical Engineering Design* (Institution of Chemical Engineers, 1961). The manufacture of acetic anhydride.
58. Westwater, J.W.: in *Advances in Chemical Engineering* (Eds T.B. Drew and J.W. Hooper) (Academic Press 1956) Boiling of Liquids.
59. Jakob, M.: *Mech. Eng.* **58** (1936) 643, 729. Heat transfer in evaporation and condensation.
60. Rohsenow, W.M. and Clark, J.A.: *Trans. Am. Soc. Mech. Eng.* **73** (1951) 609. A study of the mechanism of boiling heat transfer.
61. Rohsenow, W.M.: *Trans. Am. Soc. Mech. Eng.* **74** (1952) 969. A method of correlating heat transfer data for surface boiling of liquids.
62. Forster, H.K.: *J. Appl. Phys.* **25** (1954) 1067. On the conduction of heat into a growing vapor bubble.
63. Griffith, P. and Wallis, J.D.: *Chem. Eng. Prog. Symposium Series* No. 30, **56** (1960) 49. The role of surface conditions in nuclear boiling.
64. Jakob, M. and Fritz, W.: *Forsch. Geb. IngWes.*, **2** (1931) 435. Versuche über den Verdampfungsvorgang.
65. Nukiyama, S.: *J. Soc. Mech. Eng. (Japan)* **37** (1934) 367. English abstract pp. S53–S54. The maximum and minimum values of the heat Q transmitted from metal to boiling water under atmospheric pressure.
66. Farber, E.A. and Scorah, R.L.: *Trans. Am. Soc. Mech. Eng.* **70** (1948) 369. Heat transfer to boiling water under pressure.
67. Bonilla, C.F. and Perry, C.H.: *Trans. Am. Inst. Chem. Eng.* **37** (1941) 685. Heat transmission to boiling binary liquid mixtures.
68. Insinger, T.H. and Bliss, H.: *Trans. Am. Inst. Chem. Eng.* **36** (1940) 491. Transmission of heat to boiling liquids.
69. Sauer, E.T., Cooper, H.B.H., Akin, G.A. and McAdams, W.H.: *Mech. Eng.* **60** (1938) 669. Heat transfer to boiling liquids.

70. CRYDER, D.S. and FINALBORGO, A.C.: *Trans. Am. Inst. Chem. Eng.* **33** (1937) 346. Heat transmission from metal surfaces to boiling liquids.

71. JAKOB, M. and LINKE, W.: *Forsch. Geb. IngWes.* **4** (1933) 75. Der Wärmeübergang von einer waagerechten Platte an siedendes Wasser.

72. CICHELLI, M.T. and BONILLA, C.F.: *Trans. Am. Inst. Chem. Eng.* **41** (1945) 755. Heat transfer to liquids boiling under pressure.

73. FORSTER, H.K. and ZUBER, N.: *J. Appl. Phys.* **25** (1954) 474. Growth of a vapor bubble in a superheated liquid.

74. FORSTER, H.K. and ZUBER, N.: *A.I.Ch.E. Jl* **1** (1955) 531. Dynamics of vapor bubbles and boiling heat transfer.

75. FRITZ, W.: *Physik Z.* **36** (1935) 379. Berechnung des Maximalvolumens von Dampfblasen.

76. MCNELLY, M.J.: *J. Imp. Coll. Chem. Eng. Soc.* **7** (1953) 18. A correlation of the rates of heat transfer to nucleate boiling liquids.

77. MOSTINSKI, I.L.: *Brit. Chem. Eng.* **8** (1963) 580. Calculation of boiling heat transfer coefficients, based on the law of corresponding states.

78. JESCHKE, D.: *Z. Ver. deut. Ing.* **69** (1925) 1526. Wärmeübergang und Druckverlust in Röhrschlangen.

79. PRATT, N.H.: *Trans. Inst. Chem. Eng.* **25** (1947), 163. The heat transfer in a reaction tank cooled by means of a coil.

80. CHILTON, T.H., DREW, T.B. and JEBENS, R.H.: *Ind. Eng. Chem.* **36** (1944), 570. Heat transfer coefficients in agitated vessels.

81. CUMMINGS, G.H. and WEST, A.S.: *Ind. Eng. Chem.* **42** (1950) 2303. Heat transfer data for kettles with jackets and coils.

82. BROWN, R.W., SCOTT, M.A. and TOYNE, C.: *Trans. Inst. Chem. Eng.* **25** (1947) 181. An investigation of heat transfer in agitated jacketed cast iron vessels.

83. FLETCHER, P.: *The Chemical Engineer, London* No 435 (1987) 33. Heat transfer coefficients for stirred batch reactor design.

84. *Standards of the Tubular Exchanger Manufacturers Association* (TEMA), 7th edn (New York, 1988).

85. BS 5500: (British Standards Institution, London) British Standard 5500 1976: Specification for unfired fusion welded pressure vessels.

86. UNDERWOOD, A.J.V.: *J. Inst. Petrol. Technol.* **20** (1934) 145. The calculation of the mean temperature difference in multipass heat exchangers.

87. BOWMAN, R.A., MUELLER, A.C. and NAGLE, W.M.: *Trans. Am. Soc. Mech. Eng.* **62** (1940) 283. Mean temperature difference in design.

88. LINNHOFF, B., TOWNSEND, D.W., BOLAND, D., HEWITT, G.F., THOMAS, B.E.A., GUY, A.R. and MARSLAND, R.H.: *A User Guide on Process Integration for the Efficient Use of Energy.* (I. Chem.E., Rugby, England, 1982).

89. LINNHOFF, B.: *Chem. Eng. Prog.* **90** (Aug. 1994) 32. Use pinch analysis to knock down capital costs and emissions.

90. WILSON, E.E.: *Trans. Am. Soc. Mech. Eng.* **37** (1915) 546. A basis for rational design of heat transfer apparatus.

91. RHODES, F.H. and YOUNGER, K.R.: *Ind. Eng. Chem.* **27** (1935) 957. Rate of heat transfer between condensing organic vapours and a metal tube.

92. COULSON, J.M. and MEHTA, R.R.: *Trans. Inst. Chem. Eng.* **31** (1953) 208. Heat transfer coefficients in a climbing film evaporator.

93. KAYS, W.M. and LONDON, A.L.: *Compact Heat Exchangers* 2nd edn (McGraw-Hill, New York, 1964).

94. NEIL, D.S.: *Processing* **11** (1984) 12. The use of superheated steam in calorifiers.

95. REDMAN, J.: *The Chemical Engineer* (London) No. 452 (1988) 12. Compact future for heat exchangers.

96. GREGORY, E.: *The Chemical Engineer* (London) No. 440 (1987) 33. Plate and fin heat exchangers.

97. JOHNSTON, A.: *The Chemical Engineer* (London) No. 431 (1986) 36. Miniaturized heat exchangers for chemical processing.

98. RAMSHAW, C.: *The Chemical Engineer* (London) No. 415 (1985) 30. Process intensification — a game for *n* players.

99. NELSON, W.L.: *Petroleum Refinery Engineering,* 4th edn (McGraw-Hill, New York, 1958).

100. HUGGINS, F.E.: *Ind. Eng. Chem.* **23** (1931) 749. Effects of scrapers on heating, cooling and mixing.

101. LAUGHLIN, H.G.: *Trans. Am. Inst. Chem. Eng.* **36** (1940) 345. Data on evaporation and drying in a jacketed kettle.

102. HOULTON, H.G.: *Ind. Eng. Chem.* **36** (1944) 522. Heat transfer in the votator.

103. BOLANOWSKI, S.P. and LINEBERRY, D.D.: *Ind. Eng. Chem.* **44** (1952) 657. Special problems of the food industry.

104. SKELLAND, A.H.P.: *Chem. Eng. Sci.* **7** (1958) 166. Correlation of scraped-film heat transfer in the votator.

105. SKELLAND, A.H.P.: *Brit. Chem. Eng.* **3** (1958) 325. Scale-up relationships for heat transfer in the votator.

106. SKELLAND, A.H.P., OLIVER, D.R. and TOOKE, S.: *Brit. Chem. Eng.* **7** (1962) 346. Heat transfer in a water-cooled scraped-surface heat exchanger.
107. BOTT, T.R.: *Brit. Chem. Eng.* **11** (1966) 339. Design of scraped-surface heat exchangers.
108. BOTT, T.R. and ROMERO, J.J.B.: *Can. J. Chem. Eng.* **41** (1963) 213. Heat transfer across a scraped-surface.
109. BOTT, T.R. and SHEIKH, M.R.: *Brit. Chem. Eng.* **9** (1964) 229. Effects of blade design in scraped-surface heat transfer.
110. LYLE, O.: *Efficient Use of Steam* (HMSO, London, 1947).

9.14. NOMENCLATURE

		Units in SI System	Dimensions in M, L, T, θ (or M, L, T, θ, H)
A	Area available for heat transfer *or* area of radiating surface	m^2	L^2
A_s	Maximum cross-flow area over tube bundle	m^2	L^2
a	Absorptivity	—	—
a_s	Absorptivity of a gas	—	—
b	Wetted perimeter of condensation surface or perimeter of fin	m	L
C	Constant	—	—
C_f	Constant for friction in flow past a tube bundle	—	—
C_h	Constant for heat transfer in flow past a tube bundle	—	—
C'	Clearance between tubes in heat exchanger	m	L
C''	Coefficient in equation 9.94 (SI units only)	J/m^{3n+1}sK^{n+1}	$\mathbf{ML}^{1-3n}\,\mathbf{T}^{-3}\theta^{-n-1}$ (*or* $\mathbf{HL}^{-3n-1}\mathbf{T}^{-1}\theta^{-n-1}$)
C_p	Specific heat at constant pressure	J/kg K	$\mathbf{L}^2\,\mathbf{T}^{-2}\theta^{-1}$ (*or* $\mathbf{HM}^{-1}\theta^{-1}$)
D	Diffusivity of vapour	m^2/s	$\mathbf{L}^2\,\mathbf{T}^{-1}$
D_H	Thermal diffusivity $(k/C_p\rho)$	m^2/s	$\mathbf{L}^2\,\mathbf{T}^{-1}$
d	Diameter (internal or of sphere)	m	L
d_1, d_2	Inner and outer diameters of annulus	m	L
d_c	Diameter of helix	m	L
d_e	Hydraulic mean diameter	m	L
d_g	Gap between turns in coil	m	L
d_m	Logarithmic mean diameter of lagging	m	L
d_o	Outside diameter of tube	m	L
d_p	Height of coil	m	L
d_r	Diameter of shaft	m	L
d_s	Outside diameter of lagging *or* inside diameter of shell	m	L
d_t	Thickness of fixed tube sheet	m	L
d_v	Internal diameter of vessel	m	L
d_w	Mean diameter of pipe wall	m	L
E	Emissive power	W/m^2	$\mathbf{MT}^{-3}$ (*or* $\mathbf{HL}^{-2}\,\mathbf{T}^{-1}$)
E'_z	Energy emitted per unit area and unit time per unit wavelength	W/m^3	$\mathbf{ML}^{-1}\,\mathbf{T}^{-3}$ (*or* $\mathbf{HL}^{-3}\,\mathbf{T}^{-1}$)
e	Emissivity	—	—
e'	$\frac{1}{2}(e_s+1)$	—	—
e_F	Emissivity of a flame	—	—
e_g	Effective emissivity of a gas	—	—
e_{sh}	Emissivity of shield	—	—
F	Geometric factor for radiation *or* correction factor for logarithmic mean temperature difference	—	—
f	Working stress	N/m^2	$\mathbf{ML}^{-1}\,\mathbf{T}^{-2}$
f'	Shell-side friction factor	—	—
G	Mass rate of flow	kg/s	$\mathbf{MT}^{-1}$
G'	Mass rate of flow per unit area	kg/m^{2}s	$\mathbf{ML}^{-2}\,\mathbf{T}^{-1}$
G'_s	Mass flow per unit area over tube bundle	kg/m^{2}s	$\mathbf{ML}^{-2}\,\mathbf{T}^{-1}$
g	Acceleration due to gravity	m/s^2	$\mathbf{LT}^{-2}$

		Units in SI System	Dimensions in M, L, T, θ (or M, L, T, θ, H)
h	Heat transfer coefficient	W/m^2 K	$MT^{-3}\theta^{-1}$ (or $HL^{-2} T^{-1}\theta^{-1}$)
h_b	Film coefficient for liquid adjacent to vessel	W/m^2 K	$MT^{-3}\theta^{-1}$ (or $HL^{-2} T^{-1}\theta^{-1}$)
h_c	Heat transfer coefficient for convection	W/m^2 K	$MT^{-3}\theta^{-1}$ (or $HL^{-2} T^{-1}\theta^{-1}$)
h_f	Fin-side film coefficient	W/m^2 K	$MT^{-3}\theta^{-1}$ (or $HL^{-2} T^{-1}\theta^{-1}$)
h_m	Mean value of h over whole surface	W/m^2 K	$MT^{-3}\theta^{-1}$ (or $HL^{-2} T^{-1}\theta^{-1}$)
h_n	Heat transfer coefficient for liquid boiling at T_n	W/m^2 K	$MT^{-3}\theta^{-1}$ (or $HL^{-2} T^{-1}\theta^{-1}$)
h_r	Heat transfer coefficient for radiation	W/m^2 K	$MT^{-3}\theta^{-1}$ (or $HL^{-2} T^{-1}\theta^{-1}$)
I	Intensity of radiation	W/m^2	MT^{-3} (or $HL^{-2} T^{-1}$)
I'	Intensity of radiation falling on body	W/m^2	MT^{-3} (or $HL^{-2} T^{-1}$)
J	For fin: $(km - h)/(km + h)$	—	—
j	Number of vertical rows of tubes	—	—
j_d	j-factor for mass transfer	—	—
j_h	j-factor for heat transfer	—	—
K	Concentration of particles in a flame	m^{-3}	L^{-3}
k	Thermal conductivity	W/m K	$MLT^{-3}\theta^{-1}$ (or $HL^{-1} T^{-1}\theta^{-1}$)
k_a	Arithmetic mean thermal conductivity	W/m K	$MLT^{-3}\theta^{-1}$ (or $HL^{-1} T^{-1}\theta^{-1}$)
k_m	Mean thermal conductivity	W/m K	$MLT^{-3}\theta^{-1}$ (or $HL^{-1} T^{-1}\theta^{-1}$)
k_o	Thermal conductivity at zero temperature	W/m K	$MLT^{-3}\theta^{-1}$ (or $HL^{-1} T^{-1}\theta^{-1}$)
k'	Constant in equation 9.13	W/m	MLT^{-3} (or $HL^{-1} T^{-1}$)
k_G	Mass transfer coefficient (mass/unit area. unit time. unit partial pressure difference)	s/m	$L^{-1} T$
L	Length of paddle, length of fin or characteristic dimension	m	L
l	Length of tube or plate, or distance apart of faces, *or* thickness of gas stream	m	L
l_B	Distance between baffles	m	L
l_v	Length of side of vessel	m	L
M	Mass rate of flow of condensate per unit length of perimeter for vertical pipe and per unit length of pipe for horizontal pipe	kg/s m	$ML^{-1} T^{-1}$
m	Mass of liquid	kg	M
m	For fin: $(hb/kA)^{0.5}$	m^{-1}	L^{-1}
n	Number of baffles	—	—
N	Number of revolutions in unit time	Hz	T^{-1}
N	Number of general term in series, *or* number of transfer units	—	—
n	An index	—	—
n_B	Number of blades on agitator	—	—
P	Pressure	N/m^2	$ML^{-1} T^{-2}$
P_{Bm}	Logarithmic mean partial pressure of inert gas $\mathbf{B}$	N/m^2	$ML^{-1} T^{-2}$
P_c	Critical pressure	N/m^2	$ML^{-1} T^{-2}$
P_R	Reduced pressure (P/P_c)	—	—
P_r	Vapour pressure at surface of radius r	N/m^2	$ML^{-1} T^{-2}$
P_s	Saturation vapour pressure	N/m^2	$ML^{-1} T^{-2}$
p	Parameter in Laplace transform	s^{-1}	T^{-1}
Q	Heat flow *or* generation per unit time	W	$ML^2 T^{-3}$ (or HT^{-1})
Q_e	Radiation emitted per unit time	W	$ML^2 T^{-3}$ (or HT^{-1})
Q_f	Heat flow to root of fin per unit time	W	$ML^2 T^{-3}$ (or HT^{-1})
Q_G	Rate of heat generation per unit volume	W/m^3	$ML^{-1} T^{-3}$ (or $HL^{-3} T^{-1}$)
Q_i	Total incident radiation per unit time	W	$ML^2 T^{-3}$ (or HT^{-1})
Q_k	Heat flow per unit time by conduction in fluid	W	$ML^2 T^{-3}$ (or HT^{-1})
Q_o	Total radiation leaving surface per unit time	W	$ML^2 T^{-3}$ (or HT^{-1})
Q_r	Total heat reflected from surface per unit time	W	$ML^2 T^{-3}$ (or HT^{-1})
q	Heat flow per unit time and unit area	W/m^2	MT^{-3} (or $HL^{-2} T^{-1}$)

		Units in SI System	Dimensions in M, L, T, θ (or M, L, T, θ, H)
q_{sh}	Value of q with shield	W/m^2	$\mathbf{MT^{-3}}$ (or $\mathbf{HL^{-2}T^{-1}}$)
R	Thermal resistance	m^2 K/W	$\mathbf{M^{-1}T^3\theta}$ (or $\mathbf{H^{-1}}$ $\mathbf{L^2T\theta}$)
R_i, R_0	Thermal resistance of scale on inside, outside of tubes	m^2 K/W	$\mathbf{M^{-1}}$ $\mathbf{T^3\theta}$ (or $\mathbf{H^{-1}L^2T\theta}$)
R'	Shear stress at free surface of condensate film	N/m^2	$\mathbf{ML^{-1}}$ $\mathbf{T^{-2}}$
r	Radius	m	$\mathbf{L}$
r_1, r_2	Radius (inner, outer) of annulus or tube	m	$\mathbf{L}$
r_a	Arithmetic mean radius	m	$\mathbf{L}$
r_m	Logarithmic mean radius	m	$\mathbf{L}$
S	Flow area for condensate film	m^2	$\mathbf{L^2}$
s	Thickness of condensate film at a point	m	$\mathbf{L}$
T	Temperature	K	θ
T_c	Temperature of free surface of condensate	K	θ
T_{cm}	Temperature of cooling medium	K	θ
T_f	Mean temperature of film	K	θ
T_G	Temperature of atmosphere surrounding fin	K	θ
T_m	Mean temperature of fluid	K	θ
T_n	Standard boiling point	K	θ
T_s	Temperature of condensing vapour	K	θ
T_{sh}	Temperature of shield	K	θ
T_w	Temperature of wall	K	θ
t	Time	s	$\mathbf{T}$
U	Overall heat transfer coefficient	W/m^2 K	$\mathbf{MT^{-3}\theta^{-1}}$ (or $\mathbf{HL^{-2}}$ $\mathbf{T^{-1}\theta^{-1}}$)
u	Velocity	m/s	$\mathbf{LT^{-1}}$
u_m	Maximum velocity in condensate film	m/s	$\mathbf{LT^{-1}}$
u_y	Velocity at distance y from surface	m/s	$\mathbf{LT^{-1}}$
W	Width of stirrer	m	$\mathbf{L}$
w	Width of fin or surface	m	$\mathbf{L}$
$w_1, w_2, \ldots$	Indices in equation for heat transfer by convection	—	—
X	Distance between centres of tubes in direction of flow	m	$\mathbf{L}$
X	Ratio of temperature differences used in calculation of mean temperature difference	—	—
x	Distance in direction of transfer or along surface	m	$\mathbf{L}$
Y	Distance between centres of tubes at right angles to flow direction	m	$\mathbf{L}$
Y	Ratio of temperature differences used in calculation of mean temperature difference	—	—
y	Distance perpendicular to surface	m	$\mathbf{L}$
Z	Wavelength	m	$\mathbf{L}$
Z_m	Wavelength at which maximum energy is emitted	m	$\mathbf{L}$
z	Distance in third principal direction	m	$\mathbf{L}$
α	Angle between normal and direction of radiation	—	—
β	Coefficient of cubical expansion	K^{-1}	θ^{-1}
ε	Coefficient relating h to $u^{0.8}$	$J/s^{0.2}m^{2.8}$ K	$\mathbf{ML^{-0.8}}$ $\mathbf{T^{-2.2}\theta^{-1}}$ (or $\mathbf{HL^{-2.8}}$ $\mathbf{T^{-0.2}\theta^{-1}}$)
λ	Latent heat of vaporisation per unit mass	J/kg	$\mathbf{L^2}$ $\mathbf{T^{-2}}$ (or $\mathbf{HM^{-1}}$)
μ	Viscosity	N s/m^2	$\mathbf{ML^{-1}}$ $\mathbf{T^{-1}}$
μ_s	Viscosity of fluid at surface	N s/m^2	$\mathbf{ML^{-1}}$ $\mathbf{T^{-1}}$
ρ	Density or density of liquid	kg/m^3	$\mathbf{ML^{-3}}$
ρ_v	Density of vapour	kg/m^3	$\mathbf{ML^{-3}}$
σ	Stefan — Boltzmann constant, or	W/m^2 K^4	$\mathbf{MT^{-3}\theta^{-4}}$ (or $\mathbf{HL^{-2}}$ $\mathbf{T^{-1}\theta^{-4}}$)
	Surface tension	J/m^2	$\mathbf{MT^{-2}}$
η	Effectiveness of heat exchanger, defined by equation 9.153	—	—
ψ	$\dfrac{G_1 C_{p1} + G_2 C_{p2}}{G_1 C_{p1} G_2 C_{p2}}$		

		Units in SI System	Dimensions in **M, L, T, θ** (or **M, L, T, θ, H**)
ϕ	Angle between surface and horizontal *or* angle of contact	—	—
θ	Temperature or temperature difference	K	θ
θ_a	Temperature difference in Schmidt method	K	θ
θ_m	Logarithmic mean temperature difference	K	θ
θ_{xt}	Temperature at $t = t$, $x = x$	K	θ
θ_o	Initial uniform temperature of body	K	θ
θ'	Temperature of source or surroundings	K	θ
$\bar{\theta}$	Laplace transform of temperature	s K	**T** θ
ω	Solid angle	—	—
Fo	Fourier number $D_H t/L^2$	—	—
Gr	Grashof number	—	—
Gz	Graetz number	—	—
Nu	Nusselt number	—	—
Nu'	Particle Nusselt number	—	—
Pr	Prandtl number	—	—
Re	Reynolds number	—	—
Re'	Particle Reynolds number	—	—
Re''	Rotational flow Reynolds number	—	—
Re_x	Reynolds number for flat plate	—	—
Δ	Finite difference in a property	—	—

Suffix, w refers to wall material
Suffixes i, o refer to inside, outside of wall or lagging; or inlet, outlet conditions
Suffixes b, l, v refer to bubble, liquid, vapour
Suffixes g, b, and s refer to gas, black body, and surface.

CHAPTER 10

Mass Transfer

10.1. INTRODUCTION

Mass transfer is the term used to denote the transference of a component in a mixture from a region in which its concentration is high to a region where the concentration is lower. The process can take place in a gas or vapour or in a liquid. It can result from the random velocities of the molecules (molecular diffusion) or from the circulating or eddy currents present in a turbulent fluid (eddy diffusion).

The simplest case to consider is that of a box consisting of two compartments, each containing a different gas, initially separated by an impermeable partition. When the partition is removed the gases start to mix and the mixing process continues at a constantly decreasing rate until eventually (theoretically after the elapse of an infinite time) the whole system acquires a uniform composition. The process is one of molecular diffusion in which the mixing is attributable solely to the random motion of the molecules. The rate of diffusion is governed by Fick's Law, first proposed by FICK[1] in 1855 which expresses the mass transfer rate as a linear function of the molar concentration gradient. In a mixture of two gases **A** and **B**, assumed ideal, Fick's Law may be written for steady state diffusion as:

$$N_A = -D_{AB} \frac{dC_A}{dy} \tag{10.1}$$

where N_A is the molar flux of **A** (moles per unit area at unit time),

$\quad C_A$ is the concentration of **A** (moles of **A** per unit volume),

$\quad D_{AB}$ is known as the diffusivity or diffusion coefficient for **A** in **B**, and

$\quad y$ is distance in the direction of transfer.

An exactly similar equation may be written for **B**:

$$N_B = -D_{BA} \frac{dC_B}{dy} \tag{10.2}$$

where D_{BA} is the diffusivity of **B** in **A**.

As indicated in the next section, for an ideal gas mixture, at constant pressure $(C_A + C_B)$, is constant (equation 10.9) and hence:

$$\frac{dC_A}{dy} = -\frac{dC_B}{dy} \tag{10.3}$$

Now the condition for the pressure or molar concentration to remain constant in such a system is that there should be no net transference of numbers of molecules; the process is then referred to as one of *equimolecular counterdiffusion*,

i.e.:
$$N_A + N_B = 0$$

This relation is satisfied only if $D_{BA} = D_{AB}$ and therefore the suffixes may be dropped and equation 10.1 may be written:

$$N_A = -D \frac{dC_A}{dy} \qquad (10.4)$$

Equation 10.4 describes the mass transfer rate arising solely from the random movement of the molecules and is applicable to a stationary medium or a fluid in streamline flow. If circulating currents or eddies are present, then the molecular mechanism will be reinforced and the total mass transfer rate may be written:

$$N_A = -(D + E_D) \frac{dC_A}{dy} \qquad (10.5)$$

Whereas D is a physical property of the system and a function only of its composition, pressure and temperature, E_D, which is known as the *eddy diffusivity*, is dependent on the flow pattern and will vary with position. Its estimation presents some difficulty, but the problem is considered in Chapter 12.

The molecular diffusivity D can be expressed in terms of the molecular velocity u_m and the mean free path of the molecules λ. In Chapter 12 it will be shown that for conditions where the kinetic theory of gases is applicable, the molecular diffusivity is proportional to the product $u_m \lambda$.

10.2. DIFFUSION IN BINARY GAS MIXTURES

10.2.1. Properties of binary mixtures

If **A** and **B** are ideal gases in a mixture, the ideal gas law (equation 2.15) may be applied to each gas separately and to the mixture:

$$\left. \begin{array}{l} P_A V = n_A \mathbf{R}T \\ P_B V = n_B \mathbf{R}T \\ PV = n \mathbf{R}T \end{array} \right\} \qquad (10.6)$$

where n_A and n_B are the number of moles of **A** and **B** and n is the total number of moles in a volume V, and P_A, P_B and P are the respective partial pressures and the total pressure.

Then:
$$P_A = \frac{n_A}{V} \mathbf{R}T = C_A \mathbf{R}T = \frac{c_A}{M_A} \mathbf{R}T \qquad (10.7a)$$

$$P_B = \frac{n_B}{V} \mathbf{R}T = C_B \mathbf{R}T = \frac{c_B}{M_B} \mathbf{R}T \qquad (10.7b)$$

and:
$$P = \frac{n}{V} \mathbf{R}T = C_T \mathbf{R}T \qquad (10.7c)$$

where c_A and c_B are mass concentrations and M_A and M_B molecular weights, and C_A, C_B, C_T are, respectively, the molar concentrations of **A** and **B** and the total molar concentration of the mixture.

From Dalton's Law of partial pressures:

$$P = P_A + P_B = \mathbf{R}T(C_A + C_B) = \mathbf{R}T \left(\frac{c_A}{M_A} + \frac{c_B}{M_B} \right) \tag{10.8}$$

Thus:
$$C_T = C_A + C_B \tag{10.9}$$

and:
$$1 = x_A + x_B \tag{10.10}$$

where x_A and x_B are the mole fractions of **A** and **B**.

Then for a system at constant pressure P and constant molar concentration C_T:

$$\frac{dP_A}{dy} = -\frac{dP_B}{dy} \tag{10.11}$$

$$\frac{dC_A}{dy} = -\frac{dC_B}{dy} \tag{10.12}$$

$$\frac{dc_A}{dy} = -\frac{dc_B}{dy}\frac{M_A}{M_B} \tag{10.13}$$

and:
$$\frac{dx_A}{dy} = -\frac{dx_B}{dy} \tag{10.14}$$

By substituting from equations 10.7a and 10.7b into equation 10.4, the mass transfer rates N_A and N_B can be expressed in terms of partial pressure gradients rather than concentration gradients. Furthermore, N_A and N_B can be expressed in terms of gradients of mole fraction.

Thus:
$$N_A = -\frac{D}{\mathbf{R}T}\frac{dP_A}{dy} \tag{10.15}$$

$$= -DC_T\frac{dx_A}{dy} \tag{10.16}$$

Similarly:
$$N_B = -\frac{D}{\mathbf{R}T}\frac{dP_B}{dy} = +\frac{D}{\mathbf{R}T}\frac{dP_A}{dy} \quad \text{(from equation 10.11)} \tag{10.17}$$

$$= -DC_T\frac{dx_B}{dy} = +DC_T\frac{dx_A}{dy} \quad \text{(from equation 10.14)} \tag{10.18}$$

10.2.2. Equimolecular counterdiffusion

When the mass transfer rates of the two components are equal and opposite the process is said to be one of *equimolecular counterdiffusion*. Such a process occurs in the case of the box with a movable partition, referred to in Section 10.1. It occurs also in a distillation column when the molar latent heats of the two components are the same. At any point in the column a falling stream of liquid is brought into contact with a rising stream of vapour with which it is *not* in equilibrium. The less volatile component is transferred from the vapour to the liquid and the more volatile component is transferred in the opposite direction. If the molar latent heats of the components are equal, the condensation of a given amount of less volatile component releases exactly the amount of latent heat required to volatilise the same molar quantity of the more volatile component. Thus at the interface, and consequently throughout the liquid and vapour phases, equimolecular counterdiffusion is taking place.

Under these conditions, the differential forms of equation for N_A (10.4, 10.15 and 10.16) may be simply integrated, for constant temperature and pressure, to give respectively:

$$N_A = -D\frac{C_{A_2} - C_{A_1}}{y_2 - y_1} = \frac{D}{y_2 - y_1}(C_{A_1} - C_{A_2}) \qquad (10.19)$$

$$N_A = -\frac{D}{RT}\frac{P_{A_2} - P_{A_1}}{y_2 - y_1} = \frac{D}{RT(y_2 - y_1)}(P_{A_1} - P_{A_2}) \qquad (10.20)$$

$$N_A = -DC_T\frac{x_{A_2} - x_{A_1}}{y_2 - y_1} = \frac{DC_T}{y_2 - y_1}(x_{A_1} - x_{A_2}) \qquad (10.21)$$

Similar equations apply to N_B which is equal to $-N_A$.

Suffixes 1 and 2 represent, respectively, the values of quantities at positions y_1 and y_2. Equation 10.19 may be written as:

$$N_A = h_D(C_{A_1} - C_{A_2}) \qquad (10.22)$$

where $h_D = D/(y_2 - y_1)$ is a *mass transfer coefficient* with the driving force expressed as a *difference in molar concentration* and it should be noted that its dimensions are those of velocity ($\mathbf{LT^{-1}}$).

Similarly, equation 10.20 may be written as:

$$N_A = k'_G(P_{A_1} - P_{A_2}) \qquad (10.23)$$

where $k'_G = D/[\mathbf{R}T(y_2 - y_1)]$ is a *mass transfer coefficient* with the driving force expressed as a *difference in partial pressure*. It should be noted that its dimensions, $\mathbf{L^{-1}T}$, are different from those of h_D. It is always necessary to be careful to use the form of mass transfer coefficient corresponding to the appropriate driving force.

Finally, equation 10.21 may be written as:

$$N_A = k_x(x_{A_1} - x_{A_2}) \qquad (10.24)$$

where $k_x = DC_T/(y_2 - y_1)$ is a *mass transfer coefficient* with the driving force in the form of a *difference in mole fraction*. Its dimensions are $\mathbf{NL^{-2}T^{-1}}$.

10.2.3. Mass transfer through a stationary gas

In several important processes, one component in a gaseous mixture will be transported relative to a fixed plane (say, a liquid interface) and the other will undergo no net movement. In gas absorption a soluble gas **A** will be transferred to the liquid surface where it dissolves, whereas the insoluble gas **B** will undergo no net movement with respect to the interface. Again, in evaporation from a free surface, the vapour will move away from the surface but the air will have no net movement. The mass transfer process therefore differs from that described in the previous section.

Considering the absorption of a soluble gas **A** from a mixture with an insoluble gas **B**, the respective diffusion rates are given by:

$$N_A = -D\frac{dC_A}{dy} \qquad \text{(equation 10.4)}$$

$$N_B = -D\frac{dC_B}{dy} = D\frac{dC_A}{dy} \qquad \text{(from equation 10.3)}$$

The total mass transfer rate of **B** is zero, and there must be a "bulk flow" of the system away from the liquid surface exactly to counterbalance the diffusional flux towards the surface (see Figure 10.1).

$$\text{Bulk flow of } \mathbf{B} = -D\frac{dC_A}{dy} \qquad (10.25)$$

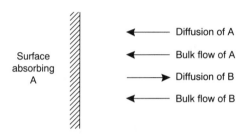

Figure 10.1.　Mass transfer through stationary gas **B**

Now the corresponding bulk flow of **A** must be C_A/C_B times that of **B**, since bulk flow implies that the gas moves en masse.

$$\text{Bulk flow of } \mathbf{A} = -D\frac{dC_A}{dy}\frac{C_A}{C_B} \qquad (10.26)$$

The total flux of **A**, N'_A, is therefore given by:

$$N'_A = -D\frac{dC_A}{dy} - D\frac{dC_A}{dy}\frac{C_A}{C_B}$$

$$= -D\frac{dC_A}{dy}\frac{C_T}{C_B} \qquad (10.27)$$

This equation is known as Stefan's Law[2]. Thus the bulk flow enhances the mass transfer rate by a factor C_T/C_B, known as the *drift factor*.

The fluxes of the components are given in Table 10.1.

Writing equation 10.27 as:

$$N'_A = D\frac{C_T}{C_B}\frac{dC_B}{dy} \qquad \text{(from equation 10.3)}$$

On integration:

$$N'_A = \frac{DC_T}{y_2 - y_1}\ln\frac{C_{B_2}}{C_{B_1}} \qquad (10.28)$$

Table 10.1.   Fluxes of components of a gas mixture

	Component A	Component B	A + B
Diffusion	$-D\dfrac{dC_A}{dy}$	$+D\dfrac{dC_A}{dy}$	0
Bulk flow	$-D\dfrac{dC_A}{dy}\cdot\dfrac{C_A}{C_B}$	$-D\dfrac{dC_A}{dy}$	$-D\dfrac{dC_A}{dy}\cdot\dfrac{C_A+C_B}{C_B}$
Total	$-D\dfrac{dC_A}{dy}\cdot\dfrac{C_A+C_B}{C_B}$	0	$-D\dfrac{dC_A}{dy}\cdot\dfrac{C_A+C_B}{C_B}$
	$=-D\dfrac{dC_A}{dy}\cdot\dfrac{C_T}{C_B}$		$=-D\dfrac{dC_A}{dy}\cdot\dfrac{C_T}{C_B}$

By definition, C_{Bm}, the logarithmic mean of C_{B_1} and C_{B_2}, is given by:

$$C_{Bm} = \frac{C_{B_2} - C_{B_1}}{\ln(C_{B_2}/C_{B_1})} \tag{10.29}$$

Thus:

$$N_A' = \frac{DC_T}{y_2 - y_1}\frac{C_{B_2} - C_{B_1}}{C_{Bm}}$$

$$= \frac{D}{y_2 - y_1}\frac{C_T}{C_{Bm}}(C_{A_1} - C_{A_2}) \tag{10.30}$$

In terms of partial pressures:

$$N_A' = \frac{D}{RT(y_2 - y_1)}\frac{P}{P_{Bm}}(P_{A_1} - P_{A_2}) \tag{10.31}$$

In terms of mole fractions:

$$N_A' = \frac{DC_T}{y_2 - y_1}\frac{1}{x_{Bm}}(x_{A_1} - x_{A_2}) \tag{10.32}$$

Equation 10.28 can be simplified when the concentration of the diffusing component A is small. Under these conditions C_A is small compared with C_T. Equation 10.28 gives:

$$N_A' = \frac{DC_T}{y_2 - y_1}\ln\left[1 - \left(\frac{C_{A_2} - C_{A_1}}{C_T - C_{A_1}}\right)\right]$$

$$= \frac{DC_T}{y_2 - y_1}\left[-\left(\frac{C_{A_2} - C_{A_1}}{C_T - C_{A_1}}\right) - \frac{1}{2}\left(\frac{C_{A_2} - C_{A_1}}{C_T - C_{A_1}}\right)^2 - \cdots\right]$$

For small values of C_A, $C_T - C_{A_1} \approx C_T$ and only the first term in the series is significant.

$$\therefore \qquad N_A' \approx \frac{D}{y_2 - y_1}(C_{A_1} - C_{A_2}) \tag{10.33}$$

This is identical to equation 10.19 for equimolecular counterdiffusion. Thus, the effects of bulk flow can be neglected at low concentrations.

Equation 10.30 can be written in terms of a mass transfer coefficient h_D:

$$N'_A = h_D(C_{A_1} - C_{A_2}) \tag{10.34}$$

where:

$$h_D = \frac{D}{y_2 - y_1} \frac{C_T}{C_{Bm}} \tag{10.35}$$

Similarly, working in terms of partial pressure difference as driving force, equation 10.31 can be written:

$$N'_A = k_G(P_{A_1} - P_{A_2}) \tag{10.36}$$

where:

$$k_G = \frac{D}{RT(y_2 - y_1)} \frac{P}{P_{Bm}} \tag{10.37}$$

and using mole fractions (equation 10.32):

$$N'_A = k_x(x_{A_1} - x_{A_2}) \tag{10.38}$$

where:

$$k_x = \frac{DC_T}{y_2 - y_1} \frac{C_T}{C_{Bm}} = \frac{DC_T}{(y_2 - y_1)x_{Bm}} \tag{10.39}$$

It will be noted that all the transfer coefficients here are greater than those for equimolecular counterdiffusion by the factor $(C_T/C_{Bm})(= P/P_{Bm})$, an integrated form of the drift factor.

When the concentration C_A of the gas being transferred is low, C_T/C_{Bm} approaches unity and the two sets of coefficients become identical.

Table 10.2. Diffusivities (diffusion coefficients) of gases and vapours in air at 298 K and atmospheric pressure[3]

Substance	D (m²/s × 10⁶)	$\frac{\mu}{\rho D}$	Substance	D (m²/s × 10⁶)	$\frac{\mu}{\rho D}$
Ammonia	28.0	0.55	Valeric acid	6.7	2.31
Carbon dioxide	16.4	0.94	i-Caproic acid	6.0	2.58
Hydrogen	71.0	0.22	Diethyl amine	10.5	1.47
Oxygen	20.6	0.75	Butyl amine	10.1	1.53
Water	25.6	0.60	Aniline	7.2	2.14
Carbon disulphide	10.7	1.45	Chlorobenzene	7.3	2.12
Ethyl ether	9.3	1.66	Chlorotoluene	6.5	2.38
Methanol	15.9	0.97	Propyl bromide	10.5	1.47
Ethanol	11.9	1.30	Propyl iodide	9.6	1.61
Propanol	10.0	1.55	Benzene	8.8	1.76
Butanol	9.0	1.72	Toluene	8.4	1.84
Pentanol	7.0	2.21	Xylene	7.1	2.18
Hexanol	5.9	2.60	Ethyl benzene	7.7	2.01
Formic acid	15.9	0.97	Propyl benzene	5.9	2.62
Acetic acid	13.3	1.16	Diphenyl	6.8	2.28
Propionic acid	9.9	1.56	n-Octane	6.0	2.58
i-Butyric acid	8.1	1.91	Mesitylene	6.7	2.31

Note: the group $(\mu/\rho D)$ in the above table is evaluated for mixtures composed largely of air.

The above figures from PERRY[4] are based on data in International Critical Tables 5 (1928) and Landolt-Börnstein, Physikalische-Chemische Tabellen (1935).

10.2.4. Diffusivities of gases and vapours

Values of diffusivities measured experimentally are given in Table 10.2 for a number of gases and vapours in air at 298K and atmospheric pressure. The table also includes values of the Schmidt number Sc, the ratio of the kinematic viscosity (μ/ρ) to the diffusivity (D) for very low concentrations of the diffusing gas or vapour. The importance of the Schmidt number in problems involving mass transfer is discussed in Chapter 12.

Experimental determination of diffusivities

Diffusivities of vapours are most conveniently determined by the method developed by WINKELMANN[4] in which liquid is allowed to evaporate in a vertical glass tube over the top of which a stream of vapour-free gas is passed, sufficiently rapidly for the vapour pressure to be maintained almost at zero (Figure 10.2). If the apparatus is maintained at a steady temperature, there will be no eddy currents in the vertical tube and mass transfer will take place from the surface by molecular diffusion alone. The rate of evaporation can be followed by the rate of fall of the liquid surface, and since the concentration gradient is known, the diffusivity can then be calculated.

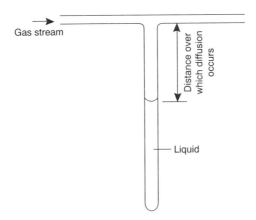

Figure 10.2. Determination of diffusivities of vapours

Example 10.1

The diffusivity of the vapour of a volatile liquid in air can be conveniently determined by Winkelmann's method in which liquid is contained in a narrow diameter vertical tube, maintained at a constant temperature, and an air stream is passed over the top of the tube sufficiently rapidly to ensure that the partial pressure of the vapour there remains approximately zero. On the assumption that the vapour is transferred from the surface of the liquid to the air stream by molecular diffusion, calculate the diffusivity of carbon tetrachloride vapour in air at 321 K and atmospheric pressure from the experimental data in Table 10.3.

The vapour pressure of carbon tetrachloride at 321 K is 37.6 kN/m² and the density of the liquid is 1540 kg/m³. Take the kilogram molecular volume as 22.4 m³.

Solution

From equation 10.30 the rate of mass transfer is given by:

$$N'_A = D \frac{C_A}{L} \frac{C_T}{C_{Bm}}$$

Table 10.3. Experimental data for diffusivity calculation

Time from commencement of experiment			Liquid level (mm)	Time from commencement of experiment			Liquid level (mm)
(h	m)	(ks)		(h	m)	(ks)	
0	0	0.0	0.0	32	38	117.5	54.7
0	26	1.6	2.5	46	50	168.6	67.0
3	5	11.1	12.9	55	25	199.7	73.8
7	36	27.4	23.2	80	22	289.3	90.3
22	16	80.2	43.9	106	25	383.1	104.8

where C_A is the saturation concentration at the interface and L is the effective distance through which mass transfer is taking place. Considering the evaporation of the liquid:

$$N'_A = \frac{\rho_L}{M} \frac{dL}{dt}$$

where ρ_L is the density of the liquid.

Thus:

$$\frac{\rho_L}{M} \frac{dL}{dt} = D \frac{C_A}{L} \frac{C_T}{C_{Bm}}$$

Integrating and putting $L = L_0$ at $t = 0$:

$$L^2 - L_0^2 = \frac{2MD}{\rho_L} \frac{C_A C_T}{C_{Bm}} t$$

L_0 will not be measured accurately nor is the effective distance for diffusion, L, at time t. Accurate values of $(L - L_0)$ are available, however, and hence:

$$(L - L_0)(L - L_0 + 2L_0) = \frac{2MD}{\rho_L} \frac{C_A C_T}{C_{Bm}} t$$

or:

$$\frac{t}{L - L_0} = \frac{\rho_L}{2MD} \frac{C_{Bm}}{C_A C_T}(L - L_0) + \frac{\rho_L C_{Bm}}{MDC_A C_T} L_0$$

If s is the slope of a graph of $t/(L - L_0)$ against $(L - L_0)$:

$$s = \frac{\rho_L C_{Bm}}{2MDC_A C_T} \quad \text{or} \quad D = \frac{\rho_L C_{Bm}}{2MC_A C_T s}$$

From a plot of $t/(L - L_0)$ against $(L - L_0)$ as in Figure 10.3:

$$s = 0.0310 \text{ ks/mm}^2 \quad \text{or} \quad 3.1 \times 10^7 \text{ s/m}^2$$

$$C_T = \frac{1}{22.4} \frac{273}{321} = 0.0380 \text{ kmol/m}^3$$

$$M = 154 \text{ kg/kmol}$$

$$C_A = \frac{37.6}{101.3} \times 0.0380 = 0.0141 \text{ kmol/m}^3$$

$$\rho_L = 1540 \text{ kg/m}^3$$

$$C_{B1} = 0.0380 \text{ kmol/m}^3, \quad C_{B2} = \frac{101.3 - 37.6}{101.3} \times 0.0380 = 0.0238 \text{ kmol/m}^3$$

$$\therefore \quad C_{Bm} = \frac{0.0380 - 0.0238}{\ln(0.0380/0.0238)} = 0.0303 \text{ kmol/m}^3$$

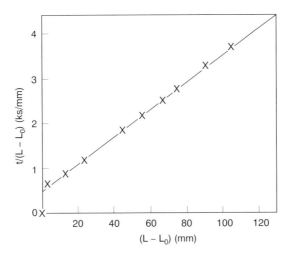

Figure 10.3. $t/(L - L_0)$ versus $(L - L_0)$ (Example 10.1)

and:
$$D = \frac{1540 \times 0.0303}{2 \times 154 \times 0.0141 \times 0.0380 \times 3.1 \times 10^7}$$

$$= 9.12 \times 10^{-6} \text{ m}^2/\text{s}$$

Prediction of diffusivities

In many cases, the diffusivity D for the transfer of one gas in another may not be known and experimental determination may not be practicable. It is then necessary to revert to one of the many predictive procedures. A commonly used method due to GILLILAND[5] is based on the "Stefan–Maxwell" hard sphere model. It takes the form:

$$D = \frac{4.3 \times 10^{-4} T^{1.5} \sqrt{(1/M_A) + (1/M_B)}}{P \left(V_A^{1/3} + V_B^{1/3} \right)^2} \tag{10.40}$$

where D is the diffusivity in m²/s, T the absolute temperature (K), M_A, M_B are the molecular masses of **A** and **B**, P is the total pressure in N/m², and $\mathbf{V}_A$, $\mathbf{V}_B$ are the molecular volumes of **A** and **B**. The molecular volume is the volume in m³ of 1 kmol of the material in the form of liquid at its boiling point, and is a measure of the volume occupied by the molecules themselves. It may not always be known, but an approximate value can be obtained, for all but simple molecules, by application of *Kopp's law* of additive volumes. Kopp gave a particular value for the equivalent atomic volume of each element. When the atomic volumes (Table 10.4) of the elements of the molecule in question are added in the appropriate proportions, the equivalent molecular volume is obtained approximately. There are certain exceptions to this rule, and additions or subtractions have to be made if the elements are combined in particular manners.

It will be noted from equation 10.40 that the diffusivity of a vapour is inversely proportional to the pressure and varies with the absolute temperature raised to the 1.5 power, though there is evidence that this underestimates the temperature dependence.

Table 10.4.   Atomic and structural diffusion volume increments $(m^3/kmol)$[6]

		Oxygen, double-bonded	0.0074
Antimony	0.0242	Coupled to two other elements:	
Arsenic	0.0305	in aldehydes and ketones	0.0074
Bismuth	0.0480	in methyl esters	0.0091
Bromine	0.0270	in ethyl esters	0.0099
Carbon	0.0148	in higher esters and ethers	0.0110
Chlorine, terminal, as in R-Cl	0.0216	in acids	0.0120
medial, as in R-CHCl-R'	0.0246	in union with S, P, N	0.0083
Chromium	0.0274	Phosphorus	0.0270
Fluorine	0.0087	Silicon	0.0320
Germanium	0.0345	Sulphur	0.0256
Hydrogen	0.0037	Tin	0.0423
Nitrogen, double-bonded	0.0156	Titanium	0.0357
in primary amines	0.0105	Vanadium	0.0320
in secondary amines	0.0120	Zinc	0.0204

Notes

For three-membered ring, as in ethylene oxide,	deduct 0.0060.
For four-membered ring, as in cyclobutane,	deduct 0.0085.
For five-membered ring, as in furane,	deduct 0.0115.
For six-membered ring, as in benzene, pyridine,	deduct 0.0150.
For anthracene ring formation,	deduct 0.0475.
For naphthalene	deduct 0.0300.

Diffusion volumes of simple molecules $(m^3/kmol)$

H_2	0.0143	CO_2	0.0340	NH_3	0.0258
O_2	0.0256	H_2O	0.0189	H_2S	0.0329
N_2	0.0312	SO_2	0.0448	Cl_2	0.0484
Air	0.0299	NO	0.0236	Br_2	0.0532
CO	0.0307	N_2O	0.0364	I_2	0.0715

A more recent method, proposed by FULLER, SCHETTLER and GIDDINGS[7], is claimed to give an improved correlation. The values of the "diffusion volume" have been *modified* to give a better correspondence with experimental values, and have then been adjusted arbitrarily to make the coefficient in the equation equal to unity. It does contain some anomalies, particularly in relation to the values of **V** for nitrogen, oxygen and air. Details of the method are given in Volume 6.

Example 10.2

Ammonia is absorbed in water from a mixture with air using a column operating at 1 bar and 295 K. The resistance to transfer can be regarded as lying entirely within the gas phase. At a point in the column, the partial pressure of the ammonia is 7.0 kN/m^2. The back pressure at the water interface is negligible and the resistance to transfer can be regarded as lying in a stationary gas film 1 mm thick. If the diffusivity of ammonia in air is $2.36 \times 10^{-5} \text{ m}^2/\text{s}$, what is the transfer rate per unit area at that point in the column? How would the rate of transfer be affected if the ammonia air mixture were compressed to double the pressure?

Solution

Concentration of ammonia in the gas

$$= \frac{1}{22.4} \frac{101.3}{101.3} \frac{273}{295} \frac{7.0}{101.3} = 0.00285 \text{ kmol/m}^3$$

$$\frac{C_T}{C_{Bm}} = \frac{101.3 \ln(101.3/94.3)}{101.3 - 94.3} = 1.036$$

From equation 10.30:

$$N'_A = \frac{D}{y_2 - y_1} \frac{C_T}{C_{Bm}} (C_{A1} - C_{A2})$$

$$= \frac{2.36 \times 10^{-5}}{1 \times 10^{-3}} \times 1.036 \times 0.00285$$

$$= \underline{\underline{6.97 \times 10^{-5} \text{ kmol/m}^2\text{s}}}$$

If the pressure is doubled, the driving force is doubled, C_T/C_{Bm} is essentially unaltered, and the diffusivity, being inversely proportional to the pressure (equation 10.40) is halved. The mass transfer rate therefore remains the same.

10.2.5. Mass transfer velocities

It is convenient to express mass transfer rates in terms of velocities for the species under consideration.

$$\text{Velocity} = \frac{\text{Flux}}{\text{Concentration}}, \quad \left(\text{e.g.} \ \frac{\text{kmol/m}^2\text{s}}{\text{kmol/m}^3} \right)$$

For diffusion according to Fick's Law:

$$u_{DA} = \frac{N_A}{C_A} = -\frac{D}{C_A}\frac{dC_A}{dy} \tag{10.41a}$$

$$u_{DB} = \frac{N_B}{C_B} = -\frac{D}{C_B}\frac{dC_B}{dy} = \frac{D}{C_B}\frac{dC_A}{dy} \tag{10.41b}$$

It should be noted that, whereas $N_B = -N_A$,

$$u_{DB} = -u_{DA}\frac{C_A}{C_B} = -u_{DA}\frac{x_A}{x_B} \tag{10.42}$$

As a result of the diffusional process, there is no net overall molecular flux arising from diffusion in a binary mixture, the two components being transferred at equal and opposite rates. In the process of equimolecular counterdiffusion, which occurs in a distillation column when the two components have equal molar latent heats, the diffusional velocities are the same as the velocities of the molecular species relative to the walls of the equipment or the phase boundary.

If the physical constraints placed upon the system result in a bulk flow, the velocities of the molecular species relative to one another remain the same, but in order to obtain the velocity relative to a fixed point in the equipment, it is necessary to add the bulk flow velocity. An example of a system in which there is a bulk flow velocity is that in which one of the components is transferred through a second component which is undergoing no net transfer, as in the absorption of a soluble gas **A** from a mixture with an insoluble gas **B**. (See Section 10.2.3). In this case, because there is no set flow of **B**, the sum of its diffusional velocity and the bulk flow velocity must be zero.

In this case:

	A	**B**
Diffusional velocity	$u_{DA} = -\dfrac{D}{C_A}\dfrac{dC_A}{dy}$	$u_{DB} = +\dfrac{D}{C_B}\dfrac{dC_A}{dy}$
Bulk flow velocity	$u_F = -\dfrac{D}{C_B}\dfrac{dC_A}{dy}$	$u_F = -\dfrac{D}{C_B}\dfrac{dC_A}{dy}$
Total velocity	$u_A = -D\dfrac{C_T}{C_A C_B}\dfrac{dC_A}{dy}$	$u_B = 0$
Flux	$N'_A = u_A C_A = -D\dfrac{C_T}{C_B}\dfrac{dC_A}{dy}$	$N'_B = 0$

The flux of **A** has been given previously (equation 10.27) as Stefan's Law.

10.2.6. General case for gas-phase mass transfer

Whatever the physical constraints placed on the system, the diffusional process causes the two components to be transferred at equal and opposite rates and the values of the diffusional velocities u_{DA} and u_{DB} given in Section 10.2.5 are always applicable. It is the bulk flow velocity u_F which changes with imposed conditions and gives rise to differences in overall mass transfer rates.

In equimolecular counterdiffusion, u_F is zero. In the absorption of a soluble gas **A** from a mixture the bulk velocity must be equal and opposite to the diffusional velocity of **B** as this latter component undergoes no net transfer.

In general, for any component:

$$\text{Total transfer} = \text{Transfer by diffusion} + \text{Transfer by bulk flow.}$$

For component **A:**

Total transfer (moles/area time) $= N'_A$

Diffusional transfer according to Fick's Law $= N_A = -D\dfrac{dC_A}{dy}$

Transfer by bulk flow $= u_F C_A$

Thus: $$N'_A = N_A + u_F C_A \tag{10.43a}$$

and for **B**: $$N'_B = N_B + u_F C_B \tag{10.43b}$$

Now the bulk flow velocity $u_F = \dfrac{\text{Total moles transferred/area time}}{\text{Total molar concentration}}$

$$= \frac{N'_A + N'_B}{C_T} \tag{10.44}$$

Substituting:

$$N'_A = N_A + \frac{C_A}{C_T}(N'_A + N'_B)$$

$$= -D\frac{dC_A}{dy} + x_A(N'_A + N'_B)$$

$$= -DC_T\frac{dx_A}{dy} + x_A(N'_A + N'_B) \qquad (10.45)$$

Similarly for **B:** $\qquad N'_B = DC_T\frac{dx_A}{dy} + (1 - x_A)(N'_A + N'_B) \qquad (10.46)$

For equimolecular counterdiffusion $N'_A = -N'_B$ and equation 10.45 reduces to Fick's Law. For a system in which **B** undergoes no net transfer, $N'_B = 0$ and equation 10.45 is identical to Stefan's Law.

For the general case: $\qquad\qquad fN'_A = -N'_B \qquad (10.47)$

As an example, if the molar latent heat of **A** is f times that of **B**, in a distillation column, the condensation of 1 mole of **A** (taken as the less volatile component) will result in the volatilisation of f moles of **B** and the mass transfer of **B** will be f times that of **A** in the opposite direction.
Substituting into equation 10.45:

$$N'_A = -DC_T\frac{dx_A}{dy} + x_A(N'_A - fN'_A) \qquad (10.48)$$

Thus: $\qquad\qquad [1 - x_A(1 - f)]N'_A = -DC_T\frac{dx_A}{dy}$

If x_A changes from x_{A_1} to x_{A_2} as y goes from y_1 to y_2:

$$N'_A\int_{y_1}^{y_2} dy = -DC_T\int_{x_{A_1}}^{x_{A_2}} \frac{dx_A}{1 - x_A(1 - f)}$$

$$\therefore \qquad N'_A(y_2 - y_1) = -DC_T\frac{1}{1 - f}\left[\ln\frac{1}{(1 - f)^{-1} - x_A}\right]_{x_{A_1}}^{x_{A_2}}$$

or: $\qquad\qquad N'_A = \frac{DC_T}{y_2 - y_1}\frac{1}{1 - f}\ln\frac{1 - x_{A_2}(1 - f)}{1 - x_{A_1}(1 - f)} \qquad (10.49)$

10.2.7. Diffusion as a mass flux

Fick's Law of diffusion is normally expressed in *molar* units as in equation 10.4:

$$N_A = -D\frac{dC_A}{dy} = -DC_T\frac{dx_A}{dy} \qquad \text{(equation 10.4)}$$

where x_A is the mole fraction of component **A.**
    It has already been seen that the corresponding equation for component **B** gives rise to an equal and opposite molar flux of that component. If each side of equation 10.4 is

multiplied by the molecular weight of A, M_A:

$$J_A = -D\frac{dc_A}{dy} = -DM_A\frac{dC_A}{dy} = -DC_TM_A\frac{dx_A}{dy} \tag{10.50}$$

where J_A is a flux in mass per unit area and unit time, e.g. kg/m^2 s, and c_A is a concentration in mass terms, e.g. kg/m^3.

Similarly, for component **B:**

$$J_B = -D\frac{dc_B}{dy} \tag{10.51}$$

Whereas the sum of the molar concentrations is constant in an ideal gas at constant pressure, the sum of the mass concentrations is not constant, and dc_A/dy and dc_B/dy are not equal and opposite,

i.e.:
$$C_A + C_B = C_T = \frac{c_A}{M_A} + \frac{c_B}{M_B} = \text{constant} \tag{10.52}$$

Thus:
$$\frac{1}{M_A}\frac{dc_A}{dy} + \frac{1}{M_B}\frac{dc_B}{dy} = 0$$

and:
$$\frac{dc_B}{dy} = -\frac{M_B}{M_A}\frac{dc_A}{dy} \tag{10.53}$$

Thus, the diffusional process does not give rise to equal and opposite mass fluxes.

10.2.8. Thermal diffusion

If a temperature gradient is maintained in a binary gaseous mixture, a concentration gradient becomes established with the light component collecting preferentially at the hot end and the heavier one at the cold end. This phenomenon, known as the *Soret effect*, may be used as the basis of a separation technique — commercially of significance in the separation of isotopes.

Conversely, when mass transfer is occurring as a result of the maintenance of a concentration gradient, a temperature gradient may be generated; this is known as the *Dufour effect*.

In a binary mixture consisting of two gaseous components A and B subject to a temperature gradient, the flux due to thermal diffusion is given by[8]:

$$(N_A)_{Th} = -D_{Th}\frac{1}{T}\frac{dT}{dy} \tag{10.54}$$

where $(N_A)_{Th}$ is the molar flux of A ($kmol/m^2$ s) in the Y-direction, and D_{Th} is the diffusion coefficient for thermal diffusion ($kmol/m$ s).

Equation 10.54, with D_{Th} positive, applies to the component which travels preferentially to the *low* temperature end of the system. For the component which moves to the high temperature end, the sign of D_{Th} is negative. In a binary mixture, the gas of higher molecular weight has the positive value of D_{Th} and therefore tends towards the lower temperature end of the system.

If two vessels each containing completely mixed gas, one at temperature T_1 and the other at a temperature T_2, are connected by a lagged non-conducting pipe in which there are no turbulent eddies (such as a capillary tube), then under steady state conditions, the rate of transfer of **A** by thermal diffusion and molecular diffusion must be equal and opposite, or:

$$(N_A)_{Th} + N_A = 0 \tag{10.55}$$

N_A is given by Fick's Law:

$$N_A = -D\frac{dC_A}{dy} = -DC_T\frac{dx_A}{dy} \tag{equation 10.50}$$

where x_A is the mole fraction of **A**, and C_T is the total molar concentration at y and will not be quite constant because the temperature varies.

Substitution of equations 10.50 and 10.54 into equation 10.55 gives:

$$-D_{Th}\frac{1}{T}\frac{dT}{dy} - DC_T\frac{dx_A}{dy} = 0 \tag{10.56}$$

The relative magnitudes of the thermal diffusion and diffusion effects are represented by the dimensionless ratio:

$$\frac{D_{Th}}{DC_T} = K_{ABT}$$

where K_{ABT} is known as the *thermal diffusion ratio*.

Then:
$$-K_{ABT}\frac{1}{T}\frac{dT}{dy} = \frac{dx_A}{dy} \tag{10.57}$$

If temperature gradients are small, C_T can be regarded as effectively constant. Furthermore, K_{ABT} is a function of composition, being approximately proportional to the product $x_A x_B$. It is therefore useful to work in terms of the *thermal diffusion factor* α, where:

$$\alpha = \frac{K_{ABT}}{x_A(1 - x_A)}$$

Substituting for α in equation 10.57 and integrating, assuming α to be constant, gives:

$$\alpha \ln\frac{T_1}{T_2} = \ln\left(\frac{x_{A_2}}{1 - x_{A_2}}\frac{1 - x_{A_1}}{x_{A_1}}\right)$$

Thus:
$$\frac{x_{A_2}\, x_{B_1}}{x_{A_1}\, x_{B_2}} = \left(\frac{T_1}{T_2}\right)^{\alpha} \tag{10.58}$$

Equation 10.58 gives the mole fraction of the two components **A** and **B** as a function of the absolute temperatures and the thermal diffusion factor.

Some values of α taken from data in GREW and IBBS[8] and HIRSCHFELDER, CURTISS and BIRD[9] are given in Table 10.5.

Table 10.5.   Values of thermal diffusion factor (α) for binary gas mixtures
            (**A** is the heavier component, which moves towards the cooler end)

| Systems | | Temperature | Mole fraction of **A** | |
A	B	(K)	x_A	α
D_2	H_2	288–373	0.48	0.17
He	H_2	273–760	0.50	0.15
N_2	H_2	288–373	0.50	0.34
Ar	H_2	258	0.53	0.26
O_2	H_2	90–294	0.50	0.19
CO	H_2	288–373	0.50	0.33
CO_2	H_2	288–456	0.50	0.28
$C_3 H_8$	H_2	230–520	0.50	0.30
N_2	He	287–373	0.50	0.36
		260	0.655	0.37
Ar	He	330	0.90	0.28
		330	0.70	0.31
		330	0.50	0.37
Ne	He	205	0.46	0.31
		330	0.46	0.315
		365	0.46	0.315
O_2	N_2	293	0.50	0.018

10.2.9. Unsteady state mass transfer

In many mass transfer processes encountered in practice, unsteady state conditions prevail. Thus, in the example given in Section 10.1, a box is divided into two compartments each containing a different gas and the partition is removed. Molecular diffusion of the gases takes place and concentrations, and concentration gradients, change with time. If a bowl of liquid evaporates into an enclosed space, the partial pressure in the gas phase progressively increases, and the concentrations and the rate of evaporation are both time-dependent.

Consider an element of gas of unit cross-sectional area and of thickness δy in the direction of mass transfer in which the concentrations C_A and C_B of the components **A** and **B** are a function of both position y and time t (Figure 10.4). Suppose that the mass transfer flux is composed of two components, one attributable to diffusion according to Fick's Law and the other to a bulk flow velocity u_F, then the fluxes of **A** and **B** at a distance y from the origin are N'_A and N'_B, respectively. These will increase to $N'_A + (\partial N'_A/\partial y)\delta y$ and $N'_B + (\partial N'_B/\partial y)\delta y$ at a distance $y + \delta y$ from the origin. At position y, the fluxes N'_A and N'_B will be made up as in Table 10.6.

At a distance $y + \delta y$ from the origin, i.e. at the further boundary of the element, these fluxes will increase by the amounts shown in the second part of Table 10.6.

Thus the difference between the inlet and the outlet flux of **A** for the element will be:

$$-\left\{ -D\frac{\partial^2 C_A}{\partial y^2} + \frac{\partial(u_F C_A)}{\partial y} \right\} \delta y$$

This must, in turn, be equal to the rate of accumulation of **A** in the element, which can be expressed as the rate of change of molar concentration with time multiplied by the volume of the element,

or:
$$\frac{\partial C_A}{\partial t} \times (\delta y \times 1)$$

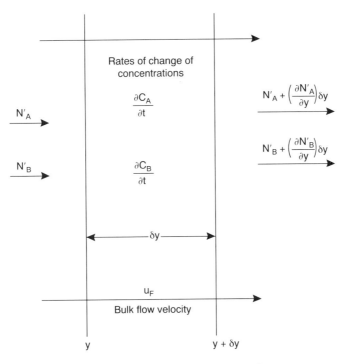

Figure 10.4.   Unsteady state mass transfer

Table 10.6.   Fluxes of a gas mixture

	Diffusional flux	Flux due to bulk flow	Total flux	
A	$-D\dfrac{\partial C_A}{\partial y}$	$u_F C_A$	$N'_A = -D\dfrac{\partial C_A}{\partial y} + u_F C_A$	(10.59)
B	$-D\dfrac{\partial C_B}{\partial y}$	$u_F C_B$	$N'_B = -D\dfrac{\partial C_B}{\partial y} + u_F C_B$	(10.60)

Changes in fluxes over distance δy:

A	$-D\left(\dfrac{\partial^2 C_A}{\partial y^2}\right)\delta y$	$\dfrac{\partial(u_F C_A)}{\partial y}\delta y$	$\left\{-D\dfrac{\partial^2 C_A}{\partial y^2} + \dfrac{\partial(u_F C_A)}{\partial y}\right\}\delta y$	
B	$-D\left(\dfrac{\partial^2 C_B}{\partial y^2}\right)\delta y$	$\dfrac{\partial(u_F C_B)}{\partial y}\delta y$	$\left\{-D\dfrac{\partial^2 C_B}{\partial y^2} + \dfrac{\partial(u_F C_B)}{\partial y}\right\}\delta y$	

Equating:
$$\frac{\partial C_A}{\partial t} = D\frac{\partial^2 C_A}{\partial y^2} - \frac{\partial(u_F C_A)}{\partial y} \tag{10.61}$$

For component **B**:
$$\frac{\partial C_B}{\partial t} = D\frac{\partial^2 C_B}{\partial y^2} - \frac{\partial(u_F C_B)}{\partial y} \tag{10.62}$$

Adding:
$$\frac{\partial(C_A + C_B)}{\partial t} = D\frac{\partial^2(C_A + C_B)}{\partial y^2} - \frac{\partial}{\partial y}[(C_A + C_B)u_F]$$

Since $C_A + C_B = C_T =$ constant for an ideal gas (equation 10.9):

$$0 = 0 - \frac{\partial}{\partial y}(u_F C_T)$$

Thus:

$$\frac{\partial u_F}{\partial y} = 0$$

and u_F is therefore independent of y.

Thus equation 10.61 can be written:

$$\frac{\partial C_A}{\partial t} = D\frac{\partial^2 C_A}{\partial y} - u_F\frac{\partial C_A}{\partial y} \tag{10.63}$$

Equimolecular counterdiffusion

For equimolecular counterdiffusion, $u_F = 0$ and equation 10.63 simplifies to:

$$\frac{\partial C_A}{\partial t} = D\frac{\partial^2 C_A}{\partial y^2} \tag{10.64}$$

Equation 10.64 is referred to as Fick's Second Law. It can also be applied when u_F is small, corresponding to conditions where C_A is always low. This equation can be solved for a number of important boundary conditions and examples will be given later. It should be compared with the corresponding equation for unsteady state heat transfer (equation 9.28).

For the more general three-dimensional case where concentration gradients are changing in the x, y and z directions, the changes taking place in each of the three principal directions must be added to give:

$$\frac{\partial C_A}{\partial t} = D\left[\frac{\partial^2 C_A}{\partial x^2} + \frac{\partial^2 C_A}{\partial y^2} + \frac{\partial^2 C_A}{\partial z^2}\right] \tag{10.65}$$

Gas absorption

In general, it is necessary to specify the physical constraints operating on the system in order to evaluate the bulk flow velocity u_F. In gas absorption, there will be no overall flux of the insoluble component **B** at the liquid interface ($y = 0$, say). However, in an unsteady state process where, by definition, concentrations will be changing with time throughout the system, the flux of **B** will be zero only at $y = 0$. At the interface ($y = 0$), total flux of **B** (from equation 10.43b) is given by:

$$N'_B = -D\left(\frac{\partial C_B}{\partial y}\right)_{y=0} + u_F(C_B)_{y=0} = 0$$

i.e.:

$$u_F = \frac{D\left(\dfrac{\partial C_B}{\partial y}\right)_{y=0}}{(C_B)_{y=0}} = \frac{-D}{(C_T - C_A)_{y=0}}\left(\frac{\partial C_A}{\partial y}\right)_{y=0} \tag{10.66}$$

Substituting in equation 10.63:

$$\frac{\partial C_A}{\partial t} = D \left[\frac{\partial^2 C_A}{\partial y^2} + \frac{1}{(C_T - C_A)_{y=0}} \left(\frac{\partial C_A}{\partial y} \right)_{y=0} \frac{\partial C_A}{\partial y} \right] \tag{10.67}$$

Thus at the interface ($y = 0$):

$$\left(\frac{\partial C_A}{\partial t} \right)_{y=0} = D \left[\left(\frac{\partial^2 C_A}{\partial y^2} \right)_{y=0} + \frac{1}{(C_T - C_A)_{y=0}} \left(\frac{\partial C_A}{\partial y} \right)^2_{y=0} \right] \tag{10.68}$$

This equation is not capable of an exact analytical solution. It has been discussed by ARNOLD[10] in relation to evaporation from a free surface.

Substituting into equation 10.59 for N_A':

$$N_A' = -D \frac{\partial C_A}{\partial y} - \frac{D C_A}{(C_T - C_A)_{y=0}} \left(\frac{\partial C_A}{\partial y} \right)_{y=0}$$

$$= -D \left[\frac{\partial C_A}{\partial y} + \frac{C_A}{(C_T - C_A)_{y=0}} \left(\frac{\partial C_A}{\partial y} \right)_{y=0} \right] \tag{10.69}$$

10.3. MULTICOMPONENT GAS-PHASE SYSTEMS

10.3.1. Molar flux in terms of effective diffusivity

For a multicomponent system, the bulk flow velocity u_F is given by:

$$u_F = \frac{1}{C_T} (N_A' + N_B' + N_C' + \cdots) \tag{10.70}$$

$$u_F = x_A u_A + x_B u_B + x_C u_C + \cdots \tag{10.71}$$

i.e.:

$$u_F = \sum x_A u_A = \frac{1}{C_T} \sum N_A' \tag{10.72}$$

Then, since $N_A' = N_A + u_F C_A$ (equation 10.43),

$$N_A' = -D' \frac{dC_A}{dy} + \left[\frac{1}{C_T} \Sigma N_A' \right] C_A$$

$$= -D' C_T \frac{dx_A}{dy} + x_A \Sigma N_A' \tag{10.73}$$

where D' is the effective diffusivity for transfer of **A** in a mixture of **B**, **C**, **D**, etc.

For the particular case, where N_B', N_C', etc., are all zero:

$$N_A' = -D' C_T \frac{dx_A}{dy} + x_A N_A'$$

or:

$$N_A' = -D' \frac{C_T}{1 - x_A} \frac{dx_A}{dy} \tag{10.74}$$

A means of calculating the effective diffusivity D' in terms of each of the binary diffusivities is given in the following section.

10.3.2. Maxwell's law of diffusion

Maxwell's law for a binary system

MAXWELL[11] postulated that the partial pressure gradient in the direction of diffusion for a constituent of a two-component gaseous mixture was proportional to:

(a) the relative velocity of the molecules in the direction of diffusion, and
(b) the product of the molar concentrations of the components.

Thus:
$$-\frac{dP_A}{dy} = FC_AC_B(u_A - u_B) \tag{10.75}$$

where u_A and u_B are the mean molecular velocities of **A** and **B** respectively in the direction of mass transfer and F is a coefficient.

Writing:
$$u_A = \frac{N'_A}{C_A} \tag{10.76}$$

$$u_B = \frac{N'_B}{C_B} \tag{10.77}$$

$$P_A = C_A\mathbf{R}T \quad \text{(for an ideal gas)} \qquad \text{(equation 10.7a)}$$

On substitution into equation 10.75:

$$-\frac{dC_A}{dy} = \frac{F}{\mathbf{R}T}(N_AC_B - N_BC_A) \tag{10.78}$$

Equimolecular counterdiffusion

By definition:
$$N'_A = -N'_B = N_A$$

Substituting in equation 10.78:

$$-\frac{dC_A}{dy} = \frac{FN_A}{\mathbf{R}T}(C_B + C_A) \tag{10.79}$$

Thus:
$$N_A = -\frac{\mathbf{R}T}{FC_T}\frac{dC_A}{dy} \tag{10.80}$$

Then, by comparison with Fick's Law (equation 10.4):

$$D = \frac{\mathbf{R}T}{FC_T} \tag{10.81}$$

or:
$$F = \frac{\mathbf{R}T}{DC_T} \tag{10.82}$$

*Transfer of **A** through stationary **B***

By definition:
$$N'_B = 0$$

$$\therefore \quad -\frac{dC_A}{dy} = \frac{F}{\mathbf{R}T}N'_AC_B \qquad \text{(from equation 10.78)}$$

or:
$$N'_A = -\frac{RT}{FC_T} \frac{C_T}{C_B} \frac{dC_A}{dy}$$
(10.83)

Substituting from equation 10.81:

$$N'_A = -D\frac{C_T}{C_B} \frac{dC_A}{dy}$$
(10.84)

It is seen that equation 10.84 is identical to equation 10.27. (Stefan's Law.)

Thus, Stefan's law can also be derived from Maxwell's law of diffusion.

Maxwell's law for multicomponent mass transfer

The previous argument can be applied to the diffusion of a constituent of a multi-component gas. Consider the transfer of component **A** through a stationary gas consisting of components **B**, **C**, ... Suppose that the total partial pressure gradient can be regarded as being made up of a series of terms each representing the contribution of the individual component gases, then from equation 10.78:

$$-\frac{dC_A}{dy} = \frac{F_{AB}N'_A C_B}{RT} + \frac{F_{AC}N'_A C_C}{RT} + \cdots$$

or:
$$-\frac{dC_A}{dy} = \frac{N'_A}{RT}(F_{AB}C_B + F_{AC}C_C + \cdots)$$
(10.85)

From equation 10.82, writing:

$$F_{AB} = \frac{RT}{D_{AB}C_T}, \quad \text{etc.}$$

where D_{AB} is the diffusivity of **A** in **B**, etc.:

$$-\frac{dC_A}{dy} = \frac{N'_A}{C_T}\left(\frac{C_B}{D_{AB}} + \frac{C_C}{D_{AC}} + \cdots\right)$$
(10.86)

$$\therefore \quad N'_A = -\frac{C_T}{\dfrac{C_B}{D_{AB}} + \dfrac{C_C}{D_{AC}} + \cdots} \frac{dC_A}{dy}$$

$$= -\frac{1}{\dfrac{C_B}{C_T - C_A}\dfrac{1}{D_{AB}} + \dfrac{C_C}{C_T - C_A}\dfrac{1}{D_{AC}} + \cdots} \frac{C_T}{C - C_A}\frac{dC_A}{dy}$$

$$= -\frac{1}{\dfrac{y'_B}{D_{AB}} + \dfrac{y'_C}{D_{AC}} + \cdots} \frac{C_T}{C_T - C_A}\frac{dC_A}{dy}$$
(10.87)

where y'_B is the mole fraction of **B** in the stationary components of the gas, etc.

By comparing equation 10.87 with Stefan's law (equation 10.27) the effective diffusivity of **A** in the mixture (D') is given by:

$$\frac{1}{D'} = \frac{y'_B}{D_{AB}} + \frac{y'_C}{D_{AC}} + \cdots$$
(10.88)

Multicomponent mass transfer is discussed in detail by TAYLOR and KRISHNA[12].

10.4. DIFFUSION IN LIQUIDS

The diffusion of solution in a liquid is governed by the same equations as for the gas phase but the diffusion coefficient D is about two orders of magnitude smaller for a liquid than for a gas. Furthermore, the diffusion coefficient is a much more complex function of the molecular properties.

For an ideal gas, the total molar concentration C_T is constant at a given total pressure P and temperature T. This approximation holds quite well for real gases and vapours, except at high pressures. For a liquid however, C_T may show considerable variations as the concentrations of the components change and, in practice, the total mass concentration (density ρ of the mixture) is much more nearly constant. Thus for a mixture of ethanol and water, the mass density will range from about 790 to 1000 kg/m^3 whereas the molar density will range from about 17 to 56 kmol/m^3. For this reason the diffusion equations are frequently written in the form of a mass flux J_A (mass/unit area, unit time) and the concentration gradients in terms of mass concentrations, for example c_A.

Thus, for component **A**:

$$J_A = -D\frac{dc_A}{dy} \tag{10.89}$$

$$= -D\rho\frac{d\omega_A}{dy} \tag{10.90}$$

where ρ is mass density (now taken as constant), and ω_A is the mass fraction of **A** in the liquid.

For component **B**:

$$J_B = -D\rho\frac{d\omega_B}{dy} \tag{10.91}$$

$$= D\rho\frac{d\omega_A}{dy} \quad \text{(since } \omega_A + \omega_B = 1\text{).} \tag{10.92}$$

Thus, the diffusional process in a liquid gives rise to a situation where the components are being transferred at approximately equal and opposite mass (rather than molar) rates.

Liquid phase diffusivities are strongly dependent on the concentration of the diffusing component; this is in strong contrast to gas phase diffusivities which are substantially independent of concentration. Normally, values of liquid phase diffusivities which are quoted apply to very dilute concentrations of the diffusing component, the only condition under which analytical solutions can be produced for the diffusion equations. For that reason, only dilute solutions will be considered in the work which follows, and in these circumstances there is no serious error involved in using Fick's first and second laws expressed in molar units.

$$N_A = -D\frac{dC_A}{dy} \tag{equation 10.4}$$

and:

$$\frac{\partial C_A}{\partial t} = D\frac{\partial^2 C_A}{\partial y^2} \tag{equation 10.64}$$

where D is now the liquid phase diffusivity and C_A is the molar concentration in the liquid phase.

Then, on integration, equation 10.4 again becomes:

$$N_A = -D\frac{C_{A_2} - C_{A_1}}{y_2 - y_1} = \frac{D}{y_2 - y_1}(C_{A_1} - C_{A_2}) \tag{10.93}$$

and $D/(y_2 - y_1)$ is now the liquid phase mass transfer coefficient.

An example of the integration of equation 10.64 is given later in Section 10.5.3.

10.4.1. Liquid phase diffusivities

Values of the diffusivities of various materials in water are given in Table 10.7. Where experimental values are not available, it is necessary to use one of the predictive methods which have been developed.

A useful equation for the calculation of liquid phase diffusivities of dilute solutions of non-electrolytes has been given by WILKE and CHANG[13]. It is not dimensionally consistent and therefore the value of the coefficient depends on the units employed. Using SI units:

$$D = \frac{1.173 \times 10^{-16}\phi_B^{1/2}M_B^{1/2}T}{\mu V_A^{0.6}} \tag{10.94}$$

where D is the diffusivity of solute **A** in solvent **B** (m^2/s),

ϕ_B is the association factor for the solvent (2.26 for water, 1.9 for methanol, 1.5 for ethanol and 1.0 for unassociated solvents such as hydrocarbons and ethers),

M_B is the molecular weight of the solvent,

μ is the viscosity of the solution $(N\,s/m^2)$,

T is temperature (K), and

V_A is the molecular volume of the solute $(m^3/kmol)$. Values for simple molecules are given in Table 10.4. For more complex molecules, V_A is calculated by summation of the atomic volume contributions given in Table 10.4.

N.B. For water a value of $0.0756\ m^3/kmol$ should be used.

Equation 10.94 applies neither to electrolytes nor to concentrated solutions. REID, PRAUSNITZ and SHERWOOD[14] discuss diffusion in electrolytes. Little information is available on diffusivities in concentrated solutions but it appears that, for ideal mixtures, the product μD is a linear function of the molar concentration.

The calculation of liquid phase diffusivities is discussed further in Volume 6.

10.5. MASS TRANSFER ACROSS A PHASE BOUNDARY

The theoretical treatment which has been developed in the preceding sections has related to mass transfer within a single phase in which no discontinuities exist. In most of the important applications of mass transfer, however, material is transferred across a phase boundary. Thus, in distillation a vapour and liquid are brought into contact in the fractionating column and the more volatile material is transferred from the liquid to the vapour while the less volatile constituent is transferred in the opposite direction; this is

Table 10.7.   Diffusivities (diffusion coefficients) in liquids at 293 K[4]

Solute	Solvent	D $(m^2/s \times 10^9)$	Sc $(\mu/\rho D)^*$
O_2	Water	1.80	558
CO_2	Water	1.50	670
N_2O	Water	1.51	665
NH_3	Water	1.76	570
Cl_2	Water	1.22	824
Br_2	Water	1.2	840
H_2	Water	5.13	196
N_2	Water	1.64	613
HCl	Water	2.64	381
H_2S	Water	1.41	712
H_2SO_4	Water	1.73	580
HNO_3	Water	2.6	390
Acetylene	Water	1.56	645
Acetic acid	Water	0.88	1140
Methanol	Water	1.28	785
Ethanol	Water	1.00	1005
Propanol	Water	0.87	1150
Butanol	Water	0.77	1310
Allyl alcohol	Water	0.93	1080
Phenol	Water	0.84	1200
Glycerol	Water	0.72	1400
Pyrogallol	Water	0.70	1440
Hydroquinone	Water	0.77	1300
Urea	Water	1.06	946
Resorcinol	Water	0.80	1260
Urethane	Water	0.92	1090
Lactose	Water	0.43	2340
Maltose	Water	0.43	2340
Glucose	Water	0.60	1680
Mannitol	Water	0.58	1730
Raffinose	Water	0.37	2720
Sucrose	Water	0.45	2230
Sodium chloride	Water	1.35	745
Sodium hydroxide	Water	1.51	665
CO_2	Ethanol	3.4	445
Phenol	Ethanol	0.8	1900
Chloroform	Ethanol	1.23	1230
Phenol	Benzene	1.54	479
Chloroform	Benzene	2.11	350
Acetic acid	Benzene	1.92	384
Ethylene dichloride	Benzene	2.45	301

*Based on $\mu/\rho = 1.005 \times 10^{-6} m^2/s$ for water, 7.37×10^{-7} for benzene, and 1.511×10^{-6} for ethanol, all at 293 K; applies only for dilute solutions.

Figures are based mainly on *International Critical Tables* **5** (1928).

an example of equimolecular counterdiffusion. In gas absorption the soluble gas diffuses to the surface, dissolves in the liquid, and then passes into the bulk of the liquid; in this case, the carrier gas is not transferred. In both of these examples, one phase is a liquid and the other a gas. However, in liquid–liquid extraction, a solute is transferred from one liquid solvent to another across a phase boundary, and in the dissolution of a crystal the solute is transferred from a solid to a liquid.

Each of the above processes is characterised by a transference of material across an interface. Because no material accumulates there, the rate of transfer on each side of the interface must be the same, and therefore the concentration gradients automatically adjust themselves so that they are proportional to the resistance to transfer in the particular phase. Furthermore, if there is no resistance to transfer at the interface, the concentrations on each side will be related to each other by the phase equilibrium relationship. The existence or otherwise of a resistance to transfer at the phase boundary is the subject of conflicting opinions[15]. It appears likely that it is not high, except in the case of crystallisation, and in the subsequent treatment equilibrium between the phases will be assumed to exist at the interface. Interfacial resistance may occur, however, if a surfactant is present as it may collect at the interface (see Section 10.5.5).

The mass transfer rate between two fluid phases will depend on the physical properties of the two phases, the concentration difference, the interfacial area, and the degree of turbulence. Mass transfer equipment is therefore designed to give a large area of contact between the phases and to promote turbulence in each of the fluids. In the majority of plants, the two phases flow continuously in a countercurrent manner. In a steady state process, therefore, though the composition of each element of fluid is changing as it passes through the equipment, conditions at any given point do not change with time. In most industrial equipment, the flow pattern is so complex that it is not capable of expression in mathematical terms, and the interfacial area is not known precisely.

A number of mechanisms have been suggested to represent conditions in the region of the phase boundary. The earliest of these is the *two-film* theory propounded by WHITMAN[16] in 1923. He suggested that the resistance to transfer in each phase could be regarded as lying in a thin film close to the interface. The transfer across these films is regarded as a steady state process of molecular diffusion following equations of the type of equation 10.19. The turbulence in the bulk fluid is considered to die out at the interface of the films. In 1935 HIGBIE[17] suggested that the transfer process was largely attributable to fresh material being brought by the eddies to the interface, where a process of unsteady state transfer took place for a fixed period at the freshly exposed surface. This theory is generally known as the *penetration theory*. DANCKWERTS[18] has since suggested a modification of this theory and considers that the material brought to the surface will remain there for varying periods of time. He later discusses the random age distribution of such elements from which the transfer is by an unsteady state process to the second phase. Subsequently, TOOR and MARCHELLO[19] have proposed a more general theory, the *film-penetration theory*, and have shown that each of the earlier theories is a particular limiting case of their own. A number of other theoretical treatments have also been given, including that of KISHINEVSKIJ[20]. The two-film theory and the penetration theory will now be considered, followed by an examination of the film-penetration theory.

10.5.1. The two-film theory

The two-film theory of WHITMAN[16] was the first serious attempt to represent conditions occurring when material is transferred from one fluid stream to another. Although it does not closely reproduce the conditions in most practical equipment, it gives expressions which can be applied to the experimental data which are generally available, and for that reason it is still extensively used.

In this theory, it is assumed that turbulence dies out at the interface and that a laminar layer exists in each of the two fluids. Outside the laminar layer, turbulent eddies supplement the action caused by the random movement of the molecules, and the resistance to transfer becomes progressively smaller. For equimolecular counterdiffusion the concentration gradient is therefore linear close to the interface, and gradually becomes less at greater distances as shown in Figure 10.5 by the full lines *ABC* and *DEF*. The basis of the theory is the assumption that the zones in which the resistance to transfer lies can be replaced by two hypothetical layers, one on each side of the interface, in which the transfer is entirely by molecular diffusion. The concentration gradient is therefore linear in each of these layers and zero outside. The broken lines *AGC* and *DHF* indicate the hypothetical concentration distributions, and the thicknesses of the two films are L_1 and L_2. Equilibrium is assumed to exist at the interface and therefore the relative positions of the points C and D are determined by the equilibrium relation between the phases. In Figure 10.5, the scales are not necessarily the same on the two sides of the interface.

Figure 10.5.   Two-film theory

The mass transfer is treated as a steady state process and therefore the theory can be applied only if the time taken for the concentration gradients to become established is very small compared with the time of transfer, or if the capacity of the films is negligible.

From equation 10.19 the rate of transfer per unit area in terms of the two-film theory for equimolecular counterdiffusion is given for the first phase as:

$$N_A = \frac{D_1}{L_1}(C_{Ao_1} - C_{Ai_1}) = h_{D1}(C_{Ao_1} - C_{Ai_1}) \tag{10.95a}$$

where L_1 is the thickness of the film, C_{Ao_1} the molar concentration outside the film, and C_{Ai_1} the molar concentration at the interface.

For the second phase, with the same notation:

$$N_A = \frac{D_2}{L_2}(C_{Ai_2} - C_{Ao_2}) = h_{D2}(C_{Ai_2} - C_{Ao_2}) \tag{10.95b}$$

Because material does not accumulate at the interface, the two rates of transfer must be the same and:

$$\frac{h_{D1}}{h_{D2}} = \frac{C_{Ai_2} - C_{Ao_2}}{C_{Ao_1} - C_{Ai_1}} \tag{10.96}$$

The relation between C_{Ai_1} and C_{Ai_2} is determined by the phase equilibrium relationship since the molecular layers and each side of the interface are assumed to be in equilibrium with one another. It will be noted that the ratio of the differences in concentrations is inversely proportional to the ratio of the mass transfer coefficients. If the bulk concentrations, C_{Ao_1} and C_{Ao_2} are fixed, the interface concentrations will adjust themselves so that equation 10.96 is satisfied. This means that if the relative value of the coefficients changes, the interface concentrations will change too. In general, if the degree of turbulence of the fluid is increased, the effective film thicknesses will be reduced and the mass transfer coefficients will be correspondingly increased.

The theory is equally applicable when bulk flow occurs. In the case, for example, of gas absorption, the mass transfer rate in terms of the concentration gradient in the gas phase will be:

$$N'_A = -D \frac{dC_A}{dy} \cdot \frac{C_T}{C_B} \qquad \text{(equation 10.27)}$$

In this case, for a steady state process, $(dC_A/dy)(C_T/C_B)$, as opposed to dC_A/dy, will be constant through the film and dC_A/dy will increase as C_A decreases. Thus lines GC and DH in Figure 10.5 will no longer be quite straight.

10.5.2. The penetration theory

The penetration theory was propounded in 1935 by HIGBIE[17] who was investigating whether or not a resistance to transfer existed at the interface when a pure gas was absorbed in a liquid. He carried out his experiments by allowing a slug-like bubble of carbon dioxide to rise through a vertical column of water in a 3 mm diameter glass tube. As the bubble rose, the displaced liquid ran back as a thin film between the bubble and the tube. Higbie assumed that each element of surface in this liquid was exposed to the gas for the time taken for the gas bubble to pass it; that is for the time given by the quotient of the bubble length and its velocity. He supposed that during this short period, which varied between 0.01 and 0.1 s in his experiments, absorption took place as the result of unsteady state molecular diffusion into the liquid; for the purposes of calculation he regarded the liquid as infinite in depth because the time of exposure was so short.

The way in which the concentration gradient builds up as a result of exposing a liquid — initially pure — to the action of a soluble gas in shown in Figure 10.6 which is based on Higbie's calculations. The percentage saturation of the liquid is plotted against distance from the surface for a number of exposure times in arbitrary units. Initially only the surface layer contains solute and the concentration changes abruptly from 100 per cent to 0 per cent at the surface. For progressively longer exposure times the concentration profile develops as shown, until after an infinite time the whole of the liquid becomes saturated. The shape of the profiles is such that at any time the effective depth of liquid which contains an appreciable concentration of solute can be specified, and hence the theory is referred to as the *Penetration Theory*. If this depth of penetration is less than the total depth of liquid, no significant error is introduced by assuming that the total depth is infinite.

The work of Higbie laid the basis of the penetration theory in which it is assumed that the eddies in the fluid bring an element of fluid to the interface where it is exposed

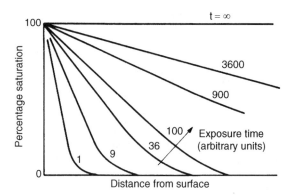

Figure 10.6.   Penetration of solute into a solvent

to the second phase for a definite interval of time, after which the surface element is mixed with the bulk again. Thus, fluid whose initial composition corresponds with that of the bulk fluid remote from the interface is suddenly exposed to the second phase. It is assumed that equilibrium is immediately attained by the surface layers, that a process of unsteady state molecular diffusion then occurs and that the element is remixed after a fixed interval of time. In the calculation, the depth of the liquid element is assumed to be infinite and this is justifiable if the time of exposure is sufficiently short for penetration to be confined to the surface layers. Throughout, the existence of velocity gradients within the fluids is ignored and the fluid at all depths is assumed to be moving at the same rate as the interface.

Diffusion of solute **A** away from the interface (Y-direction) is thus given by equation 10.64:

$$\frac{\partial C_A}{\partial t} = D \frac{\partial^2 C_A}{\partial y^2} \qquad \text{(equation 10.64)}$$

for conditions of equimolecular counterdiffusion, or when the concentrations of diffusing materials are sufficiently low for the bulk flow velocity to be negligible. Because concentrations of **A** are low, there is no objection to using molar concentration for calculation of mass transfer rates in the liquid phase (see section 10.4).

The following boundary conditions apply:

$$t = 0 \quad 0 < y < \infty \quad C_A = C_{Ao}$$

$$t > 0 \qquad y = 0 \qquad C_A = C_{Ai}$$

$$t > 0 \qquad y = \infty \quad C_A = C_{Ao}$$

where C_{Ao} is the concentration in the bulk of the phase, and C_{Ai} the equilibrium value at the interface.

It is convenient to work in terms of a "deviation" variable C' as opposed to C_A. C' is the amount by which the concentration of **A** exceeds the initial uniform concentration C_{Ao}. This change effects some simplification of the algebra.

With the substitution:

$$C' = C_A - C_{Ao} \qquad (10.97)$$

equation 10.64 becomes:

$$\frac{\partial C'}{\partial t} = D\frac{\partial^2 C'}{\partial y^2} \tag{10.98}$$

because C_{Ao} is a constant with respect to both t and y; the boundary conditions are then:

$$t = 0 \quad 0 < y < \infty \quad C' = 0$$

$$t > 0 \qquad y = 0 \qquad C' = C'_i = C_{Ai} - C_{Ao}$$

$$t > 0 \qquad y = \infty \quad C' = 0$$

These boundary conditions are necessary and sufficient for the solution of equation 10.98 which is first order with respect to t and second order with respect to y.

The equation is most conveniently solved by the method of Laplace transforms, used for the solution of the unsteady state thermal conduction problem in Chapter 9.

By definition, the Laplace transform $\overline{C'}$ of C' is given by:

$$\overline{C'} = \int_0^\infty e^{-pt}C'\,dt \tag{10.99}$$

Then:
$$\frac{\overline{\partial C'}}{\partial t} = \int_0^\infty e^{-pt}\frac{\partial C'}{\partial t}\,dt \tag{10.100}$$

$$= \left[e^{-pt}C'\right]_0^\infty + p\int_0^\infty e^{-pt}C'\,dt$$

$$= p\overline{C'} \tag{10.101}$$

Since the Laplace transform operation is independent of y:

$$\frac{\overline{\partial^2 C'}}{\partial y^2} = \frac{\partial^2 \overline{C'}}{\partial y^2} \tag{10.102}$$

Thus, taking Laplace transforms of both sides of equation 10.98:

$$p\overline{C'} = D\frac{\partial^2 \overline{C'}}{\partial y^2}$$

$$\therefore \qquad \frac{\partial^2 C'}{\partial y^2} - \frac{p}{D}\overline{C'} = 0$$

Equation 10.98 has therefore been converted from a partial differential equation in C' to an ordinary second order linear differential equation in $\overline{C'}$.

Thus:
$$\overline{C'} = B_1 e^{\sqrt{(p/D)}y} + B_2 e^{-\sqrt{(p/D)}y} \tag{10.103}$$

When:
$$y = 0, \; C_A = C_{Ai}, \quad \text{and} \quad C' = C_{Ai} - C_{Ao} = C'_i$$

and when:
$$y = \infty, \; C_A = C_{Ao}, \quad \text{and} \quad C' = 0$$

Hence:
$$B_1 = 0$$

and:
$$\overline{C'} = B_2 e^{-\sqrt{(p/D)}y} = \overline{C'_i}e^{-\sqrt{(p/D)}y} \tag{10.104}$$

Now:
$$B_2 = \int_0^\infty (C_{Ai} - C_{Ao})e^{-pt}\, dt$$

$$= \frac{1}{p}(C_{Ai} - C_{Ao})$$

Thus:
$$\overline{C'} = \frac{1}{p}(C_{Ai} - C_{Ao})e^{-\sqrt{(p/D)}y} \qquad (10.105)$$

Taking the inverse of the transform (see Appendix, Table 12, No. 83):

$$C' = C_A - C_{Ao} = (C_{Ai} - C_{Ao})\operatorname{erfc}\frac{y}{2\sqrt{Dt}}$$

or:
$$\frac{C_A - C_{Ao}}{C_{Ai} - C_{Ao}} = \operatorname{erfc}\frac{y}{2\sqrt{Dt}} = 1 - \operatorname{erf}\frac{y}{2\sqrt{Dt}} \qquad (10.106)$$

This expression gives concentration C_A as a function of position y and of time t.

erf X is known as the error function and values are tabulated as for any other function of X; erfc X is the *complementary error function* $(1 - \operatorname{erf} X)$.

By definition:
$$\operatorname{erfc} X = \frac{2}{\sqrt{\pi}}\int_X^\infty e^{-X^2}\, dX \qquad (10.107)$$

Since:
$$\int_0^\infty e^{-X^2}\, dX = \frac{\sqrt{\pi}}{2}, \qquad (10.108)$$

erfc X goes from 1 to 0 as X goes from 0 to 1.

The concentration gradient is then obtained by differentiation of equation 10.106 with respect to y.

Thus:

$$\frac{1}{C_{Ai} - C_{Ao}}\frac{\partial C_A}{\partial y} = \frac{\partial}{\partial y}\left[\frac{2}{\sqrt{\pi}}\int_{(y/2\sqrt{Dt})}^\infty e^{-y^2/4Dt}\, d\left(\frac{y}{2\sqrt{Dt}}\right)\right]$$

$$\therefore \qquad \frac{\partial C_A}{\partial y} = -(C_{Ai} - C_{Ao})\frac{2}{\sqrt{\pi}}\frac{1}{2\sqrt{Dt}}(e^{-y^2/4Dt})$$

$$= -(C_{Ai} - C_{Ao})\frac{1}{\sqrt{\pi Dt}}e^{-y^2/4Dt} \qquad (10.109)$$

The mass transfer rate at any position y at time t is given by:

$$(N_A)_t = -D\frac{\partial C_A}{\partial y}$$

$$= (C_{Ai} - C_{Ao})\sqrt{\frac{D}{\pi t}}e^{-y^2/4Dt} \qquad (10.110)$$

The mass transfer rate at the surface is then given by:

$$(N_A)_{t,y=0} = -D\left(\frac{\partial C_A}{\partial y}\right)_{y=0}$$

$$= (C_{Ai} - C_{Ao})\sqrt{\frac{D}{\pi t}} \tag{10.111}$$

The point value of the mass transfer coefficient is therefore $\sqrt{D/\pi t}$.

Regular surface renewal

It is important to note that the mass transfer rate falls off progressively during the period of exposure, theoretically from infinity at $t = 0$ to zero at $t = \infty$.

In a practical situation, the average mass transfer rate over a time of exposure t_e is needed. This is obtained by integrating to obtain the total number of moles transferred during the period and dividing by the time. Assuming that all the surface elements are exposed for the same time t_e (Higbie's assumption):

$$N_A = (C_{Ai} - C_{Ao})\sqrt{\frac{D}{\pi}}\frac{1}{t_e}\int_0^{t_e}\frac{dt}{\sqrt{t}} \tag{10.112}$$

$$= 2(C_{Ai} - C_{Ao})\sqrt{\frac{D}{\pi t_e}} \tag{10.113}$$

i.e. the average rate over the time interval $t = 0$ to $t = t_e$ is twice the point value at $t = t_e$.

Thus, the shorter the time of exposure the greater is the rate of mass transfer. No precise value can be assigned to t_e in any industrial equipment, but its value will clearly become less as the degree of agitation of the fluid is increased.

If it is assumed that each element resides for the same time interval t_e in the surface, equation 10.113 gives the overall mean rate of transfer. It will be noted that the rate is a linear function of the driving force expressed as a concentration difference, as in the two-film theory, but that it is proportional to the diffusivity raised to the power of 0.5 instead of unity.

When mass transfer rates are very high, limitations may be placed on the rate at which a component may be transferred, by virtue of the frequency with which the molecules collide with the surface. For a gas, the collision rate can be calculated from the Kinetic Theory and allowance must then be made for the fact that only a fraction of these molecules may be absorbed, the remainder being reflected. Thus, when even a pure gas is brought suddenly into contact with a fresh solvent, the initial mass transfer rate may be controlled by the rate at which gas molecules can reach the surface. However, the resistance to transfer rapidly builds up in the liquid phase to a level where this effect can be neglected. The point is well illustrated in Example 10.4.

Example 10.3

In an experimental wetted wall column, pure carbon dioxide is absorbed in water. The mass transfer rate is calculated using the penetration theory, application of which is limited by the fact that the concentration should not reach more than 1 per cent of the saturation value at a depth below the surface at which the velocity is 95 per cent of the surface velocity. What is the maximum length of column to which the theory can be applied if the flowrate of water is 3 cm^3/s per cm of perimeter?

$$\text{Viscosity of water} \qquad\qquad\qquad = 10^{-3} \text{ N s/m}^2$$

$$\text{Diffusivity of carbon dioxide in water} = 1.5 \times 10^{-9} \text{ m}^2/\text{s}$$

Solution

For the flow of a vertical film of fluid, the mean velocity of flow is governed by equation 3.87:

$$u_m = \frac{\rho g s^2}{3\mu}$$

where s is the thickness of the film.

Flowrate per unit perimeter $= \rho g s^3/3\mu = 3 \times 10^{-4}$ m^2/s

Thus:
$$s = \left(\frac{3 \times 10^{-4} \times 10^{-3} \times 3}{1000 \times 9.81}\right)^{1/3}$$

$$= 4.51 \times 10^{-4} \text{ m}$$

The velocity u_x at a distance y' from the column wall is given by equation 3.85 (using y' in place of y) as:

$$u_x = \frac{\rho g(s y' - \frac{1}{2} y'^2)}{\mu}$$

The free surface velocity u_s is given by putting $y' = s$:

$$u_s = \frac{\rho g s^2}{2\mu}$$

$$\therefore \qquad \frac{u_x}{u_s} = 2\left(\frac{y'}{s}\right) - \left(\frac{y'}{s}\right)^2 = 1 - \left(1 - \frac{y'}{s}\right)^2$$

When $u_x/u_s = 0.95$ (velocity is 95 per cent of surface velocity):

$$1 - \frac{y'}{s} = 0.224$$

$$\therefore \qquad y = s - y' = 1.010 \times 10^{-4} \text{ m (distance below free surface)}$$

Equation 10.106 gives the relationship between concentration C_A, time and depth as:

$$\frac{C_A - C_{Ao}}{C_{Ai} - C_{Ao}} = \text{erfc}\left(\frac{y}{2\sqrt{Dt}}\right)$$

The time at which concentration reaches 0.01 of saturation value at a depth of 1.010×10^{-4} m is given by:

$$0.01 = \text{erfc}\left(\frac{1.010 \times 10^{-4}}{2\sqrt{1.5 \times 10^{-9}t}}\right)$$

Thus:
$$\text{erf}\left(\frac{1.305}{\sqrt{t}}\right) = 0.99$$

Using tables of error functions (Appendix, Table 12):

$$\frac{1.305}{\sqrt{t}} = 1.822$$

$$t = 0.51s$$

Surface velocity $\qquad\qquad u_s = \dfrac{\rho g s^2}{2\mu}$

$$= \frac{1000 \times 9.81 \times (4.51 \times 10^{-4})^2}{2 \times 10^{-3}}$$

$$= 1 \text{ m/s}$$

Maximum length of column $\qquad = 1 \times 0.51 = \underline{\underline{0.51 \text{ m}}}$

Example 10.4

In a gas–liquid contactor, a pure gas is absorbed in a solvent and the Penetration Theory provides a reasonable model by which to describe the transfer mechanism. As fresh solvent is exposed to the gas, the transfer rate is initially limited by the rate at which the gas molecules can reach the surface. If at $20°C$ and a pressure of 1 bar the maximum possible rate of transfer of gas is 50 m^3/m^2s, express this as an equivalent resistance, when the gas solubility is 0.04 $kmol/m^3$.

If the diffusivity in the liquid phase is 1.8×10^{-9} m^2/s, at what time after the initial exposure will the resistance attributable to access of gas be equal to about 10 per cent of the total resistance to transfer?

Solution

Bulk gas concentration $\quad = \dfrac{1}{22.4} \times \dfrac{273}{293} = 0.0416 \text{ kmol/m}^3$

Initial mass transfer rate $= 50 \text{ m}^3/\text{m}^2 \text{ s}$

$$= 50 \times 0.0416 = 2.08 \text{ kmol/m}^2 \text{ s}$$

Concentration driving force in liquid phase

$$= 0.04 - 0 = 0.04 \text{ kmol/m}^3$$

Effective mass transfer coefficient initially

$$= \frac{2.08}{0.04} = 52.0 \text{ m/s}$$

Equivalent resistance $\quad = 1/52.0 = 0.0192 \text{ s/m}$

When this constitutes 10 per cent of the total resistance,

liquid phase resistance $\quad = 0.0192 \times 9 = 0.173 \text{ s/m}$

liquid phase coefficient $\ = 5.78 \text{ m/s}$

From equation 10.111, point value of liquid phase mass transfer coefficient $= \sqrt{\dfrac{D}{\pi t}}$

$$= \sqrt{\frac{1.8 \times 10^{-9}}{\pi t}} = 2.394 \times 10^{-5} t^{-1/2} \text{ m/s}$$

Resistance $= 4.18 \times 10^4 t^{1/2}$ s/m

Thus: $\qquad\qquad\qquad\qquad 4.18 \times 10^4 t^{1/2} = 0.173$

and: $\qquad\qquad\qquad\qquad\qquad t = \underline{\underline{1.72 \times 10^{-11} \text{ s}}}$

Thus the limited rate of access of gas molecules is not likely to be of any significance.

Example 10.5

A deep pool of ethanol is suddenly exposed to an atmosphere of pure carbon dioxide and unsteady state mass transfer, governed by Fick's Law, takes place for 100 s. What proportion of the absorbed carbon dioxide will have accumulated in the 1 mm layer closest to the surface in this period?

$$\text{Diffusivity of carbon dioxide in ethanol} = 4 \times 10^{-9} \text{ m}^2/\text{s.}$$

Solution

The accumulation in the 1 mm layer near the surface will be equal to the total amount of CO_2 entering the layer from the surface ($y = 0$) less that leaving ($y = 10^{-3}$ m) in the course of 100 s.

The mass rate of transfer at any position y and time t is given by equation 10.110:

$$(N_A)_t = C_{Ai} \sqrt{\frac{D}{\pi t}} e^{-y^2/4Dt}$$

Here N_A is expressed in mols per unit area and unit time and C_{Ao} is zero because the solvent is pure ethanol.

Considering unit area of surface, the mols transferred in time t_e at depth y

$$= C_{Ai} \sqrt{\frac{D}{\pi}} \int_0^{t_e} t^{-1/2} e^{-y^2/4Dt} \, dt$$

Putting $y^2/4Dt = X^2$:

$$t^{-1/2} = \frac{2\sqrt{D}X}{y}$$

$$dt = \frac{y^2}{4Dt} \frac{-2}{X^3} \, dX$$

Thus:

$$\text{Integral} = \int_\infty^{X_e} \frac{y^2}{4D} \frac{-2}{X^3} \frac{2\sqrt{D}X}{y} e^{-X^2} \, dX$$

$$= -\frac{y}{\sqrt{D}} \int_\infty^{X_e} X^{-2} e^{-X^2} \, dX$$

$$\text{Molar transfer per unit area} = C_{Ai} \sqrt{\frac{D}{\pi}} \left(\frac{-y}{\sqrt{D}}\right) \left\{ \left[e^{-X^2}(-X^{-1}) \right]_\infty^{X_e} - \int_\infty^{X_e} \left[-2X e^{-X^2}(-X^{-1}) \right] dX \right\}$$

$$= C_{Ai} \left(\frac{-y}{\sqrt{\pi}}\right) \left\{ -X_e^{-1} e^{-X_e^2} + 2 \int_{X_e}^\infty e^{-X^2} \, dX \right\}$$

$$= C_{Ai} \left(\frac{y}{\sqrt{\pi}}\right) \left\{ \frac{2\sqrt{Dt_e}}{y} e^{-y^2/4Dt_e} - \sqrt{\pi} \text{ erfc } \frac{y}{2\sqrt{Dt_e}} \right\}$$

$$= C_{Ai} \left\{ 2\sqrt{\frac{Dt_e}{\pi}} e^{-y^2/4Dt_e} - y \text{ erfc } \frac{y}{2\sqrt{Dt_e}} \right\}$$

Putting $D = 4 \times 10^{-9}$ m^2/s and $t = 100$ s:

At $y = 0$: $\text{moles transferred} = 2C_{Ai} \sqrt{\dfrac{4 \times 10^{-9} \times 100}{\pi}} = 7.14 \times 10^{-4} C_{Ai}$

At $y = 10^{-3}$ m: $\text{moles transferred} = C_{Ai}\{7.14 \times 10^{-4} e^{-0.626} - 10^{-3} \text{ erfc } 0.791\}$

$$= C_{Ai}\{3.82 \times 10^{-4} - 2.63 \times 10^{-4}\}$$

$$= 1.19 \times 10^{-4} C_{Ai}$$

Proportion retained in layer $= (7.14 - 1.19)/7.14 = \underline{\underline{0.83}}$

or 83 per cent has been retained

Random surface renewal

DANCKWERTS[18] suggested that each element of surface would not be exposed for the same time, but that a random distribution of ages would exist. It was assumed that the probability of any element of surface becoming destroyed and mixed with the bulk of the fluid was independent of the age of the element and, on this basis, the age distribution of the surface elements was calculated as follows:

Supposing that the rate of production of fresh surface per unit total area of surface is s, and that s is independent of the age of the element in question, the area of surface of age between t and $t + dt$ will be a function of t and may be written as $f(t)\,dt$. This will be equal to the area passing in time dt from the age range $[(t - dt)$ to $t]$ to the age range $[t$ to $(t + dt)]$. Further, this in turn will be equal to the area in the age group $[(t - dt)$ to $t]$, less that replaced by fresh surface in time dt:

$$f(t)dt = f(t - dt)dt - [f(t - dt)dt]s\,dt \qquad (10.114)$$

$$\therefore \qquad \frac{f(t) - f(t - dt)}{dt} = -sf(t - dt)$$

$$\therefore \qquad f'(t) + sf(t) = 0 \text{ (as } dt \to 0) \qquad (10.115)$$

$$\therefore \qquad e^{st}f(t) = \text{constant}$$

$$\therefore \qquad f(t) = \text{constant } e^{-st}$$

The total area of surface considered was unity:

$$\therefore \qquad \int_0^\infty f(t)dt = \text{constant} \int_0^\infty e^{-st}\,dt = 1 \qquad (10.116)$$

$$\therefore \qquad \text{constant} \times \frac{1}{s} = 1$$

$$\therefore \qquad f(t) = s\,e^{-st} \qquad (10.117)$$

Thus the age distribution of the surface is of an exponential form. Now from equation 10.111 the mass transfer rate at unit area of surface of age t is given by:

$$(N_A)_t = (C_{Ai} - C_{Ao})\sqrt{\frac{D}{\pi t}} \qquad \text{(equation 10.111)}$$

Thus, the overall rate of transfer per unit area when the surface is renewed in a random manner is:

$$N_A = (C_{Ai} - C_{Ao})\int_{t=0}^{t=\infty} \sqrt{\frac{D}{\pi t}}\,s\,e^{-st}\,dt$$

$$= (C_{Ai} - C_{Ao})\,s\sqrt{\frac{D}{\pi}}\int_0^\infty t^{-1/2}e^{-st}\,dt \qquad (10.118)$$

Put $st = \beta^2$. Then $s\,dt = 2\beta\,d\beta$.

$$\therefore \qquad N_A = (C_{Ai} - C_{Ao})s\sqrt{\frac{D}{\pi}}\frac{2}{s^{1/2}}\int_0^\infty e^{-\beta^2}\,d\beta$$

Then, since the value of the integral is $\sqrt{\pi}/2$:

$$N_A = (C_{Ai} - C_{Ao})\sqrt{Ds} \qquad\qquad (10.119)$$

Equation 10.119 might be expected to underestimate the mass transfer rate because, in any practical equipment, there will be a finite upper limit to the age of any surface element. However, the proportion of the surface in the older age group is very small and the overall rate is largely unaffected. It will be seen that the mass transfer rate is again proportional to the concentration difference and to the square root of the diffusivity. The numerical value of s is difficult to estimate, but will clearly increase as the fluid becomes more turbulent. In a packed column, s will be of the same order as the velocity of the liquid flowing over the packing, divided by the length of packing.

Varying interface composition

In previous sections the penetration theory was used to calculate the rate of mass transfer across an interface for conditions where the concentration C_{Ai} of solute **A** in the interfacial layers ($y = 0$) remained constant throughout the process. When there is no resistance to mass transfer in the other phase, for instance when this consists of pure solute **A**, there will be no concentration gradient in that phase and the composition at the interface will therefore at all times be the same as the bulk composition. Since the composition of the interfacial layers of the *penetration* phase is determined by the phase equilibrium relationship, it, too, will remain constant and the conditions necessary for the penetration theory to apply will hold. If, however, the other phase offers a significant resistance to transfer this condition will not, in general, be fulfilled.

As an example, suppose that in phase 1 there is a constant finite resistance to mass transfer which can in effect be represented as a resistance in a laminar film, and in phase 2 the penetration model is applicable. Immediately after surface renewal has taken place, the mass transfer resistance in phase 2 will be negligible and therefore the whole of the concentration driving force will lie across the film in phase 1. The interface compositions will therefore correspond to the bulk value in phase 2 (the penetration phase). As the time of exposure increases, the resistance to mass transfer in phase 2 will progressively increase and an increasing proportion of the total driving force will lie across this phase. Thus the interface composition, initially determined by the bulk composition in phase 2 (the penetration phase) will progressively approach the bulk composition in phase 1 as the time of exposure increases.

Because the boundary condition at $y = \infty$ ($C' = 0$) is unaltered by the fact that the concentration at the interface is a function of time, equation 10.104 is still applicable, but the evaluation of the constant B_2 is more complicated because $(C')_{y=0}$ is not now constant.

$$\overline{C'} = B_2\,e^{-\sqrt{(p/D)}y} \qquad\qquad \text{(equation 10.104)}$$

$$\frac{d\overline{C'}}{dy} = -\sqrt{\frac{p}{D}} B_2 \, e^{-\sqrt{(p/D)}y}$$

so that:
$$(\overline{C'})_{y=0} = B_2 \tag{10.120}$$

and:
$$\left(\frac{d\overline{C'}}{dy}\right)_{y=0} = -\sqrt{\frac{p}{D}} B_2 \tag{10.121}$$

In order to evaluate B_2 it is necessary to consider the mass transfer rates on each side of the interface and to equate these.

The mass transfer rate per unit area across the film at any time t is given by:

$$(N_A)_t = -\frac{D_f}{L_f}(C''_{Ai} - C''_{Ao}) \tag{10.122}$$

where D_f is the diffusivity in the film of thickness L_f, and C''_{Ai} and C''_{Ao} are the concentrations of **A** at the interface and in the bulk.

The capacity of the film will be assumed to be small so that the hold-up of solute is negligible. If Henry's law is applicable, the interface concentration in the second (penetration) phase is given by:

$$C_{Ai} = \frac{1}{\mathcal{H}} C''_{Ai} \tag{10.123}$$

Here C_A is used to denote concentration in the penetration phase. Thus, by substitution, the interface composition C_{Ai} is obtained in terms of the mass transfer rate $(N_A)_t$.

However, $(N_A)_t$ must also be given by applying Fick's law to the interfacial layers of phase 2 (the penetration phase).

Thus:
$$(N_A)_t = -D\left(\frac{\partial C_A}{\partial y}\right)_{y=0} \qquad \text{(from equation 10.4)} \tag{10.124}$$

Combining equations 10.122, 10.123, and 10.124:

$$(C_A)_{y=0} = C_{Ai} = \frac{C''_{Ao}}{\mathcal{H}} + \frac{DL_f}{D_f\mathcal{H}}\left(\frac{\partial C_A}{\partial y}\right)_{y=0} \tag{10.125}$$

Replacing C_A by $C' + C_{Ao}$:

$$(C')_{y=0} = \left(\frac{C''_{Ao}}{\mathcal{H}} - C_{Ao}\right) + \frac{DL_f}{D_f\mathcal{H}}\left(\frac{\partial C'}{\partial y}\right)_{y=0}$$

Taking Laplace transforms of each side, noting that the first term on the RHS is constant:

$$(\overline{C'})_{y=0} = \frac{1}{p}\left(\frac{C''_{Ao}}{\mathcal{H}} - C_{Ao}\right) + \frac{DL_f}{D_f\mathcal{H}}\left(\frac{d\overline{C'}}{dy}\right)_{y=0} \tag{10.126}$$

Substituting into equation 10.126 from equations 10.120 and 10.121:

$$B_2 = \frac{1}{p}\left(\frac{C''_{Ao}}{\mathcal{H}} - C_{Ao}\right) + \frac{DL_f}{D_f\mathcal{H}}\left(-\sqrt{\frac{p}{D}} B_2\right)$$

This leads to:

$$B_2 = \frac{D_f}{\sqrt{DL_f}}[C''_{Ao} - \mathcal{H}C_{Ao}] \bigg/ p\left\{\frac{D_f\mathcal{H}}{\sqrt{DL_f}} + \sqrt{p}\right\} \qquad (10.127)$$

On substitution into equation 10.121:

$$\left(\frac{d\overline{C'}}{dy}\right)_{y=0} = \left(\frac{d\overline{C'}}{dy}\right)_{y=0} = -\frac{D_f}{DL_f}(C''_{Ao} - \mathcal{H}C_{Ao}) \bigg/ \sqrt{p}\left\{\frac{D_f\mathcal{H}}{\sqrt{DL_f}} + \sqrt{p}\right\} \qquad (10.128)$$

On inversion (See Appendix, Table 12, No. 43):

$$\left(\frac{dC'}{dy}\right)_{y=0} = -\frac{D_f}{DL_f}[C''_{Ao} - \mathcal{H}C_{Ao}]e^{[(D_f^2\mathcal{H}^2)/(DL_f^2)]t} \operatorname{erfc}\sqrt{\frac{D_f^2\mathcal{H}^2 t}{DL_f^2}} \qquad (10.129)$$

Then the mass transfer rate at time t at the interface is given by:

$$(N_A)_t = -D\left(\frac{\partial C_A}{\partial y}\right)_{y=0} = -D\left(\frac{\partial C'}{\partial y}\right)_{y=0}$$

or:

$$(N_A)_t = (C''_{Ao} - \mathcal{H}C_{Ao})\frac{D_f}{L_f}\left(e^{[(D_f^2\mathcal{H}^2)/(DL_f^2)]t} \operatorname{erfc}\sqrt{\frac{D_f^2\mathcal{H}^2 t}{DL_f^2}}\right) \qquad (10.130)$$

Average rates of mass transfer can then be obtained, as previously, by using either the Higbie or the Danckwerts model for surface renewal.

Penetration model with laminar film at interface

HARRIOTT[21] suggested that, as a result of the effects of interfacial tension, the layers of fluid in the immediate vicinity of the interface would frequently be unaffected by the mixing process postulated in the penetration theory. There would then be a thin laminar layer unaffected by the mixing process and offering a constant resistance to mass transfer. The overall resistance can then be calculated in a manner similar to that used in the previous section where the total resistance to transfer was made up of two components — a film resistance in one phase and a penetration model resistance in the other. In equation 10.130, it will be necessary to put the Henry's law constant equal to unity and the diffusivity D_f in the film equal to that in the remainder of the fluid D. The driving force will be $C_{Ai} - C_{Ao}$ in place of $C'_{Ao} - \mathcal{H}C_{Ao}$. Thus the mass transfer rate at time t is given for a film thickness L by:

$$(N_A)_t = (C_{Ai} - C_{Ao})\frac{D}{L}\left(e^{Dt/L^2} \operatorname{erfc}\sqrt{\frac{Dt}{L^2}}\right) \qquad (10.131)$$

The average transfer rate according to the Higbie model for surface age distribution then becomes:

$$N_A = 2(C_{Ai} - C_{Ao})\sqrt{\frac{D}{\pi t_e}}\left[1 + \frac{1}{2}\sqrt{\frac{\pi L^2}{Dt_e}}\left(e^{Dt_e/L^2} \operatorname{erfc}\sqrt{\frac{Dt_e}{L^2}} - 1\right)\right] \qquad (10.132)$$

Using the Danckwerts model:

$$N_A = (C_{Ai} - C_{Ao})\sqrt{Ds}\left(1 + \sqrt{\frac{L^2 s}{D}}\right)^{-1} \tag{10.133}$$

10.5.3. The film–penetration theory

A theory which incorporates some of the principles of both the two-film theory and the penetration theory has subsequently been suggested by TOOR and MARCHELLO[19]. The whole of the resistance to transfer is regarded as lying within a laminar film at the interface, as in the two-film theory, but the mass transfer is regarded as an unsteady state process. It is assumed that fresh surface is formed at intervals from fluid which is brought from the bulk of the fluid to the interface by the action of the eddy currents. Mass transfer then takes place as in the penetration theory, with the exception that the resistance is confined to the finite film, and material which traverses the film is immediately completely mixed with the bulk of the fluid. For short times of exposure, when none of the diffusing material has reached the far side of the layer, the process is identical to that postulated in the penetration theory. For prolonged periods of exposure when a steady concentration gradient has developed, conditions are similar to those considered in the two-film theory.

The mass transfer process is again governed by equation 10.64, but the third boundary condition is applied at $y = L$, the film thickness, and not at $y = \infty$. The Laplace transform is then as before:

$$\overline{C'} = B_1 e^{\sqrt{(p/D)}y} + B_2 e^{-\sqrt{(p/D)}y} \tag{equation 10.103}$$

$t > 0 \quad y = 0 \quad C_A = C_{Ai} \quad C' = C_{Ai} - C_{Ao} = C'_i \quad \overline{C'} = (1/p)C'_i$

$t > 0 \quad y = L \quad C_A = C_{Ao} \quad C' = 0 \quad\quad\quad\quad\quad \overline{C'} = 0$

Thus: $\quad\quad \dfrac{C'_i}{p} = B_1 + B_2$

$\therefore \quad\quad 0 = B_1 e^{\sqrt{(p/D)}L} + B_2 e^{-\sqrt{(p/D)}L}$

$\therefore \quad\quad B_1 = -B_2 e^{-2\sqrt{(p/D)}L}$

$\therefore \quad\quad B_2 = \dfrac{C'_i}{p}(1 - e^{-2\sqrt{(p/D)}L})^{-1}$

$\quad\quad B_1 = -\dfrac{C'_i}{p}e^{-2\sqrt{(p/D)}L}(1 - e^{-2\sqrt{(p/D)}L})^{-1}$

$\therefore \quad \overline{C'} = \dfrac{C'_i}{p}(1 - e^{-2\sqrt{(p/D)}L})^{-1}(e^{-\sqrt{(p/D)}y} - e^{-\sqrt{(p/D)}(2L-y)}) \tag{10.134}$

Since there is no inverse of equation 10.134 in its present form, it is necessary to expand using the binomial theorem to give:

$$\overline{C'} = \frac{C'_i}{p}(e^{-\sqrt{(p/D)}y} - e^{-\sqrt{(p/D)}(2L-y)}) \sum_{n=0}^{n=\infty} e^{-2n\sqrt{(p/D)}L}$$

$$= C'_i \left[\sum_{n=0}^{n=\infty} \frac{1}{p} e^{-\sqrt{(p/D)}(2nL+y)} - \sum_{n=0}^{n=\infty} \frac{1}{p} e^{-\sqrt{(p/D)}\{2(n+1)L-y\}} \right] \qquad (10.135)$$

On inversion of equation 10.135:

$$\frac{C_A - C_{Ao}}{C_{Ai} - C_{Ao}} = \frac{C'}{C'_i} = \sum_{n=0}^{n=\infty} \text{erfc} \frac{(2nL + y)}{2\sqrt{Dt}} - \sum_{n=0}^{n=\infty} \text{erfc} \frac{2(n + 1)L - y}{2\sqrt{Dt}} \qquad (10.136)$$

Differentiating with respect to y:

$$\frac{1}{C_{Ai} - C_{Ao}} \frac{\partial C_A}{\partial y} = \sum_{n=0}^{n=\infty} -\frac{2}{\sqrt{\pi}} \frac{1}{2\sqrt{Dt}} e^{-(2nL+y)^2/(4Dt)}$$

$$- \sum_{n=0}^{n=\infty} \frac{2}{\sqrt{\pi}} \frac{1}{2\sqrt{Dt}} e^{-[2(n+1)L-y]^2/4(Dt)}$$

At the free surface, $y = 0$ and:

$$\frac{1}{C_{Ai} - C_{Ao}} \left(\frac{\partial C_A}{\partial y} \right)_{y=0} = -\frac{1}{\sqrt{\pi Dt}} \left(\sum_{n=0}^{n=\infty} e^{-(n^2 L^2)/(Dt)} + \sum_{n=0}^{n=\infty} e^{-[(n+1)^2 L^2]/(Dt)} \right)$$

$$= -\frac{1}{\sqrt{\pi Dt}} \left(1 + 2 \sum_{n=1}^{n=\infty} e^{-(n^2 L^2)/(Dt)} \right)$$

The mass transfer rate across the interface per unit area is given by:

$$(N_A)_t = -D \left(\frac{\partial C_A}{\partial y} \right)_{y=0}$$

$$= (C_{Ai} - C_{Ao}) \sqrt{\frac{D}{\pi t}} \left(1 + 2 \sum_{n=1}^{n=\infty} e^{-(n^2 L^2)/(Dt)} \right) \qquad (10.137)$$

Equation 10.137 converges rapidly for high values of L^2/Dt. For low values of L^2/Dt, it is convenient to use an alternative form by using the identity[22]:

$$\sqrt{\frac{\alpha}{\pi}} \left(1 + 2 \sum_{n=1}^{n=\infty} e^{-n^2 \alpha} \right) \equiv 1 + 2 \sum_{n=1}^{n=\infty} e^{-(n^2 \pi^2)/\alpha} \qquad (10.138)$$

Taking $\alpha = L^2/Dt$:

$$(N_A)_t = (C_{Ai} - C_{Ao}) \frac{D}{L} \left(1 + 2 \sum_{n=1}^{n=\infty} e^{-(n^2 \pi^2 Dt)/L^2} \right) \qquad (10.139)$$

It will be noted that equations 10.137 and 10.139 become identical in form and in convergence when $L^2/Dt = \pi$.

Then: $(N_A)_t = (C_{Ai} - C_{Ao})\dfrac{D}{L}\left(1 + 2\displaystyle\sum_{n=1}^{n=\infty} e^{-n^2\pi}\right)$

$$= (C_{Ai} - C_{Ao})\dfrac{D}{L}[1 + 2(e^{-\pi} + e^{-4\pi} + e^{-9\pi} + \cdots)]$$

$$= (C_{Ai} - C_{Ao})\dfrac{D}{L}(1 + 0.0864 + 6.92 \times 10^{-6} + 1.03 \times 10^{-12} + \cdots)$$

Thus, provided the rate of convergence is not worse than that for $L^2/Dt = \pi$, all terms other than the first in the series may be neglected. Now equation 10.137 will converge more rapidly than this for $L^2/Dt > \pi$, and equation 10.139 will converge more rapidly for $L^2/Dt < \pi$.

Thus: $\qquad \pi \leqslant \dfrac{L^2}{Dt} < \infty \qquad (N_A)_t = (C_{Ai} - C_{Ao})\sqrt{\dfrac{D}{\pi t}}(1 + 2e^{-L^2/Dt}) \qquad$ (10.140)

$$0 < \dfrac{L^2}{Dt} \leqslant \pi \qquad (N_A)_t = (C_{Ai} - C_{Ao})\dfrac{D}{L}(1 + 2e^{-(\pi^2 Dt)/L^2}) \qquad (10.141)$$

Now it will be noted that the second terms in equations 10.140 and 10.141 never exceeds 8.64 per cent of the first term. Thus, with an error not exceeding 8.64 per cent,

$$\pi \leqslant \dfrac{L^2}{Dt} < \infty \qquad (N_A)_t = (C_{Ai} - C_{Ao})\sqrt{\dfrac{D}{\pi t}} \qquad (10.142)$$

$$0 < \dfrac{L^2}{Dt} \leqslant \pi \qquad (N_A)_t = (C_{Ai} - C_{Ao})\dfrac{D}{L} \qquad (10.143)$$

Thus either the penetration theory or the film theory (equation 10.142 or 10.143) respectively can be used to describe the mass transfer process. The error will not exceed about 9 per cent provided that the appropriate equation is used, equation 10.142 for $L^2/Dt > \pi$ and equation 10.143 for $L^2/Dt < \pi$. Equation 10.143 will frequently apply quite closely in a wetted-wall column or in a packed tower with large packings. Equation 10.142 will be likely to apply when one of the phases is dispersed in the form of droplets, as in a spray tower, or in a packed tower with small packing elements.

Equations 10.140 and 10.141 give the point value of N_A at time t. The average values N_A can then be obtained by applying the age distribution functions obtained by Higbie and by Danckwerts, respectively, as in the earlier section on the penetration theory.

10.5.4. Other theories of mass transfer

KISHINEVSKIJ[20] has developed a model for mass transfer across an interface in which molecular diffusion is assumed to play no part. He considers that fresh material is continuously brought to the interface as a result of turbulence within the fluid and that, after exposure to the second phase, the fluid element attains equilibrium with it and then becomes mixed again with the bulk of the phase. The model thus presupposes surface renewal without penetration by diffusion and therefore the effect of diffusivity should not be important. No reliable experimental results are available to test the theory adequately.

10.5.5. Interfacial turbulence

An important feature of the behaviour of the interface which has not been taken into account in the preceding treatment is the possibility of turbulence being generated by a means other than that associated with the fluid dynamics of the process. This *interfacial turbulence* may arise from local variations in the interfacial tension set up during the course of the mass transfer process. It can occur at both gas–liquid and liquid–liquid interfaces, but the latter case has commanded the greater attention. It may give rise to violent intermittent eruptions at localised regions in the interface, as a result of the rapid mixing which occurs, the mass transfer rate may be considerably enhanced.

The effect, which arises in cases where the interfacial tension is strongly dependent on the concentration of diffusing solute, will generally be dependent on the direction (sense) in which mass transfer is taking place.

If mass transfer causes interfacial tension to decrease, localised regions of low interfacial tension will form and, as a result, surface spreading will take place. If the gradient of interfacial tension is very high, the surface spreading will be rapid enough to give rise to intense ripples at the interface and a rapid increase in the mass transfer rate. On the other hand, if the mass transfer gives rise to increased interfacial tension, the surface shows no tendency to spread and tends to be stable.

The phenomenon, frequently referred to as the *Marangoni Effect*, explains some of the anomalously high mass transfer rates reported in the literature.

The effect may be reduced by the introduction of surfactants which tend to concentrate at the interface where they exert a stabilising influence. However, they may introduce an interface resistance and substantially reduce the mass transfer rate. Thus, for instance, hexadecanol when added to open ponds of water will collect at the interface and substantially reduce the rate of evaporation.

The effects referred to above are described in more detail by SHERWOOD, PIGFORD and WILKE[23].

10.5.6. Mass transfer coefficients

On the basis of each of the theories discussed, the rate of mass transfer in the absence of bulk flow is directly proportional to the driving force, expressed as a molar concentration difference, and, therefore:

$$N_A = h_D(C_{Ai} - C_{Ao}) \qquad (10.144)$$

where h_D is a mass transfer coefficient (see equation 10.95). In the two-film theory, h_D is directly proportional to the diffusivity and inversely proportional to the film thickness. According to the penetration theory it is proportional to the square root of the diffusivity and, when all surface elements are exposed for an equal time, it is inversely proportional to the square root of time of exposure; when random surface renewal is assumed, it is proportional to the square root of the rate of renewal. In the film-penetration theory, the mass transfer coefficient is a complex function of the diffusivity, the film thickness, and either the time of exposure or the rate of renewal of surface.

In most cases, the value of the transfer coefficient cannot be calculated from the first principles. However, the way in which the coefficient will vary as operating conditions are altered can frequently be predicted by using the theory which is most nearly applicable to the problem in question.

The penetration and film-penetration theories have been developed for conditions of equimolecular counterdiffusion only; the equations are too complex to solve explicitly for transfer through a stationary carrier gas. For gas absorption, therefore, they apply only when the concentration of diffusing material is low. On the other hand, in the two-film theory the additional contribution to the mass transfer which is caused by bulk flow is easily calculated and h_D is equal to $(D/L)(C_T/C_{Bm})$ instead of D/L.

In a process where mass transfer takes place across a phase boundary, the same type of theoretical approach can be applied to each of the phases, though it does not follow that the same theory is best applied to both phases. For example, the film model might be applicable to one phase and the penetration model to the other. This problem has already been discussed earlier.

When the film theory is applicable to each phase (the two-film theory), the process is steady state throughout and the interface composition does not then vary with time. For this case the two film coefficients can readily be combined. Because material does not accumulate at the interface, the mass transfer rate on each side of the phase boundary will be the same and for two phases it follows that:

$$N_A = h_{D1}(C_{Ao1} - C_{Ai1}) = h_{D2}(C_{Ai2} - C_{Ao2}) \tag{10.145}$$

If there is no resistance to transfer at the interface C_{Ai1} and C_{Ai2} will be corresponding values in the phase-equilibrium relationship.

Usually, the values of the concentration at the interface are not known and the mass transfer coefficient is considered for the overall process. Overall transfer coefficients are then defined by the relations:

$$N_A = K_1(C_{Ao1} - C_{Ae1}) = K_2(C_{Ae2} - C_{Ao2}) \tag{10.146}$$

where C_{Ae1} is the concentration in phase 1 in equilibrium with C_{Ao2} in phase 2, and C_{Ae2} is the concentration in phase 2 in equilibrium with C_{Ao1} in phase 1. If the equilibrium relationship is linear:

$$\mathscr{H} = \frac{C_{Ai1}}{C_{Ai2}} = \frac{C_{Ae1}}{C_{Ao2}} = \frac{C_{Ao1}}{C_{Ae2}} \tag{10.147}$$

where $\mathscr{H}$ is a proportionality constant.

The relationships between the various transfer coefficients are obtained as follows: From equations 10.145 and 10.146:

$$\frac{1}{K_1} = \frac{1}{h_{D1}} \frac{C_{Ae1} - C_{Ao1}}{C_{Ai1} - C_{Ao1}} = \frac{1}{h_{D1}} \frac{C_{Ae1} - C_{Ai1}}{C_{Ai1} - C_{Ao1}} + \frac{1}{h_{D1}} \frac{C_{Ai1} - C_{Ao1}}{C_{Ai1} - C_{Ao1}}$$

But: $$\frac{1}{h_{D1}} = \frac{1}{h_{D2}} \frac{C_{Ai1} - C_{Ao1}}{C_{Ao2} - C_{Ai2}} \qquad \text{(from equation 10.145)}$$

$\therefore$ $$\frac{1}{K_1} = \frac{1}{h_{D1}} + \frac{1}{h_{D2}} \left(\frac{C_{Ae1} - C_{Ai1}}{C_{Ai1} - C_{Ao1}} \right) \left(\frac{C_{Ai1} - C_{Ao1}}{C_{Ao2} - C_{Ai2}} \right)$$

But from equation 10.147: $$\frac{C_{Ae1} - C_{Ai1}}{C_{Ao2} - C_{Ai2}} = \mathscr{H}$$

$\therefore$

$$\frac{1}{K_1} = \frac{1}{h_{D1}} + \frac{\mathcal{H}}{h_{D2}} \tag{10.148}$$

Similarly:

$$\frac{1}{K_2} = \frac{1}{\mathcal{H}h_{D1}} + \frac{1}{h_{D2}} \tag{10.149}$$

and hence:

$$\frac{1}{K_1} = \frac{\mathcal{H}}{K_2} \tag{10.150}$$

It follows, therefore, that when h_{D1} is large compared with h_{D2}, K_2 and h_{D2} are approximately equal, and, when h_{D2} is large compared with h_{D1}, K_1 and h_{D1} are almost equal.

The above relations between the various coefficients are valid provided that the transfer rate is linearly related to the driving force and that the equilibrium relationship is a straight line. They are therefore applicable for the two-film theory, and for any instant of time for the penetration and film-penetration theories. In general, application to time-averaged coefficients obtained from the penetration and film-penetration theories is not permissible because the condition at the interface will be time-dependent unless all of the resistance lies in one of the phases.

10.5.7. Countercurrent mass transfer and transfer units

Mass transfer processes involving two fluid streams are frequently carried out continuously using either countercurrent or cocurrent flow in a column apparatus. The former is the more usual, but cocurrent flow has advantages in special cases. Three common examples of processes of countercurrent mass transfer are as follows:

1. *In a packed distillation column.* Here a vapour stream is rising against the downward flow of a liquid reflux, and a state of dynamic equilibrium is set up in a steady state process. The more volatile constituent is transferred under the action of a concentration gradient from the liquid to the interface where it evaporates and then diffuses into the vapour stream. The less volatile component is transferred in the opposite direction and, if the molar latent heats of the components are the same, equimolecular counterdiffusion takes place.

2. *In a packed absorption column.* The flow pattern is similar to that in a packed distillation column but the vapour stream is replaced by a carrier gas together with a solute gas. The solute diffuses through the gas phase to the liquid surface where it dissolves and is then transferred to the bulk of the liquid. In this case there is no mass transfer in the opposite direction and the transfer rate is supplemented by bulk flow.

3. *In a liquid-liquid extraction column.* Here the process is similar to that occurring in an absorption column with the exception that both streams are liquid, and the lighter liquid rises through the denser one.

The mass transfer relations do differ in a number of important respects in distillation, gas absorption and liquid–liquid extraction. Thus in distillation, equimolecular counterdiffusion frequently takes place and the molar rate of flow of the two phases remains approximately constant throughout the whole height of the column. On the other hand, in gas absorption the mass transfer process is increased as a result of bulk flow and, when the solute gas is present in high concentrations, the molar rate of flow at the top of the column will be less than that at the bottom. For low concentrations of the solute gas however, conditions are similar in distillation and gas absorption. Consideration will now be given to mass transfer in a column under these conditions.

In Figure 10.7 the conditions existing in a column during the steady state operation of a countercurrent process are illustrated. The molar rates of flow of the two streams are G_1 and G_2, which are assumed constant over the whole column. Suffixes 1 and 2 will be used to denote the two phases, and t and b to denote the top and bottom of the column.

Figure 10.7.   Countercurrent mass transfer in a column

The height of the column is Z, its total cross-sectional area is S, and a is the interfacial area between the two phases per unit volume of column. Then the rate of transfer of the diffusing component in a height dZ of column:

$$G_1 \frac{1}{C_T} dC_{Ao1} = h_{D1}(C_{Ai1} - C_{Ao1}) Sa\, dZ \qquad (10.151)$$

giving:

$$\frac{dC_{Ao1}/dZ}{C_{Ai1} - C_{Ao1}} = \frac{h_{D1}aSC_T}{G_1} = \frac{h_{D1}aC_T}{G'_1} \qquad (10.152)$$

where G'_1 is the molar rate of flow per unit cross-section of column. The exact interfacial area cannot normally be determined independently of the transfer coefficient, and therefore values of the product $h_{D1}a$ are usually quoted for any particular system.

The left-hand side of equation 10.152 is the rate of change of concentration with height for unit driving force, and is therefore a measure of the efficiency of the column. Thus a high value of $(h_{D1}aC_T)/G'_1$ will be obtained in a column of high efficiency. The reciprocal of this quantity is $G'_1/(h_{D1}aC_T)$ which has linear dimensions and is known as the *height of the transfer unit* **H**$_1$ (HTU).

If equation 10.152 is rearranged and integrated:

$$\int_{(C_{Aol_b})}^{(C_{Aol_t})} \frac{dC_{Aol}}{C_{Ai1} - C_{Aol}} = \frac{h_{D1}aC_T}{G_1'}Z = \frac{Z}{G_1'/(h_{D1}aC_T)} = \frac{Z}{\mathbf{H}_1} \tag{10.153}$$

The right-hand side of equation 10.153 is the height of the column divided by the HTU and is known as the *number of transfer units* $\mathbf{N}_1$. It is obtained by evaluating the integral on the left-hand side of the equation.

Therefore:

$$\mathbf{N}_1 = \frac{Z}{\mathbf{H}_1} \tag{10.154}$$

In some cases, such as the evaporation of a liquid at an approximately constant temperature or the dissolving of a highly soluble gas in a liquid, the interface concentration may be either substantially constant or negligible in comparison with that of the bulk. In such cases, the integral on the left-hand side of equation 10.153 may be evaluated directly.

Then:

$$\int_{(C_{Aol})_b}^{(C_{Aol})_t} \frac{dC_{Aol}}{C_{Ai1} - C_{Aol}} = \ln \frac{(C_{Ai1} - C_{Aol})_b}{(C_{Ai1} - C_{Aol})_t} = \frac{Z}{\mathbf{H}_1} \tag{10.155}$$

where b and t represent values at the bottom and the top, as shown in Figure 10.7.

i.e.:

$$\frac{(C_{Ai1} - C_{Aol})_b}{(C_{Ai1} - C_{Aol})_t} = e^{Z/\mathbf{H}_1} \tag{10.156}$$

Thus $\mathbf{H}_1$ is the height of column over which the *driving force changes by a factor of* e.

Equation 10.151 could have been written in terms of the film coefficient for the second phase (h_{D2}) or either of the overall transfer coefficients (K_1, K_2). Transfer units based on either film coefficient or overall coefficient can therefore be defined. The following equations are analogous to equation 10.154:

The number of transfer units based on phase 2, $\mathbf{N}_2 = Z/\mathbf{H}_2$ (10.157)

Number of overall transfer units based on phase 1, $\mathbf{N}_{o1} = Z/\mathbf{H}_{o1}$ (10.158)

Number of overall transfer units based on phase 2, $\mathbf{N}_{o2} = Z/\mathbf{H}_{o2}$ (10.159)

In this notation the introduction of o into the suffix indicates an overall transfer unit. The values of $\mathbf{H}_1$, $\mathbf{H}_2$, $\mathbf{H}_{o1}$, $\mathbf{H}_{o2}$ are of the following form:

$$\mathbf{H}_1 = \frac{G_1'}{h_{D1}aC_T} \tag{10.160}$$

If one phase is a gas, as in gas absorption, it is often more convenient to express concentrations as partial pressures so that:

$$\mathbf{H}_1 = \mathbf{H}_G = \frac{G_1'}{k_Gap} \tag{10.161}$$

where, in SI units, G_1' is expressed in kmol/m^2 s, k_G in kmol/[m^2 s (N/m^2)], P in N/m^2, and a in m^2/m^3.

The overall values of the HTU can be expressed in terms of the film values by using equation 10.148 which gives the relation between the coefficients.

$$\frac{1}{K_1} = \frac{1}{h_{D1}} + \frac{\mathscr{H}}{h_{D2}}$$ (equation 10.148)

Substituting for the coefficients in terms of the HTU:

$$\mathbf{H}_{o1}\frac{aC_T}{G_1'} = \mathbf{H}_1\frac{aC_T}{G_1'} + \frac{\mathscr{H}aC_T}{G_2'}\mathbf{H}_2$$ (10.162)

i.e.:
$$\mathbf{H}_{o1} = \mathbf{H}_1 + \mathscr{H}\frac{G_1'}{G_2'}\mathbf{H}_2$$ (10.163)

Similarly:
$$\mathbf{H}_{o2} = \mathbf{H}_2 + \frac{1}{\mathscr{H}}\frac{G_2'}{G_1'}\mathbf{H}_1$$ (10.164)

The advantage of using the transfer unit in preference to the transfer coefficient is that the former remains much more nearly constant as flow conditions are altered. This is particularly important in problems of gas absorption where the concentration of the solute gas is high and the flow pattern changes in the column because of the change in the total rate of flow of gas at different sections. In most cases the coefficient is proportional to the flowrate raised to a power slightly less than unity and therefore the HTU is substantially constant.

As already seen, for equimolecular counterdiffusion, the film transfer coefficients, and hence the corresponding HTUs, can be expressed in terms of the physical properties of the system and the assumed film thickness or exposure time, using the two-film, the penetration, or the film-penetration theories. For conditions where bulk flow is important, however, the transfer rate of constituent $\mathbf{A}$ is increased by the factor C_T/C_{Bm} and the diffusion equations can be solved only on the basis of the two-film theory. In the design of equipment it is usual to work in terms of transfer coefficients or HTUs and not to endeavour to evaluate them in terms of properties of the system.

10.5.8. Other applications of theories of interphase transfer

The two-film, penetration, and film-penetration theories can be applied satisfactorily to problems involving mass transfer between a fluid and the surface of a solid. In this case the basic assumption that turbulence dies out at the interface is fully justified. The layer of fluid immediately in contact with the solid is in laminar flow and constitutes the laminar sub-layer. However, in the vicinity of a solid surface, appreciable velocity gradients exist within the fluid and the calculation of transfer rates become complex if they are taken into account.

The preceding treatment has related to processes of mass transfer. When a temperature difference exists between two phases, heat will be transferred from the hotter fluid to the cooler, and that proportion of the heat which is transferred by forced convection can be calculated by the application of the theories discussed here. Furthermore, if there is a change of phase at the interface, as in a humidifying or water cooling tower, the heat transfer rate in the two phases will not, in general, be the same.

Example 10.6

Ammonia is absorbed at 1 bar from an ammonia-air stream by passing it up a vertical tube, down which dilute sulphuric acid is flowing. The following laboratory data are available:

$$\text{Length of tube} = 825 \text{ mm}$$

$$\text{Diameter of tube} = 15 \text{ mm}$$

Partial pressures of ammonia: at inlet $= 7.5$ kN/m^2;   at outlet $= 2.0$ kN/m^2

$$\text{Air rate} = 2 \times 10^{-5} \text{ kmol/s}$$

What is the overall transfer coefficient K_G based on the gas phase?

Solution

Driving force at inlet $= 7500$ N/m^2

Driving force at outlet $= 2000$ N/m^2

$$\text{Mean driving force} = \frac{7500 - 2000}{\ln(7.5/2.0)} = 4200 \text{ N/m}^2$$

$$\text{Ammonia absorbed} = 2 \times 10^{-5} \left[\frac{7.5}{93.8} - \frac{2.0}{99.3} \right] = 1.120 \times 10^{-6} \text{ kmol/s}$$

$$\text{Wetted surface} = \pi \times 0.015 \times 0.825 = 0.0388 \text{ m}^2$$

Hence $K_G = \dfrac{1.120 \times 10^{-6}}{0.0388 \times 4200} = 6.87 \times 10^{-9}$ kmol/[m^2 s(N/m^2)]

10.6. MASS TRANSFER TO A SPHERE

In the problems considered previously, the transfer has been taking place in a single direction of a rectangular coordinate system. In many applications of mass transfer, one of the fluids is injected as approximately spherical droplets into a second immiscible fluid, and transfer of the solute occurs as the droplet passes through the continuous medium.

The problem of a sphere of pure liquid of radius b which is suddenly immersed in a gas, the whole of the mass transfer resistance lying within the liquid, may be considered. On the assumption that mass transfer is governed by Fick's law and that angular symmetry exists:

At radius r within sphere, mass transfer rate $= -(4\pi r^2)D \left(\dfrac{\partial C_A}{\partial r} \right)$

Change in mass transfer rate over a distance dr

$$= \frac{\partial}{\partial r} \left(-4\pi r^2 D \frac{\partial C_A}{\partial r} \right) dr$$

Then a material balance over a shell gives:

$$-\frac{\partial}{\partial r} \left(-4\pi r^2 D \frac{\partial C_A}{\partial r} \right) dr = \frac{\partial C_A}{\partial t} (4\pi r^2 dr)$$

$$\therefore \quad r^2 \frac{\partial C_A}{\partial t} = D \frac{\partial}{\partial r} \left(r^2 \frac{\partial C_A}{\partial r} \right) \tag{10.165}$$

This equation may be solved by taking Laplace transforms.

The concentration gradient at the surface of the sphere $(r = b)$ is then found to be given by:

$$\left(\frac{\partial C_A}{\partial r}\right)_{r=b} = \frac{-C_{Ai}}{b} + \frac{C_{Ai}}{\sqrt{\pi Dt}}\left(1 + 2\sum_{n=1}^{n=\infty} e^{-(n^2 b^2)/Dt}\right) \tag{10.166}$$

The mass transfer rate at surface of sphere at time t

$$= 4\pi b^2(-D)\left(\frac{\partial C_A}{\partial r}\right)_{r=b}$$

$$= 4\pi b D C_{Ai}\left[1 - \frac{b}{\sqrt{\pi Dt}}\left(1 + 2\sum_{n=1}^{n=\infty} e^{-(n^2 b^2)/Dt}\right)\right] \tag{10.167}$$

The total mass transfer during the passage of a drop is therefore obtained by integration of equation 10.167 over the time of exposure.

10.7. MASS TRANSFER AND CHEMICAL REACTION

In many applications of mass transfer the solute will react with the medium as in the case, for example, of the absorption of carbon dioxide in an alkaline solution. The mass transfer rate will then decrease in the direction of diffusion as a result of the reaction. Consider the unidirectional molecular diffusion of a component **A** through a distance dy over unit area. Then, neglecting the effects of bulk flow, a material balance for an irreversible reaction of order n gives:

$$\frac{\partial C_A}{\partial t}dy = -\frac{\partial}{\partial y}\left(-D\frac{\partial C_A}{\partial y}\right)dy - kC_A^n dy$$

or:

$$\frac{\partial C_A}{\partial t} = D\frac{\partial^2 C_A}{\partial y^2} - kC_A^n \tag{10.168}$$

where k is the reaction rate constant. This equation has no analytical solution for the general case.

10.7.1. Steady-state process

For a steady-state process:

$$D\frac{\partial^2 C_A}{\partial y^2} - kC_A^n = 0 \tag{10.169}$$

Equation 10.169 may be integrated using the appropriate boundary conditions. The most important case is for a first-order reaction. Putting $n = 1$ in equation 10.169:

$$D\frac{\partial^2 C_A}{\partial y^2} - kC_A = 0 \tag{10.170}$$

The solution of equation 10.170 is:

$$C_A = B_1' e^{\sqrt{(k/D)}y} + B_2' e^{-\sqrt{(k/D)}y} \tag{10.171}$$

B_1' and B_2' must then be evaluated using the appropriate boundary conditions.

Example 10.7

In a gas absorption process, the solute gas **A** diffuses into a solvent liquid with which it reacts. The mass transfer is one of steady state unidirectional molecular diffusion and the concentration of **A** is always sufficiently small for bulk flow to be negligible. Under these conditions the reaction is first order with respect to the solute **A**.

At a depth l below the liquid surface, the concentration of **A** has fallen to one-half of the value at the surface. What is the ratio of the mass transfer rate at this depth l to the rate at the surface? Calculate the numerical value of the ratio when $l\sqrt{k/D} = 0.693$, where D is the molecular diffusivity and k the first-order rate constant.

Solution

The above process is described by equation 10.171:

$$C_A = B'_1 e^{\sqrt{(k/D)}y} + B'_2 e^{-\sqrt{(k/D)}y} \qquad \text{(equation 10.171)}$$

If C_{Ai} is the surface concentration ($y = 0$):

$$C_{Ai} = B'_1 + B'_2$$

At $y = l$, $C_A = C_{Ai}/2$.

$$\therefore \qquad \frac{C_{Ai}}{2} = B'_1 e^{\sqrt{(k/D)}l} + B'_2 e^{-\sqrt{(k/D)}l}$$

Solving for B'_1 and B'_2:

$$B'_1 = \frac{C_{Ai}}{2}(1 - 2e^{-\sqrt{(k/D)}l})(e^{\sqrt{(k/D)}l} - e^{-\sqrt{(k/D)}l})^{-1}$$

$$B'_2 = -\frac{C_{Ai}}{2}(1 - 2e^{\sqrt{(k/D)}l})(e^{\sqrt{(k/D)}l} - e^{-\sqrt{(k/D)}l})^{-1}$$

$$\frac{(N_A)_{y=l}}{(N_A)_{y=0}} = \frac{-D(dC_A/dy)_{y=l}}{-D(dC_A/dy)_{y=0}} = \frac{(dC_A/dy)_{y=l}}{(dC_A/dy)_{y=0}}$$

$$\frac{dC_A}{dy} = \sqrt{\frac{k}{D}}(B'_1 e^{\sqrt{(k/D)}y} - B'_2 e^{-\sqrt{(k/D)}y})$$

$$\frac{(N_A)_{y=l}}{(N_A)_{y=0}} = \frac{B'_1 e^{\sqrt{(k/D)}l} - B'_2 e^{-\sqrt{(k/D)}l}}{B'_1 - B'_2}$$

$$= \frac{(1 - 2e^{-\sqrt{(k/D)}l})e^{\sqrt{(k/D)}l} + (1 - 2e^{\sqrt{(k/D)}l})e^{-\sqrt{(k/D)}l}}{(1 - 2e^{-\sqrt{(k/D)}l}) + (1 - 2e^{\sqrt{(k/D)}l})}$$

$$= \frac{e^{\sqrt{(k/D)}l} + e^{-\sqrt{(k/D)}l} - 4}{2(1 - e^{-\sqrt{(k/D)}l} - e^{\sqrt{(k/D)}l})}$$

When:

$$l\sqrt{\frac{k}{D}} = 0.693, \quad e^{\sqrt{(k/D)}l} = 2, \quad e^{-\sqrt{(k/D)}l} = 0.5$$

$$\frac{(N_A)_{y=1}}{(N_A)_{y=0}} = \frac{2 + \frac{1}{2} - 4}{2(1 - 2 - \frac{1}{2})} = \underline{\underline{0.5}}$$

10.7.2. Unsteady-state process

For an unsteady-state process, equation 10.168 may be integrated only in the case of a first-order reaction ($n = 1$). It then becomes:

$$\frac{\partial C_A}{\partial t} = D\frac{\partial^2 C_A}{\partial y^2} - kC_A \tag{10.172}$$

The solution of this equation has been discussed by DANCKWERTS[24]. It will now be solved using the Laplace transform method for a semi-infinite liquid initially free of solute. On the assumption that the liquid is in contact with pure solute gas, the concentration C_{Ai} at the liquid interface will be constant and equal to the saturation value. The boundary conditions will be those applicable to the penetration theory:

$$t = 0 \quad 0 < y < \infty \quad C_A = 0$$

$$t > 0 \qquad y = 0 \qquad C_A = C_{Ai}$$

$$t > 0 \qquad y = \infty \quad C_A = 0$$

Then, from equations 10.101 and 10.102, taking Laplace transforms of both sides of equation 10.172:

$$p\overline{C}_A = D\frac{d^2\overline{C}_A}{dy^2} - k\overline{C}_A$$

or:
$$\frac{d^2\overline{C}_A}{dy^2} - \frac{p+k}{D}\overline{C}_A = 0$$

Thus:
$$\overline{C}_A = B_1 e^{\sqrt{(p+k)/D}\,y} + B_2 e^{-\sqrt{(p+k)/D}\,y} \tag{10.173}$$

When $y = \infty$, $C_A = 0$ and therefore $\overline{C}_A = 0$, from which $B_1 = 0$.
When $y = 0$, $C_A = C_{Ai}$ and $\overline{C}_A = C_{Ai}/p$, from which $B_2 = C_{Ai}/p$
  Equation 10.173 therefore becomes:

$$\overline{C}_A = \frac{C_{Ai}}{p} e^{-\sqrt{(p+k)/D}\,y} \tag{10.174}$$

The mass transfer rate N_A at the interface must be evaluated in order to obtain the rate at which gas is transferred to the liquid from the gas.
  Differentiating equation 10.174 with respect to y:

$$\frac{d\overline{C}_A}{dy} = \frac{d\overline{C}_A}{dy} = -\sqrt{\frac{p+k}{D}}\frac{C_{Ai}}{p} e^{-\sqrt{(p+k)/D}\,y} \tag{10.175}$$

At the interface ($y = 0$):

$$\left(\frac{d\overline{C}_A}{dy}\right)_{y=0} = -\frac{C_{Ai}}{\sqrt{D}}\frac{\sqrt{p+k}}{p} \tag{10.176}$$

Now it is not possible to invert equation 10.176 directly using the transforms listed in Table 12 in the Appendix. However, entry number 38 gives:

$$\text{Inverse of } \frac{\sqrt{p}}{p - a^2} = \frac{1}{\sqrt{\pi t}} + a e^{a^2 t} \text{ erf}(a\sqrt{t})$$

or putting $a = \sqrt{k}$:

$$\text{Inverse of } \frac{\sqrt{p}}{p - k} = \frac{1}{\sqrt{\pi t}} + \sqrt{k} e^{kt} \text{ erf}(\sqrt{kt})$$

Now, from the shift theorem;
  If Laplace transform of f(t) is $\bar{f}(p)$,
  then Laplace transform of $f(t)e^{-kt}$

$$= \int_0^\infty f(t)e^{-kt}e^{-pt}dt = \int_0^\infty f(t)e^{-(p+k)t}dt = \bar{f}(p + k)$$

Thus, inverse of $\dfrac{\sqrt{p + k}}{p}$ is $\dfrac{1}{\sqrt{\pi t}}e^{-kt} + \sqrt{k} \text{ erf}(\sqrt{kt})$

Hence, inverting equation 10.176:

$$\left(\frac{dC_A}{dy}\right)_{y=0} = -\frac{C_{Ai}}{\sqrt{D}}\left\{\frac{1}{\sqrt{\pi t}}e^{-kt} + \sqrt{k} \text{ erf } \sqrt{kt}\right\} \tag{10.177}$$

and:
$$(N_A)_t = -D\left(\frac{dC_A}{dy}\right)_{y=0} = C_{Ai}\sqrt{D}\left\{\frac{1}{\sqrt{\pi t}}e^{-kt} + \sqrt{k} \text{ erf } \sqrt{kt}\right\} \tag{10.178}$$

Now equation 10.178 gives the instantaneous value of $(N_A)_t$ at time t.
  The average value N_A for mass transfer over an exposure time t_e is given by:

$$N_A = \frac{1}{t_e}\int_0^{t_e} N_A \, dt = \frac{1}{t_e}C_{Ai}\sqrt{D}\int_0^{t_e}\left\{\frac{1}{\sqrt{\pi t}}e^{-kt} + \sqrt{k} \text{ erf } \sqrt{kt}\right\} dt$$

$$= \frac{1}{t_e}C_{Ai}\sqrt{\frac{D}{k}}\int_0^{t_e}\left\{\left(k \text{ erf } \sqrt{kt} + kt\sqrt{\frac{k}{\pi t}}e^{-kt}\right)\right.$$

$$\left. + \left(-kt\sqrt{\frac{k}{\pi t}}\right)e^{-kt} + \frac{1}{2}\sqrt{\frac{k}{\pi t}}e^{-kt}\right) + \frac{1}{2}\sqrt{\frac{k}{\pi t}}e^{-kt}\right\} dt$$

Now $\dfrac{d}{dt}(\text{erf } \sqrt{kt}) = \dfrac{d}{dt}\left\{\dfrac{2}{\sqrt{\pi}}\displaystyle\int_0^{\sqrt{kt}} e^{-kt} \, d(\sqrt{kt})\right\}$

$$= \frac{d}{dt}\left\{\frac{2}{\sqrt{\pi}}\int_0^t e^{-kt}\left(\frac{1}{2}\sqrt{\frac{k}{t}}\right) dt\right\}$$

$$= \frac{k}{\sqrt{\pi t}}e^{-kt}$$

Thus:

$$NA = \frac{1}{t_e}C_{Ai}\sqrt{\frac{D}{k}}\int_0^{t_e}\left\{\frac{d}{dt}(kt \text{ erf }\sqrt{kt}) + \frac{d}{dt}\left(\sqrt{\frac{kt}{\pi}}e^{-kt}\right) + \frac{1}{2}\frac{d}{dt}(\text{erf }\sqrt{kt})\right\} dt$$

$$= \frac{1}{t_e}C_{Ai}\sqrt{\frac{D}{k}}\left\{kt_e \text{ erf }\sqrt{kt_e} + \sqrt{\frac{kt_e}{\pi}}e^{-kt_e} + \frac{1}{2}\text{ erf }\sqrt{kt_e}\right\}$$

$$= C_{Ai}\sqrt{\frac{D}{k}}\left\{\left(k + \frac{1}{2t_e}\right)\text{ erf }\sqrt{kt_e} + \sqrt{\frac{k}{\pi t_e}}e^{-kt_e}\right\} \tag{10.179}$$

Thus the mass transfer coefficient, enhanced by chemical reaction, h'_D is given by:

$$h'_D = \frac{N_A}{C_{Ai}} = \sqrt{\frac{D}{k}}\left\{\left(k + \frac{1}{2t_e}\right)\text{ erf }\sqrt{kt_e} + \sqrt{\frac{k}{\pi t_e}}e^{-kt_e}\right\} \tag{10.180}$$

Two special cases will be considered:

(1) When the reaction rate is very low and $k \to 0$.

$$\text{Lim}\left[\left(k + \frac{1}{2t_e}\right)\text{erf}\sqrt{kt_e}\right]_{k\to0} = \frac{1}{2t_e}\left[\frac{d}{d(\sqrt{kt_e})}\left\{\frac{2}{\sqrt{\pi}}\int_0^{t_e}e^{-kt}d(\sqrt{kt})\right\}(\sqrt{kt_e})\right]_{k\to0}$$

$$= \frac{1}{2t_e}\left[\frac{2}{\sqrt{\pi}}e^{-kt}(\sqrt{kt_e})\right]_{k\to0}$$

$$= \sqrt{\frac{k}{\pi t_e}}$$

$$\text{Lim}\left[\sqrt{\frac{k}{\pi t_e}}e^{-kt_e}\right]_{k\to0} = \sqrt{\frac{k}{\pi t_e}}$$

$$\text{Then } h'_D \to \sqrt{\frac{D}{k}}2\sqrt{\frac{k}{\pi t_e}} = 2\sqrt{\frac{D}{\pi t_e}} \tag{10.181}$$

as for mass transfer without chemical reaction (equation 10.113)

(2) When the reaction rate is very high.

$$h'_D = \sqrt{\frac{D}{k}}[k \times 1 + 0] = \sqrt{Dk} \tag{10.182}$$

and is independent of exposure time.

10.8 PRACTICAL STUDIES OF MASS TRANSFER

The principal applications of mass transfer are in the fields of distillation, gas absorption and the other separation processes involving mass transfer which are discussed in Volume 2, to which reference should be made. In particular, mass transfer coefficients and heights of transfer units in distillation are discussed in Volume 2, Section 11.11, and for gas absorption in Volume 2, Section 12.7. In this section an account is given of some of the experimental studies of mass transfer in equipment of simple geometry, in order to provide a historical perspective.

10.8.1. The j-factor of Chilton and Colburn for flow in tubes

Heat transfer

Because the mechanisms governing mass transfer are similar to those involved both in heat transfer by conduction and convection and in momentum transfer (fluid flow), quantitative relations exist between the three processes; these are discussed in Chapter 12. There is generally more published information available on heat transfer than on mass transfer, and these relationships often therefore provide a useful means of estimating mass transfer coefficients.

Results of experimental studies of heat transfer may be conveniently represented by means of the j-factor method developed by COLBURN[25] and by CHILTON and COLBURN[26] for representing data on heat transfer between a turbulent fluid and the wall of a pipe. From equation 9.57 (equation 9.54 with $n = 0.33$):

$$Nu = 0.023Re^{0.8}Pr^{0.33} \qquad \text{(equation 9.57)}$$

where the viscosity is measured at the mean film temperature, and Nu, Re and Pr denote the Nusselt, Reynolds and Prandtl numbers, respectively.

If both sides of the equation are divided by the product $Re\ Pr$:

$$St = 0.023Re^{-0.2}Pr^{-0.67} \qquad (10.183)$$

where $St(= h/C_p\rho u)$ is the Stanton number.

Equation 10.183 can be rearranged to give:

$$j_h = St\ Pr^{0.67} = 0.023Re^{-0.2} \qquad (10.184)$$

The left-hand side of equation 10.184 is referred to as the j-factor for heat transfer (j_h). Chilton and Colburn found that a plot of j_h against Re gave approximately the same curve as the friction chart (ϕ versus Re) for turbulent flow of a fluid in a pipe.

The right-hand side of equation 10.184 gives numerical values which are very close to those obtained from the Blasius equation for the friction factor ϕ for the turbulent flow of a fluid through a smooth pipe at Reynolds numbers up to about 10^6.

$$\phi = 0.0396Re^{-0.25} \qquad \text{(equation 3.10)}$$

Re	3×10^3	10^4	3×10^4	10^5	3×10^5	10^6
ϕ (equation 3.10)	0.0054	0.0040	0.0030	0.0022	0.0017	0.0013
ϕ (equation 10.184)	0.0046	0.0037	0.0029	0.0023	0.0019	0.0015

Mass transfer

Several workers have measured the rate of transfer from a liquid flowing down the inside wall of a tube to a gas passing countercurrently upwards. GILLILAND and SHERWOOD[27] have vaporised a number of liquids including water, toluene, aniline and propyl, amyl and butyl alcohols into an air stream flowing up the tube. They worked with a small tube, diameter $d = 25$ mm and length $= 450$ mm, fitted with calming sections at the top and bottom, and varied the pressure from 14 to about 300 kN/m^2.

They correlated their results using an equation of the form:

$$\frac{h_D}{u} \frac{C_{Bm}}{C_T} \left(\frac{\mu}{\rho D} \right)^{0.56} = 0.023 Re^{-0.17} \qquad (10.186)$$

The index of the Schmidt group Sc is less than the value of 0.67 for the Prandtl group for heat transfer but the range of values of Sc used was very small.

There has for long been uncertainty concerning the appropriate value to be used for the exponent of the Schmidt number in equation 10.173. SHERWOOD, PIGFORD and WILKE[23] have analysed experimental results obtained by a number of workers for heat transfer to the walls of a tube for a wide range of gases and liquids (including water, organic liquids, oils and molten salts). They give a logarithmic plot of the Stanton number ($St = h/C_p \rho u$) against Prandtl number (Pr) for a constant Reynolds number of 10,000. Superimposed on the graph are results for mass transfer obtained from experiments on the dissolution of solute from the walls of tubes composed of solid organics into liquids, and on the evaporation of liquid films from the walls of tubes to turbulent air streams using a wetted-wall column, again all at a Reynolds number of 10,000: these were plotted as Stanton number for mass transfer (h_D/u) against Schmidt number (Sc). There is very close agreement between the results for heat transfer and for mass transfer, with a line a slope of about -0.67 correlating all the results in a satisfactory manner. The range of values of Prandtl and Schmidt numbers was from about 0.4 to 10,000. This established that the exponents for both the Prandtl and Schmidt numbers in the j-factors should be the same, namely 0.67. These conclusions are consistent with the experimental results of LINTON and SHERWOOD[28] who measured the rates of dissolution of cast tubes of benzoic acid, cinnamic acid and β-naphthol into water; their Schmidt numbers ranged from 1000 to 3000.

In defining a j-factor (j_d) for mass transfer there is therefore good experimental evidence for modifying the exponent of the Schmidt number in Gilliland and Sherwood's correlation (equation 10.186). Furthermore, there is no very strong case for maintaining the small differences in the exponent of Reynolds number. On this basis, the j-factor for mass transfer may be defined as follows:

$$j_d = \frac{h_D}{u} \frac{C_{Bm}}{C_T} Sc^{0.67} = 0.023 Re^{-0.2} \qquad (10.187)$$

The term C_{Bm}/C_T (the ratio of the logarithmic mean concentration of the insoluble component to the total concentration) is introduced because $h_D(C_{Bm}/C_T)$ is less dependent than h_D on the concentrations of the components. This reflects the fact that the analogy between momentum, heat and mass transfer relates only to that part of the mass transfer which is *not* associated with the *bulk flow* mechanism; this relevant component is a fraction C_{Bm}/C_T of the total mass transfer. For equimolecular counterdiffusion, as in binary distillation when the molar latent heats of the components are equal, the term C_{Bm}/C_T should be omitted as there is no bulk flow contributing to the mass transfer.

SHERWOOD and PIGFORD[7] have shown that if the data of GILLILAND and SHERWOOD[27] and others[26,29,30] are plotted with the Schmidt group raised to this power of 0.67, as shown in Figure 10.8, a reasonably good correlation is obtained. Although the points are rather more scattered than with heat transfer, it is reasonable to assume that both j_d and j_h are approximately equal to ϕ. Equations 10.184 and 10.187 apply in the absence of ripples which can be responsible for a very much increased rate of mass transfer. The constant of 0.023 in the equations will then have a higher value.

Figure 10.8.   Mass transfer in wetted-wall columns

By equating j_h and j_d (equations 10.184 and 10.187), the mass transfer coefficient may be expressed in terms of the heat transfer coefficient, giving:

$$h_D = \frac{h}{C_p \rho} \left(\frac{C_T}{C_{Bm}} \right) \left(\frac{Pr}{Sc} \right)^{0.67} \qquad (10.188)$$

10.8.2. Mass transfer at plane surfaces

Many of the earlier studies of mass transfer involved measuring the rate of vaporisation of liquids by passing a turbulent air stream over a liquid surface. In addition, some experiments have been carried out in the absence of air flow, under what have been termed still air conditions. Most of these experiments have been carried out in some form of wind tunnel where the rate of flow of air and its temperature and humidity could be controlled and measured. In these experiments it was found to be important to keep the surface of the liquid level with the rim of the pan in order to avoid the generation of eddies at the leading edge.

HINCHLEY and HIMUS[31] measured the rate of evaporation from heated rectangular pans fitted flush with the floor of the wind tunnel (0.46m wide by 0.23m high). They showed that the rate of vaporisation was proportional to the difference between the saturation vapour pressure of the water P_s and the partial pressure of water in the air P_w. They represented their results for the mass rate of evaporation W by an empirical equation:

$$W = \text{constant } (P_s - P_w) \tag{10.189}$$

where the constant varies with the geometry of the pan and the air velocity.

This early work showed that the driving force in the process was the pressure difference $(P_s - P_w)$. Systematic work in more elaborate equipment by POWELL and GRIFFITHS[32], WADE[33], and PASQUILL[34] then followed. Wade, who vaporised a variety of organic liquids, including acetone, benzene and tri-chloroethylene at atmospheric pressure, used a small pan 88mm square in a wind tunnel. Powell and Griffiths stretched canvas sheeting over rectangular pans and, by keeping the canvas wet at all times, ensured that it behaved as a free water surface. In all of these experiments the rate of vaporisation showed a similar form of dependence on the partial pressure difference and the rate of flow of the air stream. Powell and Griffiths found that the vaporisation rate per unit area decreased in the downwind direction. For rectangular pans, of length L, the vaporisation rate was proportional to $L^{0.77}$. This can be explained in terms of the thickening of the boundary layer (see Chapter 11) and the increase in the partial pressure of vapour in the air stream arising from the evaporation at upstream positions.

In these experiments, it would be anticipated that, with high concentrations of vapour in the air, the rate of evaporation would no longer be linearly related to the partial pressure difference because of the contribution of bulk flow to the mass transfer process (see Section 10.2.3); however, there is no evidence of this even at mol fractions of vapour at the surface as high as 0.5. Possibly the experimental measurements were not sufficiently sensitive to detect this effect.

SHERWOOD and PIGFORD[7] have plotted the results of several workers[32,33,34,35,36] in terms of the Reynolds number Re_L, using the length L of the pan as the characteristic linear dimension. Figure 10.9, taken from their work, shows j_d plotted against Re_L for a number of liquids evaporating into an air stream. Although the individual points show some scatter, j_d is seen to follow the same general trend as j_h in this work. The Schmidt number was varied over such a small range that the correlation was not significantly poorer if it was omitted from the correlation.

MAISEL and SHERWOOD[37] also carried out experiments in a wind tunnel in which water was evaporated from a wet porous surface; this was preceded by a dry surface of length L_0. Thus, a velocity boundary layer had become established in the air before it came into contact with the evaporating surface. Their results were correlated by the equation:

$$j_d = 0.0415 Re_L^{-0.2} \left[1 - \left(\frac{L_0}{L} \right)^{0.8} \right]^{-0.11} \tag{10.190}$$

where L is the total length of the surface (dry + wet).

Figure 10.9.   Evaporation from plane surfaces

10.8.3. Effect of surface roughness and form drag

The results discussed above give reasonably good support to the treatment of heat, mass, and momentum transfer by the j-factor, but it is important to remember that, in all cases considered, the drag is almost entirely in the form of skin friction (i.e. viscous drag at the surface). As soon as an attempt is made to apply the relation to cases where form drag (i.e. additional drag caused by the eddies set up as a result of the fluid impinging on an obstruction) is important, such as beds of granular solids or evaporation from cylinders or spheres, the j-factor and the friction factor are found no longer to be equal. This problem will receive further consideration in Volume 2. SHERWOOD[38,39] carried out experiments where the form drag was large compared with the skin friction which he calculated approximately by subtracting the form drag from the total drag force. He obtained reasonable agreement between the corresponding value of the friction factor ϕ and j_h and j_d.

GAMSON et al.[40] have successfully used the j-factor method to correlate their experimental results for heat and mass transfer between a bed of granular solids and a gas stream.

PRATT[41] has examined the effect of using artificially roughened surfaces and of introducing "turbulence promoters", which increase the amount of form drag. He found that the values of ϕ and the heat and mass transfer coefficients were a minimum for smooth surfaces and all three quantities increased as the surface roughness was increased. ϕ increased far more rapidly than either of the other two quantities however, and the heat and mass transfer coefficients were found to reach a limiting value whereas ϕ could be increased almost indefinitely. Pratt has suggested that these limiting values are reached when the velocity gradient at the surface corresponds with that in the turbulent part of the fluid; that is, at a condition where the buffer layer ceases to exist (Chapter 11).

10.8.4. Mass transfer to single spheres

Mass transfer from a single spherical drop to still air is controlled by molecular diffusion and, at low concentrations when bulk flow is negligible, the problem is analogous to that of heat transfer by conduction from a sphere, which has been considered in Chapter 9, Section 9.3.5. Thus, for steady-state radial diffusion into a large expanse of stationary fluid in which the partial pressure falls off to zero over an infinite distance, the equation for mass transfer will take the same form as that for heat transfer (equation 9.26):

i.e.:
$$Sh' = \frac{h_D d'}{D} = 2 \tag{10.191}$$

Here Sh' is the Sherwood number which, for mass transfer, is the counterpart of the Nusselt number $Nu' (= hd/k)$ for heat transfer to a sphere. This value of 2 for the Sherwood number is the theoretical minimum in any continuous medium and will be increased if the concentration difference occurs over a finite, as opposed to an infinite, distance and if there is turbulence in the fluid.

For conditions of forced convection, FRÖSSLING[42] studied the evaporation of drops of nitrobenzene, aniline and water, and of spheres of naphthalene, into an air stream. The drops were mainly small and of the order of 1 mm diameter. POWELL[35] measured the evaporation of water from the surfaces of wet spheres up to 150 mm diameter and from spheres of ice.

The experimental results of Frössling can be represented by the equation:

$$Sh' = \frac{h_D d'}{D} = 2.0(1 + 0.276Re'^{0.5} Sc^{0.33}) \tag{10.192}$$

SHERWOOD and PIGFORD[7] found that the effect of the Schmidt group was also influenced by the Reynolds group and that the available data were fairly well correlated as shown in Figure 10.10, in which $(h_D d')/D$ is plotted against $Re' Sc^{0.67}$.

GARNER and KEEY[43,44] dissolved pelleted spheres of organic acids in water in a low-speed water tunnel at particle Reynolds numbers between 2.3 and 255 and compared their results with other data available at Reynolds numbers up to 900. Natural convection was found to exert some influence at Reynolds numbers up to 750. At Reynolds numbers greater than 250, the results are correlated by equation 10.191:

$$Sh' = 0.94Re'^{0.5} Sc^{0.33} \tag{10.193}$$

They also studied mass transfer under conditions of natural convection.

ROWE et al.[45] have reviewed the literature on heat and mass transfer between spherical particles and a fluid. For heat transfer, their results which are discussed in Chapter 9, Section 9.4.6., are generally well represented by equation 9.90:

$$Nu' = 2 + \beta'' Re'^{0.5} Pr^{0.33} \quad (0.4 < \beta'' < 0.8) \tag{10.194}$$

For mass transfer:

$$Sh' = \alpha + \beta' Re'^{0.5} Sc^{0.33} \quad (0.3 < \beta' < 1.0) \tag{10.195}$$

The constant α appears to be a function of the Grashof number, but approaches a value of about 2 as the Grashof number approaches zero.

Figure 10.10.   Mass transfer to single spheres

In an experimental investigation[45] they confirmed that equations 10.192 and 10.193 can be used to represent the results obtained for transfer from both air and water to spheres. The constants β', β'' varied from 0.68 to 0.79.

10.9. FURTHER READING

BENNETT, C.O. and MYERS, J.E.: *Momentum, Heat, and Mass Transfer*, 3rd edn (McGraw-Hill, New York, 1983).

BIRD, R.B., STEWART, W.E. and LIGHTFOOT, E.N.: *Transport Phenomena* (Wiley, New York, 1960).

DANCKWERTS, P.V.: *Gas-liquid Reactions* (McGraw-Hill, New York, 1970).

ECKERT, E.R.G. and DRAKE, R.M.: *Analysis of Heat and Mass Transfer* (McGraw-Hill, New York, 1972).

EDWARDS, D. A., BRENNER, H. and WASAN, D. T. *Interface Transport Processes and Rheology* (Butterworth-Heinemann, 1991).

SHERWOOD, T.K.: *Chemical Engineering Education* (Fall, 1974), 204. A review of the development of mass transfer theory.

SHERWOOD, T.K., PIGFORD, R.L. and WILKE, C.R.: *Mass Transfer* (McGraw-Hill, New York, 1975).

TAYLOR, R., and KRISHNA, R.: *Multicomponent Mass Transfer* (Wiley, New York, 1993).

TREYBAL, R.E.: *Mass-Transfer Operations*, 3rd edn (McGraw-Hill, New York, 1980).

10.10. REFERENCES

1. FICK, A.: *Ann. Phys.* **94** (1855) 59. Ueber Diffusion.

2. STEFAN, J.: *Wiener Akad. Wissensch.* **68** (1873) 385; **79** (1879) 169; **98** (1889) 1418; *Ann. Physik* **41** (1890) 723. Versuche über die Verdampfung.

3. PERRY, R.H. and GREEN, D.W. (eds): *Perry's Chemical Engineers' Handbook*. 6th edn (McGraw-Hill, New York, 1984).

4. WINKELMANN, A.: *Ann. Physik*. **22** (1884) 1, 152. Ueber die Diffusion von Gasen und Dämpfen.

5. GILLILAND, E.R.: *Ind. Eng. Chem.* **26** (1934) 681. Diffusion coefficients in gaseous systems.

6. SHERWOOD, T.K. and PIGFORD, R.L.: *Absorption and Extraction* (McGraw-Hill, New York, 1952).

7. FULLER, E.N., SCHETTLER, P.D. and GIDDINGS, J.C.: *Ind. Eng. Chem.* **58** (1966) 19. A new method for prediction of binary gas-phase diffusion coefficients.

8. GREW, K.E. and IBBS, T.L.: *Thermal Diffusion in Gases* (Cambridge University Press, Cambridge, 1952).

9. HIRSCHFELDER, J.O., CURTISS, C.F. and BIRD, R.B.: *Molecular Theory of Gases and Liquids*, pp. 584–585 (Wiley, New York, 1954).

10. ARNOLD, J.H.: *Trans. Am. Inst. Chem. Eng.* **40** (1944) 361. Studies in diffusion: III, Steady-state vaporization and absorption.

11. MAXWELL, J.C.: *Phil. Trans. Roy. Soc.* **157** (1867) 49. The dynamical theory of gases.

12. TAYLOR, R. and KRISHNA, R.: *Multicomponent Mass Transfer* (Wiley, New York, 1993).

13. WILKE C.R. and CHANG, P.: *A.I.Ch.E.Jl* **1** (1955) 264. Correlation of diffusion coefficients in dilute solutions.

14. REID, R.C., PRAUSNITZ, J.M. and SHERWOOD, T.K.: *The Properties of Gases and Liquids* (McGraw-Hill, New York, 1977).

15. GOODRIDGE, F. and BRICKNELL, D.J.: *Trans. Inst. Chem. Eng.* **40** (1962) 54. Interfacial resistance in the carbon dioxide-water system.

16. WHITMAN, W.G.: *Chem. and Met. Eng.* **29** (1923) 147. The two-film theory of absorption.

17. HIGBIE, R.: *Trans. Am. Inst. Chem. Eng.* **31** (1935) 365. The rate of absorption of pure gas into a still liquid during short periods of exposure.

18. DANCKWERTS, P.V.: *Ind. Eng. Chem.* **43** (1951) 1460. Significance of liquid film coefficients in gas absorption.

19. TOOR, H.L. and MARCHELLO, J.M.: *A.I.Ch.E.Jl* **4** (1958) 97. Film-penetration model for mass and heat transfer.

20. KISHINEVSKIJ, M.K.: *Zhur. Priklad. Khim.* **24** (1951) 542; *J. Appl. Chem. U.S.S.R.* **24** 593. The kinetics of absorption under intense mixing.

21. HARRIOTT, P.: *Chem. Eng. Sci.* **17** (1962) 149. A random eddy modification of the penetration theory.

22. DWIGHT, H.B.: *Tables of Integrals and Other Mathematical Data* (Macmillan, New York, 1957).

23. SHERWOOD, T.K., PIGFORD, R.L. and WILKE, C.R.: *Mass Transfer* (McGraw-Hill, New York, 1975).

24. DANCKWERTS, P. V.: *Trans. Faraday Soc.* **46** (1950) 300. Absorption by simultaneous diffusion and chemical reaction.

25. COLBURN, A.P.: *Trans. Am. Inst. Chem. Eng.* **29** (1933) 174. A method of correlating forced convection heat transfer data and a comparison with fluid friction.

26. CHILTON, T.H. and COLBURN, A.P.: *Ind. Eng. Chem.* **26** (1934) 1183. Mass transfer (absorption) coefficients — production from data on heat transfer and fluid friction.

27. GILLILAND, E.R. and SHERWOOD, T.K.: *Ind. Eng. Chem.* **26** (1934) 516. Diffusion of vapours into air streams.

28. LINTON, W.H. and SHERWOOD, T.K.: *Chem. Eng. Prog.* **46** (1950) 258. Mass transfer from solid shapes to water in streamline and turbulent flow.

29. BARNET, W.I. and KOBE, K.A.: *Ind. Eng. Chem.* **33** (1941) 436. Heat and vapour transfer in a wetted-wall column.

30. CHAMBERS, F.S. and SHERWOOD, T.K.: *Ind. Eng. Chem.* **29** (1937) 579. *Trans. Am. Inst. Chem. Eng.* **33** (1937) 579. Absorption of nitrogen dioxide by aqueous solutions.

31. HINCHLEY, J.W. and HIMUS, G.W.: *Trans. Inst. Chem. Eng.* **2** (1924) 57. Evaporation in currents of air.

32. POWELL, R.W. and GRIFFITHS, E.: *Trans. Inst. Chem. Eng.* **13** (1935) 175. The evaporation of water from plane and cylindrical surfaces.

33. WADE, S.H.: *Trans. Inst. Chem. Eng.* **20** (1942) 1. Evaporation of liquids in currents of air.

34. PASQUILL, F.: *Proc. Roy. Soc.* **A 182** (1943) 75. Evaporation from a plane free liquid surface into a turbulent air stream.

35. POWELL, R.W.: *Trans. Inst. Chem. Eng.* **18** (1940) 36. Further experiments on the evaporation of water from saturated surfaces.

36. LURIE, M. and MICHAILOFF, M.: *Ind. Eng. Chem.* **28** (1936) 345. Evaporation from free water surfaces.

37. MAISEL, D.S. and SHERWOOD, T.K.: *Chem. Eng. Prog.* **46** (1950) 131. Evaporation of liquids into turbulent gas streams.

38. SHERWOOD, T.K.: *Trans. Am. Inst. Chem. Eng.* **36** (1940) 817. Mass transfer and friction in turbulent flow.

39. SHERWOOD, T.K.: *Ind. Eng. Chem.* **42** (1950) 2077. Heat transfer, mass transfer, and fluid friction.

40. GAMSON, B.W., THODOS, G. and HOUGEN, O.A.: *Trans Am. Inst. Chem. Eng.* **39** (1943) 1. Heat mass and momentum transfer in the flow of gases through granular solids.

41. PRATT, H.R.C.: *Trans. Inst. Chem. Eng.* **28** (1950) 77. The application of turbulent flow theory to transfer processes in tubes containing turbulence promoters and packings.

42. FRÖSSLING, N.: *Gerlands Beitr. Geophys.* **52** (1938) 170. Über die Verdunstung fallender Tropfen.

43. GARNER, F.H. and KEEY, R.B.: *Chem. Eng. Sci.* **9** (1958) 119. Mass transfer from single solid spheres — I. Transfer at low Reynolds numbers.

44. GARNER, F.H. and KEEY, R.B.: *Chem. Eng. Sci.* **9** (1959) 218. Mass transfer from single solid spheres — II. Transfer in free convection.

45. ROWE, P.N., CLAXTON, K.T., and LEWIS, J.B.: *Trans. Inst. Chem. Eng.* **41** (1965) T14. Heat and mass transfer from a single sphere in an extensive flowing fluid.

10.11. NOMENCLATURE

		Units in SI system	Dimensions in in $\mathbf{M}, \mathbf{N}, \mathbf{L}, \mathbf{T}, \theta$
a	Interfacial area per unit volume	m^2/m^3	$\mathbf{L}^{-1}$
B_1, B_2	Integration constants	kmol s/m^3	$\mathbf{NL}^{-3}\mathbf{T}$
B_1', B_2'	Integration constants	kmol/m^3	$\mathbf{NL}^{-3}$
b	Radius of sphere	m	$\mathbf{L}$
C	Molar concentration	kmol/m^3	$\mathbf{NL}^{-3}$
C_A, C_B	Molar concentration of $\mathbf{A}, \mathbf{B}$	kmol/m^3	$\mathbf{NL}^{-3}$
C_p	Specific heat at constant pressure	J/kg K	$\mathbf{L}^2\mathbf{T}^{-2}\theta^{-1}$
C_T	Total molar concentration	kmol/m^3	$\mathbf{NL}^{-3}$
C_{Bm}	Logarithmic mean value of C_B	kmol/m^3	$\mathbf{NL}^{-3}$
C'	$C_A - C_{Ao}$	kmol/m^3	$\mathbf{NL}^{-3}$
C_A'', C_B''	Molar concentration of $\mathbf{A}, \mathbf{B}$ in film	kmol/m^3	$\mathbf{NL}^{-3}$
$\overline{C'}$	Laplace transform of C'	kmol s/m^3	$\mathbf{NL}^{-3}\mathbf{T}$
c	Mass concentration	kg/m^3	$\mathbf{ML}^{-3}$
D	Diffusivity	m^2/s	$\mathbf{L}^2\mathbf{T}^{-1}$
D_f	Diffusivity of fluid in film	m^2/s	$\mathbf{L}^2\mathbf{T}^{-1}$
D_L	Liquid phase diffusivity	m^2/s	$\mathbf{L}^2\mathbf{T}^{-1}$
D_{AB}	Diffusivity of $\mathbf{A}$ in $\mathbf{B}$	m^2/s	$\mathbf{L}^2\mathbf{T}^{-1}$
D_{Th}	Coefficient of thermal diffusion	kmol/ms	$\mathbf{NL}^{-1}\mathbf{T}^{-1}$
D_{BA}	Diffusivity of $\mathbf{B}$ in $\mathbf{A}$	m^2/s	$\mathbf{L}^2\mathbf{T}^{-1}$
D'	Effective diffusivity in multicomponent system	m^2/s	$\mathbf{L}^2\mathbf{T}^{-1}$
d	Pipe diameter	m	$\mathbf{L}$
d'	Diameter of sphere	m	$\mathbf{L}$
E_D	Eddy diffusivity	m^2/s	$\mathbf{L}^2\mathbf{T}^{-1}$
F	Coefficient in Maxwell's law of diffusion	m^3/kmols	$\mathbf{N}^{-1}\mathbf{L}^3\mathbf{T}^{-1}$
f	Ratio $-N_B'/N_A'$	—	—
G	Molar flow of stream	kmol/s	$\mathbf{NT}^{-1}$
G'	Molar flow per unit area	kmol/m^{2}s	$\mathbf{NL}^{-2}\mathbf{T}^{-1}$
$\mathcal{H}$	Ratio of equilibrium values of concentrations in two phases	—	—
$\mathbf{H}$	Height of transfer unit	m	$\mathbf{L}$
h	Heat transfer coefficient	W/m^2 K	$\mathbf{MT}^{-3}\theta^{-1}$
h_D	Mass transfer coefficient	m/s	$\mathbf{LT}^{-1}$
h_D'	Mass transfer coefficient enhanced by chemical reaction	m/s	$\mathbf{LT}^{-1}$
j_d	j-factor for mass transfer	—	—
j_h	j-factor for heat transfer	—	—
K	Overall mass transfer coefficient	m/s	$\mathbf{LT}^{-1}$
J	Flux as *mass* per unit area and unit time	kg/m^{2}s	$\mathbf{ML}^{-2}\mathbf{T}^{-1}$
K_{ABT}	Ratio of transport rate by thermal diffusion to that by Fick's law	—	—
k_G	Mass transfer coefficient for transfer through stationary fluid	N s/kg m	$\mathbf{NM}^{-1}\mathbf{L}^{-1}\mathbf{T}$
k_G'	Mass transfer coefficient for equimolecular counterdiffusion	N s/kg m	$\mathbf{NM}^{-1}\mathbf{L}^{-1}\mathbf{T}$
k_x	Mass transfer coefficient (mole fraction driving force)	kmol/m^{2}s	$\mathbf{NL}^{-2}\mathbf{T}^{-1}$
k	Reaction rate constant	*	*
L	Length of surface, or film thickness	m	$\mathbf{L}$
L_f	Thickness of film	m	$\mathbf{L}$
L_0	Length of surface, unheated	m	$\mathbf{L}$
M	Molecular weight	kg/kmol	$\mathbf{MN}^{-1}$
N	Molar rate of diffusion per unit area (average value)	kmol/m^{2}s	$\mathbf{NL}^{-2}\mathbf{T}^{-1}$
$(N)_t$	Molar rate of diffusion per unit area at time t	kmol/m^{2}s	$\mathbf{NL}^{-2}\mathbf{T}^{-1}$
N'	Total molar rate of transfer per unit area	kmol/m^{2}s	$\mathbf{NL}^{-2}\mathbf{T}^{-1}$
$(N)_{Th}$	Molar flux due to thermal diffusion	kmol/m^{2}s	$\mathbf{NL}^{-2}\mathbf{T}^{-1}$
$\mathbf{N}$	Number of transfer units	—	—
n	Number of moles of gas	kmol	$\mathbf{N}$

		Units in SI system	Dimensions in in $\mathbf{M, N, L, T}, \theta$
n	Order of reaction, or number of term in series	—	—
P	Total pressure	N/m^2	$\mathbf{ML^{-1}T^{-2}}$
P_A, P_B	Partial pressure of $\mathbf{A, B}$	N/m^2	$\mathbf{ML^{-1}T^{-2}}$
P_s	Vapour pressure of water	N/m^2	$\mathbf{ML^{-1}T^{-2}}$
P_w	Partial pressure of water in gas stream	N/m^2	$\mathbf{ML^{-1}T^{-2}}$
P_{Bm}	Logarithmic mean value of P_B	N/m^2	$\mathbf{ML^{-1}T^{-2}}$
p	Parameter in Laplace Transform	s^{-1}	$\mathbf{T^{-1}}$
R	Shear stress acting on surface	N/m^2	$\mathbf{ML^{-1}T^{-2}}$
$\mathbf{R}$	Universal gas constant	(8314) J/kmol K	$\mathbf{MN^{-1}L^2T^{-2}\theta^{-1}}$
r	Radius within sphere	m	$\mathbf{L}$
S	Cross-sectional area of flow	m^2	$\mathbf{L^2}$
s	Rate of production of fresh surface per unit area	s^{-1}	$\mathbf{T^{-1}}$
T	Absolute temperature	K	θ
t	Time	s	$\mathbf{T}$
t_e	Time of exposure of surface element	s	$\mathbf{T}$
u	Mean velocity	m/s	$\mathbf{LT^{-1}}$
u_A, u_B	Mean molecular velocity in direction of transfer	m/s	$\mathbf{LT^{-1}}$
u_{DA}, u_{DB}	Diffusional velocity of transfer	m/s	$\mathbf{LT^{-1}}$
u_F	Velocity due to bulk flow	m/s	$\mathbf{LT^{-1}}$
u_m	Molecular velocity	m/s	$\mathbf{LT^{-1}}$
u_s	Stream velocity	m/s	$\mathbf{LT^{-1}}$
V	Volume	m^3	$\mathbf{L^3}$
$\mathbf{V}$	Molecular volume	m^3/kmol	$\mathbf{N^{-1}L^3}$
$\mathbf{V}_o$	Correction term in equation 10.94	m^3/kmol	$\mathbf{N^{-1}L^3}$
W	Mass rate of evaporation	kg/s	$\mathbf{MT^{-1}}$
x	Distance from leading edge of surface or in X-direction or mole fraction	m —	$\mathbf{L}$ —
y	Distance from surface or in direction of diffusion	m	$\mathbf{L}$
y'	Mol fraction in stationary gas	—	—
Z	Height of column	m	$\mathbf{L}$
z	Distance in Z-direction	m	$\mathbf{L}$
α	Thermal diffusion factor	—	—
λ	Mean free path of molecules	m	$\mathbf{L}$
μ	Viscosity of fluid	N s/m^2	$\mathbf{ML^{-1}T^{-1}}$
ρ	Density of fluid	kg/m^3	$\mathbf{ML^{-3}}$
ω	Mass fraction	—	—
Nu	Nusselt number hd/k	—	—
Nu'	Nusselt number for sphere hd'/k	—	—
Pr	Prandtl number $C_p\mu/k$	—	—
Re	Reynolds number $ud\rho/\mu$	—	—
Re'	Reynolds number for sphere $ud'\rho/\mu$	—	—
Re_L	Reynolds number for flat plate $u_s L\rho/\mu$	—	—
Sc	Schmidt number $\mu/\rho D$	—	—
Sh'	Sherwood number for sphere $h_D d'/D$	—	—
St	Stanton number $h/C_p\rho u$	—	—

* Dimensions depend on order of reaction.

Suffixes

0	Value in bulk of phase
1	Phase 1
2	Phase 2
A	Component $\mathbf{A}$
B	Component $\mathbf{B}$
AB	Of $\mathbf{A}$ in $\mathbf{B}$
b	Bottom of column

e	Value in equilibrium with bulk of other phase
G	Gas phase
i	Interface value
L	Liquid phase
o	Overall value (for height and number of transfer units)
	or value in bulk of phase
t	Top of column

CHAPTER 11

The Boundary Layer

11.1. INTRODUCTION

When a fluid flows over a surface, that part of the stream which is close to the surface suffers a significant retardation, and a velocity profile develops in the fluid. The velocity gradients are steepest close to the surface and become progressively smaller as distance from the surface increases. Although theoretically there is no outer bound at which the velocity gradient becomes zero, it is convenient to divide the flow into two parts for practical purposes.

(1) A *boundary layer* close to the surface in which the velocity increases from zero at the surface itself to a near constant stream velocity at its outer bound.

(2) A region outside the boundary layer in which the velocity gradient in a direction perpendicular to the surface is negligibly small and in which the velocity is everywhere equal to the stream velocity.

The thickness of the boundary layer may be arbitrarily defined as the distance from the surface at which the velocity reaches some proportion (0.9, 0.99, 0.999) of the undisturbed stream velocity. Alternatively, it may be possible to approximate to the velocity profile by means of an equation which is then used to give the distance from the surface at which the velocity gradient is zero and the velocity is equal to the stream velocity. Difficulties arise in comparing the thicknesses obtained using these varied definitions, because velocity is changing so slowly with distance that a small difference in the criterion used for the selection of velocity will account for a very large difference in the corresponding thickness of the boundary layer.

The flow conditions in the boundary layer are of considerable interest to chemical engineers because these influence, not only the drag effect of the fluid on the surface, but also the heat or mass transfer rates if a temperature or concentration gradient exists.

It is convenient first to consider the flow over a thin plate inserted parallel to the flow of a fluid with a constant stream velocity u_s. It will be assumed that the plate is sufficiently wide for conditions to be constant across any finite width w of the plate which is being considered.

Furthermore, the extent of the fluid in a direction perpendicular to the surface is sufficiently large for the velocity of the fluid remote from the surface to be unaffected and to remain constant at the value of u_s. It may be taken that part of the fluid flows on one side of the flat plate and part on the other. The change in flow on only one side will be considered.

On the assumption that there is no slip at the surface (see Section 11.3), the fluid layers at all points on the surface ($y = 0$) will be zero. At some position a distance x from the leading edge, the velocity will increase from zero at the surface to approach the stream velocity u_s asymptotically. At the leading edge ($x = 0$), the fluid will have been influenced by the surface for only an infinitesimal time and therefore only the molecular layer of fluid at the surface will have been retarded. At progressively greater distances (x) along the surface, the fluid will have been retarded for a greater time and the effects will be felt to greater depths in the fluid. Thus the thickness (δ) of the boundary layer will increase, starting from a zero value at the leading edge. Furthermore, the velocity gradient at the surface $(du_x/dy)_{y=0}$ (where u_x is the velocity in the X-direction at a distance y from the surface) will becomes less because the velocity will change by the same amount (from $u_x = 0$ at $y = 0$ to $u_x = u_s$ at $y = \delta$) over a greater distance. The development of the boundary layer is illustrated in Figure 11.1.

Figure 11.1.   Development of boundary layer

Near the leading edge of the surface where the boundary layer thickness is small, the flow will be streamline, or laminar, and the shear stresses will arise solely from viscous shear effects. However, when the boundary layer thickness exceeds a critical value, streamline flow ceases to be stable and turbulence sets in. The transition from streamline to turbulent flow is not as sharply defined as in pipe flow (Chapter 3) and may be appreciably influenced by small disturbances generated at the leading edge of the surface and by roughness and imperfections of the surface in itself, all of which lead to an early development of turbulence. However, for equivalent conditions, the important flow parameter is the Reynolds number $u_s\delta\rho/\mu(= Re_\delta)$. Because δ can be expressed as a function of x, the distance from the leading edge of the surface, the usual criterion is taken as the value of the Reynolds number $u_s x\rho/\mu$ ($= Re_x$). If x_c gives the location of the transition point, $u_s x_c\rho/\mu$ ($= Re_{x_c}$) is of the order of 10^5.

The transition from streamline to turbulent flow in the neighbourhood of a surface has been studied by F. N. M. BROWN[1] with the aid of a three-dimensional smoke tunnel with transparent faces. Photographs were taken using short-interval flash lamps giving effective exposure times of about 20 μs. Figure 11.2 refers to the flow over an aerofoil and shows the manner in which vortices are created. Figure 11.3 shows the wake of a cascade of flat plates with sharpened leading edges and blunt trailing edges.

When the flow in the boundary layer is turbulent, streamline flow persists in a thin region close to the surface called the *laminar sub-layer*. This region is of particular importance because, in a heat or mass transfer problem, it is where the greater part of the resistance to transfer lies. High heat and mass transfer rates therefore depend on the laminar sub-layer being thin. Separating the laminar sub-layer from the turbulent

Figure 11.2. Formation of vortices in flow over an aerofoil

Figure 11.3.   Wake produced by cascade of flat plates with sharpened leading edges and blunt trailing edges

part of the boundary layer is the *buffer layer* (Figure 11.1) in which the contributions of the viscous effects and of the turbulent eddies are of comparable magnitudes. This phenomenon is discussed in more detail in Chapter 12. However, the general picture is as follows.

The average size of the eddies, or turbulent circulating currents, becomes progressively smaller as the surface is approached. The eddies are responsible for transference of momentum from the faster moving fluid remote from the surface to slower moving fluid near the surface. Thus momentum is transferred by a progression of eddies of diminishing dimensions down to about 1 mm in size, at which point they tend to die out and viscous forces (due to transfer by molecular-scale movement) predominate near the surface. Although the laminar sub-layer is taken to be the region where eddies are completely absent, there is evidence that occasionally eddies do penetrate right up to the surface. The situation is therefore highly complex and any attempt to model the process involves simplifying assumptions.

The treatment that follows is based on the simplified approach suggested by PRANDTL[2]. It involves the following assumptions:

(1) That the flow can be considered essentially as unidirectional (X-direction) and that the effects of velocity components perpendicular to the surface *within* the boundary layer can be neglected (i.e. $u_y \ll u_x$). This condition will not be met at very low Reynolds numbers where the boundary layer thickens very rapidly.

(2) That the existence of the buffer layer can be neglected and that in turbulent flow the boundary layer can be considered as consisting of a turbulent region adjacent to a laminar sub-layer.

(3) That the stream velocity does not change in the direction of flow. On this basis, from Bernoulli's theorem, the pressure then does not change (i.e. $\partial P/\partial x = 0$). In practice, $\partial P/\partial x$ may be positive or negative. If positive, it will tend to result in a greater retardation of the fluid and the boundary layer will thicken more rapidly. If negative, the converse will be true.

For flow against a pressure gradient ($\partial P/\partial x$ positive in the direction of flow) the combined force due to pressure gradient and friction may be sufficient to bring the fluid

Figure 11.4.   Boundary layer separation

completely to rest and to cause some backflow close to the surface. When this occurs the fluid velocity will be zero, not only at the surface, but also at a second position a small distance away. In these circumstances, the boundary layer is said to *separate* and circulating currents are set up as shown in Figure 11.4. An example of this phenomenon is described in Volume 2, Chapter 3, in which the flow relative to a cylinder or sphere is considered.

The procedure which will be adopted in the following section consists of taking a momentum balance on an element of fluid. The resulting *Momentum Equation* involves no assumptions concerning the nature of the flow. However, it includes an integral, the evaluation of which requires a knowledge of the velocity profile $u_x = f(y)$. At this stage assumptions must be made concerning the nature of the flow in order to obtain realistic expressions for the velocity profile.

11.2. THE MOMENTUM EQUATION

A fluid of density ρ and viscosity μ flows over a plane surface and the velocity of flow outside the boundary layer is u_s. A boundary layer of thickness δ forms near the surface, and at a distance y from the surface the velocity of the fluid is reduced to a value u_x.

Consider the equilibrium of an element of fluid bounded by the planes 1–2 and 3–4 at distances x and $x + dx$ respectively from the leading edge; the surface and the plane 2–4 parallel to the surface and at a distance l from it; and by two planes parallel to the plane of the paper and distance w apart. The distance l is greater than the boundary layer thickness δ (Figure 11.5); conditions are constant over the width w. Consider the velocities and forces in the X-direction.

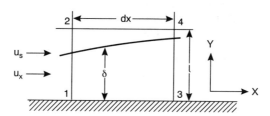

Figure 11.5.   Element of boundary layer

At plane 1–2, mass rate of flow through a strip of thickness dy at distance y from the surface

$$= \rho u_x w \, dy$$

Total flow through plane 1–2

$$= w \int_0^l \rho u_x \, dy$$

Rate of transfer of momentum through the elementary strip

$$= \rho u_x w \, dy \, u_x = w \rho u_x^2 \, dy$$

Total rate of transfer of momentum through plane 1–2

$$= w \int_0^l \rho u_x^2 \, dy$$

In passing from plane 1–2 to plane 3–4, the mass flow changes by:

$$w \frac{\partial}{\partial x} \left(\int_0^l \rho u_x \, dy \right) dx$$

and the momentum flux changes by:

$$w \frac{\partial}{\partial x} \left(\int_0^l \rho u_x^2 \, dy \right) dx$$

A mass flow of fluid equal to the difference between the flows at planes 3–4 and 1–2 must therefore occur through plane 2–4, as it is assumed that there is symmetry over the width of the element.

Since plane 2–4 lies outside the boundary layer, this entering fluid must have a velocity u_s in the X-direction. Because the fluid in the boundary layer is being retarded, there will be a smaller flow at plane 3–4 than at 1–2, and hence the amount of fluid entering through plane 2–4 is negative i.e. fluid actually leaves the element.

Thus the rate of transfer of momentum through plane 2–4 into the element

$$= w u_s \frac{\partial}{\partial x} \left(\int_0^l \rho u_x \, dy \right) dx$$

The net rate of change of momentum in the X-direction on the element must be equal to the momentum added from outside (through plane 2–4) together with the net force acting on it.

The forces in the X-direction acting on the element of fluid are:

(1) A shear force resulting from the shear stress R_0 acting at the surface. This is a retarding force and therefore R_0 is negative.
(2) The force produced as a result of any difference in pressure dP between the planes 3–4 and 1–2. However, if the velocity u_s outside the boundary layer remains constant, from Bernoulli's theorem, there can be no pressure gradient in the X-direction (i.e. $\partial P / \partial x = 0$).

Thus, the net force acting $= R_0 w \, dx$.
Hence:

$$w \frac{\partial}{\partial x} \left(\int_0^l \rho u_x^2 \, dy \right) dx = w u_s \frac{\partial}{\partial x} \left(\int_0^l \rho u_x \, dy \right) dx + R_0 w \, dx$$

Since u_s does not vary with x:

$$\frac{\partial}{\partial x} \int_0^l \rho (u_s - u_x) u_x \, dy = -R_0 \qquad (11.1)$$

This expression, known as the *momentum equation*, can be integrated provided that the relation between u_x and y is known.

With the velocity of the main stream remaining constant at u_s, the density ρ may be taken as constant and equation 11.1 therefore becomes:

$$\rho \frac{\partial}{\partial x} \int_0^l (u_s - u_x)u_x \, dy = -R_0 \qquad (11.2)$$

It will be noted that no assumptions have been made concerning the nature of the flow within the boundary layer and therefore this relation is applicable to both the streamline and the turbulent regions. The relation between u_x and y will now be derived for streamline and turbulent flow over a plane surface and the integral in equation 11.2 will be evaluated.

11.3. THE STREAMLINE PORTION OF THE BOUNDARY LAYER

In the streamline boundary layer the only forces acting within the fluid are pure viscous forces and no transfer of momentum takes place by eddy motion.

Assuming that the relation between u_x and y can be expressed approximately by:

$$u_x = u_0 + ay + by^2 + cy^3 \qquad (11.3)$$

the coefficients a, b, c and u_0 can be evaluated because the boundary conditions which the relation must satisfy are known (Figure 11.6).

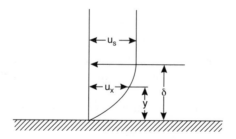

Figure 11.6.   Velocity distribution in streamline boundary layer

In fluid dynamics it is generally assumed that the velocity of flow at a solid boundary, such as a pipe wall, is zero. This is referred to as the *no-slip condition*. If the fluid "wets" the surface, this assumption can be justified in physical terms in that the molecules are small compared with the irregularities on the surface of even the smallest pipe, and therefore a fluid layer is effectively held stationary at the wall. All gases and the majority of liquids will have a sufficiently low contact angle for this condition to be met.

If the fluid does not wet the wall, the no-slip condition no longer applies and the pressure gradient at a given flowrate will be lower. This effect is particularly important during the flow of molten polymers, but it does not seem to be significant in other applications.

It will be assumed here that the fluid in contact with the surface is at rest and therefore u_0 must be zero. Furthermore, all the fluid close to the surface is moving at very low velocity and therefore any changes in its momentum as it flows parallel to the surface must be extremely small. Consequently, the net shear force acting on any element of fluid near the surface is negligible, the retarding force at its lower boundary being balanced by the accelerating force at its upper boundary. Thus the shear stress R_0 in the fluid near the surface must approach a constant value.

Since $R_0 = -\mu (\partial u_x / \partial y)_{y=0}$, $\partial u_x / \partial y$ must also be constant at small values of y and:

$$\left(\frac{\partial^2 u_x}{\partial y^2} \right)_{y=0} = 0.$$

At the distant edge of the boundary layer it is assumed that the velocity just equals the main stream velocity and that there is no discontinuity in the velocity profile.

Thus, when $y = \delta$: $\qquad\qquad u_x = u_s \quad$ and $\quad \dfrac{\partial u_x}{\partial y} = 0$

Now with $u_0 = 0$, equation 11.3 becomes:

$$u_x = ay + by^2 + cy^3$$

$$\frac{\partial u_x}{\partial y} = a + 2by + 3cy^2$$

and: $\qquad\qquad\qquad\qquad \dfrac{\partial^2 u_x}{\partial y^2} = 2b + 6cy$

At $y = 0$: $\qquad\qquad\qquad \dfrac{\partial^2 u_x}{\partial y^2} = 0$

Thus: $\qquad\qquad\qquad\qquad\quad b = 0$

At $y = \delta$: $\qquad\qquad\qquad u_x = a\delta + c\delta^3 = u_s$

and: $\qquad\qquad\qquad\qquad \dfrac{\partial u_x}{\partial y} = a + 3c\delta^2 = 0$

Thus: $\qquad\qquad\qquad\qquad\quad a = -3c\delta^2$

Hence: $\qquad\qquad\qquad c = -\dfrac{u_s}{2\delta^3} \quad$ and $\quad a = \dfrac{3u_s}{2\delta}$

The equation for the velocity profile is therefore:

$$u_x = \frac{3u_s}{2} \frac{y}{\delta} - \frac{u_s}{2} \left(\frac{y}{\delta} \right)^3 \tag{11.4}$$

$$\frac{u_x}{u_s} = \frac{3}{2} \left(\frac{y}{\delta} \right) - \frac{1}{2} \left(\frac{y}{\delta} \right)^3 \tag{11.5}$$

Equation 11.5 corresponds closely to experimentally determined velocity profiles in a laminar boundary layer.

This relation applies over the range $0 < y < \delta$.

When $y > \delta$: $\qquad\qquad\qquad\qquad u_x = u_s \qquad\qquad\qquad\qquad\qquad (11.6)$

The integral in the momentum equation (11.2) can now be evaluated for the streamline boundary layer by considering the ranges $0 < y < \delta$ and $\delta < y < l$ separately.

Thus:

$$\int_0^l (u_s - u_x) u_x \, dy = \int_0^\delta u_s^2 \left(1 - \frac{3}{2}\frac{y}{\delta} + \frac{y^3}{2\delta^3}\right)\left(\frac{3}{2}\frac{y}{\delta} - \frac{y^3}{2\delta^3}\right) dy + \int_\delta^l (u_s - u_s) u_s \, dy$$

$$= u_s^2 \int_0^\delta \left(\frac{3}{2}\frac{y}{\delta} - \frac{9}{4}\frac{y^2}{\delta^2} - \frac{1}{2}\frac{y^3}{\delta^3} + \frac{3}{2}\frac{y^4}{\delta^4} - \frac{1}{4}\frac{y^6}{\delta^6}\right) dy$$

$$= u_s^2 \delta \left(\tfrac{3}{4} - \tfrac{3}{4} - \tfrac{1}{8} + \tfrac{3}{10} - \tfrac{1}{28}\right)$$

$$= \frac{39}{280}\delta u_s^2 \tag{11.7}$$

Now:

$$R_0 = -\mu \left(\frac{\partial u_x}{\partial y}\right)_{y=0} = -\frac{3}{2}\mu\frac{u_s}{\delta} \tag{11.8}$$

Substitution from equations 11.7 and 11.8 in equation 11.2, gives:

$$\rho \frac{\partial}{\partial x}\left(\frac{39}{280}\delta u_s^2\right) = \frac{3}{2}\mu\frac{u_s}{\delta}$$

∴

$$\delta \, d\delta = \frac{140}{13}\frac{\mu}{\rho}\frac{1}{u_s}dx$$

∴

$$\frac{\delta^2}{2} = \frac{140}{13}\frac{\mu x}{\rho u_s} \quad \text{(since } \delta = 0 \text{ when } x = 0) \tag{11.9}$$

Thus:

$$\delta = 4.64\sqrt{\frac{\mu x}{\rho u_s}}$$

and:

$$\frac{\delta}{x} = 4.64\sqrt{\frac{\mu}{x \rho u_s}} = 4.64 Re_x^{-1/2} \tag{11.10}$$

The rate of thickening of the boundary layer is obtained by differentiation of equation 11.10:

$$\frac{d\delta}{dx} = 2.32 Re_x^{-1/2} \tag{11.11}$$

This relation for the thickness of the boundary layer has been obtained on the assumption that the velocity profile can be described by a polynomial of the form of equation 11.3 and that the main stream velocity is reached at a distance δ from the surface, whereas, in fact, the stream velocity is approached asymptotically. Although equation 11.4 gives the velocity u_x accurately as a function of y, it does not provide a means of calculating accurately the distance from the surface at which u_x has a particular value when u_x is near u_s, because $\partial u_x/\partial y$ is then small. The thickness of the boundary layer as calculated is therefore a function of the particular approximate relation which is taken to represent the velocity profile. This difficulty can be overcome by introducing a new concept, the *displacement thickness* δ^*.

When a viscous fluid flows over a surface it is retarded and the overall flowrate is therefore reduced. A non-viscous fluid, however, would not be retarded and therefore a boundary layer would not form. The displacement thickness δ^* is defined as the distance the surface would have to be moved in the Y-direction in order to obtain the same rate

of flow with this non-viscous fluid as would be obtained for the viscous fluid with the
surface retained at $x = 0$.

The mass rate of flow of a frictionless fluid between $y = \delta^*$ and $y = \infty$

$$= \rho \int_{\delta^*}^{\infty} u_s \, dy$$

Mass rate of flow of the real fluid between $y = 0$ and $y = \infty$

$$= \rho \int_{0}^{\infty} u_x \, dy$$

Then, by definition of the displacement thickness:

$$\rho \int_{\delta^*}^{\infty} u_s \, dy = \rho \int_{0}^{\infty} u_x \, dy$$

or:

$$\int_{0}^{\infty} u_s \, dy - \int_{0}^{\delta^*} u_s \, dy = \int_{0}^{\infty} u_x \, dy$$

and:

$$\delta^* = \int_{0}^{\infty} \left(1 - \frac{u_x}{u_s} \right) dy \tag{11.12}$$

Using equation 11.5 to give the velocity profile:

$$\delta^* = \int_{0}^{\delta} \left(1 - \frac{3}{2} \frac{y}{\delta} + \frac{1}{2} \frac{y^3}{\delta^3} \right) dy$$

(since equation 11.5 applies only over the limits $0 < y < \delta$; outside this region $u_x = u_s$
and the integral is zero):

Thus:

$$\delta^* = \delta(1 - \tfrac{3}{4} + \tfrac{1}{8})$$

and:

$$\frac{\delta^*}{\delta} = 0.375 \tag{11.13}$$

11.3.1. Shear stress at the surface

The shear stress in the fluid at the surface is given by:

$$R_0 = -\mu \left(\frac{\partial u_x}{\partial y} \right)_{y=0}$$

$$= -\frac{3}{2} \mu \frac{u_s}{\delta} \quad \text{(from equation 11.8)}$$

$$= -\frac{3}{2} \mu u_s \frac{1}{x} \frac{1}{4.64} \sqrt{\frac{x \rho u_s}{\mu}}$$

$$= -0.323 \rho u_s^2 \sqrt{\frac{\mu}{x \rho u_s}} = -0.323 \rho u_s^2 Re_x^{-1/2}$$

The shear stress R acting on the surface itself is equal and opposite to the shear stress on the fluid at the surface; that is $R = -R_0$.

Thus:
$$\frac{R}{\rho u_s^2} = 0.323 Re_x^{-1/2} \tag{11.14}$$

Equation 11.14 gives the point values of R and $R/\rho u_s^2$ at $x = x$. In order to calculate the total frictional force acting at the surface, it is necessary to multiply the average value of R between $x = 0$ and $x = x$ by the area of the surface.

The average value of $R/\rho u_s^2$ denoted by the symbol $(R/\rho u_s^2)_m$ is then given by:

$$\left(\frac{R}{\rho u_s^2}\right)_m x = \int_0^x \frac{R}{\rho u_s^2} dx$$

$$= \int_0^x 0.323 \sqrt{\frac{\mu}{x \rho u_s}} dx \quad \text{(from equation 11.14)}$$

$$= 0.646 x \sqrt{\frac{\mu}{x \rho u_s}}$$

$$\left(\frac{R}{\rho u_s^2}\right)_m = 0.646 \sqrt{\frac{\mu}{x \rho u_s}}$$

$$= 0.646 Re_x^{-0.5} \approx 0.65 Re_x^{-0.5} \tag{11.15}$$

11.4. THE TURBULENT PORTION OF THE BOUNDARY LAYER

Equation 11.5 does not fit velocity profiles measured in a turbulent boundary layer and an alternative approach must be used. In the simplified treatment of the flow conditions within the turbulent boundary layer the existence of the buffer layer (Figure 11.7) is neglected and it is assumed that the boundary layer consists of a laminar sublayer, in which momentum transfer is by molecular motion alone, outside which there is a turbulent region in which transfer is effected entirely by eddy motion. The method is based on the assumption that the shear stress at a plane surface can be calculated from the simple power law developed by Blasius, already referred to in Chapter 3.

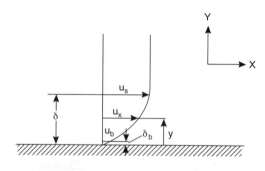

Figure 11.7.   Turbulent boundary layer

BLASIUS[3] has given an approximate expression for the shear stress at a plane smooth surface over which a fluid is flowing with a velocity u_s, for conditions where $Re_x < 10^7$. His equation is as follows:

$$\frac{R}{\rho u_s^2} = 0.0228 \left(\frac{\mu}{u_s \delta \rho}\right)^{0.25} \tag{11.16}$$

Thus, the shear stress is expressed as a function of the boundary layer thickness δ and it is therefore implicitly assumed that a certain velocity profile exists in the fluid. As a first assumption, it may be assumed that a simple power relation exists between the velocity and the distance from the surface in the boundary layer,

or:
$$\frac{u_x}{u_s} = \left(\frac{y}{\delta}\right)^f \tag{11.17}$$

Hence
$$R = 0.0228 \rho u_s^2 \left(\frac{\mu}{u_s \delta \rho}\right)^{0.25}$$

$$= 0.0228 \rho^{0.75} \mu^{0.25} \delta^{-0.25} u_s^{1.75}$$

$$= 0.0228 \rho^{0.75} \mu^{0.25} \delta^{-0.25} u_x^{1.75} \left(\frac{\delta}{y}\right)^{1.75 f} \qquad \text{(from equation 11.17)}$$

$$= 0.0228 \rho^{0.75} \mu^{0.25} u_x^{1.75} y^{-1.75 f} \delta^{1.75 f - 0.25} \tag{11.18}$$

If the velocity profile is the same for all stream velocities, the shear stress must be defined by specifying the velocity u_x at any distance y from the surface. The boundary layer thickness (determined by the velocity profile) is then no longer an independent variable so that the index of δ in equation 11.18 must be zero i.e:

$$1.75 f - 0.25 = 0$$

and:
$$f = \tfrac{1}{7}$$

Thus:
$$\frac{u_x}{u_s} = \left(\frac{y}{\delta}\right)^{1/7} \tag{11.19}$$

Equation 11.19 is sometimes known as the *Prandtl seventh power law*.
Differentiating equation 11.19 with respect to y:

$$\frac{\partial u_x}{\partial y} = \frac{1}{7} u_s \delta^{-1/7} y^{-6/7} = \frac{1}{7} \frac{u_s}{y} \left(\frac{y}{\delta}\right)^{1/7} \tag{11.20}$$

This relation is not completely satisfactory in that it gives an infinite velocity gradient at the surface (where the laminar sub-layer exists) and a finite velocity gradient at the outer edge of the boundary layer. This is in contradiction to the conditions which must exist in the stream. However, little error is introduced by using this relation for the whole of the boundary layer in the momentum equation because, firstly both the velocities and hence the momentum fluxes near the surface are very low, and secondly it gives the correct value of the velocity at the edge of the boundary layer. Accepting equation 11.19 for the limits $0 < y < \delta$, the integral in equation 11.2 becomes:

$$\int_0^l (u_s - u_x)u_x\,dy = u_s^2\left\{\int_0^\delta \left[1 - \left(\frac{y}{\delta}\right)^{1/7}\right]\left(\frac{y}{\delta}\right)^{1/7}dy\right\} + \int_\delta^l (u_s - u_s)u_s\,dy$$

$$= u_s^2\int_0^\delta \left[\left(\frac{y}{\delta}\right)^{1/7} - \left(\frac{y}{\delta}\right)^{2/7}\right]dy$$

$$= u_s^2\delta\left(\frac{7}{8} - \frac{7}{9}\right)$$

$$= \frac{7}{72}u_s^2\delta \qquad\qquad (11.21)$$

From the Blasius equation:

$$-R_0 = R = 0.0228\rho u_s^2\left(\frac{\mu}{u_s\delta\rho}\right)^{1/4} \qquad \text{(from equation 11.16)}$$

Substituting from equations 11.16 and 11.21 in equation 11.2:

$$\rho\frac{\partial}{\partial x}\left[\frac{7}{72}u_s^2\delta\right] = 0.0228\rho u_s^2\left(\frac{\mu}{u_s\delta\rho}\right)^{1/4}$$

$$\delta^{1/4}\,d\delta = 0.235\left(\frac{\mu}{u_s\rho}\right)^{1/4}dx$$

and:

$$\frac{4}{5}\delta^{5/4} = 0.235x\left(\frac{\mu}{u_s\rho}\right)^{1/4} + \text{constant}$$

Putting the constant equal to zero, implies that $\delta = 0$ when $x = 0$, i.e. that the turbulent boundary layer extends to the leading edge of the surface. An error is introduced by this assumption, but it is found to be small except where the surface is only slightly longer than the critical distance x_c for the laminar-turbulent transition.

Thus:
$$\delta = 0.376x^{0.8}\left(\frac{\mu}{u_s\rho}\right)^{0.2} \qquad\qquad (11.22)$$

$$= 0.376x\left(\frac{\mu}{u_s\rho x}\right)^{0.2}$$

or:
$$\frac{\delta}{x} = 0.376Re_x^{-0.2} \qquad\qquad (11.23)$$

The rate of thickening of the boundary layer is obtained by differentiation of equation 11.23:

$$\frac{d\delta}{dx} = 0.301Re_x^{-0.2} \qquad\qquad (11.24)$$

The displacement thickness δ^* is given by equation 11.12:

$$\delta^* = \int_0^\infty \left(1 - \frac{u_x}{u_s}\right)dy \qquad\qquad \text{(equation 11.12)}$$

$$= \int_0^\delta \left(1 - \left(\frac{y}{\delta}\right)^{1/7}\right) dy \quad \left(\text{since } 1 - \frac{u_x}{u_s} = 0 \text{ when } y > \delta\right)$$

$$= \tfrac{1}{8}\delta \tag{11.25}$$

As already explained δ^* is independent of the particular approximation used for the velocity profile.

It is of interest to compare the rates of thickening of the streamline and turbulent boundary layers at the transition point. Taking a typical value of $Re_{xc} = 10^5$:

For the streamline boundary layer, from equation 11.11, $\dfrac{d\delta}{dx} = 0.0073$

For the turbulent boundary layer, from equation 11.24, $\dfrac{d\delta}{dx} = 0.0301$

Thus the turbulent boundary layer is thickening at about 4 times the rate of the streamline boundary layer at the transition point.

11.4.1. The laminar sub-layer

Supposing that at a distance x from the leading edge the laminar sub-layer is of thickness δ_b and that the total thickness of the boundary layer is δ, the properties of the laminar sub-layer can be found by equating the shear stress at the surface as given by the Blasius equation (11.16) to that obtained from the velocity gradient near the surface.

It has already been explained that the shear stress and hence the velocity gradient are almost constant near the surface. Since the laminar sub-layer is very thin, the velocity gradient within it can therefore be taken as constant.

Thus the shear stress in the fluid at the surface

$$R_0 = -\mu \left(\frac{\partial u_x}{\partial y}\right)_{y=0} = -\mu \frac{u_x}{y}, \quad \text{where } y < \delta_b$$

Equating this to the value obtained from equation 11.16:

$$0.0228\rho u_s^2 \left(\frac{\mu}{u_s \delta \rho}\right)^{1/4} = \mu \frac{u_x}{y}$$

and:

$$u_x = 0.0228\rho u_s^2 \frac{1}{\mu} \left(\frac{\mu}{u_s \delta \rho}\right)^{1/4} y$$

If the velocity at the edge of the laminar sub-layer is u_b, i.e. if $u_x = u_b$ when $y = \delta_b$:

$$u_b = 0.0228\rho u_s^2 \frac{1}{\mu} \left(\frac{\mu}{u_s \delta \rho}\right)^{1/4} \delta_b$$

$$= 0.0228 \frac{\rho u_s^2}{\mu} \frac{\mu}{u_s \delta \rho} \delta_b \left(\frac{\mu}{u_s \delta \rho}\right)^{-3/4}$$

Thus:

$$\frac{\delta_b}{\delta} = \frac{1}{0.0228} \left(\frac{u_b}{u_s}\right) \left(\frac{\mu}{u_s \delta \rho}\right)^{3/4} \tag{11.26}$$

Now the velocity at the inner edge of the turbulent region must also be given by the equation for the velocity distribution in the turbulent region.

Hence:
$$\left(\frac{\delta_b}{\delta}\right)^{1/7} = \frac{u_b}{u_s} \qquad \text{(from equation 11.19)}$$

Thus:
$$\left(\frac{u_b}{u_s}\right)^7 = \frac{1}{0.0228}\left(\frac{u_b}{u_s}\right)\left(\frac{\mu}{u_s\delta\rho}\right)^{3/4} \qquad \text{(from equation 11.26)}$$

or:
$$\frac{u_b}{u_s} = 1.87\left(\frac{\mu}{u_s\delta\rho}\right)^{1/8}$$

$$= 1.87 Re_\delta^{-1/8} \qquad (11.27)$$

Since:
$$\delta = 0.376x^{0.8}\left(\frac{\mu}{u_s\rho}\right)^{0.2}, \qquad \text{(equation 11.22)}$$

$$\frac{u_b}{u_s} = 1.87\left(\frac{u_s\rho}{\mu}0.376\frac{x^{0.8}\mu^{0.2}}{u_s^{0.2}\rho^{0.2}}\right)^{-1/8}$$

$$= \frac{1.87}{0.376^{1/8}}\left(\frac{u_s^{0.8}x^{0.8}\rho^{0.8}}{\mu^{0.8}}\right)^{-1/8}$$

$$= 2.11 Re_x^{-0.1} \approx 2.1 Re_x^{-0.1} \qquad (11.28)$$

The thickness of the laminar sub-layer is given by:

$$\frac{\delta_b}{\delta} = \left(\frac{u_b}{u_s}\right)^7 = \frac{190}{Re_x^{0.7}} \qquad \text{(from equations 11.19 and 11.28)}$$

or:
$$\frac{\delta_b}{x} = \frac{190}{Re_x^{0.7}}\frac{0.376}{Re_x^{0.2}} \qquad \text{(from equation 11.23)}$$

$$= 71.5 Re_x^{-0.9} \qquad (11.29)$$

Thus $\delta_b \propto x^{0.1}$; that is δ_b it increases very slowly as x increases. Further, $\delta_b \propto u_s^{-0.9}$ and therefore decreases rapidly as the velocity is increased, and heat and mass transfer coefficients are therefore considerably influenced by the velocity.

The shear stress at the surface, at a distance x from the leading edge, is given by:

$$R_0 = -\mu\frac{u_b}{\delta_b}$$

Then, since $R_0 = -R$:

$$R = \mu\, 2.11\, u_s\, Re_x^{-0.1}\frac{1}{x}\frac{1}{71.5}Re_x^{0.9} \qquad \text{(from equations 11.28 and 11.29)}$$

$$= 0.0296 Re_x^{0.8}\frac{\mu u_s}{x}$$

$$= 0.0296\rho u_s^2 Re_x^{-0.2} \approx 0.03\rho u_s^2 Re_x^{-0.2} \qquad (11.30)$$

or:
$$\frac{R}{\rho u_s^2} = 0.0296 Re_x^{-0.2} \tag{11.31}$$

or approximately:
$$\frac{R}{\rho u_s^2} = 0.03 Re_x^{-0.2} \tag{11.32}$$

The mean value of $R/\rho u_s^2$ over the range $x = 0$ to $x = x$ is given by:

$$\left(\frac{R}{\rho u_s^2}\right)_m x = \int_0^x \left(\frac{R}{\rho u_s^2}\right) dx$$

$$= \int_0^x 0.0296 \left(\frac{\mu}{u_s x \rho}\right)^{0.2} dx$$

$$= 0.0296 \left(\frac{\mu}{u_s x \rho}\right)^{0.2} \frac{x}{0.8}$$

or:
$$\left(\frac{R}{\rho u_s^2}\right)_m = 0.037 Re_x^{-0.2} \tag{11.33}$$

The total shear force acting on the surface is found by adding the forces acting in the streamline ($x < x_c$) and turbulent ($x > x_c$) regions. This can be done provided the critical value Re_{x_c}, is known.

In the streamline region:
$$\left(\frac{R}{\rho u_s^2}\right)_m = 0.646 Re_x^{-0.5} \qquad \text{(equation 11.15)}$$

In the turbulent region:
$$\left(\frac{R}{\rho u_s^2}\right)_m = 0.037 Re_x^{-0.2} \qquad \text{(equation 11.33)}$$

In calculating the mean value of $(R/\rho u_s^2)_m$ in the turbulent region, it was assumed that the turbulent boundary layer extended to the leading edge. A more accurate value for the mean value of $(R/\rho u_s^2)_m$ over the whole surface can be obtained by using the expression for streamline conditions over the range from $x = 0$ to $x = x_c$ (where x_c is the critical distance from the leading edge) and the expression for turbulent conditions in the range $x = x_c$ to $x = x$.

$$\left(\frac{R}{\rho u_s^2}\right)_m = \frac{1}{x}(0.646 Re_{x_c}^{-0.5} x_c + 0.037 Re_x^{-0.2} x - 0.037 Re_{x_c}^{-0.2} x_c)$$

$$= 0.646 Re_{x_c}^{-0.5} \frac{Re_{x_c}}{Re_x} + 0.037 Re_x^{-0.2} - 0.037 Re_{x_c}^{-0.2} \frac{Re_{x_c}}{Re_x}$$

$$= 0.037 Re_x^{-0.2} + Re_x^{-1}(0.646 Re_{x_c}^{0.5} - 0.037 Re_{x_c}^{0.8}) \tag{11.34}$$

Example 11.1

Water flows at a velocity of 1 m/s over a plane surface 0.6 m wide and 1 m long. Calculate the total drag force acting on the surface if the transition from streamline to turbulent flow in the boundary layer occurs when the Reynolds group $Re_{x_c} = 10^5$.

Solution

Taking $\mu = 1$ mN s/m$^2 = 10^{-3}$ Ns/m^2,
at the far end of the surface, $Re_x = 1 \times 1 \times 10^3/10^{-3} = 10^6$
Mean value of $R/\rho u_s^2$ from equation 11.34

$$= 0.037(10^6)^{-0.2} + (10^6)^{-1}[0.646(10^5)^{0.5} - 0.037(10^5)^{0.8}]$$

$$= 0.00214$$

$$\text{Total drag force} = \frac{R}{\rho u_s^2}(\rho u_s^2) \times (\text{area of surface})$$

$$= 0.00214 \times 1000 \times 1^2 \times 1 \times 0.6$$

$$= \underline{\underline{1.28 \text{ N}}}$$

Example 11.2

Calculate the thickness of the boundary layer at a distance of 150 mm from the leading edge of a surface over which oil, of viscosity 0.05 N s/m^2 and density 1000 kg/m^3 flows with a velocity of 0.3 m/s. What is the displacement thickness of the boundary layer?

Solution

$Re_x = 0.150 \times 0.3 \times 1000/0.05 = 900$

$$\frac{\delta}{x} = \frac{4.64}{Re_x^{0.5}} \quad \text{for streamline flow (from equation 11.10)}$$

$$= \frac{4.64}{900^{0.5}} = 0.1545$$

Hence: $\delta = 0.1545 \times 0.150 = 0.0232$ m $= \underline{\underline{23.2 \text{ mm}}}$

Displacement thickness $= 0.375 \times 23.2 = \underline{\underline{8.7 \text{ mm}}}$ (from equation 11.13)

11.5. BOUNDARY LAYER THEORY APPLIED TO PIPE FLOW

11.5.1. Entry conditions

When a fluid flowing with a uniform velocity enters a pipe, a boundary layer forms at the walls and gradually thickens as the distance from the entry point increases. Since the fluid in the boundary layer is retarded and the total flow remains constant, the fluid in the central stream will be accelerated. At a certain distance from the inlet, the boundary layers, which have formed in contact with the walls, join at the axis of the pipe, and, from that point onwards, occupy the whole cross-section and consequently remain of constant thickness. *Fully developed flow* then exists. If the boundary layers are still streamline when fully developed flow commences, the flow in the pipe remains streamline. On the other hand, if the boundary layers are already turbulent, turbulent flow will persist (Figure 11.8).

An approximate experimental expression for the inlet length L_e for laminar flow is:

$$\frac{L_e}{d} = 0.0575 Re \qquad (11.35)$$

Figure 11.8.   Conditions at entry to pipe

where d is the diameter of the pipe and Re is the Reynolds group with respect to pipe diameter, and based on the mean velocity of flow in the pipe. This expression is only approximate, and is inaccurate for Reynolds numbers in the region of 2500 because the boundary layer thickness increases very rapidly in this region. An average value of L_e at a Reynolds number of 2500 is about $100d$. The inlet length is somewhat arbitrary as steady conditions in the pipe are approached asymptotically — the boundary layer thickness being a function of the assumed velocity profile.

At the inlet to the pipe the velocity across the whole section is constant. The velocity at the pipe axis will progressively increase in the direction of flow and reach a maximum value when the boundary layers join. Beyond this point the velocity profile, and the velocity at the axis, will not change. Since the fluid at the axis has been accelerated, its kinetic energy per unit mass will increase and therefore there must be a corresponding fall in its pressure energy.

Under streamline conditions, the velocity at the axis u_s will increase from a value u at the inlet to a value $2u$ where fully developed flow exists (Figure 11.9) (see equation 3.36), where u is the mean velocity of flow in the pipe which is half of the axial velocity.

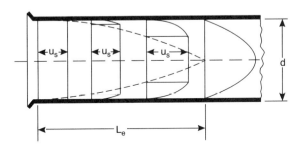

Figure 11.9.   Development of laminar velocity profile at entry to pipe

Thus the kinetic energy per unit mass of the fluid at the axis inlet $= \frac{1}{2}u^2$

Corresponding kinetic energy at the end of the inlet length $= \frac{1}{2}(2u)^2 = 2u^2$.

The increase in the kinetic energy per unit mass $= \frac{3}{2}u^2$.

Thus the fall in pressure due to the increase of velocity of the fluid $= \frac{3}{2}\rho u^2$.

If the flow in the pipe is turbulent, the velocity at the axis increases from u to only about $u/0.817$ (see equation 3.60).

Under these conditions, the fall in pressure

$$= \frac{1}{2}\rho u^2 \left(\frac{1}{0.817^2} - 1 \right)$$

$$\approx \tfrac{1}{4}\rho u^2 \tag{11.36}$$

If the fluid enters the pipe from a duct of larger cross-section, the existence of a radial velocity component will give rise to the formation of a *vena contracta* near the entry to the pipe but this has been neglected here.

11.5.2. Application of boundary layer theory

The velocity distribution and frictional resistance have already been calculated from purely theoretical considerations for the streamline flow of a fluid in a pipe. The boundary layer theory can now be applied in order to calculate, approximately, the conditions when the fluid is turbulent. For this purpose it is assumed that the boundary layer expressions can be applied to flow over a cylindrical surface and that the flow conditions in the region of fully developed flow are the same as those when the boundary layers first join. The thickness of the boundary layer is thus taken to be equal to the radius of the pipe and the velocity at the outer edge of the boundary layer is assumed to be the velocity at the axis. Such assumptions are valid very close to the walls but significant errors will arise near the centre of the pipe.

The velocity of the fluid can be assumed to obey the Prandtl one-seventh power law (equation 11.19). If the boundary layer thickness δ is replaced by the pipe radius r this becomes:

$$\frac{u_x}{u_s} = \left(\frac{y}{r} \right)^{1/7} \tag{11.37}$$

The relation between the mean velocity and the velocity at the axis has already been derived using this expression in Chapter 3. The mean velocity u was found to be 0.82 times the velocity u_s at the axis, but in this calculation the thickness of the laminar sub-layer was neglected and the Prandtl velocity distribution assumed to apply over the whole cross-section. The result therefore is strictly applicable only at very high Reynolds numbers where the thickness of the laminar sub-layer is very small. At lower Reynolds numbers the mean velocity will be rather less than 0.82 times the velocity at the axis.

The expressions for the shear stress at the walls, the thickness of the laminar sub-layer, and the velocity at the outer edge of the laminar sub-layer can be applied to the turbulent flow of a fluid in a pipe. It is convenient to express these relations in terms of the mean velocity in the pipe, the pipe diameter, and the Reynolds group with respect to the mean velocity and diameter.

The shear stress at the walls is given by the Blasius equation (11.16):

$$\frac{R}{\rho u_s^2} = 0.0228 \left(\frac{\mu}{u_s r \rho} \right)^{1/4}$$

Writing $u = 0.817u_s$ and $d = 2r$:

$$\frac{R}{\rho u^2} = 0.0386 \left(\frac{\mu}{ud\rho}\right)^{1/4} = 0.0386 Re^{-1/4} \tag{11.38}$$

This equation is more usually written:

$$\frac{R}{\rho u^2} = 0.0396 Re^{-1/4} \quad \text{(See equation 3.10)} \tag{11.39}$$

The discrepancy between the coefficients in equations 11.38 and 11.39 is attributable to the fact that the effect of the curvature of the pipe wall has not been taken into account in applying the equation for flow over a plane surface to flow through a pipe.

Equation 11.39 is applicable for Reynolds numbers up to 10^5.

The velocity at the edge of the laminar sub-layer is given by:

$$\frac{u_b}{u_s} = 1.87 \left(\frac{\mu}{u_s r\rho}\right)^{1/8} \quad \text{(equation 11.27)}$$

which becomes:

$$\frac{u_b}{u} = 2.49 \left(\frac{\mu}{ud\rho}\right)^{1/8}$$

$$= 2.49 Re^{-1/8} \tag{11.40}$$

and:

$$\frac{u_b}{u_s} = 2.0 Re^{-1/8} \tag{11.41}$$

The thickness of the laminar sub-layer is given by:

$$\frac{\delta_b}{r} = \left(\frac{u_b}{u_s}\right)^7 \quad \text{(from equation 11.19)}$$

$$= (1.87)^7 \left(\frac{\mu}{u_s r\rho}\right)^{7/8} \quad \text{(from equation 11.27)}$$

Thus:

$$\frac{\delta_b}{d} = 62 \left(\frac{\mu}{ud\rho}\right)^{7/8}$$

$$= 62 Re^{-7/8} \tag{11.42}$$

The thickness of the laminar sub-layer is therefore nearly inversely proportional to the Reynolds number and hence to the velocity.

Example 11.3

Calculate the thickness of the laminar sub-layer when benzene flows through a pipe 50 mm in diameter at 2000 cm^3/s. What is the velocity of the benzene at the edge of the laminar sub-layer? Assume that fully developed flow exists within the pipe and for benzene, $\rho = 870$ kg/m^3 and $\mu = 0.7$ mN s/m^2.

Solution

$$2000 \text{ cm}^3/\text{s} = (2000 \times 10^{-6} \times 870)$$

$$= 1.74 \text{ kg/s}$$

$$\text{Reynolds number} = \frac{4G}{\mu\pi D} = \frac{4 \times 1.74}{0.7 \times 10^{-3}\pi \times 0.050} \qquad \text{(equation 4.52)}$$

$$= 63,290$$

From equation 11.42:

$$\frac{\delta_b}{d} = 62Re^{-7/8}$$

$$\delta_b = \frac{62 \times 0.050}{63,290^{7/8}}$$

$$= 1.95 \times 10^{-4} \text{ m}$$

or:

$$\underline{0.195 \text{ mm}}$$

The mean velocity

$$= \frac{1.74}{870 \times (\pi/4)0.050^2}$$

$$= 1.018 \text{ m/s}$$

From equation 11.40:

$$\frac{u_b}{u} = \frac{2.49}{Re^{1/8}}$$

$$u_b = \frac{2.49 \times 1.018}{63,290^{1/8}}$$

$$= \underline{\underline{0.637 \text{ m/s}}}$$

11.6. THE BOUNDARY LAYER FOR HEAT TRANSFER

If fluid flows over a surface which is at a different temperature, heat transfer will occur and a temperature profile will be established in the vicinity of the surface. A number of possible conditions may be considered. At the outset, the heat transfer rate may be sufficient to change the temperature of the fluid stream significantly or it may remain at a substantially constant temperature. Furthermore, a variety of conditions may apply at the surface. Thus the surface may be maintained at a constant temperature, particularly if it is in good thermal conduct with a heat source or sink of high thermal capacity. Alternatively, the heat flux at the surface may be maintained constant, or conditions may be intermediate between the constant temperature and constant heat flux condition. In general, there is likely to be a far greater possible variety of conditions as compared with the momentum transfer problem previously considered. Temperature gradients are likely to be highest in the vicinity of the surface and it is useful to develop the concept of a *thermal boundary layer*, analogous to the velocity boundary layer already considered, within which the whole of the temperature gradient may be regarded as existing.

Thus, a velocity boundary layer and a thermal boundary layer may develop simultaneously. If the physical properties of the fluid do not change significantly over the temperature range to which the fluid is subjected, the velocity boundary layer will not be affected by the heat transfer process. If physical properties are altered, there will be an interactive effect between the momentum and heat transfer processes, leading to a comparatively complex situation in which numerical methods of solution will be necessary.

In general, the thermal boundary layer will not correspond with the velocity boundary layer. In the following treatment, the simplest non-interacting case will be considered with physical properties assumed to be constant. The stream temperature will be taken as constant (θ_s). In the first case which will be considered the wall temperature will also be taken as a constant, and then by choosing the temperature scale so that the wall temperature is zero, the boundary conditions are similar to those for the momentum transfer problem.

It will be shown, however, that the momentum and thermal boundary layers coincide only if the Prandtl number is unity, implying equal values for the kinematic viscosity (μ/ρ) and the thermal diffusivity $(D_H = k/C_p\rho)$.

The condition of constant heat flux at the surface, as opposed to constant surface temperature, is then considered in a later section.

11.6.1. The energy balance

The procedure is similar to that adopted previously. A heat balance, as opposed to a momentum balance, is taken over an element which extends beyond the limits of both the velocity and thermal boundary layers. In this way, any fluid entering or leaving the element through the face distant from the surface is at the stream velocity u_s and stream temperature θ_s. A heat balance is made therefore on the element shown in Figure 11.10 in which the length l is greater than the velocity boundary layer thickness δ and the thermal boundary layer thickness δ_t.

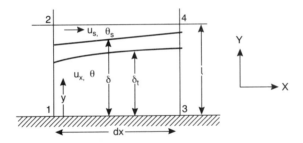

Figure 11.10.   The thermal boundary layer

The heat transferred through an element of width w of the plane $1-2$, of thickness dy at a distance y from the surface

$$= C_p \rho\, \theta u_x w\, dy$$

where C_p is the specific heat capacity of the fluid at constant pressure, ρ the density of the fluid, and θ and u_x are the temperature and velocity at a distance y from the surface.
The total rate of transfer of heat through the plane, $1-2$

$$= C_p \rho w \int_0^l \theta u_x\, dy$$

assuming that the physical properties of the fluid are independent of temperature. In the distance dx this heat flow changes by an amount:

$$C_p \rho w \frac{\partial}{\partial x} \left(\int_0^l \theta\, u_x\, dy \right) dx$$

It has already been shown (Section 11.2) that there is a mass rate of flow of fluid through plane, 2–4, into the element equal to $\rho w (\partial/\partial x)(\int_0^l u_x \, dy) \, dx$.

Since the plane, 2–4, lies outside the boundary layers, the heat introduced into the element through the plane as a result of this flow is:

$$C_p \rho \theta_s w \frac{\partial}{\partial x} \left(\int_0^l u_x \, dy \right) dx$$

where θ_s is the temperature outside the thermal boundary layer.

The heat transferred by thermal conduction into the element through plane, 1–3,

$$= -k w \, dx \left(\frac{\partial \theta}{\partial y} \right)_{y=0}$$

If the temperature θ_s of the main stream is unchanged, a heat balance on the element gives:

$$C_p \rho w \left(\frac{\partial}{\partial x} \int_0^l \theta u_x \, dy \right) dx = C_p \rho \theta_s w \left(\frac{\partial}{\partial x} \int_0^l u_x \, dy \right) dx - k \left(\frac{\partial \theta}{\partial y} \right)_{y=0} w \, dx$$

or:

$$\frac{\partial}{\partial x} \int_0^l u_x (\theta_s - \theta) \, dy = D_H \left(\frac{\partial \theta}{\partial y} \right)_{y=0} \tag{11.43}$$

where $D_H \ (= k/C_p \rho)$ is the thermal diffusivity of the fluid.

The relations between u_x and y have already been obtained for both streamline and turbulent flow. A relation between θ and y for streamline conditions in the boundary layer will now be derived, but it is not possible to define the conditions in the turbulent boundary layer sufficiently precisely to derive a similar expression for that case.

11.6.2. Heat transfer for streamline flow over a plane surface — constant surface temperature

The flow of fluid over a plane surface, which is heated at distances greater than x_0 from the leading edge, is now considered. The velocity boundary layer will start at the leading edge and the thermal boundary layer at a distance x_0 from it. If the temperature of the heated portion of the plate remains constant, this may be taken as the datum temperature (Figure 11.11). It is assumed that the temperature at a distance y from the surface may be represented by a polynomial of the form:

$$\theta = a_0 y + b_0 y^2 + c_0 y^3 \tag{11.44}$$

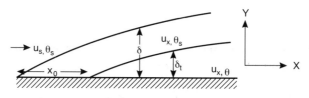

Figure 11.11.   Thermal boundary layer — streamline flow

If the fluid layer in contact with the surface is assumed to be at rest, any heat flow in the vicinity of the surface must be by pure thermal conduction. Thus the heat transferred per unit area and unit time q_0 is given by:

$$q_0 = -k \left(\frac{\partial \theta}{\partial y} \right)_{y=0}$$

If the temperature of the fluid element in contact with the wall is to remain constant, the heat transfer rate into and out of the element must be the same, or:

$$\left(\frac{\partial \theta}{\partial y} \right)_{y=0} = \text{a constant} \quad \text{and} \quad \left(\frac{\partial^2 \theta}{\partial y^2} \right)_{y=0} = 0$$

At the outer edge of the thermal boundary layer, the temperature is θ_s and the temperature gradient $(\partial \theta / \partial y) = 0$ if there is to be no discontinuity in the temperature profile.

Thus the conditions for the thermal boundary layer, with respect to temperature, are the same as those for the velocity boundary layer with respect to velocity. Then, if the thickness of the thermal boundary layer is δ_t, the temperature distribution is given by:

$$\frac{\theta}{\theta_s} = \frac{3}{2} \left(\frac{y}{\delta_t} \right) - \frac{1}{2} \left(\frac{y}{\delta_t} \right)^3 \tag{11.45}$$

(which may be compared with equation 11.5)

and:

$$\left(\frac{\partial \theta}{\partial y} \right)_{y=0} = \frac{3 \theta_s}{2 \delta_t} \tag{11.46}$$

It will now be assumed that the velocity boundary layer is everywhere thicker than the thermal boundary layer, so that $\delta > \delta_t$ (Figure 11.11). Thus the velocity distribution everywhere within the thermal boundary layer is given by equation 11.5. The implications of this assumption will be discussed later.

The integral in equation 11.43 clearly has a finite value within the thermal boundary layer but is zero outside it. When the expression for the temperature distribution in the boundary layer is inserted therefore, the upper limit of integration must be altered from l to δ_t.

Then: $\displaystyle \int_0^l (\theta_s - \theta) u_x \, dy = \theta_s u_s \int_0^{\delta_t} \left[1 - \frac{3}{2} \frac{y}{\delta_t} + \frac{1}{2} \left(\frac{y}{\delta_t} \right)^3 \right] \left[\frac{3y}{2\delta} - \frac{1}{2} \left(\frac{y}{\delta} \right)^3 \right] dy$

$$= \theta_s u_s \left[\frac{3}{4} \frac{\delta_t^2}{\delta} - \frac{3}{4} \frac{\delta_t^2}{\delta} - \frac{1}{8} \frac{\delta_t^4}{\delta^3} + \frac{3}{20} \left(\frac{\delta_t^2}{\delta} + \frac{\delta_t^4}{\delta^3} \right) - \frac{1}{28} \frac{\delta_t^4}{\delta^3} \right]$$

$$= \theta_s u_s \left(\frac{3}{20} \frac{\delta_t^2}{\delta} - \frac{3}{280} \frac{\delta_t^4}{\delta^3} \right)$$

$$= \theta_s u_s \delta \left(\frac{3}{20} \sigma^2 - \frac{3}{280} \sigma^4 \right) \tag{11.47}$$

where $\sigma = \delta_t / \delta$.

Since $\delta_t < \delta$, the second term is small compared with the first, then:

$$\int_0^l (\theta_s - \theta)u_x \, dy = \frac{3}{20}\theta_s u_s \delta\sigma^2 \qquad (11.48)$$

Substituting from equations 11.46 and 11.48 in equation 11.43:

$$\frac{\partial}{\partial x}\left(\frac{3}{20}\theta_s u_s \delta\sigma^2\right) = D_H \frac{3\theta_s}{2\delta_t} = D_H \frac{3\theta_s}{2\delta\sigma}$$

$$\therefore \qquad \frac{1}{10}u_s \delta\sigma \frac{\partial}{\partial x}(\delta\sigma^2) = D_H$$

$$\therefore \qquad \frac{1}{10}u_s\left(\delta\sigma^3\frac{\partial\delta}{\partial x} + 2\delta^2\sigma^2\frac{\partial\sigma}{\partial x}\right) = D_H \qquad (11.49)$$

It has already been shown that:

$$\delta^2 = \frac{280\mu x}{13\rho u_s} = 21.5\frac{\mu x}{\rho u_s} \qquad \text{(equation 11.9)}$$

and hence:

$$\delta\frac{\partial\delta}{\partial x} = \frac{140}{13}\frac{\mu}{\rho u_s}$$

Substituting in equation 11.49:

$$\frac{u_s}{10}\frac{\mu}{\rho u_s}\left(\frac{140}{13}\sigma^3 + \frac{560}{13}x\sigma^2\frac{\partial\sigma}{\partial x}\right) = D_H$$

$$\therefore \qquad \frac{14}{13}\frac{\mu}{\rho D_H}\left(\sigma^3 + 4x\sigma^2\frac{\partial\sigma}{\partial x}\right) = 1$$

$$\therefore \qquad \sigma^3 + \frac{4x}{3}\frac{\partial\sigma^3}{\partial x} = \frac{13}{14}Pr^{-1}$$

(where the Prandtl number $Pr = C_p\mu/k = \mu/\rho D_H$)

$$\therefore \qquad \frac{3}{4}x^{-1}\sigma^3 + \frac{\partial\sigma^3}{\partial x} = \frac{13}{14}Pr^{-1}\frac{3}{4}x^{-1}$$

$$\therefore \qquad \frac{3}{4}x^{-1/4}\sigma^3 + x^{3/4}\frac{\partial\sigma^3}{\partial x} = \frac{13}{14}Pr^{-1}\frac{3}{4}x^{-1/4}$$

Integrating:

$$x^{3/4}\sigma^3 = \frac{13}{14}Pr^{-1}x^{3/4} + \text{constant}$$

or:

$$\sigma^3 = \frac{13}{14}Pr^{-1} + \text{constant } x^{-3/4}$$

When $x = x_0$

$$\sigma = 0.$$

so that:

$$\text{constant} = -\frac{13}{14}Pr^{-1}x_0^{3/4}$$

Hence:

$$\sigma^3 = \frac{13}{14}Pr^{-1}\left[1 - \left(\frac{x_0}{x}\right)^{3/4}\right]$$

and:
$$\sigma = 0.976 Pr^{-1/3} \left[1 - \left(\frac{x_0}{x} \right)^{3/4} \right]^{1/3} \tag{11.50}$$

If the whole length of the plate is heated, $x_0 = 0$ and:
$$\sigma = 0.976 Pr^{-1/3} \tag{11.51}$$

In the above derivation, it has been assumed that $\sigma < 1$.

For all liquids other than molten metals, $Pr > 1$ and hence $\sigma < 1$ (from equation 11.51). For gases, $Pr \neq 0.6$, so that $\sigma \neq 1.18$.

Thus only a small error is introduced when the above expression is applied to gases. The only serious deviations occur for molten metals, which have very low Prandtl numbers.

If h is the heat transfer coefficient:
$$q_0 = -h\theta_s$$

and:
$$-h\theta_s = -k \left(\frac{\partial \theta}{\partial y} \right)_{y=0}$$

or, from equation 11.46:
$$h = \frac{k}{\theta_s} \frac{3}{2} \frac{\theta_s}{\delta_t}$$
$$= \frac{3}{2} \frac{k}{\delta_t} = \frac{3}{2} \frac{k}{\delta\sigma} \tag{11.52}$$

Substituting for δ from equation 11.10, and σ from equation 11.50:
$$h = \frac{3k}{2} \frac{1}{4.64} \sqrt{\frac{\rho u_s}{\mu x}} \frac{Pr^{1/3}}{0.976[1 - (x_0/x)^{3/4}]^{1/3}}$$

or:
$$\frac{hx}{k} = 0.332 Pr^{1/3} Re_x^{1/2} \frac{1}{[1 - (x_0/x)^{3/4}]^{1/3}} \tag{11.53}$$

If the surface is heated over its entire length, so that $x_0 = 0$, then:
$$Nu_x = \frac{hx}{k} = 0.332 Pr^{1/3} Re_x^{1/2} \tag{11.54}$$

It is seen from equation 11.54 that the heat transfer coefficient theoretically has an infinite value at the leading edge, where the thickness of the thermal boundary layer is zero, and that it decreases progressively as the boundary layer thickens. Equation 11.54 gives the point value of the heat transfer coefficient at a distance x from the leading edge. The mean value between $x = 0$ and $x = x$ is given by:
$$h_m = x^{-1} \int_0^x h \, dx$$
$$= x^{-1} \int_0^x \psi x^{-1/2} \, dx$$

(where ψ is not a function of x).
$$= x^{-1} \left[2\psi x^{1/2} \right]_0^x = 2h$$

Thus the mean value of the heat transfer coefficient between $x = 0$ and $x = x$ is equal to twice the point value at $x = x$. The mean value of the Nusselt group is given by:

$$(Nu_x)_m = 0.662Pr^{1/3}Re_x^{1/2} \tag{11.55}$$

11.6.3. Heat transfer for streamline flow over a plane surface — constant surface heat flux

The second important case which will be considered is where the heat flux at the surface, as opposed to the temperature, is constant. An example of this is where the surface is electrically heated. In this case, the surface temperature θ_0 will increase in the direction of flow (x-direction) as the value of the heat transfer coefficient decreases. The equation for the temperature profile in the boundary layer becomes:

$$\frac{\theta - \theta_0}{\theta_s - \theta_0} = \frac{3}{2}\left(\frac{y}{\delta_t}\right) - \frac{1}{2}\left(\frac{y}{\delta_t}\right)^3 \tag{11.56}$$

(which may be compared with equation 11.45) and the temperature gradient at the walls is given by:

$$\left(\frac{\partial\theta}{\partial y}\right)_{y=0} = \frac{3(\theta_s - \theta_0)}{2\delta_t} \tag{11.57}$$

(which may be compared with equation 11.46)

The value of the integral in the energy balance (equation 11.43) is again given by equation 11.48 [substituting $(\theta_s - \theta_0)$ for θ_s]. Furthermore, the heat flux q_0 at the surface is constant and the right hand side of equation 11.43 may be expressed as $(-q_0/C_p\rho)$. Thus, for constant heat flux, equation 11.43 becomes:

$$\frac{\partial}{\partial x}\left(\frac{3}{20}(\theta_s - \theta_0)u_s\delta\sigma^2\right) = -\frac{q_0}{C_p\rho}$$

Integrating:

$$\frac{3}{20}(\theta_s - \theta_0)u_s\delta\sigma^2 = -\frac{q_0 x}{C_p\rho} \tag{11.58}$$

Because $\delta_t = 0$ when $x = 0$, the integration constant in equation 11.58 is zero. Now, from equation 11.57:

$$q_0 = -\frac{3(\theta_s - \theta_0)k}{2\delta_t}$$

Thus:

$$\frac{3}{20}(\theta_s - \theta_0)u_s\delta\sigma^2 = \frac{3}{2}(\theta_s - \theta_0)\frac{k}{\delta_t}\frac{x}{C_p\rho}$$

or:

$$\sigma^3 = 10\frac{x^2}{\delta^2}\frac{\mu}{u_s x\rho}\frac{k}{C_p\mu} \tag{11.59}$$

Now:

$$\frac{\delta^2}{x^2} = \frac{280}{13}\frac{\mu}{u_s x\rho} \qquad \text{(equation 11.9)}$$

Substituting from equation 11.9 into equation 11.59:

$$\sigma^3 = \frac{13}{28}Pr^{-1}$$

or:
$$\sigma = 0.774Pr^{-1/3} \tag{11.60}$$

Equation 11.60 may be compared with equation 11.51 for constant surface temperature. Now:

$$\frac{\delta_t}{x} = \sigma\frac{\delta}{x} = (0.774Pr^{-1/3})\left(\frac{280}{13}\frac{\mu}{u_s x\rho}\right)^{1/2}$$

$$= 3.59Pr^{-1/3}Re_x^{-1/2} \tag{11.61}$$

The heat transfer coefficient at a distance x from the leading edge of the surface is given by:

$$h = \frac{-q_0}{\theta_s - \theta_0} = \frac{k\,(\partial\theta/\partial y)_{y=0}}{\theta_s - \theta_0}$$

or:
$$h = \frac{k(\theta_s - \theta_0)/\delta_t}{\theta_s - \theta_0}\frac{3}{2}$$

$$= \frac{3}{2}\frac{k}{\delta_t} \tag{11.62}$$

$$= \frac{3}{2}k\frac{1}{x}\frac{1}{3.59}Pr^{1/3}Re_x^{1/2}$$

Thus:
$$\frac{hx}{k} = Nu_x = 0.417Pr^{1/3}Re_x^{1/2} \tag{11.63}$$

By comparison with equation 11.54, it is seen that the local Nusselt number and the heat transfer coefficient are both some 26 per cent higher for a constant surface heat flux as compared with a constant surface temperature.

Equation 11.63 is based on the energy integral equation which neglects the effect of the heat transference by the convective effect of the velocity components perpendicular to the surface. If these effects are taken into account the coefficient in equations 11.61 becomes 0.453[4], i.e. 8.6 per cent higher.

For constant surface temperature (equation 11.54) this convective effect is less significant[4].

11.7. THE BOUNDARY LAYER FOR MASS TRANSFER

If a concentration gradient exists within a fluid flowing over a surface, mass transfer will take place. Again the whole of the resistance to transfer can be regarded as lying within a *diffusion boundary layer* in the vicinity of the surface. If the concentration gradients, and hence the mass transfer rates, are small, variations in physical properties may be neglected and it can be shown that the velocity and thermal boundary layers are unaffected[5]. For low concentrations of the diffusing component, the effects of bulk flow will be small and the mass balance equation for component **A** is:

$$\frac{\partial}{\partial x} \int_0^l (C_{As} - C_A)u_x \, dy = D \left(\frac{\partial C_A}{\partial y} \right)_{y=0}$$ (11.64)

where C_A and C_{As} are the molar concentrations of **A** at a distance y from the surface and outside the boundary layer respectively; and l is a distance at right angles to the surface which is greater than the thickness of any of the three boundary layers. Equation 11.56 is obtained in exactly the same manner as equation 11.43 for heat transfer.

Again, the form of the concentration profile in the diffusion boundary layer depends on the conditions which are assumed to exist at the surface and in the fluid stream. For the conditions corresponding to those used in consideration of the thermal boundary layer, that is constant concentrations both in the stream outside the boundary layer and at the surface, the concentration profile will be of similar form to that given by equation 11.56:

$$\frac{C_A - C_{A0}}{C_{As} - C_{A0}} = \frac{3}{2} \left(\frac{y}{\delta_D} \right) - \frac{1}{2} \left(\frac{y}{\delta_D} \right)^3$$ (11.65)

where δ_D is the thickness of the concentration boundary layer,

$\quad C_A$ is the concentration of **A** at $y = y$,

$\quad C_{A0}$ is the concentration of **A** at the surface ($y = 0$), and

$\quad C_{As}$ is the concentration of **A** outside the boundary layer

Substituting from equation 11.65 to evaluate the integral in equation 11.64, assuming that mass transfer takes place over the whole length of the surface (i.e. $x_0 = 0$), by analogy with equation 11.51:

$$\frac{\delta_D}{\delta} \approx 0.976 Sc^{-1/3}$$ (11.66)

where $Sc = \mu/\rho D$ is the Schmidt number. Equation 11.66 is applicable provided that $Sc > 1$. If Sc is only slightly less than 1, little error is introduced and it is therefore applicable to most mixtures of gases, as seen in Table 10.1 (Chapter 10). The arguments are identical to those relating to the validity of equation 11.51 for heat transfer.

The point values of the Sherwood number Sh_x and mass transfer coefficient h_D are then given by:

$$Sh_x = \frac{h_D x}{D} = 0.331 Sc^{1/3} Re_x^{1/2}$$ (11.67)

The mean value of the coefficient between $x = 0$ and $x = x$ is then given by:

$$(Sh_x)_m = 0.662 Sc^{1/3} Re_x^{1/2}$$ (11.68)

In equations 11.67 and 11.68 Sh_x and $(Sh_x)_m$ represent the point and mean values respectively of the Sherwood numbers.

11.8. FURTHER READING

KAYS, W. M. and CRAWFORD, M. E.: *Convective Heat and Mass Transfer* (McGraw-Hill, New York, 1980).
SCHLICHTING, H.: *Boundary Layer Theory* (trans. by KESTIN, J.) 6th edn (McGraw-Hill, New York, 1968).
WHITE, F. M.: *Viscous Fluid Flow* (McGraw-Hill, New York, 1974).

11.9. REFERENCES

1. BROWN, F. N. M.: *Proc. Midwest Conf. Fluid Mechanics* (Sept. 1959) 331. The organized boundary layer.
2. PRANDTL, L.: *The Essentials of Fluid Dynamics* (Hafner, New York, 1949).
3. BLASIUS, H.: *Forsch. Ver. deut. Ing.* **131** (1913). Das Ähnlichkeitsgesetz bei Reibungsvorgängen in Flüssigkeiten.
4. KAYS, W. M. and CRAWFORD, M. E.: *Convective Heat and Mass Transfer* (McGraw-Hill, New York, 1980).
5. ECKERT, E. R. G. and DRAKE, R. M. Jr.: *Analysis of Heat and Mass Transfer* (McGraw-Hill, New York, 1972).

11.10. NOMENCLATURE

		Units in SI system	Dimensions in M, N, L, T, θ
a	Coefficient of y	K/m	$L^{-1}\theta$
a_0	Coefficient of y	s^{-1}	T^{-1}
b	Coefficient of y^2	$m^{-1}\,s^{-1}$	$L^{-1}T^{-1}$
b_0	Coefficient of y^2	K/m^2	$L^{-2}\theta$
C_A	Molar concentration of **A**	$kmol/m^3$	NL^{-3}
C_{A0}	Molar concentration of **A** at surface ($y=0$)	$kmol/m^3$	NL^{-3}
C_{AS}	Molar concentration of **A** outside boundary layer	$kmol/m^3$	NL^{-3}
C_p	Specific heat at constant pressure	J/kg K	$L^2T^{-2}\theta^{-1}$
c	Coefficient of y^3	$m^{-2}\,s^{-1}$	$L^{-2}T^{-1}$
c_0	Coefficient of y^3	K/m^3	$L^{-3}\theta$
D	Molecular diffusivity	m^2/s	L^2T^{-1}
D_H	Thermal diffusivity	m^2/s	L^2T^{-1}
d	Pipe diameter	m	L
f	Index	—	—
h	Heat transfer coefficient	W/m^2 K	$MT^{-3}\theta^{-1}$
h_D	Mass transfer coefficient	$kmol/[(m^2)(s)(kmol/m^3)]$	LT^{-1}
h_m	Mean value of heat transfer coefficient	W/m^2 K	$MT^{-3}\theta^{-1}$
k	Thermal conductivity	W/m K	$MLT^{-3}\theta^{-1}$
L_e	Inlet length of pipe	m	L
l	Thickness of element of fluid	m	L
P	Total pressure	N/m^2	$ML^{-1}T^{-2}$
q_0	Rate of transfer of heat per unit area at walls	W/m^2	MT^{-3}
R	Shear stress acting on surface	N/m^2	$ML^{-1}T^{-2}$
R_0	Shear stress acting on fluid at surface	N/m^2	$ML^{-1}T^{-2}$
r	Radius of pipe	m	L
t	Time	s	T
u	Mean velocity	m/s	LT^{-1}
u_b	Velocity at edge of laminar sub-layer	m/s	LT^{-1}
u_0	Velocity of fluid at surface	m/s	LT^{-1}
u_s	Velocity of fluid outside boundary layer, or at pipe axis	m/s	LT^{-1}
u_x	Velocity in X-direction at $y=y$	m/s	LT^{-1}
u_y	Velocity in Y-direction (perpendicular to surface)	m/s	LT^{-1}
u_z	Velocity in Z-direction	m/s	LT^{-1}
w	Width of surface	m	L
x	Distance from leading edge of surface in X-direction	m	L
x_c	Value of x at which flow becomes turbulent	m	L
x_0	Unheated length of surface	m	L
y	Distance from surface	m	L
δ	Thickness of boundary layer	m	L
δ_b	Thickness of laminar sub-layer	m	L
δ_D	Diffusion boundary layer thickness	m	L

		Units in SI system	Dimensions in M, N, L, T, θ
δ_t	Thickness of thermal boundary layer	m	L
δ^*	Displacement thickness of boundary layer	m	L
μ	Viscosity of fluid	$\mathrm{N\,s/m^2}$	$\mathbf{ML^{-1}T^{-1}}$
ρ	Density of fluid	$\mathrm{kg/m^3}$	$\mathbf{ML^{-3}}$
σ	Ratio of δ_t to δ	—	—
θ	Temperature at $y = y$	K	θ
θ_s	Temperature outside boundary layer, or at pipe axis	K	θ
Nu_x	Nusselt number hx/k	—	—
Re	Reynolds number $ud\rho/\mu$	—	—
Re_x	Reynolds number $u_s x\rho/\mu$	—	—
Re_{x_c}	Reynolds number $u_s x_c\rho/\mu$	—	—
Re_δ	Reynolds number $u_s\delta\rho/\mu$	—	—
Pr	Prandtl number $C_p\mu/k$	—	—
Sc	Schmidt number $\mu/\rho D$	—	—
Sh_x	Sherwood number $h_D x/D$	—	—

CHAPTER 12

Momentum, Heat, and Mass Transfer

12.1. INTRODUCTION

In the previous chapters the stresses arising from relative motion within a fluid, the transfer of heat by conduction and convection, and the mechanism of mass transfer have all been discussed. These three major processes of momentum, heat, and mass transfer have, however, been regarded as independent problems.

In most of the unit operations encountered in the chemical and process industries, one or more of the processes of momentum, heat, and mass transfer is involved. Thus, in the flow of a fluid under adiabatic conditions through a bed of granular particles, a pressure gradient is set up in the direction of flow and a velocity gradient develops approximately perpendicularly to the direction of motion in each fluid stream; momentum transfer then takes place between the fluid elements which are moving at different velocities. If there is a temperature difference between the fluid and the pipe wall or the particles, heat transfer will take place as well, and the convective component of the heat transfer will be directly affected by the flow pattern of the fluid. Here, then, is an example of a process of simultaneous momentum and heat transfer in which the same fundamental mechanism is affecting both processes. Fractional distillation and gas absorption are frequently carried out in a packed column in which the gas or vapour stream rises countercurrently to a liquid. The function of the packing in this case is to provide a large interfacial area between the phases and to promote turbulence within the fluids. In a very turbulent fluid, the rates of transfer per unit area of both momentum and mass are high; and as the pressure drop rises the rates of transfer of both momentum and mass increase together. In some cases, momentum, heat, and mass transfer all occur simultaneously as, for example, in a water-cooling tower (see Chapter 13), where transfer of sensible heat and evaporation both take place from the surface of the water droplets. It will now be shown not only that the process of momentum, heat, and mass transfer are physically related, but also that quantitative relations between them can be developed.

Another form of interaction between the transfer processes is responsible for the phenomenon of *thermal diffusion* in which a component in a mixture moves under the action of a temperature gradient. Although there are important applications of thermal diffusion, the magnitude of the effect is usually small relative to that arising from concentration gradients.

When a fluid is flowing under streamline conditions over a surface, a forward component of velocity is superimposed on the random distribution of velocities of the molecules, and movement at right angles to the surface occurs solely as a result of the random motion of the molecules. Thus if two adjacent layers of fluid are moving at different velocities, there will be a tendency for the faster moving layer to be retarded and the slower moving layer to be accelerated by virtue of the continuous passage of molecules in each direction. There will therefore be a net transfer of momentum from the fast- to the slow-moving stream. Similarly, the molecular motion will tend to reduce any temperature gradient or any concentration gradient if the fluid consists of a mixture of two or more components. At the boundary the effects of the molecular transfer are balanced by the drag forces at the surface.

If the motion of the fluid is turbulent, the transfer of fluid by eddy motion is superimposed on the molecular transfer process. In this case, the rate of transfer to the surface will be a function of the degree of turbulence. When the fluid is highly turbulent, the rate of transfer by molecular motion will be negligible compared with that by eddy motion. For small degrees of turbulence the two may be of the same order.

It was shown in the previous chapter that when a fluid flows under turbulent conditions over a surface, the flow can conveniently be divided into three regions:

(1) At the surface, the laminar sub-layer, in which the only motion at right angles to the surface is due to molecular diffusion.
(2) Next, the buffer layer, in which molecular diffusion and eddy motion are of comparable magnitude.
(3) Finally, over the greater part of the fluid, the turbulent region in which eddy motion is large compared with molecular diffusion.

In addition to momentum, both heat and mass can be transferred either by molecular diffusion alone or by molecular diffusion combined with eddy diffusion. Because the effects of eddy diffusion are generally far greater than those of the molecular diffusion, the main resistance to transfer will lie in the regions where only molecular diffusion is occurring. Thus the main resistance to the flow of heat or mass to a surface lies within the laminar sub-layer. It has been shown in Chapter 11 that the thickness of the laminar sub-layer is almost inversely proportional to the Reynolds number for fully developed turbulent flow in a pipe. Thus the heat and mass transfer coefficients are much higher at high Reynolds numbers.

There are strict limitations to the application of the analogy between momentum transfer on the one hand, and heat and mass transfer on the other. Firstly, it must be borne in mind that momentum is a vector quantity, whereas heat and mass are scalar quantities. Secondly, the quantitative relations apply only to that part of the momentum transfer which arises from *skin friction*. If *form drag* is increased there is little corresponding increase in the rates at which heat transfer and mass transfer will take place.

Skin friction is the drag force arising from shear stress attributable to the viscous force in the laminar region in the neighbourhood of a surface. *Form drag* is the inertial component arising from vortex formation arising from the presence of an obstruction of flow by, for instance, a baffle or a roughness element on the surface of the pipe. Thus, in the design of contacting devices such as column packings, it is important that they are so shaped that the greater part of the pressure drop is attributable to skin friction rather than to form drag.

12.2. TRANSFER BY MOLECULAR DIFFUSION

12.2.1. Momentum transfer

When the flow characteristics of the fluid are *Newtonian*, the shear stress R_y in a fluid is proportional to the velocity gradient and to the viscosity.

Thus: $$R_y = -\mu \frac{du_x}{dy} = -\frac{\mu}{\rho} \frac{d(\rho u_x)}{dy} \quad \text{(for constant density)} \quad (12.1)$$
(cf. equation 3.3)

where u_x is the velocity of the fluid parallel to the surface at a distance y from it.

The shear stress R_y within the fluid, at a distance y from the boundary surface, is a measure of the rate of transfer of momentum per unit area at right angles to the surface.

Since (ρu_x) is the momentum per unit volume of the fluid, the rate of transfer of momentum per unit area is proportional to the gradient in the Y-direction of the momentum per unit volume. The negative sign indicates that momentum is transferred from the fast- to the slow-moving fluid and the shear stress acts in such a direction as to oppose the motion of the fluid.

12.2.2. Heat transfer

From the definition of thermal conductivity, the heat transferred per unit time through unit area at a distance y from the surface is given by:

$$q_y = -k \frac{d\theta}{dy} = -\frac{k}{C_p \rho} \frac{d(C_p \rho \theta)}{dy} \quad \text{(cf. equation 9.11)} \quad (12.2)$$

where C_p is the specific heat of the fluid at constant pressure, θ the temperature, and k the thermal conductivity. C_p and ρ are both assumed to be constant.

The term $(C_p \rho \theta)$ represents the heat content per unit volume of fluid and therefore the flow of heat is proportional to the gradient in the Y-direction of the heat content per unit volume. The proportionality constant $k/C_p \rho$ is called the thermal diffusivity D_H.

12.2.3. Mass transfer

It is shown in Chapter 10, from Fick's Law of diffusion, that the rate of diffusion of a constituent **A** in a mixture is proportional to its concentration gradient.

Thus, from equation 10.4:

$$N_A = -D \frac{dC_A}{dy} \quad (12.3)$$

where N_A is the molar rate of diffusion of constituent **A** per unit area, C_A the molar concentration of constituent **A** and D the diffusivity.

The essential similarity between the three processes is that the rates of transfer of momentum, heat, and mass are all proportional to the concentration gradients of these quantities. In the case of gases the proportionality constants μ/ρ, D_H, and D, all of which have the dimensions length^2/time, all have a physical significance. For liquids the constants cannot be interpreted in a similar manner. The viscosity, thermal conductivity, and diffusivity of a gas will now be considered.

12.2.4. Viscosity

Consider the flow of a gas parallel to a solid surface and the movement of molecules at right angles to this direction through a plane a–a of unit area, parallel to the surface and sufficiently close to it to be within the laminar sublayer (Figure 12.1). During an interval of time dt, molecules with an average velocity $i_1 u_m$ in the Y-direction will pass through the plane (where u_m is the root mean square velocity and i_1 is some fraction of it, depending on the actual distribution of velocities).

Figure 12.1.   Transfer of momentum near a surface

If all these molecules can be considered as having the same component of velocity in the Y-direction, molecules from a volume $i_1 u_m$ dt will pass through the plane in time dt.

If N is the numerical concentration of molecules close to the surface, the number of molecules passing $= i_1 u_m \mathbf{N}$ dt.

Thus the rate of passage of molecules $= i_1 u_m \mathbf{N}$.

These molecules have a mean velocity u_x (say) in the X-direction.

Thus the rate at which momentum is transferred across the plane away from the surface

$$= i_1 \mathbf{N} u_m m u_x$$

where m is the mass of each molecule.

By similar reasoning there must be an equivalent stream of molecules also passing through the plane in the opposite direction; otherwise there would be a resultant flow perpendicular to the surface.

If this other stream of molecules has originated at a distance $j\lambda$ from the previous ones, and the mean component of their velocities in the X-direction is u'_x (where λ is the mean free path of the molecules and j is some fraction of the order of unity) then:

The net rate of transfer of momentum away from the surface

$$= i_1 \mathbf{N} u_m m (u_x - u'_x)$$

The gradient of the velocity with respect to the Y-direction

$$= \frac{\mathrm{d}u_x}{\mathrm{d}y} = \frac{(u'_x - u_x)}{j\lambda}$$

since λ is small.

Thus the rate of transfer of momentum per unit area which can be written as:

$$R_y = -i_1 N u_m m j \lambda \frac{du_x}{dy}$$

$$= -i_1 j \rho u_m \lambda \frac{du_x}{dy} \qquad (12.4)$$

(since $Nm = \rho$, the density of the fluid).

But:
$$R_y = -\mu \frac{du_x}{dy} \quad \text{(from equation 12.1)}$$

$$\therefore \qquad \frac{\mu}{\rho} = i_1 j u_m \lambda \qquad (12.5)$$

The value of the product $i_1 j$ has been variously given by different workers, from statistical treatment of the velocities of the molecules;[1] a value of 0.5 will be taken.

Thus:
$$\frac{\mu}{\rho} = \frac{1}{2} u_m \lambda \qquad (12.6)$$

It is now possible to give a physical interpretation to the Reynolds number:

$$Re = \frac{ud\rho}{\mu} = ud \frac{2}{u_m \lambda} = 2 \frac{u}{u_m} \frac{d}{\lambda} \qquad (12.7)$$

or Re is proportional to the product of the ratio of the flow velocity to the molecular velocity and the ratio of the characteristic linear dimension of the system to the mean free path of the molecules.

From the kinetic theory,[1] $u_m = \sqrt{(8RT/\pi M)}$ and is independent of pressure, and $\rho \lambda$ is a constant.

Thus, the viscosity of a gas would be expected to be a function of temperature but not of pressure.

12.2.5. Thermal conductivity

Considering now the case where there is a temperature gradient in the Y-direction, the rate of passage of molecules through the unit plane a–a $= i_2 u_m N$ (where i_2 is some fraction of the order of unity). If the temperature difference between two planes situated a distance $j\lambda$ apart is $(\theta - \theta')$, the net heat transferred as one molecule passes in one direction and another molecule passes in the opposite direction is $c_m(\theta - \theta')$, where c_m is the heat capacity per molecule.

The net rate of heat transfer per unit area $= i_2 u_m N c_m (\theta - \theta')$.
The temperature gradient $d\theta/dy = (\theta' - \theta)/j\lambda$ since λ is small.
Thus the net rate of heat transfer per unit area

$$q = -i_2 j u_m N c_m \lambda \frac{d\theta}{dy}$$

$$= -i_2 j u_m C_v \rho \lambda \frac{d\theta}{dy} \qquad (12.8)$$

(since $N c_m = \rho C_v$, the specific heat per unit volume of fluid)

and:
$$q = -k\frac{d\theta}{dy} \quad \text{(from equation 12.2)}$$

Thus the thermal diffusivity:

$$\frac{k}{C_p\rho} = i_2 j u_m \lambda \frac{C_v}{C_p} \tag{12.9}$$

From statistical calculations[1] the value of $i_2 j$ has been given as $(9\gamma - 5)/8$ (where $\gamma = C_p/C_v$, the ratio of the specific heat at constant pressure to the specific heat at constant volume).

Thus:
$$\frac{k}{C_p\rho} = u_m\lambda\frac{9\gamma - 5}{8\gamma} \tag{12.10}$$

The Prandtl number Pr is defined as the ratio of the kinematic viscosity to the thermal diffusivity.

Thus:
$$Pr = \frac{\mu/\rho}{k/C_p\rho} = \frac{C_p\mu}{k} = \frac{\frac{1}{2}u_m\lambda}{u_m\lambda(9\lambda - 5)/8\gamma}$$

$$= \frac{4\gamma}{9\gamma - 5} \tag{12.11}$$

Values of Pr calculated from equation 12.11 are in close agreement with practical figures.

12.2.6. Diffusivity

Considering the diffusion, in the Y-direction, of one constituent $\mathbf{A}$ of a mixture across the plane a–a, if the numerical concentration is $\mathbf{N}_A$ on one side of the plane and $\mathbf{N}'_A$ on the other side at a distance of $j\lambda$:

The net rate of passage of molecules per unit area
$$= i_3 u_m(\mathbf{N}_A - \mathbf{N}'_A)$$

where i_3 is an appropriate fraction of the order of unity.

The rate of mass transfer per unit area
$$= i_3 u_m(\mathbf{N}_A - \mathbf{N}'_A)m$$

The concentration gradient of $\mathbf{A}$ in the Y-direction
$$= \frac{dC_A}{dy} = \frac{(\mathbf{N}'_A - \mathbf{N}_A)m}{j\gamma}$$

Thus the rate of mass transfer per unit area
$$= -i_3 j\lambda u_m\frac{dC_A}{dy} \tag{12.12}$$

$$= -D\frac{dC_A}{dy} \quad \text{(from equation 12.3)}$$

Thus: $$D = i_3 j u_m \lambda \qquad (12.13)$$

There is, however, no satisfactory evaluation of the product $i_3 j$.

The ratio of the kinematic viscosity to the diffusivity is the Schmidt number, Sc, where:

$$Sc = \frac{(\mu/\rho)}{D} = \frac{\mu}{\rho D} \qquad (12.14)$$

It is thus seen that the kinematic viscosity, the thermal diffusivity, and the diffusivity for mass transfer are all proportional to the product of the mean free path and the root mean square velocity of the molecules, and that the expressions for the transfer of momentum, heat, and mass are of the same form.

For liquids the same qualitative forms of relationships exist, but it is not possible to express the physical properties of the liquids in terms of molecular velocities and distances.

12.3. EDDY TRANSFER

In the previous section, the molecular basis for the processes of momentum transfer, heat transfer and mass transfer has been discussed. It has been shown that, in a fluid in which there is a momentum gradient, a temperature gradient or a concentration gradient, the consequential momentum, heat and mass transfer processes arise as a result of the random motion of the molecules. For an ideal gas, the kinetic theory of gases is applicable and the physical properties μ/ρ, $k/C_p\rho$ and D, which determine the transfer rates, are all seen to be proportional to the product of a molecular velocity and the mean free path of the molecules.

A fluid in turbulent flow is characterised by the presence of circulating or eddy currents, and these are responsible for fluid mixing which, in turn, gives rise to momentum, heat or mass transfer when there is an appropriate gradient of the "property" in question. The following simplified analysis of the transport processes in a turbulent fluid is based on the work and ideas of Prandtl. By analogy with the kinetic theory, it is suggested that the relationship between transfer rate and driving force should depend on quantities termed the *eddy kinematic viscosity* E, the *eddy thermal diffusivity* E_H and the *eddy diffusivity* E_D analogous to μ/ρ, $k/C_p\rho$ and D for molecular transport. Extending the analogy further, E, E_H and E_D might be expected to be proportional to the product of a velocity term and a length term, each of which is characteristic of the eddies in the fluid. Whereas μ/ρ, $k/C_p\rho$ and D are all physical properties of the fluid and, for a material of given composition at a specified temperature at pressure have unique values, the eddy terms E, E_H and E_D all depend on the intensity of the eddies. In general, therefore, they are a function of the flow pattern and vary from point to point within the fluid.

In Chapter 11 the concept of a *boundary layer* is discussed. It is suggested that, when a fluid is in turbulent flow over a surface, the eddy currents tend to die out in the region very close to the surface, giving rise to a *laminar sub-layer* in which E, E_H and E_D are all very small. With increasing distance from the surface these quantities become progressively greater, rising from zero in the laminar sub-layer to values considerably in excess of μ/ρ, $k/C_p\rho$ and D in regions remote from the surface. Immediately outside the laminar sub-layer is a buffer zone in which the molecular and eddy terms are of comparable magnitudes. At its outer edge, the eddy terms have become much larger than the molecular terms and the latter can then be neglected — in what can now be regarded as the fully turbulent region.

12.3.1. The nature of turbulent flow

In turbulent flow there is a complex interconnected series of circulating or eddy currents in the fluid, generally increasing in scale and intensity with increase of distance from any boundary surface. If, for steady-state turbulent flow, the velocity is measured at any fixed point in the fluid, both its magnitude and direction will be found to vary in a random manner with time. This is because a random velocity component, attributable to the circulation of the fluid in the eddies, is superimposed on the steady state mean velocity. No net motion arises from the eddies and therefore their time average in any direction must be zero. The instantaneous magnitude and direction of velocity at any point is therefore the vector sum of the steady and fluctuating components.

If the magnitude of the fluctuating velocity component is the same in each of the three principal directions, the flow is termed *isotropic*. If they are different the flow is said to be *anisotropic*. Thus, if the root mean square values of the random velocity components in the X, Y and Z directions are respectively $\sqrt{\overline{u_{Ex}^2}}$, $\sqrt{\overline{u_{Ey}^2}}$ and $\sqrt{\overline{u_{Ez}^2}}$; for isotropic turbulence:

$$\sqrt{\overline{u_{Ex}^2}} = \sqrt{\overline{u_{Ey}^2}} = \sqrt{\overline{u_{Ez}^2}} \tag{12.15}$$

There are two principal characteristics of turbulence. One is the *scale* which is a measure of the mean size of the eddies, and the other is the *intensity* which is a function of the circulation velocity ($\sqrt{\overline{u_E^2}}$) within the eddies. Both the scale and the intensity increase as the distance from a solid boundary becomes greater. During turbulent flow in a pipe, momentum is transferred from large eddies in the central core through successively smaller eddies as the walls are approached. Eventually, when the laminar sub-layer is reached the eddies die out completely. However, the laminar sub-layer should not be regarded as a completely discrete region, because there is evidence that from time to time eddies do penetrate and occasionally completely disrupt it.

The intensity of turbulence I is defined as the ratio of the mean value of the fluctuating component of velocity to the steady state velocity. For flow in the X-direction parallel to a surface this may be written as:

$$I = \frac{\sqrt{\frac{1}{3}(\overline{u_{Ex}^2} + \overline{u_{Ey}^2} + \overline{u_{Ez}^2})}}{u_x} \tag{12.16}$$

For isotropic turbulence, from equation 12.15, this becomes:

$$I = \frac{\sqrt{\overline{u_E^2}}}{u_x} \tag{12.17}$$

The intensity of turbulence will vary with the geometry of the flow system. Typically, for a fluid flowing over a plane surface or through a pipe, it may have a value of between

0.005 and 0.02. In the presence of packings and turbulence promoting grids, very much higher values (0.05 to 0.1) are common.

The scale of turbulence is given approximately by the diameter of the eddy, or by the distance between the centres of successive eddies. The scale of turbulence is related to the dimensions of the system through which the fluid is flowing. The size of largest eddies is clearly limited by the diameter of the pipe or duct. As the wall is approached their average size becomes less and momentum transfer takes place by interchange through a succession of eddies of progressively smaller size (down to about 1 mm) until they finally die out as the laminar sub-layer is approached near the walls.

An idea of the scale of turbulence can be obtained by measuring instantaneous values of velocities at two different points within the fluid and examining how the correlation coefficient for the two sets of values changes as the distance between the points is increased.

When these are close together, most of the simultaneously measured velocities will relate to fluid in the same eddy and the correlation coefficient will be high. When the points are further apart the correlation coefficient will fall because in an appreciable number of the pairs of measurements the two velocities will relate to different eddies. Thus, the distance apart of the measuring stations at which the correlation coefficient becomes very poor is a measure of scale of turbulence. Frequently, different scales of turbulence can be present simultaneously. Thus, when a fluid in a tube flows past an obstacle or suspended particle, eddies may form in the wake of the particles and their size will be of the same order as the size of the particle; in addition, there will be larger eddies limited in size only by the diameter of the pipe.

12.3.2. Mixing length and eddy kinematic viscosity

PRANDTL[2,3] and TAYLOR[4] both developed the concept of a *mixing length* as a measure of the distance which an element of fluid must travel before it loses its original identity and becomes fully assimilated by the fluid in its new position. Its magnitude will be of the same order as the scale of turbulence or the eddy size. The mixing length is analogous in concept to the *mean free path* of gas molecules which, according to the kinetic theory is the mean distance a molecule travels before it collides with another molecule and loses its original identity.

In turbulent flow over a surface, a velocity gradient, and hence a momentum gradient, exists within the fluid. Any random movement perpendicular to the surface gives rise to a momentum transfer. Elements of fluid with high velocities are brought from remote regions towards the surface and change places with slower moving fluid elements. This mechanism is essentially similar to that involved in the random movement of molecules in a gas. It is therefore suggested that an *eddy kinematic viscosity E* for eddy transport may be defined which is analogous to the kinematic viscosity μ/ρ for molecular transport. Then for isotropic turbulence:

$$E \propto \lambda_E u_E \qquad (12.18)$$

where λ_E is the mixing length, and

u_E is some measure of the linear velocity of the fluid in the eddies.

On this basis, the momentum transfer rate per unit area in a direction perpendicular to the surface at some position y is given by:

$$R_y = -E\frac{d(\rho u_x)}{dy} \tag{12.19}$$

For constant density:

$$R_y = -E\rho\frac{du_x}{dy} \tag{12.20}$$

Prandtl has suggested that u_E is likely to increase as both the mixing length λ_E and the modulus of the velocity gradient $|du_x/dy|$ increase. The simplest form of relation between the three quantities is that:

$$u_E \propto \lambda_E \left|\frac{du_x}{dy}\right| \tag{12.21}$$

This is tantamount to saying that the velocity change over a distance equal to the mixing length approximates to the eddy velocity. This cannot be established theoretically but is probably a reasonable assumption.

Combining equations 12.18 and 12.21:

$$E \propto \lambda_E \left\{\lambda_E \left|\frac{du_x}{dy}\right|\right\} \tag{12.22}$$

Arbitrarily putting the proportionality constant equal to unity:

$$E = \lambda_E^2 \left|\frac{du_x}{dy}\right| \tag{12.23}$$

Equation 12.23 implies a small change in the definition of λ_E.

In the neighbourhood of a surface, the velocity gradient will be positive and the modulus sign in equation 12.23 may be dropped. On substitution into equation 12.20:

$$R_y = -\rho\lambda_E^2 \left(\frac{du_x}{dy}\right)^2 \tag{12.24}$$

or:

$$\sqrt{\frac{-R_y}{\rho}} = \lambda_E \frac{du_x}{dy} \tag{12.25}$$

In equations 12.19 and 12.20, R_y represents the momentum transferred per unit area and unit time. This momentum transfer *tends* to accelerate the slower moving fluid close to the surface and to retard the faster-moving fluid situated at a distance from the surface. It gives rise to a stress R_y at a distance y from the surface since, from Newton's Law of Motion, force equals rate of change of momentum. Such stresses, caused by the random motion in the eddies, are sometimes referred to as *Reynolds Stresses*.

The problem can also be approached in a slightly different manner. In Figure 12.2, the velocity profile is shown near a surface. At point 1, the velocity is u_x and at point 2, the velocity is u_x'. For an eddy velocity u_{Ey} in the direction perpendicular to the surface, the fluid is transported away from the surface at a mass rate per unit area equal to $u_{Ey}\rho$; this fluid must be replaced by an equal mass of fluid which is transferred in the opposite

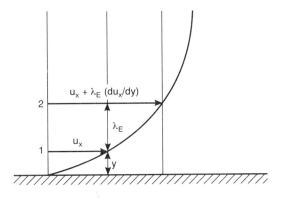

Figure 12.2.   Prandtl mixing length

direction. The momentum transferred *away* from the surface per unit and unit time is
given by:

$$R_y = \rho u_{Ey}(u_x - u'_x)$$

If the distance between the two locations is approximately equal to the mixing length
λ_E, and if the velocity gradient is nearly constant over that distance:

$$\frac{u'_x - u_x}{\lambda_E} \approx \frac{du_x}{dy}$$

Again assuming that:

$$u'_x - u_x \approx u_{Ey}$$

$$R = -\rho \lambda_E^2 \left(\frac{du_x}{dy}\right)^2 \qquad\qquad \text{(equation 12.24)}$$

It is assumed throughout that no mixing takes place with the intervening fluid when an
eddy transports fluid elements over a distance equal to the mixing length.

Close to a surface $R_y \to R_0$, the value at the surface.

The shear stress R acting on the surface must be equal and opposite to that in the fluid
in contact with the surface (i.e. $R = -R_0$).

$$\therefore \qquad\qquad \sqrt{\frac{R}{\rho}} = \lambda_E \frac{du_x}{dy} \qquad\qquad (12.26)$$

$\sqrt{R/\rho}$ is known as the *shearing stress velocity* or *friction velocity* and is usually denoted
by u^*.

In steady state flow over a plane surface, or close to the wall for flow in a pipe, u^* is
constant and equation 12.26 can be integrated provided that the relation between λ_E and
y is known. λ_E will increase with y and, if a linear relation is assumed:

$$\lambda_E = Ky \qquad\qquad (12.27)$$

This is the simplest possible form of relation; its use is justified only if it leads to results which are in conformity with experimental results for velocity profiles.

Then:
$$u^* = Ky\frac{du_x}{dy}$$
(12.28)

On integration:
$$\frac{u_x}{u^*} = \frac{1}{K}\ln y + B \qquad \text{where } B \text{ is a constant.}$$

or:
$$\frac{u_x}{u^*} = \frac{1}{K}\ln\frac{yu^*\rho}{\mu} + B'$$
(12.29)

Since $(u^*\rho/\mu)$ is constant, B' will also be constant.

Writing the dimensionless velocity term $u_x/u^* = u^+$ and the dimensionless derivative of y $(yu^*\rho/\mu) = y^+$:

$$u^+ = \frac{1}{K}\ln y^+ + B'$$
(12.30)

If equation 12.29 is applied to the outer edge of the boundary layer when $y = \delta$ (boundary layer thickness) and $u_x = u_s$ (the stream velocity).

$$\frac{u_s}{u^*} = \frac{1}{K}\ln\frac{\delta u^*\rho}{\mu} + B'$$
(12.31)

Subtracting equation 12.29 from equation 12.31:

$$\frac{u_s - u_x}{u^*} = \frac{1}{K}\ln\frac{\delta}{y}$$
(12.32)

Using experimental results for flow of fluids over both smooth and rough surfaces, NIKURADSE[5,6] found K to have a value of 0.4.

Thus:
$$\frac{u_s - u_x}{u^*} = 2.5\ln\frac{\delta}{y}$$
(12.33)

For fully developed flow in a pipe, $\delta = r$ and u_s is the velocity at the axis. Then:

$$\frac{u_s - u_x}{u^*} = 2.5\ln\frac{r}{y}$$
(12.34)

Equation 12.34 is known as the *velocity defect law* (Figure 12.3).

The application to pipe flow is not strictly valid because $u^* (= \sqrt{R/\rho})$ is constant only in regions close to the wall. However, equation 12.34 appears to give a reasonable approximation to velocity profiles for turbulent flow, except near the pipe axis. The errors in this region can be seen from the fact that on differentiation of equation 12.34 and putting $y = r$, the velocity gradient on the centre line is $2.5u^*/r$ instead of zero.

Inserting $K = 0.4$ in equation 12.27 gives the relation between mixing length (λ_E) and distance (y) from the surface:

$$\frac{\lambda_E}{y} = 0.4$$
(12.35)

Equation 12.35 applies only in those regions where eddy transfer dominates, i.e. outside both the laminar sub-layer and the buffer layer (see below).

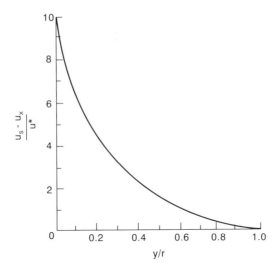

Figure 12.3.   Velocity defect law

12.4. UNIVERSAL VELOCITY PROFILE

For fully developed turbulent flow in a pipe, the whole of the flow may be regarded as lying within the boundary layer. The cross-section can then conveniently be divided into three regions:

(a) The *turbulent core* in which the contribution of eddy transport is so much greater than that of molecular transport that the latter can be neglected.
(b) The *buffer layer* in which the two mechanisms are of comparable magnitude.
(c) The *laminar sub-layer* in which turbulent eddies have effectively died out so that only molecular transport need be considered.

It is now possible to consider each of these regions in turn and to develop a series of equations to represent the velocity over the whole cross section of a pipe. Together, they constitute the *Universal Velocity Profile*.

12.4.1. The turbulent core

Equation 12.30 applies in the turbulent core, except near the axis of the pipe where the shear stress is markedly different from that at the walls. Inserting the value of 0.4 for K:

$$u^+ = 2.5 \ln y^+ + B' \qquad (12.36)$$

On plotting experimental data on velocity profiles as u^+ against log y^+ (see Figure 12.4), a series of parallel straight lines is obtained of slope 2.5 and with intercepts at ln $y^+ = 0$ varying with the relative roughness of the surface (e/d). For smooth surfaces $(e/d = 0)$, $B' = 5.5$. B' becomes progressively smaller as the relative roughness increases.
Thus for a smooth pipe:

$$u^+ = 2.5 \ln y^+ + 5.5 \qquad (12.37)$$

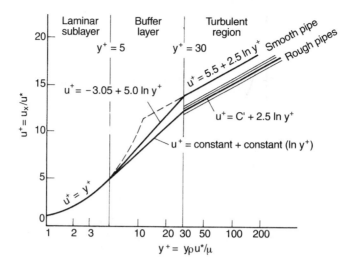

Figure 12.4.   The universal velocity profile

and for a rough pipe:

$$u^+ = 2.5 \ln y^+ + B' \tag{12.38}$$

where B' is a function of e/d and is less than 5.5.

Equations 12.37 and 12.38 correlate experimental data well for values of y^+ exceeding 30.

12.4.2. The laminar sub-layer

In the laminar sub-layer, turbulence has died out and momentum transfer is attributable solely to viscous shear. Because the layer is thin, the velocity gradient is approximately linear and equal to u_b/δ_b where u_b is the velocity at the outer edge of a laminar sub-layer of thickness δ_b (see Chapter 11).

$$R = \mu \frac{u_b}{\delta_b} = \mu \frac{u_x}{y}$$

Then:

$$u^{*2} = \frac{R}{\rho} = \frac{\mu u_x}{\rho y}$$

$\therefore$

$$\frac{u_x}{u^*} = \frac{y u^* \rho}{\mu} \tag{12.39}$$

or:

$$u^+ = y^+ \tag{12.40}$$

This relationship holds reasonably well for values of y^+ up to about 5, and it applies to both rough and smooth surfaces.

12.4.3. The buffer layer

The buffer layer covers the intermediate range $5 < y^+ < 30$. A straight line may be drawn to connect the curve for the laminar sub-layer (equation 12.40) at $y^+ = 5$ with the line for the turbulent zone for flow over a smooth surface at $y^+ = 30$ (equation 12.37) (see Figure 12.4). The data for the intermediate region are well correlated by this line whose equation must be of the form:

$$u^+ = a \ln y^+ + a' \qquad (12.41)$$

The line passes through the points ($y^+ = 5$, $u^+ = 5$) and ($y^+ = 30$, $u^+ = 2.5 \ln 30 + 5.5$) and therefore a and a' may be evaluated to give:

$$u^+ = 5.0 \ln y^+ - 3.05 \qquad (12.42)$$

An equation, similar in form to equation 12.42, will be applicable for rough surfaces but the values of the two constants will be different.

12.4.4. Velocity profile for all regions

For a smooth pipe, therefore, the complete Universal Velocity Profile is given by:

$$0 < y^+ < 5 \qquad u^+ = y^+ \qquad \text{(equation 12.40)}$$

$$5 < y^+ < 30 \qquad u^+ = 5.0 \ln y^+ - 3.05 \qquad \text{(equation 12.42)}$$

$$y^+ > 30 \qquad u^+ = 2.5 \ln y^+ + 5.5 \qquad \text{(equation 12.37)}$$

A simplified form of velocity profile is obtained by neglecting the existence of the buffer layer and assuming that there is a sudden transition from the laminar sub-layer to an eddy-dominated turbulent regime. The transition will occur at the point of intersection of the curves (shown by broken lines in Figure 12.4) representing equations 12.40 and 12.37. Solving the equations simultaneously gives $y^+ = 11.6$ as the point of intersection.

For flow in a pipe, the dimensionless distance y^+ from the walls and the corresponding velocity u^+ may be expressed in terms of three dimensionless quantities; the pipe Reynolds number $ud\rho/\mu$, the pipe friction factor $\phi \ (= R/\rho u^2)$ and the ratio y/d or y/r.

Since:

$$u^* = \sqrt{\frac{R}{\rho}} = \sqrt{\frac{R}{\rho u^2}} u = \phi^{1/2} u$$

$$u^+ = \frac{u_x}{u^*} = \frac{u_x}{u} \phi^{1/2} \qquad (12.43)$$

$$y^+ = \frac{y u^* \rho}{\mu} - \frac{y u^*}{d} \frac{u d \rho}{u} \frac{}{\mu} = \frac{y}{d} \phi^{1/2} Re \qquad (12.44)$$

12.4.5. Velocity gradients

By differentiation of equations 12.40, 12.42 and 12.37 respectively, the corresponding values of the gradient du^+/dy^+ are obtained:

$$0 < y^+ < 5 \qquad \frac{du^+}{dy^+} = 1 \qquad (12.45)$$

$$5 < y^+ < 30 \qquad \frac{du^+}{dy^+} = \frac{5.0}{y^+} \qquad (12.46)$$

$$y^+ > 30 \qquad \frac{du^+}{dy^+} = \frac{2.5}{y^+} \qquad (12.47)$$

Equations 12.45 and 12.47 are applicable to both rough and smooth surfaces; equation 12.46 is valid only for a smooth surface ($e/d \to 0$).

Velocity gradients are directly related to du^+/dy^+,

since:
$$\frac{du_x}{dy} = \frac{du^+}{dy^+} \frac{\rho u^{*2}}{\mu} \tag{12.48}$$

Thus:

$$0 < y^+ < 5 \qquad \frac{du_x}{dy} = \frac{\rho u^{*2}}{\mu} = \frac{R}{\mu} \tag{12.49}$$

$$5 < y^+ < 30 \qquad \frac{du_x}{dy} = \frac{5.0\rho u^{*2}}{y^+ \mu} = \frac{5.0\,R}{y^+\,\mu} \tag{12.50a}$$

$$\frac{du_x}{dy} = 5.0\frac{u^*}{y} = 5.0\phi^{1/2}\frac{u}{y} \tag{12.50b}$$

$$y^+ > 30 \qquad \frac{du_x}{dy} = \frac{2.5\rho u^{*2}}{y^+ \mu} = \frac{2.5\,R}{y^+\,\mu} \tag{12.51a}$$

$$\frac{du_x}{dy} = 2.5\frac{u^*}{y} = 2.5\phi^{1/2}\frac{u}{y} \tag{12.51b}$$

12.4.6. Laminar sub-layer and buffer layer thicknesses

On the basis of the Universal Velocity Profile, the laminar sub-layer extends from $y^+ = 0$ to $y^+ = 5$ and the buffer layer from $y^+ = 5$ to $y^+ = 30$.

From the definition of y^+:

$$y = y^+ \frac{\mu}{u^* \rho}$$

$$= y^+ \frac{\mu}{ud\rho}\frac{u}{u^*} d \tag{12.52}$$

Thus:
$$\frac{y}{d} = Re^{-1} \left(\frac{R}{\rho u^2}\right)^{-1/2} y^+ \tag{12.53}$$

$$= Re^{-1}\phi^{-1/2}y^+ \tag{12.53a}$$

Putting $y^+ = 5$, the laminar sub-layer thickness (δ_b) is given by:

$$\frac{\delta_b}{d} = 5Re^{-1}\phi^{-1/2} \tag{12.54}$$

The buffer layer extends to $y^+ = 30$, where:

$$\frac{y}{\delta} = 30Re^{-1}\phi^{-1/2} \tag{12.55}$$

If the buffer layer is neglected, it has been shown (Section 12.4.4) that the laminar sub-layer will extend to $y^+ = 11.6$ giving:

$$\frac{\delta_b}{d} = 11.6Re^{-1}\phi^{-1/2} \tag{12.56}$$

Using the Blasius equation (equation 11.39) to give an approximate value for $R/\rho u^2$ for a smooth pipe:

$$\phi = 0.0396Re^{-1/4} \qquad \text{(from equation 11.39)}$$

$$\frac{\delta_b}{d} = 58Re^{-7/8} \qquad (12.57)$$

The equation should be compared with equation 11.42 obtained using Prandtl's simplified approach to boundary layer theory which also disregards the existence of the buffer layer:

$$\frac{\delta_b}{d} = 62Re^{-7/8} \qquad \text{(equation 11.42)}$$

Similarly, the velocity u_b at the edge of the laminar sub-layer is given by:

$$u^+ = \frac{u_b}{u^*} = 11.6 \qquad (12.58)$$

(since $u^+ = y^+$ in the laminar sub-layer).
   Thus using the Blasius equation:

$$\frac{u_b}{u} = 11.6\phi^{1/2} \qquad (12.59)$$

Substituting for ϕ in terms of Re:

$$\frac{u_b}{u} = 2.32Re^{-1/8} \qquad (12.60)$$

   Again, equation 12.60 can be compared with equation 11.40, derived on the basis of the Prandtl approach:

$$\frac{u_b}{u} = 2.49Re^{-1/8} \qquad \text{(equation 11.40)}$$

12.4.7. Variation of eddy kinematic viscosity

Since the buffer layer is very close to the wall, R_y can be replaced by R_0.

and:

$$R_0 = -(\mu + E\rho)\frac{du_x}{dy} \qquad (12.61)$$

Thus:

$$\frac{-R_0}{\rho} = u^{*2} = \left(\frac{\mu}{\rho} + E\right)\frac{du_x}{dy} \qquad (12.62)$$

For $5 < y^+ < 30$, substituting for du_x/dy from equation 12.50a:

$$u^{*2} = \left(\frac{\mu}{\rho} + E\right)\frac{5\rho u^{*2}}{\mu y^+}$$

giving:

$$E = \frac{\mu}{\rho}\left(\frac{y^+}{5} - 1\right) \qquad (12.63)$$

   Thus E varies from zero at $y^+ = 5$, to $(5\mu/\rho)$ at $y^+ = 30$.
   For values y^+ greater than 30, μ/ρ is usually neglected in comparison with E. The error is greatest at $y^+ = 30$ where $E/(\mu/\rho) = 5$, but it rapidly becomes smaller at larger distances from the surface.

Thus, for $y^+ > 30$, from equation 12.62:

$$u^{*2} = E\frac{du_x}{dy} \tag{12.64}$$

Substituting for du_x/dy from equation 12.51a:

$$u^{*2} = E\frac{2.5\rho u^{*2}}{\mu y^+}$$

Thus:
$$E = 0.4y^+ \frac{\mu}{\rho} \tag{12.65a}$$

$$= 0.4u^* y. \tag{12.65b}$$

It should be noted that equation 12.65a gives $E = 12(\mu/\rho)$ at $y^+ = 30$, compared with $5(\mu/\rho)$ from equation 12.63. This arises because of the discontinuity in the Universal Velocity Profile at $y^+ = 30$.

12.4.8. Approximate form of velocity profile in turbulent region

A simple approximate form of the relation between u^+ and y^+ for the turbulent flow of a fluid in a pipe of circular cross-section may be obtained using the Prandtl one-seventh power law and the Blasius equation. These two equations have been shown (Section 11.4) to be mutually consistent.

The Prandtl one-seventh power law gives:

$$\frac{u_x}{u_{CL}} = \left(\frac{y}{r}\right)^{1/7} \tag{equation 3.56}$$

Then:
$$u^+ = \frac{u_x}{u^*} = \frac{u_{CL}}{u^*}\left(\frac{y}{r}\right)^{1/7} = \frac{u_{CL}}{r^{1/7}}\frac{1}{u^*}\left(\frac{y^+\mu}{\rho u^*}\right)^{1/7} \tag{12.66}$$

The Blasius relation between friction factor and Reynolds number for turbulent flow is:

$$\phi = \frac{R}{\rho u^2} = 0.0396Re^{-1/4} \tag{equation 3.10}$$

Thus:
$$\frac{R}{\rho} = u^{*2} = 0.0396\rho^{-1/4}\mu^{1/4}(2r)^{-1/4}u^{1.75}$$

Again, from the Prandtl one-seventh power law:

$$\frac{u}{u_{CL}} = \frac{49}{60} \tag{equation 3.60}$$

Thus:

$$\frac{u_{CL}}{r^{1/2}} = (0.0396)^{-1/1.75}2^{1/7}\rho^{-1/7}\mu^{1/7}u^{*8/7}\frac{60}{49}$$

$$= 8.56\rho^{-1/7}\mu^{1/7}u^{*8/7}$$

Substituting in equation 12.66:

$$u^+ = (8.56\rho^{-1/7}\mu^{1/7}u^{*8/7})\frac{1}{u^*}\left(\frac{\mu}{\rho u^*}\right)^{1/7} y^{+1/7}$$

i.e.: $$u^+ = 8.56y^{+1/7}$$ (12.67)

In Table 12.1, the values of u^+ calculated from equation 12.67 are compared with those given by the universal velocity profile (equations 12.37, 12.40 and 12.42). It will be seen that there is almost exact correspondence at $y^+ = 1000$ and differences are less than 6 per cent in the range $30 < y^+ < 4000$.

Table 12.1.   Comparison of values of u^+ calculated from equation 12.67 with those given by the universal velocity profile

y^+	u^+ from equation 12.67	u^+ from UVP	% Difference (based on column 3)
10	11.89	8.46	+40.5
15	12.60	10.49	+20.1
20	13.13	11.93	+10.1
25	13.56	14.65	−7.4
30	13.92	14.00	−0.0
100	16.53	17.01	−2.9
300	19.33	19.76	−2.2
1,000	22.96	22.76	+0.0
2,000	25.35	24.50	+3.5
3,000	26.87	25.52	+5.3
4,000	27.99	26.23	+6.7
5,000	28.90	26.79	+7.9
10,000	31.91	28.52	+11.9
100,000	44.34	34.28	+29.3

12.4.9. Effect of curvature of pipe wall on shear stress

Close to the wall of a pipe, the effect of the curvature of the wall has been neglected and the shear stress in the fluid has been taken to be independent of the distance from the wall. However, this assumption is not justified near the centre of the pipe.

As shown in Chapter 3, the shear stress varies linearly over the cross-section rising from zero at the axis of the pipe to a maximum value at the walls.

A force balance taken over the whole cross section gives:

$$-\frac{dP}{dx}\pi r^2 = -R_0 2\pi r$$

Thus: $$-R_0 = -\frac{dP}{dx}\frac{r}{2}$$ (12.68)

Taking a similar force balance over the central core of fluid lying at distances greater than y from the wall; that is for a plug of radius $(r - y)$:

$$-R_y = -\frac{dP}{dx}\frac{r - y}{2}$$ (12.69)

Thus:
$$\frac{R_y}{R_0} = 1 - \frac{y}{r} \qquad (12.70)$$

At radius $(r - y)$, equation 12.61 becomes:
$$R_y = -(\mu + E\rho)\frac{du_x}{dy} \qquad (12.71)$$

Neglecting μ/ρ compared with E, and substituting for R_y from equation 12.70:
$$\frac{-R_0}{\rho}\left(1 - \frac{y}{r}\right) = E\frac{du_x}{dy}$$

This leads to:
$$E = 0.4y + \frac{\mu}{\rho}\left(1 - \frac{y}{r}\right) \qquad (12.72a)$$

$$= 0.4u^*y\left(1 - \frac{y}{r}\right) \qquad (12.72b)$$

$$= 0.4\phi^{1/2}uy\left(1 - \frac{y}{r}\right) \qquad (12.72c)$$

12.5. FRICTION FACTOR FOR A SMOOTH PIPE

Equation 12.37 can be used in order to calculate the friction factor $\phi = R/\rho u^2$ for the turbulent flow of fluid in a pipe. It is first necessary to obtain an expression for the mean velocity u of the fluid from the relation:

$$u = \frac{\displaystyle\int_0^r [2\pi(r - y)\,dy\,u_x]}{\pi r^2}$$

$$= 2\int_0^1 u_x\left(1 - \frac{y}{r}\right)d\left(\frac{y}{r}\right) \qquad (12.73)$$

Now the velocity at the pipe axis u_s is obtained by putting $y = r$ into equation 12.37.

Thus:
$$u_s = u^*\left(2.5\ln\frac{r\rho u^*}{\mu} + 5.5\right) \qquad (12.74)$$

This is not strictly justified because equation 12.74 gives a finite, instead of zero, velocity gradient when applied at the centre of the pipe.

Substituting for u_x in equation 12.73 from equation 12.34:

$$u = 2\int_0^1\left(u_s - 2.5u^*\ln\frac{r}{y}\right)\left(1 - \frac{y}{r}\right)d\left(\frac{y}{r}\right)$$

$$\frac{u}{u_s} = 2\int_0^1\left(1 + 2.5\frac{u^*}{u_s}\ln\frac{y}{r}\right)\left(1 - \frac{y}{r}\right)d\left(\frac{y}{r}\right)$$

$$= 2\left[\frac{y}{r} - \frac{1}{2}\left(\frac{y}{r}\right)^2\right]_0^1 + 5.0\frac{u^*}{u_s}\left\{\left[\left(\ln\frac{y}{r}\right)\left[\frac{y}{r} - \frac{1}{2}\left(\frac{y}{r}\right)^2\right]\right]_0^1\right.$$

$$\left. - \int_0^1\left(\frac{y}{r}\right)^{-1}\left[\frac{y}{r} - \frac{1}{2}\left(\frac{y}{r}\right)^2\right]d\left(\frac{y}{r}\right)\right\}$$

$$= 1 + 5.0\frac{u^*}{u_s} \left\{ 0 - \left[\left(\frac{y}{r}\right) - \frac{1}{4}\left(\frac{y}{r}\right)^2 \right]_0^1 \right\}$$

$$= 1 + 5\frac{u^*}{u_s}\left(-\frac{3}{4}\right)$$

$$= 1 - 3.75\frac{u^*}{u_s} \qquad\qquad (12.75)$$

Substituting into equation 12.74:

$$u + 3.75u^* = u^* \left\{ 2.5\ln\left[\left(\frac{d\rho u}{\mu}\right)\left(\frac{r}{d}\right)\left(\frac{u^*}{u}\right)\right] + 5.5 \right\}$$

$$\frac{u}{u^*} = 2.5\ln\left\{ (Re)\frac{u^*}{u} \right\}$$

Now:
$$\phi = \frac{R}{\rho u^2} = \left(\frac{u^*}{u}\right)^2$$

$$\therefore \qquad \phi^{-1/2} = 2.5\ln[(Re)\phi^{1/2}] \qquad\qquad (12.76)$$

The experimental results for ϕ as a function of Re closely follow equation 12.74 modified by a correction term of 0.3 to give:

$$\phi^{-1/2} = 2.5\ln[(Re)\phi^{1/2}] + 0.3 \qquad\qquad (12.77)$$

The correction is largely associated with the errors involved in using equation 12.74 at the pipe axis.

Equation 12.77 is identical to equation 3.11.

Example 12.1

Air flows through a smooth circular duct of internal diameter 250 mm at an average velocity of 15 m/s. Calculate the fluid velocity at points 50 mm and 5 mm from the wall. What will be the thickness of the laminar sub-layer if this extends to $u^+ = y^+ = 5$? The density and viscosity of air may be taken as 1.10 kg/m^3 and 20×10^{-6} N s/m^2 respectively.

Solution

Reynolds number: $Re = \dfrac{0.250 \times 15 \times 1.10}{20 \times 10^{-6}} = 2.06 \times 10^5$

Hence, from Figure 3.7: $\dfrac{R}{\rho u^2} = 0.0018$

$$u_s = \frac{u}{0.817} = \frac{15}{0.817} = 18.4 \text{ m/s}$$

$$u^* = u\sqrt{\frac{R}{\rho u^2}} = 15\sqrt{0.0018} = 0.636 \text{ m/s}$$

At 50 mm from the wall: $\dfrac{y}{r} = \dfrac{0.050}{0.125} = 0.40$

Hence, from equation 12.34:

$$u_x = u_s + 2.5u^* \ln\left(\frac{y}{r}\right)$$

$$= 18.4 + 2.5 \times 0.636 \ln 0.4$$

$$= 16.9 \text{ m/s}$$

At 5 mm from the wall: $y/r = 0.005/0.125 = 0.04$

Hence: $u_x = 18.4 + 2.5 \times 0.636 \ln 0.04$

$$= 13.3 \text{ m/s}$$

The thickness of the laminar sub-layer is given by equation 12.54:

$$\delta_b = \frac{5d}{Re\sqrt{(R/\rho u^2)}}$$

$$= \frac{5 \times 0.250}{2.06 \times 10^5 \sqrt{(0.0018)}}$$

$$= 1.43 \times 10^{-4} \text{ m}$$

or: 0.143 mm

12.6. EFFECT OF SURFACE ROUGHNESS ON SHEAR STRESS

Experiments have been carried out on artificially roughened surfaces in order to determine the effect of obstructions of various heights[5,6]. Experimentally, it has been shown that the shear force is not affected by the presence of an obstruction of height e unless:

$$\frac{u_e e\rho}{\mu} > 40 \tag{12.78}$$

where u_e is the velocity of the fluid at a distance e above the surface.

If the obstruction lies entirely within the laminar sub-layer the velocity u_e is given by:

$$R = \mu\left(\frac{du_x}{dy}\right)_{y=0}$$

$$= \mu\left(\frac{u_e}{e}\right), \quad \text{approximately}$$

The shearing stress velocity:

$$u^* = \sqrt{\frac{R}{\rho}} = \sqrt{\frac{\mu u_e}{\rho e}}$$

so that: $$u_e = \frac{\rho e}{\mu} u^{*2}$$

Thus: $$\frac{u_e e\rho}{\mu} = \frac{e\rho}{\mu} u^{*2} \frac{e\rho}{\mu} = \left(\frac{e\rho u^*}{\mu}\right)^2 \tag{12.79}$$

$e\rho u^*/\mu$ is known as the *roughness Reynolds number*, Re_r.

For the flow of a fluid in a pipe:

$$u^* = \sqrt{\frac{R}{\rho}} = u\phi^{1/2}$$

where u is the mean velocity over the whole cross-section. Re_r will now be expressed in terms of the three dimensionless groups used in the friction chart.

Thus:
$$Re_r = \frac{e\rho u^*}{\mu}$$

$$= \left(\frac{d\rho u}{\mu}\right)\left(\frac{e}{d}\right)\phi^{1/2}$$

$$= Re\left(\frac{e}{d}\right)\phi^{1/2} \tag{12.80}$$

The shear stress should then be unaffected by the obstruction if $Re_r < \sqrt{40}$, that is if $Re_r < 6.5$.

If the surface has a number of closely spaced obstructions, however, all of the same height e, the shear stress is affected when $Re_r >$ about 3. If the obstructions are of varying heights, with e as the arithmetic mean, the shear stress is increased if $Re_r >$ about 0.3, because the effect of one relatively large obstruction is greater than that of several small ones.

For hydrodynamically smooth pipes, through which fluid is flowing under turbulent conditions, the shear stress is given approximately by the Blasius equation:

$$\phi = \frac{R}{\rho u^2} \propto Re^{-1/4} \qquad \text{(from equation 11.39)}$$

so that:
$$R \propto u^{1.75}$$

and is independent of the roughness.

For smooth pipes, the frictional drag at the surface is known as *skin friction*. With rough pipes, however, an additional drag known as *form drag* results from the eddy currents caused by impact of the fluid on the obstructions and, when the surface is very rough, it becomes large compared with the skin friction. Since form drag involves dissipation of kinetic energy, the losses are proportional to the square of the velocity of the fluid, so that $R \propto u^2$. This applies when $Re_r > 50$.

Thus, when: $\qquad\qquad\qquad Re_r < 0.3, \quad R \propto u^{1.75}$

when: $\qquad\qquad\qquad\qquad Re_r > 50, \quad R \propto u^2$

and when: $\qquad\quad 0.3 < Re_r < 50, R \propto u^w \quad$ where $1.75 < w < 2$.

When the thickness of the laminar sub-layer is large compared with the height of the obstructions, the pipe behaves as a smooth pipe (when $e < \delta_b/3$). Since the thickness of the laminar sub-layer decreases as the Reynolds number is increased, a surface which is hydrodynamically smooth at low Reynolds numbers may behave as a rough surface at higher values. This explains the shapes of the curves obtained for ϕ plotted against Reynolds number (Figure 3.7). The curves, for all but the roughest of pipes, follow the curve for the smooth pipe at low Reynolds numbers and then diverge at higher values. The greater the roughness of the surface, the lower is the Reynolds number at which the curve starts to diverge. At high Reynolds numbers, the curves for rough pipes become parallel to the Reynolds number axis, indicating that skin friction is negligible and $R \propto u^2$. Under these conditions, the shear stress can be calculated from equation 3.13.

Nikuradse's data[5,6] for rough pipes gives:

$$u^+ = 2.5 \ln \left(\frac{y}{e} \right) + 8.5 \tag{12.81}$$

12.7. SIMULTANEOUS MOMENTUM, HEAT AND MASS TRANSFER

It has been seen that when there is a velocity gradient in a fluid, the turbulent eddies are responsible for transferring momentum from the regions of high velocity to those of low velocity; this gives rise to shear stresses within the field. It has been suggested that the eddy kinematic viscosity E can be written as the product of an eddy velocity u_E and a mixing length λ_E giving:

$$E \propto \lambda_E u_E \tag{equation 12.18}$$

If u_E is expressed as the product of the mixing length and the modulus of velocity gradient and if the proportionality constant is equal to unity:

$$E = \lambda_E^2 \left| \frac{du_x}{dy} \right| \tag{equation 12.23}$$

If there is a temperature gradient within the fluid, the eddies will be responsible for heat transfer and an eddy thermal diffusivity E_H may be defined in a similar way. It is suggested that, since the mechanism of transfer of heat by eddies is essentially the same as that for transfer of momentum, E_H is related to mixing length and velocity gradient in a similar manner.

Thus:
$$E_H = \lambda_E^2 \left| \frac{du_x}{dy} \right| \tag{12.82}$$

On a similar basis an eddy diffusivity for mass transfer E_D can be defined for systems in which concentration gradients exist as:

$$E_D = \lambda_E^2 \left| \frac{du_x}{dy} \right| \tag{12.83}$$

E/E_H is termed the *Turbulent Prandtl Number* and E/E_D the *Turbulent Schmidt Number*.

E, E_H and E_D are all nearly equal and therefore both the dimensionless numbers are approximately equal to unity.

Thus for the eddy transfer of heat:

$$q_y = -E_H \frac{d(C_p \rho \theta)}{dy} = -\lambda_E^2 \left| \frac{du_x}{dy} \right| \frac{d(C_p \rho \theta)}{dy} \tag{12.84}$$

and similarly for mass transfer:

$$N_A = -E_D \frac{dC_A}{dy} = -\lambda_E^2 \left| \frac{du_x}{dy} \right| \frac{dC_A}{dy} \tag{12.85}$$

In the neighbourhood of a surface du_x/dy will be positive and thus:

$$R_y = -\lambda_E^2 \frac{du_x}{dy} \frac{d(\rho u_x)}{dy} \tag{12.86}$$

$$q_y = -\lambda_E^2 \frac{du_x}{dy} \frac{d(C_p \rho \theta)}{dy} \tag{12.87}$$

and:

$$N_A = -\lambda_E^2 \frac{du_x}{dy} \frac{dC_A}{dy} \tag{12.88}$$

For conditions of constant density, equation 12.25 gives:

$$\sqrt{\frac{-R_y}{\rho}} = \lambda_E \frac{du_x}{dy} \qquad \text{(equation 12.25)}$$

When molecular and eddy transport both contribute significantly, it may be assumed as a first approximation that their effects are additive. Then:

$$R_y = -\left(\frac{\mu}{\rho} + E\right)\frac{d(\rho u_x)}{dy} = -\mu \frac{du_x}{dy} - \lambda_E^2 \rho \left(\frac{du_x}{dy}\right)^2 \tag{12.89}$$

$$q_y = -\left(\frac{k}{C_p \rho} + E_H\right)\frac{d(C_p \rho \theta)}{dy} = -k\frac{d\theta}{dy} - \lambda_E^2 \rho C_p \left(\frac{du_x}{dy}\right)\left(\frac{d\theta}{dy}\right) \tag{12.90}$$

$$N_A = -(D + E_D)\frac{dC_A}{dy} = -D\frac{dC_A}{dy} - \lambda_E^2 \left(\frac{du_x}{dy}\right)\left(\frac{dC_A}{dy}\right) \tag{12.91}$$

and:

$$\frac{\lambda_E}{y} \approx 0.4 \qquad \text{(equation 12.35)}$$

Whereas the kinematic viscosity μ/ρ, the thermal diffusivity $k/C_p\rho$, and the diffusivity D are physical properties of the system and can therefore be taken as constant provided that physical conditions do not vary appreciably, the eddy coefficients E, E_H, and E_D will be affected by the flow pattern and will vary throughout the fluid. Each of the eddy coefficients is proportional to the square of the mixing length. The mixing length will

Table 12.2. Relations between physical properties

	Molecular processes only	Molecular and eddy transfer together	Eddy transfer predominating
Momentum transfer	$R_y = -\dfrac{\mu}{\rho}\dfrac{d(\rho u_x)}{dy}$	$R_y = -\left(\dfrac{\mu}{\rho} + E\right)\dfrac{d(\rho u_x)}{dy}$	$R_y = -E\dfrac{d(\rho u_x)}{dy}$
Heat transfer	$q_y = -\dfrac{k}{C_p\rho}\dfrac{d(C_p\rho\theta)}{dy}$	$q_y = -\left(\dfrac{k}{C_p\rho} + E_H\right)\dfrac{d(C_p\rho\theta)}{dy}$	$q_y = -E_H\dfrac{d(C_p\rho\theta)}{dy}$
Mass transfer	$N_A = -D\dfrac{dC_A}{dy}$	$N_A = -(D + E_D)\dfrac{dC_A}{dy}$	$N_A = -E_D\dfrac{dC_A}{dy}$

where $E \approx E_H \approx E_D \approx \lambda_E^2 \left|\dfrac{du_x}{dy}\right|$ and $\lambda_E \approx 0.4y$

normally increase with distance from a surface, and the eddy coefficients will therefore increase rapidly with position.

The relations are summarised in Table 12.2.

12.8. REYNOLDS ANALOGY

12.8.1. Simple form of analogy between momentum, heat and mass transfer

The simple concept of the Reynolds Analogy was first suggested by REYNOLDS[7] to relate heat transfer rates to shear stress, but it is also applicable to mass transfer. It is assumed that elements of fluid are brought from remote regions to the surface by the action of the turbulent eddies; the elements do not undergo any mixing with the intermediate fluid through which they pass, and they instantaneously reach equilibrium on contact with the interfacial layers. An equal volume of fluid is, at the same time, displaced in the reverse direction. Thus in a flowing fluid there is a transference of momentum and a simultaneous transfer of heat if there is a temperature gradient, and of mass if there is a concentration gradient. The turbulent fluid is assumed to have direct access to the surface and the existence of a buffer layer and laminar sub-layer is neglected. Modification of the model has been made by TAYLOR[4] and PRANDTL[8,9] to take account of the laminar sub-layer. Subsequently, the effect of the buffer layer has been incorporated by applying the *universal velocity profile*.

Consider the equilibrium set up when an element of fluid moves from a region at high temperature, lying outside the boundary layer, to a solid surface at a lower temperature if no mixing with the intermediate fluid takes place. Turbulence is therefore assumed to persist right up to the surface. The relationship between the rates of transfer of momentum and heat can then be deduced as follows (Figure 12.5).

Figure 12.5.   The Reynolds analogy — momentum, heat and mass transfer

Consider the fluid to be flowing in a direction parallel to the surface (X-direction) and for momentum and heat transfer to be taking place in a direction at right angles to the surface (Y-direction positive away from surface).

Suppose a mass M of fluid situated at a distance from the surface to be moving with a velocity u_s in the X-direction. If this element moves to the surface where the velocity is zero, it will give up its momentum Mu_s in time t, say. If the temperature difference between the element and the surface is θ_s and C_p is the specific heat of the fluid, the heat transferred to the surface will be $\mathbf{M} \, C_p \theta_s$. If the surface is of area A, the rate of heat transfer is given by:

$$\frac{\mathbf{M} C_p \theta_s}{At} = -q_0 \qquad (12.92)$$

where $-q_0$ is the heat transferred to the surface per unit area per unit time (**NB**—the negative sign has been introduced as the positive direction is away from the surface).

If the shear stress at the surface is R_0, it will equal the rate of change in momentum, per unit area.

Thus:
$$\frac{\mathbf{M} u_s}{At} = -R_0. \qquad (12.93)$$

A similar argument can be applied to the mass transfer process when a concentration gradient exists. Thus, if the molar concentration of $\mathbf{A}$ remote from the surface is C_{As} and at the surface it is C_{Aw}, the moles of $\mathbf{A}$ transferred to the surface will be $(\mathbf{M}/\rho)(C_{As} - C_{Aw})$, if the density ($\rho$) can be assumed to be constant over the range of concentrations encountered. Thus the moles of $\mathbf{A}$ transferred to the surface per unit area and unit time $(-N_A)_{y=0}$ is given by:

$$\frac{1}{At} \frac{\mathbf{M}}{\rho}(C_{As} - C_{Aw}) = (-N_A)_{y=0} \qquad (12.94)$$

Dividing equation 12.92 by 12.93:

$$\frac{C_p \theta_s}{u_s} = \frac{-q_0}{-R_0} \qquad (12.95)$$

or:
$$\frac{-R_0}{u_s} = \frac{-q_0}{C_p \theta_s} \qquad (12.96)$$

Again dividing equation 12.94 by equation 12.93:

$$\frac{C_{As} - C_{Aw}}{\rho u_s} = \frac{(-N_A)_{y=0}}{-R_0}$$

or:
$$\frac{-R_0}{\rho u_s} = \frac{(-N_A)_{y=0}}{C_{As} - C_{Aw}} \qquad (12.97)$$

Now R_0 (the shear stress in the fluid at the surface) is equal and opposite to R, the shear stress acting on the surface, $-q_0/\theta_s$ is by definition the heat transfer coefficient at the surface (h), and $(-N_A)_{y=0}/(C_{As} - C_{Aw})$ is the mass transfer coefficient (h_D). Then dividing both sides of equation 12.96 by ρu_s and of equation 12.97 by u_s to make them dimensionless:

$$\frac{R}{\rho u_s^2} = \frac{h}{C_p \rho u_s} = St \qquad (12.98)$$

where St denotes the Stanton Number ($h/C_p \rho u_s$),

and:
$$\frac{R}{\rho u_s^2} = \frac{h_D}{u_s} \qquad (12.99)$$

h_D/u_s is sometimes referred to as the Stanton number for mass transfer.

Thus:
$$\frac{h_D}{u_s} = \frac{h}{C_p \rho u_s} \tag{12.100}$$

or:
$$h_D = \frac{h}{C_p \rho} \tag{12.101}$$

Equation 12.101 is often referred to as the *Lewis Relation*. It provides an approximate method for evaluating a mass transfer coefficient if the heat transfer coefficient is known. The assumption that the turbulent eddies can penetrate right up to the surface is justified however only in special circumstances and the problem is considered further in the next section.

For flow in a smooth pipe, the friction factor for turbulent flow is given approximately by the Blasius equation and is proportional to the Reynolds number (and hence the velocity) raised to a power of $-\frac{1}{4}$. From equations 12.98 and 12.99, therefore, the heat and mass transfer coefficients are both proportional to $u_s^{0.75}$.

The application of the analogies to the problems of heat and mass transfer to plane surfaces and to pipe walls for fully developed flow is discussed later.

12.8.2. Mass transfer with bulk flow

When the mass transfer process deviates significantly from equimolecular counter-diffusion, allowance must be made for the fact that there may be a very large difference in the molar rates of transfer of the two components. Thus, in a gas absorption process, there will be no transfer of the insoluble component **B** across the interface and only the soluble component **A** will be transferred. This problem will now be considered in relation to the Reynolds Analogy. However, it gives manageable results only if physical properties such as density are taken as constant and therefore results should be applied with care.

Consider the movement of an element of fluid consisting of n molar units of a mixture of two constituents **A** and **B** from a region outside the boundary layer, where the molecular concentrations are C_{As} and C_{Bs}, to the surface where the corresponding concentrations are C_{Aw} and C_{Bw}. The total molar concentration is everywhere C_T. The transfer is effected in a time t and takes place at an area A of surface.

There is no net transference of the component **B**. When n molar units of material are transferred from outside the boundary layer to the surface:

$$\text{Transfer of } \mathbf{A} \text{ towards surface} = n \frac{C_{As}}{C_T}$$

$$\text{Transfer of } \mathbf{B} \text{ towards surface} = n \frac{C_{Bs}}{C_T}$$

In this case the molar rate of transfer of **B** away from the surface is equal to the transfer towards the surface.

$$\text{Transfer of } \mathbf{B} \text{ away from surface} = n \frac{C_{Bs}}{C_T}$$

Associated transfer of **A** away from surface

$$= n \frac{C_{Bs}}{C_T} \frac{C_{Aw}}{C_{Bw}}$$

Thus the net transfer of **A** towards the surface

$$-N'_A At = n \left(\frac{C_{As}}{C_T} - \frac{C_{Bs}}{C_T} \frac{C_{Aw}}{C_{Bw}} \right)$$

$$= n \left(\frac{C_T C_{As} - C_{Aw} C_{As} - C_{Aw} C_T + C_{Aw} C_{As}}{C_T(C_T - C_{Aw})} \right)$$

$$= n \frac{(C_{As} - C_{Aw})}{C_{Bw}}$$

It is assumed that the total molar concentration is everywhere constant. Thus the rate of transfer per unit area and unit time is given by:

$$-N'_A = \frac{n(C_{As} - C_{Aw})}{C_{Bw} At} \tag{12.102}$$

The net transfer of momentum per unit time

$$= -R_0 A = \frac{n \rho u_s}{C_T t} \tag{12.103}$$

ρ is taken as the mean mass density of the fluid.

$$\therefore \qquad -R_0 = \frac{n \rho u_s}{C_T t A} \tag{12.104}$$

Dividing equations 12.102 and 12.104:

$$\frac{N'_A}{R_0} = \frac{C_{As} - C_{Aw}}{\rho u_s} \frac{C_T}{C_{Bw}} \tag{12.105}$$

Writing $R_0 = -R$ and defining the mass transfer coefficient by the relation:

$$\frac{-N'_A}{C_{As} - C_{Aw}} = h_D \tag{12.106}$$

$$\frac{h_D}{u_s} \frac{C_{Bw}}{C_T} = \frac{R}{\rho u_s^2} \tag{12.107}$$

Thus, there is a direct proportionality between the momentum transfer and that portion of the mass transfer which is *not* attributable to bulk flow.

For heat transfer:

$$\frac{R}{\rho u_s^2} = \frac{h}{C_p \rho u_s} \tag{equation 12.98}$$

Thus for simultaneous heat transfer and mass transfer with bulk flow, combination of equations 12.107 and 12.98 gives:

$$\frac{h_D}{u_s} \frac{C_{Bw}}{C_T} = \frac{h}{C_p \rho u_s}$$

or:

$$h_D \frac{C_{Bw}}{C_T} = \frac{h}{C_p \rho} \tag{12.108}$$

Equation 12.108 is the form of the *Lewis Relation* which is applicable to mass transfer with bulk flow.

12.8.3. Taylor–Prandtl modification of Reynolds analogy for heat transfer and mass transfer

The original Reynolds analogy involves a number of simplifying assumptions which are justifiable only in a limited range of conditions. Thus it was assumed that fluid was transferred from outside the boundary layer to the surface without mixing with the intervening fluid, that it was brought to rest at the surface, and that thermal equilibrium was established. Various modifications have been made to this simple theory to take account of the existence of the laminar sub-layer and the buffer layer close to the surface.

TAYLOR[4] and PRANDTL[8,9] allowed for the existence of the laminar sub-layer but ignored the existence of the buffer layer in their treatment and assumed that the simple Reynolds analogy was applicable to the transfer of heat and momentum from the main stream to the edge of the laminar sub-layer of thickness δ_b. Transfer through the laminar sub-layer was then presumed to be attributable solely to molecular motion.

If αu_s and $b\theta_s$ are the velocity and temperature, respectively, at the edge of the laminar sub-layer (see Figure 12.5), applying the Reynolds analogy (equation 12.95) for transfer across the turbulent region:

$$\frac{-q_0}{-R_0} = \frac{C_p(\theta_s - b\theta_s)}{u_s - \alpha u_s} \tag{12.109}$$

The rate of transfer of heat by conduction through the laminar sub-layer from a surface of area A is given by:

$$-q_0 A = \frac{k b \theta_s A}{\delta_b} \tag{12.110}$$

The rate of transfer of momentum is equal to the shearing force and therefore:

$$-R_0 A = \frac{\mu \alpha u_s A}{\delta_b} = RA \tag{12.111}$$

Dividing equations 12.110 and 12.111:

$$\frac{-q_0}{-R_0} = \frac{k b \theta_s}{\mu \alpha u_s} \tag{12.112}$$

Thus from equations 12.109 and 12.112:

$$\frac{k b \theta_s}{\mu \alpha u_s} = \frac{C_p(1 - b)\theta_s}{(1 - \alpha)u_s}$$

$$\therefore \quad \frac{Pr(1 - b)}{b} = \frac{(1 - \alpha)}{\alpha}$$

$$\therefore \quad \frac{b}{\alpha} = \frac{1}{\alpha + (1 - \alpha)Pr^{-1}}$$

Substituting in equation 12.112:

$$\frac{-q_0}{-R_0} = \frac{C_p \theta_s}{u_s} \frac{1}{1 + \alpha(Pr - 1)} = \frac{h\theta_s}{R}$$

or:

$$St = \frac{h}{C_p \rho u_s} = \frac{R/\rho u_s^2}{1 + \alpha(Pr - 1)} \qquad (12.113)$$

The quantity α, which is the ratio of the velocity at the edge of the laminar sub-layer to the stream velocity, was evaluated in Chapter 11 in terms of the Reynolds number for flow over the surface. For flow over a plane surface, from Chapter 11:

$$\alpha = 2.1 Re_x^{-0.1} \qquad \text{(equation 11.28)}$$

where Re_x is the Reynolds number $u_s x \rho / \mu$, x being the distance from the leading edge of the surface.

For flow through a pipe of diameter d (see Chapter 11):

$$\alpha = 2.0 Re^{-1/8} \qquad \text{(equation 11.41)}$$

where Re is the Reynolds number $ud\rho/\mu$.

Alternatively, from equation 12.59:

$$\alpha = \frac{u_b}{u} = 11.6\phi^{1/2} \approx 11.6 \left(\frac{R}{\rho u_s^2}\right)^{1/2}$$

For mass transfer to a surface, a similar relation to equation 12.113 can be derived for equimolecular counterdiffusion except that the Prandtl number is replaced by the Schmidt number. It follows that:

$$\frac{h_D}{u_s} = \frac{R/\rho u_s^2}{1 + \alpha(Sc - 1)} \qquad (12.114)$$

where Sc is the Schmidt number $(\mu/\rho D)$. It is possible also to derive an expression to take account of bulk flow, but many simplifying assumptions must be made and the final result is not very useful.

It is thus seen that by taking account of the existence of the laminar sub-layer, correction factors arc introduced into the simple Reynolds analogy.

For heat transfer, the factor is $[1 + \alpha(Pr - 1)]$ and for mass transfer it is $[1 + \alpha(Sc - 1)]$.

There are two sets of conditions under which the correction factor approaches unity:

(i) For gases both the Prandtl and Schmidt groups are approximately unity, and therefore the simple Reynolds analogy is closely followed. Furthermore, the Lewis relation which is based on the simple analogy would be expected to hold closely for gases. The Lewis relation will also hold for any system for which the Prandtl and Schmidt numbers are equal. In this case, the correction factors for heat and mass transfer will be approximately equal.

(ii) When the fluid is highly turbulent, the laminar sub-layer will become very thin and the velocity at the edge of the laminar sub-layer will be small. In these circumstances again, the correction factor will approach unity.

Equations 12.113 and 12.114 provide a means of expressing the mass transfer coefficient in terms of the heat transfer coefficient.

From equation 12.113:

$$\frac{R}{\rho u_s^2} = \frac{h}{C_p \rho u_s}[1 + \alpha(Pr - 1)] \tag{12.115}$$

From equation 12.114:

$$\frac{R}{\rho u_s^2} = \frac{h_D}{u_s}[1 + \alpha(Sc - 1)] \tag{12.116}$$

Thus:

$$h_D = \frac{h}{C_p \rho} \frac{1 + \alpha(Pr - 1)}{1 + \alpha(Sc - 1)} \tag{12.117}$$

Equation 12.117 is a modified form of the *Lewis Relation*, which takes into account the resistance to heat and mass transfer in the laminar sub-layer.

12.8.4. Use of universal velocity profile in Reynolds analogy

In the Taylor–Prandtl modification of the theory of heat transfer to a turbulent fluid, it was assumed that the heat passed directly from the turbulent fluid to the laminar sub-layer and the existence of the buffer layer was neglected. It was therefore possible to apply the simple theory for the boundary layer in order to calculate the heat transfer. In most cases, the results so obtained are sufficiently accurate, but errors become significant when the relations are used to calculate heat transfer to liquids of high viscosities. A more accurate expression can be obtained if the temperature difference across the buffer layer is taken into account. The exact conditions in the buffer layer are difficult to define and any mathematical treatment of the problem involves a number of assumptions. However, the conditions close to the surface over which fluid is flowing can be calculated approximately using the *universal velocity profile*.[10]

The method is based on the calculation of the total temperature difference between the fluid and the surface, by adding the components attributable to the laminar sub-layer, the buffer layer and the turbulent region. In the steady state, the heat flux (q_0) normal to the surface will be constant if the effects of curvature are neglected.

Laminar sub-layer $(0 < y^+ < 5)$

Since the laminar sub-layer is thin, the temperature gradient may be assumed to be approximately linear (see also Section 11.6.1).

Thus:

$$q_0 = -k\frac{\theta_5}{y_5} \tag{12.118}$$

where y_5 and θ_5 are respectively the values of y and θ at $y^+ = 5$. As before, the temperature scale is so chosen that the surface temperature is zero.

Now, by definition:

$$y^+ = \frac{y u^* \rho}{\mu}$$

Thus:
$$y_5 = \frac{5\mu}{u^*\rho}$$

Substituting in equation 12.118:
$$\theta_5 = \frac{-q_0}{k} \frac{5\mu}{u^*\rho} \qquad (12.119)$$

Buffer layer $(5 < y^+ < 30)$

It has already been shown that it is reasonable to assume that the eddy kinematic viscosity E and the eddy thermal diffusivity E_H are equal. Now the variation of E through the buffer zone is given by:
$$E = \frac{\mu}{\rho}\left(\frac{y^+}{5} - 1\right) \qquad \text{(equation 12.63)}$$

The heat transfer rate in the buffer zone is given by:
$$q_0 = -(k + E_H C_p \rho)\frac{d\theta}{dy} \qquad \text{(from equation 12.90)}$$

Substituting from equation 12.63 and putting $E_H = E$:
$$-q_0 = \left[k + C_p \rho \frac{\mu}{\rho}\left(\frac{y^+}{5} - 1\right)\right]\frac{d\theta}{dy} \qquad (12.120)$$

From the definition of y^+:
$$\frac{d\theta}{dy} = \frac{u^*\rho}{\mu}\frac{d\theta}{dy^+} \qquad (12.121)$$

Substituting from equation 12.121 into equation 12.120:
$$-q_0 = \rho u^*\left[\frac{k}{\mu} + C_p\left(\frac{y^+}{5} - 1\right)\right]\frac{d\theta}{dy^+}$$
$$= \frac{C_p \rho u^*}{5}\left[\frac{5k}{C_p \mu} - 5 + y^+\right]\frac{d\theta}{dy^+}$$

$$\therefore \qquad \frac{d\theta}{dy^+} = \frac{-5q_0}{C_p \rho u^*[5(Pr^{-1} - 1) + y^+]} \qquad (12.122)$$

Integrating between the limits of $y^+ = 5$ and $y^+ = 30$:
$$\theta_{30} - \theta_5 = \frac{-5q_0}{C_p \rho u^*} \ln(5Pr + 1). \qquad (12.123)$$

Turbulent zone $(y^+ > 30)$

In this region the Reynolds analogy can be applied. Equation 12.109 becomes:
$$\frac{-q_0}{-R_0} = \frac{C_p(\theta_s - \theta_{30})}{u_s - u_{30}}$$

The velocity at $y^+ = 30$ for a smooth surface is given by:

$$\frac{u_x}{u^*} = u^+ = 5.0 \ln y^+ - 3.05 \qquad \text{(equation 12.42)}$$

$$\therefore \qquad u_{30} = u^*(5.0 \ln 30 - 3.05)$$

and:
$$\theta_s - \theta_{30} = \frac{-q_0}{C_p \rho u^{*2}}(u_s - 5.0 u^* \ln 30 + 3.05 u^*) \qquad (12.124)$$

The overall temperature difference θ_s is obtained by addition of equations 12.119, 12.123, and 12.124:

$$\text{Thus} \quad \theta_s = \frac{-q_0}{C_p \rho u^{*2}} \left[\frac{5 C_p \mu}{k} u^* + 5 u^* \ln(5 Pr + 1) + u_s - 5.0 u^* \ln 30 + 3.05 u^* \right]$$

$$= \frac{-q_0}{C_p \rho u^{*2}} \left[5 Pr\, u^* + 5 u^* \ln \left(\frac{Pr}{6} + \frac{1}{30} \right) + u_s + 3.05 u^* \right] \qquad (12.125)$$

i.e.:

$$\frac{u^{*2}}{u_s^2} = \frac{-q_0}{C_p \rho u_s \theta_s} \left\{ 1 + 5 \frac{u^*}{u_s} \left[Pr + \ln \left(\frac{Pr}{6} + \frac{1}{30} \right) + \ln 5 - 1 \right] \right\} \qquad \text{(since } 5 \ln 5 = 8.05\text{)}$$

$$\therefore \qquad \frac{R}{\rho u_s^2} = \frac{h}{C_p \rho u_s} \left\{ 1 + 5 \sqrt{\frac{R}{\rho u_s^2}} \left[(Pr - 1) + \ln \left(\frac{5}{6} Pr + \frac{1}{6} \right) \right] \right\} \qquad (12.126)$$

where:
$$\frac{h}{C_p \rho u_s} = St \quad \text{(the Stanton number)}$$

A similar expression can also be derived for mass transfer in the absence of bulk flow:

$$\frac{R}{\rho u_s^2} = \frac{h_D}{u_s} \left\{ 1 + 5 \sqrt{\frac{R}{\rho u_s^2}} \left[(Sc - 1) + \ln \left(\frac{5}{6} Sc + \frac{1}{6} \right) \right] \right\} \qquad (12.127)$$

12.8.5. Flow over a plane surface

The simple Reynolds analogy gives a relation between the friction factor $R/\rho u_s^2$ and the Stanton number for heat transfer:

$$\frac{R}{\rho u_s^2} = \frac{h}{C_p \rho u_s} \qquad \text{(equation 12.98)}$$

This equation can be used for calculating the point value of the heat transfer coefficient by substituting for $R/\rho u_s^2$ in terms of the Reynolds group Re_x using equation 11.32:

$$\frac{R}{\rho u_s^2} = 0.03 Re_x^{-0.2} \qquad \text{(equation 11.32)}$$

$$\therefore \qquad St = \frac{h}{C_p \rho u_s} = 0.03 Re_x^{-0.2} \qquad (12.128)$$

Equation 12.128 gives the point value of the heat transfer coefficient. If the whole surface is effective for heat transfer, the mean value is given by:

$$St = \frac{h}{C_p \rho u_s} = \frac{1}{x} \int_0^x 0.03 Re_x^{-0.2} \, dx$$

$$= 0.037 Re_x^{-0.2} \qquad\qquad (12.129)$$

These equations take no account of the existence of the laminar sub-layer and therefore give unduly high values for the transfer coefficient, especially with liquids. The effect of the laminar sub-layer is allowed for by using the Taylor–Prandtl modification:

$$St = \frac{h}{C_p \rho u_s} = \frac{R/\rho u_s^2}{1 + \alpha(Pr - 1)} \qquad\qquad \text{(equation 12.113)}$$

where:

$$\alpha = \frac{u_b}{u_s} = 2.1 Re_x^{-0.1} \qquad\qquad \text{(equation 11.28)}$$

Thus:

$$St = Nu_x Re_x^{-1} Pr^{-1} = \frac{0.03 Re_x^{-0.2}}{1 + 2.1 Re_x^{-0.1}(Pr - 1)} \qquad\qquad (12.130)$$

This expression will give the point value of the Stanton number and hence of the heat transfer coefficient. The mean value over the whole surface is obtained by integration. No general expression for the mean coefficient can be obtained and a graphical or numerical integration must be carried out after the insertion of the appropriate values of the constants.

Similarly, substitution may be made from equation 11.32 into equation 12.126 to give the point values of the Stanton number and the heat transfer coefficient, thus:

$$St = \frac{h}{C_p \rho u_s} = \frac{0.03 Re_x^{-0.2}}{1 + 0.87 Re_x^{-0.1}[(Pr - 1) + \ln(\frac{5}{6}Pr + \frac{1}{6})]} \qquad\qquad (12.131)$$

Mean values may be obtained by graphical or numerical integration.

The same procedure may be used for obtaining relationships for mass transfer coefficients, for equimolecular counterdiffusion or where the concentration of the non-diffusing constituent is small:

$$\frac{R}{\rho u_s^2} = \frac{h_D}{u_s} \qquad\qquad \text{(equation 12.99)}$$

For flow over a plane surface, substitution from equation 11.32 gives:

$$\frac{h_D}{u_s} = 0.03 Re_x^{-0.2} \qquad\qquad (12.132)$$

Equation 12.132 gives the point value of h_D. The mean value over the surface is obtained in the same manner as equation 12.129 as:

$$\frac{h_D}{u_s} = 0.037 Re_x^{-0.2} \qquad\qquad (12.133)$$

For mass transfer through a stationary second component:

$$\frac{R}{\rho u_s^2} = \frac{h_D}{u_s} \frac{C_{Bw}}{C_T} \qquad\qquad \text{(equation 12.107)}$$

The correction factor C_{Bw}/C_T must then be introduced into equations 12.132 and 12.133.

The above equations are applicable only when the Schmidt number Sc is very close to unity or where the velocity of flow is so high that the resistance of the laminar sub-layer is small. The resistance of the laminar sub-layer can be taken into account, however, for equimolecular counterdiffusion or for low concentration gradients by using equation 12.114.

$$\frac{h_D}{u_s} = \frac{R/\rho u_s^2}{1 + \alpha(Sc - 1)} \qquad \text{(equation 12.114)}$$

Substitution for $R/\rho u_s^2$ and α using equations 11.32 and 11.28 gives:

$$\frac{h_D}{u_s} = Sh_x Re_x^{-1} Sc^{-1} = \frac{0.03 Re_x^{-0.2}}{1 + 2.1 Re_x^{-0.1}(Sc - 1)} \qquad (12.134)$$

12.8.6. Flow in a pipe

For the inlet length of a pipe in which the boundary layers are forming, the equations in the previous section will give an approximate value for the heat transfer coefficient. It should be remembered, however, that the flow in the boundary layer at the entrance to the pipe may be streamline and the point of transition to turbulent flow is not easily defined. The results therefore are, at best, approximate.

In fully developed flow, equations 12.98 and 12.113 can be used, but it is preferable to work in terms of the mean velocity of flow and the ordinary pipe Reynolds number Re. Furthermore, the heat transfer coefficient is generally expressed in terms of a driving force equal to the difference between the bulk fluid temperature and the wall temperature. If the fluid is highly turbulent, however, the bulk temperature will be quite close to the temperature θ_s at the axis.

Into equations 12.95 and 12.113, equation 3.60 may then be substituted:

$$u = 0.817 u_s \qquad \text{(equation 3.60)}$$

$$\frac{u_b}{u} = 2.49 Re^{-1/8} \qquad \text{(equation 11.40)}$$

$$\frac{R}{\rho u^2} = 0.0396 Re^{-1/4} \qquad \text{(equation 11.39)}$$

Firstly, using the simple Reynolds analogy (equation 12.98):

$$\frac{h}{C_p \rho u} = \left(\frac{h}{C_p \rho u_s}\right)\left(\frac{u_s}{u}\right)$$

$$= \left(\frac{R}{\rho u_s^2}\right)\left(\frac{u_s}{u}\right)$$

$$= \left(\frac{R}{\rho u^2}\right)\left(\frac{u}{u_s}\right)$$

$$= 0.032 Re^{-1/4} \qquad (12.135)$$

Then, using the Taylor–Prandtl modification (equation 12.113):

$$\frac{h}{C_p \rho u} = \frac{0.032 Re^{-1/4}}{1 + (u_b/u)(u/u_s)(Pr - 1)}$$

$$= \frac{0.032 Re^{-1/4}}{1 + 2.0 Re^{-1/8}(Pr - 1)} \tag{12.136}$$

Finally, using equation 12.131:

$$St = \frac{h}{C_p \rho u} = \frac{0.817(R/\rho u^2)}{1 + 0.817\sqrt{(R/\rho u^2)}\, 5[(Pr - 1) + \ln(\frac{5}{6}Pr + \frac{1}{6})]}$$

$$= \frac{0.032 Re^{-1/4}}{1 + 0.817 Re^{-1/8}[(Pr - 1) + \ln(\frac{5}{6}Pr + \frac{1}{6})]} \tag{12.137}$$

For mass transfer, the equations corresponding to equations 12.133 and 12.134 are obtained in the same way as the analogous heat transfer equations.

Thus, using the simple Reynolds analogy for equimolecular counterdiffusion:

$$\frac{h_D}{u} = 0.032 Re^{-1/4} \tag{12.138}$$

and for diffusion through a stationary gas:

$$\left(\frac{h_D}{u}\right)\left(\frac{C_{Bw}}{C_T}\right) = 0.032 Re^{-1/4} \tag{12.139}$$

Using the Taylor–Prandtl form for equimolecular counterdiffusion or low concentration gradients:

$$\frac{h_D}{u} = \frac{0.032 Re^{-1/4}}{1 + 2.0 Re^{-1/8}(Sc - 1)} \tag{12.140}$$

Example 12.2

Water flows at 0.50 m/s through a 20 mm tube lined with β-naphthol. What is the mass transfer coefficient if the Schmidt number is 2330?

Solution

Reynolds number: $Re = (0.020 \times 0.50 \times 1000/1 \times 10^{-3}) = 10,000$
    From equation 12.140:

$$\frac{h_D}{u} = 0.032 Re^{-1/4}[1 + 2.0(Sc - 1)Re^{-1/8}]^{-1}$$

$$\therefore \qquad h_D = 0.032 \times 0.50 \times 0.1[1 + 2 \times 2329 \times 0.316]^{-1}$$

$$= 1.085 \times 10^{-6} \text{ kmol/m}^2 \text{ s (kmol/m}^3) = \underline{\underline{1.085 \times 10^{-6}}} \text{ m/s}$$

Example 12.3

Calculate the rise in temperature of water which is passed at 3.5 m/s through a smooth 25 mm diameter pipe, 6 m long. The water enters at 300 K and the tube wall may be assumed constant at 330 K. The following methods may be used:

(a) the simple Reynolds analogy (equation 12.135);
(b) the Taylor–Prandtl modification (equation 12.136);
(c) the universal velocity profile (equation 12.137);
(d) $Nu = 0.023Re^{0.8}Pr^{0.33}$ (equation 9.57).

Solution

Taking the fluid properties at 310 K and assuming that fully developed flow exists, an approximate solution will be obtained neglecting the variation of properties with temperature.

$$Re = \frac{0.025 \times 3.5 \times 1000}{0.7 \times 10^{-3}} = 1.25 \times 10^5$$

$$Pr = \frac{4.18 \times 10^3 \times 0.7 \times 10^{-3}}{0.65} = 4.50$$

(a) *Reynolds analogy*

$$\frac{h}{C_p \rho u} = 0.032Re^{-0.25} \qquad\qquad\qquad \text{(equation 12.135)}$$

$$h = 4.18 \times 1000 \times 1000 \times 3.5 \times 0.032(1.25 \times 10^5)^{-0.25}$$

$$= 24{,}902 \text{ W/m}^2\text{ K} \quad \text{or} \quad \underline{\underline{24.9 \text{ kW/m}^2\text{ K}}}$$

Heat transferred per unit time in length dL of pipe $= h \pi 0.025 \, \mathrm{d}L \, (330 - \theta)$ kW, where θ is the temperature at a distance L m from the inlet.

Rate of increase of heat content of fluid $= \dfrac{\pi}{4}0.025^2 \times 3.5 \times 1000 \times 4.18 \, \mathrm{d}\theta$ kW

Outlet temperature θ' is then given by:

$$\int_{300}^{\theta'} \frac{\mathrm{d}\theta}{(330 - \theta)} = 0.0109h \int_0^6 \mathrm{d}L$$

where h is in kW/m^2 K.

$$\therefore \qquad\qquad \log_{10}(330 - \theta') = \log_{10} 30 - \frac{0.0654h}{2.303} = 1.477 - 0.0283h$$

In this case: $\qquad\qquad\qquad\qquad h = 24.9 \text{ kW/m}^2\text{ K}$

$$\therefore \qquad\qquad \log_{10}(330 - \theta') = 1.477 - 0.705 = 0.772$$

$$\therefore \qquad\qquad\qquad\qquad \underline{\underline{\theta' = 324.1 \text{ K}}}$$

(b) *Taylor–Prandtl equation*

$$\frac{h}{C_p \rho u} = 0.032Re^{-1/4}[1 + 2.0Re^{-1/8}(Pr - 1)]^{-1} \qquad \text{(equation 12.136)}$$

$$\therefore \qquad\qquad h = \frac{24.9}{1 + 2.0 \times 3.5/4.34}$$

$$= \underline{\underline{9.53 \text{ kW/m}^2\text{ K}}}$$

$$\log_{10}(330 - \theta') = 1.477 - (0.0283 \times 9.53) = 1.207$$

$$\theta' = 313.9 \text{ K}$$

(c) *Universal velocity profile equation*

$$\frac{h}{C_p \rho u} = 0.032 Re^{-1/4}\{1 + 0.82 Re^{-1/8}[(Pr - 1) + \ln(0.83 Pr + 0.17)]\}^{-1} \quad \text{(equation 12.137)}$$

$$= \frac{24.9}{1 + (0.82/4.34)(3.5 + 2.303 \times 0.591)}$$

$$= 12.98 \text{ kW/m}^2 \text{ K}$$

$$\log_{10}(330 - \theta') = 1.477 - (0.0283 \times 12.98) = 1.110$$

$$\theta' = 317.1 \text{ K}$$

(d) $Nu = 0.023 Re^{0.8} Pr^{0.33}$

$$h = \frac{0.023 \times 0.65}{0.0250}(1.25 \times 10^5)^{0.8}(4.50)^{0.33} \qquad \text{(equation 9.57)}$$

$$= 0.596 \times 1.195 \times 10^4 \times 1.64$$

$$= 1.168 \times 10^4 \text{ W/m}^2 \text{ K} \quad \text{or} \quad 11.68 \text{ kW/m}^2 \text{ K}$$

$$\log_{10}(330 - \theta') = 1.477 - (0.0283 \times 11.68) = 1.147$$

$$\theta' = 316.0 \text{ K}$$

Comparing the results:

Method	h(kW/m^2 K)	θ'(K)
(a)	24.9	324.1
(b)	9.5	313.9
(c)	13.0	317.1
(d)	11.7	316.0

It is seen that the simple Reynolds analogy is far from accurate in calculating heat transfer to a liquid.

Example 12.4

The tube in Example 12.3 is maintained at 350 K and air is passed through it at 3.5 m/s, the initial temperature of the air being 290 K. What is the outlet temperature of the air for the four cases used in Example 12.3?

Solution

Taking the physical properties of air at 310 K and assuming that fully developed flow exists in the pipe:

$$Re = 0.0250 \times 3.5 \times \frac{(29/22.4)(273/310)}{0.018 \times 10^{-3}} = 5535$$

$$Pr = \frac{1.003 \times 1000 \times 0.018 \times 10^{-3}}{0.024} = 0.75$$

The heat transfer coefficients and final temperatures are then calculated as in Example 12.3.

Method	$h(\text{W/m}^2\text{ K})$	$\theta(\text{K})$
(a)	15.5	348.1
(b)	18.3	349.0
(c)	17.9	348.9
(d)	21.2	349.4

In this case the result obtained using the Reynolds analogy agrees much more closely with the other three methods.

12.9. FURTHER READING

BENNETT, C. O. and MYERS, J. E.: Momentum, Heat and Mass Transfer, 3rd edn (McGraw-Hill, New York, 1983).
BRODKEY, R. S. and HERSHEY, H. C.: Transport Phenomena (McGraw-Hill, New York, 1988).
HINZE, J. O.: Turbulence, 2nd edn (McGraw-Hill, New York, 1975).

12.10. REFERENCES

1. JEANS, J. H.: Kinetic Theory of Gases (Cambridge University Press, 1940).
2. PRANDTL, L.: Z. Angew. Math. u. Mech. 5 (1925) 136. Untersuchungen zur ausgebildeten Turbulenz.
3. PRANDTL, L.: Z. Ver. deut. Ing. 77 (1933) 105. Neuere Ergebnisse der Turbulenzforschung.
4. TAYLOR, G. I.: N.A.C.A. Rep. and Mem. No. 272 (1916) 423. Conditions at the surface of a hot body exposed to the wind.
5. NIKURADSE, J.: Forsch. Ver. deut. Ing. 356 (1932). Gesetzmässigkeiten der turbulenten Strömung in glatten Röhren.
6. NIKURADSE, J.: Forsch. Ver. deut. Ing. 361 (1933). Strömungsgesetze in rauhen Röhren.
7. REYNOLDS, O.: Proc. Manchester Lit. Phil. Soc. 14 (1874) 7. On the extent and action of the heating surface for steam boilers.
8. PRANDTL, L.: Physik. Z. 11 (1910) 1072. Eine Beziehung zwischen Wärmeaustausch und Strömungswiderstand der Flüssigkeiten.
9. PRANDTL, L.: Physik. Z. 29 (1928) 487. Bemerkung über den Wärmeübergang im Röhr.
10. MARTINELLI, R. C.: Trans. Am. Soc. Mech. Eng. 69 (1947) 947. Heat transfer to molten metals.

12.11. NOMENCLATURE

		Units in SI system	Dimensions in $\mathbf{M, N, L, T}, \theta$
A	Area of surface	m^2	$\mathbf{L}^2$
a	Constant in equation 12.41	—	—
a'	Constant in equation 12.41	—	—
B	Integration constant	m/s	$\mathbf{LT}^{-1}$
B'	B/u^*	—	—
b	Ratio of θ_b to θ_s	—	—
C	Molar concentration	kmol/m^3	$\mathbf{NL}^{-3}$
C_A, C_B	Molar concentration of $\mathbf{A}$, $\mathbf{B}$	kmol/m^3	$\mathbf{NL}^{-3}$
C_{As}, C_{Bs}	Molar concentration of $\mathbf{A}$, $\mathbf{B}$ outside boundary layer	kmol/m^3	$\mathbf{NL}^{-3}$
C_{Aw}, C_{Bw}	Molar concentration of $\mathbf{A}$, $\mathbf{B}$ at wall	kmol/m^3	$\mathbf{NL}^{-3}$
C_p	Specific heat at constant pressure	J/kg K	$\mathbf{L}^2\mathbf{T}^{-2}\theta^{-1}$
C_T	Total molar concentration	kmol/m^3	$\mathbf{NL}^{-3}$
C_v	Specific heat at constant volume	J/kg K	$\mathbf{L}^2\mathbf{T}^{-2}\theta^{-1}$
c_m	Heat capacity of one molecule	J/K	$\mathbf{ML}^2\mathbf{T}^{-2}\theta^{-1}$

		Units in SI system	Dimensions in $\mathbf{M, N, L, T}, \theta$
D	Diffusivity	m²/s	$\mathbf{L^2T^{-1}}$
D_H	Thermal diffusivity	m²/s	$\mathbf{L^2T^{-1}}$
d	Pipe or particle diameter	m	$\mathbf{L}$
E	Eddy kinematic viscosity	m²/s	$\mathbf{L^2T^{-1}}$
E_D	Eddy diffusivity	m²/s	$\mathbf{L^2T^{-1}}$
E_H	Eddy thermal diffusivity	m²/s	$\mathbf{L^2T^{-1}}$
e	Surface roughness	m	$\mathbf{L}$
h	Heat transfer coefficient	W/m² K	$\mathbf{MT^{-3}\theta^{-1}}$
h_D	Mass transfer coefficient	m/s	$\mathbf{LT^{-1}}$
I	Intensity of turbulence	—	—
i	Fraction of root mean square velocity of molecules	—	—
j	Fraction of mean free path of molecules	—	—
K	Ratio of mixing length to distance from surface	—	—
k	Thermal conductivity	W/m K	$\mathbf{MLT^{-3}\theta^{-1}}$
M	Molecular weight	kg/kmol	$\mathbf{MN^{-1}}$
$\mathbf{M}$	Mass of fluid	kg	$\mathbf{M}$
m	Mass of gas molecule	kg	$\mathbf{M}$
N	Molar rate of diffusion per unit area	kmol/m²s	$\mathbf{NL^{-2}T^{-1}}$
N'	Total molar rate of transfer per unit area	kmol/m²s	$\mathbf{NL^{-2}T^{-1}}$
$\mathbf{N}$	Number of molecules per unit volume at $y = y$	m⁻³	$\mathbf{L^{-3}}$
$\mathbf{N}'$	Number of molecules per unit volume $y = y + j\lambda$	m⁻³	$\mathbf{L^{-3}}$
n	Number of molar units	kmol	$\mathbf{N}$
q_0	Rate of transfer of heat per unit area at walls	W/m²	$\mathbf{MT^{-3}}$
q_y	Rate of transfer of heat per unit area at $y = y$	W/m²	$\mathbf{MT^{-3}}$
R	Shear stress acting on surface	N/m²	$\mathbf{ML^{-1}T^{-2}}$
R_0	Shear stress acting on fluid at surface	N/m²	$\mathbf{ML^{-1}T^{-2}}$
R_y	Shear stress in fluid at $y = y$	N/m²	$\mathbf{ML^{-1}T^{-2}}$
$\mathbf{R}$	Universal gas constant	(8314)J/kmol K	$\mathbf{MN^{-1}L^2T^{-2}\theta^{-1}}$
r	Radius of pipe	m	$\mathbf{L}$
T	Absolute temperature	K	θ
t	Time	s	$\mathbf{T}$
u	Mean velocity	m/s	$\mathbf{LT^{-1}}$
u_b	Velocity at edge of laminar sub-layer	m/s	$\mathbf{LT^{-1}}$
u_E	Mean velocity in eddy	m/s	$\mathbf{LT^{-1}}$
u_{Ex}	Mean component of eddy velocity in X-direction	m/s	$\mathbf{LT^{-1}}$
u_{Ey}	Mean component of eddy velocity in Y-direction	m/s	$\mathbf{LT^{-1}}$
u_{Ez}	Mean component of eddy velocity in Z-direction	m/s	$\mathbf{LT^{-1}}$
u_e	Velocity at distance e from surface	m/s	$\mathbf{LT^{-1}}$
u_m	Root mean square velocity of molecules	m/s	$\mathbf{LT^{-1}}$
u_s	Velocity of fluid outside boundary layer, or at pipe axis	m/s	$\mathbf{LT^{-1}}$
u_x	Velocity in X-direction at $y = y$	m/s	$\mathbf{LT^{-1}}$
u'_x	Velocity in X-direction at $y = y + j\lambda$ or $y + \lambda_E$	m/s	$\mathbf{LT^{-1}}$
u^+	Ratio of u_y to u^*	—	—
u^*	Shearing stress velocity, $\sqrt{R/\rho}$	m/s	$\mathbf{LT^{-1}}$
w	Exponent of velocity	—	—
y	Distance from surface	m	$\mathbf{L}$
y^+	Ratio of y to $\mu/\rho u^*$	—	—
α	Ratio of u_b to u_s	—	—
γ	Ratio of C_p to C_v	—	—
δ	Thickness of boundary layer	m	$\mathbf{L}$
δ_b	Thickness of laminar sub-layer	m	$\mathbf{L}$
λ	Mean free path of molecules	m	$\mathbf{L}$
λ_E	Mixing length	m	$\mathbf{L}$

		Units in SI system	Dimensions in M, N, L, T, θ
μ	Viscosity of fluid	N s/m^2	$\mathbf{ML^{-1}T^{-1}}$
ρ	Density of fluid	kg/m^3	$\mathbf{ML^{-3}}$
ϕ	Friction factor $R/\rho u^2$	—	—
θ	Temperature at $y = y$	K	θ
θ_s	Temperature outside boundary layer, or at pipe axis	K	θ
θ'	Temperature at $y = y + j\lambda$	K	θ
Nu	Nusselt number hd/k	—	—
Nu_x	Nusselt number hx/k	—	—
Re	Reynolds number $ud\rho/\mu$	—	—
Re_r	Roughness Reynolds number $u^*e\rho/\mu$	—	—
Re_x	Reynolds number $u_s x\rho/\mu$	—	—
Pr	Prandtl number $C_p\mu/k$	—	—
Sc	Schmidt number $\mu/\rho D$	—	—
Sh_x	Sherwood number h_D/D	—	—
St	Stanton number $h/C_p\rho u$ or $h/C_p\rho u_s$	—	—

Subscripts

A, B   For component **A, B**

CHAPTER 13

Humidification and Water Cooling

13.1. INTRODUCTION

In the processing of materials it is often necessary either to increase the amount of vapour present in a gas stream, an operation known as *humidification*; or to reduce the vapour present, a process referred to as *dehumidification*. In humidification, the vapour content may be increased by passing the gas over a liquid which then evaporates into the gas stream. This transfer into the main stream takes place by diffusion, and at the interface simultaneous heat and mass transfer take place according to the relations considered in previous chapters. In the reverse operation, that is dehumidification, partial condensation must be effected and the condensed vapour removed.

The most widespread application of humidification and dehumidification involves the air–water system, and a discussion of this system forms the greater part of the present chapter. Although the drying of wet solids is an example of a humidification operation, the reduction of the moisture content of the solids is the main aim and the humidification of the air stream is a secondary effect. Much of the present chapter is, however, of vital significance in any drying operation. Air conditioning and gas drying also involve humidification and dehumidification operations. For example, moisture must be removed from wet chlorine so that the gas can be handled in steel equipment which otherwise would be severely corroded. Similarly, the gases used in the manufacture of sulphuric acid must be dried or dehumidified before entering the converters, and this is achieved by passing the gas through a dehydrating agent such as sulphuric acid, in essence an absorption operation, or by an alternative dehumidification process discussed later.

In order that hot condenser water may be re-used in a plant, it is normally cooled by contact with an air stream. The equipment usually takes the form of a tower in which the hot water is run in at the top and allowed to flow downwards over a packing against a countercurrent flow of air which enters at the bottom of the cooling tower. The design of such towers forms an important part of the present chapter, though at the outset it is necessary to consider basic definitions of the various quantities involved in humidification, in particular *wet-bulb* and *adiabatic saturation temperatures*, and the way in which humidity data are presented on charts and graphs. While the present discussion is devoted to the very important air-water system, which is in some ways unique, the same principles may be applied to other liquids and gases, and this topic is covered in a final section.

13.2. HUMIDIFICATION TERMS

13.2.1. Definitions

The more important terms used in relation to humidification are defined as follows:

Humidity ($\mathscr{H}$)	mass of vapour associated with unit mass of dry gas
Humidity of saturated gas ($\mathscr{H}_0$)	humidity of the gas when it is saturated with vapour at a given temperature
Percentage humidity	$100(\mathscr{H}/\mathscr{H}_0)$
Humid heat (s)	heat required to raise unit mass of dry gas and its associated vapour through unit temperature difference at constant pressure.

That is:

$$s = C_a + \mathscr{H} C_w$$

where C_a and C_w are the specific heat capacities of the gas and the vapour, respectively. (For the air-water system, the humid heat is approximately

$$s = 1.00 + 1.9\mathscr{H} \text{ kJ/kg K.})$$

Humid volume	volume occupied by unit mass of dry gas and its associated vapour
Saturated volume	humid volume of saturated gas
Dew point	temperature at which the gas is saturated with vapour. As a gas is cooled, the dew point is the temperature at which condensation will first occur.
Percentage relative humidity	$\left(\dfrac{\text{partial pressure of vapour in gas}}{\text{partial pressure of vapour in saturated gas}}\right) \times 100$

The above nomenclature conforms with the recommendations of BS1339[1], although there are some ambiguities in the standard.

The relationship between the partial pressure of the vapour and the humidity of a gas may be derived as follows. In unit volume of gas:

$$\text{mass of vapour} \quad = \frac{P_w M_w}{RT}$$

$$\text{and mass of non-condensable gas} \quad = \frac{(P - P_w)M_A}{RT}$$

The humidity is therefore given by:

$$\mathscr{H} = \frac{P_w}{P - P_w}\left(\frac{M_w}{M_A}\right) \tag{13.1}$$

and the humidity of the saturated gas is:

$$\mathscr{H}_0 = \frac{P_{w0}}{P - P_{w0}}\left(\frac{M_w}{M_A}\right) \tag{13.2}$$

where P_w is the partial pressure of vapour in the gas, P_{w0} the partial pressure of vapour in the saturated gas at the same temperature, M_A the mean molecular weight of the dry gas, M_w the molecular mass of the vapour, P the total pressure, **R** the gas constant (8.314 kJ/kmol K in SI units), and T the absolute temperature.

For the air–water system, P_w is frequently small compared with P and hence, substituting for the molecular masses:

$$\mathscr{H} = \frac{18}{29}\left(\frac{P_w}{P}\right)$$

The relationship between the percentage humidity of a gas and the percentage relative humidity may be derived as follows:

The percentage humidity, by definition $= 100\mathscr{H}/\mathscr{H}_0$

Substituting from equations 13.1 and 13.2 and simplifying:

$$\text{Percentage humidity} = \frac{P - P_{w0}}{P - P_w} \cdot \frac{P_w}{P_{w0}} \times 100$$

$$= \frac{P - P_{w0}}{P - P_w} \times (\text{percentage relative humidity}) \qquad (13.3)$$

When $(P - P_{w0})/(P - P_w) \approx 1$, the percentage relative humidity and the percentage humidity are equal. This condition is approached when the partial pressure of the vapour is only a small proportion of the total pressure or when the gas is almost saturated, that is as $P_w \to P_{w0}$.

Example 13.1

In a process in which it is used as a solvent, benzene is evaporated into dry nitrogen. At 297 K and 101.3 kN/m², the resulting mixture has a percentage relative humidity of 60. It is required to recover 80 per cent of the benzene present by cooling to 283 K and compressing to a suitable pressure. What should this pressure be? The vapour pressure of benzene is 12.2 kN/m² at 297 K and 6.0 kN/m² at 283 K.

Solution

From the definition of percentage relative humidity (RH):

$$P_w = P_{w0}\frac{RH}{100}$$

At 297 K:
$$P_w = \frac{12.2 \times 60}{100} = 7.32 \text{ kN/m}^2$$

In the benzene–nitrogen mixture:

$$\text{mass of benzene} = \frac{P_w M_w}{RT} = \frac{7.32 \times 78}{8.314 \times 297} = 0.231 \text{ kg}$$

$$\text{mass of nitrogen} = \frac{(P - P_w)M_A}{RT} = \frac{(101.3 - 7.32)28}{8.314 \times 297} = 1.066 \text{ kg}$$

Hence the humidity is:
$$\mathscr{H} = \frac{0.231}{1.066} = 0.217 \text{ kg/kg}$$

In order to recover 80 per cent of the benzene, the humidity must be reduced to 20 per cent of the initial value. As the vapour will be in contact with liquid benzene, the nitrogen will be saturated with benzene vapour and hence at 283 K:

$$\mathscr{H}_0 = \frac{0.217 \times 20}{100} = 0.0433 \text{ kg/kg}$$

Thus in equation 13.2:

$$0.0433 = \frac{6.0}{P - 6.0} \frac{78}{28}$$

from which: $P = 392 \text{ kN/m}^2$

Example 13.2

In a vessel at 101.3 kN/m² and 300 K, the percentage relative humidity of the water vapour in the air is 25. If the partial pressure of water vapour when air is saturated with vapour at 300 K is 3.6 kN/m², calculate:

(a) the partial pressure of the water vapour in the vessel;
(b) the specific volumes of the air and water vapour;
(c) the humidity of the air and humid volume; and
(d) the percentage humidity.

Solution

(a) From the definition of percentage relative humidity:

$$P_w = P_{w0}\frac{RH}{100} = \frac{3.6 \times 25}{100} = 0.9 \text{ kN/m}^2$$

(b) In 1 m³ of air:

$$\text{mass of water vapour} = \frac{0.9 \times 18}{8.314 \times 300} = 0.0065 \text{ kg}$$

$$\text{mass of air} = \frac{(101.3 - 0.9)29}{8.314 \times 300} = 1.167 \text{ kg}$$

Hence: specific volume of water vapour at 0.9 kN/m² $= \dfrac{1}{0.0065} = 154 \text{ m}^3/\text{kg}$

specific volume of air at 100.4 kN/m² $= \dfrac{1}{1.167} = 0.857 \text{ m}^3/\text{kg}$

(c) Humidity: $\mathscr{H} = \dfrac{0.0065}{1.1673} = 0.0056 \text{ kg/kg}$

(Using the approximate relationship:

$$\mathscr{H} = \frac{18 \times 0.9}{29 \times 101.3} = 0.0055 \text{ kg/kg.})$$

∴   Humid volume = volume of 1 kg air + associated vapour = specific volume of air at 100.4 kN/m²

$$= 0.857 \text{ m}^3/\text{kg}$$

(d) From equation 13.3:

$$\text{percentage humidity} = \frac{101.3 - 3.6}{101.3 - 0.9}25$$

$$= 24.3 \text{ per cent}$$

13.2.2. Wet-bulb temperature

When a stream of unsaturated gas is passed over the surface of a liquid, the humidity of the gas is increased due to evaporation of the liquid. The temperature of the liquid falls below that of the gas and heat is transferred from the gas to the liquid. At equilibrium the rate of heat transfer from the gas just balances that required to vaporise the liquid and the liquid is said to be at the *wet-bulb temperature*. The rate at which this temperature is reached depends on the initial temperatures and the rate of flow of gas past the liquid surface. With a small area of contact between the gas and the liquid and a high gas flowrate, the temperature and the humidity of the gas stream remain virtually unchanged.

The rate of transfer of heat from the gas to the liquid can be written as:

$$Q = hA(\theta - \theta_w) \tag{13.4}$$

where Q is the heat flow, h the coefficient of heat transfer, A the area for transfer, and θ and θ_w are the temperatures of the gas and liquid phases.

The liquid evaporating into the gas is transferred by diffusion from the interface to the gas stream as a result of a concentration difference $(c_0 - c)$, where c_0 is the concentration of the vapour at the surface (mass per unit volume) and c is the concentration in the gas stream. The rate of evaporation is then given by:

$$W = h_D A(c_0 - c) = h_D A \frac{M_w}{RT}(P_{w0} - P_w) \tag{13.5}$$

where h_D is the coefficient of mass transfer.

The partial pressures of the vapour, P_w and P_{w0}, may be expressed in terms of the corresponding humidities $\mathscr{H}$ and $\mathscr{H}_w$ by equations 13.1 and 13.2.

If P_w and P_{w0} are small compared with P, $(P - P_w)$ and $(P - P_{w0})$ may be replaced by a mean partial pressure of the gas P_A and:

$$W = h_D A \frac{(\mathscr{H}_w - \mathscr{H})M_w}{RT} \cdot \left(P_A \frac{M_A}{M_w}\right)$$

$$= h_D A \rho_A (\mathscr{H}_w - \mathscr{H}) \tag{13.6}$$

where ρ_A is the density of the gas at the partial pressure P_A.

The heat transfer required to maintain this rate of evaporation is:

$$Q = h_D A \rho_A (\mathscr{H}_w - \mathscr{H})\lambda \tag{13.7}$$

where λ is the latent heat of vaporisation of the liquid.

Thus, equating equations 13.4 and 13.7:

$$(\mathscr{H} - \mathscr{H}_w) = -\frac{h}{h_D \rho_A \lambda}(\theta - \theta_w) \tag{13.8}$$

Both h and h_D are dependent on the equivalent gas film thickness, and thus any decrease in the thickness, as a result of increasing the gas velocity for example, increases both h and h_D. At normal temperatures, (h/h_D) is virtually independent of the gas velocity provided this is greater than about 5 m/s. Under these conditions, heat transfer by convection from the gas stream is large compared with that from the surroundings by radiation and conduction.

The wet-bulb temperature θ_w depends only on the temperature and the humidity of the gas and values normally quoted are determined for comparatively high gas velocities, such that the condition of the gas does not change appreciably as a result of being brought into contact with the liquid and the ratio (h/h_D) has reached a constant value. For the air–water system, the ratio $(h/h_D\rho_A)$ is about 1.0 kJ/kg K and varies from 1.5 to 2.0 kJ/kg K for organic liquids.

Example 13.3

Moist air at 310 K has a wet-bulb temperature of 300 K. If the latent heat of vaporisation of water at 300 K is 2440 kJ/kg, estimate the humidity of the air and the percentage relative humidity. The total pressure is 105 kN/m^2 and the vapour pressure of water vapour at 300 K is 3.60 kN/m^2 and 6.33 kN/m^2 at 310 K.

Solution

The humidity of air saturated at the wet-bulb temperature is given by equation 13.2:

$$\mathscr{H}_w = \frac{P_{w0}}{P - P_{w0}} \frac{M_w}{M_A}$$

$$= \left(\frac{3.6}{105.0 - 3.6}\right) \frac{18}{29} = 0.0220 \text{ kg/kg}$$

Therefore, taking $(h/h_D\rho_A)$ as 1.0 kJ/kg K, in equation 13.8:

$$(0.0220 - \mathscr{H}) = \frac{1.0}{2440}(310 - 300)$$

or:

$$\mathscr{H} = 0.018 \text{ kg/kg}$$

At 310 K,

$$P_{w0} = 6.33 \text{ kN/m}^2$$

In equation 13.2:

$$0.0780 = \frac{18P_w}{(105.0 - P_w)29}$$

$$\therefore \quad P_w = 2.959 \text{ kN/m}^2$$

and the percentage relative humidity

$$= \frac{100 \times 2.959}{6.33} = 46.7 \text{ per cent}$$

13.2.3. Adiabatic saturation temperature

In the system just considered, neither the humidity nor the temperature of the gas is appreciably changed. If the gas is passed over the liquid at such a rate that the time of contact is sufficient for equilibrium to be established, the gas will become saturated and both phases will be brought to the same temperature. In a thermally insulated system, the total sensible heat falls by an amount equal to the latent heat of the liquid evaporated. As a result of continued passage of the gas, the temperature of the liquid gradually approaches an equilibrium value which is known as the *adiabatic saturation temperature*.

These conditions are achieved in an infinitely tall thermally insulated humidification column through which gas of a given initial temperature and humidity flows countercurrently to the liquid under conditions where the gas is completely saturated at the top of the column. If the liquid is continuously circulated round the column, and if any fresh liquid which is added is at the same temperature as the circulating liquid, the temperature of the liquid at the top and bottom of the column, and of the gas at the top, approach the adiabatic saturation temperature. Temperature and humidity differences are a maximum at the bottom and zero at the top, and therefore the rates of transfer of heat and mass decrease progressively from the bottom to the top of the tower. This is illustrated in Figure 13.1.

Figure 13.1.   Adiabatic saturation temperature θ_S

Making a heat balance over the column, it is seen that the heat of vaporisation of the liquid must come from the sensible heat in the gas. The temperature of the gas falls from θ to the adiabatic saturation temperature θ_s, and its humidity increases from $\mathscr{H}$ to $\mathscr{H}_s$ (the saturation value at θ_s). Then working on the basis of unit mass of dry gas:

$$(\theta - \theta_s)s = (\mathscr{H}_s - \mathscr{H})\lambda$$

or:
$$(\mathscr{H} - \mathscr{H}_s) = -\frac{s}{\lambda}(\theta - \theta_s) \qquad (13.9)$$

where s is the humid heat of the gas and λ the latent heat of vaporisation at θ_s. s is almost constant for small changes in $\mathscr{H}$.

Equation 13.9 indicates an approximately linear relationship between humidity and temperature for all mixtures of gas and vapour having the same adiabatic saturation temperature θ_s. A curve of humidity versus temperature for gases with a given adiabatic saturation temperature is known as an *adiabatic cooling line*. For a range of adiabatic saturation temperatures, a family of curves, approximating to straight lines of slopes equal to $-(s/\lambda)$, is obtained. These lines are not exactly straight and parallel because of variations in λ and s.

Comparing equations 13.8 and 13.9, it is seen that the adiabatic saturation temperature is equal to the wet-bulb temperature when $s = h/h_D\rho_A$. This is the case for most water vapour systems and accurately so when $\mathcal{H} = 0.047$. The ratio $(h/h_D\rho_A s) = b$ is sometimes known as the *psychrometric ratio* and, as indicated, b is approximately unity for the air–water system. For most systems involving air and an organic liquid, $b = 1.3 - 2.5$ and the wet-bulb temperature is higher than the adiabatic saturation temperature. This was confirmed in 1932 by SHERWOOD and COMINGS[2] who worked with water, ethanol, n-propanol, n-butanol, benzene, toluene, carbon tetrachloride, and n-propyl acetate, and found that the wet-bulb temperature was always higher than the adiabatic saturation temperature except in the case of water.

In Chapter 12 it is shown that when the Schmidt and Prandtl numbers for a mixture of gas and vapour are approximately equal to unity, the *Lewis relation* applies, or:

$$h_D = \frac{h}{C_p\rho} \qquad \text{(equation 12.101)}$$

where C_p and ρ are the mean specific heat and density of the vapour phase.

Therefore:
$$\frac{h}{h_D\rho_A} = \frac{C_p\rho}{\rho_A} \qquad (13.10)$$

Where the humidity is relatively low, $C_p \approx s$ and $\rho \approx \rho_A$ and hence:

$$s \approx \frac{h}{h_D\rho_A} \qquad (13.11)$$

For systems containing vapour other than that of water, s is only approximately equal to $h/h_D\rho_A$ and the difference between the two quantities may be as high as 50 per cent.

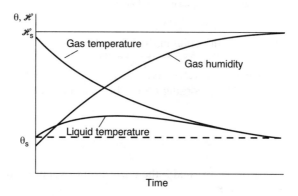

Figure 13.2.   Saturation of gas with liquid other than water at adiabatic saturation temperature

If an unsaturated gas is brought into contact with a liquid which is at the adiabatic saturation temperature of the gas, a simultaneous transfer of heat and mass takes place. The temperature of the gas falls and its humidity increases (Figure 13.2). The temperature of the liquid at any instant tends to change and approach the wet-bulb temperature corresponding to the particular condition of the gas at that moment. For a liquid other than water, the adiabatic saturation temperature is less than the wet-bulb temperature and therefore in the initial stages, the temperature of the liquid rises. As the gas becomes humidified, however, its wet-bulb temperature falls and consequently the temperature to which the liquid is tending decreases as evaporation takes place. In due course, therefore, a point is reached where the liquid actually reaches the wet-bulb temperature of the gas in contact with it. It does not remain at this temperature, however, because the gas is not then completely saturated, and further humidification is accompanied by a continued lowering of the wet-bulb temperature. The temperature of the liquid therefore starts to fall and continues to fall until the gas is completely saturated. The liquid and gas are then both at the adiabatic saturation temperature.

The air–water system is unique, however, in that the Lewis relation holds quite accurately, so that the adiabatic saturation temperature is the same as the wet-bulb temperature. If, therefore, an unsaturated gas is brought into contact with water at the adiabatic saturation temperature of the gas, there is no tendency for the temperature of the water to change, and it remains in a condition of dynamic equilibrium through the whole of the humidification process (Figure 13.3). In this case, the adiabatic cooling line represents the conditions of gases of constant wet-bulb temperatures as well as constant adiabatic saturation temperatures. The change in the condition of a gas as it is humidified with water vapour is therefore represented by the adiabatic cooling line and the intermediate conditions of the gas during the process are readily obtained. This is particularly useful because only partial humidification is normally obtained in practice.

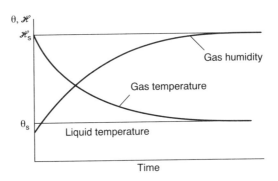

Figure 13.3.   Saturation of air with water at adiabatic saturation temperature

13.3. HUMIDITY DATA FOR THE AIR–WATER SYSTEM

To facilitate calculations, various properties of the air–water system are plotted on a *psychrometric* or *humidity chart*. Such a chart is based on either the temperature or the enthalpy of the gas. The temperature–humidity chart is the more commonly used though the enthalpy–humidity chart is particularly useful for determining the effect of mixing two gases or of mixing a gas and a liquid. Each chart refers to a particular total pressure of the system. A humidity–temperature chart for the air–water system at atmospheric pressure is given in Figure 13.4 (based on the original chart by GROSVENOR[3]) and the corresponding humidity–enthalpy chart is given in Figure 13.5.

Figure 13.4.   Humidity–temperature chart

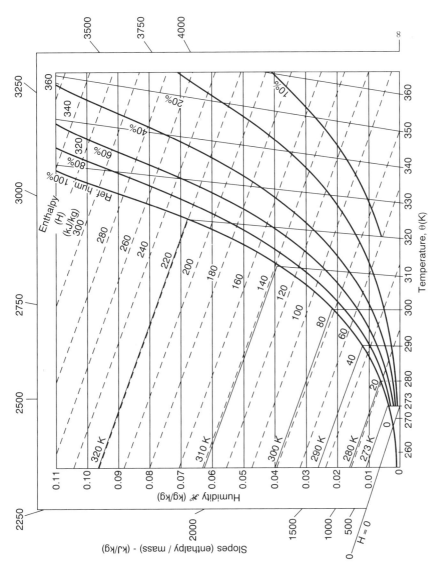

Figure 13.5. Humidity–enthalpy diagram for air–water vapour system at atmospheric pressure

13.3.1. Temperature–humidity chart

In Figure 13.4 it will be seen that the following quantities are plotted against temperature:
(i) The *humidity* $\mathscr{H}$ for various values of the percentage humidity.

For saturated gas:
$$\mathscr{H}_0 = \frac{P_{w0}}{P - P_{w0}} \left(\frac{M_w}{M_A} \right)$$
(equation 13.2)

From equation 13.1 for a gas with a humidity less than the saturation value:

$$\mathscr{H} = \frac{P_w}{P - P_w} \left(\frac{M_w}{M_A} \right) = \mathscr{H}_0 \frac{P_w}{P_{w0}} \frac{P - P_{w0}}{P - P_w}$$
(13.12)

(ii) *The specific volume of dry gas.* This is a linear function of temperature.

(iii) *The saturated volume.* This increases more rapidly with temperature than the specific volume of dry gas because both the quantity and the specific volume of vapour increase with temperature. At a given temperature the humid volume varies linearly with humidity and hence the humid volume of unsaturated gas can be found by interpolation.

(iv) *The latent heat of vaporisation*

In addition, the *humid heat* is plotted as the abscissa in Figure 13.4 with the humidity as the ordinate.

Adiabatic cooling lines are included in the diagram and, as already discussed, these have a slope of $-(s/\lambda)$ and they are slightly curved since s is a function of $\mathscr{H}$. On the chart they appear as straight lines, however, since the inclination of the axis has been correspondingly adjusted. Each adiabatic cooling line represents the composition of all gases whose adiabatic saturation temperature is given by its point of intersection with the 100 per cent humidity curve. For the air–water system, the adiabatic cooling lines represent conditions of constant wet-bulb temperature as well and, as previously mentioned, enable the change in composition of a gas to be followed as it is humidified by contact with water at the adiabatic saturation temperature of the gas.

Example 13.4

Air containing 0.005 kg water vapour per kg of dry air is heated to 325 K in a dryer and passed to the lower shelves. It leaves these shelves at 60 per cent humidity and is reheated to 325 K and passed over another set of shelves, again leaving at 60 per cent humidity. This is again repeated for the third and fourth sets of shelves, after which the air leaves the dryer. On the assumption that the material on each shelf has reached the wet-bulb temperature and that heat losses from the dryer may be neglected, determine:

(a) the temperature of the material on each tray;
(b) the amount of water removed in kg/s, if 5 m³/s moist air leaves the dryer;
(c) the temperature to which the inlet air would have to be raised to carry out the drying in a single stage.

Solution

For each of the four sets of shelves, the condition of the air is changed to 60 per cent humidity along an adiabatic cooling line.

Initial condition of air: $\theta = 325$ K, $\mathscr{H} = 0.005$ kg/kg

On humidifying to 60 per cent humidity:

$$\theta = 301 \text{ K}, \quad \mathscr{H} = 0.015 \text{ kg/kg and } \theta_w = 296 \text{ K}$$

At the end of the second pass: $\qquad \theta = 308 \text{ K}, \quad \mathscr{H} = 0.022 \text{ kg/kg and } \theta_w = 301 \text{ K}$

At the end of the third pass: $\qquad \theta = 312 \text{ K}, \quad \mathscr{H} = 0.027 \text{ kg/kg and } \theta_w = 305 \text{ K}$

At the end of the fourth pass: $\qquad \theta = 315 \text{ K}, \quad \mathscr{H} = 0.032 \text{ kg/kg and } \theta_w = 307 \text{ K}$

Thus the temperatures of the material on each of the trays are:

$$\underline{\underline{296 \text{ K}, 301 \text{ K}, 305 \text{ K, and } 307 \text{ K}}}$$

Total increase in humidity $\qquad = (0.032 - 0.005) = 0.027 \text{ kg/kg}$

The air leaving the system is at 315 K and 60 per cent humidity.

From Figure 13.4, specific volume of dry air $\quad = 0.893 \text{ m}^3/\text{kg}$

Specific volume of saturated air

$$(\text{saturated volume}) = 0.968 \text{ m}^3/\text{kg}$$

Therefore, by interpolation, the humid volume of air of 60 per cent humidity $= 0.937 \text{ m}^3/\text{kg}$

Mass of air passing through the dryer $= \dfrac{5}{0.937} = 5.34 \text{ kg/s}$

Mass of water evaporated $= (5.34 \times 0.027) = \underline{\underline{0.144 \text{ kg/s}}}$

If the material is to be dried by air in a single pass, the air must be heated before entering the dryer such that its wet-bulb temperature is 307 K.
For air with a humidity of 0.005 kg/kg, this corresponds to a dry bulb temperature of $\underline{\underline{370 \text{ K}}}$.

The various steps in this calculation are shown in Figure 13.6.

Figure 13.6. Humidification stages for Example 13.4 (schematic)

13.3.2. Enthalpy–humidity chart

In the calculation of enthalpies it is necessary to define some standard reference state at which the enthalpy is taken as zero. It is most convenient to take the melting point of the material constituting the vapour as the reference temperature, and the liquid state of the material as its standard state.

If H is the enthalpy of the humid gas per unit mass of dry gas, H_a the enthalpy of the dry gas per unit mass, H_w the enthalpy of the vapour per unit mass, C_a the specific heat of the gas at constant pressure, C_w the specific heat of the vapour at constant pressure, θ the temperature of the humid gas, θ_0 the reference temperature, λ the latent heat of vaporisation of the liquid at θ_0 and $\mathscr{H}$ the humidity of the gas,

then for an unsaturated gas: $H = H_a + H_w \mathscr{H}$ (13.13)

where: $$H_a = C_a(\theta - \theta_0)$$ (13.14)

and: $$H_w = C_w(\theta - \theta_0) + \lambda$$ (13.15)

Thus, in equation 13.13: $$H = (C_a + \mathscr{H}C_w)(\theta - \theta_0) + \mathscr{H}\lambda$$ (13.16)
$$= (\theta - \theta_0)(s + \mathscr{H}\lambda)$$

If the gas contains more liquid or vapour than is required to saturate it at the temperature in question, either the gas will be supersaturated or the excess material will be present in the form of liquid or solid according to whether the temperature θ is greater or less than the reference temperature θ_0. The supersaturated condition is unstable and will not be considered further.

If the temperature θ is greater than θ_0 and if the humidity $\mathscr{H}$ is greater than the humidity $\mathscr{H}_0$ of saturated gas, the enthalpy H per unit mass of dry gas is given by:

$$H = C_a(\theta - \theta_0) + \mathscr{H}_0[C_w(\theta - \theta_0) + \lambda] + C_L(\mathscr{H} - \mathscr{H}_0)(\theta - \theta_0)$$ (13.17)

where C_L is the specific heat of the liquid.

If the temperature θ is less than θ_0, the corresponding enthalpy H is given by:

$$H = C_a(\theta - \theta_0) + \mathscr{H}_0[C_w(\theta - \theta_0) + \lambda] + (\mathscr{H} - \mathscr{H}_0)[C_s(\theta - \theta_0) + \lambda_f]$$ (13.18)

where C_s is the specific heat of the solid and λ_f is the latent heat of freezing of the liquid, a negative quantity.

Equations 13.16 to 13.18 give the enthalpy in terms of the temperature and humidity of the humid gas for the three conditions: $\theta = \theta_0$, $\theta > \theta_0$, and $\theta < \theta_0$ respectively. Thus, given the percentage humidity and the temperature, the humidity may be obtained from Figure 13.4, the enthalpy calculated from equations 13.16, 13.17 or 13.18 and plotted against the humidity, usually with enthalpy as the abscissa. Such a plot is shown in Figure 13.7 for the air–water system, which includes the curves for 100 per cent humidity and for some lower value, say Z per cent.

Considering the nature of the isothermals for the three conditions dealt with previously, at constant temperature θ the relation between enthalpy and humidity for an unsaturated gas is:

$$H = \text{constant} + [C_w(\theta - \theta_0) + \lambda]\mathscr{H}$$ (13.19)

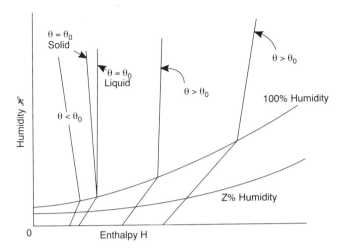

Figure 13.7.   Humidity–enthalpy diagram for air-water system — rectangular axes

Thus, the isothermal is a straight line of slope $[C_w(\theta - \theta_0) + \lambda]$ with respect to the humidity axis. At the reference temperature θ_0, the slope is λ; at higher temperatures, the slope is greater than λ, and at lower temperatures it is less than λ. Because the latent heat is normally large compared with the sensible heat, the slope of the isothermals remains positive down to very low temperatures. Since the humidity is plotted as the ordinate, the slope of the isothermal relative to the X-axis decreases with increase in temperature. When $\theta > \theta_0$ and $\mathscr{H} > \mathscr{H}_0$, the saturation humidity, the vapour phase consists of a saturated gas with liquid droplets in suspension. The relation between enthalpy and humidity at constant temperature θ is:

$$H = \text{constant} + C_L(\theta - \theta_0)\mathscr{H} \qquad (13.20)$$

The isothermal is therefore a straight line of slope $C_L(\theta - \theta_0)$. At the reference temperature θ_0, the slope is zero and the isothermal is parallel to the humidity axis. At higher temperatures, the slope has a small positive value. When $\theta < \theta_0$ and $\mathscr{H} > \mathscr{H}_0$, solid particles are formed and the equation of the isothermal is:

$$H = \text{constant} + [C_s(\theta - \theta_0) + \lambda_f]\mathscr{H} \qquad (13.21)$$

This represents a straight line of slope $[C_s(\theta - \theta_0) + \lambda_f]$. Both $C_s(\theta - \theta_0)$ and λ_f are negative and therefore the slopes of all these isothermals are negative. When $\theta = \theta_0$, the slope is λ_f. In the supersaturated region therefore, there are two distinct isothermals at temperature θ_0; one corresponds to the condition where the excess vapour is present in the form of liquid droplets and the other to the condition where it is present as solid particles. The region between these isothermals represents conditions where a mixture of liquid and solid is present in the saturated gas at the temperature θ_0.

The shape of the humidity-enthalpy line for saturated air is such that the proportion of the total area of the diagram representing saturated, as opposed to supersaturated, air is small when rectangular axes are used. In order to enable greater accuracy to be obtained in the use of the diagram, oblique axes are normally used, as in Figure 13.5, so that the isothermal for unsaturated gas at the reference temperature θ_0 is parallel to the humidity axis.

It should be noted that the curves of humidity plotted against either temperature or enthalpy have a discontinuity at the point corresponding to the freezing point of the humidifying material. Above the temperature θ_0 the lines are determined by the vapour–liquid equilibrium and below it by the vapour–solid equilibrium.

Two cases may be considered to illustrate the use of enthalpy–humidity charts. These are the mixing of two streams of humid gas and the addition of liquid or vapour to a gas.

Mixing of two streams of humid gas

Consider the mixing of two gases of humidities $\mathcal{H}_1$ and $\mathcal{H}_2$, temperatures θ_1 and θ_2, and enthalpies H_1 and H_2 to give a mixed gas of temperature θ, enthalpy H, and humidity $\mathcal{H}$. If the masses of dry gas concerned are m_1, m_2, and m respectively, then taking a balance on the dry gas, vapour, and enthalpy:

$$m_1 + m_2 = m \tag{13.22}$$

$$m_1\mathcal{H}_1 + m_2\mathcal{H}_2 = m\mathcal{H} \tag{13.23}$$

and:

$$m_1 H_1 + m_2 H_2 = mH \tag{13.24}$$

Elimination of m gives:

$$m_1(\mathcal{H} - \mathcal{H}_1) = m_2(\mathcal{H}_2 - \mathcal{H}) \tag{13.25}$$

and:

$$m_1(H - H_1) = m_2(H_2 - H)$$

Dividing these two equations:

$$\frac{\mathcal{H} - \mathcal{H}_1}{H - H_1} = \frac{\mathcal{H} - \mathcal{H}_2}{H - H_2} \tag{13.26}$$

The condition of the resultant gas is therefore represented by a point on the straight line joining $(\mathcal{H}_1, H_1)$ and $(\mathcal{H}_2, H_2)$. The humidity $\mathcal{H}$ is given, from equation 13.25, by:

$$\frac{\mathcal{H} - \mathcal{H}_1}{\mathcal{H}_2 - \mathcal{H}} = \frac{m_2}{m_1} \tag{13.27}$$

The gas formed by mixing two unsaturated gases may be either unsaturated, saturated, or supersaturated. The possibility of producing supersaturated gas arises because the 100 per cent humidity line on the humidity–enthalpy diagram is concave towards the humidity axis.

Example 13.5

In an air-conditioning system, 1 kg/s air at 350 K and 10 per cent humidity is mixed with 5 kg/s air at 300 K and 30 per cent humidity. What is the enthalpy, humidity, and temperature of the resultant stream?

Solution

From Figure 13.4:

$$\text{at } \theta_1 = 350 \text{ K and humidity} = 10 \text{ per cent;} \quad \mathcal{H}_1 = 0.043 \text{ kg/kg}$$

$$\text{at } \theta_2 = 300 \text{ K and humidity} = 30 \text{ per cent;} \quad \mathcal{H}_2 = 0.0065 \text{ kg/kg}$$

Thus, in equation 13.23:

$$(1 \times 0.043) + (5 \times 0.0065) = (1 + 5)\mathscr{H}$$

and:

$$\mathscr{H} = 0.0125 \text{ kg/kg}$$

From Figure 13.5:

$$\text{at } \theta_1 = 350 \text{ K and } \mathscr{H}_1 = 0.043 \text{ kg/kg;} \quad H_1 = 192 \text{ kJ/kg}$$

$$\text{at } \theta_2 = 300 \text{ K and } \mathscr{H}_2 = 0.0065 \text{ kg/kg;} \quad H_2 = 42 \text{ kJ/kg}$$

Thus, in equation 13.25:

$$1(H - 192) = 5(42 - H)$$

and:

$$H = 67 \text{ kJ/kg}$$

From Figure 13.5:

$$\text{at } H = 67 \text{ kJ/kg and } \mathscr{H} = 0.0125 \text{ kg/kg}$$

$$\theta = 309 \text{ K}$$

The data used in this example are shown in Figure 13.8.

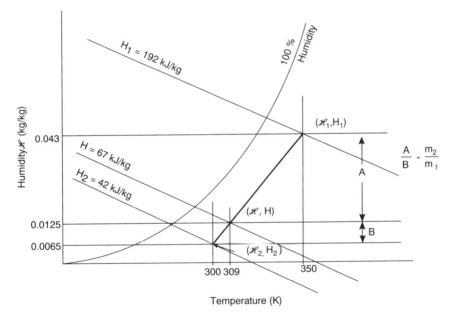

Figure 13.8.   Data used in Example 13.5

Addition of liquid or vapour to a gas

If a mass m_3 of liquid or vapour of enthalpy H_3 is added to a gas of humidity $\mathscr{H}_1$ and enthalpy H_1 and containing a mass m_1 of dry gas, then:

$$m_1(\mathscr{H} - \mathscr{H}_1) = m_3 \tag{13.28}$$

$$m_1(H - H_1) = m_3 H_3 \tag{13.29}$$

Thus:
$$\frac{H - H_1}{\mathcal{H} - \mathcal{H}_1} = H_3 \qquad (13.30)$$

where $\mathcal{H}$ and H are the humidity and enthalpy of the gas produced on mixing.

The composition and properties of the mixed stream are therefore represented by a point on the straight line of slope H_3, relative to the humidity axis, which passes through the point $(H_1, \mathcal{H}_1)$. In Figure 13.5 the edges of the plot are marked with points which, when joined to the origin, give a straight line of the slope indicated. Thus in using the chart, a line of slope H_3 is drawn through the origin and a parallel line drawn through the point $(H_1, \mathcal{H}_1)$. The point representing the final gas stream is then given from equation 13.28:

$$\mathcal{H} - \mathcal{H}_1 = \frac{m_3}{m_1}$$

It can be seen from Figure 13.5 that for the air-water system a straight line, of slope equal to the enthalpy of dry saturated steam (2675 kJ/kg), is almost parallel to the isothermals, so that the addition of live steam has only a small effect on the temperature of the gas. The addition of water spray, even if the water is considerably above the temperature of the gas, results in a lowering of the temperature after the water has evaporated. This arises because the latent heat of vaporisation of the liquid constitutes the major part of the enthalpy of the vapour. Thus when steam is added it gives up a small amount of sensible heat to the gas, whereas when hot liquid is added a small amount of sensible heat is given up and a very much larger amount of latent heat is absorbed from the gas.

Example 13.6

0.15 kg/s steam at atmospheric pressure and superheated to 400 K is bled into an air stream at 320 K and 20 per cent relative humidity. What is the temperature, enthalpy, and relative humidity of the mixed stream if the air is flowing at 5 kg/s? How much steam would be required to provide an exit temperature of 330 K and what would be the humidity of this mixture?

Solution

Steam at atmospheric pressure is saturated at 373 K at which the latent heat

$$= 2258 \text{ kJ/kg}$$

Taking the specific heat of superheated steam as 2.0 kJ/kg K;

$$\text{enthalpy of the steam:} H_3 = 4.18(373 - 273) + 2258 + 2.0(400 - 373)$$

$$= 2730 \text{ kJ/kg}$$

From Figure 13.5:

at $\theta_1 = 320$ K and 20 per cent relative humidity; $\mathcal{H}_1 = 0.013$ kg/kg   and $H_1 = 83$ kJ/kg

The line joining the axis and slope $H_3 = 2730$ kJ/kg at the edge of the chart is now drawn in and a parallel line is drawn through $(H_1, \mathcal{H}_1)$.

Now:
$$(\mathcal{H} - \mathcal{H}_1) = \frac{m_3}{m_1} = \frac{0.15}{5} = 0.03 \text{ kg/kg}$$

and:
$$\mathcal{H} = (0.03 + 0.013) = 0.043 \text{ kg/kg}$$

At the intersection of $\mathscr{H} = 0.043$ kg/kg and the line through $(\mathscr{H}_1, H_1)$

$$H = 165 \text{ kJ/kg and } \theta = 324 \text{ K}$$

When $\theta = 330$ K the intersection of this isotherm and the line through $(\mathscr{H}_1, H_1)$ gives an outlet stream in which $\mathscr{H} = 0.094$ kg/kg (83 per cent relative humidity) and $H = 300$ kJ/kg.

Thus, in equation 13.28:

$$m_3 = 5(0.094 - 0.013) = 0.41 \text{ kg/s}$$

The data used in this example are shown in Figure 13.9.

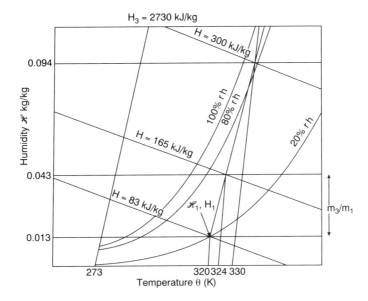

Figure 13.9. Data used in Example 13.6 (schematic)

13.4. DETERMINATION OF HUMIDITY

The most important methods for determining humidity are as follows:

(1) *Chemical methods.* A known volume of the gas is passed over a suitable absorbent, the increase in mass of which is measured. The efficiency of the process can be checked by arranging a number of vessels containing absorbent in series and ascertaining that the increase in mass in the last of these is negligible. The method is very accurate but is laborious. Satisfactory absorbents for water vapour are phosphorus pentoxide dispersed in pumice, and concentrated sulphuric acid.

(2) *Determination of the wet-bulb temperature.* Equation 13.8 gives the humidity of a gas in terms of its temperature, its wet-bulb temperature, and various physical properties of the gas and vapour. The wet-bulb temperature is normally determined as the temperature attained by the bulb of a thermometer which is covered with a piece of material which is maintained saturated with the liquid. The gas should be passed over the surface of the wet bulb at a high enough velocity (> 5 m/s) (a) for the condition of the gas stream not to be affected appreciably by the evaporation of liquid, (b) for the heat transfer by convection to be large compared with that by radiation and conduction from the surroundings, and

(c) for the ratio of the coefficients of heat and mass transfer to have reached a constant value. The gas should be passed long enough for equilibrium to be attained and, for accurate work, the liquid should be cooled nearly to the wet-bulb temperature before it is applied to the material.

The stream of gas over the liquid surface may be produced by a small fan or other similar means (Figure 13.10a). The crude forms of wet-bulb thermometer which make

(a)

Figure 13.10.   (a) Wet-bulb thermometer (b) Dew-point meter (c) Hygrometer indicating percentage relative humidity and temperature

(b)

(c)

Figure 13.10. (*continued*)

no provision for the rapid passage of gas cannot be used for accurate determinations of humidity.

(3) *Determination of the dew point.* The dew point is determined by cooling a highly polished surface (Figure 13.10*b*) in the gas and observing the highest temperature at which condensation takes place. The humidity of the gas is equal to the humidity of saturated gas at the dew-point.

(4) *Measurement of the change in length of a hair or fibre.* The length of a hair or fibre is influenced by the humidity of the surrounding atmosphere. Many forms of apparatus for automatic recording of humidity depend on this property. The method has the disadvantage that the apparatus needs frequent calibration because the zero tends to shift. This difficulty is most serious when the instrument is used over a wide range of humidities. A typical hair hygrometer is shown in Figure 13.10*c*.

(5) *Measurement of conductivity of a fibre.* If a fibre is impregnated with an electrolyte, such as lithium chloride, its electrical resistance will be governed by its moisture content, which in turn depends on the humidity of the atmosphere in which it is situated. In a lithium chloride cell, a skein of very fine fibres is wound on a plastic frame carrying the electrodes and the current flowing at a constant applied voltage gives a direct measure of the relative humidity.

(6) *Measurement of heat of absorption on to a surface.*

(7) *Electrolytic hygrometry* in which the quantity of electricity required to electrolyse water absorbed from the atmosphere on to a thin film of desiccant is measured.

(8) *Piezo-electric hygrometry* employing a quartz crystal with a hygroscopic coating in which moisture is alternately absorbed from a wet-gas and desorbed in a dry-gas stream; the dynamics is a function of the gas humidity.

(9) *Capacitance meters* in which the electrical capacitance is a function of the degree of deposition of moisture from the atmosphere.

(10) *Observation of colour changes* in active ingredients, such as cobaltous chloride.

Further details of instruments for the measurement of humidity are given in Volume 3, Chapter 6. Reference should also be made to standard works on psychrometry[4,5,6].

13.5. HUMIDIFICATION AND DEHUMIDIFICATION

13.5.1. Methods of increasing humidity

The following methods may be used for increasing the humidity of a gas:

(1) Live steam may be added directly in the required quantity. It has been shown that this produces only a slight increase in the temperature, but the method is not generally favoured because any impurities that are present in the steam may be added at the same time.

(2) Water may be sprayed into the gas at such a rate that, on complete vaporisation, it gives the required humidity. In this case, the temperature of the gas will fall as the latent heat of vaporisation must be supplied from the sensible heat of the gas and liquid.

(3) The gas may be mixed with a stream of gas of higher humidity. This method is frequently used in laboratory work when the humidity of a gas supplied to an apparatus is controlled by varying the proportions in which two gas streams are mixed.

(4) The gas may be brought into contact with water in such a way that only part of the liquid is evaporated. This is perhaps the most common method and will now be considered in more detail.

In order to obtain a high rate of humidification, the area of contact between the air and the water is made as large as possible by supplying the water in the form of a fine spray; alternatively, the interfacial area is increased by using a packed column. Evaporation occurs if the humidity at the surface is greater than that in the bulk of the air; that is, if the temperature of the water is above the dew point of the air.

When humidification is carried out in a packed column, the water which is not evaporated can be recirculated so as to reduce the requirements of fresh water. As a result of continued recirculation, the temperature of the water will approach the adiabatic saturation temperature of the air, and the air leaving the column will be cooled — in some cases to within 1 deg K of the temperature of the water. If the temperature of the air is to be maintained constant, or raised, the water must be heated.

Two methods of changing the humidity and temperature of a gas from $A(\theta_1, \mathscr{H}_1)$ to $B(\theta_2, \mathscr{H}_2)$ can be traced on the humidity chart as shown in Figure 13.11. The first method consists of saturating the air by water artificially maintained at the dew point of air of humidity $\mathscr{H}_2$ (line AC) and then heating at constant humidity to θ_2 (line CB). In the second method, the air is heated (line AD) so that its adiabatic saturation temperature corresponds with the dew point of air of humidity $\mathscr{H}_2$. It is then saturated by water at the adiabatic saturation temperature (line DC) and heated at constant humidity to θ_2 (line CB). In this second method, an additional operation — the preliminary heating — is carried out on the air, but the water temperature automatically adjusts itself to the required value.

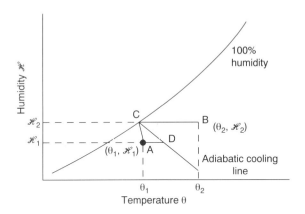

Figure 13.11.   Two methods of changing conditions of gas from $(\theta_1, \mathscr{H}_1)$ to $(\theta_2, \mathscr{H}_2)$

Since complete humidification is not always obtained, an allowance must be made when designing air humidification cycles. For example, if only 95 per cent saturation is obtained the adiabatic cooling line should be followed only to the point corresponding to that degree of saturation, and therefore the gas must be heated to a slightly higher temperature before adiabatic cooling is commenced.

Example 13.7

Air at 300 K and 20 per cent humidity is to be heated in two stages with intermediate saturation with water to 90 per cent humidity so that the final stream is at 320 K and 20 per cent humidity. What is the humidity of the exit stream and the conditions at the end of each stage?

Solution

At $\theta_1 = 300$ K and 20 per cent humidity: $\mathcal{H}_1 = 0.0045$ kg/kg, from Figure 13.4, and
at $\theta_2 = 320$ K and 20 per cent humidity: $\mathcal{H}_2 = 0.0140$ kg/kg

When $\mathcal{H}_2 = 0.0140$ kg/kg, air is saturated at 292 K and has a humidity of 90 per cent at 293 K.
The adiabatic cooling line corresponding to 293 K intersects with $\mathcal{H} = 0.0045$ kg/kg at a temperature, $\theta = 318$ K.
Thus the stages are:

(i) Heat the air at $\mathcal{H} = 0.0045$ from 300 to 318 K.
(ii) Saturate with water at an adiabatic saturation temperature of 293 K until 90 per cent humidity is attained. At the end of this stage:

$$\mathcal{H} = 0.0140 \text{ kg/kg} \quad \text{and} \quad \theta = 294.5 \text{ K}$$

(iii) Heat the saturated air at $\mathcal{H} = 0.0140$ kg/kg from 294.5 to 320 K

13.5.2. Dehumidification

Dehumidification of air can be effected by bringing it into contact with a cold surface, either liquid or solid. If the temperature of the surface is lower than the dew point of the gas, condensation takes place and the temperature of the gas falls. The temperature of the surface tends to rise because of the transfer of latent and sensible heat from the air. It would be expected that the air would cool at constant humidity until the dew point was reached, and that subsequent cooling would be accompanied by condensation. It is found, in practice, that this occurs only when the air is well mixed. Normally the temperature and humidity are reduced simultaneously throughout the whole of the process. The air in contact with the surface is cooled below its dew point, and condensation of vapour therefore occurs before the more distant air has time to cool. Where the gas stream is cooled by cold water, countercurrent flow should be employed because the temperature of the water and air are changing in opposite directions.

The humidity can be reduced by compressing air, allowing it to cool again to its original temperature, and draining off the water which has condensed. During compression, the partial pressure of the vapour is increased and condensation takes place as soon as it reaches the saturation value. Thus, if air is compressed to a high pressure, it becomes saturated with vapour, but the partial pressure is a small proportion of the total pressure. Compressed air from a cylinder therefore has a low humidity. Town gas is frequently compressed before it is circulated so as to prevent condensation in the mains.

Many large air-conditioning plants incorporate automatic control of the humidity and temperature of the issuing air. Temperature control is effected with the aid of a thermo-couple or resistance thermometer, and humidity control by means of a thermocouple recording the difference between the wet- and dry-bulb temperatures.

13.6. WATER COOLING

13.6.1. Cooling towers

Cooling of water can be carried out on a small scale either by allowing it to stand in an open pond or by the spray pond technique in which it is dispersed in spray form and then collected in a large, open pond. Cooling takes place both by the transference of sensible heat and by evaporative cooling as a result of which sensible heat in the water provides the latent heat of vaporisation.

On the large scale, air and water are brought into countercurrent contact in a cooling tower which may employ either natural draught or mechanical draught. The water flows down over a series of wooden slats which give a large interfacial area and promote turbulence in the liquid. The air is humidified and heated as it rises, while the water is cooled mainly by evaporation.

The natural draught cooling tower depends on the chimney effect produced by the presence in the tower of air and vapour of higher temperature and therefore of lower density than the surrounding atmosphere. Thus atmospheric conditions and the temperature and quantity of the water will exert a very important effect on the operation of the tower. Not only will these factors influence the quantity of air drawn through the tower, but they will also affect the velocities and flow patterns and hence the transfer coefficients between gas and liquid. One of the prime considerations in design therefore is to construct a tower in such a way that the resistance to air flow is low. Hence the packings and distributors must be arranged in open formation. The draught of a cooling tower at full load is usually only about 50 N/m^2,[7] and the air velocity in the region of 1.2–1.5 m/s, so that under the atmospheric conditions prevailing in this country the air usually leaves the tower in a saturated condition. The density of the air stream at outlet is therefore determined by its temperature. Calculation of conditions within the tower is carried out in the manner described in the following pages. It is, however, necessary to work with a number of assumed air flowrates and to select the one which fits both the transfer conditions and the relationship between air rate and pressure difference in the tower.

The *natural draught cooling tower* consists of an empty shell constructed either of timber or ferroconcrete; the upper portion is empty and merely serves to increase the draught. The lower portion, amounting to about 10–12 per cent of the total height, is usually fitted with grids on to which the water is fed by means of distributors or sprays as shown in Figure 13.12. The shells of cooling towers are now generally constructed in ferroconcrete in a shape corresponding approximately to a hyperboloid of revolution. The shape is chosen mainly for constructional reasons, but it does take account of the fact that the entering air will have a radial velocity component; the increase in cross-section towards the top causes a reduction in the outlet velocity and there is a small recovery of kinetic energy into pressure energy. Ferroconcrete towers are now made up to about 100–140 m in height and 80–100 m in diameter.

The *mechanical draught cooling tower* may employ forced draught with the fan at the bottom, or induced draught with the fan driving the moist air out at the top. The air velocity can be increased appreciably above that in the natural draught tower, and a greater depth of packing can be used. The tower will extend only to the top of the packing unless atmospheric conditions are such that a chimney must be provided in order to prevent recirculation of the moist air. The danger of recirculation is considerably

Figure 13.12.   Water-cooling tower. View of spray distribution system

less with the induced-draught type because the air is expelled with a higher velocity. Mechanical draught towers are generally confined to small installations and to conditions where the water must be cooled to as low a temperature as possible. In some cases it is possible to cool the water to within 1 deg K of the wet-bulb temperature of the air. Although the initial cost of the tower is less, maintenance and operating costs are of course higher than in natural draught towers which are now used for all large installations. A typical steel-framed mechanical draught cooling tower is shown in Figure 13.13.

The operation of the conventional water cooling tower is often characterised by the discharge of a plume consisting of a suspension of minute droplets of water in air. This is formed when the hot humid air issuing from the top of the tower mixes with the ambient atmosphere, and precipitation takes place as described earlier (Section 13.3.2). In the *hybrid* (or wet/dry) cooling tower[8], mist formation is avoided by cooling *part* of the water in a finned-tube exchanger bundle which thus generates a supply of warm dry air which is then blended with the air issuing from the evaporative section. By adjusting the proportion of the water fed to the heat exchanger, the plume can be completely eliminated.

In the cooling tower the temperature of the liquid falls and the temperature and humidity of the air rise, and its action is thus similar to that of an air humidifier. The limiting temperature to which the water can be cooled is the wet-bulb temperature corresponding to the condition of the air at inlet. The enthalpy of the air stream does not remain

1 Fan assembly
2 Gearbox
3 Fan stack
4 Drive shaft assembly
5 Motor
6 Fan deck
7 Mechanical equipment supports
8 Drift eliminators (PVC or Timber - Timber shown)
9 Cooling tower packing (plastic plate or wooden lath)
10 Inlet water distribution pipe
11 Open type distribution system
12 Timber laths for even water distribution
13 Cladding
14 Cladding extended to form handrail
15 Access ladder
16 Internal access ladder to distribution system and drift
   eliminators
17 Diagonal wind baffles
18 Air inlet louvres
19 Steel structures with horizontal and diagonal ties
20 Cold water sump

Some structural members have been omitted for clarity

Figure 13.13.   Visco 2000 series steel framed, mechanical draught, water cooling tower constant since the temperature of the liquid changes rapidly in the upper portion of the tower. Towards the bottom, however, the temperature of the liquid changes less rapidly because the temperature differences are smaller. At the top of the tower, the temperature falls from the bulk of the liquid to the interface and then again from the interface to the bulk of the gas. Thus the liquid is cooled by transfer of sensible heat and by evaporation at the surface. At the bottom of a tall tower, however, the temperature gradient in the liquid is in the same direction, though smaller, but the temperature gradient in the gas is in the opposite direction. Transfer of sensible heat to the interface therefore takes place from the bulk of the liquid and from the bulk of the gas, and all the cooling is caused by the evaporation at the interface. In most cases, about 80 per cent of the heat loss from the water is accounted for by evaporative cooling.

13.6.2. Design of natural-draught towers

The air flow through a natural-draught or hyperbolic-type tower is due largely to the difference in density between the warm air in the tower and the external ambient air; thus a draught is created in the stack by a chimney effect which eliminates the need for mechanical fans. It has been noted by McKELVEY and BROOKE[9] that natural-draught towers commonly operate at a pressure difference of some 50 N/m² under full load, and above the packing the mean air velocity is typically 1–2 m/s. The performance of a natural-draught tower differs from that of a mechanical-draught installation in that the cooling achieved depends upon the relative humidity as well as the wet-bulb temperature. It is important therefore, at the design stage, to determine correctly, and to specify, the density of the inlet and exit air streams in addition to the usual tower design conditions of water temperature range, how closely the water temperature should approach the wet bulb temperature of the air, and the quantity of water to be handled. Because the performance depends to a large extent on atmospheric humidity, the outlet water temperature is difficult to control with natural-draught towers.

In the design of natural-draught towers, a ratio of height z_t to base diameter of 3:2 is normally used and a design method has been proposed by CHILTON[10]. Chilton showed that the duty coefficient D_t of a tower is approximately constant over the normal range of operation and is related to tower size by an efficiency factor or performance coefficient C_t given by:

$$D_t = \frac{19.50 A_b z_t^{0.5}}{C_t^{1.5}} \tag{13.31}$$

where for water loadings in excess of 1 kg/m²s, C_t is usually about 5.2 though lower values are obtained with new packings which are being developed.

The duty coefficient is given by the following equation (in which SI units must be used as it is not dimensionally consistent):

$$\frac{W_L}{D_t} = 0.00369 \frac{\Delta H'}{\Delta T} (\Delta T' + 0.0752 \Delta H')^{0.5} \tag{13.32}$$

where W_L (kg/s) is the water load in the tower, $\Delta H'$ (kJ/kg) the change in enthalpy of the air passing through the tower, ΔT(K) the change in water temperature in passing through the tower and $\Delta T'$(K), the difference between the temperature of the air leaving the packing and the dry-bulb temperature of the inlet air. The air leaving the packing inside the tower is assumed to be saturated at the mean of the inlet and outlet water temperatures. Any divergence between theory and practice of a few degrees in this respect does not significantly affect the results as the draught component depends on the ratio of the change of density to change in enthalpy and not on change in temperature alone.[11] The use of equations 13.31 and 13.32 is illustrated in the following example.

Example 13.8

What are the diameter and height of a hyperbolic natural-draught cooling tower handling 4810 kg/s of water with the following temperature conditions:

$$\text{water entering the tower} = 301 \text{ K}$$

$$\text{water leaving the tower} = 294 \text{ K}.$$

$$\text{air: dry bulb} = 287 \text{ K}$$

$$\text{wet bulb} = 284 \text{ K}$$

Solution

Temperature range for the water, $\Delta T = (301 - 294) = 7$ K.
At a mean water temperature of $0.5(301 + 294) = 297.5$ K, the enthalpy $= 92.6$ kJ/kg
At a dry bulb temperature of 287 K, the enthalpy $= 49.5$ kJ/kg

$$\therefore \qquad \Delta T' = (297.5 - 287) = 10.5 \text{ K}$$

and:
$$\Delta H' = (92.6 - 49.5) = 43.1 \text{ kJ/kg}$$

In equation 13.32:

$$\frac{4810}{D_t} = 0.00369 \frac{43.1}{7}[10.5 + (0.0752 \times 43.1)]^{0.5}$$

and:
$$D_t = 57,110$$

Taking C_t as 5.0 and assuming as a first estimate a tower height of 100 m, in equation 13.31:

$$57,110 = 19.50 A_b \frac{100^{0.5}}{5.0^{1.5}}$$

and:
$$A_b = 3274 \text{ m}^2$$

Thus the internal diameter of the column at sill level $= \left(\frac{3274 \times 4}{\pi}\right)^{0.5}$

$$= \underline{\underline{64.6 \text{ m}}}$$

Since this gives a height: diameter ratio of approximately $(100 : 64.6) = 3 : 2$, the design is acceptable.

13.6.3. Height of packing for both natural and mechanical draught towers[12]

The height of a water-cooling tower can be determined by setting up a material balance on the water, an enthalpy balance, and rate equations for the transfer of heat in the liquid and gas and for mass transfer in the gas phase. There is no concentration gradient in the liquid and therefore there is no resistance to mass transfer in the liquid phase.

Considering the countercurrent flow of water and air in a tower of height z (Figure 13.14), the mass rate of flow of air per unit cross-section G' is constant throughout the whole height of the tower and, because only a small proportion of the total supply of water is normally evaporated (1–5 per cent), the liquid rate per unit area L' can be taken as constant. The temperature, enthalpy, and humidity will be denoted by the symbols θ, H, and $\mathcal{H}$ respectively, suffixes G, L, 1, 2, and f being used to denote conditions in the gas and liquid, at the bottom and top of the column, and of the air in contact with the water.

Figure 13.14.   Flow in water-cooling tower

The five basic equations for an incremental height of column, dz, are as follows:
(1) Water balance:

$$dL' = G'\, d\mathcal{H} \tag{13.33}$$

(2) Enthalpy balance:

$$G'\, dH_G = L'\, dH_L \tag{13.34}$$

since only a small proportion of the liquid is evaporated.

Now: $\qquad\qquad\qquad\qquad H_G = s(\theta_G - \theta_0) + \lambda\mathcal{H} \tag{13.35}$

and: $\qquad\qquad\qquad\qquad H_L = C_L(\theta_L - \theta_0) \tag{13.36}$

Thus: $\qquad\qquad\qquad G'\, dH_G = L'C_L\, d\theta_L \tag{13.37}$

and: $\qquad\qquad\qquad dH_G = s\, d\theta_G + \lambda\, d\mathcal{H} \tag{13.38}$

Integration of this expression over the whole height of the column, on the assumption that the physical properties of the materials do not change appreciably, gives:

$$G'(H_{G2} - H_{G1}) = L'C_L(\theta_{L2} - \theta_{L1}) \tag{13.39}$$

(3) Heat transfer from the body of the liquid to the interface:

$$h_L a \, dz(\theta_L - \theta_f) = L'C_L \, d\theta_L \tag{13.40}$$

where h_L is the heat transfer coefficient in the liquid phase and a is the interfacial area per unit volume of column. It will be assumed that the area for heat transfer is equal to that available for mass transfer, though it may be somewhat greater when the packing is not completely wetted.

Rearranging equation 13.40:

$$\frac{d\theta_L}{\theta_L - \theta_f} = \frac{h_L a}{L'C_L} \, dz \tag{13.41}$$

(4) Heat transfer from the interface to the bulk of the gas:

$$h_G a \, dz(\theta_f - \theta_G) = G's \, d\theta_G \tag{13.42}$$

where h_G is the heat transfer coefficient in the gas phase.

Rearranging:
$$\frac{d\theta_G}{\theta_f - \theta_G} = \frac{h_G a}{G's} \, dz \tag{13.43}$$

(5) Mass transfer from the interface to the gas:

$$h_D \rho a \, dz(\mathcal{H}_f - \mathcal{H}) = G' \, d\mathcal{H} \tag{13.44}$$

where h_D is the mass transfer coefficient for the gas and ρ is the mean density of the air (see equation 13.6).

Rearranging:
$$\frac{d\mathcal{H}}{\mathcal{H}_f - \mathcal{H}} = \frac{h_D a \rho}{G'} \, dz \tag{13.45}$$

These equations cannot be integrated directly since the conditions at the interface are not necessarily constant; nor can they be expressed directly in terms of the corresponding property in the bulk of the gas or liquid.

If the Lewis relation (equation 13.11) is applied, it is possible to obtain workable equations in terms of enthalpy instead of temperature and humidity. Thus, writing h_G as $h_D \rho s$, from equation 13.42:

$$G's \, d\theta_G = h_D \rho a \, dz(s\theta_f - s\theta_G) \tag{13.46}$$

and from equation 13.44:

$$G'\lambda \, d\mathcal{H} = h_D \rho a \, dz(\lambda \mathcal{H}_f - \lambda \mathcal{H}) \tag{13.47}$$

Adding these two equations gives:

$$G'(s \, d\theta_G + \lambda \, d\mathcal{H}) = h_D \rho a \, dz[(s\theta_f + \lambda \mathcal{H}_f) - (s\theta_G + \lambda \mathcal{H})]$$

$$G' \, dH_G = h_D \rho a \, dz(H_f - H_G) \quad \text{(from equation 13.35)} \tag{13.48}$$

or:
$$\frac{dH_G}{H_f - H_G} = \frac{h_D a \rho}{G'} \, dz \tag{13.49}$$

The use of an enthalpy driving force, as in equation 13.48, was first suggested by MERKEL[13], and the following development of the treatment was proposed by MICKLEY[12].

Combination of equations 13.37, 13.40, and 13.48 gives:

$$\frac{H_G - H_f}{\theta_L - \theta_f} = -\frac{h_L}{h_D \rho} \tag{13.50}$$

From equations 13.46 and 13.48:

$$\frac{H_G - H_f}{\theta_G - \theta_f} = \frac{dH_G}{d\theta_G} \tag{13.51}$$

and from equations 13.46 and 13.44:

$$\frac{\mathcal{H} - \mathcal{H}_f}{\theta_G - \theta_f} = \frac{d\mathcal{H}}{d\theta_G} \tag{13.52}$$

These equations will now be employed in the determination of the required height of a cooling tower for a given duty. The method consists of the graphical evaluation of the relation between the enthalpy of the body of gas and the enthalpy of the gas at the interface with the liquid. The required height of the tower is then obtained by integration of equation 13.49.

Suppose water is to be cooled at a mass rate L' per unit area from a temperature θ_{L2} to θ_{L1}. The air will be assumed to have a temperature θ_{G1}, a humidity $\mathcal{H}_1$, and an enthalpy H_{G1} (which can be calculated from the temperature and humidity), at the inlet point at the bottom of the tower, and its mass flow per unit area will be taken as G'. The change in the condition of the liquid and gas phases will now be followed on an enthalpy–temperature diagram (Figure 13.15). The enthalpy–temperature curve PQ for saturated air is plotted either using calculated data or from the humidity chart (Figure 13.4). The region below this line relates to unsaturated air and the region above it to supersaturated air. If it is

Figure 13.15.   Construction for height of water-cooling tower

assumed that the air in contact with the liquid surface is saturated with water vapour, this curve represents the relation between air enthalpy H_f and temperature θ_f at the interface.

The curve connecting air enthalpy and water temperature is now drawn using equation 13.39. This is known as the operating line and is a straight line of slope $(L'C_L/G')$, passing through the points $A(\theta_{L1}, H_{G1})$ and $B(\theta_{L2}, H_{G2})$. Since θ_{L1}, H_{G1} are specified, the procedure is to draw a line through (θ_{L1}, H_{G1}) of slope $L'C_L/G'$ and to produce it to a point whose abscissa is equal to θ_{L2}. This point B then corresponds to conditions at the top of the tower and the ordinate gives the enthalpy of the air leaving the column.

Equation 13.50 gives the relation between liquid temperature, air enthalpy, and conditions at the interface, for any position in the tower, and is represented by a family of straight lines of slope $-(h_L/h_D\rho)$. The line for the bottom of the column passes through the point $A(\theta_{L1}, H_{G1})$ and cuts the enthalpy-temperature curve for saturated air at the point C, representing conditions at the interface. The difference in ordinates of points A and C is the difference in the enthalpy of the air at the interface and that of the bulk air at the bottom of the column.

Similarly, line $A'C'$, parallel to AC, enables the difference in the enthalpies of the bulk air and the air at the interface to be determined at some other point in the column. The procedure can be repeated for a number of points and the value of $(H_f - H_G)$ obtained as a function of H_G for the whole tower.

Now:
$$\frac{dH_G}{H_f - H_G} = \frac{h_D a\rho}{G'} \, dz \qquad \text{(equation 13.49)}$$

On integration:
$$z = \int_1^2 dz = \frac{G'}{h_D a\rho} \int_1^2 \frac{dH_G}{H_f - H_G} \qquad (13.53)$$

assuming h_D to remain approximately constant.

Since $(H_f - H_G)$ is now known as a function of H_G, $1/(H_f - H_G)$ can be plotted against H_G and the integral evaluated between the required limits. The height of the tower is thus determined.

The integral in equation 13.53 cannot be evaluated by taking a logarithmic mean driving force because the saturation line PQ is far from linear. CAREY and WILLIAMSON[14] have given a useful approximate method of evaluating the integral. They assume that the enthalpy difference $(H_f - H_G) = \Delta H$ varies in a parabolic manner. The three fixed points taken to define the parabola are at the bottom and top of the column (ΔH_1 and ΔH_2 respectively) and ΔH_m, the value at the mean water temperature in the column. The effective mean driving force is $f \Delta H_m$, where f is a factor for converting the driving force at the mean water temperature to the effective value. In Figure 13.16, $(\Delta H_m/\Delta H_1)$ is plotted against $(\Delta H_m/\Delta H_2)$ and contours representing constant values of f are included.

Using the mean driving force, integration of equation 13.53 gives:
$$\frac{H_{G2} - H_{G1}}{f \Delta H_m} = \frac{h_D a\rho}{G'} z \qquad (13.54)$$

or:
$$z = \frac{G'}{h_D a\rho} \frac{H_{G2} - H_{G1}}{f \Delta H_m}$$

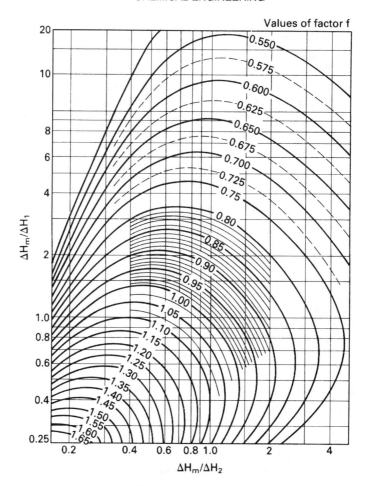

Figure 13.16. Correction factor f for obtaining mean effective driving force in column

13.6.4. Change in air condition

The change in the humidity and temperature of the air is obtained as follows. The enthalpy and temperature of the air are known only at the bottom of the tower, where fresh air is admitted. Here the condition of the air can be represented by a point E with coordinates (H_{G1}, θ_{G1}). Thus the line AE (Figure 13.15) is parallel to the temperature axis.

Since:
$$\frac{H_G - H_f}{\theta_G - \theta_f} = \frac{dH_G}{d\theta_G} \tag{13.51}$$

the slope of the line EC is $(dH_G/d\theta_G)$ and represents the rate of change of air enthalpy with air temperature at the bottom of the column. If the gradient $(dH_G/d\theta_G)$ is taken as constant over a small section, the point F, on EC, will represent the condition of the gas at a small distance from the bottom. The corresponding liquid temperature is found by drawing through F a line parallel to the temperature axis. This cuts the operating line at some point H, which indicates the liquid temperature. The corresponding value of the temperature and enthalpy of the gas at the interface is then obtained by drawing a line through H, parallel to AC. This line then cuts the curve for saturated air at a point J, which represents the conditions of the gas at the interface. The rate of change of enthalpy with temperature for the gas is then given by the slope of the line FJ. Again, this slope can be considered to remain constant over a small height of the column, and the condition of the gas is thus determined for the next point in the tower. The procedure is then repeated until the curve representing the condition of the gas has been extended to a point whose ordinate is equal to the enthalpy of the gas at the top of the column. This point is obtained by drawing a straight line through B, parallel to the temperature axis. The final point on the line then represents the condition of the air which leaves the top of the water-cooling tower.

The size of the individual increments of height which are considered must be decided for the particular problem under consideration and will depend, primarily, on the rate of change of the gradient $(\mathrm{d}H_G/\mathrm{d}\theta_G)$. It should be noted that, for the gas to remain unsaturated throughout the whole of the tower, the line representing the condition of the gas must be below the curve for saturated gas. If at any point in the column, the air has become saturated, it is liable to become supersaturated as it passes further up the column and comes into contact with hotter liquid. It is difficult to define precisely what happens beyond this point as partial condensation may occur, giving rise to a mist. Under these conditions the preceding equations will no longer be applicable. However, an approximate solution is obtained by assuming that once the air stream becomes saturated it remains so during its subsequent contact with the water through the column.

13.6.5. Temperature and humidity gradients in a water cooling tower

In a water cooling tower, the temperature profiles depend on whether the air is cooler or hotter than the surface of the water. Near the top, hot water makes contact with the exit air which is at a lower temperature, and sensible heat is therefore transferred both from the water to the interface and from the interface to the air. The air in contact with the water is saturated at the interface temperature and humidity therefore falls from the interface to the air. Evaporation followed by mass transfer of water vapour therefore takes place and latent heat is carried away from the interface in the vapour. The sensible heat removed from the water is then equal to the sum of the latent and sensible heats transferred to the air. Temperature and humidity gradients are then as shown in Figure 13.17a.

If the tower is sufficiently tall, the interface temperature can fall below the dry bulb temperature of the air (but not below its wet bulb temperature), and sensible heat will then be transferred from both the air and the water to the interface. The corresponding temperature and humidity profiles are given in Figure 13.17b. In this part of the tower, therefore, the sensible heat removed from the water will be that transferred as latent heat *less* the sensible heat transferred from the air.

13.6.6. Evaluation of heat and mass transfer coefficients

In general, coefficients of heat and mass transfer in the gas phase and the heat transfer coefficient for the liquid phase are not known. They can be determined, however, by

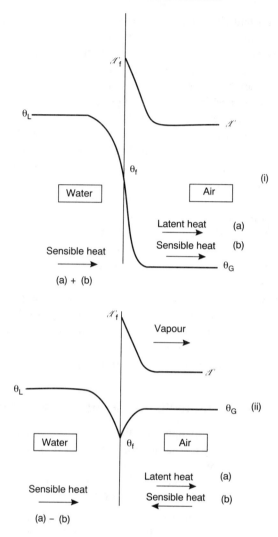

Figure 13.17. Temperature and humidity gradients in a water cooling tower (*i*) upper sections of tower
(*ii*) bottom of tower

carrying out tests in the laboratory or pilot scale using the same packing. If, for the air-
water system, a small column is operated at steady water and air rates and the temperature
of the water at the top and bottom and the initial and final temperatures and humidities of
the air stream are noted, the operating line for the system is obtained. Assuming a value
of the ratio $-(h_L/h_D\rho)$, for the slope of the tie-lines AC, etc., the graphical construction
is carried out, starting with the conditions at the bottom of the tower. The condition
of the gas at the top of the tower is thus calculated and compared with the measured
value. If the difference is significant, another value of $-(h_L/h_D\rho)$ is assumed and the
procedure repeated. Now that the slope of the tie line is known, the value of the integral
of $dH_G/(H_f - H_G)$ over the whole column can be calculated. Since the height of the

column is known, the product $h_D a$ is found by solution of equation 13.49. $h_G a$ can then be calculated using the Lewis relation.

The values of the three transfer coefficients are therefore obtained at any given flow rates from a single experimental run. The effect of liquid and gas rate can be found by means of a series of similar experiments.

Several workers have measured heat and mass transfer coefficients in water-cooling towers and in humidifying towers. THOMAS and HOUSTON[15], using a tower 2 m high and 0.3 m square in cross-section, fitted with wooden slats, give the following figures for heat and mass transfer coefficients for packed heights greater than 75 mm:

$$h_G a = 3.0 L'^{0.26} G'^{0.72} \qquad (13.55)$$

$$h_L a = 1.04 \times 10^4 L'^{0.51} G'^{1.00} \qquad (13.56)$$

$$h_D a = 2.95 L'^{0.26} G'^{0.72} \qquad (13.57)$$

In the above equations, L' and G' are expressed in kg/m^{2}s, s in J/kg K, $h_G a$ and $h_L a$ in W/m^3 K, and $h_D a$ in s^{-1}. A comparison of the gas and liquid film coefficients can then be made for a number of gas and liquid rates; taking the humid heat s as 1.17×10^3 J/kg K:

	$L' = G' = 0.5$ kg/m^{2}s	$L' = G' = 1.0$ kg/m^{2}s	$L' = G' = 2.0$ kg/m^{2}s
$h_G a$	1780	3510	6915
$h_L a$	3650	10,400	29,600
$h_L a / h_G a$	2.05	2.96	4.28

CRIBB[16] quotes values of the ratio h_L / h_G ranging from 2.4 to 8.5.

It will be seen that the liquid film coefficient is generally considerably higher than the gas film coefficient, but that it is not always safe to ignore the resistance to transfer in the liquid phase.

LOWE and CHRISTIE[17] used a 1.3 m square experimental column fitted with a number of different types of packing and measured heat and mass transfer coefficients and pressure drops. They showed that in most cases:

$$h_D a \propto L'^{1-n} G'^n \qquad (13.58)$$

The index n was found to vary from about 0.4 to 0.8 according to the type of packing. It will be noted that when $n \approx 0.75$, there is close agreement with the results of Thomas and Houston (equation 13.57).

The heat-transfer coefficient for the liquid is often large compared with that for the gas phase. As a first approximation, therefore, it can be assumed that the whole of the resistance to heat transfer lies within the gas phase and that the temperature at the water-air interface is equal to the temperature of the bulk of the liquid. Thus, everywhere in the tower, $\theta_f = \theta_L$. This simplifies the calculations, because the lines AC, HJ, etc., have a slope of $-\infty$, i.e. they become parallel to the enthalpy axis.

Some workers have attempted to base the design of humidifiers on the overall heat transfer coefficient between the liquid and gas phases. This treatment is not satisfactory since the quantities of heat transferred through the liquid and through the gas are not the same, as some of the heat is utilised in effecting evaporation at the interface. In fact, at

the bottom of a tall tower, the transfer of heat in both the liquid and the gas phases may
be towards the interface, as already mentioned. A further objection to the use of overall
coefficients is that the Lewis relation can be applied only to the heat and mass transfer
coefficients in the gas phase.

In the design of commercial units, nomographs[18,19] are available which give a performance characteristic (KaV/L'), where K is a mass transfer coefficient (kg water/m²s) and
V is the active cooling volume (m³/m² plan area), as a function of θ, θ_w and (L'/G'). For
a given duty (KaV/L') is calculated from:

$$\frac{KaV}{C_LL'} = \int_{\theta_1}^{\theta_2} \frac{d\theta}{H_f - H_G} \tag{13.59}$$

and then a suitable tower with this value of (KaV/L') is sought from performance
curves[20,21]. In normal applications the performance characteristic varies between 0.5–2.5.

Example 13.9

Water is to be cooled from 328 to 293 K by means of a countercurrent air stream entering at 293 K with a
relative humidity of 20 per cent. The flow of air is 0.68 m³/m²s and the water throughput is 0.26 kg/m²s. The
whole of the resistance to heat and mass transfer may be assumed to be in the gas phase and the product (h_Da)
may be taken as 0.2 (m/s)(m²/m³), that is 0.2 s⁻¹.

What is the required height of packing and the condition of the exit air stream?

Solution

Assuming the latent heat of water at 273 K = 2495 kJ/kg
specific heat of air = 1.003 kJ/kg K
and specific heat of water vapour = 2.006 kJ/kg K

the enthalpy of the inlet air stream

$$H_{G1} = 1.003(293 - 273) + \mathcal{H}[2495 + 2.006(293 - 273)]$$

From Figure 13.4:

at $\theta = 293$ K and 20 per cent RH, $\mathcal{H} = 0.003$ kg/kg, and hence

$$H_{G1} = (1.003 \times 20) + 0.003[2495 + (2.006 \times 20)]$$

$$= 27.67 \text{ kJ/kg}$$

In the inlet air, water vapour = 0.003 kg/kg dry air

or: $\dfrac{(0.003/18)}{(1/29)} = 0.005$ kmol/kmol dry air

Thus flow of dry air $= (1 - 0.005)0.68 = 0.677$ m³/m²s

Density of air at 293 K $= \dfrac{29}{22.4}\dfrac{273}{293} = 1.206$ kg/m³

and mass flow of dry air $= (1.206 \times 0.677) = 0.817$ kg/m²s

Slope of operating line: $(L'C_L/G') = \dfrac{0.26 \times 4.18}{0.817} = 1.33$

The coordinates of the bottom of the operating line are:

$$\theta_{L1} = 293 \text{ K}, \quad H_{G1} = 27.67 \text{ kJ/kg}$$

Hence on an enthalpy–temperature diagram, the operating line of slope 1.33 is drawn through the point $(293, 27.67) = (\theta_{L1}, H_{G1})$.

The top point of the operating line is given by $\theta_{L2} = 328$ K, and H_{G2} is found to be 76.5 kJ/kg (Figure 13.18).

Figure 13.18. Calculation of height of water-cooling tower

From Figures 13.4 and 13.5 the curve representing the enthalpy of saturated air as a function of temperature is obtained and drawn in. Alternatively, this plot may be calculated from:

$$H_F = C_a(\theta_f - 273) + \mathcal{H}_0[C_w(\theta_f - 273) + \lambda] \text{ kJ/kg}$$

The curve represents the relation between enthalpy and temperature at the interface, that is H_f as a function of θ_f.

It now remains to evaluate the integral $\int dH_G/(H_f - H_G)$ between limits, $H_{G1} = 27.7$ kJ/kg and $H_{G2} = 76.5$ kJ/kg. Various values of H_G between these limits are selected and the value of θ obtained from the operating line. At this value of θ_1, now θ_f, the corresponding value of H_f is obtained from the curve for saturated air. The working is as follows:

H_G	$\theta = \theta_f$	H_f	$(H_f - H_G)$	$1/(H_f - H_G)$
27.7	293	57.7	30	0.0330
30	294.5	65	35	0.0285
40	302	98	58	0.0172
50	309	137	87	0.0114
60	316	190	130	0.0076
70	323	265	195	0.0051
76.5	328	355	279	0.0035

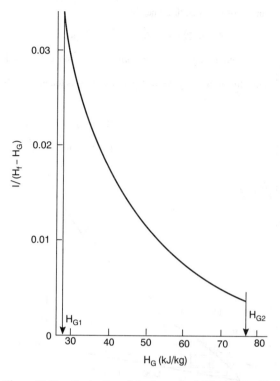

Figure 13.19.   Evaluation of the integral of $dH_G/(H_f - H_G)$

A plot of $1/(H_f - H_G)$ and H_G is now made as in Figure 13.19 from which the area under the curve $= 0.65$. This value may be checked using the approximate solution of CAREY and WILLIAMSON[14]. At the bottom of the column:

$$H_{G1} = 27.7 \text{ kJ/kg}, \quad H_{f1} = 57.7 \text{ kJ/kg} \quad \therefore \, \Delta H_1 = 30 \text{ kJ/kg}$$

At the top of the column:

$$H_{G2} = 76.5 \text{ kJ/kg}, \quad H_{f2} = 355 \text{ kJ/kg} \quad \therefore \, \Delta H_2 = 279 \text{ kJ/kg}$$

At the mean water temperature of $0.5(328 + 293) = 310.5$ K:

$$H_{Gm} = 52 \text{ kJ/kg}, \quad H_f = 145 \text{ kJ/kg} \quad \therefore \, \Delta H_m = 93 \text{ kJ/kg}$$

$$\frac{\Delta H_m}{\Delta H_1} = 3.10, \qquad \frac{\Delta H_m}{\Delta H_2} = 0.333,$$

and from Figure 13.16: $f = 0.79$

Thus:
$$\frac{H_{G2} - H_{G1}}{f \Delta H_m} = \frac{76.5 - 27.7}{0.79 \times 93} = 0.66$$

which agrees well with the value (0.65) obtained by graphical integration.
    Thus, in equation 13.53:

$$\text{height of packing, } z = \int_{H_{G1}}^{H_{G2}} \frac{dH_G}{H_f - H_G} \frac{G'}{h_D a \rho}$$

$$= \frac{0.65 \times 0.817}{0.2 \times 1.206}$$

$$= \underline{\underline{2.20 \text{ m}}}$$

Assuming that the resistance to mass transfer lies entirely within the gas phase, the lines connecting θ_L and θ_f are parallel with the enthalpy axis.

In Figure 13.18 a plot of H_G and θ_G is obtained using the construction given in Section 13.6.4 and shown in Figure 13.15. From this curve, the value of θ_{G2} corresponding to $H_{G2} = 76.5$ kJ/kg is 300 K. From Figure 13.5, under these conditions, the exit air has a humidity of 0.019 kg/kg which from Figure 13.4 corresponds to a relative humidity of 83 per cent.

13.6.7. Humidifying towers

If the main function of the tower is to produce a stream of humidified air, the final temperature of the liquid will not be specified but the humidity of the gas leaving the top of the tower will be given instead. It is therefore not possible to fix any point on the operating line, though its slope can be calculated from the liquid and gas rates. In designing a humidifier, therefore, it is necessary to calculate the temperature and enthalpy, and hence the humidity, of the gas leaving the tower for a number of assumed water outlet temperatures and thereby determine the outlet water temperature resulting in the air leaving the tower with the required humidity. The operating line for this water-outlet temperature is then used in the calculation of the height of the tower required to effect this degree of humidification. The calculation of the dimensions of a humidifier is therefore rather more tedious than that for the water-cooling tower.

In a humidifier in which the make-up liquid is only a small proportion of the total liquid circulating, its temperature approaches the adiabatic saturation temperature θ_s, and remains constant, so that there is no temperature gradient in the liquid. The gas in contact with the liquid surface is approximately saturated and has a humidity $\mathcal{H}_s$.

Thus:
$$\mathrm{d}\theta_L = 0$$

and:
$$\theta_{L1} = \theta_{L2} = \theta_L = \theta_f = \theta_s$$

Hence:
$$-G's\,\mathrm{d}\theta_G = h_G a\,\mathrm{d}z(\theta_G - \theta_s) \quad \text{(from equation 13.42)}$$

and:
$$-G'\,\mathrm{d}\mathcal{H} = h_D \rho a\,\mathrm{d}z(\mathcal{H} - \mathcal{H}_s) \quad \text{(from equation 13.44)}$$

Integration of these equations gives:

$$\ln\frac{\theta_{G1} - \theta_s}{\theta_{G2} - \theta_s} = \frac{h_G a}{G's}z \tag{13.60}$$

and:
$$\ln\frac{\mathcal{H}_s - \mathcal{H}_1}{\mathcal{H}_s - \mathcal{H}_2} = \frac{h_D a \rho}{G'}z \tag{13.61}$$

assuming h_G, h_D, and s to remain approximately constant.

From these equations the temperature θ_{G2} and the humidity $\mathcal{H}_2$ of the gas leaving the humidifier can be calculated in terms of the height of the tower. Rearrangement of equation 13.61 gives:

$$\ln\left(1 + \frac{\mathcal{H}_1 - \mathcal{H}_2}{\mathcal{H}_s - \mathcal{H}_1}\right) = -\frac{h_D a \rho}{G'}z$$

i.e.:
$$\frac{\mathcal{H}_2 - \mathcal{H}_1}{\mathcal{H}_s - \mathcal{H}_1} = 1 - \mathrm{e}^{-h_D a \rho z/G'} \tag{13.62}$$

Thus the ratio of the actual increase in humidity produced in the saturator to the maximum possible increase in humidity (i.e. the production of saturated gas) is equal to $(1 - e^{-h_D a p z/G'})$, and complete saturation of the gas is reached exponentially. A similar relation exists for the change in the temperature of the gas stream, namely:

$$\frac{\theta_{G1} - \theta_{G2}}{\theta_{G2} - \theta_s} = 1 - e^{-h_G az/G's} \qquad (13.63)$$

Further, the relation between the temperature and the humidity of the gas at any stage in the adiabatic humidifier is given by:

$$\frac{d\mathcal{H}}{d\theta_G} = \frac{\mathcal{H} - \mathcal{H}_s}{\theta_G - \theta_s} \qquad \text{(from equation 13.52)}$$

On integration:
$$\ln \frac{(\mathcal{H}_s - \mathcal{H}_2)}{(\mathcal{H}_s - \mathcal{H}_1)} = \ln \frac{(\theta_{G2} - \theta_s)}{(\theta_{G1} - \theta_s)} \qquad (13.64)$$

or:
$$\frac{(\mathcal{H}_s - \mathcal{H}_2)}{(\mathcal{H}_s - \mathcal{H}_1)} = \frac{(\theta_{G2} - \theta_s)}{(\theta_{G1} - \theta_s)} \qquad (13.65)$$

13.7. SYSTEMS OTHER THAN AIR–WATER

Calculations relating to systems where the Lewis relation is not applicable are very much more complicated because the adiabatic saturation temperature and the wet-bulb temperature do not coincide. Thus the significance of the adiabatic cooling lines on the psychrometric chart is very much restricted. They no longer represent the changes which take place in a gas as it is humidified by contact with liquid initially at the adiabatic saturation temperature of the gas, but simply give the compositions of all gases with the same adiabatic saturation temperature.

Calculation of the change in the condition of the liquid and the gas in a humidification tower is rendered more difficult since equation 13.49, which was derived for the air-water system, is no longer applicable. LEWIS and WHITE[22] have developed a method of calculation based on the use of a *modified enthalpy* in place of the true enthalpy of the system.

For the air–water system, from equation 13.11:

$$h_G = h_D \rho s \qquad (13.66)$$

This relationship applies quite closely for the conditions normally encountered in practice. For other systems, the relation between the heat and mass transfer coefficients in the gas phase is given by:

$$h_G = b h_D \rho s \qquad (13.67)$$

where b is approximately constant and generally has a value greater than unity. For these systems, therefore, equation 13.46 becomes:

$$G's\, d\theta_G = b h_D \rho a\, dz(s\theta_f - s\theta_G) \qquad (13.68)$$

Adding equations 13.68 and 13.47 to obtain the relationship corresponding to equation 13.48:

$$G'(s\,d\theta_G + \lambda\,d\mathscr{H}) = h_D\rho a\,dz[(bs\theta_f + \lambda\mathscr{H}_f) - (bs\theta_G + \lambda\mathscr{H})] \tag{13.69}$$

Lewis and White use a *modified latent heat of vaporisation* λ' defined by:

$$b = \frac{\lambda}{\lambda'} \tag{13.70}$$

and a *modified enthalpy* per unit mass of dry gas defined by:

$$H'_G = s(\theta_G - \theta_0) + \lambda'\mathscr{H} \tag{13.71}$$

Substituting in equation 13.67; from equations 13.38, 13.70, and 13.71:

$$G'\,dH_G = bh_D\rho a\,dz(H'_f - H'_G) \tag{13.72}$$

and:

$$\frac{dH_G}{H'_f - H'_G} = \frac{bh_D\rho a}{G'}\,dz \tag{13.73}$$

Combining equations 13.37, 13.40, and 13.72:

$$\frac{H'_G - H'_f}{\theta_L - \theta_f} = -\frac{h_L}{h_D\rho b} \quad \text{(cf. equation 13.50)} \tag{13.74}$$

From equations 13.66 and 13.72:

$$\frac{H'_G - H'_f}{\theta_G - \theta_f} = \frac{dH_G}{d\theta_G} \quad \text{(cf. equation 13.51)} \tag{13.75}$$

From equations 13.44 and 13.67:

$$\frac{\mathscr{H} - \mathscr{H}_f}{\theta_G - \theta_f} = b\frac{d\mathscr{H}}{d\theta_G} \quad \text{(cf. equation 13.52)} \tag{13.76}$$

The calculation of conditions within a countercurrent column operating with a system other than air–water is carried out in a similar manner to that already described by applying equations 13.73, 13.74, and 13.75 in conjunction with equation 3.39:

$$G'(H_{G2} - H_{G1}) = L'C_L(\theta_{L2} - \theta_{L1}) \quad \text{(equation 13.39)}$$

On an enthalpy–temperature diagram (Figure 13.20) the enthalpy of saturated gas is plotted against its temperature. If equilibrium between the liquid and gas exists at the interface, this curve PQ represents the relation between gas enthalpy and temperature at the interface (H_f v. θ_f). The modified enthalpy of saturated gas is then plotted against temperature (curve RS) to give the relation between H'_f and θ_f. Since b is greater than unity, RS will lie below PQ. By combining equations 13.35, 13.70, and 13.72, H'_G is obtained in terms of H_G.

$$H'_G = \frac{1}{b}[H_G + (b - 1)s(\theta_G - \theta_0)] \tag{13.77}$$

Figure 13.20.   Construction for height of column for vapour other than water

H'_G can conveniently be plotted against H_G for a number of constant temperatures. If b and s are constant, a series of straight lines is obtained. The operating line AB given by equation 13.39 is drawn in Figure 13.20. Point A has coordinates (θ_{L1}, H_{G1}) corresponding to the bottom of the column. Point a has coordinates (θ_{L1}, H'_{G1}), H'_{G1} being obtained from equation 13.77.

From equation 13.72, a line through a, of slope $-(h_L/h_D\rho b)$, will intersect curve RS at c, (θ_{f1}, H'_{f1}) to give the interface conditions at the bottom of the column. The corresponding air enthalpy is given by C, (θ_{f1}, H_{f1}). The difference between the ordinates of c and a then gives the driving force in terms of modified enthalpy at the bottom of the column $(H'_{f1} - H'_{G1})$. A similar construction at other points, such as A', enables the driving force to be calculated at any other point. Hence $(H'_f - H'_G)$ is obtained as a function of H_G throughout the column. The height of column corresponding to a given change in air enthalpy can be obtained from equation 13.71 since the left-hand side can now be evaluated.
Thus:

$$\int_{H_{G1}}^{H_{G2}} \frac{dH_G}{H'_f - H'_G} = \frac{bh_D a\rho}{G'}z \tag{13.78}$$

The change in the condition of the gas stream is obtained as follows: E, with coordinates (θ_{G1}, H_{G1}), represents the condition of the inlet gas. The modified enthalpy of this gas is given by $e(\theta_{G1}, H'_{G1})$. From equation 13.75 it is seen that ec gives the rate of change of gas enthalpy with temperature $(dH_G/d\theta_G)$ at the bottom of the column. Thus ED, parallel to ec, describes the way in which gas enthalpy changes at the bottom of the column. At some arbitrary small distance from the bottom, F represents the condition of the gas and H gives the corresponding liquid temperature. In exactly the same way the next small change is obtained by drawing through h a line hj parallel to ac. The slope of fj gives the new value of $(dH_G/d\theta_G)$ and therefore the gas condition at a higher point in the column is obtained by drawing FT parallel to fj. In this way the change in the condition of the gas through the column can be followed by continuing the procedure until the gas enthalpy reaches the value H_{G2} corresponding to the top of the column.

A detailed description of the method of construction of psychrometric charts is given by SHALLCROSS and LOW[23]. They illustrate their method by producing charts for three systems; air–water, air–benzene and air–toluene at pressures of 1 and 2 bar.

Example 13.10

In a countercurrent packed column, n-butanol flows down at a rate of 0.25 kg/m² s and is cooled from 330 to 295 K. Air at 290 K, initially free of n-butanol vapour, is passed up the column at the rate of 0.7 m³/m² s. Calculate the required height of tower and the condition of the exit air.

Data:

Mass transfer coefficient per unit volume, $h_D a = 0.1$ s^{-1}
Psychrometric ratio, $h_G/(h_D \rho_A s) = b = 2.34$
Heat transfer coefficients, $h_L = 3h_G$
Latent heat of vaporisation of n-butanol, $\lambda = 590$ kJ/kg
Specific heat of liquid n-butanol, $C_L = 2.5$ kJ/kg K
Humid heat of gas, $s = 1.05$ kJ/kg K

Temperature (K)	Vapour pressure (kN/m²) of butanol
295	0.59
300	0.86
305	1.27
310	1.75
315	2.48
320	3.32
325	4.49
330	5.99
335	7.89
340	10.36
345	14.97
350	17.50

Solution

The first stage is to calculate the enthalpy of the saturated gas by way of the saturated humidity, $\mathscr{H}_0$ given by:

$$\mathscr{H}_0 = \frac{P_{w0}}{P - P_{w0}} \frac{M_w}{M_A} = \frac{P_{w0}}{101.3 - P_{w0}} \frac{74}{29}$$

The enthalpy is then:

$$H_f = \frac{1}{1 + \mathscr{H}_0} \times 1.001(\theta_f - 273) + \mathscr{H}_0[2.5(\theta_f - 273) + 590] \text{ kJ/kg}$$

where 1.001 kJ/kg K is the specific heat of dry air.

$$H_f = \frac{1.001\theta_f - 273.27}{1 + \mathscr{H}_0} + \mathscr{H}_0(2.5\theta_f - 92.5) \text{ kJ/kg moist air}$$

The results of this calculation are presented in the following table and H_f is plotted against θ_f in Figure 13.21.

The modified enthalpy at saturation H'_f is given by:

$$H'_f = \frac{1.001\theta_f - 273.27}{1 + \mathcal{H}_0} + \mathcal{H}_0[2.5(\theta_f - 273) + \lambda']$$

where from equation 13.70: $\lambda' = \lambda/b = 590/2.34$ or 252 kJ/kg

$$\therefore \qquad H'_f = \frac{1.001\theta_f - 273.27}{1 + \mathcal{H}_0} + \mathcal{H}_0(2.5\theta_f - 430.5) \text{ kJ/kg moist air}$$

These results are also given in the following Table and plotted as H'_f against θ_f in Figure 13.21.

θ_f (K)	P_{w0} (kN/m^2)	$\mathcal{H}_0$ (kg/kg)	$(1.001\theta_f - 273.27)/(1 + \mathcal{H}_0)$ (kJ/kg)	$\mathcal{H}_0(2.5\theta_f - 92.5)$ (kJ/kg)	H_f (kJ/kg)	$\mathcal{H}_0(2.5\theta_f - 430.5)$ (kJ/kg)	H'_f (kJ/kg)
295	0.59	0.0149	21.70	9.61	31.31	4.57	26.28
300	0.86	0.0218	24.45	14.33	40.78	6.97	33.42
305	1.27	0.0324	31.03	21.71	52.74	10.76	41.79
310	1.75	0.0448	35.45	30.58	66.03	15.43	50.88
315	2.48	0.0640	39.52	44.48	84.00	22.85	62.37
320	3.32	0.0864	43.31	61.13	104.44	31.92	75.23
325	4.49	0.1183	46.55	85.18	131.73	45.19	91.74
330	5.99	0.1603	49.18	117.42	166.60	63.23	112.41
335	7.89	0.2154	51.07	160.47	211.54	87.67	138.73
340	10.36	0.2905	51.97	220.05	272.02	121.87	173.83
345	14.97	0.4422	49.98	340.49	390.47	191.03	241.01
350	17.50	0.5325	50.30	416.68	466.98	236.70	287.00

The bottom of the operating line (point a) has coordinates, $\theta_{L1} = 295$ K and H_{G1}, where:

$$H_{G1} = 1.05(290 - 273) = 17.9 \text{ kJ/kg.}$$

At a mean temperature of, say, 310 K, the density of air is:

$$\frac{29}{22.4} \frac{273}{310} = 1.140 \text{ kg/m}^3$$

and: $$G' = (0.70 \times 1.140) = 0.798 \text{ kg/m}^2 \text{ s}$$

Thus, the slope of the operating line becomes:

$$\frac{L'C_L}{G'} = \frac{0.25 \times 2.5}{0.798} = 0.783 \text{ kJ/kg K}$$

and this is drawn in as AB in Figure 13.21 and at $\theta_{L2} = 330$ K, $H_{G2} = 46$ kJ/kg.

From equation 13.77: $H'_G = [H_G + (b - 1)s(\theta_G - \theta_0)]/b$

$$\therefore \qquad H'_{G1} = \frac{17.9 + (2.34 - 1)1.05(290 - 273)}{2.34} = 17.87 \text{ kJ/kg}$$

Point a coincides with the bottom of the column.

A line is drawn through a of slope $-\dfrac{h_L}{h_D \rho b} = -\dfrac{3h_G}{h_D \rho b} \cdot \dfrac{h_D \rho s}{h_G}$

$$= -3s = -3.15 \text{ kJ/kg K}$$

It meets curve RS at c (θ_{f1}, H'_{f1}) to give the interface conditions at the bottom of the column. The corresponding air enthalpy is given by point C whose co-ordinates are:

$$\theta_{f1} = 293 \text{ K} \qquad H_{f1} = 29.0 \text{ kJ/kg}$$

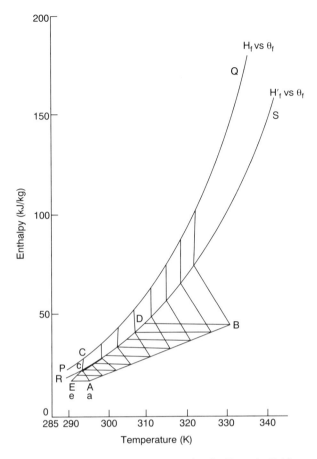

Figure 13.21. Graphical construction for Example 13.10

The difference between the ordinates of c and a gives the driving force in terms of the modified enthalpy at the bottom of the column, or:

$$(H'_{f1} - H'_{G1}) = (23.9 - 17.9) = 6.0 \text{ kJ/kg}$$

A similar construction is made at other points along the operating line with the results shown in the following table.

θ_f (K)	H_G (kJ/kg)	H'_G (kJ/kg)	H'_f (kJ/kg)	$(H'_f - H'_G)$ (kJ/kg)	$1/(H'_f - H'_G)$ (kg/kJ)	Mean value in interval	Interval	Value of integral over interval
295	17.9	17.9	23.9	6.0	0.167			
300	22.0	22.0	29.0	7.0	0.143	0.155	4.1	0.636
305	26.0	26.0	35.3	9.3	0.108	0.126	4.0	0.504
310	30.0	30.0	42.1	12.1	0.083	0.096	4.0	0.384
315	34.0	34.0	50.0	16.0	0.063	0.073	4.0	0.292
320	38.1	38.1	57.9	19.8	0.051	0.057	4.1	0.234
325	42.0	42.0	66.7	24.7	0.041	0.046	3.9	0.179
330	46.0	46.0	75.8	29.8	0.034	0.0375	4.0	0.150
						Value of integral = 2.379		

from which:

$$\int_{H_{G1}}^{H_{G2}} \frac{dH_G}{H'_f - H'_G} = 2.379$$

Substituting in equation 13.78:

$$\frac{bh_D \rho a z}{G'} = 2.379$$

and:

$$z = \frac{2.379 \times 0.798}{2.34 \times 0.1} = \underline{\underline{8.1 \ m}}$$

It remains to evaluate the change in gas conditions.

Point e, $(\theta_{G1} = 290 \ K, H_{G1} = 17.9 \ kJ/kg)$ represents the condition of the inlet gas. ec is now drawn in, and from equation 13.75, this represents $dH_G/d\theta_G$. As for the air-water system, this construction is continued until the gas enthalpy reaches H_{G2}. The final point is given by D at which $\underline{\underline{\theta_{G2} = 308 \ K}}$.

It is fortuitous in this problem that $H'_G = H_G$. This is not always the case and reference should be made to Section 13.7 for elaboration on this point.

13.8. FURTHER READING

BACKHURST, J. R., HARKER, J. H., and PORTER, J. E.: *Problems in Heat and Mass Transfer* (Edward Arnold, London, 1974).

BURGER, R.: *Chem. Eng. Progr.*, **71**(7) (1975) 73. Cooling tower drift elimination.

DEMONBRUN, J. R.: *Chem. Eng.*, **75**(19) (1968) 106. Factors to consider in selecting a cooling tower.

DONOHUE, J. M. and NATHAN, C. C.: *Chem. Eng. Progr.*, **71**(7) (1975) 88. Unusual problems in cooling water treatment.

ECKERT, E. R. G. and DRAKE, R. M.: *Analysis of Heat and Mass Transfer* (McGraw-Hill, New York, 1972).

ELGAWHARY, A. W.: *Chem. Eng. Progr.*, **71**(7) (1975) 83. Spray cooling system design.

FRIAR, F.: *Chem. Eng.*, **81**(15) (1974) 122. Cooling-tower basin design.

GUTHRIE, K. M.: *Chem. Eng.*, **76**(6) (1969) 114. Capital cost estimating.

HALL, W. A.: *Chem. Eng. Progr.*, **67**(7) (1971) 52. Cooling tower plume abatement.

HANSEN, E. P., and PARKER, J. J.: *Power Eng.*, **71**(5) (1967) 38. Status of big cooling towers.

HOLZHAUER, R.: *Plant. Eng.*, **29**(15) (1975) 60. Industrial cooling towers.

INCROPERO, F. P. and DE WITT, D. P.: *Fundamentals of Heat and Mass Transfer* (Wiley, New York, 1985).

INDUSTRIAL WATER SOCIETY: *Guide to Mechanical Draught Evaporative Cooling Towers: Selection, Operation and Maintenance* (London, 1987).

JACKSON, J.: *Cooling Towers* (Butterworths, London, 1951).

JORDAN, D. R., BEARDEN, M. D. and MCILHENNY, W. F.: *Chem. Eng. Progr.*, **71** (1975) 77. Blowdown concentration by electrodialysis.

JUONG, J. F.: *Hydrocarbon Process.*, **48**(7) (1969) 200. How to estimate cooling tower costs.

KELLY, G. M.: ASME Paper, 75-IPWR-9, (1975). Cooling tower design and evaluation parameters.

KOLFLAT, T. D.: *Power Eng.*, **78**(1) (1974) 32. Cooling tower practices.

MAZE, R. W.: *Hydrocarbon Process.*, **46**(2) (1967) 123. Practical tips on cooling tower sizing.

MAZE, R. W.: *Chem. Eng.*, **83**(1) (1975) 106. Air cooler or water tower — which for heat disposal?

MCCABE, W. L., SMITH, J. C. and HARRIOTT, P.: *Unit Operations of Chemical Engineering* (McGraw-Hill, New York, 1985).

MCKELVEY, K. K. and BROOKE, M.: *The Industrial Cooling Tower* (Elsevier, New York, 1959).

MEYTSAR, J.: *Hydrocarbon Process.*, **57**(11) (1978) 238. Estimate cooling tower requirements easily.

NELSON, W. L.: *Oil & Gas J.*, **65**(47) (1967) 182. What is cost of cooling towers?

NORMAN, W. S.: *Absorption, Distillation and Cooling Towers* (Longmans, London, 1961).

OLDS, F. C.: *Power Eng.*, **76**(12) (1972) 30. Cooling towers.

PAIGE, P. M.: *Chem. Eng.*, **74**(14) (1967) 93. Costlier cooling towers require a new approach to water-systems design.

PARK, J. E. and VANCE, J. M.: *Chem. Eng. Progr.*, **67** (1971) 55. Computer model of crossflow towers.

PICCIOTTI, M.: *Hydrocarbon Process.*, **56**(6) (1977) 163. Design quench water towers.

PICCIOTTI, M.: *Hydrocarbon Process.*, **56**(9) (1977) 179. Optimize quench water systems.

RABB, A.: *Hydrocarbon Process.*, **47**(2) (1968) 122. Are dry cooling towers economical?

UCHIYAMA, T.: *Hydrocarbon Process.*, **55**(12) (1976) 93. Cooling tower estimates made easy.

WALKER, R.: *Water Supply, Treatment and Distribution* (Prentice-Hall, New York, 1978).
WRINKLE, R. B.: *Chem. Eng. Progr.*, **67** (1971) 45. Performance of counterflow cooling tower cells.

13.9. REFERENCES

1. BS 1339: 1965(1981): British Standard 1339 (British Standards Institution, London) Definitions, formulae and constants relating to the humidity of the air.
2. SHERWOOD, T.K. and COMINGS, E.W.: *Trans. Am. Inst. Chem. Eng.* **28** (1932) 88. An experimental study of the wet bulb hygrometer.
3. GROSVENOR, M. M.: *Trans. Am. Inst. Chem. Eng.* **1** (1908) 184. Calculations for dryer design.
4. WEXLER, A.: In *Humidity and Moisture. Measurements and Control in Science and Industry*. Volume 1 Principles and Methods of Humidity Measurement in Gases, (Ed. R. E. Ruskin) (Reinhold, New York, 1965).
5. HICKMAN, M. J.: *Measurement of Humidity*, 4th edn (National Physical Laboratory. Notes on Applied Science No. 4). (HMSO, 1970).
6. MEADOWCROFT, D. B.: In *Instrumentation Reference Book*, (Ed. B. E. Noltingk), Chapter 6. Chemical analysis — moisture measurement (Butterworth, 1988).
7. WOOD, B. and BETTS, P.: *Proc. Inst. Mech. Eng. (Steam Group)* **163** (1950) 54. A contribution to the theory of natural draught cooling towers.
8. CLARK, R.: *The Chemical Engineer* (London) No. 529 (1992) 22. Cutting the fog.
9. McKELVEY, K. K. and BROOKE, M.: *The Industrial Cooling Tower* (Elsevier, New York, 1959).
10. CHILTON, C. H.: *Proc. Inst. Elec. Engrs.* **99** (1952), 440. Performance of natural-draught cooling towers.
11. PERRY, R. H. and GREEN, D. W. (eds): *Perry's Chemical Engineers' Handbook.* 6th edn (McGraw-Hill, New York, 1984).
12. MICKLEY, H. S.: *Chem. Eng. Prog.* **45** (1949) 739. Design of forced draught air conditioning equipment.
13. MERKEL, F.: *Ver. deut. Ing. Forschungsarb.* No. 275 (1925). Verdunstungs-Kühlung.
14. CAREY, W. F. and WILLIAMSON, G. J.: *Proc. Inst. Mech. Eng. (Steam Group)* **163** (1950) 41. Gas cooling and humidification: design of packed towers from small scale tests.
15. THOMAS, W.J. and HOUSTON, P.: *Brit. Chem. Eng.* **4** (1959) 160, 217. Simultaneous heat and mass transfer in cooling towers.
16. CRIBB, G.: *Brit. Chem. Eng.* **4** (1959) 264. Liquid phase resistance in water cooling.
17. LOWE, H. J. and CHRISTIE, D. G.: *Inst. Mech. Eng. Symposium on Heat Transfer* (1962), Paper 113, 933. Heat transfer and pressure drop data on cooling tower packings, and model studies of the resistance of natural-draught towers to airflow.
18. WOOD, B. and BETTS, P.: *Engineer* **189** (1950) (4912) 337, (4913) 349. A total heat–temperature diagram for cooling tower calculations.
19. ZIVI, S. M. and BRAND, B. B.: *Refrig. Eng.* **64** (1956) 8, 31, 90. Analysis of cross-flow cooling towers.
20. *Counter-flow Cooling Tower Performance* (J. F. Pritchard & Co., Kansas City, 1957).
21. COOLING TOWER INSTITUTE: *Performance Curves* (Houston, 1967).
22. LEWIS, J. G. and WHITE, R. R.: *Ind. Eng. Chem.* **45** (1953), 486. Simplified humidification calculations.
23. SHALLCROSS, D. C. and LOW, S. L.: *Ch.E.R.D.* **72** (1994) 763. Construction of psychrometric charts for systems other than water vapour in air. Errata: *Ch.E.R.D.* **73** (1995) 865.

13.10. NOMENCLATURE

		Units in SI system	Dimensions in M, N, L, T, θ
A	Interfacial area	m^2	$\mathbf{L}^2$
A_b	Base area of hyperbolic tower	m^2	$\mathbf{L}^2$
a	Interfacial area per unit volume of column	m^2/m^3	$\mathbf{L}^{-1}$
b	Psychrometric ratio ($h/h_D \rho_A s$)	—	—
C_a	Specific heat of gas at constant pressure	J/kg K	$\mathbf{L}^2\,\mathbf{T}^{-2}\theta^{-1}$
C_L	Specific heat of liquid	J/kg K	$\mathbf{L}^2\,\mathbf{T}^{-2}\theta^{-1}$
C_p	Specific heat of gas and vapour mixture at constant pressure	J/kg K	$\mathbf{L}^2\,\mathbf{T}^{-2}\theta^{-1}$
C_s	Specific heat of solid	J/kg K	$\mathbf{L}^2\,\mathbf{T}^{-2}\theta^{-1}$
C_t	Performance coefficient or efficiency factor	—	—

		Units in SI system	Dimensions in $\mathbf{M, N, L, T}, \theta$
C_w	Specific heat of vapour at constant pressure	J/kg K	$\mathbf{L^2\ T^{-2}\theta^{-1}}$
c	Mass concentration of vapour	kg/m^3	$\mathbf{ML^{-3}}$
c_0	Mass concentration of vapour in saturated gas	kg/m^3	$\mathbf{ML^{-3}}$
D_t	Duty coefficient of tower (equation 13.31)	—	—
f	Correction factor for mean driving force	—	—
G'	Mass rate of flow of gas per unit area	kg/m^2 s	$\mathbf{ML^{-2}\ T^{-1}}$
H	Enthalpy of humid gas per unit mass of dry gas	J/kg	$\mathbf{L^2\ T^{-2}}$
H_a	Enthalpy per unit mass, of dry gas	J/kg	$\mathbf{L^2\ T^{-2}}$
H_w	Enthalpy per unit mass, of vapour	J/kg	$\mathbf{L^2\ T^{-2}}$
H_1	Enthalpy of stream of gas, per unit mass of dry gas	J/kg	$\mathbf{L^2\ T^{-2}}$
H_2	Enthalpy of another stream of gas, per unit mass of dry gas	J/kg	$\mathbf{L^2\ T^{-2}}$
H_3	Enthalpy per unit mass of liquid or vapour	J/kg	$\mathbf{L^2\ T^{-2}}$
H'	Modified enthalpy of humid gas defined by (13.69)	J/kg	$\mathbf{L^2\ T^{-2}}$
ΔH	Enthalpy driving force ($H_f - H_G$)	J/kg	$\mathbf{L^2\ T^{-2}}$
$\Delta H'$	Change in air enthalpy on passing through tower	J/kg	$\mathbf{L^2\ T^{-2}}$
h	Heat transfer coefficient	W/m^2 K	$\mathbf{MT^{-3}\theta^{-1}}$
h_D	Mass transfer coefficient	kmol/(kmol/m^3)m^2 s	$\mathbf{LT^{-1}}$
h_G	Heat transfer coefficient for gas phase	W/m^2 K	$\mathbf{MT^{-3}\theta^{-1}}$
h_L	Heat transfer coefficient for liquid phase	W/m^2 K	$\mathbf{MT^{-3}\theta^{-1}}$
$\mathscr{H}$	Humidity	kg/kg	—
$\mathscr{H}_s$	Humidity of gas saturated at the adiabatic saturation temperature	kg/kg	—
$\mathscr{H}_w$	Humidity of gas saturated at the wet-bulb temperature	kg/kg	—
$\mathscr{H}_0$	Humidity of saturated gas	kg/kg	—
$\mathscr{H}_1$	Humidity of a gas stream	kg/kg	—
$\mathscr{H}_2$	Humidity of second gas stream	kg/kg	—
L'	Mass rate of flow of liquid per unit area	kg/m^2 s	$\mathbf{ML^{-2}\ T^{-1}}$
M_A	Molecular weight of gas	kg/kmol	$\mathbf{MN^{-1}}$
M_w	Molecular weight of vapour	kg/kmol	$\mathbf{MN^{-1}}$
m, m_1, m_2	Masses of dry gas	kg	$\mathbf{M}$
m_3	Mass of liquid or vapour	kg	$\mathbf{M}$
P	Total pressure	N/m^2	$\mathbf{ML^{-1}\ T^{-2}}$
P_A	Mean partial pressure of gas	N/m^2	$\mathbf{ML^{-1}\ T^{-2}}$
P_w	Partial pressure of vapour	N/m^2	$\mathbf{ML^{-1}\ T^{-2}}$
P_{w0}	Partial pressure of vapour in saturated gas	N/m^2	$\mathbf{ML^{-1}\ T^{-2}}$
Q	Rate of transfer of heat to liquid surface	W	$\mathbf{ML^2\ T^{-3}}$
$\mathbf{R}$	Universal gas constant	(8314)J/kmol K	$\mathbf{MN^{-1}\ L^2\ T^{-2}\theta^{-1}}$
s	Humid heat of gas	J/kg K	$\mathbf{L^2\ T^{-2}\theta^{-1}}$
T	Absolute temperature	K	θ
ΔT	Change in water temperature in passing through the tower	K	θ
$\Delta T'$	(Temperature of air leaving packing — ambient dry-bulb temperature)	K	θ
V	Active volume per plan area of column	m^3/m^2	$\mathbf{L}$
W_L	Water loading on tower	kg/s	$\mathbf{MT^{-1}}$
w	Rate of evaporation	kg/s	$\mathbf{MT^{-1}}$
Z	Percentage humidity	—	—
z	Height from bottom of tower	m	$\mathbf{L}$
z_t	Height of cooling tower	m	$\mathbf{L}$
θ	Temperature of gas stream	K	θ

		Units in SI system	Dimensions in M, N, L, T, θ
θ_0	Reference temperature, taken as the melting point of the material	K	θ
θ_s	Adiabatic saturation temperature	K	θ
θ_w	Wet bulb temperature	K	θ
λ	Latent heat of vaporisation per unit mass, at datum temperature	J/kg	$L^2\,T^{-2}$
λ_f	Latent heat of freezing per unit mass, at datum temperature	J/kg	$L^2\,T^{-2}$
λ'	Modified latent heat of vaporisation per unit mass defined by (13.68)	J/kg	$L^2\,T^{-2}$
ρ	Mean density of gas and vapour	kg/m^3	ML^{-3}
ρ_A	Mean density of gas at partial pressure P_A	kg/m^3	ML^{-3}
Pr	Prandtl number	—	—
Sc	Schmidt number	—	—

Suffixes 1, 2, f, L, G denote conditions at the bottom of the tower, the top of the tower, the interface, the liquid, and the gas, respectively.

Suffix m refers to the mean water temperature.

Appendix

A1. CONVERSION FACTORS FOR SOME COMMON SI UNITS

An asterisk (*) denotes an exact relationship.

Length	*1 in.	: 25.4 mm
	*1 ft	: 0.304 8 m
	*1 yd	: 0.914 4 m
	1 mile	: 1.609 3 km
	*1 Å (angstrom)	: 10^{-10} m
Time	*1 min	: 60 s
	*1 h	: 3.6 ks
	*1 day	: 86.4 ks
	1 year	: 31.5 Ms
Area	*1 in.^{2}	: 645.16 mm^2
	1 ft^2	: 0.092 903 m^2
	1 yd^2	: 0.836 13 m^2
	1 acre	: 4046.9 m^2
	1 mile^2	: 2.590 km^2
Volume	1 in.^{3}	: 16.387 cm^3
	1 ft^3	: 0.028 32 m^3
	1 yd^3	: 0.764 53 m^3
	1 UK gal	: 4546.1 cm^3
	1 US gal	: 3785.4 cm^3
Mass	1 oz	: 28.352 g
	*1 lb	: 0.453 592 37 kg
	1 cwt	: 50.802 3 kg
	1 ton	: 1016.06 kg
Force	1 pdl	: 0.138 26 N
	1 lbf	: 4.448 2 N
	1 kgf	: 9.806 7 N
	1 tonf	: 9.964 0 kN
	*1 dyn	: 10^{-5} N
Temperature difference	*1 deg F (deg R)	: $\frac{5}{9}$ deg C (deg K)
Energy (work, heat)	1 ft lbf	: 1.355 8 J
	1 ft pdl	: 0.042 14 J
	*1 cal (internat. table)	: 4.186 8 J
	1 erg	: 10^{-7} J
	1 Btu	: 1.055 06 kJ
	1 hp h	: 2.684 5 MJ
	*1 kW h	: 3.6 MJ
	1 therm	: 105.51 MJ
	1 thermie	: 4.185 5 MJ
Calorific value (volumetric)	1 Btu/ft^3	: 37.259 kJ/m^3
Velocity	1 ft/s	: 0.304 8 m/s
	1 mile/h	: 0.447 04 m/s
Volumetric flow	1 ft^3/s	: 0.028 316 m^3/s
	1 ft^3/h	: 7.865 8 cm^3/s
	1 UK gal/h	: 1.262 8 cm^3/s
	1 US gal/h	: 1.051 5 cm^3/s

Mass flow	1 lb/h	: 0.126 00 g/s
	1 ton/h	: 0.282 24 kg/s
Mass per unit area	1 lb/in.^{2}	: 703.07 kg/m^2
	1 lb/ft^2	: 4.882 4 kg/m^2
	1 ton/sq mile	: 392.30 kg/km^2
Density	1 lb/in^3	: 27.680 g/cm^3
	1 lb/ft^3	: 16.019 kg/m^3
	1 lb/UK gal	: 99.776 kg/m^3
	1 lb/US gal	: 119.83 kg/m^3
Pressure	1 lbf/in.^{2}	: 6.894 8 kN/m^2
	1 tonf/in.^{2}	: 15.444 MN/m^2
	1 lbf/ft^2	: 47.880 N/m^2
	*1 standard atm	: 101.325 kN/m^2
	*1 atm	
	(1 kgf/cm^2)	: 98.066 5 kN/m^2
	*1 bar	: 10^5 N/m^2
	1 ft water	: 2.989 1 kN/m^2
	1 in. water	: 249.09 N/m^2
	1 in. Hg	: 3.386 4 kN/m^2
	1 mm Hg (1 torr)	: 133.32 N/m^2
Power (heat flow)	1 hp (British)	: 745.70 W
	1 hp (metric)	: 735.50 W
	1 erg/s	: 10^{-7} W
	1 ft lbf/s	: 1.355 8 W
	1 Btu/h	: 0.293 07 W
	1 ton of refrigeration	: 3516.9 W
Moment of inertia	1 lb ft^2	: 0.042 140 kg m^2
Momentum	1 lb ft/s	: 0.138 26 kg m/s
Angular momentum	1 lb ft^2/s	: 0.042 140 kg m^2/s
Viscosity, dynamic	*1 P (Poise)	: 0.1 N s/m^2
	1 lb/ft h	: 0.413 38 mN s/m^2
	1 lb/ft s	: 1.488 2 Ns/m^2
Viscosity, kinematic	*1 S (Stokes)	: 10^{-4} m^2/s
	1 ft^2/h	: 0.258 06 cm^2/s
Surface energy (surface tension)	1 erg/cm^2 (1 dyn/cm)	: 10^{-3} J/m^2 : (10^{-3} N/m)
Mass flux density	1 lb/h ft^2	: 1.356 2 g/s m^2
Heat flux density	1 Btu/h ft^2	: 3.154 6 W/m^2
	*1 kcal/h m^2	: 1.163 W/m^2
Heat transfer coefficient	1 Btu/h ft^2 °F	: 5.678 3 W/m^{2}K
Specific enthalpy (latent heat, etc.)	*1 Btu/lb	: 2.326 kJ/kg
Specific heat capacity	*1 Btu/lb °F	: 4.186 8 kJ/kgK
Thermal conductivity	1 Btu/h ft °F	: 1.730 7 W/mK
	1 kcal/h m °C	: 1.163 W/mK

(Taken from MULLIN, J.W.: *The Chemical Engineer* No. 211 (Sept. 1967), 176. SI units in Chemical Engineering.)

A2. TABLES OF PHYSICAL PROPERTIES

Table 1. Temperature conversion chart

General formula: $°F = (°C \times \frac{9}{5}) + 32$; $°C = (°F - 32) \times \frac{5}{9}$; $K = °C + 273.15$

°C	val	°F	°C	val	°F	°C	val	°F	°C	val	°F	°C	val	°F	°C	val	°F	°C	val	°F	°C	val	°F	°C	val	°F
-273.1	-459.7		-17.8	0	32	10.0	50	122.0	38	100	212	260	500	932	538	1000	1832	816	1500	2732	1093	2000	3632	1371	2500	4532
-268	-450		-17.2	1	33.8	10.6	51	123.8	43	110	230	266	510	950	543	1010	1850	821	1510	2750	1099	2010	3650	1377	2510	4550
-262	-440		-16.7	2	35.6	11.1	52	125.6	49	120	248	271	520	968	549	1020	1868	827	1520	2768	1104	2020	3668	1382	2520	4568
-257	-430		-16.1	3	37.4	11.7	53	127.4	54	130	266	277	530	986	554	1030	1886	832	1530	2786	1110	2030	3686	1388	2530	4586
-251	-420		-15.6	4	39.2	12.2	54	129.2	60	140	284	282	540	1004	560	1040	1904	838	1540	2804	1116	2040	3704	1393	2540	4604
-246	-410		-15.0	5	41.0	12.8	55	131.0	66	150	302	288	550	1022	566	1050	1922	843	1550	2822	1121	2050	3722	1399	2550	4622
-240	-400		-14.4	6	42.8	13.3	56	132.8	71	160	320	293	560	1040	571	1060	1940	849	1560	2840	1127	2060	3740	1404	2560	4640
-234	-390		-13.9	7	44.6	13.9	57	134.6	77	170	338	299	570	1058	577	1070	1958	854	1570	2858	1132	2070	3758	1410	2570	4658
-229	-380		-13.3	8	46.4	14.4	58	136.4	82	180	356	304	580	1076	582	1080	1976	860	1580	2876	1138	2080	3776	1416	2580	4676
-223	-370		-12.8	9	48.2	15.0	59	138.2	88	190	374	310	590	1094	588	1090	1994	866	1590	2894	1143	2090	3794	1421	2590	4694
-218	-360		-12.2	10	50.0	15.6	60	140.0	93	200	392	316	600	1112	593	1100	2012	871	1600	2912	1149	2100	3812	1427	2600	4712
-212	-350		-11.7	11	51.8	16.1	61	141.8	99	210	410	321	610	1130	599	1110	2030	877	1610	2930	1154	2110	3830	1432	2610	4730
-207	-340		-11.1	12	53.6	16.7	62	143.6	100	212	413	327	620	1148	604	1120	2048	882	1620	2948	1160	2120	3848	1438	2620	4748
-201	-330		-10.6	13	55.4	17.2	63	145.4	104	220	428	332	630	1166	610	1130	2066	888	1630	2966	1166	2130	3866	1443	2630	4766
-196	-320		-10.0	14	57.2	17.8	64	147.2	110	230	446	338	640	1184	616	1140	2084	893	1640	2984	1171	2140	3884	1449	2640	4784
-190	-310		-9.44	15	59.0	18.3	65	149.0	116	240	464	343	650	1202	621	1150	2102	899	1650	3002	1177	2150	3902	1454	2650	4802
-184	-300		-8.89	16	60.8	18.9	66	150.8	121	250	482	349	660	1220	627	1160	2120	904	1660	3020	1182	2160	3920	1460	2660	4820
-179	-290		-8.33	17	62.6	19.4	67	152.6	127	260	500	354	670	1238	632	1170	2138	910	1670	3038	1188	2170	3938	1466	2670	4838
-173	-280		-7.78	18	64.4	20.0	68	154.4	132	270	518	360	680	1256	638	1180	2156	916	1680	3056	1193	2180	3956	1471	2680	4856
-169.5	-273.1	-459.7	-7.22	19	66.2	20.6	69	156.2	138	280	536	366	690	1274	643	1190	2174	921	1690	3074	1199	2190	3974	1477	2690	4874
-168	-270	-454	-6.67	20	68.0	21.1	70	158.0	143	290	554	371	700	1292	649	1200	2192	927	1700	3092	1204	2200	3992	1482	2700	4892
-162	-260	-436	-6.11	21	69.8	21.7	71	159.8	149	300	572	377	710	1310	654	1210	2210	932	1710	3110	1210	2210	4010	1488	2710	4910
-157	-250	-418	-5.56	22	71.6	22.2	72	161.6	154	310	590	382	720	1328	660	1220	2228	938	1720	3128	1216	2220	4028	1493	2720	4928
-151	-240	-400	-5.00	23	73.4	22.8	73	163.4	160	320	608	388	730	1346	666	1230	2246	943	1730	3146	1221	2230	4046	1499	2730	4946
-146	-230	-382	-4.44	24	75.2	23.3	74	165.2	166	330	626	393	740	1364	671	1240	2264	949	1740	3164	1227	2240	4064	1504	2740	4964
-140	-220	-364	-3.89	25	77.0	23.9	75	167.0	171	340	644	399	750	1382	677	1250	2282	954	1750	3182	1232	2250	4082	1510	2750	4982
-134	-210	-346	-3.33	26	78.8	24.4	76	168.8	177	350	662	404	760	1400	682	1260	2300	960	1760	3200	1238	2260	4100	1516	2760	5000
-129	-200	-328	-2.78	27	80.6	25.0	77	170.6	182	360	680	410	770	1418	688	1270	2318	966	1770	3218	1243	2270	4118	1521	2770	5018
-123	-190	-310	-2.22	28	82.4	25.6	78	172.4	188	370	698	416	780	1436	693	1280	2336	971	1780	3236	1249	2280	4136	1527	2780	5036
-118	-180	-292	-1.67	29	84.2	26.1	79	174.2	193	380	716	421	790	1454	699	1290	2354	977	1790	3254	1254	2290	4154	1532	2790	5054

Table 1. (continued)

C.	F.	F.
-112	**-170**	-274
-107	**-160**	-256
-101	**-150**	-238
-95.6	**-140**	-220
-90.0	**-130**	-202
-84.4	**-120**	-184
-78.9	**-110**	-166
-73.3	**-100**	-148
-67.8	**-90**	-130
-62.2	**-80**	-112
-56.7	**-70**	-94
-51.1	**-60**	-76
-45.6	**-50**	-58
-40.0	**-40**	-40
-34.4	**-30**	-22
-28.9	**-20**	-4
-23.3	**-10**	14
-17.8	**0**	32

C.	C.	F.
-1.11	**30**	86.0
-0.56	**31**	87.8
0	**32**	89.6
0.56	**33**	91.4
1.11	**34**	93.2
1.67	**35**	95.0
2.22	**36**	96.8
2.78	**37**	98.6
3.33	**38**	100.4
3.89	**39**	102.2
4.44	**40**	104.0
5.00	**41**	105.8
5.56	**42**	107.6
6.11	**43**	109.4
6.67	**44**	111.2
7.22	**45**	113.0
7.78	**46**	114.8
8.33	**47**	116.6
8.89	**48**	118.4
9.44	**49**	120.2

C.	F.	F.
26.7	**80**	176.0
27.2	**81**	177.8
27.8	**82**	179.6
28.3	**83**	181.4
28.9	**84**	183.2
29.4	**85**	185.0
30.0	**86**	186.8
30.6	**87**	188.6
31.1	**88**	190.4
31.7	**89**	192.2
32.2	**90**	194.0
32.8	**91**	195.8
33.3	**92**	197.6
33.9	**93**	199.4
34.4	**94**	201.2
35.0	**95**	203.0
35.6	**96**	204.8
36.1	**97**	206.6
36.7	**98**	208.4
37.2	**99**	210.2
37.8	**100**	212.0

C.	F.	F.
199	**390**	734
204	**400**	752
210	**410**	770
216	**420**	788
221	**430**	806
227	**440**	824
232	**450**	842
238	**460**	860
243	**470**	878
249	**480**	896
254	**490**	914

C.	C.	F.
427	**800**	1472
432	**810**	1490
438	**820**	1508
443	**830**	1526
449	**840**	1544
454	**850**	1562
460	**860**	1580
466	**870**	1598
471	**880**	1616
477	**890**	1634
482	**900**	1652
488	**910**	1670
493	**920**	1688
499	**930**	1706
504	**940**	1724
510	**950**	1742
516	**960**	1760
521	**970**	1778
527	**980**	1796
532	**990**	1814

C.	C.	F.
704	**1300**	2372
710	**1310**	2390
716	**1320**	2408
721	**1330**	2426
727	**1340**	2444
732	**1350**	2462
738	**1360**	2480
743	**1370**	2498
749	**1380**	2516
754	**1390**	2534
760	**1400**	2552
766	**1410**	2570
771	**1420**	2588
777	**1430**	2606
782	**1440**	2624
788	**1450**	2642
793	**1460**	2660
799	**1470**	2678
804	**1480**	2696
810	**1490**	2714

C.	C.	F.
982	**1800**	3272
988	**1810**	3290
993	**1820**	3308
999	**1830**	3326
1004	**1840**	3344
1010	**1850**	3362
1016	**1860**	3380
1021	**1870**	3398
1027	**1880**	3416
1032	**1890**	3434
1038	**1900**	3452
1043	**1910**	3470
1049	**1920**	3488
1054	**1930**	3506
1060	**1940**	3524
1066	**1950**	3542
1071	**1960**	3560
1077	**1970**	3578
1082	**1980**	3596
1088	**1990**	3614
1093	**2000**	3632

C.	C.	F.
1260	**2300**	4172
1266	**2310**	4190
1271	**2320**	4208
1277	**2330**	4226
1282	**2340**	4244
1288	**2350**	4262
1293	**2360**	4280
1299	**2370**	4298
1304	**2380**	4316
1310	**2390**	4334
1316	**2400**	4352
1321	**2410**	4370
1327	**2420**	4388
1332	**2430**	4406
1338	**2440**	4424
1343	**2450**	4442
1349	**2460**	4460
1354	**2470**	4478
1360	**2480**	4496
1366	**2490**	4514

C.	C.	F.
1538	**2800**	5072
1543	**2810**	5090
1549	**2820**	5108
1554	**2830**	5126
1560	**2840**	5144
1566	**2850**	5162
1571	**2860**	5180
1577	**2870**	5198
1582	**2880**	5216
1588	**2890**	5234
1593	**2900**	5252
1599	**2910**	5270
1604	**2920**	5288
1610	**2930**	5306
1616	**2940**	5324
1621	**2950**	5342
1627	**2960**	5360
1632	**2970**	5378
1638	**2980**	5396
1643	**2990**	5414
1649	**3000**	5432

Interpolation factors

C.	F.	C.	F.
0.56	**1**	1.8	
1.11	**2**	3.6	
1.67	**3**	5.4	
2.22	**4**	7.2	
2.78	**5**	9.0	
3.33	**6**	10.8	
3.89	**7**	12.6	
4.44	**8**	14.4	
5.00	**9**	16.2	
5.56	**10**	18.0	

Note.—The numbers in bold-face type refer to the temperature (in either Centigrade or Fahrenheit degrees) which it is desired to convert into the other scale. If converting from Fahrenheit degrees to Centigrade degrees the equivalent temperature is in the left column, while if converting from degrees Centigrade to degrees Fahrenheit, the equivalent temperature is in the column on the right. This table, made by Albert Sauveur, is published by permission of Mme. Albert Sauveur.

Table 2.   Thermal conductivities of liquids*

A linear variation with temperature may be assumed. The extreme values given constitute also the temperature limits over which the data are recommended.

Liquid	k (W/m K)	(K)	k (Btu/h ft °F)
Acetic acid 100 per cent	0.171	293	0.099
50 per cent	0.35	293	0.20
Acetone	0.177	303	0.102
	0.164	348	0.095
Allyl alcohol	0.180	298 to 303	0.104
Ammonia	0.50	258 to 303	0.29
Ammonia, aqueous	0.45	293	0.261
	0.50	333	0.29
Amyl acetate	0.144	283	0.083
Amyl alcohol (n-)	0.163	303	0.094
	0.154	373	0.089
Amyl alcohol (iso-)	0.152	303	0.088
	0.151	348	0.087
Aniline	0.173	273 to 293	0.100
Benzene	0.159	303	0.092
	0.151	333	0.087
Bromobenzene	0.128	303	0.074
	0.121	373	0.070
Butyl acetate (n-)	0.147	298 to 303	0.085
Butyl alcohol (n-)	0.168	303	0.097
	0.164	348	0.095
Butyl alcohol (iso-)	0.157	283	0.091
Calcium chloride brine			
30 per cent	0.55	303	0.32
15 per cent	0.59	303	0.34
Carbon disulphide	0.161	303	0.093
	0.152	348	0.088
Carbon tetrachloride	0.185	273	0.107
Chlorobenzene	0.163	341	0.094
Chloroform	0.144	283	0.083
	0.138	303	0.080

Liquid	k (W/m K)	(K)	k (Btu/h ft °F)
Hexane (n-)	0.138	303	0.080
	0.135	333	0.078
Heptyl alcohol (n-)	0.163	303	0.094
	0.157	348	0.091
Hexyl alcohol (n-)	0.161	303	0.093
	0.156	348	0.090
Kerosene	0.149	293	0.086
	0.140	348	0.081
Mercury	8.36	301	4.83
Methyl alcohol 100 per cent	0.215	293	0.124
80 per cent	0.267	293	0.154
60 per cent	0.329	293	0.190
40 per cent	0.405	293	0.234
20 per cent	0.492	293	0.284
100 per cent	0.197	323	0.114
Methyl chloride	0.192	258	0.111
	0.154	303	0.089
Nitrobenzene	0.164	303	0.095
	0.152	373	0.088
Nitromethane	0.216	303	0.125
	0.208	333	0.120
Nonane (n-)	0.145	303	0.084
	0.142	333	0.082
Octane (n-)	0.144	303	0.083
	0.140	333	0.081
Oils, Petroleum	0.138 to 0.156	273	0.08 to 0.09
Oil, Castor	0.180	293	0.104
	0.173	373	0.100
	0.168	293	0.097

Table 2. (continued)

Liquid	k (W/m K)	(K)	k (Btu/h ft °F)
Cymene (para)	0.135	303	0.078
	0.137	333	0.079
Decane (n-)	0.147	303	0.085
	0.144	333	0.083
Dichlorodifluoromethane	0.099	266	0.057
	0.092	289	0.053
	0.083	311	0.048
	0.074	333	0.043
	0.066	355	0.038
Dichloroethane	0.142	323	0.082
Dichloromethane	0.192	258	0.111
	0.166	303	0.096
Ethyl acetate	0.175	293	0.101
Ethyl alcohol 100 per cent	0.182	293	0.105
80 per cent	0.237	293	0.137
60 per cent	0.305	293	0.176
40 per cent	0.388	293	0.224
20 per cent	0.486	293	0.281
100 per cent	0.151	323	0.087
Ethyl benzene	0.149	303	0.086
	0.142	333	0.082
Ethyl bromide	0.121	293	0.070
Ethyl ether	0.138	303	0.080
	0.135	348	0.078
Ethyl iodide	0.111	313	0.064
	0.109	348	0.063
Ethylene glycol	0.265	273	0.153
Gasoline	0.135	303	0.078
Glycerol 100 per cent	0.284	293	0.164
80 per cent	0.327	293	0.189
60 per cent	0.381	293	0.220
40 per cent	0.448	293	0.259
20 per cent	0.481	293	0.278
100 per cent	0.284	373	0.164
Heptane (n-)	0.140	303	0.081
	0.137	333	0.079

k (W/m K)	(K)	k (Btu/h ft °F)	Liquid
0.164	373	0.095	Oil, Olive
0.145	303	0.084	Paraldehyde
0.135	373	0.078	
0.135	303	0.078	Pentane (n-)
0.128	348	0.074	
0.159	323	0.092	Perchloroethylene
0.130	303	0.075	Petroleum ether
0.126	348	0.073	
0.171	303	0.099	Propyl alcohol (n-)
0.164	348	0.095	
0.157	303	0.091	Propyl alcohol (iso-)
0.155	333	0.090	
0.85	373	49	Sodium
0.80	483	46	
0.57	303	0.33	Sodium chloride brine 25.0 per cent
0.59	303	0.34	12.5 per cent
0.36	303	0.21	Sulphuric acid 90 per cent
0.43	303	0.25	60 per cent
0.52	303	0.30	30 per cent
0.22	258	0.128	Sulphur dioxide
0.192	303	0.111	
0.149	303	0.086	Toluene
0.145	348	0.084	
0.133	323	0.077	β-trichloroethane
0.138	323	0.080	Trichloroethylene
0.128	288	0.074	Turpentine
0.184	288	0.106	Vaseline
0.57	273	0.330	Water
0.615	303	0.356	
0.658	333	0.381	
0.688	353	0.398	
0.155	293	0.090	Xylene (ortho-)
0.155	293	0.090	(meta-)

*By permission from *Heat Transmission*, by W. H. McAdams, copyright 1942, McGraw-Hill

Table 3.   Latent heats of vaporisation*

Example: For water at 373 K, $\theta_c - \theta = (647 - 373) = 274$ K, and the latent heat of vaporisation is 2257 kJ/kg

No.	Compound	Range $\theta_c - \theta$ (°F)	θ_c (°F)	Range $\theta_c - \theta$ (K)	θ_c (K)	No.	Compound	Range $\theta_c - \theta$ (°F)	θ_c (°F)	Range $\theta_c - \theta$ (K)	θ_c (K)
18	Acetic acid	180–405	610	100–225	594	2	Freon-12 (CCl_2F_2)	72–360	232	40–200	384
22	Acetone	216–378	455	120–210	508	5	Freon-21 ($CHCl_2F$)	126–450	354	70–250	451
29	Ammonia	90–360	271	50–200	406	6	Freon-22 ($CHClF_2$)	90–306	205	50–170	369
13	Benzene	18–720	552	10–00	562	1	Freon-113 (CCl_2F-$CClF_2$)	162–450	417	90–250	487
16	Butane	162–360	307	90–200	426	10	Heptane	36–540	512	20–300	540
21	Carbon dioxide	18–180	88	10–100	304	11	Hexane	90–450	455	50–225	508
4	Carbon disulphide	252–495	523	140–275	546	15	Isobutane	144–360	273	80–200	407
2	Carbon tetrachloride	54–450	541	30–250	556	27	Methanol	72–450	464	40–250	513
7	Chloroform	252–495	505	140–275	536	20	Methyl chloride	126–450	289	70–250	416
8	Dichloromethane	270–450	421	150–250	489	19	Nitrous oxide	45–270	97	25–150	309
3	Diphenyl	315–720	981	175–400	800	9	Octane	54–540	565	30–300	569
25	Ethane	45–270	90	25–150	305	12	Pentane	36–360	387	20–200	470
26	Ethyl alcohol	36–252	469	20–140	516	23	Propane	72–360	205	40–200	369
28	Ethyl alcohol	252–540	469	140–300	516	24	Propyl alcohol	36–360	507	20–200	537
17	Ethyl chloride	180–450	369	100–250	460	14	Sulphur dioxide	162–288	314	90–160	430
13	Ethyl ether	18–720	381	10–00	467	30	Water	180–900	705	100–500	647
2	Freon-11 (CCl_3F)	126–450	389	70–250	471						

*By permission from *Heat Transmission*, by W. H. McAdams, Copyright 1942, McGraw-Hill.

Latent heats of vaporisation

Table 4.　Specific heats of liquids*

No.	Liquid	Range (K)
29	Acetic acid, 100 per cent	273–353
32	Acetone	293–323
52	Ammonia	203–323
37	Amyl alcohol	223–298
26	Amyl acetate	273–373
30	Aniline	273–403
23	Benzene	283–353
27	Benzyl alcohol	253–303
10	Benzyl chloride	243–303
49	Brine, 25 per cent CaCl₂	233–293
51	Brine, 25 per cent NaCl	233–293
44	Butyl alcohol	273–373
2	Carbon disulphide	173–298
3	Carbon tetrachloride	283–333
8	Chlorobenzene	273–373
4	Chloroform	273–323
21	Decane	193–298
6A	Dichloroethane	243–333
5	Dichloromethane	233–323
15	Diphenyl	353–393
22	Diphenylmethane	303–373
16	Diphenyl oxide	273–473
16	Dowtherm A	273–473
24	Ethyl acetate	223–298
42	Ethyl alcohol, 100 per cent	303–353
46	Ethyl alcohol, 95 per cent	293–353
50	Ethyl alcohol, 50 per cent	293–353
25	Ethyl benzene	273–373
1	Ethyl bromide	278–298
13	Ethyl chloride	243–313
36	Ethyl ether	173–298
7	Ethyl iodide	273–373
39	Ethylene glycol	233–473
2A	Freon-11 (CCl₃F)	253–343
6	Freon-12 (CCl₂F₂)	233–288
4A	Freon-21 (CHCl₂F)	253–343
7A	Freon-22 (CHClF₂)	253–333
3A	Freon-113 (CCl₂F-CClF₂)	253–343
38	Glycerol	233–293
28	Heptane	273–333
35	Hexane	193–293
48	Hydrochloric acid, 30 per cent	293–373
41	Isoamyl alcohol	283–373
43	Isobutyl alcohol	273–373
47	Isopropyl alcohol	253–323
31	Isopropyl ether	193–293
40	Methyl alcohol	233–293
13A	Methyl chloride	193–293
14	Naphthalene	363–473
12	Nitrobenzene	273–373
34	Nonane	223–298
33	Octane	223–298
3	Perchloroethylene	243–413
45	Propyl alcohol	253–373
20	Pyridine	223–298
9	Sulphuric acid, 98 per cent	283–318
11	Sulphur dioxide	253–373
23	Toluene	273–333
53	Water	283–473
19	Xylene (ortho)	273–373
18	Xylene (meta)	273–373
17	Xylene (para)	273–373

*By permission from *Heat Transmission*, by W. H. McAdams, copyright 1942, McGraw-Hill.

Specific heats of liquids

Temperature: °F, K scales. Specific heat (kJ/kg K) (Btu/lb °F).

Table 5. Specific heats of gases and vapours at 101.3 kN/m²*

No.	Gas	Range (K)	No.	Gas	Range (K)
10	Acetylene	273–473	1	Hydrogen	273–873
15	Acetylene	473–673	2	Hydrogen	873–1673
16	Acetylene	673–1673	35	Hydrogen bromide	273–1673
27	Air	273–1673	30	Hydrogen chloride	273–1673
12	Ammonia	273–873	20	Hydrogen fluoride	273–1673
14	Ammonia	873–1673	36	Hydrogen iodide	273–1673
18	Carbon dioxide	273–673	19	Hydrogen sulphide	273–973
24	Carbon dioxide	673–1673	21	Hydrogen sulphide	973–1673
26	Carbon monoxide	273–1673	5	Methane	273–573
32	Chlorine	273–473	6	Methane	573–973
34	Chlorine	473–1673	7	Methane	973–1673
3	Ethane	273–473	25	Nitric oxide	273–973
9	Ethane	473–873	28	Nitric oxide	973–1673
8	Ethane	873–1673	26	Nitrogen	273–1673
4	Ethylene	273–473	23	Oxygen	273–773
11	Ethylene	473–873	29	Oxygen	773–1673
13	Ethylene	873–1673	33	Sulphur	573–1673
17B	Freon-11 (CCl₃F)	273–423	22	Sulphur dioxide	273–673
17C	Freon-21 (CHCl₂F)	273–423	31	Sulphur dioxide	673–1673
17A	Freon-22 (CHClF₂)	273–423	17	Water	273–1673
17D	Freon-113 (CCl₂F-CClF₂)	273–423			

*By permission from *Heat Transmission*, by W. H. McAdams, copyright, 1942, McGraw-Hill.

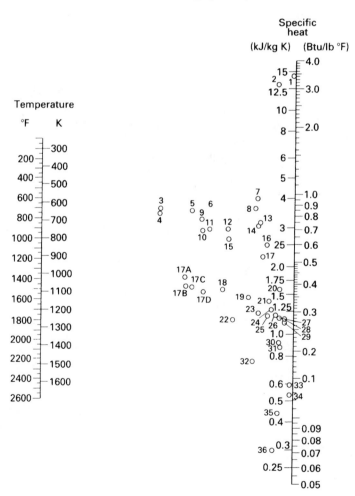

Table 6.   Viscosity of water*

Temperature (K)	Viscosity (mN s/m²)	Temperature (K)	Viscosity (mN s/m²)	Temperature (K)	Viscosity (mN s/m²)
273	1.7921	306	0.7523	340	0.4233
274	1.7313	307	0.7371	341	0.4174
275	1.6728	308	0.7225	342	0.4117
276	1.6191	309	0.7085	343	0.4061
277	1.5674	310	0.6947	344	0.4006
278	1.5188	311	0.6814	345	0.3952
279	1.4728	312	0.6685	346	0.3900
280	1.4284	313	0.6560	347	0.3849
281	1.3860	314	0.6439	348	0.3799
282	1.3462	315	0.6321	349	0.3750
283	1.3077	316	0.6207	350	0.3702
284	1.2713	317	0.6097	351	0.3655
285	1.2363	318	0.5988	352	0.3610
286	1.2028	319	0.5883	353	0.3565
287	1.1709	320	0.5782	354	0.3521
288	1.1404	321	0.5683	355	0.3478
289	1.1111	322	0.5588	356	0.3436
290	1.0828	323	0.5494	357	0.3395
291	1.0559	324	0.5404	358	0.3355
292	1.0299	325	0.5315	359	0.3315
293	1.0050	326	0.5229	360	0.3276
293.2	1.0000	327	0.5146	361	0.3239
294	0.9810	328	0.5064	362	0.3202
295	0.9579	329	0.4985	363	0.3165
296	0.9358	330	0.4907	364	0.3130
297	0.9142	331	0.4832	365	0.3095
298	0.8937	332	0.4759	366	0.3060
299	0.8737	333	0.4688	367	0.3027
300	0.8545	334	0.4618	368	0.2994
301	0.8360	335	0.4550	369	0.2962
302	0.8180	336	0.4483	370	0.2930
303	0.8007	337	0.4418	371	0.2899
304	0.7840	338	0.4355	372	0.2868
305	0.7679	339	0.4293	373	0.2838

*Calculated by the formula:

$$1/\mu = 2.1482[(\theta - 281.435) + \sqrt{(8078.4 + (\theta - 281.435)^2)}] - 120$$

(By permission from *Fluidity and Plasticity*, by E.C. Bingham. Copyright 1922, McGraw-Hill Book Company Inc.)

Table 7.   Thermal conductivities of gases and vapours

The extreme temperature values given constitute the experimental range. For extrapolation to other temperatures, it is suggested that the data given be plotted as log k vs. log T, or that use be made of the assumption that the ratio $C_p\mu/k$ is practically independent of temperature (and of pressure, within moderate limits).

Substance	k (W/m K)	(K)	k (Btu/h ft °F)	Substance	k (W/m K)	(K)	k (Btu/h ft °F)
Acetone	0.0098	273	0.0057	Chlorine	0.0074	273	0.0043
	0.0128	319	0.0074	Chloroform	0.0066	273	0.0038
	0.0171	373	0.0099		0.0080	319	0.0046
	0.0254	457	0.0147		0.0100	373	0.0058
Acetylene	0.0118	198	0.0068		0.0133	457	0.0077
	0.0187	273	0.0108	Cyclohexane	0.0164	375	0.0095
	0.0242	323	0.0140				
	0.0298	373	0.0172	Dichlorodifluoromethane	0.0083	273	0.0048
Air	0.0164	173	0.0095		0.0111	323	0.0064
	0.0242	273	0.0140		0.0139	373	0.0080
	0.0317	373	0.0183		0.0168	423	0.0097
	0.0391	473	0.0226				
	0.0459	573	0.0265	Ethane	0.0114	203	0.0066
Ammonia	0.0164	213	0.0095		0.0149	239	0.0086
	0.0222	273	0.0128		0.0183	273	0.0106
	0.0272	323	0.0157		0.0303	373	0.0175
	0.0320	373	0.0185	Ethyl acetate	0.0125	319	0.0072
					0.0166	373	0.0096
Benzene	0.0090	273	0.0052		0.0244	457	0.0141
	0.0126	319	0.0073	alcohol	0.0154	293	0.0089
	0.0178	373	0.0103		0.0215	373	0.0124
	0.0263	457	0.0152	chloride	0.0095	273	0.0055
	0.0305	485	0.0176		0.0164	373	0.0095
Butane (n-)	0.0135	273	0.0078		0.0234	457	0.0135
	0.0234	373	0.0135		0.0263	485	0.0152
(iso-)	0.0138	273	0.0080	ether	0.0133	273	0.0077
	0.0241	373	0.0139		0.0171	319	0.0099
					0.0227	373	0.0131
Carbon dioxide	0.0118	223	0.0068		0.0327	457	0.0189
	0.0147	273	0.0085		0.0362	485	0.0209
	0.0230	373	0.0133	Ethylene	0.0111	202	0.0064
	0.0313	473	0.0181		0.0175	273	0.0101
	0.0396	573	0.0228		0.0267	323	0.0131
disulphide	0.0069	273	0.0040		0.0279	373	0.0161
	0.0073	280	0.0042				
monoxide	0.0071	84	0.0041	Heptane (n-)	0.0194	473	0.0112
	0.0080	94	0.0046		0.0178	373	0.0103
	0.0234	213	0.0135	Hexane (n-)	0.0125	273	0.0072
tetrachloride	0.0071	319	0.0041		0.0138	293	0.0080
	0.0090	373	0.0052	Hexene	0.0106	273	0.0061
	0.0112	457	0.0065		0.0109	373	0.0189
Hydrogen	0.0113	173	0.065		0.0225	457	0.0130
	0.0144	223	0.083		0.0256	485	0.0148
	0.0173	273	0.100	Methylene chloride	0.0067	273	0.0039
	0.0199	323	0.115		0.0085	319	0.0049
	0.0223	373	0.129		0.0109	373	0.0063
	0.0308	573	0.178		0.0164	485	0.0095

Table 7. (*continued*)

Substance	k (W/m K)	(K)	k (Btu/h ft °F)	Substance	k (W/m K)	(K)	k (Btu/h ft °F)
Hydrogen and carbon dioxide		273		Nitric oxide	0.0178	203	0.0103
0 per cent H_2	0.0144		0.0083		0.0239	273	0.0138
20 per cent	0.0286		0.0165	Nitrogen	0.0164	173	0.0095
40 per cent	0.0467		0.0270		0.0242	273	0.0140
60 per cent	0.0709		0.0410		0.0277	323	0.0160
80 per cent	0.1070		0.0620		0.0312	373	0.0180
100 per cent	0.173		0.10	Nitrous oxide	0.0116	201	0.0067
Hydrogen and nitrogen		273			0.0157	273	0.0087
0 per cent H_2	0.0230		0.0133		0.0222	373	0.0128
20 per cent	0.0367		0.0212				
40 per cent	0.0542		0.0313	Oxygen	0.0164	173	0.0095
60 per cent	0.0758		0.0438		0.0206	223	0.0119
80 per cent	0.1098		0.0635		0.0246	273	0.0142
Hydrogen and nitrous oxide		273			0.0284	323	0.0164
0 per cent H_2	0.0159		0.0092		0.0321	373	0.0185
20 per cent	0.0294		0.0170				
40 per cent	0.0467		0.0270	Pentane (n-)	0.0128	273	0.0074
60 per cent	0.0709		0.0410		0.0144	293	0.0083
80 per cent	0.112		0.0650	(iso-)	0.0125	273	0.0072
Hydrogen sulphide	0.0132	273	0.0076		0.0220	373	0.0127
				Propane	0.0151	273	0.0087
Mercury	0.0341	473	0.0197		0.0261	373	0.0151
Methane	0.0173	173	0.0100				
	0.0251	223	0.0145	Sulphur dioxide	0.0087	273	0.0050
	0.0302	273	0.0175		0.0119	373	0.0069
	0.0372	323	0.0215				
Methyl alcohol	0.0144	273	0.0083	Water vapour	0.0208	319	0.0120
	0.0222	373	0.0128		0.0237	373	0.0137
acetate	0.0102	273	0.0059		0.0324	473	0.0187
	0.0118	293	0.0068		0.0429	573	0.0248
chloride	0.0092	273	0.0053		0.0545	673	0.0315
	0.0125	319	0.0072		0.0763	773	0.0441
	0.0163	373	0.0094				

*By permission from *Heat Transmission*, by W. H. McAdams, copyright 1942, McGraw-Hill.

Table 8.   Viscosities of gases*

Co-ordinates for use with graph on facing page

No.	Gas	X	Y
1	Acetic acid	7.7	14.3
2	Acetone	8.9	13.0
3	Acetylene	9.8	14.9
4	Air	11.0	20.0
5	Ammonia	8.4	16.0
6	Argon	10.5	22.4
7	Benzene	8.5	13.2
8	Bromine	8.9	19.2
9	Butene	9.2	13.7
10	Butylene	8.9	13.0
11	Carbon dioxide	9.5	18.7
12	Carbon disulphide	8.0	16.0
13	Carbon monoxide	11.0	20.0
14	Chlorine	9.0	18.4
15	Chloroform	8.9	15.7
16	Cyanogen	9.2	15.2
17	Cyclohexane	9.2	12.0
18	Ethane	9.1	14.5
19	Ethyl acetate	8.5	13.2
20	Ethyl alcohol	9.2	14.2
21	Ethyl chloride	8.5	15.6
22	Ethyl ether	8.9	13.0
23	Ethylene	9.5	15.1
24	Fluorine	7.3	23.8
25	Freon-11 (CCl_8F)	10.6	15.1
26	Freon-12 (CCl_2F_2)	11.1	16.0
27	Freon-21 ($CHCl_2F$)	10.8	15.3
28	Freon-22 ($CHClF_2$)	10.1	17.0
29	Freon-113 ($CCl_2F\text{-}CClF_3$)	11.3	14.0
30	Helium	10.9	20.5
31	Hexane	8.6	11.8
32	Hydrogen	11.2	12.4
33	$3H_2 + 1N_2$	11.2	17.2
34	Hydrogen bromide	8.8	20.9
35	Hydrogen chloride	8.8	18.7
36	Hydrogen cyanide	9.8	14.9
37	Hydrogen iodide	9.0	21.3
38	Hydrogen sulphide	8.6	18.0
39	Iodine	9.0	18.4
40	Mercury	5.3	22.9
41	Methane	9.9	15.5
42	Methyl alcohol	8.5	15.6
43	Nitric oxide	10.9	20.5
44	Nitrogen	10.6	20.0
45	Nitrosyl chloride	8.0	17.6
46	Nitrous oxide	8.8	19.0
47	Oxygen	11.0	21.3
48	Pentane	7.0	12.8
49	Propane	9.7	12.9
50	Propyl alcohol	8.4	13.4
51	Propylene	9.0	13.8
52	Sulphur dioxide	9.6	17.0
53	Toluene	8.6	12.4
54	2, 3, 3-trimethylbutane	9.5	10.5
55	Water	8.0	16.0
56	Xenon	9.3	23.0

[To convert to lb/ft h multiply by 2.42.]

Viscosities of gases

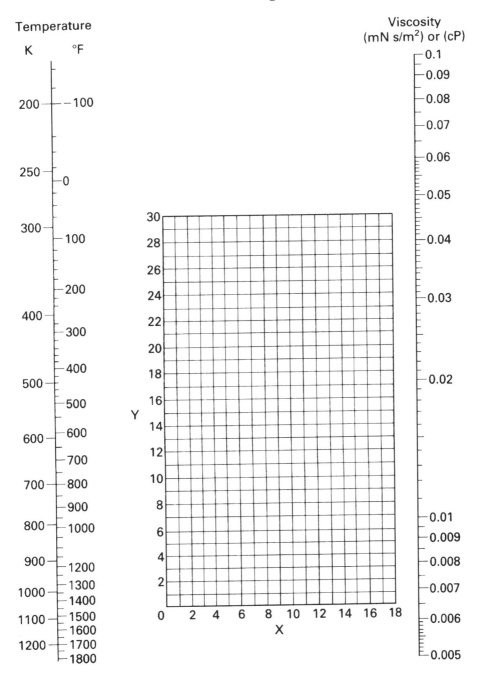

Table 9.   Viscosities and densities of liquids*

Co-ordinates for graph on following page

No.	Liquid	X	Y	Density at 293 K (kg/m³)	No.	Liquid	X	Y	Density at 293 K (kg/m³)
1	Acetaldehyde	15.2	4.8	783 (291 K)	57	Freon-113 (CCl₂F-CClF₂)	12.5	11.4	1576
2	Acetic acid, 100 per cent	12.1	14.2	1049	58	Glycerol, 100 per cent	2.0	30.0	1261
3	Acetic acid, 70 per cent	9.5	17.0	1069	59	Glycerol, 50 per cent	6.9	19.6	1126
4	Acetic anhydride	12.7	12.8	1083	60	Heptane	14.1	8.4	684
5	Acetone, 100 per cent	14.5	7.2	792	61	Hexane	14.7	7.0	659
6	Acetone, 35 per cent	7.9	15.0	948	62	Hydrochloric acid,			
7	Allyl alcohol	10.2	14.3	854		31.5 per cent	13.0	16.6	1157
8	Ammonia, 100 per cent	12.6	2.0	817 (194 K)	63	Isobutyl alcohol	7.1	18.0	779 (299 K)
9	Ammonia, 26 per cent	10.1	13.9	904	64	Isobutyric acid	12.2	14.4	949
10	Amyl acetate	11.8	12.5	879	65	Isopropyl alcohol	8.2	16.0	789
11	Amyl alcohol	7.5	18.4	817	66	Kerosene	10.2	16.9	780–820
12	Aniline	8.1	18.7	1022	67	Linseed oil, raw	7.5	27.2	934 ± 4 (288 K)
13	Anisole	12.3	13.5	990	68	Mercury	18.4	16.4	13546
14	Arsenic trichloride	13.9	14.5	2163	69	Methanol, 100 per cent	12.4	10.5	792
15	Benzene	12.5	10.9	880	70	Methanol, 90 per cent	12.3	11.8	820
16	Brine, CaCl₂, 25 per cent	6.6	15.9	1228	71	Methanol, 40 per cent	7.8	15.5	935
17	Brine, NaCl, 25 per cent	10.2	16.6	1186 (298 K)	72	Methyl acetate	14.2	8.2	924
18	Bromine	14.2	13.2	3119	73	Methyl chloride	15.0	3.8	952 (273 K)
19	Bromotoluene	20.0	15.9	1410	74	Methyl ethyl ketone	13.9	8.6	805
20	Butyl acetate	12.3	11.0	882	75	Naphthalene	7.9	18.1	1145
21	Butyl alcohol	8.6	17.2	810	76	Nitric acid, 95 per cent	12.8	13.8	1493
22	Butyric acid	12.1	15.3	964	77	Nitric acid, 60 per cent	10.8	17.0	1367
23	Carbon dioxide	11.6	0.3	1101 (236 K)	78	Nitrobenzene	10.6	16.2	1205 (291 K)
24	Carbon disulphide	16.1	7.5	1263	79	Nitrotoluene	11.0	17.0	1160
25	Carbon tetrachloride	12.7	13.1	1595	80	Octane	13.7	10.0	703
26	Chlorobenzene	12.3	12.4	1107	81	Octyl alcohol	6.6	21.1	827
27	Chloroform	14.4	10.2	1489	82	Pentachloroethane	10.9	17.3	1671 (298 K)
28	Chlorosulphonic acid	11.2	18.1	1787 (298 K)	83	Pentane	14.9	5.2	630 (291 K)
29	Chlorotoluene, ortho	13.0	13.3	1082	84	Phenol	6.9	20.8	1071 (298 K)
30	Chlorotoluene, meta	13.3	12.5	1072	85	Phosphorus tribromide	13.8	16.7	2852 (288 K)
31	Chlorotoluene, para	13.3	12.5	1070	86	Phosphorus trichloride	16.2	10.9	1574
32	Cresol, meta	2.5	20.8	1034	87	Propionic acid	12.8	13.8	992
33	Cyclohexanol	2.9	24.3	962	88	Propyl alcohol	9.1	16.5	804
34	Dibromoethane	12.7	15.8	2495	89	Propyl bromide	14.5	9.6	1353
35	Dichloroethane	13.2	12.2	1256	90	Propyl chloride	14.4	7.5	890
36	Dichloromethane	14.6	8.9	1336	91	Propyl iodide	14.1	11.6	1749
37	Diethyl oxalate	11.0	16.4	1079	92	Sodium	16.4	13.9	970
38	Dimethyl oxalate	12.3	15.8	1148 (327 K)	93	Sodium hydroxide, 50%	3.2	25.8	1525
39	Diphenyl	12.0	18.3	992 (346 K)	94	Stannic chloride	13.5	12.8	2226
40	Dipropyl oxalate	10.3	17.7	1038 (273 K)	95	Sulphur dioxide	15.2	7.1	1434 (273 K)
41	Ethyl acetate	13.7	9.1	901	96	Sulphuric acid, 110 per cent	7.2	27.4	1980
42	Ethyl alcohol, 100 per cent	10.5	13.8	789	97	Sulphuric acid, 98 per cent	7.0	24.8	1836
43	Ethyl alcohol, 95 per cent	9.8	14.3	804	98	Sulphuric acid, 60 per cent	10.2	21.3	1498
44	Ethyl alcohol, 40 per cent	6.5	16.6	935	99	Sulphuryl chloride	15.2	12.4	1667
45	Ethyl benzene	13.2	11.5	867	100	Tetrachloroethane	11.9	15.7	1600
46	Ethyl bromide	14.5	8.1	1431	101	Tetrachloroethylene	14.2	12.7	1624 (288 K)
47	Ethyl chloride	14.8	6.0	917 (279 K)	102	Titanum tetrachloride	14.4	12.3	1726
48	Ethyl ether	14.5	5.3	708 (298 K)	103	Toluene	13.7	10.4	866
49	Ethyl formate	14.2	8.4	923	104	Trichloroethylene	14.8	10.5	1466
50	Ethyl iodide	14.7	10.3	1933	105	Turpentine	11.5	14.9	861–867
51	Ethylene glycol	6.0	23.6	1113	106	Vinyl acetate	14.0	8.8	932
52	Formic acid	10.7	15.8	1220	107	Water	10.2	13.0	998
53	Freon-11 (CCl₃F)	14.4	9.0	1494 (290 K)	108	Xylene, ortho	13.5	12.1	881
54	Freon-12 (CCl₂F₂)	16.8	5.6	1486 (293 K)	109	Xylene, meta	13.9	10.6	867
55	Freon-21 (CHCl₂F)	15.7	7.5	1426 (273 K)	110	Xylene, para	13.9	10.9	861
56	Freon-22 (CHClF₂)	17.2	4.7	3870 (273 K)					

*By permission from *Perry's Chemical Engineers' Handbook*, by Perry, R. H. and Green, D. W. (eds), 6th edn. Copyright 1984, McGraw-Hill.

Viscosities of liquids

Temperature

K   °F

Viscosity
(mN s/m²) or (cP)

Table 10. Critical constants of gases*

	Critical temperature T_c(K)	Critical pressure P_c, (MN/m²)	Compressibility constant in critical state Z_c
Paraffins			
Methane	191	4.64	0.290
Ethane	306	4.88	0.284
Propane	370	4.25	0.276
n-Butane	425	3.80	0.274
Isobutane	408	3.65	0.282
n-Pentane	470	3.37	0.268
Isopentane	461	3.33	0.268
Neopentane	434	3.20	0.268
n-Hexane	508	3.03	0.264
n-Heptane	540	2.74	0.260
n-Octane	569	2.49	0.258
Mono-olefins			
Ethylene	282	5.07	0.268
Propylene	365	4.62	0.276
1-Butene	420	4.02	0.276
1-Pentene	474	4.05	
Miscellaneous organic compounds			
Acetic acid	595	5.78	0.200
Acetone	509	4.72	0.237
Acetylene	309	6.24	0.274
Benzene	562	4.92	0.274
1,3-Butadiene	425	4.33	0.270
Cyclohexane	553	4.05	0.271
Dichlorodifluoromethane (Freon-12)	385	4.01	0.273
Diethyl ether	467	3.61	0.261
Ethyl alcohol	516	6.38	0.249
Ethylene oxide	468	7.19	0.25
Methyl alcohol	513	7.95	0.220
Methyl chloride	416	6.68	0.276
Methyl ethyl ketone	533	4.00	0.26
Toluene	594	4.21	0.27
Trichlorofluoromethane (Freon-11)	471	4.38	0.277
Trichlorotrifluoroethane (Freon-113)	487	3.41	0.274
Elementary gases			
Bromine	584	10.33	0.307
Chlorine	417	7.71	0.276
Helium	5.3	0.23	0.300
Hydrogen	33.3	1.30	0.304
Neon	44.5	2.72	0.307
Nitrogen	126	3.39	0.291
Oxygen	155	5.08	0.29
Miscellaneous inorganic compounds			
Ammonia	406	11.24	0.242
Carbon dioxide	304	7.39	0.276
Carbon monoxide	133	3.50	0.294
Hydrogen chloride	325	8.26	0.266
Hydrogen sulphide	374	9.01	0.284
Nitric oxide (NO)	180	6.48	0.25
Nitrous oxide (N₂O)	310	7.26	0.271
Sulphur	1313	11.75	
Sulphur dioxide	431	7.88	0.268
Sulphur trioxide	491	8.49	0.262
Water	647	22.1	0.23

*Selected values from K. A. Kobe and R. E. Lynn, Jr., *Chem. Rev.*, **52**, 117 (1953). By permission.

A3. STEAM TABLES

Tables 11A, 11B, 11C and 11D are adapted from the
Abridged Callendar Steam Tables
by permission of Messrs Edward Arnold (Publishers) Ltd.

Table 11A. Properties of saturated steam (S.I. units)

Absolute pressure (kN/m²)	Temperature ($^\circ$C) θ_s	Temperature (K) T_s	Enthalpy per unit mass (H_s) (kJ/kg) water	latent	steam	Entropy per unit mass (S_s) (kJ/kg K) water	latent	steam	Specific volume (v) (m³/kg) water	steam
Datum: Triple point of water										
0.611	0.01	273.16	0.0	2501.6	2501.6	0	9.1575	9.1575	0.0010002	206.16
1.0	6.98	280.13	29.3	2485.0	2514.4	0.1060	8.8706	8.9767	0.001000	129.21
2.0	17.51	290.66	73.5	2460.2	2533.6	0.2606	8.4640	8.7246	0.001001	67.01
3.0	24.10	297.25	101.0	2444.6	2545.6	0.3543	8.2242	8.5785	0.001003	45.67
4.0	28.98	302.13	121.4	2433.1	2554.5	0.4225	8.0530	8.4755	0.001004	34.80
5.0	32.90	306.05	137.8	2423.8	2561.6	0.4763	7.9197	8.3960	0.001005	28.19
6.0	36.18	309.33	151.5	2416.0	2567.5	0.5209	7.8103	8.3312	0.001006	23.74
7.0	39.03	312.18	163.4	2409.2	2572.6	0.5591	7.7176	8.2767	0.001007	20.53
8.0	41.54	314.69	173.9	2403.2	2577.1	0.5926	7.6370	8.2295	0.001008	18.10
9.0	43.79	316.94	183.3	2397.9	2581.1	0.6224	7.5657	8.1881	0.001009	16.20
10.0	45.83	318.98	191.8	2392.9	2584.8	0.6493	7.5018	8.1511	0.001010	14.67
12.0	49.45	322.60	206.9	2384.2	2591.2	0.6964	7.3908	8.0872	0.001012	12.36
14.0	52.58	325.73	220.0	2376.7	2596.7	0.7367	7.2966	8.0333	0.001013	10.69
16.0	55.34	328.49	231.6	2370.0	2601.6	0.7721	7.2148	7.9868	0.001015	9.43
18.0	57.83	330.98	242.0	2363.9	2605.9	0.8036	7.1423	7.9459	0.001016	8.45
20.0	60.09	333.24	251.5	2358.4	2609.9	0.8321	7.0773	7.9094	0.001017	7.65
25.0	64.99	338.14	272.0	2346.4	2618.3	0.8933	6.9390	7.8323	0.001020	6.20
30.0	69.13	342.28	289.3	2336.1	2625.4	0.9441	6.8254	7.7695	0.001022	5.23
35.0	72.71	345.86	304.3	2327.2	2631.5	0.9878	6.7288	7.7166	0.001025	4.53
40.0	75.89	349.04	317.7	2319.2	2636.9	1.0261	6.6448	7.6709	0.001027	3.99
45.0	78.74	351.89	329.6	2312.0	2641.7	1.0603	6.5703	7.6306	0.001028	3.58
50.0	81.35	354.50	340.6	2305.4	2646.0	1.0912	6.5035	7.5947	0.001030	3.24
60.0	85.95	359.10	359.9	2293.6	2653.6	1.1455	6.3872	7.5327	0.001033	2.73
70.0	89.96	363.11	376.8	2283.3	2660.1	1.1921	6.2883	7.4804	0.001036	2.37
80.0	93.51	366.66	391.7	2274.0	2665.8	1.2330	6.2022	7.4352	0.001039	2.09
90.0	96.71	369.86	405.2	2265.6	2670.9	1.2696	6.1258	7.3954	0.001041	1.87
100.0	99.63	372.78	417.5	2257.9	2675.4	1.3027	6.0571	7.3598	0.001043	1.69
101.325	100.00	373.15	419.1	2256.9	2676.0	1.3069	6.0485	7.3554	0.0010437	1.6730
105	101.00	374.15	423.3	2254.3	2677.6	1.3182	6.0252	7.3434	0.001045	1.618
110	102.32	375.47	428.8	2250.8	2679.6	1.3330	5.9947	7.3277	0.001046	1.549
115	103.59	376.74	434.2	2247.4	2681.6	1.3472	5.9655	7.3127	0.001047	1.486
120	104.81	377.96	439.4	2244.1	2683.4	1.3609	5.9375	7.2984	0.001048	1.428
125	105.99	379.14	444.4	2240.9	2685.2	1.3741	5.9106	7.2846	0.001049	1.375
130	107.13	380.28	449.2	2237.8	2687.0	1.3868	5.8847	7.2715	0.001050	1.325
135	108.24	381.39	453.9	2234.8	2688.7	1.3991	5.8597	7.2588	0.001050	1.279
140	109.32	382.47	458.4	2231.9	2690.3	1.4109	5.8356	7.2465	0.001051	1.236
145	110.36	383.51	462.8	2229.0	2691.8	1.4225	5.8123	7.2347	0.001052	1.196
150	111.37	384.52	467.1	2226.2	2693.4	·1.4336	5.7897	7.2234	0.001053	1.159
155	112.36	385.51	471.3	2223.5	2694.8	1.4445	5.7679	7.2123	0.001054	1.124
160	113.32	386.47	475.4	2220.9	2696.2	1.4550	5.7467	7.2017	0.001055	1.091
165	114.26	387.41	479.4	2218.3	2697.6	1.4652	5.7261	7.1913	0.001056	1.060
170	115.17	388.32	483.2	2215.7	2699.0	1.4752	5.7061	7.1813	0.001056	1.031
175	116.06	389.21	487.0	2213.3	2700.3	1.4849	5.6867	7.1716	0.001057	1.003
180	116.93	390.08	490.7	2210.8	2701.5	1.4944	5.6677	7.1622	0.001058	0.977
185	117.79	390.94	494.3	2208.5	2702.8	1.5036	5.6493	7.1530	0.001059	0.952
190	118.62	391.77	497.9	2206.1	2704.0	1.5127	5.6313	7.1440	0.001059	0.929
195	119.43	392.58	501.3	2203.8	2705.1	1.5215	5.6138	7.1353	0.001060	0.907
200	120.23	393.38	504.7	2201.6	2706.3	1.5301	5.5967	7.1268	0.001061	0.885

Table 11A. (*continued*)

Absolute pressure (kN/m²)	Temperature (°C) θ_s	Temperature (K) T_s	Enthalpy per unit mass (H_s) (kJ/kg) water	latent	steam	Entropy per unit mass (S_s) (kJ/kg K) water	latent	steam	Specific volume (v) (m³/kg) water	steam
210	121.78	394.93	511.3	2197.2	2708.5	1.5468	5.5637	7.1105	0.001062	0.846
220	123.27	396.42	517.6	2193.0	2710.6	1.5628	5.5321	7.0949	0.001064	0.810
230	124.71	397.86	523.7	2188.9	2712.6	1.5781	5.5018	7.0800	0.001065	0.777
240	126.09	399.24	529.6	2184.9	2714.5	1.5929	5.4728	7.0657	0.001066	0.746
250	127.43	400.58	535.4	2181.0	2716.4	1.6072	5.4448	7.0520	0.001068	0.718
260	128.73	401.88	540.9	2177.3	2718.2	1.6209	5.4179	7.0389	0.001069	0.692
270	129.99	403.14	546.2	2173.6	2719.9	1.6342	5.3920	7.0262	0.001070	0.668
280	131.21	404.36	551.5	2170.1	2721.5	1.6471	5.3669	7.0140	0.001071	0.646
290	132.39	405.54	556.5	2166.6	2723.1	1.6596	5.3427	7.0022	0.001072	0.625
300	133.54	406.69	561.4	2163.2	2724.7	1.6717	5.3192	6.9909	0.001074	0.606
320	135.76	408.91	570.9	2156.7	2727.6	1.6948	5.2744	6.9692	0.001076	0.570
340	137.86	411.01	579.9	2150.4	2730.3	1.7168	5.2321	6.9489	0.001078	0.538
360	139.87	413.02	588.5	2144.4	2732.9	1.7376	5.1921	6.9297	0.001080	0.510
380	141.79	414.94	596.8	2138.6	2735.3	1.7575	5.1541	6.9115	0.001082	0.485
400	143.63	416.78	604.7	2132.9	2737.6	1.7764	5.1179	6.8943	0.001084	0.462
420	145.39	418.54	612.3	2127.5	2739.8	1.7946	5.0833	6.8779	0.001086	0.442
440	147.09	420.24	619.6	2122.3	2741.9	1.8120	5.0503	6.8622	0.001088	0.423
460	148.73	421.88	626.7	2117.2	2743.9	1.8287	5.0186	6.8473	0.001089	0.405
480	150.31	423.46	633.5	2112.2	2745.7	1.8448	4.9881	6.8329	0.001091	0.389
500	151.85	425.00	640.1	2107.4	2747.5	1.8604	4.9588	6.8192	0.001093	0.375
520	153.33	426.48	646.5	2102.7	2749.3	1.8754	4.9305	6.8059	0.001095	0.361
540	154.77	427.92	652.8	2098.1	2750.9	1.8899	4.9033	6.7932	0.001096	0.348
560	156.16	429.31	658.8	2093.7	2752.5	1.9040	4.8769	6.7809	0.001098	0.337
580	157.52	430.67	664.7	2089.3	2754.0	1.9176	4.8514	6.7690	0.001100	0.326
600	158.84	431.99	670.4	2085.0	2755.5	1.9308	4.8267	6.7575	0.001101	0.316
620	160.12	433.27	676.0	2080.8	2756.9	1.9437	4.8027	6.7464	0.001102	0.306
640	161.38	434.53	681.5	2076.7	2758.2	1.9562	4.7794	6.7356	0.001104	0.297
660	162.60	435.75	686.8	2072.7	2759.5	1.9684	4.7568	6.7252	0.001105	0.288
680	163.79	436.94	692.0	2068.8	2760.8	1.9803	4.7348	6.7150	0.001107	0.280
700	164.96	438.11	697.1	2064.9	2762.0	1.9918	4.7134	6.7052	0.001108	0.272
720	166.10	439.25	702.0	2061.1	2763.2	2.0031	4.6925	6.6956	0.001109	0.266
740	167.21	440.36	706.9	2057.4	2764.3	2.0141	4.6721	6.6862	0.001110	0.258
760	168.30	441.45	711.7	2053.7	2765.4	2.0249	4.6522	6.6771	0.001112	0.252
780	169.37	442.52	716.3	2050.1	2766.4	2.0354	4.6328	6.6683	0.001114	0.246
800	170.41	443.56	720.9	2046.5	2767.5	2.0457	4.6139	6.6596	0.001115	0.240
820	171.44	444.59	725.4	2043.0	2768.5	2.0558	4.5953	6.6511	0.001116	0.235
840	172.45	445.60	729.9	2039.6	2769.4	2.0657	4.5772	6.6429	0.001118	0.229
860	173.43	446.58	734.2	2036.2	2770.4	2.0753	4.5595	6.6348	0.001119	0.224
880	174.40	447.55	738.5	2032.8	2771.3	2.0848	4.5421	6.6269	0.001120	0.220
900	175.36	448.51	742.6	2029.5	2772.1	2.0941	4.5251	6.6192	0.001121	0.215
920	176.29	449.44	746.8	2026.2	2773.0	2.1033	4.5084	6.6116	0.001123	0.210
940	177.21	450.36	750.8	2023.0	2773.8	2.1122	4.4920	6.6042	0.001124	0.206
960	178.12	451.27	754.8	2019.8	2774.6	2.1210	4.4759	6.5969	0.001125	0.202
980	179.01	452.16	758.7	2016.7	2775.4	2.1297	4.4602	6.5898	0.001126	0.198
1000	179.88	453.03	762.6	2013.6	2776.2	2.1382	4.4447	6.5828	0.001127	0.194
1100	184.06	457.21	781.1	1998.6	2779.7	2.1786	4.3712	6.5498	0.001133	0.177
1200	187.96	461.11	798.4	1984.3	2782.7	2.2160	4.3034	6.5194	0.001139	0.163
1300	191.60	464.75	814.7	1970.7	2785.4	2.2509	4.2404	6.4913	0.001144	0.151
1400	195.04	468.19	830.1	1957.7	2787.8	2.2836	4.1815	6.4651	0.001149	0.141
1500	198.28	471.43	844.6	1945.3	2789.9	2.3144	4.1262	6.4406	0.001154	0.132
1600	201.37	474.52	858.5	1933.2	2791.7	2.3436	4.0740	6.4176	0.001159	0.124

(*continued overleaf*)

Table 11A.   (*continued*)

Absolute pressure (kN/m²)	Temperature		Enthalpy per unit mass (H_s) (kJ/kg)			Entropy per unit mass (S_s) (kJ/kg K)			Specific volume (v) (m³/kg)	
	(°C)	(K)	water	latent	steam	water	latent	steam	water	steam
	θ_s	T_s								
1700	204.30	477.45	871.8	1921.6	2793.4	2.3712	4.0246	6.3958	0.001163	0.117
1800	207.11	480.26	884.5	1910.3	2794.8	2.3976	3.9776	6.3751	0.001168	0.110
1900	209.79	482.94	896.8	1899.3	2796.1	2.4227	3.9327	6.3555	0.001172	0.105
2000	212.37	485.52	908.6	1888.7	2797.2	2.4468	3.8899	6.3367	0.001177	0.0996
2200	217.24	490.39	930.9	1868.1	2799.1	2.4921	3.8094	6.3015	0.001185	0.0907
2400	221.78	494.93	951.9	1848.5	2800.4	2.5342	3.7348	6.2690	0.001193	0.0832
2600	226.03	499.18	971.7	1829.7	2801.4	2.5736	3.6652	6.2388	0.001201	0.0769
3000	233.84	506.99	1008.3	1794.0	2802.3	2.6455	3.5383	6.1838	0.001216	0.0666
3500	242.54	515.69	1049.7	1752.2	2802.0	2.7252	3.3976	6.1229	0.001235	0.0570
4000	250.33	523.48	1087.4	1712.9	2800.3	2.7965	3.2720	6.0685	0.001252	0.0498
4500	257.41	530.56	1122.1	1675.6	2797.7	2.8612	3.1579	6.0191	0.001269	0.0440
5000	263.92	537.07	1154.5	1639.7	2794.2	2.9207	3.0528	5.9735	0.001286	0.0394
6000	275.56	548.71	1213.7	1571.3	2785.0	3.0274	2.8633	5.8907	0.001319	0.0324
7000	285.80	558.95	1267.5	1506.0	2773.4	3.1220	2.6541	5.8161	0.001351	0.0274
8000	294.98	568.13	1317.2	1442.7	2759.9	3.2077	2.5393	5.7470	0.001384	0.0235
9000	303.31	576.46	1363.8	1380.8	2744.6	3.2867	2.3952	5.6820	0.001418	0.0205
10000	310.96	584.11	1408.1	1319.7	2727.7	3.3606	2.2592	5.6198	0.001453	0.0180
11000	318.04	591.19	1450.6	1258.8	2709.3	3.4304	2.1292	5.5596	0.001489	0.0160
12000	324.64	597.79	1491.7	1197.5	2698.2	3.4971	2.0032	5.5003	0.001527	0.0143
14000	336.63	609.78	1571.5	1070.9	2642.4	3.6241	1.7564	5.3804	0.0016105	0.01150
16000	347.32	620.47	1650.4	934.5	2584.9	3.7470	1.5063	5.2533	0.0017102	0.00931
18000	356.96	630.11	1734.8	779.0	2513.9	3.8766	1.2362	5.1127	0.0018399	0.007497
20000	365.71	638.86	1826.6	591.6	2418.2	4.0151	0.9259	4.9410	0.0020374	0.005875
22000	373.68	646.83	2010.3	186.3	2196.6	4.2934	0.2881	4.5814	0.0026675	0.003735
22120	374.15	647.30	2107.4	0	2107.4	4.4429	0	4.4429	0.0031700	0.003170

Table 11B. Properties of saturated steam (Centigrade and Fahrenheit units)

Pressure		Temperature		Enthalpy per unit mass						Entropy (Btu/lb°F)		Specific volume (ft³/lb)
				Centigrade units (kcal/kg)			Fahrenheit units (Btu/lb)					
Absolute (lb/in.²)	Vacuum (in. Hg)	(°C)	(°F)	Water	Latent	Steam	Water	Latent	Steam	Water	Steam	Steam
0.5	28.99	26.42	79.6	26.45	582.50	608.95	47.6	1048.5	1096.1	0.0924	2.0367	643.0
0.6	28.79	29.57	85.3	29.58	580.76	610.34	53.2	1045.4	1098.6	0.1028	2.0214	540.6
0.7	28.58	32.28	90.1	32.28	579.27	611.55	58.1	1042.7	1100.8	0.1117	2.0082	466.6
0.8	28.38	34.67	94.4	34.66	577.95	612.61	62.4	1040.3	1102.7	0.1196	1.9970	411.7
0.9	28.17	36.80	98.2	36.80	576.74	613.54	66.2	1038.1	1104.3	0.1264	1.9871	368.7
1.0	27.97	38.74	101.7	38.74	575.60	614.34	69.7	1036.1	1105.8	0.1326	1.9783	334.0
1.1	27.76	40.52	104.9	40.52	574.57	615.09	72.9	1034.3	1107.2	0.1381	1.9702	305.2
1.2	27.56	42.17	107.9	42.17	573.63	615.80	75.9	1032.5	1108.4	0.1433	1.9630	281.1
1.3	27.35	43.70	110.7	43.70	572.75	616.45	78.7	1030.9	1109.6	0.1484	1.9563	260.5
1.4	27.15	45.14	113.3	45.12	571.94	617.06	81.3	1029.5	1110.8	0.1527	1.9501	243.0
1.5	26.95	46.49	115.7	46.45	571.16	617.61	83.7	1028.1	1111.8	0.1569	1.9442	228.0
1.6	26.74	47.77	118.0	47.73	570.41	618.14	86.0	1026.8	1112.8	0.1609	1.9387	214.3
1.7	26.54	48.98	120.2	48.94	569.71	618.65	88.2	1025.5	1113.7	0.1646	1.9336	202.5
1.8	26.33	50.13	122.2	50.08	569.06	619.14	90.2	1024.4	1114.6	0.1681	1.9288	191.8
1.9	26.13	51.22	124.2	51.16	568.47	619.63	92.1	1023.3	1115.4	0.1715	1.9243	182.3
2.0	25.92	52.27	126.1	52.22	567.89	620.11	94.0	1022.2	1116.2	0.1749	1.9200	173.7
3.0	23.88	60.83	141.5	60.78	562.89	623.67	109.4	1013.2	1122.6	0.2008	1.8869	118.7
4.0	21.84	67.23	153.0	67.20	559.29	626.49	121.0	1006.7	1127.7	0.2199	1.8632	90.63
5.0	19.80	72.38	162.3	72.36	556.24	628.60	130.2	1001.6	1131.8	0.2348	1.8449	73.52
6.0	17.76	76.72	170.1	76.71	553.62	630.33	138.1	996.6	1134.7	0.2473	1.8299	61.98
7.0	15.71	80.49	176.9	80.52	551.20	631.72	144.9	992.2	1137.1	0.2582	1.8176	53.64
8.0	13.67	83.84	182.9	83.89	549.16	633.05	151.0	988.5	1139.5	0.2676	1.8065	47.35
9.0	11.63	86.84	188.3	86.88	547.42	634.30	156.5	985.2	1141.7	0.2762	1.7968	42.40
10.0	9.59	89.58	193.2	89.61	545.82	635.43	161.3	982.5	1143.8	0.2836	1.7884	38.42
11.0	7.55	92.10	197.8	92.15	544.26	636.41	165.9	979.6	1145.5	0.2906	1.7807	35.14
12.0	5.50	94.44	202.0	94.50	542.75	637.25	170.1	976.9	1147.0	0.2970	1.7735	32.40
13.0	3.46	96.62	205.9	96.69	541.34	638.03	173.9	974.6	1148.5	0.3029	1.7672	30.05
14.0	1.42	98.65	209.6	98.73	540.06	638.79	177.7	972.2	1149.9	0.3086	1.7613	28.03
14.696	Gauge (lb/in.²)	100.00	212.0	100.06	539.22	639.28	180.1	970.6	1150.7	0.3122	1.7574	26.80
15	0.3	100.57	213.0	100.65	538.9	639.5	181.2	970.0	1151.2	0.3137	1.7556	26.28
16	1.3	102.40	216.3	102.51	537.7	640.2	184.5	967.9	1152.4	0.3187	1.7505	24.74
17	2.3	104.13	219.5	104.27	536.5	640.8	187.6	965.9	1153.5	0.3231	1.7456	23.38
18	3.3	105.78	222.4	105.94	535.5	641.4	190.6	964.0	1154.6	0.3276	1.7411	22.17
19	4.3	107.36	225.2	107.53	534.5	642.0	193.5	962.2	1155.7	0.3319	1.7368	21.07
20	5.3	108.87	228.0	109.05	533.6	642.6	196.3	960.4	1156.7	0.3358	1.7327	20.09
21	6.3	110.32	230.6	110.53	532.6	643.1	198.9	958.8	1157.7	0.3396	1.7287	19.19
22	7.3	111.71	233.1	111.94	531.7	643.6	201.4	957.2	1158.6	0.3433	1.7250	18.38
23	8.3	113.05	235.5	113.30	530.8	644.1	203.9	955.6	1159.5	0.3468	1.7215	17.63
24	9.3	114.34	237.8	114.61	530.0	644.6	206.3	954.0	1160.3	0.3502	1.7181	16.94
25	10.3	115.59	240.1	115.87	529.2	645.1	208.6	952.5	1161.1	0.3534	1.7148	16.30
26	11.3	116.80	242.2	117.11	528.4	645.5	210.8	951.1	1161.9	0.3565	1.7118	15.72
27	12.3	117.97	244.4	118.31	527.6	645.9	212.9	949.7	1162.6	0.3595	1.7089	15.17
28	13.3	119.11	246.4	119.47	526.8	646.3	215.0	948.3	1163.3	0.3625	1.7060	14.67
29	14.3	120.21	248.4	120.58	526.1	646.7	217.0	947.0	1164.0	0.3654	1.7032	14.19
30	15.3	121.3	250.3	121.7	525.4	647.1	219.0	945.6	1164.6	0.3682	1.7004	13.73
32	17.3	123.3	254.0	123.8	524.1	647.9	222.7	943.1	1165.8	0.3735	1.6952	12.93
34	19.3	125.3	257.6	125.8	522.8	648.6	226.3	940.7	1167.0	0.3785	1.6905	12.21

(*continued overleaf*)

Table 11B.   (*continued*)

Pressure		Temperature		Enthalpy per unit mass						Entropy (Btu/lb°F)		Specific volume (ft³/lb)
				Centigrade units (kcal/kg)			Fahrenheit units (Btu/lb)					
Absolute (lb/in.²)	Vacuum (in. Hg)	(°C)	(°F)	Water	Latent	Steam	Water	Latent	Steam	Water	Steam	Steam
36	21.3	127.2	260.9	**127.7**	521.5	**649.2**	**229.7**	938.5	**1168.2**	0.3833	1.6860	11.58
38	23.3	128.9	264.1	**129.5**	520.3	**649.8**	**233.0**	936.4	**1169.4**	0.3879	1.6817	11.02
40	25.3	130.7	267.2	**131.2**	519.2	**650.4**	**236.1**	934.4	**1170.5**	0.3923	1.6776	10.50
42	27.3	132.3	270.3	**132.9**	518.0	**650.9**	**239.1**	932.3	**1171.4**	0.3964	1.6737	10.30
44	29.3	133.9	273.1	**134.5**	516.9	**651.4**	**242.0**	930.3	**1172.3**	0.4003	1.6700	9.600
46	31.3	135.4	275.8	**136.0**	515.9	**651.9**	**244.9**	928.3	**1173.2**	0.4041	1.6664	9.209
48	33.3	136.9	278.5	**137.5**	514.8	**652.3**	**247.6**	926.4	**1174.0**	0.4077	1.6630	8.848
50	35.3	138.3	281.0	**139.0**	513.8	**652.8**	**250.2**	924.6	**1174.8**	0.4112	1.6597	8.516
52	37.3	139.7	283.5	**140.4**	512.8	**653.2**	**252.7**	922.9	**1175.6**	0.4146	1.6566	8.208
54	39.3	141.0	285.9	**141.8**	511.8	**653.6**	**255.2**	921.1	**1176.3**	0.4179	1.6536	7.922
56	41.3	142.3	288.3	**143.1**	510.9	**654.0**	**257.6**	919.4	**1177.0**	0.4211	1.6507	7.656
58	43.3	143.6	290.5	**144.4**	510.0	**654.4**	**259.9**	917.8	**1177.7**	0.4242	1.6478	7.407
60	45.3	144.9	292.7	**145.6**	509.2	**654.8**	**262.2**	916.2	**1178.4**	0.4272	1.6450	7.175
62	47.3	146.1	294.9	**146.8**	508.4	**655.2**	**264.4**	914.6	**1179.0**	0.4302	1.6423	6.957
64	49.3	147.3	296.9	**148.0**	507.6	**655.6**	**266.5**	913.1	**1179.6**	0.4331	1.6398	6.752
66	51.3	148.4	299.0	**149.2**	506.7	**655.9**	**268.6**	911.6	**1180.2**	0.4359	1.6374	6.560
68	53.3	149.5	301.0	**150.3**	505.9	**656.2**	**270.7**	910.1	**1180.8**	0.4386	1.6350	6.378
70	55.3	150.6	302.9	**151.5**	505.0	**656.5**	**272.7**	908.7	**1181.4**	0.4412	1.6327	6.206
72	57.3	151.6	304.8	**152.6**	504.2	**656.8**	**274.6**	907.4	**1182.0**	0.4437	1.6304	6.044
74	59.3	152.6	306.7	**153.6**	503.4	**657.0**	**276.5**	906.0	**1182.5**	0.4462	1.6282	5.890
76	61.3	153.6	308.5	**154.7**	502.6	**657.3**	**278.4**	904.6	**1183.0**	0.4486	1.6261	5.743
78	63.3	154.6	310.3	**155.7**	501.8	**657.5**	**280.3**	903.2	**1183.5**	0.4510	1.6240	5.604
80	65.3	155.6	312.0	**156.7**	501.1	**657.8**	**282.1**	901.9	**1184.0**	0.4533	1.6219	5.472
82	67.3	156.5	313.7	**157.7**	500.3	**658.0**	**283.9**	900.6	**1184.5**	0.4556	1.6199	5.346
84	69.3	157.5	315.4	**158.6**	499.6	**658.2**	**285.6**	899.4	**1185.0**	0.4579	1.6180	5.226
86	71.3	158.4	317.1	**159.6**	498.9	**658.5**	**287.3**	898.1	**1185.4**	0.4601	1.6161	5.110
88	73.3	159.4	318.7	**160.5**	498.3	**658.8**	**289.0**	896.8	**1185.8**	0.4622	1.6142	5.000
90	75.3	160.3	320.3	**161.5**	497.6	**659.1**	**290.7**	895.5	**1186.2**	0.4643	1.6124	4.896
92	77.3	161.2	321.9	**162.4**	496.9	**659.3**	**292.3**	894.3	**1186.6**	0.4664	1.6106	4.796
94	79.3	162.0	323.3	**163.3**	496.3	**659.6**	**293.9**	893.1	**1187.0**	0.4684	1.6088	4.699
96	81.3	162.8	324.8	**164.1**	495.7	**659.8**	**295.5**	891.9	**1187.4**	0.4704	1.6071	4.607
98	83.3	163.6	326.6	**165.0**	495.0	**660.0**	**297.0**	890.8	**1187.8**	0.4723	1.6054	4.519
100	85.3	164.4	327.8	**165.8**	494.3	**660.1**	**298.5**	889.7	**1188.2**	0.4742	1.6038	4.434
105	90.3	166.4	331.3	**167.9**	492.7	**660.6**	**302.2**	886.9	**1189.1**	0.4789	1.6000	4.230
110	95.3	168.2	334.8	**169.8**	491.2	**661.0**	**305.7**	884.2	**1189.9**	0.4833	1.5963	4.046
115	100.3	170.0	338.1	**171.7**	489.8	**661.5**	**309.2**	881.5	**1190.7**	0.4876	1.5927	3.880
120	105.3	171.8	341.3	**173.6**	488.3	**661.9**	**312.5**	878.9	**1191.4**	0.4918	1.5891	3.729
125	110.3	173.5	344.4	**175.4**	486.9	**662.3**	**315.7**	876.4	**1192.1**	0.4958	1.5856	3.587
130	115.3	175.2	347.3	**177.1**	485.6	**662.7**	**318.8**	874.0	**1192.8**	0.4997	1.5823	3.456
135	120.3	176.8	350.2	**178.8**	484.2	**663.0**	**321.9**	871.5	**1193.4**	0.5035	1.5792	3.335
140	125.3	178.3	353.0	**180.5**	482.9	**663.4**	**324.9**	869.1	**1194.0**	0.5071	1.5763	3.222
145	130.3	179.8	355.8	**182.1**	481.6	**663.7**	**327.8**	866.8	**1194.6**	0.5106	1.5733	3.116
150	135.3	181.3	358.4	**183.7**	480.3	**664.0**	**330.6**	864.5	**1195.1**	0.5140	1.5705	3.015

Table 11C. Enthalpy of superheated steam, H (kJ/kg)

Pressure P (kN/m²)	Saturation T_s (K)	S_s (kJ/kg)	Temperature, θ (°C) Temperature, T(K)	100 373.15	200 473.15	300 573.15	400 673.15	500 773.15	600 873.15	700 973.15	800 1073.15
100	372.78	2675.4		2676.0	2875.4	3074.6	3278.0	3488.0	3705.0	3928.0	4159.0
200	393.38	2706.3			2870.4	3072.0	3276.4	3487.0	3704.0	3927.0	4158.0
300	406.69	2724.7			2866.0	3069.7	3275.0	3486.0	3703.1	3927.0	4158.0
400	416.78	2737.6			2861.3	3067.0	3273.5	3485.0	3702.0	3926.0	4157.0
500	425.00	2747.5			2856.0	3064.8	3272.1	3484.0	3701.2	3926.0	4156.8
600	431.99	2755.5			2850.7	3062.0	3270.0	3483.0	3701.0	3925.0	4156.2
700	438.11	2762.0			2845.5	3059.5	3269.0	3482.6	3700.2	3924.0	4156.0
800	443.56	2767.5			2839.7	3057.0	3266.8	3480.4	3699.0	3923.8	4155.0
900	448.56	2772.1			2834.0	3055.0	3266.2	3479.5	3698.6	3923.0	4155.0
1000	453.03	2776.2			2828.7	3051.7	3264.3	3478.0	3697.5	3922.8	4154.0
2000	485.59	2797.2				3024.8	3248.0	3467.0	3690.0	3916.0	4150.0
3000	506.98	2802.3				2994.8	3231.7	3456.0	3681.6	3910.4	4145.0
4000	523.49	2800.3				2962.0	3214.8	3445.0	3673.4	3904.0	4139.6
5000	537.09	2794.2				2926.0	3196.9	3433.8	3665.4	3898.0	4135.5
6000	548.71	2785.0				2886.0	3178.0	3421.7	3657.0	3891.8	4130.0
7000	558.95	2773.4				2840.0	3159.1	3410.0	3648.8	3886.0	4124.8
8000	568.13	2759.9				2785.0	3139.5	3398.0	3640.4	3880.8	4121.0
9000	576.46	2744.6					3119.0	3385.5	3632.0	3873.6	4116.0
10000	584.11	2727.7					3097.7	3373.6	3624.0	3867.2	4110.8
11000	591.19	2709.3					3075.6	3361.0	3615.5	3862.0	4106.0
12000	597.79	2698.2					3052.9	3349.0	3607.0	3855.3	4101.2

Table 11D. Entropy of superheated steam, S (kJ/kg K)

Pressure P (kN/m²)	Saturation T_s (K)	H_s (kJ/kg)	Temperature, θ (°C) Temperature, T(K)	100 373.15	200 473.15	300 573.15	400 673.15	500 773.15	600 873.15	700 973.15	800 1073.15
100	372.78	7.3598		7.362	7.834	8.216	8.544	8.834	9.100	9.344	9.565
200	393.38	7.1268			7.507	7.892	8.222	8.513	8.778	9.020	9.246
300	406.69	6.9909			7.312	7.702	8.033	8.325	8.591	8.833	9.057
400	416.78	6.8943			7.172	7.566	7.898	8.191	8.455	8.700	8.925
500	425.00	6.8192			7.060	7.460	7.794	8.087	8.352	8.596	8.820
600	431.99	6.7575			6.968	7.373	7.708	8.002	8.268	8.510	8.738
700	438.11	6.7052			6.888	7.298	7.635	7.930	8.195	8.438	8.665
800	443.56	6.6596			6.817	7.234	7.572	7.867	8.133	8.375	8.602
900	448.56	6.6192			6.753	7.176	7.515	7.812	8.077	8.321	8.550
1000	453.03	6.5828			6.695	7.124	7.465	7.762	8.028	8.272	8.502
2000	485.59	6.3367				6.768	7.128	7.431	7.702	7.950	8.176
3000	506.98	6.1838				6.541	6.922	7.233	7.508	7.756	7.985
4000	523.49	6.0685				6.364	6.770	7.090	7.368	7.620	7.850
5000	537.07	5.9735				6.211	6.647	6.977	7.258	7.510	7.744
6000	548.71	5.8907				6.060	6.542	6.880	7.166	7.422	7.655
7000	558.95	5.8161				5.933	6.450	6.798	7.088	7.345	7.581
8000	568.13	5.7470				5.792	6.365	6.724	7.020	7.280	7.515
9000	576.46	5.6820					6.288	6.659	6.958	7.220	7.457
10000	584.11	5.6198					6.215	6.598	6.902	7.166	7.405
11000	591.19	5.5596					6.145	6.540	6.850	7.117	7.357
12000	597.79	5.5003					6.077	6.488	6.802	7.072	7.312

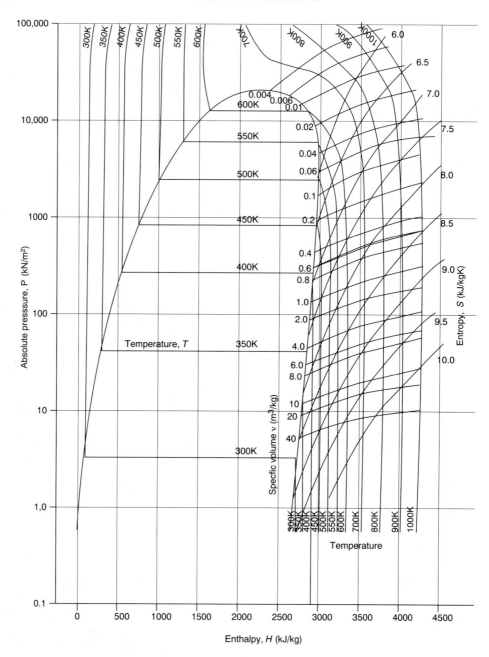

Figure 11A. Pressure–enthalpy diagram for water and steam

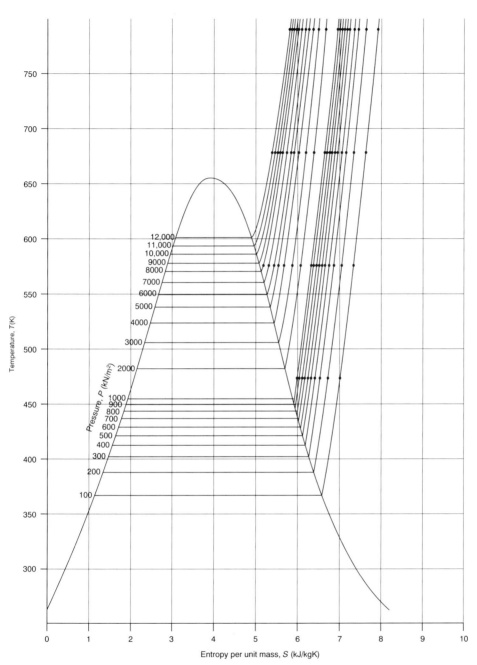

Figure 11B.   Temperature–entropy diagram for water and steam

A4. MATHEMATICAL TABLES

Table 12. Laplace transforms[a]

No.	Transform $\bar{f}(p) = \int_0^\infty e^{-pt} f(t)\,dt$	Function $f(t)$
1	$\dfrac{1}{p}$	1
2	$\dfrac{1}{p^2}$	t
3	$\dfrac{1}{p^n} \quad n = 1, 2, 3, \ldots$	$\dfrac{t^{n-1}}{(n-1)!}$
4	$\dfrac{1}{\sqrt{p}}$	$\dfrac{1}{\sqrt{(\pi t)}}$
5	$\dfrac{1}{p^{3/2}}$	$2\sqrt{\dfrac{t}{\pi}}$
6	$\dfrac{1}{p^{n+\frac{1}{2}}} \quad n = 1, 2, 3, \ldots$	$\dfrac{2^n t^{n-\frac{1}{2}}}{[1.3.5 \ldots (2n-1)]\sqrt{\pi}}$
7	$\dfrac{\Gamma(k)}{p^k} \quad k > 0$	t^{k-1}
8	$\dfrac{1}{p-a}$	e^{at}
9	$\dfrac{1}{(p-a)^2}$	$t e^{at}$
10	$\dfrac{1}{(p-a)^n} \quad n = 1, 2, 3, \ldots$	$\dfrac{1}{(n-1)!} t^{n-1} e^{at}$
11	$\dfrac{\Gamma(k)}{(p-a)^k} \quad k > 0$	$t^{k-1} e^{at}$
12	$\dfrac{1}{(p-a)(p-b)} \quad a \neq b$	$\dfrac{1}{a-b}(e^{at} - e^{bt})$
13	$\dfrac{p}{(p-a)(p-b)} \quad a \neq b$	$\dfrac{1}{a-b}(ae^{at} - be^{bt})$
14	$\dfrac{1}{(p-a)(p-b)(p-c)}$ $a \neq b \neq c$	$-\dfrac{(b-c)e^{at} + (c-a)e^{bt} + (a-b)e^{ct}}{(a-b)(b-c)(c-a)}$
15	$\dfrac{1}{p^2 + a^2}$	$\dfrac{1}{a}\sin at$
16	$\dfrac{p}{p^2 + a^2}$	$\cos at$
17	$\dfrac{1}{p^2 - a^2}$	$\dfrac{1}{a}\sinh at$

[a] By permission from *Operational Mathematics* by R.V. Churchill, McGraw-Hill 1958.

Table 12. (*continued*)

No.	Transform $\bar{f}(p) = \int_0^\infty e^{-pt} f(t)dt$	Function $f(t)$
18	$\dfrac{p}{p^2 - a^2}$	$\cosh at$
19	$\dfrac{1}{p(p^2 + a^2)}$	$\dfrac{1}{a^2}(1 - \cos at)$
20	$\dfrac{1}{p^2(p^2 + a^2)}$	$\dfrac{1}{a^3}(at - \sin at)$
21	$\dfrac{1}{(p^2 + a^2)^2}$	$\dfrac{1}{2a^3}(\sin at - at \cos at)$
22	$\dfrac{p}{(p^2 + a^2)^2}$	$\dfrac{t}{2a}\sin at$
23	$\dfrac{p^2}{(p^2 + a^2)^2}$	$\dfrac{1}{2a}(\sin at + at \cos at)$
24	$\dfrac{p^2 - a^2}{(p^2 + a^2)^2}$	$t \cos at$
25	$\dfrac{p}{(p^2 + a^2)(p^2 + b^2)} \quad a^2 \ne b^2$	$\dfrac{\cos at - \cos bt}{b^2 - a^2}$
26	$\dfrac{1}{(p - a)^2 + b^2}$	$\dfrac{1}{b}e^{at}\sin bt$
27	$\dfrac{p - a}{(p - a)^2 + b^2}$	$e^{at}\cos bt$
28	$\dfrac{3a^2}{p^3 + a^3}$	$e^{-at} - e^{at/2}\left(\cos\dfrac{at\sqrt{3}}{2} - \sqrt{3}\sin\dfrac{at\sqrt{3}}{2}\right)$
29	$\dfrac{4a^3}{p^4 + 4a^4}$	$\sin at \cosh at - \cos at \sinh at$
30	$\dfrac{p}{p^4 + 4a^4}$	$\dfrac{1}{2a^2}\sin at \sinh at$
31	$\dfrac{1}{p^4 - a^4}$	$\dfrac{1}{2a^3}(\sinh at - \sin at)$
32	$\dfrac{p}{p^4 - a^4}$	$\dfrac{1}{2a^2}(\cosh at - \cos at)$
33	$\dfrac{8a^3 p^2}{(p^2 + a^2)^3}$	$(1 + a^2 t^2)\sin at - at \cos at$
34	$\dfrac{1}{p}\left(\dfrac{p - 1}{p}\right)^n$	$\dfrac{e^t}{n!}\dfrac{d^n}{dt^n}(t^n e^{-t}) = $ Laguerre polynomial of degree n

(*continued overleaf*)

Table 12.   (*continued*)

No.	Transform $\bar{f}(p) = \int_0^\infty e^{-pt}f(t)dt$	Function $f(t)$
35	$\dfrac{p}{(p-a)^{3/2}}$	$\dfrac{1}{\sqrt{(\pi t)}} - e^{at}(1+2at)$
36	$\sqrt{(p-a)} - \sqrt{(p-b)}$	$\dfrac{1}{2\sqrt{(\pi t^3)}}(e^{bt} - e^{at})$
37	$\dfrac{1}{\sqrt{p}+a}$	$\dfrac{1}{\sqrt{(\pi t)}} - ae^{a^2 t} \, \text{erfc}\,(a\sqrt{t})$
38	$\dfrac{\sqrt{p}}{p-a^2}$	$\dfrac{1}{\sqrt{(\pi t)}} + ae^{a^2 t} \text{erf}(a\sqrt{t})$
39	$\dfrac{\sqrt{p}}{p+a^2}$	$\dfrac{1}{\sqrt{(\pi t)}} - \dfrac{2ae^{-a^2 t}}{\sqrt{\pi}} \int_0^{a\sqrt{t}} e^{\lambda^2} 2d\lambda$
40	$\dfrac{1}{\sqrt{p(p-a^2)}}$	$\dfrac{1}{a}e^{a^2 t}\text{erf}(a\sqrt{t})$
41	$\dfrac{1}{\sqrt{p(p+a^2)}}$	$\dfrac{2e^{-a^2 t}}{a\sqrt{\pi}} \int_0^{a\sqrt{t}} e^{\lambda^2} d\lambda$
42	$\dfrac{b^2-a^2}{(p-a^2)(b+\sqrt{p})}$	$e^{a^2 t}[b - a\text{erf}(a\sqrt{t})] - be^{b^2 t} \, \text{erfc}\,(b\sqrt{t})$
43	$\dfrac{1}{\sqrt{p}(\sqrt{p}+a)}$	$e^{a^2 t} \, \text{erfc}\,(a\sqrt{t})$
44	$\dfrac{1}{(p+a)\sqrt{(p+b)}}$	$\dfrac{1}{\sqrt{(b-a)}}e^{-at}\text{erf}[\sqrt{(b-a)}\sqrt{t}]$
45	$\dfrac{b^2-a^2}{\sqrt{p(p-a^2)}(\sqrt{p}+b)}$	$e^{a^2 t}\left[\dfrac{b}{a}\text{erf}(a\sqrt{t}) - 1\right] + e^{b^2 t} \, \text{erfc}\,(b\sqrt{t})$
46	$\dfrac{(1-p)^n}{p^{n+1/2}}$	$\dfrac{n!}{(2n)!\sqrt{(\pi t)}}H_{2n}(\sqrt{t})$ where $H_n(t) = e^{t^2}\dfrac{d^n}{dt^n}e^{-t^2}$ is the Hermite polynomial
47	$\dfrac{(1-p)^n}{p^{n+3/2}}$	$-\dfrac{n!}{\sqrt{\pi}(2n+1)!}H_{2n+1}(\sqrt{t})$
48	$\dfrac{\sqrt{(p+2a)}}{\sqrt{p}} - 1$	$ae^{-at}[I_1(at) + I_0(at)]$
49	$\dfrac{1}{\sqrt{(p+a)}\sqrt{(p+b)}}$	$e^{-\frac{1}{2}(a+b)t}I_0\left(\dfrac{a-b}{2}t\right)$

Table 12. (*continued*)

No.	Transform $\bar{f}(p) = \int_0^\infty e^{-pt}f(t)dt$	Function f(t)
50	$\dfrac{\Gamma(k)}{(p+a)^k(p+b)^k} \quad k > 0$	$\sqrt{\pi}\left(\dfrac{t}{a-b}\right)^{k-\frac{1}{2}}e^{-\frac{1}{2}(a+b)t}I_{k-\frac{1}{2}}\left(\dfrac{a-b}{2}t\right)$
51	$\dfrac{1}{\sqrt{(p+a)}(p+b)^{3/2}}$	$te^{-\frac{1}{2}(a+b)t}\left[I_0\left(\dfrac{a-b}{2}t\right) + I_1\left(\dfrac{a-b}{2}t\right)\right]$
52	$\dfrac{\sqrt{(p+2a)} - \sqrt{p}}{\sqrt{(p+2a)} + \sqrt{p}}$	$\dfrac{1}{t}e^{-at}I_1(at)$
53	$\dfrac{(a-b)^k}{[\sqrt{(p+a)} + \sqrt{(p+b)}]^{2k}} \quad k > 0$	$\dfrac{k}{t}e^{-\frac{1}{2}(a+b)t}I_k\left(\dfrac{a-b}{2}t\right)$
54	$\dfrac{[\sqrt{(p+a)} + \sqrt{p}]^{-2j}}{\sqrt{p}\sqrt{(p+a)}} \quad j > -1$	$\dfrac{1}{a^j}e^{-\frac{1}{2}at}I_j\left(\dfrac{1}{2}at\right)$
55	$\dfrac{1}{\sqrt{(p^2+a^2)}}$	$J_0(at)$
56	$\dfrac{[\sqrt{(p^2+a^2)} - p]^j}{\sqrt{(p^2+a^2)}} \quad j > 1$	$a^jJ_j(at)$
57	$\dfrac{1}{(p^2+a^2)^k} \quad k > 0$	$\dfrac{\sqrt{\pi}}{\Gamma(k)}\left(\dfrac{t}{2a}\right)^{k-\frac{1}{2}}J_{k-\frac{1}{2}}(at)$
58	$[\sqrt{(p^2+a^2)} - p]^k \quad k > 0$	$\dfrac{ka^k}{t}J_k(at)$
59	$\dfrac{[p - \sqrt{(p^2-a^2)}]^j}{\sqrt{(p^2-a^2)}} \quad j > -1$	$a^jI_j(at)$
60	$\dfrac{1}{(p^2-a^2)^k} \quad k > 0$	$\dfrac{\sqrt{\pi}}{\Gamma(k)}\left(\dfrac{t}{2a}\right)^{k-\frac{1}{2}}I_{k-\frac{1}{2}}(at)$
61	$\dfrac{e^{-kp}}{p}$	$S_k(t) = \begin{cases} 0 & \text{when } 0 < t < k \\ 1 & \text{when } t > k \end{cases}$
62	$\dfrac{e^{-kp}}{p^2}$	$\begin{cases} 0 & \text{when } 0 < t < k \\ t - k & \text{when } t > k \end{cases}$
63	$\dfrac{e^{-kp}}{p^j} \quad j > 0$	$\begin{cases} 0 & \text{when } 0 < t < k \\ \dfrac{(t-k)^{j-1}}{\Gamma(j)} & \text{when } t > k \end{cases}$
64	$\dfrac{1 - e^{-kp}}{p}$	$\begin{cases} 1 & \text{when } 0 < t < k \\ 0 & \text{when } t > k \end{cases}$

(*continued overleaf*)

Table 12. (*continued*)

No.	Transform $\bar{f}(p) = \int_0^\infty e^{-pt} f(t)dt$	Function $f(t)$
65	$\dfrac{1}{p(1 - e^{-kp})} = \dfrac{1 + \coth \frac{1}{2}kp}{2p}$	$S(k, t) = n$ when $(n-1)k < t < nk$ $\quad n = 1, 2, 3, \ldots$
66	$\dfrac{1}{p(e^{kp} - a)}$	$\begin{cases} 0 \quad \text{when } 0 < t < k \\ 1 + a + a^2 + \ldots + a^{n-1} \\ \quad \text{when } nk < t < (n+1)k \quad n = 1, 2, 3, \ldots \end{cases}$
67	$\dfrac{1}{p} \tanh kp$	$M(2k, t) = (-1)^{n-1}$ $\quad$ when $2k(n-1) < t < 2kn$ $n = 1, 2, 3, \ldots$
68	$\dfrac{1}{p(1 + e^{-kp})}$	$\dfrac{1}{2}M(k, t) + \dfrac{1}{2} = \dfrac{1 - (-1)^n}{2}$ $\quad$ when $(n-1)k < t < nk$
69	$\dfrac{1}{p^2} \tanh kp$	$H(2k, t) = \begin{cases} t & \text{when } 0 < t < 2k \\ 4k - t & \text{when } 2k < t < 4k \end{cases}$
70	$\dfrac{1}{p \sinh kp}$	$2S(2k, t + k) - 2 = 2(n - 1)$ $\quad$ when $(2n - 3)k < t < (2n - 1)k \quad t > 0$
71	$\dfrac{1}{p \cosh kp}$	$M(2k, t + 3k) + 1 = 1 + (-1)^n$ $\quad$ when $(2n - 3)k < t < (2n - 1)k \quad t > 0$
72	$\dfrac{1}{p} \coth kp$	$2S(2k, t) - 1 = 2n - 1$ $\quad$ when $2k(n - 1) < t < 2kn$
73	$\dfrac{k}{p^2 + k^2} \coth \dfrac{\pi p}{2k}$	$|\sin kt|$
74	$\dfrac{1}{(p^2 + 1)(1 - e^{-\pi p})}$	$\begin{cases} \sin t & \text{when } (2n - 2)\pi < t < (2n - 1)\pi \\ 0 & \text{when } (2n - 1)\pi < t < 2n\pi \end{cases}$
75	$\dfrac{1}{p} e^{-k/p}$	$J_0[2\sqrt{(kt)}]$
76	$\dfrac{1}{\sqrt{p}} e^{-k/p}$	$\dfrac{1}{\sqrt{(\pi t)}} \cos 2\sqrt{(kt)}$
77	$\dfrac{1}{\sqrt{p}} e^{k/p}$	$\dfrac{1}{\sqrt{(\pi t)}} \cosh 2\sqrt{(kt)}$
78	$\dfrac{1}{p^{3/2}} e^{-k/p}$	$\dfrac{1}{\sqrt{(\pi k)}} \sin 2\sqrt{(kt)}$
79	$\dfrac{1}{p^{3/2}} e^{k/p}$	$\dfrac{1}{\sqrt{(\pi k)}} \sinh 2\sqrt{(kt)}$
80	$\dfrac{1}{p^j} e^{k/p} \quad j > 0$	$\left(\dfrac{t}{k}\right)^{(j-1)/2} J_{j-1}[2\sqrt{(kt)}]$

Table 12. (*continued*)

No.	Transform $\bar{f}(p) = \int_0^\infty e^{-pt}f(t)dt$	Function f(t)
81	$\dfrac{1}{p^j}e^{k/p}\quad j>0$	$\left(\dfrac{t}{k}\right)^{(j-1)/2}I_{j-1}[2\sqrt{(kt)}]$
82	$e^{-k\sqrt{p}}\quad k>0$	$\dfrac{k}{2\sqrt{(\pi t^3)}}\exp\left(-\dfrac{k^2}{4t}\right)$
83	$\dfrac{1}{p}e^{-k\sqrt{p}}\quad k\geq 0$	$\text{erfc}\left(\dfrac{k}{2\sqrt{t}}\right)$
84	$\dfrac{1}{\sqrt{p}}e^{-k\sqrt{p}}\quad k\geq 0$	$\dfrac{1}{\sqrt{(\pi t)}}\exp\left(-\dfrac{k^2}{4t}\right)$
85	$p^{-3/2}e^{-k\sqrt{p}}\quad k\geq 0$	$2\sqrt{\dfrac{t}{\pi}}\left[\exp\left(-\dfrac{k^2}{4t}\right)\right]-k\,\text{erfc}\left(\dfrac{k}{2\sqrt{t}}\right)$
86	$\dfrac{ae^{-k\sqrt{p}}}{p(a+\sqrt{p})}\quad k\geq 0$	$-\exp(ak)\exp(a^2t)\,\text{erfc}\left(a\sqrt{t}+\dfrac{k}{2\sqrt{t}}\right)+$ $\text{erfc}\left(\dfrac{k}{2\sqrt{t}}\right)$
87	$\dfrac{e^{-k\sqrt{p}}}{\sqrt{p}(a+\sqrt{p})}\quad k\geq 0$	$\exp(ak)\exp(a^2t)\,\text{erfc}\left(a\sqrt{t}+\dfrac{k}{2\sqrt{t}}\right)$
88	$\dfrac{e^{-k\sqrt{[p(p+a)]}}}{\sqrt{[p(p+a)]}}$	$\begin{cases}0\quad\text{when }0<t<k\\ \exp(-\tfrac{1}{2}at)I_0\left[\tfrac{1}{2}a\sqrt{(t^2-k^2)}\right]\text{ when }t>k\end{cases}$
89	$\dfrac{e^{-k\sqrt{(p^2+a^2)}}}{\sqrt{(p^2+a^2)}}$	$\begin{cases}0\qquad\qquad\text{when }0<t<k\\ J_0[a\sqrt{(t^2-k^2)}]\quad\text{when }t>k\end{cases}$
90	$\dfrac{e^{-k\sqrt{(p^2-a^2)}}}{\sqrt{(p^2-a^2)}}$	$\begin{cases}0\qquad\qquad\text{when }0<t<k\\ I_0[a\sqrt{(t^2-k^2)}]\quad\text{when }t>k\end{cases}$
91	$\dfrac{e^{-k[\sqrt{(p^2+a^2)}-p]}}{\sqrt{(p^2+a^2)}}\quad k\geq 0$	$J_0[a\sqrt{(t^2+2kt)}]$
92	$e^{-kp}-e^{-k\sqrt{(p^2+a^2)}}$	$\begin{cases}0\quad\text{when }0<t<k\\ \dfrac{ak}{\sqrt{(t^2-k^2)}}J_1[a\sqrt{(t^2-k^2)}]\\ \quad\text{when }t>k\end{cases}$
93	$e^{-k\sqrt{(p^2-a^2)}}-e^{-kp}$	$\begin{cases}0\quad\text{when }0<t<k\\ \dfrac{ak}{\sqrt{(t^2-k^2)}}J_1[a\sqrt{(t^2-k^2)}]\\ \quad\text{when }t>k\end{cases}$

(*continued overleaf*)

Table 12. (*continued*)

No.	Transform $\bar{f}(p) = \int_0^\infty e^{-pt} f(t)dt$	Function $f(t)$
94	$\dfrac{a^j e^{-k\sqrt{(p^2+a^2)}}}{\sqrt{(p^2+a^2)}[\sqrt{(p^2+a^2)}+p]^j}$ $\quad j > -1$	$\begin{cases} 0 \quad \text{when } 0 < t < k \\ \left(\dfrac{t-k}{t+k}\right)^{(1/2)j} J_j[a\sqrt{(t^2-k^2)}] \\ \quad \text{when } t > k \end{cases}$
95	$\dfrac{1}{p} \ln p$	$\lambda - \ln t \quad \lambda = -0.5772\ldots$
96	$\dfrac{1}{p^k} \ln p \quad k > 0$	$t^{k-1}\left\{\dfrac{\lambda}{[\Gamma(k)]^2} - \dfrac{\ln t}{\Gamma(k)}\right\}$
97[b]	$\dfrac{\ln p}{p-a} \quad a > 0$	$(\exp at)[\ln a - Ei(-at)]$
98[c]	$\dfrac{\ln p}{p^2+1}$	$\cos t Si(t) - \sin t Ci(t)$
99[c]	$\dfrac{p \ln p}{p^2+1}$	$-\sin t Si(t) - \cos t Ci(t)$
100[b]	$\dfrac{1}{p} \ln(1+kp) \quad k > 0$	$-Ei\left(-\dfrac{t}{k}\right)$
101	$\ln \dfrac{p-a}{p-b}$	$\dfrac{1}{t}(e^{bt} - e^{at})$
102[c]	$\dfrac{1}{p} \ln(1+k^2p^2)$	$-2Ci\left(\dfrac{t}{k}\right)$
103[c]	$\dfrac{1}{p} \ln(p^2+a^2) \quad a > 0$	$2\ln a - 2Ci(at)$
104[c]	$\dfrac{1}{p^2} \ln(p^2+a^2) \quad a > 0$	$\dfrac{2}{a}[at \ln a + \sin at - atCi(at)]$
105	$\ln \dfrac{p^2+a^2}{p^2}$	$\dfrac{2}{t}(1 - \cos at)$
106	$\ln \dfrac{p^2-a^2}{p^2}$	$\dfrac{2}{t}(1 - \cosh at)$
107	$\tan^{-1} \dfrac{k}{p}$	$\dfrac{1}{t} \sin kt$

[b] $Ei(-t) = -\int_t^\infty \dfrac{e^{-x}}{x} dx$ (for $t > 0$) = exponential integral function.

[c] $Si(t) = \int_0^t \dfrac{\sin x}{x} dx$ = sine integral function.

$Ci(t) = -\int_t^\infty \dfrac{\cos x}{x} dx$ = cosine integral function.

These functions are tabulated in M. ABRAMOWITZ and I.A. STEGUN, *Handbook of Mathematical Functions*, Dover Publications, New York 1965.

Table 12. (*continued*)

No.	Transform $\bar{f}(p) = \int_0^\infty e^{-pt} f(t) dt$	Function $f(t)$
108^c	$\dfrac{1}{p} \tan^{-1} \dfrac{k}{p}$	Si (kt)
109	$\exp(k^2 p^2)$ erfc (kp) $k > 0$	$\dfrac{1}{k\sqrt{\pi}} \exp\left(-\dfrac{t^2}{4k^2}\right)$
110	$\dfrac{1}{p} \exp(k^2 p^2)$ erfc (kp) $k > 0$	erf $\left(\dfrac{t}{2k}\right)$
111	$\exp(kp)$ erfc $[\sqrt{(kp)}]$ $k > 0$	$\dfrac{\sqrt{k}}{\pi\sqrt{t(t+k)}}$
112	$\dfrac{1}{\sqrt{p}}$ erfc $[\sqrt{(kp)}]$	$\begin{cases} 0 & \text{when } 0 < t < k \\ (\pi t)^{-\frac{1}{2}} & \text{when } t > k \end{cases}$
113	$\dfrac{1}{\sqrt{p}} \exp(kp)$ erfc $[\sqrt{(kp)}]$ $k > 0$	$\dfrac{1}{\sqrt{[\pi(t+k)]}}$
114	erf $\left(\dfrac{k}{\sqrt{p}}\right)$	$\dfrac{1}{\pi t} \sin(2k\sqrt{t})$
115	$\dfrac{1}{\sqrt{p}} \exp\left(\dfrac{k^2}{p}\right)$ erfc $\left(\dfrac{k}{\sqrt{p}}\right)$	$\dfrac{1}{\sqrt{(\pi t)}} \exp(-2k\sqrt{t})$
116^d	$K_0(kp)$	$\begin{cases} 0 & \text{when } 0 < t < k \\ (t^2 - k^2)^{-\frac{1}{2}} & \text{when } t > k \end{cases}$
117^d	$K_0(k\sqrt{p})$	$\dfrac{1}{2t} \exp\left(-\dfrac{k^2}{4t}\right)$
118^d	$\dfrac{1}{p} \exp(kp)K_1(kp)$	$\dfrac{1}{k} \sqrt{[t(t+2k)]}$
119^d	$\dfrac{1}{\sqrt{p}} K_1(k\sqrt{p})$	$\dfrac{1}{k} \exp\left(-\dfrac{k^2}{4t}\right)$
120^d	$\dfrac{1}{\sqrt{p}} \exp\left(\dfrac{k}{p}\right) K_0\left(\dfrac{k}{p}\right)$	$\dfrac{2}{\sqrt{(\pi t)}} K_0[2\sqrt{(2kt)}]$
121^e	$\pi \exp(-kp)I_0(kp)$	$\begin{cases} [t(2k-t)]^{-\frac{1}{2}} & \text{when } 0 < t < 2k \\ 0 & \text{when } t > 2k \end{cases}$
122^e	$\exp(-kp)I_1(kp)$	$\begin{cases} \dfrac{k-t}{\pi k\sqrt{[t(2k-t)]}} & \text{when } 0 < t < 2k \\ 0 & \text{when } t > 2k \end{cases}$
123	unity	unit impulse

$^d K_n(x)$ denotes the Bessel function of the second kind for the imaginary argument.
$^e I_n(x)$ denotes the Bessel function of the first kind for the imaginary argument.

CHEMICAL ENGINEERING

Table 13. Error function and its derivative

x	$\frac{2}{\sqrt{\pi}}e^{-x^2}$	erf$x$	x	$\frac{2}{\sqrt{\pi}}e^{-x^2}$	erf$x$
0.00	1.12837	0.00000	0.50	0.87878	0.52049
0.01	1.12826	0.01128	0.51	0.86995	0.52924
0.02	1.12792	0.02256	0.52	0.86103	0.53789
0.03	1.12736	0.03384	0.53	0.85204	0.54646
0.04	1.12657	0.04511	0.54	0.84297	0.55493
0.05	1.12556	0.05637	0.55	0.83383	0.56332
0.06	1.12432	0.06762	0.56	0.82463	0.57161
0.07	1.12286	0.07885	0.57	0.81536	0.57981
0.08	1.12118	0.09007	0.58	0.80604	0.58792
0.09	1.11927	0.10128	0.59	0.79666	0.59593
0.10	1.11715	0.11246	0.60	0.78724	0.60385
0.11	1.11480	0.12362	0.61	0.77777	0.61168
0.12	1.11224	0.13475	0.62	0.76826	0.61941
0.13	1.10946	0.14586	0.63	0.75872	0.62704
0.14	1.10647	0.15694	0.64	0.74914	0.63458
0.15	1.10327	0.16799	0.65	0.73954	0.64202
0.16	1.09985	0.17901	0.66	0.72992	0.64937
0.17	1.09623	0.18999	0.67	0.72027	0.65662
0.18	1.09240	0.20093	0.68	0.71061	0.66378
0.19	1.08837	0.21183	0.69	0.70095	0.67084
0.20	1.08413	0.22270	0.70	0.69127	0.67780
0.21	1.07969	0.23352	0.71	0.68159	0.68466
0.22	1.07506	0.24429	0.72	0.67191	0.69143
0.23	1.07023	0.25502	0.73	0.66224	0.69810
0.24	1.06522	0.26570	0.74	0.65258	0.70467
0.25	1.06001	0.27632	0.75	0.64293	0.71115
0.26	1.05462	0.28689	0.76	0.63329	0.71753
0.27	1.04904	0.29741	0.77	0.62368	0.72382
0.28	1.04329	0.30788	0.78	0.61408	0.73001
0.29	1.03736	0.31828	0.79	0.60452	0.73610
0.30	1.03126	0.32862	0.80	0.59498	0.74210
0.31	1.02498	0.33890	0.81	0.58548	0.74800
0.32	1.01855	0.34912	0.82	0.57601	0.75381
0.33	1.01195	0.35927	0.83	0.56659	0.75952
0.34	1.00519	0.36936	0.84	0.55720	0.76514
0.35	0.99828	0.37938	0.85	0.54786	0.77066
0.36	0.99122	0.38932	0.86	0.53858	0.77610
0.37	0.98401	0.39920	0.87	0.52934	0.78143
0.38	0.97665	0.40900	0.88	0.52016	0.78668
0.39	0.96916	0.41873	0.89	0.51103	0.79184
0.40	0.96154	0.42839	0.90	0.50196	0.79690
0.41	0.95378	0.43796	0.91	0.49296	0.80188
0.42	0.94590	0.44746	0.92	0.48402	0.80676
0.43	0.93789	0.45688	0.93	0.47515	0.81156
0.44	0.92977	0.46622	0.94	0.46635	0.81627
0.45	0.92153	0.47548	0.95	0.45761	0.82089
0.46	0.91318	0.48465	0.96	0.44896	0.82542
0.47	0.90473	0.49374	0.97	0.44037	0.82987
0.48	0.89617	0.50274	0.98	0.43187	0.83423
0.49	0.88752	0.51166	0.99	0.42345	0.83850

Table 13. (*continued*)

x	$\dfrac{2}{\sqrt{\pi}}e^{-x^2}$	erf$x$	x	$\dfrac{2}{\sqrt{\pi}}e^{-x^2}$	erf$x$
1.00	0.41510	0.84270	1.50	0.11893	0.96610
1.01	0.40684	0.84681	1.51	0.11540	0.96727
1.02	0.39867	0.85083	1.52	0.11195	0.96841
1.03	0.39058	0.85478	1.53	0.10859	0.96951
1.04	0.38257	0.85864	1.54	0.10531	0.97058
1.05	0.37466	0.86243	1.55	0.10210	0.97162
1.06	0.36684	0.86614	1.56	0.09898	0.97262
1.07	0.35911	0.86977	1.57	0.09593	0.97360
1.08	0.35147	0.87332	1.58	0.09295	0.97454
1.09	0.34392	0.87680	1.59	0.09005	0.97546
1.10	0.33647	0.88020	1.60	0.08722	0.97634
1.11	0.32912	0.88353	1.61	0.08447	0.97720
1.12	0.32186	0.88678	1.62	0.08178	0.97803
1.13	0.31470	0.88997	1.63	0.07917	0.97884
1.14	0.30764	0.89308	1.64	0.07662	0.97962
1.15	0.30067	0.89612	1.65	0.07414	0.98037
1.16	0.29381	0.89909	1.66	0.07173	0.98110
1.17	0.28704	0.90200	1.67	0.06938	0.98181
1.18	0.28037	0.90483	1.68	0.06709	0.98249
1.19	0.27381	0.90760	1.69	0.06487	0.98315
1.20	0.26734	0.91031	1.70	0.06271	0.98379
1.21	0.26097	0.91295	1.71	0.06060	0.98440
1.22	0.25471	0.91553	1.72	0.05856	0.98500
1.23	0.24854	0.91805	1.73	0.05657	0.98557
1.24	0.24248	0.92050	1.74	0.05464	0.98613
1.25	0.23652	0.92290	1.75	0.05277	0.98667
1.26	0.23065	0.92523	1.76	0.05095	0.98719
1.27	0.22489	0.92751	1.77	0.04918	0.98769
1.28	0.21923	0.92973	1.78	0.04747	0.98817
1.29	0.21367	0.93189	1.79	0.04580	0.98864
1.30	0.20820	0.93400	1.80	0.04419	0.98909
1.31	0.20284	0.93606	1.81	0.04262	0.98952
1.32	0.19757	0.93806	1.82	0.04110	0.98994
1.33	0.19241	0.94001	1.83	0.03963	0.99034
1.34	0.18734	0.94191	1.84	0.03820	0.99073
1.35	0.18236	0.94376	1.85	0.03681	0.99111
1.36	0.17749	0.94556	1.86	0.03547	0.99147
1.37	0.17271	0.94731	1.87	0.03417	0.99182
1.38	0.16802	0.94901	1.88	0.03292	0.99215
1.39	0.16343	0.95067	1.89	0.03170	0.99247
1.40	0.15894	0.95228	1.90	0.03052	0.99279
1.41	0.15453	0.95385	1.91	0.02938	0.99308
1.42	0.15022	0.95537	1.92	0.02827	0.99337
1.43	0.14600	0.95685	1.93	0.02721	0.99365
1.44	0.14187	0.95829	1.94	0.02617	0.99392
1.45	0.13783	0.95969	1.95	0.02517	0.99417
1.46	0.13387	0.96105	1.96	0.02421	0.99442
1.47	0.13001	0.96237	1.97	0.02328	0.99466
1.48	0.12623	0.96365	1.98	0.02237	0.99489
1.49	0.12254	0.96489	1.99	0.02150	0.99511
			2.00	0.02066	0.99532

Problems

1.1. 98% sulphuric acid of viscosity 0.025 Ns/m^2 and density 1840 kg/m^3 is pumped at 685 cm^3/s through a 25 mm line. Calculate the value of the Reynolds number.

1.2. Compare the costs of electricity at 1 p per kWh and gas at 15 p per therm.

1.3. A boiler plant raises 5.2 kg/s of steam at 1825 kN/m^2 pressure, using coal of calorific value 27.2 MJ/kg. If the boiler efficiency is 75%, how much coal is consumed per day? If the steam is used to generate electricity, what is the power generation in kilowatts, assuming a 20% conversion efficiency of the turbines and generators?

1.4. The power required by an agitator in a tank is a function of the following four variables:

   (a) Diameter of impeller.
   (b) Number of rotations of impeller per unit time.
   (c) Viscosity of liquid.
   (d) Density of liquid.

From a dimensional analysis, obtain a relation between the power and the four variables. The power consumption is found, experimentally, to be proportional to the square of the speed of rotation. By what factor would the power be expected to increase if the impeller diameter were doubled?

1.5. It is found experimentally that the terminal settling velocity u_0 of a spherical particle in a fluid is a function of the following quantities:

          particle diameter d,
          buoyant weight of particle (weight of particle-weight of displaced fluid) W,
          fluid density ρ,
          fluid viscosity μ.

Obtain a relationship for u_0 using dimensional analysis.
   Stokes established, from theoretical considerations, that for small particles which settle at very low velocities, the settling velocity is independent of the density of the fluid except in so far as this affects the buoyancy. Show that the settling velocity *must* then be inversely proportional to the viscosity of the fluid.

1.6. A drop of liquid spreads over a horizontal surface. What are the factors which will influence:

   (a) the rate at which the liquid spreads, and
   (b) the final shape of the drop?

Obtain dimensionless groups involving the physical variables in the two cases.

1.7. Liquid is flowing at a volumetric flowrate Q per unit width down a vertical surface. Obtain from dimensional analysis the form of the relationship between flowrate and film thickness. If the flow is streamline, show that the volumetric flowrate is directly proportional to the density of the liquid.

1.8. Obtain, by dimensional analysis, a functional relationship for the heat transfer coefficient for forced convection at the inner wall of an annulus through which a cooling liquid is flowing.

1.9. Obtain by dimensional analysis a functional relationship for the wall heat transfer coefficient for a fluid flowing through a straight pipe of circular cross-section. Assume that the effects of natural convection can be neglected in comparison with those of forced convection.

It is found by experiment that, when the flow is turbulent, increasing the flowrate by a factor of 2 always results in a 50% increase in the coefficient. How would a 50% increase in density of the fluid be expected to affect the coefficient, all other variables remaining constant?

1.10. A stream of droplets of liquid is formed rapidly at an orifice submerged in a second, immiscible liquid. What physical properties would be expected to influence the mean size of droplet formed? Using dimensional analysis obtain a functional relation between the variables.

1.11. Liquid flows under steady-state conditions along an open channel of fixed inclination to the horizontal. On what factors will the depth of liquid in the channel depend? Obtain a relationship between the variables using dimensional analysis.

1.12. Liquid flows down an inclined surface as a film. On what variables will the thickness of the liquid film depend? Obtain the relevant dimensionless groups. It may be assumed that the surface is sufficiently wide for edge effects to be negligible.

1.13. A glass particle settles under the action of gravity in a liquid. Upon which variables would you expect the terminal velocity of the particle to depend? Obtain a relevant dimensionless grouping of the variables. The falling velocity is found to be proportional to the square of the particle diameter when other variables are kept constant. What will be the effect of doubling the viscosity of the liquid? What does this suggest regarding the nature of the flow?

1.14. Heat is transferred from condensing steam to a vertical surface and the resistance to heat transfer is attributable to the thermal resistance of the condensate layer on the surface.
What variables will be expected to affect the film thickness at a point?
Obtain the relevant dimensionless groups.
For streamline flow it is found that the film thickness is proportional to the one third power of the volumetric flowrate per unit width. Show that the heat transfer coefficient would be expected to be inversely proportional to the one third power of viscosity.

1.15. A spherical particle settles in a liquid contained in a narrow vessel. Upon what variables would you expect the falling velocity of the particle to depend? Obtain the relevant dimensionless groups.
For particles of a given density settling in a vessel of large diameter, the settling velocity is found to be inversely proportional to the viscosity of the liquid. How would you expect it to depend on particle size?

1.16. A liquid is in steady state flow in an open trough of rectangular cross-section inclined at an angle θ to the horizontal. On what variables would you expect the mass flow per unit time to depend? Obtain the dimensionless groups which are applicable to this problem.

1.17. The resistance force on a spherical particle settling in a fluid in given by Stokes' Law. Obtain an expression for the terminal falling velocity of the particle. It is convenient to express the results of experiments in the form of a dimensionless group which may be plotted against a Reynolds group with respect to the particle. Suggest a suitable form for this dimensionless group.
Force on particle from Stokes' Law $= 3\pi\mu du$; where μ equals fluid viscosity, d equals particle diameter and u is equal to the velocity of the particle relative to the fluid.
What will be the terminal falling velocity of a particle of diameter 10 μm and of density 1600 kg/m^3 settling in a liquid of density 1000 kg/m^3 and of viscosity 0.001 Ns/m^2?
If Stokes' Law applies for particle Reynolds numbers up to 0.2, what is the diameter of the largest particle whose behaviour is governed by Stokes' Law for this solid and liquid?

1.18. A sphere, initially at a constant temperature, is immersed in a liquid whose temperature is maintained constant. The time t taken for the temperature of the centre of the sphere to reach a given temperature θ_c is a function of the following variables:

Diameter of sphere d
Thermal conductivity of sphere k
Density of sphere ρ
Specific heat capacity of sphere C_p
Temperature of fluid in which it is immersed θ_s

Obtain relevant dimensionless groups for this problem.

1.19. Upon what variables would you expect the rate of filtration of a suspension of fine solid particles to depend? Consider the flow through unit area of filter medium and express the variables in the form of dimensionless groups.

It is found that the filtration rate is doubled if the pressure difference is doubled. What effect would you expect from raising the temperature of filtration from 293 to 313 K?

The viscosity of the liquid is given by:

$$\mu = \mu_0(1 - 0.015(T - 273))$$

where: μ is the viscosity at a temperature TK
and: μ_0 is the viscosity at 273 K.

2.1. Calculate the ideal available energy produced by the discharge to atmosphere through a nozzle of air stored in a cylinder of capacity 0.1 m^3 at a pressure of 5 MN/m^2. The initial temperature of the air is 290 K and the ratio of the specific heats is 1.4.

2.2. Obtain expressions for the variation of:

   (a) internal energy with change of volume,
   (b) internal energy with change of pressure, and
   (c) enthalpy with change of pressure.

all at constant temperature, for a gas whose equation of state is given by van der Waals' law.

2.3. Calculate the energy stored in 1000 cm^3 of gas at 80 MN/m^2 and 290 K using a datum of STP.

2.4. Compressed gas is distributed from a works in cylinders which are filled to a pressure P by connecting them to a large reservoir of gas which remains at a steady pressure P and temperature T. If the small cylinders are initially at a temperature T and pressure P_0, what is the final temperature of the gas in the cylinders if heat losses can be neglected and if the compression can be regarded as reversible? Assume that the ideal gas laws are applicable.

3.1. Calculate the hydraulic mean diameter of the annular space between a 40 mm and a 50 mm tube.

3.2. 0.015 m^3/s of acetic acid is pumped through a 75 mm diameter horizontal pipe 70 m long. What is the pressure drop in the pipe?

$$\text{Viscosity of acid} = 2.5 \text{ mN s/m}^2,$$

$$\text{Density of acid} = 1060 \text{ kg/m}^3.$$

$$\text{Roughness of pipe surface} = 6 \times 10^{-5} \text{ m}.$$

3.3. A cylindrical tank, 5 m in diameter, discharges through a mild steel pipe 90 m long and 230 mm diameter connected to the base of the tank. Find the time taken for the water level in the tank to drop from 3 m to 1 m above the bottom. Take the viscosity of water as 1 mN s/m^2.

3.4. Two storage tanks A and B containing a petroleum product discharge through pipes each 0.3 m in diameter and 1.5 km long to a junction at D. From D the product is carried by a 0.5 m diameter pipe to a third storage tank C, 0.8 km away. The surface of the liquid in A is initially 10 m above that in C and the liquid level in B is 7 m higher than that in A. Calculate the initial rate of discharge of the liquid if the pipes are of mild steel. Take the density of the petroleum product as 870 kg/m^3 and the viscosity as 0.7 mN s/m^2.

3.5. Find the drop in pressure due to friction in a pipe 300 m long and 150 mm diameter when water is flowing at the rate of 0.05 m^3/s. The pipe is of glazed porcelain.

3.6. Two tanks, the bottoms of which are at the same level, are connected with one another by a horizontal pipe 75 mm diameter and 300 m long. The pipe is bell-mouthed at each end so that losses on entry and exit are negligible. One tank is 7 m diameter and contains water to a depth of 7 m. The other tank is 5 m diameter and contains water to a depth of 3 m. If the tanks are connected to each other by means of the pipe, how long will it take before the water level in the larger tank has fallen to 6 m? Assume the pipe to be an old mild steel pipe.

3.7. Two immiscible fluids A and B, of viscosities μ_A and μ_B, flow under streamline conditions between two horizontal parallel planes of width b, situated a distance $2a$ apart (where a is much less than b), as two distinct parallel layers one above the other, each of depth a. Show that the volumetric rate of flow of A is:

$$\frac{-\Delta P a^3 b}{12\mu_A l} \times \frac{7\mu_A + \mu_B}{\mu_A + \mu_B},$$

where $-\Delta P$ is the pressure drop over a length l in the direction of flow.

3.8. Glycerol is pumped from storage tanks to rail cars through a single 50 mm diameter main 10 m long, which must be used for all grades of glycerol. After the line has been used for commercial material, how much pure glycerol must be pumped before the issuing liquid contains not more than 1% of the commercial material? The flow in the pipeline is streamline and the two grades of glycerol have identical densities and viscosities.

3.9. A viscous fluid flows through a pipe with slightly porous walls so that there is a leakage of kP m³/m² s, where P is the local pressure measured above the discharge pressure and k is a constant. After a length L the liquid is discharged into a tank. If the internal diameter of the pipe is D m and the volumetric rate of flow at the inlet is Q m³/s, show that the pressure drop in the pipe is given by:

$$-\Delta P = \frac{Q}{\pi k D}a \tanh aL,$$

where:

$$a = \left(\frac{128 k \mu}{D^3}\right)^{0.5}$$

Assume a fully developed flow with $(R/\rho\mu^2) = 8Re^{-1}$.

3.10. A petroleum product of viscosity 0.5 mN s/m² and specific gravity 0.7 is pumped through a pipe of 0.15 m diameter to storage tanks situated 100 m away. The pressure drop along the pipe is 70 kN/m². The pipeline has to be repaired and it is necessary to pump the liquid by an alternative route consisting of 70 m of 20 cm pipe followed by 50 m of 10 cm pipe. If the existing pump is capable of developing a pressure of 300 kN/m², will it be suitable for use during the period required for the repairs? Take the roughness of the pipe surface as 0.00005 m.

3.11. Explain the phenomenon of hydraulic jump which occurs during the flow of a liquid in an open channel.
    A liquid discharges from a tank into an open channel under a gate so that the liquid is initially travelling at a velocity of 1.5 m/s and a depth of 75 mm. Calculate, from first principles, the corresponding velocity and depth after the jump.

3.12. What is a non-Newtonian fluid? Describe the principal types of behaviour exhibited by these fluids. The viscosity of a non-Newtonian fluid changes with the rate of shear according to the approximate relationship:

$$\mu_a = k\left(-\frac{du_x}{dr}\right)^{-0.5}$$

where μ_a is the apparent viscosity, and du_x/dr is the velocity gradient normal to the direction of motion.
    Show that the volumetric rate of streamline flow through a horizontal tube of radius a is:

$$\frac{\pi}{5}a^5\left(\frac{-\Delta P}{2kl}\right)^2$$

where $-\Delta P$ is the pressure drop over length l of the tube.

3.13. Calculate the pressure drop when 3 kg/s of sulphuric acid flows through 60 m of 25 mm pipe ($\rho = 1840$ kg/m³, $\mu = 0.025$ Ns/m²).

3.14. The relation between cost per unit length C of a pipeline installation and its diameter d is given by

$$C = a + bd,$$

where a and b are independent of pipe size. Annual charges are a fraction β of the capital cost. Obtain an expression for the optimum pipe diameter on a minimum cost basis for a fluid of density ρ and viscosity μ flowing at a mass rate of G. Assume that the fluid is in turbulent flow and that the Blasius equation is applicable, i.e. the friction factor is proportional to the Reynolds number to the power of minus one quarter. Indicate clearly how the optimum diameter depends on flowrate and fluid properties.

3.15. A heat exchanger is to consist of a number of tubes each 25 mm diameter and 5 m long arranged in parallel. The exchanger is to be used as a cooler with a rating of 4 MW and the temperature rise in the water feed to the tubes is to be 20 K.

If the pressure drop over the tubes is not to exceed 2 kN/m², calculate the minimum number of tubes that are required. Assume that the tube walls are smooth and that entrance and exit effects can be neglected.

$$\text{Viscosity of water} = 1 \text{ mN s/m}^2.$$

3.16. Sulphuric acid is pumped at 3 kg/s through a 60 m length of smooth 25 mm pipe. Calculate the drop in pressure. If the pressure drop falls by one half, what will the new flowrate be?

$$\text{Density of acid} = 1840 \text{ kg/m}^3.$$

$$\text{Viscosity of acid} = 25 \text{ mN s/m}^2.$$

3.17. A Bingham plastic material is flowing under streamline conditions in a pipe of circular cross-section. What are the conditions for one half of the total flow to be within the central core across which the velocity profile is flat? The shear stress acting within the fluid R_y varies with velocity gradient du_x/dy according to the relation

$$R_y - R_c = -k\frac{du_x}{dy},$$

where R_c and k are constants for the material.

3.18. Oil of viscosity 10 mN s/m² and density 950 kg/m³ is pumped 8 km from an oil refinery to a distribution depot through a 75 mm diameter pipeline and is then despatched to customers at a rate of 500 tonne/day. Allowance must be made for periods of maintenance which may interrupt the supply from the refinery for up to 72 hours. If the maximum permissible pressure drop over the pipeline is 3450 kN/m², what is the shortest time in which the storage tanks can be completely recharged after a 72 hour shutdown? Take the roughness of the pipe surface as 0.05 mm.

3.19. Water is pumped at 1.4 m³/s from a tank at a treatment plant to a tank at a local works through two parallel pipes, 0.3 m and 0.6 m diameter respectively. What is the velocity in each pipe and, if a single pipe is used, what diameter will be needed if this flow of water is to be transported, the pressure drop being the same? Assume turbulent flow, with the friction factor inversely proportional to the one quarter power of the Reynolds number.

3.20. Oil of viscosity 10 mN s/m² and specific gravity 0.90, flows through 60 m of 100 mm diameter pipe and the pressure drop is 13.8 kN/m². What will be the pressure drop for a second oil of viscosity 30 mN s/m² and specific gravity 0.95 flowing at the same rate through the pipe? Assume the pipe wall to be smooth.

3.21. Crude oil is pumped from a terminal to a refinery through a foot diameter pipeline. As a result of frictional heating, the temperature of the oil is 20 deg K higher at the refinery end than at the terminal end of the pipe and the viscosity has fallen to one half its original value. What is the ratio of the pressure gradient in the pipeline at the refinery end to that at the terminal end?

$$\text{Viscosity of oil at terminal} = 90 \text{ mN s/m}^2$$

$$\text{Density of oil (approximately constant)} = 960 \text{ kg/m}^3$$

$$\text{Flowrate of oil} = 20,000 \text{ tonne/day}$$

Outline a method for calculating the temperature of the oil as a function of distance from the inlet for a given value of the heat transfer coefficient between the pipeline and the surroundings.

3.22. Oil with a viscosity of 10 mN s/m^2 and density 900 kg/m^3 is flowing through a 500 mm diameter pipe 10 km long. The pressure difference between the two ends of the pipe is 10^6N/m^2. What will the pressure drop be at the same flowrate if it is necessary to replace the pipe by one only 300 mm diameter? Assume the pipe surface to be smooth.

3.23. Oil of density 950 kg/m^3 and viscosity 10^{-2} Ns/m^2 is to be pumped 10 km through a pipeline and the pressure drop must not exceed 2×10^5N/m^2. What is the minimum diameter of pipe which will be suitable, if a flowrate of 50 tonne/h is to be maintained? Assume the pipe wall to be smooth. *Either* use pipe friction chart *or* use Blasius equation ($R/\rho u^2 = 0.0396Re^{-1/4}$).

3.24. On the assumption that the velocity profile in a fluid in turbulent flow is given by the Prandtl one-seventh power law, calculate the radius at which the flow between it and the centre is equal to that between it and the wall, for a pipe 100 mm in diameter.

3.25. A pipeline 0.5 m diameter and 1200 m long is used for transporting an oil of density 950 kg/m^3 and of viscosity 0.01 Ns/m^2 at 0.4 m^3/s. If the roughness of the pipe surface is 0.5 mm, what is the pressure drop? With the same pressure drop, what will be the flowrate of a second oil of density 980 kg/m^3 and of viscosity 0.02 Ns/m^2? (Friction charts provided)

3.26. Water (density 1000 kg/m^3, viscosity 1 mN s/m^2) is pumped through a 50 mm diameter pipeline at 4 kg/s and the pressure drop is 1 MN/m^2. What will be the pressure drop for a solution of glycerol in water (density 1050 kg/m^3, viscosity 10 mN s/m^2) when pumped at the same rate? Assume the pipe to be smooth.

3.27. A liquid is pumped in streamline flow through a pipe of diameter d. At what distance from the centre of the pipe will the fluid be flowing at the average velocity?

3.28. Cooling water is supplied to a heat exchanger and flows through 25 mm diameter tubes each 5 m long arranged in parallel. If the pressure drop over the heat exchanger is not to exceed 8000 N/m^2, how many tubes must be included for a total flowrate of water of 110 tonne/h?

$$\text{Density of water} \quad 1000 \text{ kg/m}^3$$

$$\text{Viscosity of water} \quad 1 \text{ mN s/m}^2$$

$$\text{Assume pipes to be smooth walled}$$

If ten per cent of the tubes became blocked, what would the new pressure drop be?

3.29. The effective viscosity of a non-Newtonian fluid can be expressed by the following relationship:

$$\mu_a = k'' \frac{du_x}{dr}$$

where k'' is constant.
  Show that the volumetric flowrate of this fluid in a horizontal pipe of radius a under isothermal laminar flow conditions with a pressure gradient $-\Delta P/l$ per unit length is:

$$Q = \frac{2\pi}{7} a^{7/2} \left(\frac{-\Delta P}{2k''l} \right)^{1/2}$$

3.30. Determine the yield stress of a Bingham fluid of density 2000 kg/m^3 which will just flow out of an open-ended vertical tube of diameter 300 mm under the influence of its own weight.

3.31. A fluid of density 1.2×10^3 kg/m^3 flows down an inclined plane at $15°$ to the horizontal. If the viscous behaviour is described by the relationship:

$$R_{yx} = -k \left(\frac{du_x}{dy} \right)^n$$

where $k = 4.0$ Ns$^{0.4}$/m^2, and
$\quad n = 0.4$

calculate the volumetric flowrate per unit width if the fluid film is 10 mm thick.

3.32. A fluid with a finite yield stress is sheared between two concentric cylinders, 50 mm long. The inner cylinder is 30 mm diameter and the gap is 20 mm. The outer cylinder is held stationary while a torque is applied to the inner. The moment required just to produce motion was 0.01 N m. Calculate the force needed to ensure all the fluid is flowing under shear if the plastic viscosity is 0.1 Ns/m^2.

3.33. Experiments with a capillary viscometer of length 100 mm and diameter 2 mm gave the following results:

Applied pressure (N/m^2)	Volumetric flowrate (m^3/s)
1×10^3	1×10^{-7}
2×10^3	2.8×10^{-7}
5×10^3	1.1×10^{-7}
1×10^4	3×10^{-6}
2×10^4	9×10^{-6}
5×10^4	3.5×10^{-5}
1×10^5	1×10^{-4}

Suggest a suitable model to describe the fluid properties.

3.34. Data obtained with a cone and plate viscometer were as follows:

cone half-angle $89°$
cone radius 50 mm

cone speed (Hz)	measured torque (Nm)
0.1	4.6×10^{-2}
0.5	7×10^{-2}
1	1.0
5	3.4
10	6.4
50	3.0×10

Suggest a suitable model to describe the fluid properties.

3.35. Tomato purée of density 1300 kg/m^3 is pumped through a 50 mm diameter factory pipeline at a flowrate of 0.00028 m^3/s. It is suggested that in order to double production either:

(a) a similar line with pump should be put in parallel to the existing one, or
(b) a large pump should force the material through the present line, or
(c) a large pump should supply the liquid through a line of twice the cross-sectional area.

Given that the flow properties of the purée can be described by the Casson equation:

$$(-R_y)^{1/2} = (-R_Y)^{1/2} + \left(-\mu_c \frac{du_x}{dy} \right)^{1/2}$$

where R_Y is a yield stress, here 20 N/m^2,

μ_c is a characteristic Casson plastic viscosity, 5 Ns/m^2, and

$\dfrac{du_x}{dy}$ is the velocity gradient.

evaluate the relative pressure drops of the three suggestions, assuming laminar flow throughout.

3.36. The rheological properties of a particular suspension can be approximated reasonably well by either a "power law" or a "Bingham plastic" model over the shear rate range of 10 to 50 s^{-1}. If the consistency K is 10 Ns^n/m^2 and the flow behaviour index n is 0.2 in the power law model, what will be the approximate values of the yield stress and of the plastic viscosity in the Bingham plastic model?

What will be the pressure drop, when the suspension is flowing under laminar conditions in a pipe 200 m long and 40 mm diameter, when the centre line velocity is 1 m/s, according to the power law model? Calculate the centre line velocity for this pressure drop for the Bingham plastic model and comment on the result.

3.37. Show how, by suitable selection of the index n, the power law may be used to describe the behaviour of both shear-thinning and shear-thickening non-Newtonian fluids over a limited range of shear rates. What are the main objections to the use of the power law? Give some examples of different types of shear-thinning fluids.

A power law fluid is flowing under laminar conditions through a pipe of circular cross-section. At what radial position is the fluid velocity equal to the mean velocity in the pipe? Where does this occur for a fluid with an n- value of 0.2?

3.38. A liquid whose rheology can be represented by the "power law" model is flowing under streamline conditions through a pipe of 5 mm diameter. If the mean velocity of flow in 1 m/s and the velocity at the pipe axis is 1.2 m/s, what is the value of the power law index n?

Water, of viscosity 1 mN s/m^2 flowing through the pipe at the same mean velocity gives rise to a pressure drop of 10^4 N/m^2 compared with 10^5 N/m^2 for the non-Newtonian fluid. What is the consistency ("k" value) of the non-Newtonian fluid?

3.39. Two liquids of equal densities, the one Newtonian and the other a non-Newtonian "power law" fluid, flow at equal volumetric rates down two wide vertical surfaces of the same widths. The non-Newtonian fluid has a power law index of 0.5 and has the same apparent viscosity as the Newtonian fluid when its shear rate is 0.01 s^{-1}. Show that, for equal surface velocities of the two fluids, the film thickness for the non-Newtonian fluid is 1.125 times that of the Newtonian fluid.

4.1. A gas, having a molecular weight of 13 kg/kmol and a kinematic viscosity of 0.25 cm^2/s, is flowing through a pipe 0.25 m internal diameter and 5 km long at the rate of 0.4 m^3/s and is delivered at atmospheric pressure. Calculate the pressure required to maintain this rate of flow under isothermal conditions.

The volume occupied by 1 kmol at 273 K and 101.3 kN/m^2 is 22.4 m^3.

What would be the effect on the required pressure if the gas were to be delivered at a height of 150 m (i) above and (ii) below its point of entry into the pipe?

4.2. Nitrogen at 12 MN/m^2 pressure is fed through a 25 mm diameter mild steel pipe to a synthetic ammonia plant at the rate of 1.25 kg/s. What will be the drop in pressure over a 30 m length of pipe for isothermal flow of the gas at 298 K?

Absolute roughness of the pipe surface = 0.005 mm.
Kilogram molecular volume = 22.4 m^3.
Viscosity of nitrogen = 0.02 mN s/m^2.

4.3. Hydrogen is pumped from a reservoir at 2 MN/m^2 pressure through a clean horizontal mild steel pipe 50 mm diameter and 500 m long. The downstream pressure is also 2 MN/m^2 and the pressure of this gas is raised to 2.6 MN/m^2 by a pump at the upstream end of the pipe. The conditions of flow are isothermal and the temperature of the gas is 293 K. What is the flowrate and what is the effective rate of working of the pump?

Viscosity of hydrogen = 0.009 mN s/m^2 at 293 K.

4.4. In a synthetic ammonia plant the hydrogen is fed through a 50 mm steel pipe to the converters. The pressure drop over the 30 m length of pipe is 500 kN/m^2, the pressure at the downstream end being 7.5 MN/m^2. What power is required in order to overcome friction losses in the pipe? Assume isothermal expansion of the gas at 298 K. What error is introduced by assuming the gas to be an incompressible fluid of density equal to that at the mean pressure in the pipe? $\mu = 0.02$ mN s/m^2.

4.5. A vacuum distillation plant operating at 7 kN/m^2 pressure at the top has a boil-up rate of 0.125 kg/s of xylene. Calculate the pressure drop along a 150 mm bore vapour pipe used to connect the column to the condenser. The pipe length may be taken as equivalent to 6 m, $e/d = 0.002$ and $\mu = 0.01$ mN s/m^2.

4.6. Nitrogen at 12 MN/m^2 pressure is fed through a 25 mm diameter mild steel pipe to a synthetic ammonia plant at the rate of 0.4 kg/s. What will be the drop in pressure over a 30 m length of pipe assuming isothermal expansion of the gas at 300 K? What is the average quantity of heat per unit area of pipe surface that must pass through the walls in order to maintain isothermal conditions? What would be the pressure drop in the pipe if it were perfectly lagged?

4.7. Air, at a pressure of 10 MN/m^2 and a temperature of 290 K, flows from a reservoir through a mild steel pipe of 10 mm diameter and 30 m long into a second reservoir at a pressure P_2. Plot the mass rate of flow of the air as a function of the pressure P_2. Neglect any effects attributable to differences in level and assume an adiabatic expansion of the air. $\mu = 0.018$ mN s/m^2, $\gamma = 1.36$.

4.8. Over a 30 m length of 150 mm vacuum line carrying air at 293 K the pressure falls from 1 kN/m^2 to 0.1 kN/m^2. If the relative roughness e/d is 0.002, what is the approximate flowrate?

4.9. A vacuum system is required to handle 10 g/s of vapour (molecular weight 56 kg/kmol) so as to maintain a pressure of 1.5 kN/m^2 in a vessel situated 30 m from the vacuum pump. If the pump is able to maintain a pressure of 0.15 kN/m^2 at its suction point, what diameter pipe is required? The temperature is 290 K, and isothermal conditions may be assumed in the pipe, whose surface can be taken as smooth. The ideal gas law is followed.

$$\text{Gas viscosity} = 0.01 \text{ mN s/m}^2.$$

4.10. In a vacuum system, air is flowing isothermally at 290 K through a 150 mm diameter pipeline 30 m long. If the relative roughness of the pipewall e/d is 0.002 and the downstream pressure is 130 N/m^2, what will the upstream pressure be if the flow rate of air is 0.025 kg/s?

Assume that the ideal gas law applies and that the viscosity of air is constant at 0.018 mN s/m^2.

What error would be introduced if the change in kinetic energy of the gas as a result of expansion were neglected?

4.11. Air is flowing at a rate of 30 kg/m^{2}s through a smooth pipe of 50 mm diameter and 300 m long. If the upstream pressure is 800 kN/m^2, what will the downstream pressure be if the flow is isothermal at 273 K? Take the viscosity of air as 0.015 mN s/m^2 and the kg molecular volume as 22.4 m^3. What is the significance of the change in kinetic energy of the fluid?

4.12. If temperature does not change with height, estimate the boiling point of water at a height of 3000 m above sea-level. The barometer reading at sea-level is 98.4 kN/m^2 and the temperature is 288.7 K. The vapour pressure of water at 288.7 K is 1.77 kN/m^2. The molecular mass of air is 29 kg/kmol.

4.13. A 150 mm gas main is used for transferring coal gas (molecular mass 13 kg/kmol and kinematic viscosity 0.25 cm^2/s) at 295 K from a plant to a storage station 100 m away, at a rate of 1 m^3/s. Calculate the pressure drop, if the pipe can be considered to be smooth.

If the maximum permissible pressure drop is 10 kN/m^2, is it possible to increase the flowrate by 25%?

5.1. It is required to transport sand of particle size 1.25 mm and density 2600 kg/m^3 at the rate of 1 kg/s through a horizontal pipe, 200 m long. Estimate the air flowrate required, the pipe diameter and the pressure drop in the pipe-line.

5.2. Sand of mean diameter 0.2 mm is to be conveyed by water flowing at 0.5 kg/s in a 25 mm ID horizontal pipe, 100 m long. What is the maximum amount of sand which may be transported in this way if the head developed by the pump is limited to 300 kN/m^2? Assume fully suspended heterogeneous flow.

5.3. Explain the various mechanisms by which particles may be maintained in suspension during hydraulic transport in a horizontal pipeline and indicate when each is likely to be important.

A highly concentrated suspension of flocculated kaolin in water behaves as a pseudo-homogeneous fluid with shear-thinning characteristics which can be represented approximately by the Ostwald–de Waele power law, with an index of 0.15. It is found that, if air is injected into the suspension when in laminar flow, the pressure gradient may be reduced even though the flowrate of suspension is kept constant. Explain how this is possible in "slug" flow and estimate the possible reduction in pressure gradient for equal volumetric flowrates of suspension and air.

6.1. Sulphuric acid of specific gravity 1.3 is flowing through a pipe of 50 mm internal diameter. A thin-lipped orifice, 10 mm diameter, is fitted in the pipe and the differential pressure shown by a mercury manometer is 10 cm. Assuming that the leads to the manometer are filled with the acid, calculate (a) the weight of acid flowing per second, and (b) the approximate loss of pressure (in kN/m^2) caused by the orifice.

The coefficient of discharge of the orifice may be taken as 0.61, the specific gravity of mercury as 13.55, and the density of water as 1000 kg/m^3.

6.2. The rate of discharge of water from a tank is measured by means of a notch for which the flowrate is directly proportional to the height of liquid above the bottom of the notch. Calculate and plot the profile of the notch if the flowrate is 0.01 m^3/s when the liquid level is 150 mm above the bottom of the notch.

6.3. Water flows at between 3000 and 4000 cm^3/s through a 50 mm pipe and is metered by means of an orifice. Suggest a suitable size of orifice if the pressure difference is to be measured with a simple water manometer. What approximately is the pressure difference recorded at the maximum flowrate?

6.4. The rate of flow of water in a 150 mm diameter pipe is measured with a venturi meter with a 50 mm diameter throat. When the pressure drop over the converging section is 100 mm of water, the flowrate is 2.7 kg/s. What is the coefficient for the converging cone of the meter at that flowrate and what is the head lost due to friction? If the total loss of head over the meter is 15 mm water, what is the coefficient for the diverging cone?

6.5. A venturi meter with a 50 mm throat is used to measure a flow of slightly salt water in a pipe of inside diameter 100 mm. The meter is checked by adding 20 cm^3/s of normal sodium chloride solution above the meter and analysing a sample of water downstream from the meter. Before addition of the salt, 1000 cm^3 of water requires 10 cm^3 of 0.1 M silver nitrate solution in a titration. 1000 cm^3 of the downstream sample required 23.5 cm^3 of 0.1 M silver nitrate. If a mercury-under-water manometer connected to the meter gives a reading of 205 mm, what is the discharge coefficient of the meter? Assume that the density of the liquid is not appreciably affected by the salt.

6.6. A gas cylinder containing 30 m^3 of air at 6 MN/m^2 pressure discharges to the atmosphere through a valve which may be taken as equivalent to a sharp-edged orifice of 6 mm diameter (coefficient of discharge = 0.6). Plot the rate of discharge against the pressure in the cylinder. How long will it take for the pressure in the cylinder to fall to (a) 1 MN/m^2, and (b) 150 kN/m^2?

Assume an adiabatic expansion of the gas through the valve and that the contents of the cylinder remain at 273 K.

6.7. Air at a pressure of 1500 kN/m^2 and a temperature of 370 K flows through an orifice of 30 mm^2 to atmospheric pressure. If the coefficient of discharge is 0.65, the critical pressure ratio is 0.527, and the ratio of the specific heats is 1.4, calculate the mass flowrate.

6.8. Water flows through an orifice of 25 mm diameter situated in a 75 mm pipe at the rate of 300 cm^3/s. What will be the difference in level on a water manometer connected across the meter? Take the viscosity of water as 1 mN s/m^2.

6.9. Water flowing at 1500 cm³/s in a 50 mm diameter pipe is metered by means of a simple orifice of diameter 25 mm. If the coefficient of discharge of the meter is 0.62, what will be the reading on a mercury-under-water manometer connected to the meter?

What is the Reynolds number for the flow in the pipe?

$$\text{Density of water} = 1000 \text{ kg/m}^3,$$
$$\text{Viscosity of water} = 1 \text{ mN s/m}^2.$$

6.10. What size of orifice would give a pressure difference of 0.3 m water gauge for the flow of a petroleum product of specific gravity 0.9 at 0.05 m³/s in a 150 mm diameter pipe?

6.11. The flow of water through a 50 mm pipe is measured by means of an orifice meter with a 40 mm aperture. The pressure drop recorded is 150 mm on a mercury-under-water manometer and the coefficient of discharge of the meter is 0.6. What is the Reynolds number in the pipe and what would you expect the pressure drop over a 30 m length of the pipe to be?

$$\text{Friction factor, } \phi = \frac{R}{\rho u^2} = 0.0025.$$
$$\text{Specific gravity of mercury} = 13.6$$
$$\text{Viscosity of water} = 1 \text{ mN s/m}^2.$$

What type of pump would you use, how would you drive it, and what material of construction would be suitable?

6.12. A rotameter has a tube 0.3 m long which has an internal diameter of 25 mm at the top and 20 mm at the bottom. The diameter of the float is 20 mm, its effective specific gravity is 4.80, and its volume 6.6 cm³. If the coefficient of discharge is 0.72, at what height will the float be when metering water at 100 cm³/s?

6.13. Explain why there is a critical pressure ratio across a nozzle at which, for a given upstream pressure, the flowrate is a maximum.

Obtain an expression for the maximum flow for a given upstream pressure for isentropic flow through a horizontal nozzle. Show that for air (ratio of specific heats $\gamma = 1.4$) the critical pressure ratio is 0.53 and calculate the maximum flow through an orifice of area 30 mm² and coefficient of discharge 0.65 when the upstream pressure is 1.5 MN/m² and the upstream temperature 293 K.

$$\text{Kilogram molecular volume} = 22.4 \text{ m}^3.$$

6.14. A gas cylinder containing air discharges to atmosphere through a valve whose characteristics may be considered similar to those of a sharp-edged orifice. If the pressure in the cylinder is initially 350 kN/m², by how much will the pressure have fallen when the flowrate has decreased to one-quarter of its initial value?

The flow through the valve may be taken as isentropic and the expansion in the cylinder as isothermal. The ratio of the specific heats at constant pressure and constant volume is 1.4.

6.15. Water discharges from the bottom outlet of an open tank 1.5 m by 1 m in cross-section. The outlet is equivalent to an orifice 40 mm diameter with a coefficient of discharge of 0.6. The water level in the tank is regulated by a float valve on the feed supply which shuts off completely when the height of water above the bottom of the tank is 1 m and which gives a flowrate which is directly proportional to the distance of the water surface below this maximum level. When the depth of water in the tank is 0.5 m the inflow and outflow are directly balanced.

As a result of a short interruption in the supply, the water level in the tank falls to 0.25 m above the bottom but is then restored again. How long will it take the level to rise to 0.45 m above the bottom?

6.16. The flowrate of air at 298 K in a 0.3 m diameter duct is measured with a pitot tube which is used to traverse the cross-section. Readings of the differential pressure recorded on a water manometer are taken with the pitot tube at ten different positions in the cross-section. These positions are so chosen as to be the midpoints of ten concentric annuli each of the same cross-sectional area. The readings are as follows:

Position	1	2	3	4	5
Manometer reading (mm water)	18.5	18.0	17.5	16.8	15.7
Position	6	7	8	9	10
Manometer reading (mm water)	14.7	13.7	12.7	11.4	10.2

The flow is also metered using a 15 cm orifice plate across which the pressure differential is 50 mm on a mercury-under-water manometer. What is the coefficient of discharge of the orifice meter?

6.17. Explain the principle of operation of the pitot tube and indicate how it can be used in order to measure the total flowrate of fluid in a duct.

If a pitot tube is inserted in a circular cross-section pipe in which a fluid is in streamline flow, calculate at what point in the cross-section it should be situated so as to give a direct reading representative of the mean velocity of flow of the fluid.

6.18. The flowrate of a fluid in a pipe is measured using a pitot tube, which gives a pressure differential equivalent to 40 mm of water when situated at the centre line of the pipe and 22.5 mm of water when midway between the axis and the wall. Show that these readings are consistent with streamline flow in the pipe.

6.19. Derive a relationship between the pressure difference recorded between the two orifices of a pitot tube and the velocity of flow of an incompressible fluid. A pitot tube is to be situated in a large circular duct in which fluid is in turbulent flow so that it gives a direct reading of the mean velocity in the duct. At what radius in the duct should it be located, if the radius of the duct is r?

The point velocity in the duct can be assumed to be proportional to the one-seventh power of the distance from the wall.

6.20. A gas of molecular mass 44 kg/kmol temperature 373 K and pressure 202.6 kN/m^2 is flowing in a duct. A pitot tube is located at the centre of the duct and is connected to a differential manometer containing water. If the differential reading is 38.1 mm water, what is the velocity at the centre of the duct?

The volume occupied by 1 kmol at 273 K and 101.3 kN/m^2 is 22.4 m^3.

6.21. Glycerol, of density 1260 kg/m^3 and viscosity 50 mN s/m^2, is flowing through a 50 mm pipe and the flowrate is measured using an orifice meter with a 38 mm orifice. The pressure differential is 150 mm as indicated on a manometer filled with a liquid of the same density as the glycerol. There is reason to suppose that the orifice meter may have become partially blocked and that the meter is giving an erroneous reading. A check is therefore made by inserting a pitot tube at the centre of the pipe. It gives a reading of 100 mm on a water manometer. What does this suggest?

6.22. The flowrate of air in a 305 mm diameter duct is measured with a pitot tube which is used to traverse the cross-section. Readings of the differential pressure recorded on a water manometer are taken with the pitot tube at ten different positions in the cross-section. These positions are so chosen as to be the mid-points of ten concentric annuli each of the same cross-sectional area. The readings are as follows:

Position	1	2	3	4	5
Manometer reading (mm water)	18.5	18.0	17.5	16.8	15.8
Position	6	7	8	9	10
Manometer reading	14.7	13.7	12.7	11.4	10.2

The flow is also metered using a 50 mm orifice plate across which the pressure differential is 150 mm on a mercury-under-water manometer. What is the coefficient of discharge of the orifice meter?

6.23. The flow of liquid in a 25 mm diameter pipe is metered with an orifice meter in which the orifice has a diameter of 19 mm. The aperture becomes partially blocked with dirt from the liquid. What fraction of the area can become blocked before the error in flowrate at a given pressure differential exceeds 15 per cent? Assume that the coefficient of discharge of the meter remains constant when calculated on the basis of the actual free area of the orifice.

6.24. Water is flowing through a 100 mm diameter pipe and its flowrate is metered by means of a 50 mm diameter orifice across which the pressure drop is 13.8 kN/m^2. A second stream, flowing through a 75 mm diameter pipe, is also metered using a 50 mm diameter orifice across which the pressure differential is 150 mm measured on a mercury-under-water manometer. The two streams join and flow through a 150 mm diameter pipe. What would you expect the pressure difference to be across a 75 mm diameter orifice, measured using a mercury-under-water manometer, inserted in this pipe?

The coefficients of discharge for all the orifice meters are equal.

(Density of mercury = 13600 kg/m^3)

6.25. Water is flowing through a 150 mm diameter pipe and its flowrate is measured by means of a 50 mm diameter orifice, across which the pressure differential is 2.27×10^4 N/m^2. The coefficient of discharge of the orifice meter is independently checked by means of a pitot tube which, when situated at the axis of the pipe, gave a reading of 15.6 mm on a differential water manometer. On the assumption that the flow in the pipe is turbulent and that the velocity distribution over the cross-section is given by the Prandtl one-seventh power law, calculate the coefficient of discharge of the orifice meter.

6.26. Air at 323 K and 152 kN/m^2 pressure flows through a duct of circular cross-section, diameter 0.5 m. In order to measure the flow rate of air, the velocity profile across a diameter of the duct is measured using a Pitot-static tube connected to a water manometer inclined at an angle of $\cos^{-1} 0.1$ to the vertical. The following results are obtained:

Distance from duct centre line (m)	Manometer Reading (mm)
0	104
0.05	100
0.10	96
0.15	86
0.175	79
0.20	68
0.225	50

From the measurements calculate the mass flow rate of air through the duct, the average velocity, the ratio of the average to the maximum velocity and the Reynolds number. Comment on these results.

Discuss the application of this method of measuring gas flow rates with particular emphasis on the best distribution of experimental points across the duct and on the accuracy of the results.

(Take the viscosity of air as 1.9×10^{-2} mN s/m^2 and the molecular weight of air as 29 kg/kmol.)

7.1. A reaction is to be carried out in an agitated vessel. Pilot-plant experiments were performed under fully turbulent conditions in a tank 0.6 m in diameter, fitted with baffles and provided with a flat-bladed turbine. It was found that the satisfactory mixing was obtained at a rotor speed of 4 Hz, when the power consumption was 0.15 kW and the Reynolds number 160,000. What should be the rotor speed in order to retain the same mixing performance if the linear scale of the equipment is increased 6 times? What will be the power consumption and the Reynolds number?

7.2. A three-bladed propeller is used to mix a fluid in the laminar region. The stirrer is 0.3 m in diameter and is rotated at 1.5 Hz. Due to corrosion, the propeller has to be replaced by a flat two-bladed paddle, 0.75 m in diameter. If the same motor is used, at what speed should the paddle rotate?

7.3. Compare the capital and operating costs of a three-bladed propeller with those of a constant speed six-bladed turbine, both constructed from mild steel. The impeller diameters are 0.3 and 0.45 m respectively and both stirrers are driven by a 1 kW motor. What is the recommended speed of rotation in each case? Assume operation for 8000 hr/year, power at £0.01/kWh and interest and depreciation at 15%/year.

7.4. In a leaching operation, the rate at which solute goes into solution is given by an equation of the form:

$$\frac{dM}{dt} = k(c_s - c) \text{ kg/s}$$

where M kg is the amount of solute dissolving in t s, k m^3/s is a constant and c_s and c are the saturation and bulk concentrations of the solute respectively in kg/m^3. In a pilot test on a vessel 1 m^3 in volume, 75% saturation was attained in 10 s. If 300 kg of a solid containing 28% by mass of a water soluble solid is agitated with 100 m^3 of water, how long will it take for all the solute to dissolve assuming conditions are the same as in the pilot unit? Water is saturated with the solute at a concentration of 2.5 kg/m^3.

7.5. For producing an oil-water emulsion, two portable three-bladed propeller mixers are available; a 0.5 m diameter impeller rotating at 1 Hz and a 0.35 m impeller rotating at 2 Hz. Assuming turbulent conditions prevail, which unit will have the lower power consumption?

7.6. A reaction is to be carried out in an agitated vessel. Pilot-plant experiments were performed under fully turbulent conditions in a tank 0.6 m in diameter, fitted with baffles and provided with a flat-bladed turbine. It was found that satisfactory mixing was obtained at a rotor speed of 4 Hz, when the power consumption was 0.15 kW and the Reynolds number 160,000. What should be the rotor speed in order to retain the same mixing performance if the linear scale of the equipment is increased 6 times? What will be the power consumption and the Reynolds number?

7.7. Tests on a small scale tank 0.3 m diameter (Rushton impeller diameter 0.1 m) have shown that a blending process between two miscible liquids (aqueous solutions, properties approximately the same as water, i.e. viscosity 1 mN s/m^2, density 1000 kg/m^3) is satisfactorily completed after 1 minute using an impeller speed of 250 rev/min. It is decided to scale up the process to a tank of 2.5 m diameter using the criterion of constant tip-speed.

(a) What speed should be chosen for the larger impeller?
(b) What power will be required?
(c) What will be the blend time in the large tank?

7.8. An agitated tank with a standard Rushton impeller is required to disperse gas in a solution of properties similar to those of water. The tank will be 3 m diameter (1 m diameter impeller). A power level of 0.8 kW/m^3 is chosen. Assuming fully turbulent conditions and that the presence of the gas does not significantly affect the relation between the Power and Reynolds numbers:

(a) What power will be required by the impeller?
(b) At what speed should the impeller be driven?
(c) If a small pilot scale tank 0.3 m diameter is to be constructed to test the process, at what speed should the impeller be driven?

8.1. A three-stage compressor is required to compress air from 140 kN/m^2 and 283 K to 4000 kN/m^2. Calculate the ideal intermediate pressures, the work required per kilogram of gas, and the isothermal efficiency of the process. Assume the compression to be adiabatic and the interstage cooling to cool the air to the initial temperature. Show qualitatively, by means of temperature-entropy diagrams, the effect of unequal work distribution and imperfect intercooling, on the performance of the compressor.

8.2. A twin-cylinder, single-acting compressor, working at 5 Hz, delivers air at 515 kN/m^2 pressure, at the rate of 0.2 m^3/s. If the diameter of the cylinder is 20 cm, the cylinder clearance ratio 5% and the temperature of the inlet air 283 K, calculate the length of stroke of the piston and the delivery temperature.

8.3. A single-stage double-acting compressor running at 3 Hz is used to compress air from 110 kN/m^2 and 282 K to 1150 kN/m^2. If the internal diameter of the cylinder is 20 cm, the length of stroke 25 cm and the piston clearance 5%, calculate (a) the maximum capacity of the machine, referred to air at the initial temperature and pressure, and (b) the theoretical power requirements under isentropic conditions.

8.4. Methane is to be compressed from atmospheric pressure to 30 MN/m^2 in four stages.
   Calculate the ideal intermediate pressures and the work required per kilogram of gas. Assume compression to be isentropic and the gas to behave as an ideal gas. Indicate on a temperature-entropy diagram the effect of imperfect intercooling on the work done at each stage.

8.5. An air-lift raises 0.01 m^3/s of water from a well 100 m deep through a 100 mm diameter pipe. The level of the water is 40 m below the surface. The air consumed is 0.1 m^3/s of free air compressed to 800 kN/m^2.
   Calculate the efficiency of the pump and the mean velocity of the mixture in the pipe.

8.6. In a single-stage compressor:

$$\text{Suction pressure} = 101.3 \text{ kN/m}^2.$$
$$\text{Suction temperature} = 283 \text{ K.}$$

$$\text{Final pressure} \ = \ 380 \ \text{kN/m}^2.$$
$$\text{Compression is adiabatic.}$$

If each new charge is heated 18 K by contact with the clearance gases, calculate the maximum temperature attained in the cylinder.

8.7. A single-acting reciprocating pump has a cylinder diameter of 115 mm and a stroke of 230 mm. The suction line is 6 m long and 50 mm in diameter, and the level of the water in the suction tank is 3 m below the cylinder of the pump. What is the maximum speed at which the pump can run without an air vessel if separation is not to occur in the suction line? The piston undergoes approximately simple harmonic motion. Atmospheric pressure is equivalent to a head of 10.4 m of water and separation occurs at a pressure corresponding to a head of 1.22 m of water.

8.8. An air-lift pump is used for raising 800 cm^3/s of a liquid of specific gravity 1.2 to a height of 20 m. Air is available at 450 kN/m^2. If the efficiency of the pump is 30%, calculate the power requirement, assuming isentropic compression of the air ($\gamma = 1.4$).

8.9. A single-acting air compressor supplies 0.1 m^3/s of air (at STP) compressed to 380 kN/m^2 from 101.3 kN/m^2 pressure. If the suction temperature is 288.5 K, the stroke is 250 mm, and the speed is 4 Hz, find the cylinder diameter. Assume the cylinder clearance is 4% and compression and re-expansion are isentropic ($\gamma = 1.4$). What is the theoretical power required for the compression?

8.10. Air at 290 K is compressed from 101.3 to 2000 kN/m^2 pressure in a two-stage compressor operating with a mechanical efficiency of 85%. The relation between pressure and volume during the compression stroke and expansion of the clearance gas is $PV^{1.25} = $ constant. The compression ratio in each of the two cylinders is the same and the interstage cooler may be taken as perfectly efficient. If the clearances in the two cylinders are 4% and 5% respectively, calculate:

(a) the work of compression per unit mass of gas compressed;
(b) the isothermal efficiency;
(c) the isentropic efficiency ($\gamma = 1.4$);
(d) the ratio of the swept volumes in the two cylinders.

8.11. Explain briefly the significance of the "specific speed" of a centrifugal or axial-flow pump.

A pump is designed to be driven at 10 Hz and to operate at a maximum efficiency when delivering 0.4 m^3/s of water against a head of 20 m. Calculate the specific speed. What type of pump does this value suggest?

A pump, built for these operating conditions, has a measured maximum overall efficiency of 70%. The same pump is now required to deliver water at 30 m head. At what speed should the pump be driven if it is to operate at maximum efficiency? What will be the new rate of delivery and the power required?

8.12. A centrifugal pump is to be used to extract water from a condenser in which the vacuum is 640 mm of mercury. At the rated discharge the net positive suction head must be at least 3 m above the cavitation vapour pressure of 710 mm mercury vacuum. If losses in the suction pipe account for a head of 1.5 m, what must be the least height of the liquid level in the condenser above the pump inlet?

8.13. What is meant by the Net Positive Suction Head (NPSH) required by a pump? Explain why it exists and how it can be made as low as possible. What happens if the necessary NPSH is not provided?

A centrifugal pump is to be used to circulate liquid (sp. gr. 0.80 and viscosity 0.5 mN s/m^2) from the reboiler of a distillation column through a vaporiser at the rate of 400 cm^3/s, and to introduce the superheated liquid above the vapour space in the reboiler which contains liquid to a depth of 0.7 m. Suggest a suitable layout if a smooth bore 25 mm pipe is to be used. The pressure of the vapour in the reboiler is 1 kN/m^2 and the NPSH required by the pump is 2 m of liquid.

8.14. 1250 cm^3/s of water is to be pumped through a steel pipe, 25 mm diameter and 30 m long, to a tank 12 m higher than its reservoir. Calculate approximately the power required. What type of pump would you install for the purpose and what power motor (in kW) would you provide?

$$\text{Viscosity of water} \ = 1.30 \ \text{mN s/m}^2.$$

$$\text{Density of water} \ = 1000 \ \text{kg/m}^3.$$

8.15. Calculate the pressure drop in, and the power required to operate, a condenser consisting of 400 tubes 4.5 m long and 10 mm internal diameter. The coefficient of contraction at the entrance of the tubes is 0.6, and 0.04 m^3/s of water is to be pumped through the condenser.

8.16. 75% sulphuric acid, of density 1650 kg/m^3 and viscosity 8.6 mN s/m^2, is to be pumped for 0.8 km along a 50 mm internal diameter pipe at the rate of 3.0 kg/s, and then raised vertically 15 m by the pump. If the pump is electrically driven and has an efficiency of 50%, what power will be required? What type of pump would you use and of what material would you construct the pump and pipe?

8.17. 60% sulphuric acid is to be pumped at the rate of 4000 cm^3/s through a lead pipe 25 mm diameter and raised to a height of 25 m. The pipe is 30 m long and includes two right-angled bends. Calculate the theoretical power required.

The specific gravity of the acid is 1.531 and its kinematic viscosity is 0.425 cm^2/s. The density of water may be taken as 1000 kg/m^3.

8.18. 1.3 kg/s of 98% sulphuric acid is to be pumped through a 25 mm diameter pipe, 30 m long, to a tank 12 m higher than its reservoir. Calculate the power required and indicate the type of pump and material of construction of the line that you would choose.

$$\text{Viscosity of acid} = 0.025 \text{ N s/m}^2.$$

$$\text{Specific gravity} = 1.84.$$

8.19. A petroleum fraction is pumped 2 km from a distillation plant to storage tanks through a mild steel pipeline, 150 mm in diameter, at the rate of 0.04 m^3/s. What is the pressure drop along the pipe and the power supplied to the pumping unit if it has an efficiency of 50%?

The pump impeller is eroded and the pressure at its delivery falls to one half. By how much is the flowrate reduced?

$$\text{Specific gravity of the liquid} = 0.705.$$

$$\text{Viscosity of the liquid} = 0.5 \text{ mN s/m}^2.$$

$$\text{Roughness of pipe surface} = 0.004 \text{ mm}.$$

8.20. Calculate the power required to pump oil of specific gravity 0.85 and viscosity 3 mN s/m^2 at 4000 cm^3/s through a 50 mm pipeline 100 m long, the outlet of which is 15 m higher than the inlet. The efficiency of the pump is 50%. What effect does the nature of the surface of the pipe have on the resistance?

8.21. 600 cm^3/s of water at 320 K is pumped in a 40 mm i.d. pipe through a length of 150 m in a horizontal direction and up through a vertical height of 10 m. In the pipe there is a control valve which may be taken as equivalent to 200 pipe diameters and other pipe fittings equivalent to 60 pipe diameters. Also in the line there is a heat exchanger across which there is a loss in head of 1.5 m of water. If the main pipe has a roughness of 0.0002 m, what power must be delivered to the pump if the unit is 60% efficient?

8.22. A pump developing a pressure of 800 kN/m^2 is used to pump water through a 150 mm pipe 300 m long to a reservoir 60 m higher. With the valves fully open, the flowrate obtained is 0.05 m^3/s. As a result of corrosion and scaling the effective absolute roughness of the pipe surface increases by a factor of 10. By what percentage is the flowrate reduced?

$$\text{Viscosity of water} = 1 \text{ mN s/m}^2.$$

9.1. Calculate the time taken for the distant face of a brick wall, of thermal diffusivity, $D_H = 0.0042$ cm^2/s and thickness $l = 0.45$ m, initially at 290 K, to rise to 470 K if the near face is suddenly raised to a temperature of $\theta' = 870$ K and maintained at that temperature. Assume that all the heat flow is perpendicular to the faces of the wall and that the distant face is perfectly insulated.

9.2. Calculate the time for the distant face to reach 470 K under the same conditions, except that the distant face is not perfectly lagged. Instead, a very large thickness of material of the same thermal properties as the brickwork is stacked against it.

9.3. Benzene vapour, at atmospheric pressure, condenses on a plane surface 2 m long and 1 m wide, maintained at 300 K and inclined at an angle of $45°$ to the horizontal. Plot the thickness of the condensate film and the point heat transfer coefficient against distance from the top of the surface.

9.4. It is desired to warm 0.9 kg/s of air from 283 to 366 K by passing it through the pipes of a bank consisting of 20 rows with 20 pipes in each row. The arrangement is in-line with centre to centre spacing, in both directions, equal to twice the pipe diameter. Flue gas, entering at 700 K and leaving at 366 K with a free flow mass velocity of 10 kg/m²s, is passed across the outside of the pipes.

Neglecting gas radiation, how long should the pipes be?

For simplicity, outer or inner pipe diameter may be taken as 12 mm.

Values of k and μ, which may be used for both air and flue gases, are given below. The specific heat of air and flue gases is 1.0 kJ/kg K.

Temperature (K)	Thermal conductivity k (W/m K)	Viscosity μ (mN s/m²)
250	0.022	0.0165
500	0.040	0.0276
800	0.055	0.0367

9.5. A cooling coil, consisting of a single length of tubing through which water is circulated, is provided in a reaction vessel, the contents of which are kept uniformly at 360 K by means of a stirrer. The inlet and outlet temperatures of the cooling water are 280 and 320 K respectively. What would the outlet water temperature become if the length of the cooling coil were increased 5 times? Assume the overall heat transfer coefficient to be constant over the length of the tube and independent of the water temperature.

9.6. In an oil cooler 216 kg/h of hot oil enters a thin metal pipe of diameter 25 mm. An equal mass of cooling water flows through the annular space between the pipe and a larger concentric pipe, the oil and water moving in opposite directions. The oil enters at 420 K and is to be cooled to 320 K. If the water enters at 290 K, what length of pipe will be required? Take coefficients of 1.6 kW/m² K on the oil side and 3.6 kW/m² K on the water side and 2.0 kJ/kg K for the specific heat of the oil.

9.7. The walls of a furnace are built up to 150 mm thickness of a refractory of thermal conductivity 1.5 W/m K. The surface temperatures of the inner and outer faces of the refractory are 1400 and 540 K respectively.

If a layer of insulating material 25 mm thick, of thermal conductivity 0.3 W/m K, is added, what temperatures will its surfaces attain assuming the inner surface of the furnace to remain at 1400 K? The coefficient of heat transfer from the outer surface of the insulation to the surroundings, which are at 290 K, may be taken as 4.2, 5.0, 6.1, and 7.1 W/m K, for surface temperatures of 370, 420, 470, and 520 K respectively. What will be the reduction in heat loss?

9.8. A pipe of outer diameter 50 mm, maintained at 1100 K, is covered with 50 mm of insulation of thermal conductivity 0.17 W/m K.

Would it be feasible to use a magnesia insulation which will not stand temperatures above 615 K and has a thermal conductivity 0.09 W/m K for an additional layer thick enough to reduce the outer surface temperature to 370 K in surroundings at 280 K? Take the surface coefficient of heat transfer by radiation and convection as 10 W/m² K.

9.9. In order to warm 0.5 kg/s of a heavy oil from 311 to 327 K. it is passed through tubes of inside diameter 19 mm and length 1.5 m, forming a bank, on the outside of which steam is condensing at 373 K. How many tubes will be needed?

In calculating Nu, Pr, and Re, the thermal conductivity of the oil may be taken as 0.14 W/m K and the specific heat as 2.1 kJ/kg K, irrespective of temperature. The viscosity is to be taken at the mean oil temperature. Viscosity of the oil at 319 and 373 K is 154 and 19.2 mN s/m² respectively.

9.10. A metal pipe of 12 mm outer diameter is maintained at 420 K. Calculate the rate of heat loss per metre run in surroundings uniformly at 290 K, (a) when the pipe is covered with 12 mm thickness of a material of thermal conductivity 0.35 W/m K and surface emissivity 0.95, and (b) when the thickness of the covering material is reduced to 6 mm, but the outer surface is treated so as to reduce its emissivity to 0.10.

The coefficients of radiation from a perfectly black surface in surroundings at 290 K are 6.25, 8.18, and 10.68 W/m^2 K at 310, 370, and 420 K respectively.

The coefficients of convection may be taken as $1.22(\theta/d)^{0.25}$ W/m^2 K, where θ (K) is the temperature difference between the surface and the surrounding air, and d (m) is the outer diameter.

9.11. A condenser consists of 30 rows of parallel pipes of outer diameter 230 mm and thickness 1.3 mm, with 40 pipes, each 2 m long, per row. Water, inlet temperature 283 K, flows through the pipes at 1 m/s, and steam at 372 K condenses on the outside of the pipes. There is a layer of scale 0.25 mm thick, of thermal conductivity 2.1 W/m K, on the inside of the pipes.

Taking the coefficients of heat transfer on the water side as 4.0, and on the steam side as 8.5 kW/m^2 K, calculate the outlet water temperature and the total weight of steam condensed per second. The latent heat of steam at 372 K is 2250 kJ/kg. 1 m^3 water weighs 1000 kg.

9.12. In an oil cooler, water flows at the rate of 360 kg/h per tube through metal tubes of outer diameter 19 mm and thickness 1.3 mm, along the outside of which oil flows in the opposite direction at the rate of 75 g/s per tube.

If the tubes are 2 m long, and the inlet temperatures of the oil and water are respectively 370 and 280 K, what will be the outlet oil temperature? The coefficient of heat transfer on the oil side is 1.7 and on the water side 2.5 kW/m^2 K, and the specific heat of the oil is 1.9 kJ/kg K.

9.13. Waste gases flowing across the outside of a bank of pipes are being used to warm air which flows through the pipes. The bank consists of 12 rows of pipes with 20 pipes, each 0.7 m long, per row. They are arranged in-line, with centre-to-centre spacing equal in both directions to one-and-a-half times the pipe diameter. Both inner and outer diameter may be taken as 12 mm. Air, with mass velocity 8 kg/m^{2}s, enters the pipes at 290 K. The initial gas temperature is 480 K and the total weight of the gases crossing the pipes per second is the same as the total weight of the air flowing through them.

Neglecting gas radiation, estimate the outlet temperature of the air. The physical constants for the waste gases may be assumed the same as for air, given below:

Temperature (K)	Thermal conductivity (W/m K)	Viscosity (mN/ s/m^2)
250	0.022	0.0165
310	0.027	0.0189
370	0.030	0.0214
420	0.033	0.0239
480	0.037	0.0260

Specific heat = 1.00 kJ/kg K.

9.14. Oil is to be warmed from 300 to 344 K by passing it at 1 m/s through the pipes of a shell-and-tube heat exchanger. Steam at 377 K condenses on the outside of the pipes, which have outer and inner diameters of 48 and 41 mm respectively; but, owing to fouling, the inside diameter has been reduced to 38 mm, and the resistance to heat transfer of the pipe wall and dirt together, based on this diameter, is 0.0009 m^2 K/W.

It is known from previous measurements under similar conditions that the oil side coefficients of heat transfer for a velocity of 1 m/s. based on a diameter of 38 mm. vary with the temperature of the oil according to the table below:

Oil temperature (K)	300	311	322	333	344
Oil side coefficient of heat transfer (W/m^2 K)	74	80	97	136	244

The specific heat and density of the oil may be assumed constant at 1.9 kJ/kg K and 900 kg/m^3 respectively, and any resistance to heat transfer on the steam side neglected.

Find the length of tube bundle required.

9.15. It is proposed to construct a heat exchanger to condense 7.5 kg/s of n-hexane at a pressure of 150 kN/m^2, involving a heat load of 4.5 MW. The hexane is to reach the condenser from the top of a fractionating column at its condensing temperature of 356 K.

From experience it is anticipated that the overall heat transfer coefficient will be 450 W/m^2 K. The available cooling water is at 289 K.

Outline the proposals that you would make for the type and size of the exchanger and explain the details of the mechanical construction that you consider require special attention.

9.16. A heat exchanger is to be mounted at the top of a fractionating column about 15 m high to condense 4 kg/s of n-pentane at 205 kN/m^2, corresponding to a condensing temperature of 333 K. Give an outline of the calculations you would make to obtain an approximate idea of the size and construction of the exchanger required.

For purposes of standardisation, the company will use 19 mm outer diameter tubes of 1.65 mm wall thickness, and these may be 2.5, 3.6, or 5 m in length. The film coefficient for condensing pentane on the outside of a horizontal tube bundle may be taken as 1.1 kW/m^2 K. The condensation is effected by pumping water through the tubes, the initial water temperature being 288 K.

The latent heat of condensation of pentane is 335 kJ/kg.

For these 19 mm tubes, a water velocity of 1 m/s corresponds to a flowrate of 200 g/s of water.

9.17. An organic liquid is boiling at 340 K on the inside of a metal surface of thermal conductivity 42 W/m K and thickness 3 mm. The outside of the surface is heated by condensing steam. Assuming that the heat transfer coefficient from steam to the outer metal surface is constant at 11 kW/m^2 K, irrespective of the steam temperature, find what value of the steam temperature would give a maximum rate of evaporation.

The coefficients of heat transfer from the inner metal surface to the boiling liquid depend upon the temperature difference as shown below:

Temperature difference metal surface to boiling liquid (K)	Heat transfer coefficient metal surface to boiling liquid (kW/m^2 K)
22.2	4.43
27.8	5.91
33.3	7.38
36.1	7.30
38.9	6.81
41.7	6.36
44.4	5.73
50.0	4.54

9.18. It is desired to warm an oil of specific heat 2.0 kJ/kg K from 300 to 325 K by passing it through a tubular heat exchanger with metal tubes of inner diameter 10 mm. Along the outside of the tubes flows water, inlet temperature 372 K and outlet temperature 361 K.

The overall heat transfer coefficient from water to oil, reckoned on the inside area of the tubes, may be assumed constant at 230 W/m^2 K, and 75 g/s of oil is to be passed through each tube.

The oil is to make two passes through the heater. The water makes one pass along the outside of the tubes. Calculate the length of the tubes required.

9.19. A condenser consists of a number of metal pipes of outer diameter 25 mm and thickness 2.5 mm. Water, flowing at 0.6 m/s, enters the pipes at 290 K, and it is not permissible that it should be discharged at a temperature in excess of 310 K.

If 1.25 kg/s of a hydrocarbon vapour is to be condensed at 345 K on the outside of the pipes, how long should each pipe be and how many pipes should be needed?

Take the coefficient of heat transfer on the water side as 2.5 and on the vapour side as 0.8 kW/m^2 K and assume that the overall coefficient of heat transfer from vapour to water, based upon these figures, is reduced 20% by the effects of the pipe walls, dirt, and scale.

The latent heat of the hydrocarbon vapour at 345 K is 315 kJ/kg.

9.20. An organic vapour is being condensed at 350 K on the outside of a nest of pipes through which water flows at 0.6 m/s, its inlet temperature being 290 K. The outer and inner diameters of the pipes are 19 mm and 15 mm respectively, but a layer of scale 0.25 mm thick and thermal conductivity 2.0 W/m K has formed on the inside of the pipes.

If the coefficients of heat transfer on the vapour and water sides respectively are 1.7 and 3.2 kW/m^2 K and it is required to condense 25 g/s of vapour on each of the pipes, how long should these be and what will be the outlet temperature of the water?

The latent heat of condensation is 330 kJ/kg.

Neglect any small resistance to heat transfer in the pipe walls.

9.21. A heat exchanger is required to cool continuously 20 kg/s of warm water from 360 to 335 K by means of 25 kg/s of cold water, inlet temperature 300 K.

Assuming that the water velocities are such as to give an overall coefficient of heat transfer of 2 kW/m^2 K. assumed constant, calculate the total area of surface required (a) in a counterflow heat exchanger, i.e. one in which the hot and cold fluids flow in opposite directions, and (b) in a multipass heat exchanger, with the cold water making two passes through the tubes and the hot water making one pass along the outside of the tubes. In case (b) assume that the hot water flows in the same direction as the inlet cold water and that its temperature over any cross-section is uniform.

9.22. Find the heat loss per m^2 of surface through a brick wall 0.5 m thick when the inner surface is at 400 K and the outside at 310 K: the thermal conductivity of the brick may be taken as 0.7 W/m K.

9.23. A furnace is constructed with 225 mm of firebrick, 120 mm of insulating brick, and 225 mm of building brick. The inside temperature is 1200 K and the outside temperature 330 K. If the thermal conductivities are 1.4, 0.2, and 0.7 W/m K, find the heat loss per unit area and the temperature at the junction of the firebrick and insulating brick.

9.24. Calculate the total heat loss by radiation and convection from an unlagged horizontal steam pipe of 50 mm outside diameter at 415 K to air at 290 K.

9.25. Toluene is continuously nitrated to mononitrotoluene in a cast-iron vessel of 1 m diameter fitted with a propeller agitator of 0.3 m diameter driven at 2 Hz. The temperature is maintained at 310 K by circulating cooling water at 0.5 kg/s through a stainless steel coil of 25 mm outside diameter and 22 mm inside diameter wound in the form of a helix of 0.81 m diameter. The conditions are such that the reacting material may be considered to have the same physical properties as 75% sulphuric acid. If the mean water temperature is 290 K, what is the overall heat transfer coefficient?

9.26. 7.5 kg/s of pure iso-butane is to be condensed at a temperature of 331.7 K in a horizontal tubular exchanger using a water inlet temperature of 301 K. It is proposed to use 19 mm outside diameter tubes of 1.6 mm wall arranged on a 25 mm triangular pitch. Under these conditions the resistance of the scale may be taken as 0.0005 m^2 K/W. It is required to determine the number and arrangement of the tubes in the shell.

9.27. 37.5 kg/s of crude oil is to be heated from 295 to 330 K by heat exchange with the bottom product from a distillation column. The bottom product, flowing at 29.6 kg/s, is to be cooled from 420 to 380 K. There is available a tubular exchanger with an inside shell diameter of 0.60 m having one pass on the shell side and two passes on the tube side. It has 324 tubes, 19 mm outside diameter with 2.1 mm wall and 3.65 m long, arranged on a 25 mm square pitch and supported by baffles with a 25% cut, spaced at 230 mm intervals. Would this exchanger be suitable?

9.28. A 150 mm internal diameter steam pipe is carrying steam at 444 K and is lagged with 50 mm of 85% magnesia. What will be the heat loss to the air at 294 K?

9.29. A refractory material which has an emissivity of 0.40 at 1500 K and 0.43 at 1420 K is at a temperature of 1420 K and is exposed to black furnace walls at a temperature of 1500 K. What is the rate of gain of heat by radiation per unit area?

9.30. The total emissivity of clean chromium as a function of surface-temperature T K is given approximately by the empirical expression:

$$e = 0.38 \left[1 - \frac{263}{T} \right].$$

Obtain an expression for the absorptivity of solar radiation as a function of surface temperature and compare the absorptivity and emissivity at 300, 400, and 1000 K.

Assume that the sun behaves as a black body at 5500 K.

9.31. Repeat question 9.30 for the case of aluminium, assuming the emissivity to be 1.25e.

9.32. Calculate the heating due to solar radiation on the flat concrete roof of a building, 8 m by 9 m, in Africa, if the surface temperature of the roof is 330 K. What would be the effect of covering the roof with a highly reflecting surface such as polished aluminium?

The total emissivity of concrete at 330 K is 0.89, whilst the total absorptivity of solar radiation (sun temperature = 5500 K) at this temperature is 0.60. Use the data of question 9.31 for aluminium.

9.33. A rectangular iron ingot 15 cm × 15 cm × 30 cm is supported at the centre of a reheating furnace. The furnace has walls of silica-brick at 1400 K, and the initial temperature of the ingot is 290 K. How long will it take to heat the ingot to 600 K?

It may be assumed that the furnace is large compared with the ingot-size, and that the ingot remains at uniform temperature throughout its volume; convection effects are negligible.

The total emissivity of the oxidised iron surface is 0.78 and both emissivity and absorptivity are independent of the surface temperature.

$$\text{Density of iron} = 7.2 \text{ Mg/m}^3.$$

$$\text{Specific heat of iron} = 0.50 \text{ kJ/kg K}.$$

9.34. A wall is made of brick, of thermal conductivity 1.0 W/m K, 230 mm thick, lined on the inner face with plaster of thermal conductivity 0.4 W/m K and of thickness 10 mm. If a temperature difference of 30 K is maintained between the two outer faces, what is the heat flow per unit area of wall?

9.35. A 50 mm diameter pipe of circular cross-section and with walls 3 mm thick is covered with two concentric layers of lagging, the inner layer having a thickness of 25 mm and a thermal conductivity of 0.08 W/m K, and the outer layer has a thickness of 40 mm and a thermal conductivity of 0.04 W/m K. What is the rate of heat loss per metre length of pipe if the temperature inside the pipe is 550 K and the outside surface temperature is 330 K?

9.36. The temperature of oil leaving a co-current flow cooler is to be reduced from 370 to 350 K by lengthening the cooler. The oil and water flow rates and inlet temperatures, and the other dimensions of the cooler, will remain constant. The water enters at 285 K and the oil at 420 K. The water leaves the original cooler at 310 K. If the original length is 1 m, what must be the new length?

9.37. In a countercurrent-flow heat exchanger, 1.25 kg/s of benzene (specific heat 1.9 kJ/kg K and specific gravity 0.88) is to be cooled from 350 to 300 K with water which is available at 290 K. In the heat exchanger, tubes of 25 mm external and 22 mm internal diameter are employed and the water passes through the tubes. If the film coefficients for the water and benzene are 0.85 and 1.70 kW/m^2 K respectively and the scale resistance can be neglected, what total length of tube will be required if the minimum quantity of water is to be used and its temperature is not to be allowed to rise above 320 K?

9.38. Calculate the rate of loss of heat from a 6 m long horizontal steam pipe of 50 mm internal diameter and 60 mm external diameter when carrying steam at 800 kN/m^2. The temperature of the atmosphere and surroundings is 290 K.

What would be the cost of steam saved by coating the pipe with a 50 mm thickness of 85% magnesia lagging of thermal conductivity 0.07 W/m K, if steam costs £0.5 per 100 kg? The emissivity of the surface of the bare pipe and of the lagging may be taken as 0.85, and the coefficient h for heat loss by natural convection can be calculated from the expression:

$$h = 1.65(\Delta T)^{0.25} (\text{W/m}^2 \text{ K}),$$

where ΔT is the temperature difference in K.

Take the Stefan-Boltzmann constant as 5.67×10^{-3} W/m^2 K^4.

9.39. A stirred reactor contains a batch of 700 kg reactants of specific heat 3.8 kJ/kg K initially at 290 K, which is heated by dry saturated steam at 170 kN/m^2 fed to a helical coil. During the heating period the steam supply rate is constant at 0.1 kg/s and condensate leaves at the temperature of the steam. If heat losses are neglected, calculate the true temperature of the reactants when a thermometer immersed in the material reads 360 K. The bulb of the thermometer is approximately cylindrical and is 100 mm long by 10 mm diameter with a water equivalent of 15 g, and the overall heat transfer coefficient to the thermometer is 300 W/m^2 K. What would a thermometer with a similar bulb of half the length and half the heat capacity indicate under these conditions?

9.40. Derive an expression relating the pressure drop for the turbulent flow of a fluid in a pipe to the heat transfer coefficient at the walls on the basis of the simple Reynolds analogy. Indicate the assumptions which are made and the conditions under which you would expect it to apply closely. Air at 320 K and atmospheric pressure is flowing through a smooth pipe of 50 mm internal diameter, and the pressure drop over a 4 m length is found to be 150 mm water gauge. By how much would you expect the air temperature to fall over the first metre if the wall temperature there is 290 K?

$$\text{Viscosity of air} = 0.018 \text{ mN s/m}^2.$$

$$\text{Specific heat } (C_p) = 1.05 \text{ kJ/kg K.}$$

$$\text{Molecular volume} = 22.4 \text{ m}^3/\text{kmol at 1 bar 273K.}$$

9.41. The radiation received by the earth's surface on a clear day with the sun overhead is 1 kW/m² and an additional 0.3 kW/m² is absorbed by the earth's atmosphere. Calculate approximately the temperature of the sun, assuming its radius to be 700,000 km and the distance between the sun and the earth to be 150,000,000 km. The sun may be assumed to behave as a black body.

9.42. A thermometer is immersed in a liquid which is heated at the rate of 0.05 K/s. If the thermometer and the liquid are both initially at 290 K, what rate of passage of liquid over the bulb of the thermometer is required if the error in the thermometer reading after 600 s is to be no more than 1 K? Take the water equivalent of the thermometer as 30 g and the heat transfer coefficient to the bulb to be given by $U = 735 \, u^{0.8}$ W/m² K, and the area of the bulb as 0.01 m², where u is the velocity in m/s.

9.43. In a shell and tube type of heat exchanger with horizontal tubes 25 mm external diameter and 22 mm internal diameter, benzene is condensed on the outside by means of water flowing through the tubes at the rate of 0.03 m³/s. If the water enters at 290 K and leaves at 300 K and the heat transfer coefficient on the water side is 850 W/m² K, what total length of tubing will be required?

9.44. In a contact sulphuric acid plant, the gases leaving the first converter are to be cooled from 845 to 675 K by means of the air required for the combustion of the sulphur. The air enters the heat exchanger at 495 K. If the flow of each of the streams is 2 m³/s at NTP, suggest a suitable design for a shell and tube type of heat exchanger employing tubes of 25 mm internal diameter.

(a) Assume parallel co-current flow of the gas streams.
(b) Assume parallel countercurrent flow.
(c) Assume that the heat exchanger is fitted with baffles giving cross-flow outside the tubes.

9.45. A large block of material of thermal diffusivity $D_H = 0.0042$ cm²/s is initially at a uniform temperature of 290 K and one face is raised suddenly to 875 K and maintained at that temperature. Calculate the time taken for the material at a depth of 0.45 m to reach a temperature of 475 K on the assumption of unidirectional heat transfer and that the material can be considered to be infinite in extent in the direction of transfer.

9.46. A 50% glycerol-water mixture is flowing at a Reynolds number of 1500 through a 25 mm diameter pipe. Plot the mean value of the heat transfer coefficient as a function of pipe length assuming that:

$$Nu = 1.62 \left(RePr\frac{d}{l} \right)^{0.33}$$

Indicate the conditions under which this is consistent with the predicted value $Nu = 4.1$ for fully developed flow.

9.47. A liquid is boiled at a temperature of 360 K using steam fed at a temperature of 380 K to a coil heater. Initially the heat transfer surfaces are clean and an evaporation rate of 0.08 kg/s is obtained from each square metre of heating surface. After a period, a layer of scale of resistance 0.0003 m² K/W is deposited by the boiling liquid on the heat transfer surface. On the assumption that the coefficient on the steam side remains unaltered and that the coefficient for the boiling liquid is proportional to its temperature difference raised to the power of 2.5, calculate the new rate of boiling.

9.48. A batch of reactants of specific heat 3.8 kJ/kg K and of mass 1000 kg is heated by means of a submerged steam coil of area 1 m² fed with steam at 390 K. If the overall heat transfer coefficient is 600 W/m² K, calculate the time taken to heat the material from 290 to 360 K, if heat losses to the surroundings are neglected.

If the external area of the vessel is 10 m^2 and the heat transfer coefficient to the surroundings at 290 K is 8.5 W/m^2 K, what will be the time taken to heat the reactants over the same temperature range and what is the maximum temperature to which the reactants can be raised?

What methods would you suggest for improving the rate of heat transfer?

9.49. What do you understand by the terms "black body" and "grey body" when applied to radiant heat transfer?

Two large, parallel plates with grey surfaces are situated 75 mm apart; one has an emissivity of 0.8 and is at a temperature of 350 K and the other has an emissivity of 0.4 and is at a temperature of 300 K. Calculate the net rate of heat exchange by radiation per square metre taking the Stefan-Boltzmann constant as 5.67 × 10^{-8} W/m^2 K^4. Any formula (other than Stefan's law) which you use must be proved.

9.50. A longitudinal fin on the outside of a circular pipe is 75 mm deep and 3 mm thick. If the pipe surface is at 400 K, calculate the heat dissipated per metre length from the fin to the atmosphere at 290 K if the coefficient of heat transfer from its surface may be assumed constant at 5 W/m^2 K. The thermal conductivity of the material of the fin is 50 W/m K and the heat loss from the extreme edge of the fin may be neglected. It should be assumed that the temperature is uniformly 400 K at the base of the fin.

9.51. Liquid oxygen is distributed by road in large spherical insulated vessels, 2 m internal diameter, well lagged on the outside. What thickness of magnesia lagging, of thermal conductivity 0.07 W/m K must be used so that not more than 1% of the liquid oxygen evaporates during a journey of 10 ks if the vessel is initially 80% full?

$$\text{Latent heat of vaporisation of oxygen} = 215 \text{ kJ/kg.}$$
$$\text{Boiling point of oxygen} = 90 \text{ K.}$$
$$\text{Density of liquid oxygen} = 1140 \text{ kg/m}^3$$
$$\text{Atmospheric temperature} = 288 \text{ K.}$$
$$\text{Heat transfer coefficient from outside}$$
$$\text{lagging to atmosphere} = 4.5 \text{ W/m}^2 \text{ K.}$$

9.52. Benzene is to be condensed at the rate of 1.25 kg/s in a vertical shell and tube type of heat exchanger fitted with tubes of 25 mm outside diameter and 2.5 m long. The vapour condenses on the outside of the tubes and the cooling water enters at 295 K and passes through the tubes at 1.05 m/s. Calculate the number of tubes required if the heat exchanger is arranged for a single pass of the cooling water. The tube wall thickness is 1.6 mm.

9.53. One end of a metal bar 25 mm in diameter and 0.3 m long is maintained at 375 K and heat is dissipated from the whole length of the bar to surroundings at 295 K. If the coefficient of heat transfer from the surface is 10 W/m^2 K, what is the rate of loss of heat? Take the thermal conductivity of the metal as 85 W/m K.

9.54. A shell and tube heat exchanger consists of 120 tubes of internal diameter 22 mm and length 2.5 m. It is operated as a single pass condenser with benzene condensing at a temperature of 350 K on the outside of the tubes and water of inlet temperature 290 K passing through the tubes. Initially there is no scale on the walls and a rate of condensation of 4 kg/s is obtained with a water velocity of 0.7 m/s through the tubes. After prolonged operation, a scale of resistance 0.0002 m^2 K/W is formed on the inner surface of the tubes. To what value must the water velocity be increased in order to maintain the same rate of condensation on the assumption that the transfer coefficient on the water side is proportional to the velocity raised to the 0.8 power, and that the coefficient for the condensing vapour is 2.25 kW/m^2 K based on the inside area? The latent heat of vaporisation of benzene is 400 kJ/kg.

9.55. Derive an expression for the radiant heat transfer rate per unit area between two large parallel planes of emissivities e_1 and e_2 and at absolute temperatures T_1 and T_2 respectively.

Two such planes are situated 2.5 mm apart in air: one has an emissivity of 0.1 and is at a temperature of 350 K and the other has an emissivity of 0.05 and is at a temperature of 300 K. Calculate the percentage change in the total heat transfer rate by coating the first surface so as to reduce its emissivity to 0.025.

$$\text{Stefan-Boltzmann constant} = 5.67 \times 10^{-8} \text{ W/m}^2 \text{ K}^4.$$
$$\text{Thermal conductivity of air} = 0.026 \text{ W/m K.}$$

9.56. Water flows at 2 m/s through a 2.5 m length of a 25 mm diameter tube. If the tube is at 320 K and the water enters and leaves at 293 and 295 K respectively, what is the value of the heat transfer coefficient? How would the outlet temperature change if the velocity were increased by 50%?

9.57. A liquid hydrocarbon is fed at 295 K to a heat exchanger consisting of a 25 mm diameter tube heated on the outside by condensing steam at atmospheric pressure. The flow rate of the hydrocarbon is measured by means of a 19 mm orifice fitted to the 25 mm feed pipe. The reading on a differential manometer containing the hydrocarbon-over-water is 450 mm and the coefficient of discharge of the meter is 0.6.

Calculate the initial rate of rise of temperature (K/s) of the hydrocarbon as it enters the heat exchanger. Outside film coefficient = 6.0 kW/m² K. Inside film coefficient h is given by:

$$\frac{hd}{k} = 0.023 \left(\frac{ud\rho}{\mu}\right)^{0.8} \left(\frac{C_p\mu}{k}\right)^{0.4}$$

where: u = linear velocity of hydrocarbon (m/s).

d = tube diameter (m).

ρ = liquid density (800 kg/m³).

μ = liquid viscosity (9×10^{-4} Ns/m²).

C_p = specific heat of liquid (1.7×10^3 J/kg K).

k = thermal conductivity of liquid (0.17 W/m K).

9.58. Water passes at 1.2 m/s through a series of 25 mm diameter tubes 5 m long maintained at 320 K. If the inlet temperature is 290 K, at what temperature would you expect it to leave?

9.59. Heat is transferred from one fluid stream to a second fluid across a heat transfer surface. If the film coefficients for the two fluids are, respectively, 1.0 and 1.5 kW/m² K, the metal is 6 mm thick (thermal conductivity 20 W/m K) and the scale coefficient is equivalent to 850 W/m² K, what is the overall heat transfer coefficient?

9.60. A pipe of outer diameter 50 mm carries hot fluid at 1100 K. It is covered with a 50 mm layer of insulation of thermal conductivity 0.17 W/m K. Would it be feasible to use magnesia insulation, which will not stand temperatures above 615 K and has a thermal conductivity of 0.09 W/m K for an additional layer thick enough to reduce the outer surface temperature to 370 K in surroundings at 280 K? Take the surface coefficient of transfer by radiation and convection as 10 W/m² K.

9.61. A jacketed reaction vessel containing 0.25 m³ of liquid of specific gravity 0.9 and specific heat 3.3 kJ/kg K is heated by means of steam fed to a jacket on the walls. The contents of the tank are agitated by a stirrer running at 3 Hz. The heat transfer area is 2.5 m² and the steam temperature is 380 K. The outside film heat transfer coefficient is 1.7 kW/m² K and the 10 mm thick wall of the tank has a thermal conductivity of 6.0 W/m K. The inside film coefficient was found to be 1.1 kW/m² K for a stirrer speed of 1.5 Hz and to be proportional to the two-thirds power of the speed of rotation.

Neglecting heat losses and the heat capacity of the tank, how long will it take to raise the temperature of the liquid from 295 to 375 K?

9.62. By dimensional analysis, derive a relationship for the heat transfer coefficient h for natural convection between a surface and a fluid on the assumption that the coefficient is a function of the following variables:

k = thermal conductivity of the fluid.
C_p = specific heat of the fluid.
ρ = density of the fluid.
μ = viscosity of the fluid.
βg = the product of the acceleration due to gravity and the coefficient of cubical expansion of the fluid.
l = a characteristic dimension of the surface.
ΔT = the temperature difference between the fluid and the surface.

Indicate why each of these quantities would be expected to influence the heat transfer coefficient and explain how the orientation of the surface affects the process.

Under what conditions is heat transfer by convection important in chemical engineering?

9.63. A shell and tube heat exchanger is used for preheating the feed to an evaporator. The liquid of specific heat 4.0 kJ/kg K and specific gravity 1.1 passes through the inside of the tubes and is heated by steam condensing at 395 K on the outside. The exchanger heats liquid at 295 K to an outlet temperature of 375 K when the flow rate is 175 cm^3/s and to 370 K when the flowrate is 325 cm^3/s. What is the heat transfer area and the value of the overall heat transfer coefficient when the flow is 175 cm^3/s?

Assume that the film heat transfer coefficient for the liquid in the tubes is proportional to the 0.8 power of the velocity that the transfer coefficient for the condensing steam remains constant at 3.4 kW/m^2 K and that the resistance of the tube wall and scale can be neglected.

9.64. 0.1 m^3 of liquid of specific heat capacity 3 kJ/kg K and density 950 kg/m^3 is heated in an agitated tank fitted with a coil, of heat transfer area 1 m^2, supplied with steam at 383 K. How long will it take to heat the liquid from 293 to 368 K, if the tank, of external area 20 m^2 is losing heat to surroundings at 293 K? To what temperature will the system fall in 1800 s if the steam is turned off?

$$\text{Overall heat transfer coefficient in coil} = 2000 \text{ W/m}^2 \text{ K}$$
$$\text{Heat transfer coefficient to surroundings} = 10 \text{ W/m}^2 \text{ K}$$

9.65. The contents of a reaction vessel are heated by means of steam at 393 K supplied to a heating coil which is totally immersed in the liquid. When the vessel has a layer of lagging 50 mm thick on its outer surfaces, it takes one hour to heat the liquid from 293 to 373 K. How long will it take if the thickness of lagging is doubled?

$$\text{Outside temperature} = 293 \text{ K}$$
$$\text{Thermal conductivity of lagging} = 0.05 \text{ W/m K}$$

Coefficient for heat loss by radiation and convection from outside surface of vessel $= 10$ W/m^2 K.

$$\text{Outside area of vessel} = 8 \text{ m}^2$$
$$\text{Coil area} = 0.2 \text{ m}^2$$

Overall heat transfer coefficient for steam coil $= 300$ W/m^2 K.

9.66. A smooth tube in a condenser which is 25 mm internal diameter and 10 m long is carrying cooling water and the pressure drop over the length of the tube is 2×10^4 N/m^2. If vapour at a temperature of 353 K is condensing on the outside of the tube and the temperature of the cooling water rises from 293 K at inlet to 333 K at outlet, what is the value of the overall heat transfer coefficient based on the inside area of the tube? If the coefficient for the condensing vapour is 15000 W/m^2 K, what is the film coefficient for the water? If the latent heat of vaporisation is 800 kJ/kg, what is the rate of condensation of vapour?

9.67. A chemical reactor, 1 m in diameter and 5 m long, operates at a temperature of 1073 K. It is covered with a 500 mm thickness of lagging of thermal conductivity 0.1 W/m K. The heat loss from the cylindrical surface to the surroundings is 3.5 kW. What is the heat transfer coefficient from the surface of the lagging to the surroundings at a temperature of 293 K? How would the heat loss be altered if the coefficient were halved?

9.68. An open cylindrical tank 500 mm diameter and 1 m deep is three quarters filled with a liquid of density 980 kg/m^3 and of specific heat capacity 3 kJ/kg K. If the heat transfer coefficient from the cylindrical walls and the base of the tank is 10 W/m^2 K and from the surface is 20 W/m^2 K, what area of heating coil, fed with steam at 383 K, is required to heat the contents from 288 K to 368 K in a half hour? The overall heat transfer coefficient for the coil may be taken as 100 W/m^2 K. The surroundings are at 288 K. The heat capacity of the tank itself may be neglected.

9.69. Liquid oxygen is distributed by road in large spherical vessels, 1.82 m in internal diameter. If the vessels were unlagged and the coefficient for heat transfer from the outside of the vessel to the atmosphere were 5 W/m^2 K, what proportion of the contents would evaporate during a journey lasting an hour? Initially the vessels are 80% full.

What thickness of lagging would be required to reduce the losses to one tenth?

$$\text{Atmospheric temperature} = 288 \text{ K}$$
$$\text{Boiling point of oxygen} = 90 \text{ K,}$$

Density of oxygen $= 1140$ kg/m^3,
Latent heat of vaporisation of oxygen $= 214$ kJ/kg,
Thermal conductivity of lagging $= 0.07$ W/m K.

9.70. Water at 293 K is heated by passing through a 6.1 m coil of 25 mm internal diameter pipe; the thermal conductivity of the pipe wall is 20 W/m K and the wall thickness is 3.2 mm. The coil is heated by condensing steam at 373 K for which the film coefficient is 8 kW/m^2 K. When the water velocity in the pipe is 1 m/s, its outlet temperature is 309 K. What will the outlet temperature be if the velocity is increased to 1.3 m/s, if the coefficient of heat transfer to the water in the tube is proportional to the velocity raised to the 0.8 power?

9.71. Liquid is heated in a vessel by means of steam which is supplied to an internal coil in the vessel. When the vessel contains 1000 kg of liquid it takes half an hour to heat the contents from 293 to 368 K if the coil is supplied with steam at 373 K. The process is modified so that liquid at 293 K is continuously fed to the vessel at the rate of 0.28 kg/s, the total contents of the vessel always being maintained at 1000 kg. What is the equilibrium temperature which the contents of the vessel will reach, if heat losses to the surroundings are neglected and the overall heat transfer coefficient remains constant?

9.72. The heat loss through a firebrick furnace wall 0.2 m thick is to be reduced by addition of a layer of insulating brick to the outside. What is the thickness of insulating brick necessary to reduce the heat loss to 400 W/m^2? The inside furnace wall temperature is 1573 K, the ambient air adjacent to the furnace exterior is at 293 K and the natural convection heat transfer coefficient at the exterior surface is given by $h_0 = 3.0\Delta T^{0.25}$ W/m^2 K, where ΔT is the temperature difference between the surface and the ambient air.

Thermal conductivity of firebrick $= 1.5$ W/m K
Thermal conductivity of insulating brick $= 0.4$ W/m K.

9.73. 2.8 kg/s of organic liquid of specific heat capacity 2.5 kJ/kg K is cooled in a heat exchanger from 363 to 313 K using water whose temperature rises from 293 to 318 K flowing countercurrently. After maintenance, the pipework is wrongly connected so that the two streams, flowing at the same rates as previously, are now in co-current flow. On the assumption that overall heat transfer coefficient is unaffected, show that the new outlet temperatures of the organic liquid and the water will be 320.6 K and 314.5 K, respectively.

9.74. An organic liquid is cooled from 353 to 328 K in a single-pass heat exchanger. When the cooling water of initial temperature 288 K flows countercurrently its outlet temperature is 333 K. With the water flowing co-currently, its feed rate has to be increased in order to give the same outlet temperature for the organic liquid, the new outlet temperature of the water is 313 K. When the cooling water is flowing countercurrently, the film heat transfer coefficient for the water is 600 W/m^2 K.

What is the coefficient when the exchanger is operating with co-current flow if its value is proportional to the 0.8 power of the water velocity?

Calculate the film coefficient from the organic liquid, on the assumptions that it remains unchanged, and that heat transfer resistances other than those attributable to the two liquids may be neglected.

9.75. A reaction vessel is heated by steam at 393 K supplied to a coil immersed in the liquid in the tank. It takes 1800 s to heat the contents from 293 K to 373 K when the outside temperature is 293 K. When the outside and initial temperatures are only 278 K, it takes 2700 s to heat the contents to 373 K. The area of the steam coil is 2.5 m^2 and of the external surface is 40 m^2. If the overall heat transfer coefficient from the coil to the liquid in the vessel is 400 W/m^2 K, show that the overall coefficient for transfer from the vessel to the surroundings is about 5 W/m^2 K.

9.76. Steam at 403 K is supplied through a pipe of 25 mm outside diameter. Calculate the heat loss per metre to surroundings at 293 K, on the assumption that there is a negligible drop in temperature through the wall of the pipe. The heat transfer coefficient h from the outside of the pipe to the surroundings is given by:

$$h = 1.22 \left(\frac{\Delta T}{d} \right)^{0.25} \text{ W/m}^2 \text{ K}$$

where d is the outside diameter of the pipe (m) and ΔT is the temperature difference (K) between the surface and surroundings.

The pipe is then lagged with a 50 mm thickness of lagging of thermal conductivity 0.1 W/m K. If the outside heat transfer coefficient is given by the same equation as for the bare pipe, by what factor is the heat loss reduced?

9.77. A vessel contains 1 tonne of liquid of specific heat capacity 4.0 kJ/kg K. It is heated by steam at 393 K which is fed to a coil immersed in the liquid; heat is lost to the surroundings at 293 K from the outside of the vessel. How long does it take to heat the liquid from 293 to 353 K and what is the maximum temperature to which the liquid can be heated? When the liquid temperature has reached 353 K, the steam supply is turned off for two hours and the vessel cools. How long will it take to reheat the material to 353 K?

Coil: Area 0.5 m^2. Overall heat transfer coefficient to liquid, 600 W/m^2 K
Outside of vessel: Area 6 m^2. Heat transfer coefficient to surroundings, 10 W/m^2 K.

9.78. A bare thermocouple is used to measure the temperature of a gas flowing through a hot pipe. The heat transfer coefficient between the gas and the thermocouple is proportional to the 0.8 power of the gas velocity and the heat transfer by radiation from the walls to the thermocouple is proportional to the temperature difference.
When the gas is flowing at 5 m/s the thermocouple reads 323 K, when it is flowing at 10 m/s it reads 313 K, and when it is flowing at 15.0 m/s it reads 309 K. Show that the gas temperature is about 298 K and calculate the approximate wall temperature. What temperature will the thermocouple indicate when the gas velocity is 20 m/s?

9.79. A hydrocarbon oil of density 950 kg/m^3 and specific heat capacity 2.5 kJ/kg K is cooled in a heat exchanger from 363 to 313 K by water flowing countercurrently. The temperature of the water rises from 293 to 323 K. If the flowrate of the hydrocarbon is 0.56 kg/s, what is the required flowrate of water?
After plant modifications, the heat exchanger is incorrectly connected so that the two streams are in co-current flow. What are the new outlet temperatures of hydrocarbon and water, if the overall heat transfer coefficient is unchanged?

9.80. A reaction mixture is heated in a vessel fitted with an agitator and a steam coil of area 10 m^2 fed with steam at 393 K. The heat capacity of the system is equal to that of 500 kg of water. The overall coefficient of heat transfer from the vessel of area 5 m^2 is 10 W/m^2 K. It takes 1800 s to heat the contents from ambient temperature of 293 to 333 K. How long will it take to heat the system to 363 K and what is the maximum temperature which can be reached?
Specific heat capacity of water = 4200 J/kg K.

9.81. A pipe, 50 mm outside diameter, is carrying steam at 413 K and the coefficient of heat transfer from its outer surface to the surroundings at 288 K is 10 W/m^2 K. What is the heat loss per metre?
It is desired to add lagging of thermal conductivity 0.03 W/m K as a thick layer to the outside of the pipe in order to cut heat losses by 90%. If the heat transfer from the outside surface of the lagging is 5 W/m^2 K, what thickness of lagging is required?

9.82. It takes 1800 s to heat a tank of liquid from 293 to 333 K using steam supplied to an immersed coil when the steam temperature is 383 K. How long will it take when the steam temperature is raised to 393 K? The overall heat transfer coefficient from the steam coil to the tank is 10 times the coefficient from the tank to surroundings at a temperature of 293 K and the area of the steam coil is equal to the outside area of the tank.

9.83. A thermometer is situated in a duct in an air stream which is at a constant temperature. The reading varies with the gas flowrate in the following way:

Air velocity (m/s)	Thermometer reading (K)
6.1	553
7.6	543
12.2	533

The wall of the duct and the gas stream are at somewhat different temperatures. If the heat transfer coefficient for radiant heat transfer from the wall to the thermometer remains constant, and the heat transfer coefficient between the gas stream and thermometer is proportional to the 0.8 power of the velocity, what is the true temperature of the air system? Neglect any other forms of heat transfer.

9.84. How long will it take to heat 0.18 m³ of liquid of density 900 kg/m³ and specific heat 2.1 kJ/kg K from 293 to 377 K in a tank fitted with a coil of area 1 m²? The coil is fed with steam at 383 K and the overall heat transfer coefficient can be taken as constant at 0.5 kW/m² K. The vessel has an external surface of 2.5 m², and the coefficient for heat transfer to the surroundings at 293 K is 5 W/m² K.

The batch system of heating is to be replaced by a continuous countercurrent heat exchanger in which the heating medium is a liquid entering at 388 K and leaving at 333 K. If the heat transfer coefficient is 250 W/m² K, what heat exchange area is required? Neglect heat losses.

10.1. Ammonia gas is diffusing at a constant rate through a layer of stagnant air 1 mm thick. Conditions are fixed so that the gas contains 50% by volume of ammonia at one boundary of the stagnant layer. The ammonia diffusing to the other boundary is quickly absorbed and the concentration is negligible at that plane. The temperature is 295 K and the pressure atmospheric, and under these conditions the diffusivity of ammonia in air is 0.18 cm²/s. Calculate the rate of diffusion of ammonia through the layer.

10.2. A simple rectifying column consists of a tube arranged vertically and supplied at the bottom with a mixture of benzene and toluene as vapour. At the top a condenser returns some of the product as a reflux which flows in a thin film down the inner wall of the tube. The tube is insulated and heat losses can be neglected. At one point in the column the vapour contains 70 mol% benzene and the adjacent liquid reflux contains 59 mol% benzene. The temperature at this point is 365 K. Assuming the diffusional resistance to vapour transfer to be equivalent to the diffusional resistance of a stagnant vapour layer 0.2 mm thick, calculate the rate of interchange of benzene and toluene between vapour and liquid. The molar latent heats of the two materials can be taken as equal. The vapour pressure of toluene at 365 K is 54.0 kN/m² and the diffusivity of the vapours is 0.051 cm²/s.

10.3. By what percentage would the rate of absorption be increased or decreased by increasing the total pressure from 100 to 200 kN/m² in the following cases?

(a) The absorption of ammonia from a mixture of ammonia and air containing 10% of ammonia by volume, using pure water as solvent. Assume that all the resistance to mass transfer lies within the gas phase.
(b) The same conditions as (a) but the absorbing solution exerts a partial vapour pressure of ammonia of 5 kN/m².

The diffusivity can be assumed to be inversely proportional to the absolute pressure.

10.4. In the Danckwerts model of mass transfer it is assumed that the fractional rate of surface renewal s is constant and independent of surface age. Under such conditions the expression for the surface age distribution function ϕ is $\phi = se^{-st}$. If the fractional rate of surface renewal were proportional to surface age (say $s = bt$, where b is a constant), show that the surface age distribution function would then assume the form:

$$\phi = \left(\frac{2b}{\pi}\right)^{1/2} e^{-bt^2/2}.$$

10.5. By consideration of the appropriate element of a sphere show that the general equation for molecular diffusion in a stationary medium and in the absence of a chemical reaction is:

$$\frac{\partial C}{\partial t} = D\left(\frac{\partial^2 C}{\partial r^2} + \frac{1}{r^2}\frac{\partial^2 C}{\partial \beta^2} + \frac{1}{r^2 \sin^2 \beta}\frac{\partial^2 C}{\partial \phi^2} + \frac{2}{r}\frac{\partial C}{\partial r} + \frac{\cot \beta}{r^2}\frac{\partial C}{\partial \beta}\right),$$

where C is the concentration of the diffusing substance, D the molecular diffusivity, t the time, and r, β, and ϕ are spherical polar coordinates, β being the latitude angle.

10.6. Prove that for equimolecular counterdiffusion from a sphere to a surrounding stationary, infinite medium, the Sherwood number based on the diameter of the sphere is equal to 2.

10.7. Show that the concentration profile for unsteady-state diffusion into a bounded medium of thickness L, when the concentration at the interface is suddenly raised to a constant value C_i and kept constant at the initial value of C_o at the other boundary is:

$$\frac{C - C_o}{C_i - C_o} = 1 - \frac{z}{L} = \frac{2}{\pi}\left[\sum_{n=1}^{n=\infty} \frac{1}{n} \exp\left(-\frac{n^2\pi^2 Dt}{L^2}\right) \sin\frac{n z \pi}{L}\right].$$

Assume the solution to be the sum of the solution for infinite time (steady-state part) and the solution of a second unsteady-state part; this simplifies the boundary conditions for the second part.

10.8. Show that under the conditions specified in Problem 10.7 and assuming the Higbie model of surface renewal, the average mass flux at the interface is given by

$$(N_A)_t = (C_i - C_o)\frac{D}{L}\left\{1 + \frac{2L^2}{\pi^2 Dt}\sum_{n=1}^{n=\infty}\left[\frac{\pi^2}{6} - \frac{1}{n^2}\exp\left(-\frac{n^2\pi^2 Dt}{L^2}\right)\right]\right\}.$$

N.B. Use the relation $\sum_{n=1}^{n=\infty}\frac{1}{n^2} = \frac{\pi^2}{6}$.

10.9. According to the simple penetration theory the instantaneous mass flux, $(N_A)_t$ is:

$$(N_A)_t = (C_i - C_o)\left(\frac{D}{\pi t}\right)^{0.5}$$

What is the equivalent expression for the instantaneous heat flux under analogous conditions?

Pure SO_2 is absorbed at 295 K and atmospheric pressure into a laminar water jet. The solubility of SO_2, assumed constant over a small temperature range, is 1.54 kmol/m^3 under these conditions and the heat of solution is 28 kJ/kmol.

Calculate the resulting jet surface temperature if the Lewis number is 90. Neglect heat transfer between the water and the gas.

10.10. In a packed column, operating at approximately atmospheric pressure and 295 K, a 10% ammonia–air mixture is scrubbed with water and the concentration is reduced to 0.1%. If the whole of the resistance to mass transfer may be regarded as lying within a thin laminar film on the gas side of the gas–liquid interface, derive from first principles an expression for the rate of absorption at any position in the column. At some intermediate point where the ammonia concentration in the gas phase has been reduced to 5%, the partial pressure of ammonia in equilibrium with the aqueous solution is 660 N/m^2 and the transfer rate is 10^{-3} kmol/m^{2}s. What is the thickness of the hypothetical gas film if the diffusivity of ammonia in air is 0.24 cm^2/s?

10.11. An open bowl, 0.3 m in diameter, contains water at 350 K evaporating into the atmosphere. If the air currents are sufficiently strong to remove the water vapour as it is formed and if the resistance to its mass transfer in air is equivalent to that of a 1 mm layer for conditions of molecular diffusion, what will be the rate of cooling due to evaporation? The water can be considered as well mixed and the water equivalent of the system is equal to 10 kg. The diffusivity of water vapour in air may be taken as 0.20 cm^2/s and the kilogram molecular volume at NTP as 22.4 m^3.

10.12. Show by substitution that when a gas of solubility C^+ is absorbed into a stagnant liquid of infinite depth, the concentration at time t and depth x is:

$$C^+ \text{ erfc } \frac{x}{2\sqrt{Dt}}.$$

Hence, on the basis of the simple penetration theory, show that the rate of absorption in a packed column will be proportional to the square root of the diffusivity.

10.13. Show that in steady-state diffusion through a film of liquid, accompanied by a first-order irreversible reaction, the concentration of solute in the film at depth z below the interface is given by:

$$\frac{C}{C_i} = \frac{\sinh\sqrt{(k/D)}(z_L - z)}{\sinh\sqrt{(k/D)}z_L}$$

if $C = 0$ at $z = z_L$ and $C = C_i$ at $z = 0$, corresponding to the interface. Hence show that according to the "film theory" of gas-absorption, the rate of absorption per unit area of interface N_A is given by:

$$N_A = K_L C_i \frac{\beta}{\tanh \beta},$$

where $\beta = \sqrt{(D\mathbf{k})}/K_L$, D is the diffusivity of the solute, $\mathbf{k}$ the rate constant of the reaction, K_L the liquid film mass transfer coefficient for physical absorption, C_i the concentration of solute at the interface, z the distance normal to the interface, and z_L the liquid film thickness.

10.14. The diffusivity of the vapour of a volatile liquid in air can be conveniently determined by Winkelmann's method, in which liquid is contained in a narrow diameter vertical tube maintained at a constant temperature, and an air stream is passed over the top of the tube sufficiently rapidly to ensure that the partial pressure of the vapour there remains approximately zero. On the assumption that the vapour is transferred from the surface of the liquid to the air stream by molecular diffusion, calculate the diffusivity of carbon tetrachloride vapour in air at 321 K and atmospheric pressure from the following experimentally obtained data:

Time from commencement of experiment (ks)	Liquid level (cm)
0	0.00
1.6	0.25
11.1	1.29
27.4	2.32
80.2	4.39
117.5	5.47
168.6	6.70
199.7	7.38
289.3	9.03
383.1	10.48

The vapour pressure of carbon tetrachloride at 321 K is 37.6 kN/m² and the density of the liquid is 1540 kg/m³. Take the kilogram molecular volume as 22.4 m³.

10.15. Ammonia is absorbed in water from a mixture with air using a column operating at atmospheric pressure and 295 K. The resistance to transfer can be regarded as lying entirely within the gas phase. At a point in the column the partial pressure of the ammonia is 6.6 kN/m². The back pressure at the water interface is negligible and the resistance to transfer can be regarded as lying in a stationary gas film 1 mm thick. If the diffusivity of ammonia in air is 0.236 cm²/s, what is the transfer rate per unit area at that point in the column? If the gas were compressed to 200 kN/m² pressure, how would the transfer rate be altered?

10.16. What are the general principles underlying the two-film, penetration and film-penetration theories for mass transfer across a phase boundary? Give the basic differential equations which have to be solved for these theories with the appropriate boundary conditions.

According to the penetration theory, the instantaneous rate of mass transfer per unit area (N_A) at some time t after the commencement of transfer is given by:

$$(N_A)_t = \Delta C \sqrt{\frac{D}{\pi t}},$$

where ΔC is the concentration force and D is the diffusivity.

Obtain expressions for the average rates of transfer on the basis of the Higbie and Danckwerts assumptions.

10.17. A solute diffuses from a liquid surface at which its molar concentration is C_i into a liquid with which it reacts. The mass transfer rate is given by Fick's law and the reaction is first order with respect to the solute. In a steady-state process the diffusion rate falls at a depth L to one half the value at the interface. Obtain an expression for the concentration C of solute at a depth z from the surface in terms of the molecular diffusivity D and the reaction rate constant k. What is the molar flux at the surface?

10.18. 4 cm³ of mixture formed by adding 2 cm³ of acetone to 2 cm³ of dibutyl phthalate is contained in a 6 mm diameter vertical glass tube immersed in a thermostat maintained at 315 K. A stream of air at 315 K and atmospheric pressure is passed over the open top of the tube to maintain a zero partial pressure of acetone vapour at that point. The liquid level is initially 11.5 mm below the top of the tube and the acetone vapour is transferred to the air stream by molecular diffusion alone. The dibutyl phthalate can be regarded as completely non-volatile and the partial pressure of acetone vapour may be calculated from Raoult's law on the assumption that the density of dibutyl phthalate is sufficiently greater than that of acetone for the liquid to be completely mixed.

Calculate the time taken for the liquid level to fall to 5 cm below the top of the tube, neglecting the effects of bulk flow in the vapour.

Kilogram molecular volume $= 22.4$ m^3

Molecular weights of acetone, dibutyl phthalate $= 58$ and 278 kg/kmol respectively.

Liquid densities of acetone, dibutyl phthalate $= 764$ and 1048 kg/m^3 respectively.

Vapour pressure of acetone at 315 K $= 60.5$ kN/m^2.

Diffusivity of acetone vapour in air at 315 K $= 1.23 \times 10^{-5}$ m^2/s.

10.19. A crystal is suspended in fresh solvent and 5% of the crystal dissolves in 300 s. How long will it take before 10% of the crystal has dissolved? Assume that the solvent can be regarded as infinite in extent, that the mass transfer in the solvent is governed by Fick's second law of diffusion and can be represented as a unidirectional process, and that changes in the surface area of the crystal may be neglected. Start your derivations using Fick's second law.

10.20. In a continuous steady state reactor, a slightly soluble gas is absorbed into a liquid in which it dissolves and reacts, the reaction being second order with respect to the dissolved gas. Calculate the reaction rate constant on the assumption that the liquid is semi-infinite in extent and that mass transfer resistance in the gas phase is negligible. The diffusivity of the gas in the liquid is 10^{-8} m^2/s, the gas concentration in the liquid falls to one half of its value in the liquid over a distance of 1 mm, and the rate of absorption at the interface is 4×10^{-6} kmol/m^2 s.

10.21. Experiments have been carried out on the mass transfer of acetone between air and a laminar water jet. Assuming that desorption produces random surface renewal with a constant fractional rate of surface renewal, s, but an upper limit on surface age equal to the life of the jet, τ, show that the surface age frequency distribution function, $\phi(t)$, for this case is given by:

$$\phi(t) = s \exp \frac{-st}{1 - \exp(-st)} \qquad \text{for} \quad 0 < t < \tau$$

$$\phi(t) = 0 \qquad \qquad \text{for} \quad t > \tau.$$

Hence, show that the enhancement, E, for the increase in value of the liquid-phase mass transfer coefficient is:

$$E = \frac{(\pi s\tau)^{1/2} \; \text{erf} \; (s\tau)^{1/2}}{2[1 - \exp(-s\tau)]}$$

where E is defined as the ratio of the mass transfer coefficient predicted by conditions described above to the mass transfer coefficient obtained from the penetration theory for a jet with an undisturbed surface. Assume that the interfacial concentration of acetone is practically constant.

10.22. Solute gas is diffusing into a stationary liquid, virtually free of solvent, and of sufficient depth for it to be regarded as semi-infinite in extent. In what depth of fluid below the surface will 90% of the material which has been transferred across the interface have accumulated in the first minute?

Diffusivity of gas in liquid $= 10^{-9}$ m^2/s.

10.23. A chamber, of volume 1 m^3, contains air at a temperature of 293 K and a pressure of 101.3 kN/m^2, with a partial pressure of water vapour of 0.8 kN/m^2. A bowl of liquid with a free surface of 0.01 m^2 and maintained at a temperature of 303 K is introduced into the chamber. How long will it take for the air to become 90% saturated at 293 K and how much water must be evaporated?

The diffusivity of water vapour in air is 2.4×10^{-5} m^2/s and the mass transfer resistance is equivalent to that of a stagnant gas film of thickness 0.25 mm. Neglect the effects of bulk flow.

Saturation vapour pressure of water $= 4.3$ kN/m^2 at 303 K

and 2.3 kN/m^2 at 293 K.

10.24. A large deep bath contains molten steel, the surface of which is in contact with air. The oxygen concentration in the bulk of the molten steel is 0.03% by mass and the rate of transfer of oxygen from the air is sufficiently high to maintain the surface layers saturated at a concentration of 0.16% by weight. The surface of the liquid is disrupted by gas bubbles rising to the surface at a frequency of 120 bubbles per m^2 of surface per second, each bubble disrupts and mixes about 15 cm^2 of the surface layer into the bulk.

On the assumption that the oxygen transfer can be represented by a surface renewal model, obtain the appropriate equation for mass transfer by starting with Fick's second law of diffusion and calculate:

(a) The mass transfer coefficient
(b) The mean mass flux of oxygen at the surface
(c) The corresponding film thickness for a film model, giving the same mass transfer rate.

Diffusivity of oxygen in steel = 1.2×10^{-8} m^2/s
Density of molten steel   = 7,100 kg/m^3

10.25. Two large reservoirs of gas are connected by a pipe of length 2L with a full-bore valve at its mid-point. Initially a gas A fills one reservoir and the pipe up to the valve and gas B fills the other reservoir and the remainder of the pipe. The valve is opened rapidly and the gases in the pipe mix by molecular diffusion.

Obtain an expression for the concentration of gas A in that half of the pipe in which it is increasing, as a function of distance y from the valve and time t after opening. The whole system is at a constant pressure and the ideal gas law is applicable to both gases. It may be assumed that the rate of mixing in the vessels is high so that the gas concentration at the two ends of the pipe do not change.

10.26. A pure gas is absorbed into a liquid with which it reacts. The concentration in the liquid is sufficiently low for the mass transfer to be governed by Fick's law and the reaction is first order with respect to the solute gas. It may be assumed that the film theory may be applied to the liquid and that the concentration of solute gas falls from the saturation value to zero across the film. Obtain an expression for the mass transfer rate across the gas-liquid interface in terms of the molecular diffusivity, D, the first order reaction rate constant k, the film thickness L and the concentration C_{AS} of solute in a saturated solution. The reaction is initially carried out at 293 K. By what factor will the mass transfer rate across the interface change, if the temperature is raised to 313 K?

Reaction rate constant at 293 K = 2.5×10^{-6} s^{-1}
Energy of activation for reaction (in Arrhenius equation)
                              = 26,430 kJ/kmol
Universal gas constant **R**   = 8314 J/kmol K
Molecular diffusivity D = 10^{-9} m^2/s
Film thickness, L = 10 mm
Solubility of gas at 313 K is 80% of solubility at 293 K.

10.27. Using Maxwell's law of diffusion obtain an expression for the effective diffusivity for a gas **A** in a binary mixture of **B** and **C**, in terms of the diffusivities of **A** in the two pure components and the molar concentrations of **A**, **B** and **C**.

Carbon dioxide is absorbed in water from a 25 per cent mixture in nitrogen. How will its absorption rate compare with that from a mixture containing 35 per cent carbon dioxide, 40 per cent hydrogen and 25 per cent nitrogen? It may be assumed that the gas-film resistance is controlling, that the partial pressure of carbon dioxide at the gas-liquid interface is negligible and that the two-film theory is applicable, with the gas film thickness the same in the two cases.

Diffusivity of CO$_2$:   in hydrogen 3.5×10^{-5} m^2/s;
                       in nitrogen 1.6×10^{-5} m^2/s.

10.28. Given that from the penetration theory for mass transfer across an interface, the instantaneous rate of mass transfer is inversely proportional to the square root of the time of exposure, obtain a relationship between exposure time in the Higbie model and surface renewal rate in the Danckwerts model which will give the same average mass transfer rate. The age distribution function and average mass transfer rate from the Danckwerts theory must be derived from first principles.

10.29. Ammonia is absorbed in a falling film of water in an absorption apparatus and the film is disrupted and mixed at regular intervals as it flows down the column. The mass transfer rate is calculated from the penetration theory on the assumption that all the relevant conditions apply. It is found from measurements that the mass transfer rate immediately before mixing is only 16 per cent of that calculated from the theory and the difference has been attributed to the existence of a surface film which remains intact and unaffected by the mixing process. If the liquid mixing process takes place every second, what thickness of surface film would account for the discrepancy?

Diffusivity of ammonia in water $= 1.76 \times 10^{-9}$ m^2/s

10.30. A deep pool of ethanol is suddenly exposed to an atmosphere consisting of pure carbon dioxide and unsteady state mass transfer, governed by Fick's Law, takes place for 100 s. What proportion of the absorbed carbon dioxide will have accumulated in the 1 mm thick layer of ethanol closest to the surface?

Diffusivity of carbon dioxide in ethanol $= 4 \times 10^{-9}$ m^2/s.

11.1. Calculate the thickness of the boundary layer at a distance of 75 mm from the leading edge of a plane surface over which water is flowing at a rate of 3 m/s. Assume that the flow in the boundary layer is streamline and that the velocity u of the fluid at a distance y from the surface can be represented by the relation $u = a + by + cy^2 + dy^3$ (where the coefficients a, b, c, and d are independent of y). Take the viscosity of water as 1 mN s/m^2.

11.2. Water flows at a velocity of 1 m/s over a plane surface 0.6 m wide and 1 m long. Calculate the total drag force acting on the surface if the transition from streamline to turbulent flow in the boundary layer occurs when the Reynolds group $Re_x = 10^5$.

11.3. Calculate the thickness of the boundary layer at a distance of 150 mm from the leading edge of a surface over which oil, of viscosity 50 mN s/m^2 and density 990 kg/m^3, flows with a velocity of 0.3 m/s. What is the displacement thickness of the boundary layer?

11.4. Calculate the thickness of the laminar sub-layer when benzene flows through a pipe of 50 mm diameter at 0.003 m^3/s. What is the velocity of the benzene at the edge of the laminar sub-layer? Assume fully developed flow exists within the pipe.

11.5. Air is flowing at a velocity of 5 m/s over a plane surface. Derive an expression for the thickness of the laminar sub-layer and calculate its value at a distance of 1 m from the leading edge of the surface.

Assume that within the boundary layer outside the laminar sub-layer the velocity of flow is proportional to the one-seventh power of the distance from the surface and that the shear stress R at the surface is given by:

$$\frac{R}{\rho u_s^2} = 0.03 \left(\frac{u_s \rho x}{\mu} \right)^{-0.2},$$

where ρ is the density of the fluid (1.3 kg/m^3 for air), μ the viscosity of the fluid (17×10^{-6} N s/m^2 for air), u_s the stream velocity (m/s), and x the distance from the leading edge (m).

11.6. Obtain the momentum equation for an element of boundary layer. If the velocity profile in the laminar region can be represented approximately by a sine function, calculate the boundary-layer thickness in terms of distance from the leading edge of the surface.

11.7. Explain the concepts of "momentum thickness" and "displacement thickness" for the boundary layer formed during flow over a plane surface. Develop a similar concept to displacement thickness in relation to heat flux across the surface for laminar flow and heat transfer by thermal conduction, for the case where the surface has a constant temperature and the thermal boundary layer is always thinner than the velocity boundary layer. Obtain an expression for this "thermal thickness" in terms of the thicknesses of the velocity and temperature boundary layers.

Similar forms of cubic equations may be used to express velocity and temperature variations with distance from the surface.

For a Prandtl number Pr less than unity, the ratio of the temperature to the velocity boundary layer thickness is equal to $Pr^{-1/3}$. Work out therefore the "thermal thickness" in terms of the thickness of the velocity boundary layer for a value of 0.7 for Pr.

11.8. Explain why it has been necessary to use concepts, such as the displacement thickness and the momentum thickness, for a boundary layer in order to obtain a boundary layer thickness which is largely independent of the approximation used for the velocity profile in the neighbourhood of the surface.

It is found that the velocity u at a distance y from the surface can be expressed as a simple power function ($u \propto y^n$) for the turbulent boundary layer at a plane surface. What is the value of n, if the ratio of the momentum thickness to the displacement thickness is 1.78?

11.9. Derive the momentum equation for the flow of a fluid over a plane surface for conditions where the pressure gradient along the surface is negligible. By assuming a sine function for the variation of velocity with distance from the surface (within the boundary layer) for streamline flow, obtain an expression for the boundary layer thickness as a function of distance from the leading edge of the surface.

11.10. Derive the momentum equation for the flow of a viscous fluid over a small plane surface.
Show that the velocity profile in the neighbourhood of the surface can be expressed as a sine function which satisfies the boundary conditions at the surface and at the outer edge of the boundary layer.
Obtain the boundary layer thickness and its displacement thickness as a function of the distance from the leading edge of the surface, when the velocity profile is expressed as a sine function.

12.1. If the temperature rise per metre length along a pipe carrying air at 12.2 m/s is 66 K, what will be the corresponding pressure drop for a pipe temperature of 420 K and an air temperature of 310 K?
The density of air at 310 K is 1.14 kg/m^3.

12.2. It is required to warm a quantity of air from 289 to 313 K by passing it through a number of parallel metal tubes of inner diameter 50 mm maintained at 373 K. The pressure drop must not exceed 250 N/m^2. How long should the individual tubes be?
The density of air at 301 K is 1.19 kg/m^3. The coefficients of heat transfer by convection from tube to air are 45, 62, and 77 W/m^{2}K for velocities of 20, 24, and 30 m/s at 301 K respectively.

12.3. Air at 330 K, flowing at 10 m/s, enters a pipe of inner diameter 25 mm, maintained at 415 K. The drop of static pressure along the pipe is 80 N/m^2 per metre length. Using the Reynolds analogy between heat transfer and friction, estimate the temperature of the air 0.6 m along the pipe.

12.4. Air flows at 12 m/s through a pipe of inside diameter 25 mm. The rate of heat transfer by convection between the pipe and the air is 60 W/m^{2}K. Neglecting the effects of temperature variation, estimate the pressure drop per metre length of pipe.

12.5. Apply Reynolds analogy to solve the following problem. Air at 320 K and atmospheric pressure is flowing through a smooth pipe of 50 mm internal diameter and the pressure drop over a 4 m length is found to be 1.5 kN/m^2. By how much would you expect the air temperature to fall over the first metre of pipe length if the wall temperature there is kept constant at 295 K?

$$\text{Viscosity of air} = 0.018 \text{ mN s/m}^2.$$
$$\text{Specific heat of air} = 1.05 \text{ kJ/kg K}.$$

12.6. Obtain an expression for the simple Reynolds analogy between heat transfer and friction. Indicate the assumptions which are made in the derivation and the conditions under which you would expect the relation to be applicable.
The Reynolds number of a gas flowing at 2.5 kg/m^{2}s through a smooth pipe is 20,000. If the specific heat of the gas at constant pressure is 1.67 kJ/kg K, what will the heat transfer coefficient be?

12.7. Explain Prandtl's concept of a "mixing length". What parallels can you draw between the mixing length and the mean free path of the molecules in a gas?
The ratio of the mixing length to the distance from the pipe wall has a constant value of 0.4 for the turbulent flow of a fluid in a pipe. What is the value of the pipe friction factor if the ratio of the mean velocity to the axial velocity is 0.8?

12.8. The velocity profile in the neighbourhood of a surface for a Newtonian fluid can be expressed in terms of a dimensionless velocity u^+ and a dimensionless distance y^+ from the surface. Obtain the relation between u^+ and y^+ in the laminar sub-layer. Outside the laminar sub-layer, the relation takes the form:

$$u^+ = 2.5 \ln y^+ + 5.5$$

At what value of y^+ does the transition from the laminar sub-layer to the turbulent zone occur?

In the "Universal Velocity Profile", the laminar sub-layer extends to values of $y^+ = 5$ and the turbulent zone starts at $y^+ = 30$ and the range $5 < y^+ < 30$, the buffer layer, is covered by a second linear relation between u^+ and $\ln y^+$. What is the *maximum* difference between the values of u^+, in the range $5 < y^+ < 30$, using the two methods of representation of the velocity profile?

$$\text{Definitions: } u^+ = \frac{u_x}{u^*}$$

$$y^+ = \frac{yu^*\rho}{\mu}$$

$$u^{*2} = \frac{R}{\rho}$$

where u_x is the velocity at distance y from surface,
       R is the wall shear stress, and
       ρ, μ are the density, the viscosity of fluid respectively.

12.9. Calculate the rise in temperature of water passed at 4 m/s through a smooth 25 mm diameter pipe, 6 m long. The water enters at 300 K and the temperature of the wall of the tube can be taken as approximately constant at 330 K. Use:

(a) The simple Reynolds analogy.
(b) The Taylor-Prandtl modification.
(c) The buffer lays equation
(d) $Nu = 0.023Re^{0.8}Pr^{0.33}$.

Comment on the differences in the results so obtained.

12.10. Calculate the rise in temperature of a stream of air, entering at 290 K and passing at 4 m/s through the tube maintained at 350 K, other conditions remaining the same as detailed in Problem 12.9.

12.11. Air flows through a smooth circular duct of internal diameter 0.25 m at an average velocity of 15 m/s. Calculate the fluid velocity at points 50 mm and 5 mm from the wall. What will be the thickness of the laminar sub-layer if this extends to $u^+ = y^+ = 5$? The density of the air may be taken as 1.12 kg/m^3 and its viscosity as 0.02 mN s/m^2.

12.12. Obtain the Taylor-Prandtl modification of the Reynolds Analogy for momentum and heat transfer, and give the corresponding relation for mass transfer (no bulk flow).

An air stream at approximately atmospheric temperature and pressure and containing a low concentration of carbon disulphide vapour is flowing at 38 m/s through a series of 50 mm diameter tubes. The inside of the tubes is covered with a thin film of liquid and both heat and mass transfer are taking place between the gas stream and the liquid film. The film heat transfer coefficient is found to be 100 W/m^{2}K. Using a pipe friction chart and assuming the tubes to behave as smooth surfaces, calculate:

(a) the film mass transfer coefficient, and
(b) the gas velocity at the interface between the laminar sub-layer and the turbulent zone of the gas.

Specific heat of air = 1.0 kJ/kg K
Viscosity of air = 0.02 mN s/m^2
Diffusivity of carbon disulphide vapour in air = 1.1×10^{-5} m^2/s
Thermal conductivity of air 0.024 = W/m K.

12.13. Obtain the Taylor-Prandtl modification of the Reynolds' analogy between momentum and heat transfer and write down the corresponding analogy for mass transfer. For a particular system a mass transfer coefficient of 8.71×10^{-8} m/s and a heat transfer coefficient of 2730 W/m^2 K were measured for similar flow conditions. Calculate the ratio of the velocity in the fluid where the laminar sub-layer terminates, to the stream velocity.

$$\text{Molecular diffusivity} = 1.5 \times 10^{-9} \text{ m}^2/\text{s}$$
$$\text{Viscosity} = 1 \text{ mN s/m}^2$$

$$\text{Density} = 1000 \text{ kg/m}^3$$
$$\text{Thermal conductivity} = 0.48 \text{ W/m K}$$
$$\text{Specific heat capacity} = 4.0 \text{ kJ/kg K}$$

12.14. Heat and mass transfer are taking place simultaneously to a surface under conditions where the Reynolds analogy between momentum, heat and mass transfer may be applied. The mass transfer is of a single component at a high concentration in a binary mixture, the other component of which undergoes no net transfer. Using the Reynolds analogy, obtain a relation between the coefficients for heat transfer and for mass transfer.

12.15. Derive the Taylor-Prandtl modification of the Reynolds Analogy between momentum and heat transfer.

In a shell and tube type condenser, water flows through the tubes which are 10 m long and 40 mm diameter. The pressure drop across the tubes is 5.6 kN/m^2 and the effects of entry and exit losses may be neglected. The tube walls are smooth and flow may be taken as fully developed. The ratio of the velocity at the edge of the laminar sub-layer to the mean velocity of flow may be taken at 2 $Re^{-0.125}$, where Re is the Reynolds number in the pipeline.

If the tube walls are at an approximately constant temperature of 393 K and the inlet temperature of the water is 293 K, estimate the outlet temperature.

Physical properties of water: density 1000 kg/m^3
$$\text{viscosity 1 mN s/m}^2$$
$$\text{thermal conductivity 0.6 W/m K}$$
$$\text{specific heat capacity 4.2 kJ/kg K}$$

12.16. Explain the importance of the universal velocity profile and derive the relation between the dimensionless derivative of velocity u^+, and the dimensionless derivative of distance from the surface y^+, using the concept of Prandtl's mixing length λ_E.

It may be assumed that the fully turbulent portion of the boundary layer starts at $y^+ = 30$, that the ratio of the mixing length λ_E to the distance y from the surface, $\lambda_E/y = 0.4$, and that for a smooth surface $u^+ = 14$ at $y^+ = 30$.

If the laminar sub-layer extends from $y^+ = 0$ to $y^+ = 5$, obtain the equation for the relation between u^+ and y^+ in the buffer zone, and show that the ratio of the eddy viscosity to the molecular viscosity increases linearly from 0 to 5 through this buffer zone.

12.17. Derive the Taylor-Prandtl modification of the Reynolds analogy between heat and momentum transfer and express it in a form in which it is applicable to pipe flow.

If the relationship between the Nusselt number Nu, Reynolds number Re and Prandtl number Pr may be expressed as:

$$Nu = 0.023 Re^{0.8} Pr^{0.33}$$

calculate the ratio of the velocity at the edge of the laminar sub-layer to the velocity at the pipe axis for water ($Pr = 10$) flowing at a Reynolds number (Re) of 10,000 in a smooth pipe. Use the pipe friction chart.

12.18. Obtain a dimensionless relation for the velocity profile in the neighbourhood of a surface for the turbulent flow of a liquid, using Prandtl's concept of a "Mixing Length" (University Velocity Profile). Neglect the existence of the buffer layer and assume that, outside the laminar sub-layer, eddy transport mechanisms dominate. Assume that in the turbulent fluid the mixing length λ_E is equal to 0.4 times the distance y from the surface and that the dimensionless velocity u^+ is equal to 5.5 when the dimensionless distance y^+ is unity.

Show that, if the Blasius relation is used for the shear stress R at the surface, the thickness of the laminar sub-layer δ_b is approximately 1.07 times that calculated on the assumption that the velocity profile in the turbulent fluid is given by Prandtl's one seventh power law.

Blasius Equation:

$$\frac{R}{\rho u_s^2} = 0.0228 \left(\frac{u_s \delta \rho}{\mu} \right)^{-0.25}$$

where ρ, μ are the density, viscosity of the fluid,
     u_s is the stream velocity, and
     δ is the total boundary layer thickness.

12.19. Obtain the Taylor-Prandtl modification of the Reynolds Analogy between momentum transfer and mass transfer (equimolecular counterdiffusion) for the turbulent flow of a fluid over a surface. Write down the corresponding analogy for heat transfer. State clearly the assumptions which are made. For turbulent flow over a surface, the film heat transfer coefficient for the fluid is measured and found to be 4 kW/m^2 K. What would the corresponding value of the mass transfer coefficient be, given the following physical properties?

$$\text{Diffusivity } D = 5 \times 10^{-9} \text{ m}^2/s$$
$$\text{Thermal conductivity } k = 0.6 \text{ W/m K}$$
$$\text{Specific heat capacity } C_p = 4 \text{ kJ/kg K}$$
$$\text{Density } \rho = 1000 \text{ kg/m}^3$$
$$\text{Viscosity } \mu = 1 \text{ mN s/m}^2$$

Assume that the ratio of the velocity at the edge of the laminar sub-layer to the stream velocity is (a) 0.2, (b) 0.6.

Comment on the difference in the two results.

12.20. By using the simple Reynolds Analogy, obtain the relation between the heat transfer coefficient and the mass transfer coefficient for the gas phase for the absorption of a soluble component from a mixture of gases. If the heat transfer coefficient is 100 W/m^2 K, what will the mass transfer coefficient be for a gas of specific heat capacity C_p equal to 1.5 kJ/kg K and density 1.5 kg/m^3? The concentration of the gas is sufficiently low for bulk flow effects to be negligible.

13.1. In a process in which benzene is used as a solvent, it is evaporated into dry nitrogen. The resulting mixture at a temperature of 297 K and a pressure of 101.3 kN/m^2 has a relative humidity of 60%. It is desired to recover 80% of the benzene present by cooling to 283 K and compressing to a suitable pressure. What must this pressure be?

Vapour pressures of benzene: at 297 K = 12.2 kN/m^2; at 283 K = 6.0 kN/m^2.

13.2. 0.6 m^3/s of gas is to be dried from a dew point of 294 K to a dew point of 277.5 K. How much water must be removed and what will be the volume of the gas after drying?

$$\text{Vapour pressure of water at 294 K} = 2.5 \text{ kN/m}^2.$$
$$\text{Vapour pressure of water at 277.5 K} = 0.85 \text{ kN/m}^2.$$

13.3. Wet material, containing 70% moisture, is to be dried at the rate of 0.15 kg/s in a countercurrent dryer to give a product containing 5% moisture (both on the wet basis). The drying medium consists of air heated to 373 K and containing water vapour equivalent to a partial pressure of 1.0 kN/m^2. The air leaves the dryer at 313 K and 70% saturated. Calculate how much air will be required to remove the moisture. The vapour pressure of water at 313 K may be taken as 7.4 kN/m^2.

13.4. 30,000 m^3 of coal gas (measured at 289 K and 101.3 kN/m^2 saturated with water vapour) is compressed to 340 kN/m^2 pressure, cooled to 289 K and the condensed water is drained off. Subsequently the pressure is reduced to 170 kN/m^2 and the gas is distributed at this pressure and 289 K. What is the percentage humidity of the gas after this treatment?

The vapour pressure of water at 289 K is 1.8 kN/m^2.

13.5. A rotary countercurrent dryer is fed with ammonium nitrate containing 5% moisture at the rate of 1.5 kg/s, and discharges the nitrate with 0.2% moisture. The air enters at 405 K and leaves at 355 K; the humidity of the entering air being 0.007 kg of moisture per kg of dry air. The nitrate enters at 294 K and leaves at 339 K.

Neglecting radiation losses, calculate the mass of dry air passing through the dryer and the humidity of the air leaving the dryer.

$$\text{Latent heat of water at 294 K} = 2450 \text{ kJ/kg.}$$
$$\text{Specific heat of ammonium nitrate} = 1.88 \text{ kJ/kg K.}$$
$$\text{Specific heat of dry air} = 0.99 \text{ kJ/kg K.}$$
$$\text{Specific heat of water vapour} = 2.01 \text{ kJ/kg K.}$$

13.6. Material is fed to a dryer at the rate of 0.3 kg/s and the moisture removed is 35% of the wet charge. The stock enters and leaves the dryer at 324 K. The air temperature falls from 341 to 310 K, its humidity rising from 0.01 to 0.02 kg/kg.

Calculate the heat loss to the surroundings.

$$\text{Latent heat of water at 324 K} = 2430 \text{ kJ/kg.}$$
$$\text{Specific heat of dry air} = 0.99 \text{ kJ/kg K.}$$
$$\text{Specific heat of water vapour} = 2.01 \text{ kJ/kg K.}$$

13.7. A rotary dryer is fed with sand at a rate of 1 kg/s. The feed is 50% wet and the sand is discharged with 3% moisture. The entering air is at 380 K and has an absolute humidity of 0.007 kg/kg. The wet sand enters at 294 K and leaves at 309 K and the air leaves at 310 K.

Calculate the mass of air passing through the dryer and the humidity of the air leaving the dryer. Allow a radiation loss of 25 kJ/kg of dry air.

$$\text{Latent heat of water at 294 K} = 2450 \text{ kJ/kg.}$$
$$\text{Specific heat of sand} = 0.88 \text{ kJ/kg K.}$$
$$\text{Specific heat of dry air} = 0.99 \text{ kJ/kg K.}$$
$$\text{Specific heat of vapour} = 2.01 \text{ kJ/kg K.}$$

13.8. Water is to be cooled in a packed tower from 330 to 295 K by means of air flowing countercurrently. The liquid flows at a rate of 275 cm^3/m^2 s and the air at 0.7 m^3/m^2 s. The entering air has a temperature of 295 K and a humidity of 20%. Calculate the required height of tower and the condition of the air leaving at the top.

The whole of the resistance to heat and mass transfer can be considered as being within the gas phase and the product of the mass transfer coefficient and the transfer surface per unit volume of column ($h_D a$) can be taken as 0.2 s^{-1}.

13.9. Water is to be cooled in a small packed column from a temperature of 330 to 285 K by means of air flowing countercurrently. The rate of flow of liquid is 1400 cm^3/m^2 s and the flow rate of the air, which enters at a temperature of 295 K and a humidity of 60%, is 3.0 m^3/m^2 s. Calculate the required height of tower if the whole of the resistance to heat and mass transfer can be considered as being in the gas phase and the product of the mass transfer coefficient and the transfer surface per unit volume of column is 2 s^{-1}.

What is the condition of the air which leaves at the top?

13.10. Air containing 0.005 kg of water vapour per kg of dry air is heated to 325 K in a dryer and passed to the lower shelves. It leaves these shelves at 60% humidity and is reheated to 325 K and passed over another set of shelves, again leaving at 60% humidity. This is again reheated for the third and fourth sets of shelves, after which the air leaves the dryer. On the assumption that the material in each shelf has reached the wet bulb temperature and that heat losses from the dryer can be neglected, determine:

(a) the temperature of the material on each tray;
(b) the rate of water removal if 5 m^3/s of moist air leaves the dryer;
(c) the temperature to which the inlet air would have to be raised to carry out the drying in a single stage.

13.11. 0.08 m^3/s of air at 305 K and 60% humidity is to be cooled to 275 K. Calculate, by use of a psychrometric chart, the amount of heat to be removed for each 10 K interval of the cooling process. What total mass of moisture will be deposited? What is the humid heat of the air at the beginning and end of the process?

13.12. A hydrogen stream at 300 K and atmospheric pressure has a dew point of 275 K. It is to be further humidified by adding to it (through a nozzle) saturated stream at 240 kN/m^2 at the rate of 1 kg steam to 30 kg of hydrogen feed. What will be the temperature and humidity of the resultant stream?

13.13. In a countercurrent packed column n-butanol flows down at a rate of 0.25 kg/m^2 s and is cooled from 330 to 295 K. Air at 290 K, initially free of n-butanol vapour, is passed up the column at the rate of 0.7 m^3/m^2 s. Calculate the required height of tower and the condition of the exit air.
Data:

Mass transfer coefficient per unit volume: $h_D a = 0.1$ s^{-1}

Psychrometric ratio: $\dfrac{h}{h_D \rho_A s} = 2.34$

Heat transfer coefficients: $h_L = 3h_G$

Latent heat of vaporisation of n-butanol, $\lambda = 590$ kJ/kg.

Specific heat of liquid n-butanol: $C_L = 2.5$ kJ/kg K.

Humid heat of gas: $s = 1.05$ kJ/kg K.

Temperature (K)	Vapour pressure of n-butanol (kN/m^2)
295	0.59
300	0.86
305	1.27
310	1.75
315	2.48
320	3.32
325	4.49
330	5.99
335	7.89
340	10.36
345	14.97
350	17.50

13.14. Estimate the height and base diameter of a natural draught hyperbolic cooling tower which will handle a flow of 5000 kg/s water entering at 300 K and leaving at 294 K. The dry-bulb air temperature is 287 K and the ambient wet bulb temperature is 284 K.

Index

This index is intended primarily to direct the reader to entries in this volume, but also contains some cross-references to material in other volumes. Such cross-references will be indicated by the volume number printed in bold Roman numerals. Thus, for example, the entry:

Adiabatic cooling line 627, 628, 629, 643, **II**

indicates that reference to that subject will be found on pages 627, 628, 629, and 643 of this volume, and that further such material will be found in Volume 2. Note that the cross-references to other volumes do not include a page number. For this see the index in the volume specified.